METHODS OF APPLIED MATHEMATICS FOR ENGINEERS AND SCIENTISTS

Based on course notes from more than twenty years of teaching engineering and physical sciences at Michigan Technological University, Tomas B. Co's engineering mathematics textbook is rich with examples, applications, and exercises. Professor Co uses analytical approaches to solve smaller problems to provide mathematical insight and understanding, and numerical methods for large and complex problems. The book emphasizes applying matrices with strong attention to matrix structure and computational issues such as sparsity and efficiency. Chapters on vector calculus and integral theorems are used to build coordinate-free physical models, with special emphasis on orthogonal coordinates. Chapters on ordinary differential equations and partial differential equations cover both analytical and numerical approaches. Topics on analytical solutions include similarity transform methods, direct formulas for series solutions, bifurcation analysis, Lagrange-Charpit formulas, and shocks/rarefaction. Topics on numerical methods include stability analysis, differential algebraic equations, high-order finite-difference formulas, and Delaunay meshes. MATLAB implementations of the methods and concepts are fully integrated.

Tomas B. Co is an associate professor of chemical engineering at Michigan Technological University. After completing his PhD in chemical engineering at the University of Massachusetts at Amherst, he was a postdoctoral researcher at Lehigh University, a visiting researcher at Honeywell Corp., and a visiting professor at Korea University. He has been teaching applied mathematics to graduate and advanced undergraduate students at Michigan Tech for more than twenty years. His research areas include advanced process control, including plantwide control, nonlinear control, and fuzzy logic. His journal publications span broad areas in such journals as *IEEE Transactions in Automatic Control*, *Automatica*, *AIChE Journal*, *Computers in Chemical Engineering*, and *Chemical Engineering Progress*. He has been nominated twice for the Distinguished Teaching Award at Michigan Tech and is a member of the Michigan Technological University Academy of Teaching Excellence.

Methods of Applied Mathematics for Engineers and Scientists

Tomas B. Co
Michigan Technological University

CAMBRIDGE
UNIVERSITY PRESS

CAMBRIDGE UNIVERSITY PRESS
Cambridge, New York, Melbourne, Madrid, Cape Town,
Singapore, São Paulo, Delhi, Mexico City

Cambridge University Press
32 Avenue of the Americas, New York, NY 10013-2473, USA

www.cambridge.org
Information on this title: www.cambridge.org/9781107004122

First published 2013

Printed in the United States of America

A catalog record for this publication is available from the British Library.

Library of Congress Cataloging in Publication Data

Co, Tomas B., 1959–
Methods of applied mathematics for engineers and scientists / Tomas B. Co.,
Michigan Technological University.
pages cm
Includes bibliographical references and index.
ISBN 978-1-107-00412-2 (hardback)
1. Matrices. 2. Differential equations – Numerical solutions. I. Title.
QA188.C63 2013
512.9′434–dc23 2012043979

ISBN 978-1-107-00412-2 Hardback

Additional resources for this publication at www.cambridge.org/co

Contents

Preface *page* xi

I MATRIX THEORY

1 **Matrix Algebra** 3
- 1.1 Definitions and Notations 4
- 1.2 Fundamental Matrix Operations 6
- 1.3 Properties of Matrix Operations 18
- 1.4 Block Matrix Operations 30
- 1.5 Matrix Calculus 31
- 1.6 Sparse Matrices 39
- 1.7 Exercises 41

2 **Solution of Multiple Equations** 54
- 2.1 Gauss-Jordan Elimination 55
- 2.2 LU Decomposition 59
- 2.3 Direct Matrix Splitting 65
- 2.4 Iterative Solution Methods 66
- 2.5 Least-Squares Solution 71
- 2.6 QR Decomposition 77
- 2.7 Conjugate Gradient Method 78
- 2.8 GMRES 79
- 2.9 Newton's Method 80
- 2.10 Enhanced Newton Methods via Line Search 82
- 2.11 Exercises 86

3 **Matrix Analysis** 99
- 3.1 Matrix Operators 100
- 3.2 Eigenvalues and Eigenvectors 107
- 3.3 Properties of Eigenvalues and Eigenvectors 113
- 3.4 Schur Triangularization and Normal Matrices 116
- 3.5 Diagonalization 117
- 3.6 Jordan Canonical Form 118
- 3.7 Functions of Square Matrices 120

3.8 Stability of Matrix Operators 124
3.9 Singular Value Decomposition 127
3.10 Polar Decomposition 132
3.11 Matrix Norms 135
3.12 Exercises 138

II VECTORS AND TENSORS

4 **Vector and Tensor Algebra and Calculus** 149
4.1 Notations and Fundamental Operations 150
4.2 Vector Algebra Based on Orthonormal Basis Vectors 154
4.3 Tensor Algebra 157
4.4 Matrix Representation of Vectors and Tensors 162
4.5 Differential Operations for Vector Functions of One Variable 164
4.6 Application to Position Vectors 165
4.7 Differential Operations for Vector Fields 169
4.8 Curvilinear Coordinate System: Cylindrical and Spherical 184
4.9 Orthogonal Curvilinear Coordinates 189
4.10 Exercises 196

5 **Vector Integral Theorems** 204
5.1 Green's Lemma 205
5.2 Divergence Theorem 208
5.3 Stokes' Theorem and Path Independence 210
5.4 Applications 215
5.5 Leibnitz Derivative Formula 224
5.6 Exercises 225

III ORDINARY DIFFERENTIAL EQUATIONS

6 **Analytical Solutions of Ordinary Differential Equations** 235
6.1 First-Order Ordinary Differential Equations 236
6.2 Separable Forms via Similarity Transformations 238
6.3 Exact Differential Equations via Integrating Factors 242
6.4 Second-Order Ordinary Differential Equations 245
6.5 Multiple Differential Equations 250
6.6 Decoupled System Descriptions via Diagonalization 258
6.7 Laplace Transform Methods 262
6.8 Exercises 263

7 **Numerical Solution of Initial and Boundary Value Problems** 273
7.1 Euler Methods 274
7.2 Runge Kutta Methods 276
7.3 Multistep Methods 282
7.4 Difference Equations and Stability 291
7.5 Boundary Value Problems 299
7.6 Differential Algebraic Equations 303
7.7 Exercises 305

8 **Qualitative Analysis of Ordinary Differential Equations** 311
8.1 Existence and Uniqueness 312
8.2 Autonomous Systems and Equilibrium Points 313
8.3 Integral Curves, Phase Space, Flows, and Trajectories 314
8.4 Lyapunov and Asymptotic Stability 317
8.5 Phase-Plane Analysis of Linear Second-Order Autonomous Systems 321
8.6 Linearization Around Equilibrium Points 327
8.7 Method of Lyapunov Functions 330
8.8 Limit Cycles 332
8.9 Bifurcation Analysis 340
8.10 Exercises 340

9 **Series Solutions of Linear Ordinary Differential Equations** 347
9.1 Power Series Solutions 347
9.2 Legendre Equations 358
9.3 Bessel Equations 363
9.4 Properties and Identities of Bessel Functions and Modified Bessel Functions 369
9.5 Exercises 371

IV PARTIAL DIFFERENTIAL EQUATIONS

10 **First-Order Partial Differential Equations and the Method of Characteristics** 379
10.1 The Method of Characteristics 380
10.2 Alternate Forms and General Solutions 387
10.3 The Lagrange-Charpit Method 389
10.4 Classification Based on Principal Parts 393
10.5 Hyperbolic Systems of Equations 397
10.6 Exercises 399

11 **Linear Partial Differential Equations** 405
11.1 Linear Partial Differential Operator 406
11.2 Reducible Linear Partial Differential Equations 408
11.3 Method of Separation of Variables 411
11.4 Nonhomogeneous Partial Differential Equations 431
11.5 Similarity Transformations 439
11.6 Exercises 443

12 **Integral Transform Methods** 450
12.1 General Integral Transforms 451
12.2 Fourier Transforms 452
12.3 Solution of PDEs Using Fourier Transforms 459
12.4 Laplace Transforms 464
12.5 Solution of PDEs Using Laplace Transforms 474
12.6 Method of Images 476
12.7 Exercises 477

13 **Finite Difference Methods** 483

13.1 Finite Difference Approximations 484
13.2 Time-Independent Equations 491
13.3 Time-Dependent Equations 504
13.4 Stability Analysis 512
13.5 Exercises 519

14 **Method of Finite Elements** 523

14.1 The Weak Form 524
14.2 Triangular Finite Elements 527
14.3 Assembly of Finite Elements 533
14.4 Mesh Generation 539
14.5 Summary of Finite Element Method 541
14.6 Axisymmetric Case 546
14.7 Time-Dependent Systems 547
14.8 Exercises 552

Bibliography B-1

Index I-1

A **Additional Details and Fortification for Chapter 1** 561

A.1 Matrix Classes and Special Matrices 561
A.2 Motivation for Matrix Operations from Solution of Equations 568
A.3 Taylor Series Expansion 572
A.4 Proofs for Lemma and Theorems of Chapter 1 576
A.5 Positive Definite Matrices 586

B **Additional Details and Fortification for Chapter 2** 589

B.1 Gauss Jordan Elimination Algorithm 589
B.2 SVD to Determine Gauss-Jordan Matrices Q and W 594
B.3 Boolean Matrices and Reducible Matrices 595
B.4 Reduction of Matrix Bandwidth 600
B.5 Block LU Decomposition 602
B.6 Matrix Splitting: Diakoptic Method and Schur Complement Method 605
B.7 Linear Vector Algebra: Fundamental Concepts 611
B.8 Determination of Linear Independence of Functions 614
B.9 Gram-Schmidt Orthogonalization 616
B.10 Proofs for Lemma and Theorems in Chapter 2 617
B.11 Conjugate Gradient Algorithm 620
B.12 GMRES Algorithm 629
B.13 Enhanced-Newton Using Double-Dogleg Method 635
B.14 Nonlinear Least Squares via Levenberg-Marquardt 639

C **Additional Details and Fortification for Chapter 3** 644

C.1 Proofs of Lemmas and Theorems of Chapter 3 644
C.2 QR Method for Eigenvalue Calculations 649
C.3 Calculations for the Jordan Decomposition 655

C.4 Schur Triangularization and SVD 658
C.5 Sylvester's Matrix Theorem 659
C.6 Danilevskii Method for Characteristic Polynomial 660

D **Additional Details and Fortification for Chapter 4** 664
D.1 Proofs of Identities of Differential Operators 664
D.2 Derivation of Formulas in Cylindrical Coordinates 666
D.3 Derivation of Formulas in Spherical Coordinates 669

E **Additional Details and Fortification for Chapter 5** 673
E.1 Line Integrals 673
E.2 Surface Integrals 678
E.3 Volume Integrals 684
E.4 Gauss-Legendre Quadrature 687
E.5 Proofs of Integral Theorems 691

F **Additional Details and Fortification for Chapter 6** 700
F.1 Supplemental Methods for Solving First-Order ODEs 700
F.2 Singular Solutions 703
F.3 Finite Series Solution of $d\mathbf{x}/dt = A\mathbf{x} + \mathbf{b}(t)$ 705
F.4 Proof for Lemmas and Theorems in Chapter 6 708

G **Additional Details and Fortification for Chapter 7** 715
G.1 Differential Equation Solvers in MATLAB 715
G.2 Derivation of Fourth-Order Runge Kutta Method 718
G.3 Adams-Bashforth Parameters 722
G.4 Variable Step Sizes for BDF 723
G.5 Error Control by Varying Step Size 724
G.6 Proof of Solution of Difference Equation, Theorem 7.1 730
G.7 Nonlinear Boundary Value Problems 731
G.8 Ricatti Equation Method 734

H **Additional Details and Fortification for Chapter 8** 738
H.1 Bifurcation Analysis 738

I **Additional Details and Fortification for Chapter 9** 745
I.1 Details on Series Solution of Second-Order Systems 745
I.2 Method of Order Reduction 748
I.3 Examples of Solution of Regular Singular Points 750
I.4 Series Solution of Legendre Equations 753
I.5 Series Solution of Bessel Equations 757
I.6 Proofs for Lemmas and Theorems in Chapter 9 761

J **Additional Details and Fortification for Chapter 10** 771
J.1 Shocks and Rarefaction 771
J.2 Classification of Second-Order Semilinear Equations: $n > 2$ 781
J.3 Classification of High-Order Semilinear Equations 784

K **Additional Details and Fortification for Chapter 11** 786
K.1 d'Alembert Solutions 786
K.2 Proofs of Lemmas and Theorems in Chapter 11 791

L **Additional Details and Fortification for Chapter 12** 795
L.1 The Fast Fourier Transform 795
L.2 Integration of Complex Functions 799
L.3 Dirichlet Conditions and the Fourier Integral Theorem 819
L.4 Brief Introduction to Distribution Theory and Delta Distributions 820
L.5 Tempered Distributions and Fourier Transforms 830
L.6 Supplemental Lemmas, Theorems, and Proofs 836
L.7 More Examples of Laplace Transform Solutions 840
L.8 Proofs of Theorems Used in Distribution Theory 846

M **Additional Details and Fortification for Chapter 13** 851
M.1 Method of Undetermined Coefficients for Finite Difference Approximation of Mixed Partial Derivative 851
M.2 Finite Difference Formulas for 3D Cases 852
M.3 Finite Difference Solutions of Linear Hyperbolic Equations 855
M.4 Alternating Direction Implicit (ADI) Schemes 863

N **Additional Details and Fortification for Chapter 14** 867
N.1 Convex Hull Algorithm 867
N.2 Stabilization via Streamline-Upwind Petrov-Galerkin (SUPG) 870

Preface

This book was written as a textbook on applied mathematics for engineers and scientist, with the expressed goal of merging both analytical and numerical methods more tightly than other textbooks. The role of applied mathematics has continued to grow increasingly important with advancement of science and technology, ranging from modeling and analysis of natural phenomenon to simulation and optimization of man-made systems. With the huge and rapid advances of computing technology, larger and more complex problems can now be tackled and analyzed in a very timely fashion. In several cases, what used to require supercomputers can now be solved using personal computers. Nonetheless, as the technological tools continue to progress, it has become even more imperative that the results can be understood and interpreted clearly and correctly, as well as the need for a deeper knowledge behind the strengths and limitations of the numerical methods used. This means that we cannot forgo the analytical techniques because they continue to provide indispensable insights on the veracity and meaning of the results. The analytical tools continue to be of prime importance for basic understanding for building mathematical models and data analysis. Still, when it comes to solving large and complex problems, numerical methods are needed.

The level of exposition in this book is aimed at graduate students, advanced undergraduate students, and researchers in the engineering and science field. Thus the topics were mostly chosen to continue several topics found in most undergraduate textbooks in applied mathematics. We have focused on advanced concepts and implementation of various mathematical tools to solve the problems that most graduate students are likely to face in their research work and other advanced courses.

The contents of the book can be divided into four main parts: matrix theory, vectors and tensors, ordinary differential equations, and partial differential equations. We begin the book with matrix theory because the tools developed in matrix theory form the crucial foundations used in the rest of the book. The next part centers on the concepts used in vector and tensor theory, including the application of tensor calculus and integral theorems to develop mathematical models of physical systems, often resulting in several differential equations. The last two parts focus on the solution of ordinary and partial differential equations. It can be argued that the primary needs of applied mathematics in engineering and the physical sciences are to obtain models for a system or phenomena in the form of differential equations and then to be able to "solve" them to predict and understand the effects of changes in model parameters, boundary conditions, or initial conditions.

Although the methods of applied mathematics are independent of computing platform and programs, we have chosen to use MATLAB as a particular platform under which we investigate the mathematical methods, techniques, and ideas so that the approaches can be tested and the results can be visualized. The supplied MATLAB codes are all included on the book's website, and the reader can modify the codes for their own use. There are several excellent MATLAB toolboxes supplied by third-party software developers, and they have been optimized for speed, efficiency, and user-friendliness. However, the unintended consequences of user-friendly tools can sometimes render the users to be "button pushers." We contend that students in applied mathematics still need to discover the mechanism and ideas behind the full-blown programs – at least to apply them to simple test problems and gain some basic understanding of the various approaches. The links to the supplemental MATLAB programs and files can be accessed through the link: `www.cambridge.org/Co`.

The appendices are collected as chapter fortifications. They include proofs, advanced topics, additional tables, and examples. The reader should be able to access these materials through the web via the link: `www.cambridge.org/Co`. The index also contains topics that can be found in the appendices, and they are given page numbers that continue the count from the main text.

Several colleagues and students have helped tremendously in the writing of this textbook. Mostly, I want to thank my best friend and wife, Faith Morrison, for the support and encouragement and the sacrifice she's made so that I could finish this extended and personally significant project. I hope the textbook will contain useful information for the readers, enough for them to share in the continued exploration of the methods and applications of mathematics to further improve the understanding and conditions of our world.

Tomas B. Co
Houghton, MI

PART I

MATRIX THEORY

Matrix theory is a powerful field of mathematics that has found applications in the solution of several real-world problems, ranging from the solution of algebraic equations to the solution of differential equations. Its importance has also been enhanced by the rapid development of several computer programs that have improved the efficiency of matrix analysis and the solution of matrix equations.

We have allotted three chapters to discussing matrix theory. Chapter 1 contains the basic notations and operations. These include conventions and notations for the various structural, algebraic, differential, and integral operations. As such, this chapter focuses on how to formulate problems in terms of matrix equations, the various approaches of matrix algebraic manipulations, and matrix partitions.

Chapter 2 then focuses on the solution of the linear equation given by $A\mathbf{x} = \mathbf{b}$, and it includes both direct and indirect methods. The most direct method is to find the inverse of A and then evaluate $\mathbf{x} = A^{-1}\mathbf{b}$. However, the major practical issue is that matrix inverses become unwieldy when the matrices are large. This chapter is concerned with finding the solutions by reformulating the problem to take advantage of available matrix properties. Direct methods use various factorizations of A based on matrices that are more easily invertible, whereas indirect methods use an iterative process starting with an initial guess of the solution. The methods can then be applied to linear least-squares problems, as well as to the solution of multivariable nonlinear equations.

Chapter 3 focuses on matrices as operators. In this case, the discussion is concerned with the analysis of matrices, for example, using eigenvalues and eigenvectors. This allows one to obtain diagonalized matrices or Jordan canonical forms. These forms provide efficient tools for evaluating matrix functions, which are also very useful for solving simultaneous differential equations. Other analysis tools such as singular values decomposition, matrix norms, and condition numbers are also included in the chapter.

The matrix theory topics are also used in the other parts of this book. In Part II, we can use matrices to represent vector coordinates and tensors. The operations and vector/tensor properties can also be evaluated and analyzed efficiently using matrix theory. For instance, the mutual orthogonalities among the principal axes of a symmetric tensor are immediate consequences of the properties of matrix eigenvectors. In Part III, matrices are also shown to be indispensable tools for solving ordinary differential equations. Specifically, the solution and analysis of a set of simultaneous linear ordinary differential equations can be represented in terms of matrix exponential functions. Moreover, numerical solution methods can now be coded in matrix forms. Finally, in Part IV of the book, both the finite difference and finite elements methods reduce partial differential equations to linear algebraic equations. Thus the tools discussed in Chapter 2 are strongly applicable because the matrices resulting from either of these methods will likely be large and sparse.

1 Matrix Algebra

In this chapter, we review some definitions and operations of matrices. Matrices play very important roles in the computation and analysis of several mathematical problems. They allow for compact notations of large sets of linear algebraic equations. Various matrix operations such as addition, multiplication, and inverses can be combined to find the required solutions in a more tractable manner. The existence of several software tools, such as MATLAB, have also made it very efficient to approach the solution by posing several problems in the form of matrix equations. Moreover, the matrices possess internal properties such as determinant, rank, trace, eigenvalues, and eigenvectors, which can help characterize the systems under consideration.

We begin with the basic notation and definitions in Section 1.1. The matrix notations introduced in this chapter are used throughout the book. Then in Section 1.2, we discuss the various matrix operations. Several matrix operations should be familiar to most readers, but some may not be as familiar, such as Kronecker products. We have classified the operations as either structural or algebraic. The structural operations are those operations that involve only the collection and arrangement of the elements. On the other hand, the algebraic operations pertain to those in which algebraic operations are implemented among the elements of a matrix or group of matrices. The properties of the different matrix operations such as associativity, commutativity, and distributivity properties are summarized in Section 1.3. In addition, we discuss the properties of determinants and include some matrix inverse formulas. The properties and formulas allow for the manipulation and simplification of matrix equations. These will be important tools used throughout this book.

In Section 1.4, we explore various block matrix operations. These operations are very useful when the structure of the matrices can be partitioned into submatrices. These block operations will also prove to be very useful when solving large sets of equations that exhibit a specific pattern.

From algebraic operations, we then move to topics involving differential and integral calculus in Section 1.5. We first define and fix various notations for the derivatives and integrals of matrices. These notations are also used throughout the book. The various properties of the matrix calculus operations are also summarized in this section. One of the applications of matrix calculus is optimization, in which the concept of positive (and negative) definiteness is needed for sufficient conditions. We devote Section A.5 in the appendix to explaining positive or negative definiteness in more detail.

Finally, in Section 1.6, we include a brief discussion on sparse matrices. These matrices often result when the problem involves a large collection of smaller elements that are connected with only few of the other elements, such as when we solve differential equations by numerical methods, for example, the finite difference methods or finite element methods.

1.1 Definitions and Notations

The primary application of matrices is in solving simultaneous linear equations. These equations can come from solving problems based on mass and energy balance of physical, chemical, and biological processes; Kirchhoff's laws in electric circuits; force and moment balances in engineering structures; and so forth. The size of the unknowns for these problems can be quite large, so the solution can become quite complicated. This is especially the case with modern engineering systems, which typically contain several stages (e.g., staged operations in chemical engineering), are highly integrated (e.g., large-scale integration in microelectronics), or are structurally large (e.g., large power grids and large buildings). Matrix methods offer techniques that allow for tractability and computational efficiency.

When solving large nonlinear problems, numerical methods become a necessary approach. The numerical computations often involve matrix formulations. For instance, several techniques for solving nonlinear equations and nonlinear optimization problems implement Newton's method and other gradient-based methods, in which the calculations include matrix operations. Matrix equations also result from finite approximations of systems of differential equations. For boundary value problems, the internal values are to be solved such that both the boundary conditions and the differential equations that describe the systems are satisfied. Here, the numerical techniques include finite element methods and finite difference methods, both of which translate the problem back to a linear set of equations.

Aside from calculating the unknowns or solving differential equations, matrix methods are also useful in operator analysis and design. In this case, matrix equations are analyzed in terms of operators, inputs, and outputs. The matrices associated with the operators can be formulated to obtain the desired behavior. For example, if we want to move a 3D point $a = (x, y, z)$ to another position, say, $b = (\widehat{x}, \widehat{y}, \widehat{z})$, in a particular way, for instance, to move it radially outward or rotate it at specified degrees counterclockwise, then we can build matrices that would produce the desired effects. Conversely, for a system (mechanical, chemical, electrical, biological, etc.) that can be written in matrix forms (both in differential equations and algebraic equations), we can often isolate the matrices associated with system operations and use matrix analysis to explore the capabilities and behavior of the system.

It is also worth mentioning that, in addition to the classical systems that are modeled with algebraic and differential equations, there are other application domains that use matrix methods extensively. These include data processing, computational geometry, and network analysis. In data processing, matrix methods help in regression analysis and statistical data analysis. These applications also include data mining in search engines, bioinformatics, and computer security. Computational geometry also uses matrix methods to handle and analyze large sets of data. Applications include computer graphics and visualization, which are also used for pattern recognition purposes. In network analysis, matrix methods are used together with graph theory to analyze the connectivity and effects of large, complex structures. Applications include the analysis of communication and control systems, as well as large power grids.

We now begin with the definition of a matrix and continue with some of the notations and conventions that are used throughout this book.

Definition 1.1. *A* **matrix** *is a collection of objects, called the* **elements** *of the matrix, arranged in rows and columns.*

These elements of the matrix could be numbers, such as

$$A = \begin{pmatrix} 1 & 0 & -0.3 \\ -2 & 3+i & \frac{1}{2} \end{pmatrix} \qquad \text{with } i = \sqrt{-1}$$

or functions, such as

$$B = \begin{pmatrix} 1 & 2x(t)+a \\ \int \sin(\omega t)dt & dy/dt \end{pmatrix}$$

The elements of matrices are restricted to a set of mathematical objects that allow algebraic binary operations such as addition, subtraction, multiplication, and division. The valid elements of the matrix are referred to as **scalars**. Note that a scalar is not the same as a matrix having only one row and one column.

We often use capital letters to denote matrices, whereas the corresponding small letters stand for the elements. Thus the elements of matrix A positioned at the ith row and jth column are denoted as a_{ij}, for example, for A having N rows and M columns,

$$A = \begin{pmatrix} a_{11} & a_{12} & \cdots & a_{1M} \\ a_{21} & a_{22} & \cdots & a_{2M} \\ \vdots & \vdots & \ddots & \vdots \\ a_{N1} & a_{N2} & \cdots & a_{NM} \end{pmatrix} \tag{1.1}$$

The size of the matrix is given by the symbol "[=]", for example, for matrix A having N rows and M columns,

$$A\,[=]\,N \times M \qquad \text{or} \qquad A_{[N\times M]} \tag{1.2}$$

A **row vector** is a matrix having one row, whereas a **column vector** is a matrix having one column. The **length** of a vector means the number of elements of the row or column vector. If the type of vector has not been specified, we take it to mean a column vector. We often use bold small letters to denote vectors. A basic vector is the ith**unit vector** of length N denoted by $\mathbf{e}_i$,

$$\mathbf{e}_i = \begin{pmatrix} 0 \\ \vdots \\ 0 \\ 1 \\ 0 \\ \vdots \\ 0 \end{pmatrix} \begin{matrix} \\ \\ \\ \leftarrow i^{\text{th}} \text{ element} \\ \\ \\ \\ \end{matrix} \tag{1.3}$$

The length N of the unit vector is determined by context.

A **square matrix** is a matrix with the same number of columns and rows. Special cases include **lower triangular**, **upper triangular**, and **diagonal** matrices. Lower triangular matrices have zero elements above the main diagonal, whereas upper triangular matrices have zero elements below the main diagonal. Diagonal matrices have zero off-diagonal elements. The diagonal matrix is also represented by

$$D = \text{diag}\left(d_{11}, d_{22}, \ldots, d_{NN}\right) \tag{1.4}$$

A special diagonal matrix in which the main diagonal elements are all 1's is known as the **identity matrix**, denoted by I. If the size of the identity matrix needs to be specified, then we use I_N to denote an $N \times N$ identity matrix. An extensive list of different matrices that have special forms such as bidiagonal, tridiagonal, Hessenberg, Toeplitz, and so forth are given in Tables A.1 through A.5 in Section A.1 as an appendix for easy reference.

1.2 Fundamental Matrix Operations

We assume that the reader is already familiar with several matrix operations. The purpose of the following sections is to summarize these operations, introduce our notations, and relate them to some of the available MATLAB commands. We can divide matrix operations into two major categories. The first category involves the restructuring or combination of matrices. The second category includes the operations that contain algebraic computations such as addition, multiplication, and inverses.

1.2.1 Matrix Restructuring Operations

A list of matrix rearrangement operations with their respective notations are summarized in Tables 1.1 and 1.2 (together with some MATLAB commands associated with the operations).

The row and column augmentation operations are designated by horizontal and vertical bars, respectively. These are used extensively throughout the book because we take advantage of block matrix operations. The reshaping operations are given by the **vectorization** operation and **reshape** operation. Both these operations are quite useful when reformulating equations such as $HX + XB + CXD = F$ into the familiar linear equation form given by $A\mathbf{x} = \mathbf{b}$.

There are two operations that involve exchanging the roles of rows and columns: the standard **transpose** operation, which we denote by superscript T, and the **conjugate transpose**, which we denote by superscript asterisk. In general, $A^T \neq A^*$, except when the elements of A are all real. When $A = A^T$, we say that A is **symmetric**, and when $A = A^*$, we say that A is **Hermitian**. The two cases are generally not the same. For instance, let

$$A = \begin{pmatrix} 1+i & 2 \\ 2 & 3 \end{pmatrix} \qquad B = \begin{pmatrix} 1 & 2+i \\ 2-i & 3 \end{pmatrix}$$

then A is symmetric but not Hermitian, whereas B is Hermitian but not symmetric. On the other hand, when $A = -A^T$, we say that A is **skew-symmetric**, and when $A = -A^*$, we say that A is **skew-Hermitian**.

Table 1.1. *Matrix restructuring operations*

	Operation	Notation	Rule
1	Column Augment	$C = \left(A \mid B \right)$ MATLAB: `C=[A,B]`	$\begin{pmatrix} c_{11} & \cdots & c_{1,M+P} \\ \vdots & \ddots & \vdots \\ c_{N1} & \cdots & c_{N,M+P} \end{pmatrix} = \begin{pmatrix} a_{11} & \cdots & a_{1M} & b_{11} & \cdots & b_{1P} \\ \vdots & \ddots & \vdots & \vdots & \ddots & \vdots \\ a_{N1} & \cdots & a_{NM} & b_{N1} & \cdots & b_{NP} \end{pmatrix}$
2	Row Augment	$C = \left(\dfrac{A}{B} \right)$ MATLAB: `C=[A;B]`	$\begin{pmatrix} c_{11} & \cdots & c_{1,M} \\ \vdots & \ddots & \vdots \\ c_{N+P,1} & \cdots & c_{N+P,M} \end{pmatrix} = \begin{pmatrix} a_{11} & \cdots & a_{1M} \\ \vdots & \ddots & \vdots \\ a_{N1} & \cdots & a_{NM} \\ b_{11} & \cdots & b_{1M} \\ \vdots & \ddots & \vdots \\ b_{P1} & \cdots & b_{PM} \end{pmatrix}$
3	Vectorize	$C = \text{vec}(A)$ MATLAB: `C=A(:)`	$\begin{pmatrix} c_1 \\ \vdots \\ c_{N\cdot M} \end{pmatrix} = \begin{pmatrix} A_{\bullet,1} \\ \hline \vdots \\ \hline A_{\bullet,M} \end{pmatrix}$ where $A_{\bullet,i}$ is the ith column of A

The **submatrix** operation is denoted by using a list of k subscript indices and ℓ superscript indices to refer to the rows and columns, respectively, extracted from a matrix. For instance,

$$A = \begin{pmatrix} 1 & 2 & 3 \\ 4 & 5 & 6 \\ 7 & 8 & 9 \end{pmatrix} \quad \rightarrow \quad A_{1,2}^{2,3} = \begin{pmatrix} 2 & 3 \\ 5 & 6 \end{pmatrix}$$

For a square matrix, if the diagonals of the submatrix are a subset of the diagonals of the original matrix, then we call it a **principal submatrix**. This happens if the superscript indices and the subscript indices of the submatrix are the same. For instance,

$$A = \begin{pmatrix} 1 & 2 & 3 \\ 4 & 5 & 6 \\ 7 & 8 & 9 \end{pmatrix} \quad \rightarrow \quad A_{1,3}^{1,3} = \begin{pmatrix} 1 & 3 \\ 7 & 9 \end{pmatrix}$$

then $A_{1,3}^{1,3}$ is a principal submatrix.

Table 1.2. *Matrix restructuring operations*

	Operation	Notation	Rule
4	Reshape	$C =$ reshape $(\mathbf{v}, N, M)$ MATLAB: `reshape(v,N,M)`	$C = \begin{pmatrix} v_1 & v_{N+1} & & v_{(M-1)N+1} \\ \vdots & \vdots & \cdots & \vdots \\ v_N & v_{2N} & & v_{MN} \end{pmatrix}$ where $\mathbf{v}$ is a vector of length NM
5	Transpose	$C = A^T$ MATLAB: `C=A.'`	$C = \begin{pmatrix} a_{11} & \cdots & a_{M1} \\ \vdots & \ddots & \vdots \\ a_{1N} & \cdots & a_{MN} \end{pmatrix}$
6	Conjugate Transpose	$C = A^*$ MATLAB: `C=A'`	$C = \begin{pmatrix} \overline{a}_{11} & \cdots & \overline{a}_{M1} \\ \vdots & \ddots & \vdots \\ \overline{a}_{1N} & \cdots & \overline{a}_{MN} \end{pmatrix}$ where $\overline{a}_{ij} =$ complex conjugate of a_{ij}

7	Submatrix	$C = A_{i_1,i_2,\ldots,i_k}^{j_1,j_2\ldots,j_\ell}$
		MATLAB:
		`rows=[i1,i2,...]`
		`cols=[j1,j2,...]`
		`C=A(rows,cols)`
			$C_{[\,k\times\ell\,]} = \begin{pmatrix} a_{i_1j_1} & \cdots & a_{i_1j_\ell} \\ \vdots & \ddots & \vdots \\ a_{i_kj_1} & \cdots & a_{i_kj_\ell} \end{pmatrix}$

8	$(ij)^{\text{th}}$ Redact	$C = A_{ij\downarrow}$
		MATLAB:
		`C=A`
		`C(i,:)=[ ]`
		`C(:,j)=[ ]`
			$C_{[\,(N-1)\times(M-1)\,]} =$
			$\left(\begin{array}{c\|c} A_{1,\ldots,i-1}^{1,\ldots,j-1} & A_{1,\ldots,i-1}^{j+1,\ldots,M} \\ \hline A_{i+1,\ldots,N}^{1,\ldots,j-1} & A_{i+1,\ldots,N}^{j+1,\ldots,M} \end{array}\right)$

Next, the operation to remove some specified rows and columns is referred to here as the $(ij)^{\text{th}}$ **redact** operation. We use $A_{ij\downarrow}$ to denote the removal of the ith row and j^{th} column. For instance,

$$A = \begin{pmatrix} 1 & 2 & 3 \\ 4 & 5 & 6 \\ 7 & 8 & 9 \end{pmatrix} \quad \rightarrow \quad A_{23\downarrow} = \begin{pmatrix} 1 & 2 \\ 7 & 8 \end{pmatrix} \tag{1.5}$$

This operation is useful in finding determinants, cofactors, and adjugates.

Table 1.3. *Matrix algebraic operations*

	Operation	Notation	Rule	MATLAB commands
1	Sum	$C = A + B$	$c_{ij} = a_{ij} + b_{ij}$	`C=A+B`
2	Scalar Product	$C = qA$	$c_{ij} = q\, a_{ij}$	`C=q*A`
3	Matrix Product	$C = AB$	$c_{ij} = \sum_{k=1}^{K} a_{ik} b_{kj}$	`C=A*B`
4	Haddamard Product	$C = A \circ B$	$c_{ij} = a_{ij} b_{ij}$	`C=A.*B`
5	Kronecker Product (tensor product)	$C = A \otimes B$	$C = \left(\begin{array}{c\|c\|c} a_{11}B & \cdots & a_{1M}B \\ \hline \vdots & & \vdots \\ \hline a_{N1}B & \cdots & a_{NM}B \end{array}\right)$	`C=kron(A,B)`
6	Determinant	$q = \det(A)$ or $q = \|A\|$	$q = \sum_{K} \epsilon(K) \left(\prod_{i=1}^{N} a_{i,k_i} \right)$ see (1.10)	`q=det(A)`
7	Cofactor	$q = \mathbf{cof}(a_{ij})$	$q = (-1)^{i+j} \left\| A_{ij\downarrow} \right\|$	
8	Adjugate	$C = \mathbf{adj}(A)$	$c_{ij} = \mathbf{cof}(a_{ji})$	
9	Inverse	$C = A^{-1}$	$C = \frac{1}{\|A\|} \mathbf{adj}(A)$	`C=inv(A)`
10	Trace	$q = \mathbf{tr}(A)$	$q = \sum_{i=1}^{N} a_{ii}$	`q=trace(A)`
11	Real Part	$C = \mathbf{Real}(A)$	$c_{ij} = \text{real}(a_{ij})$	`C=Real(A)`
12	Imag Part	$C = \mathbf{Imag}(A)$	$c_{ij} = \text{imag}(a_{ij})$	`C=Imag(A)`
13	Complex Congugate	$C = \overline{A}$	$c_{ij} = \overline{a_{ij}}$	`C=Conj(A)`

1.2.2 Matrix Algebraic Operations

The matrix algebraic operations can be classified further as either binary or unary. For binary operations, the algebraic operations require two inputs, either a scalar and a matrix or two matrices of appropriate sizes. For unary operations, the input is a matrix, and the algebraic operations are applied on the elements of the matrix. The matrix algebraic operations are given in Table 1.3, together with their corresponding MATLAB commands.

1.2.2.1 Binary Algebraic Operations

The most basic matrix binary computational operations are matrix sums, scalar products, and matrix products, which are quite familiar to most readers. To see how these operations seem to be the natural consequences of solving simultaneous equations, we refer the reader to Section A.2 included in the appendices.

Matrix products of A and B are denoted simply by $C = AB$, which requires $A[=]N \times K, B[=]K \times M$ and $C[=]N \times M$ (i.e., the columns of A must be equal to the rows of B). If this is the case, we say that A and B are **conformable** for the operation AB. Furthermore, based on the sizes of the matrices, $A_{[N\times K]}B_{[K\times M]} = C_{[N\times M]}$, we see that dropping the common value K leaves the size of C to be $N \times M$. For the matrix product AB, we say that A **premultiplies** B, or B **postmultiplies** A. For instance, let

$$A = \begin{pmatrix} 1 & 1 \\ 2 & 1 \\ -1 & 0 \end{pmatrix} \quad \text{and} \quad B = \begin{pmatrix} -2 & 1 \\ -1 & 3 \end{pmatrix} \quad \text{then} \quad C = AB = \begin{pmatrix} -3 & 4 \\ -5 & 5 \\ 2 & -1 \end{pmatrix}$$

However, B and A is not conformable for the product BA.

In several cases, $AB \neq BA$, even if the reversed order is conformable, and thus one needs to be clear whether a matrix premultiplies or postmultiplies another matrix. For the special case in which switching the order yields the same product (i.e., $AB = BA$), then we say that A and B **commutes**. It is necessary that commuting matrices are square and have the same sizes.

We list a few key results regarding matrix products:

1. For matrix products between a matrix $A[=]N \times M$ and the appropriately sized identity matrix, we have

$$AI_M = I_N A = A$$

   where I_M and I_N are identity matrices of size M and size N, respectively.
2. Based on the definition of matrix products, when B premultiplies A, the row elements of B are pairwise multiplied with the column elements of A, and the results are then summed together. This fact implies that to scale the ith row of A by a factor d_i, we can simply premultiply A by a diagonal matrix $D = \text{diag}(d_1, \ldots, d_N)$. For instance,

$$DA = \begin{pmatrix} 2 & 0 & 0 \\ 0 & 1 & 0 \\ 0 & 0 & -1 \end{pmatrix} \begin{pmatrix} 1 & 2 & 3 \\ 4 & 5 & 6 \\ 7 & 8 & 9 \end{pmatrix} = \begin{pmatrix} 2 & 4 & 6 \\ 4 & 5 & 6 \\ -7 & -8 & -9 \end{pmatrix}$$

   Likewise, to scale the jth column of A by a factor d_j, we can simply postmultiply A by a diagonal matrix $D = \text{diag}(d_1, \ldots, d_N)$. For instance,

$$AD = \begin{pmatrix} 1 & 2 & 3 \\ 4 & 5 & 6 \\ 7 & 8 & 9 \end{pmatrix} \begin{pmatrix} 2 & 0 & 0 \\ 0 & 1 & 0 \\ 0 & 0 & -1 \end{pmatrix} = \begin{pmatrix} 2 & 2 & -3 \\ 8 & 5 & -6 \\ 14 & 8 & -9 \end{pmatrix}$$

3. Premultiplying A by a row vector of 1's yields a row vector containing the sums of each column, whereas postmultiplying by a column vector of 1's yields a column vector containing the sum of each row. For instance,

$$\begin{pmatrix} 1 & 1 & 1 \end{pmatrix}\begin{pmatrix} 1 & 2 & 3 \\ 4 & 5 & 6 \\ 7 & 8 & 9 \end{pmatrix} = \begin{pmatrix} 12 & 15 & 18 \end{pmatrix} \qquad \begin{pmatrix} 1 & 2 & 3 \\ 4 & 5 & 6 \\ 7 & 8 & 9 \end{pmatrix}\begin{pmatrix} 1 \\ 1 \\ 1 \end{pmatrix} = \begin{pmatrix} 6 \\ 15 \\ 24 \end{pmatrix}$$

4. Let T be an identity matrix, but with additional nonzero nondiagonal elements in the jth column. Then $B = TA$ is a matrix whose i^{th} row $(i \neq j)$ is given by the sum of the ith row of A and $t_{ij} \cdot (j^{\text{th}}$ row of $A)$. The jth row of B remains to be the jth row of A. For instance,

$$\begin{pmatrix} 1 & 0 & -1 \\ 0 & 1 & 2 \\ 0 & 0 & 1 \end{pmatrix}\begin{pmatrix} 1 & 2 & 3 \\ 4 & 5 & 6 \\ 7 & 8 & 9 \end{pmatrix} = \begin{pmatrix} -6 & -6 & -6 \\ 18 & 21 & 24 \\ 7 & 8 & 9 \end{pmatrix}$$

Likewise, let G be an identity matrix, but with nondiagonal elements in the ith row. Then $C = AG$ is a matrix whose j^{th} column $(j \neq i)$ is given by the sum of the jth column of A and $g_{ij} \cdot (j^{\text{th}}$ column of $A)$. The ith column of G remains to be the ith column of A. For instance,

$$\begin{pmatrix} 1 & 2 & 3 \\ 4 & 5 & 6 \\ 7 & 8 & 9 \end{pmatrix}\begin{pmatrix} 1 & 0 & 0 \\ 0 & 1 & 0 \\ -1 & 2 & 1 \end{pmatrix} = \begin{pmatrix} -2 & 8 & 3 \\ -2 & 17 & 6 \\ -2 & 26 & 9 \end{pmatrix}$$

5. A square matrix P is known as a **row permutation matrix** if it is a matrix obtained by permuting the rows of an identity matrix. If P is a row permutation matrix, then PA is a matrix obtained by permuting the rows of A in the same sequence as P. For instance, let $P[=]3 \times 3$ be obtained by permuting the rows of the identity matrix according to the sequence [3, 1, 2], then

$$PA = \begin{pmatrix} 0 & 0 & 1 \\ 1 & 0 & 0 \\ 0 & 1 & 0 \end{pmatrix}\begin{pmatrix} 1 & 2 & 3 \\ 4 & 5 & 6 \\ 7 & 8 & 9 \end{pmatrix} = \begin{pmatrix} 7 & 8 & 9 \\ 1 & 2 & 3 \\ 4 & 5 & 6 \end{pmatrix}$$

Likewise, a square matrix $\widehat{P}$ is also a **column permutation matrix** if it is a matrix obtained by permuting the columns of an identity matrix. If $\widehat{P}$ is a column permutation matrix, then $A\widehat{P}$ is obtained by permuting the columns A in the same sequence as $\widehat{P}$. For instance, let $\widehat{P}[=]3 \times 3$ be obtained by permuting the columns of the identity matrix according to the sequence [3, 1, 2], then

$$A\widehat{P} = \begin{pmatrix} 1 & 2 & 3 \\ 4 & 5 & 6 \\ 7 & 8 & 9 \end{pmatrix}\begin{pmatrix} 0 & 1 & 0 \\ 0 & 0 & 1 \\ 1 & 0 & 0 \end{pmatrix} = \begin{pmatrix} 3 & 1 & 2 \\ 6 & 4 & 5 \\ 9 & 7 & 8 \end{pmatrix}$$

Remark: Matrices D, T, and P described in items 2, 4, and 5 are known as the **scaling**, **pairwise combination**, and **permutation** row operators, respectively. Collectively, they are known as the **elementary row operators**. All three operations show that premultiplication (left multiplication) is a row operation. On the other hand, D, G, and $\widehat{P}$ are **elementary column operators**, and they operate on matrices via postmultiplication (right multiplication).[1] All these matrix operations are used extensively in the Gauss-Jordan elimination method for solving linear equations.

Aside from scalar and matrix products, there are two more matrix operations involving multiplication. The **Hadamard product**, also known as **element-wise product**, is defined as follows:

$$Q = A \circ B \qquad \longleftrightarrow \qquad q_{ij} = a_{ij} b_{ij} \quad i = 1, \ldots, N; \; j = 1, \ldots, M \tag{1.6}$$

For instance,

$$\begin{pmatrix} 1 & -1 \\ -2 & 2 \end{pmatrix} \circ \begin{pmatrix} 1 & 2 \\ 3 & 4 \end{pmatrix} = \begin{pmatrix} 1 & -2 \\ -6 & 8 \end{pmatrix}$$

The **Kronecker product**, also known as the **Tensor product**, is defined as follows:

$$\Psi = A \otimes B = \left(\begin{array}{c|c|c} a_{11}B & \cdots & a_{1M}B \\ \hline \vdots & \ddots & \vdots \\ \hline a_{N1}B & \cdots & a_{NM}B \end{array} \right) \tag{1.7}$$

where the matrix blocks $a_{ij}B$ are scalar products of a_{ij} and B. For instance,

$$\begin{pmatrix} 1 & -1 \\ -2 & 2 \end{pmatrix} \otimes \begin{pmatrix} 1 & 2 \\ 3 & 4 \end{pmatrix} = \begin{pmatrix} 1 & 2 & -1 & -2 \\ 3 & 4 & -3 & -4 \\ -2 & -4 & 2 & 4 \\ -6 & -8 & 6 & 8 \end{pmatrix}$$

Both the Hadamard product and Kronecker product are useful when solving general matrix equations, some of which result from the finite difference methods.

1.2.2.2 Unary Algebraic Operations

We first look at the set of unary operations applicable only to square matrices. The first set of unary operations to consider are highly related to each other. These operations are the determinant, cofactors, adjugates, and inverses. As before, we refer the reader to Section A.2 to see how these definitions naturally developed from the application to the solution of simultaneous linear algebraic equations.

Of these unary operations, the matrix inverse can easily be defined independent of computation.

Definition 1.2. *The* **matrix inverse** *of a square matrix A is a matrix of the same size, denoted by A^{-1}, that satisfies*

$$A^{-1}A = AA^{-1} = I \tag{1.8}$$

Unfortunately, except for some special classes of matrices, the evaluation of the inverse is not straightforward in general. Instead, the computation of matrix inverses requires the definition of three other operations: the determinant, the cofactor, and the adjugate.

First, we need another function called the **permutation sign function**.

[1] We suggest the use of the mnemonics **LR** and **RC** to stand for **L**eft operation acts on **R**ows and **R**ight operation acts on **C**olumns, respectively.

Definition 1.3. *Let $K = \{k_1, k_2, \ldots, k_N\}$ be a sequence of distinct indices ranging from 1 to N. Let $\pi(K)$ be the number of pairwise exchanges among the indices in the sequence K needed to reorder the sequence in K into an ascending order given by $\{1, 2, \ldots, N\}$. Then the* **permutation sign function**, *denoted by $\epsilon(K)$, is defined by*

$$\epsilon(K) = (-1)^{\pi(K)} \tag{1.9}$$

which means it takes on the value of $+1$ or -1, depending on whether $\pi(K)$ is even or odd, respectively.

For example, we have

$$\epsilon(1,2,3) = +1 \quad \epsilon(5,1,2,4,3) = -1 \quad \epsilon(2,1,3,4) = -1 \quad \epsilon(6,2,1,5,3,4) = +1$$

Definition 1.4. *The* **determinant** *of a square matrix A of size N, denoted by either $|A|$ or $\mathbf{det}(A)$, is given by*

$$\mathbf{det}(A) = \sum_{\substack{k_i \neq k_j \\ i,j=1,\ldots,N}} \epsilon\,(k_1, \ldots, k_N)\, a_{1,k_1} a_{2,k_2} \cdots a_{N,k_N} \tag{1.10}$$

where the summation is over all nonrepeated combinations of indices $1, 2, \ldots, N$.

Definition 1.5. *The* **cofactor** *of an element a_{ij} of a square matrix A of size N, denoted by $\mathbf{cof}\,(a_{ij})$, is defined as*

$$\mathbf{cof}\,(a_{ij}) = (-1)^{i+j}\,\mathbf{det}\,(A_{ij\downarrow}) \tag{1.11}$$

where $A_{ij\downarrow}$ is the $(ij)^{\text{th}}$ redact.

Using cofactors, we can compute the determinant in a recursive manner.

LEMMA 1.1. *Let A be a square matrix of size N, then $\mathbf{det}\,(A) = a_{11}$ if $N = 1$. Otherwise, for any j*

$$\mathbf{det}\,(A) = \sum_{k=1}^{N} a_{kj}\;\mathbf{cof}\,(a_{kj}) \tag{1.12}$$

Likewise, for any i

$$\mathbf{det}\,(A) = \sum_{k=1}^{N} a_{ik}\;\mathbf{cof}\,(a_{ik}) \tag{1.13}$$

PROOF. By induction, one can show that either the column expansion formula given in (1.12) or the row expansion formula given in (1.13) will yield the same result as given in (1.10).

We refer to A as **singular** if $\mathbf{det}(A) = 0$; otherwise, A is **nonsingular**. As we show next, only nonsingular matrices can have matrix inverses.

Definition 1.6. *The* **adjugate**[2] *of a square matrix A is a matrix of the same size, denoted by* **adj** (A), *consisting of the cofactors of each element in A but collected in a transposed arrangement, that is,*

$$\textbf{adj}\left(A\right) = \begin{pmatrix} \textbf{cof}\,(a_{11}) & \cdots & \textbf{cof}\,(a_{N1}) \\ \vdots & \ddots & \vdots \\ \textbf{cof}\,(a_{1N}) & \cdots & \textbf{cof}\,(a_{NN}) \end{pmatrix} \tag{1.14}$$

Using adjugates, we arrive at one key result for the computation of matrix inverses, if they exist.

LEMMA 1.2. *Let A be any square matrix, then*

$$A\,\textbf{adj}(A) = \left(\textbf{det}(A)\right)I \qquad \textit{and} \qquad \textbf{adj}(A)\,A = \left(\textbf{det}(A)\right)I \tag{1.15}$$

Assuming matrix A is nonsingular, the inverse is given by

$$A^{-1} = \frac{1}{\textbf{det}(A)}\,\textbf{adj}(A) \tag{1.16}$$

PROOF. (See section A.4.3)

Note that matrix adjugates always exist, whereas matrix inverses A^{-1} exist only if $\textbf{det}(A) \neq 0$.

EXAMPLE 1.1. Let

$$A = \begin{pmatrix} 1 & 2 & 3 \\ 4 & 5 & 6 \\ 7 & 8 & 0 \end{pmatrix}$$

then

$$\textbf{cof}(a_{11}) = +\begin{vmatrix} 5 & 6 \\ 8 & 0 \end{vmatrix}; \quad \textbf{cof}(a_{21}) = -\begin{vmatrix} 2 & 3 \\ 8 & 0 \end{vmatrix}; \quad \textbf{cof}(a_{31}) = +\begin{vmatrix} 2 & 3 \\ 5 & 6 \end{vmatrix}$$

$$\textbf{cof}(a_{12}) = -\begin{vmatrix} 4 & 6 \\ 7 & 0 \end{vmatrix}; \quad \textbf{cof}(a_{22}) = +\begin{vmatrix} 1 & 3 \\ 7 & 0 \end{vmatrix}; \quad \textbf{cof}(a_{32}) = -\begin{vmatrix} 1 & 3 \\ 4 & 6 \end{vmatrix}$$

$$\textbf{cof}(a_{13}) = +\begin{vmatrix} 4 & 5 \\ 7 & 8 \end{vmatrix}; \quad \textbf{cof}(a_{23}) = -\begin{vmatrix} 1 & 2 \\ 7 & 8 \end{vmatrix}; \quad \textbf{cof}(a_{33}) = +\begin{vmatrix} 1 & 2 \\ 4 & 5 \end{vmatrix}$$

then

$$\textbf{adj}(A) = \begin{pmatrix} -48 & 24 & -3 \\ 42 & -21 & 6 \\ -3 & 6 & -3 \end{pmatrix} \rightarrow \textbf{adj}(A)\,A = A\,\textbf{adj}(A) \begin{pmatrix} 27 & 0 & 0 \\ 0 & 27 & 0 \\ 0 & 0 & 27 \end{pmatrix}$$

[2] In other texts, the term "**adjoint**" is used instead of **adjugate**. We chose to use the latter, because the term "adjoint" is also used to refer to another matrix in linear operator theory.

and

$$A^{-1} = \frac{1}{27} \begin{pmatrix} -48 & 24 & -3 \\ 42 & -21 & 6 \\ -3 & 6 & -3 \end{pmatrix}$$

Although (1.16) is a general method for computing the inverse, there are more efficient ways to find the matrix inverse that take advantage of special structures and properties. For instance, the inverse of diagonal matrices is another diagonal matrix consisting of the reciprocals of the diagonal elements. Another example is when the transpose happens to also be its inverse. These matrices are known as **orthogonal** matrices. To determine whether a given matrix is indeed orthogonal, we can just compute A^TA and AA^T and check whether both products yield identity matrices.

The other unary operations include the trace, real component, imaginary component, and the complex conjugate operations. The **trace** of a square matrix A, denoted $\mathbf{tr}(A)$, is defined as the sum of the diagonals.

EXAMPLE 1.2. Let $A[=]2 \times 2$, then for $M = \lambda I - A$, where λ is a scalar parameter, we have the following results:

$$\begin{aligned} \mathbf{det}\,(\lambda I - A) &= \lambda^2 - \left(\mathbf{tr}\,(A)\right)\lambda + \mathbf{det}\left(A\right) \\ \mathbf{adj}\,(\lambda I - A) &= \begin{pmatrix} \lambda - a_{22} & -a_{12} \\ -a_{21} & \lambda - a_{11} \end{pmatrix} \\ (\lambda I - A)^{-1} &= \frac{1}{\lambda^2 - \left(\mathbf{tr}\,(A)\right)\lambda + \mathbf{det}\left(A\right)} \begin{pmatrix} \lambda - a_{22} & -a_{12} \\ -a_{21} & \lambda - a_{11} \end{pmatrix} \end{aligned}$$

Note that when $\mathbf{det}\,(\lambda I - A) = 0$, the inverse will no longer exist, but $\mathbf{adj}\,(\lambda I - A)$ will still be valid.

We now show some examples in which the matrices can be used to represent the indexed equations. The first example involves the matrix formulation of the finite difference approximation of a partial differential equation. The second involves the matrix formulation of a quadratic equation.

EXAMPLE 1.3. Consider the heat equation of a $L \times W$ flat rectangular plate given by

$$\frac{\partial T}{\partial t} = \alpha \left(\frac{\partial^2 T}{\partial x^2} + \frac{\partial^2 T}{\partial y^2} \right) \tag{1.17}$$

with stationary boundary conditions

$$T(0, y, t) = f_0(y) \qquad T(x, 0, t) = g_0(x)$$

$$T(L, y, t) = f_L(y) \qquad T(x, W, t) = g_W(x)$$

and initial condition, $T(x, y, 0) = h(x, y)$. We can introduce a uniform finite time increment Δt and finite differences for x and y given by $\Delta x = L/(N+1)$ and $\Delta y = W/(M+1)$, respectively, so that $t_k = k\Delta t$, $x_n = n\Delta x$, and $y_m = m\Delta y$,

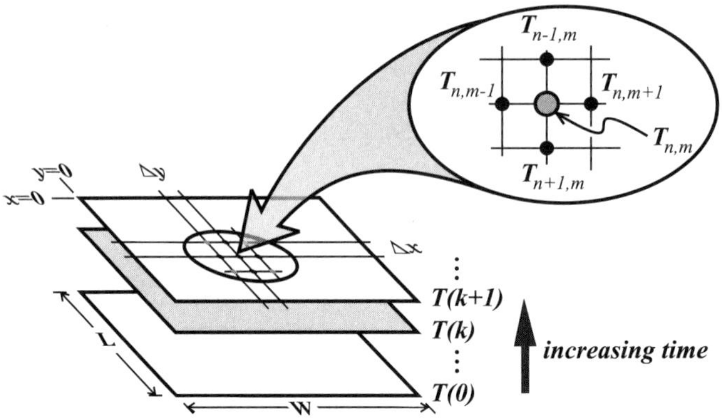

Figure 1.1. A schematic of the finite difference approximation of the temperature distribution T of a flat plate in Example 1.3.

with $k = 0, 1, \ldots$, $n = 0, \ldots, N+1$, and $m = 0, \ldots, M+1$. The points corresponding to $n = 0$, $n = N+1$, $m = 0$, and $m = M+1$ represent the boundary values. We can then let $[T(k)]$ be a $N \times M$ matrix that represents the temperature distribution of the specific internal points of the plate at time t_k (see Figure 1.1).

Using the finite difference approximation of the partial derivatives at point $x = n\Delta x$ and $y = m\Delta y$, and time $t = k\Delta t$:[3]

$$\begin{aligned}\frac{\partial T}{\partial t} &= \frac{T_{n,m}(k+1) - T_{n,m}(k)}{\Delta t}\\ \frac{\partial^2 T}{\partial x^2} &= \frac{T_{n+1,m}(k) - 2T_{n,m}(k) + T_{n-1,m}(k)}{\Delta x^2}\\ \frac{\partial^2 T}{\partial y^2} &= \frac{T_{n,m+1}(k) - 2T_{n,m}(k) + T_{n,m-1}(k)}{\Delta y^2}\end{aligned}$$

then (1.17) is approximated by the following indexed equations:

$$T_{n,m}(k+1) = \begin{array}{l}\mu_x\left(T_{n-1,m}(k) + \left(\dfrac{1}{2\mu_x} - 2\right)T_{n,m}(k) + T_{n+1,m}(k)\right)\\ +\,\mu_y\left(T_{n,m-1}(k) + \left(\dfrac{1}{2\mu_y} - 2\right)T_{n,m}(k) + T_{n,m+1}(k)\right)\end{array} \tag{1.18}$$

where

$$\mu_x = \frac{\alpha\Delta t}{(\Delta x)^2} \qquad ; \qquad \mu_y = \frac{\alpha\Delta t}{(\Delta y)^2}$$

$T_{n,m}(k)$ is the temperature at time $t = k\Delta t$ located at $(x, y) = (n\Delta x, m\Delta y)$.

The first group of terms in (1.18) involves only $T_{n-1,m}$, $T_{n,m}$, and $T_{n+1,m}$, that is, only a combination of row elements at fixed m. This means that the

[3] The finite difference methods are discussed in more detail in Chapter 13.

first group of terms can be described by the product AT for some constant $N \times N$ matrix A. Conversely, the second group of terms in (1.18) involves only acombination of column elements at fixed n, which means a product TB for some matrix $B[=]M \times M$. In anticipation of boundary conditions, we need an extra matrix $C[=]N \times M$. Thus we should be able to represent (1.18) using a matrix formulation given by

$$T(k+1) = AT(k) + T(k)B + C \tag{1.19}$$

where A and B and C are constant matrices.[4]

When formulating general matrix equations, it is often advisable to apply it to smaller matrices first. Thus let us start with a case in which $N = 4$ and $M = 3$. We can show that (1.18) can be represented by

$$\begin{aligned}\left[T(k+1)\right] &= \mu_x \begin{pmatrix} \beta_x & 1 & 0 & 0 \\ 1 & \beta_x & 1 & 0 \\ 0 & 1 & \beta_x & 1 \\ 0 & 0 & 1 & \beta_x \end{pmatrix} \begin{pmatrix} T_{11}(k) & T_{12}(k) & T_{13}(k) \\ T_{21}(k) & T_{22}(k) & T_{23}(k) \\ T_{31}(k) & T_{32}(k) & T_{33}(k) \\ T_{41}(k) & T_{42}(k) & T_{43}(k) \end{pmatrix} \\ &+ \mu_y \begin{pmatrix} T_{11}(k) & T_{12}(k) & T_{13}(k) \\ T_{21}(k) & T_{22}(k) & T_{23}(k) \\ T_{31}(k) & T_{32}(k) & T_{33}(k) \\ T_{41}(k) & T_{42}(k) & T_{43}(k) \end{pmatrix} \begin{pmatrix} \beta_y & 1 & 0 \\ 1 & \beta_y & 1 \\ 0 & 1 & \beta_y \end{pmatrix} \\ &+ \mu_x \begin{pmatrix} T_{01}(k) & T_{02}(k) & T_{03}(k) \\ 0 & 0 & 0 \\ 0 & 0 & 0 \\ T_{51}(k) & T_{52}(k) & T_{53}(k) \end{pmatrix} + \mu_y \begin{pmatrix} T_{10}(k) & 0 & T_{14}(k) \\ T_{20}(k) & 0 & T_{24}(k) \\ T_{30}(k) & 0 & T_{34}(k) \\ T_{40}(k) & 0 & T_{44}(k) \end{pmatrix}\end{aligned}$$

where $\beta_x = 1/(2\mu_x) - 2$ and $\beta_y = 1/(2\mu_y) - 2$. Generalizing, we have

$$A = \mu_x \begin{pmatrix} \beta_x & 1 & & 0 \\ 1 & \ddots & \ddots & \\ & \ddots & \beta_x & 1 \\ 0 & & 1 & \beta_x \end{pmatrix} [=] N \times N$$

$$B = \mu_y \begin{pmatrix} \beta_y & 1 & & 0 \\ 1 & \ddots & \ddots & \\ & \ddots & \beta_y & 1 \\ 0 & & 1 & \beta_y \end{pmatrix} [=] M \times M$$

$$C = \mu_x \begin{pmatrix} p_1 & \cdots & p_M \\ 0 & \cdots & 0 \\ & \vdots & \\ 0 & \cdots & 0 \\ q_1 & \cdots & q_M \end{pmatrix} + \mu_y \begin{pmatrix} r_1 & 0 & & 0 & s_1 \\ \vdots & \vdots & \cdots & \vdots & \vdots \\ r_N & 0 & & 0 & s_N \end{pmatrix}$$

[4] More generally, if the boundary conditions are time-varying, then $C = C(k)$. Also, if the coefficient $\alpha = \alpha(t)$, then A and B will need to be replaced by $A(k)$ and $B(k)$, respectively.

where $p_m = f_0(m\Delta y)$, $q_m = f_L(m\Delta y)$, $r_n = g_0(n\Delta x)$, and $s_n = g_W(n\Delta x)$. The initial matrix is obtained using the initial condition, that is, $T_{nm}(0) = h(n\Delta x, m\Delta y)$. Starting with $T(0)$, one can then march iteratively through time using (1.19). (A specific example is given in exercise **E1.21**.)

EXAMPLE 1.4. The general second-order polynomial equation in N variables is given by

$$\beta = \sum_{i=1}^{N}\sum_{j=1}^{N} a_{ij}x_ix_j$$

One could write this equation as

$$[\beta] = \mathbf{x}^T A\mathbf{x}$$

where

$$A = \begin{pmatrix} a_{11} & \dots & a_{1N} \\ \vdots & \ddots & \vdots \\ a_{N1} & \dots & a_{NN} \end{pmatrix} \quad \text{and} \quad \mathbf{x} = \begin{pmatrix} x_1 \\ \vdots \\ x_N \end{pmatrix}$$

Note that $[\beta]$ is a 1×1 matrix in this formulation. The right-hand side is known as the **quadratic form**. However, because $x_ix_j = x_jx_i$, three alternative forms are possible:

$$[\beta] = \mathbf{x}^T Q\mathbf{x} \qquad [\beta] = \mathbf{x}^T L\mathbf{x} \qquad \text{or} \qquad [\beta] = \mathbf{x}^T U\mathbf{x}$$

where

$$Q = \begin{pmatrix} q_{11} & \dots & q_{1N} \\ \vdots & \ddots & \vdots \\ q_{N1} & \dots & q_{NN} \end{pmatrix} \quad U = \begin{pmatrix} u_{11} & \dots & u_{1N} \\ & \ddots & \vdots \\ 0 & & u_{NN} \end{pmatrix} \quad L = \begin{pmatrix} \ell_{11} & & 0 \\ \vdots & \ddots & \\ \ell_{N1} & \dots & \ell_{NN} \end{pmatrix}$$

and

$$q_{ij} = \frac{a_{ij} + a_{ji}}{2}; \quad u_{ij} = \begin{cases} a_{ij} + a_{ji} & \text{if } i < j \\ a_{ii} & \text{if } i = j \\ 0 & \text{if } i > j \end{cases}; \quad \ell_{ij} = \begin{cases} a_{ij} + a_{ji} & \text{if } i > j \\ a_{ii} & \text{if } i = j \\ 0 & \text{if } i < j \end{cases}$$

(The proof that all three forms are equivalent is left as an exercise in **E1.34**.)

This example shows that more than one matrix formulation is possible in some cases. Matrix Q is symmetric, whereas L is lower triangular, and U is upper triangular. The most common formulation is to use the symmetric matrix Q.

1.3 Properties of Matrix Operations

In this section, we discuss the different properties of matrix operations. With these properties, one could manipulate matrix equations to either simplify equations, generate efficient algorithms, or analyze the problem, before actual matrix computations are implemented. We first discuss the basic properties involving addition, multiplications, and inverses. Next is a separate subsection on the properties of

Table 1.4. *Properties of matrix operations*

Commutative Operations	
$A \circ B = B \circ A$	$A + B = B + A$
$\alpha A = A\alpha$	$AA^{-1} = A^{-1}A$
Associativity of Sums and Products	
$A + (B + C) = (A + B) + C$	$A \circ (B \circ C) = (A \circ B) \circ C$
$A(BC) = (AB)C$	$A \otimes (B \otimes C) = (A \otimes B) \otimes C$
Distributivity of Products	
$A(B + C) = AB + AC$	$A \otimes (B + C) = A \otimes B + A \otimes C$
$(A + B)C = AC + BC$	$(A + B) \otimes C = A \otimes C + B \otimes C$
$A \circ (B + C) = A \circ B + A \circ C = B \circ A + C \circ A = (B + C) \circ A$	$(AB) \otimes (CD) = (A \otimes C)(B \otimes D)$
Transpose of Products	
$(AB)^T = B^T A^T$	$(A \circ B)^T = B^T \circ A^T = A^T \circ B^T$
$(A \otimes B)^T = A^T \otimes B^T$	
Inverse of Matrix Products and Kronecker Products	
$(AB)^{-1} = B^{-1}A^{-1}$	$(A \otimes B)^{-1} = (A)^{-1} \otimes (B)^{-1}$
Reversible Operations	
$\left(A^T\right)^T = A$	$\left(A^{-1}\right)^{-1} = A$
$(A^*)^* = A$	
Vectorization of Sums and Products	
$\operatorname{vec}(A + B) = \operatorname{vec}(A) + \operatorname{vec}(B)$	$\operatorname{vec}(BAC) = \left(C^T \otimes B\right) \operatorname{vec}(A)$
	$\operatorname{vec}(A \circ B) = \operatorname{vec}(A) \circ \operatorname{vec}(B)$

determinants. Finally, we include a subsection of the formulas that involve matrix inverses.

1.3.1 Basic Properties

A list of some basic properties of matrix operations is given in Table 1.4. Most of the properties can be derived by directly using the definitions given in Tables 1.1, 1.2, and 1.3. The proofs are given in Section A.4.1 as an appendix. The properties of the matrix operations allow for the manipulation of matrix equations before

Table 1.5. *Definition of vectors used in Example 1.5*

Vector	Description of elements
$\mathbf{x}$	x_k is the annual supply rate (kg/year) of material from source k.
$\mathbf{y}$	y_k is the annual production rate (kg/year) of product k.
$\mathbf{z}$	z_k is the sale price per kg of product k.
$\mathbf{w}$	w_k is the production cost per kg of the material from source k.

actual computations. They help in simplifying expressions that often yield important insights about the data or the system being investigated.

The first group of properties list the commutativity, associativity, and distributivity properties of various sums and products. One general rule is to choose associations of products that would improve computations. For instance, let **a**, **b**, **c**, **d**, **e**, and **f** be column vectors of the same length; we should use the following associations

$$\mathbf{a}\mathbf{b}^T\mathbf{c}\mathbf{d}^T\mathbf{e}\mathbf{f}^T = \mathbf{a}\left(\mathbf{b}^T\mathbf{c}\right)\left(\mathbf{d}^T\mathbf{e}\right)\mathbf{f}^T$$

because both $\left(\mathbf{b}^T\mathbf{c}\right)$ and $\left(\mathbf{d}^T\mathbf{e}\right)$ are 1×1. A similar rule holds for using the distributive properties. For example, we can use distributivity to rearrange the following equation:

$$AD + ABCD = A(D + BCD) = A(I + BC)D$$

More importantly, these properties allow for manipulations of matrix equations to help simplify the equations, as shown in the example that follows.

EXAMPLE 1.5. Consider a processing facility that can take raw material from M different sources to produce N different products. The yield of product j per kilogram of material coming from source i can be collected in matrix form as $F = (f_{ij})$. In addition, define the cost, price, supply rates, and production rates by the column vectors given in Table 1.5. We simplify the situation by assuming that all the products are sold immediately after production without need for inventory. Let S, C, and $P = (S - C)$ be the annual sale, annual cost, and annual net profit, respectively. We want to obtain a vector $\mathbf{g}$ where the kth element is the annual net profit per kilogram of material from source k, that is, $P = \mathbf{g}^T\mathbf{x}$.

Using matrix representation, we have

$$\begin{aligned} \mathbf{y} &= F\mathbf{x} \\ S &= \mathbf{z}^T\mathbf{y} \\ C &= \mathbf{w}^T\mathbf{x} \end{aligned}$$

then the net profit can be represented by

$$P = S - C = \mathbf{z}^T F\mathbf{x} - \mathbf{w}^T\mathbf{x} = \left(\mathbf{z}^T F - \mathbf{w}^T\right)\mathbf{x} = \mathbf{g}^T\mathbf{x}$$

where $\mathbf{g}$ is given by

$$\mathbf{g} = F^T\mathbf{z} - \mathbf{w}$$

More generally, the problem of maximizing net profit by adjusting the supply rates are formulated as a typical **linear programming** problem:

$$\max_{\mathbf{x}} \quad \mathbf{g}^T\mathbf{x} \qquad \text{(objective function)}$$

subject to

$$\begin{aligned} 0 \leq & \quad \mathbf{x} & \leq \mathbf{x}_{\max} & \qquad \text{(availability constraints)} \\ \mathbf{y}_{\min} \leq & \quad \mathbf{y}(= F\mathbf{x}) & \leq \mathbf{y}_{\max} & \qquad \text{(demand constraints)} \end{aligned}$$

The transposes of matrix products turn out to be equal to the matrix products of the transposes but in the reversed sequence. Together with the associative property, this can be extended to the following results:

$$\begin{aligned} (ABC\cdots EFG)^T &= G^T F^T E^T \cdots C^T D^T A^T \\ \left(A^k\right)^T &= \left(A^T\right)^k \\ \left(AA^{-1}\right)^T &= \left(A^{-1}\right)^T A^T = I = \left(A^T\right)^{-1} A^T \end{aligned}$$

The last result shows that $\left(A^T\right)^{-1} = \left(A^{-1}\right)^T$. Thus we often use the shorthand A^{-T} to mean either $\left(A^T\right)^{-1}$ or $\left(A^{-1}\right)^T$.

Similarly, the inverse of a matrix product is a product of the matrix inverses in the reverse sequence. This can be generalized to be[5]

$$\begin{aligned} (ABC\cdots)^{-1} &= \cdots C^{-1}B^{-1}A^{-1} \\ \left(A^k\right)^{-1} &= A^{-1}\cdots A^{-1} = \left(A^{-1}\right)^k \\ A^k A^\ell &= A^{k+\ell} \end{aligned}$$

Thus we can use A^{-k} to denote either $\left(A^k\right)^{-1}$ or $\left(A^{-1}\right)^k$. Note that these results are still consistent with $A^0 = I$.

EXAMPLE 1.6. Consider a resistive electrical network consisting of junction points or **nodes** that are connected to each other by **links** where the links contain three types of electrical components: resistors, current sources, and voltage sources. We simplify our network to contain only two types of links. One type of link contains only either one resistor or one voltage source, or both connected in series.[6]

[5] Note that this is not the case for Kronecker products that is,

$$(A \otimes B \otimes C \otimes \cdots)^{-1} = A^{-1} \otimes B^{-1} \otimes C^{-1} \otimes \cdots$$

[6] If multiple resistors with resistance $R_{j1}, R_{j2}, \ldots$ are connected in series the jth link, then they can be replaced by one resistor with resistance $R_j = \sum_k R_{jk}$. Likewise, if multiple voltage sources with signed voltages $s_{j1}, s_{j2}, \ldots$ are connected in series the jth link, then they can be replaced by one voltage source with signed voltage $s_j = \sum_k s_{jk}$, where the sign is positive if the polarity goes from positive to negative along the current flow.

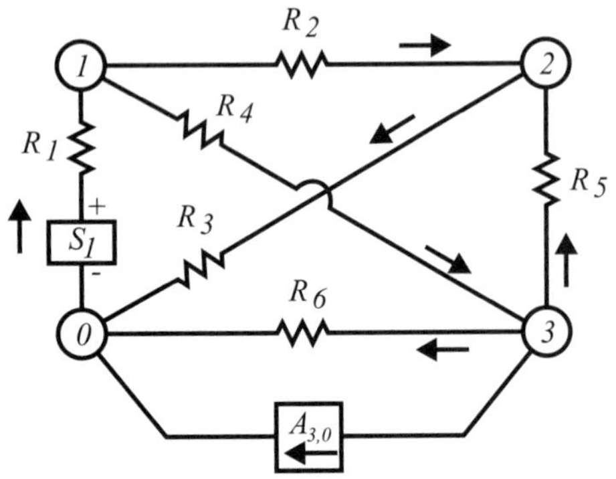

Figure 1.2. An electrical network with resistors R_j in link j, voltage sources s_j in link j, and current sources $A_{k,\ell}$ from node k to node ℓ.

The other type of link contains only a current source. One such network is shown in Figure 1.2.

Suppose there are $n+1$ nodes and $m \geq (n+1)$ links. By setting one of the nodes as having zero potential (the ground node), we want to determine the potentials of the remaining n nodes as well as the current flowing through each link and the voltages across each of the resistors. To obtain the required equations, we need to first propose the directions of each link, select the ground node (node 0), and label the remaining nodes (nodes 1 to n). Based on the choices of current flow and node labels, we can form the **node-link incidence matrix** $\Psi[=]n \times m$, which is a matrix composed of only 0, 1, and -1. The ith row of Ψ refers to the ith node, whereas the jth column refers to the jth link. Note that the links containing only current sources are not included during the formulation of incidence matrix. (Instead, these links are involved only during the implementation of Kirchhoff's current laws.) We set $\psi_{ij} = 1$ if the current is flowing into node i along the jth link, and $\psi_{ij} = -1$ if the current is flowing out of node i along the jth link. For the network shown in Figure 1.2, the incidence matrix is given by

$$\Psi = \begin{pmatrix} +1 & -1 & 0 & -1 & 0 & 0 \\ 0 & +1 & -1 & 0 & +1 & 0 \\ 0 & 0 & 0 & +1 & -1 & -1 \end{pmatrix}$$

Let p_i be the potential of node i with respect to ground and let e_j be the potential difference along link j between nodes k and ℓ, that is, where $\psi_{kj} \neq 0$ and $\psi_{\ell j} \neq 0$. Because the current flows from high to low potential,

$$\mathbf{e} = -\Psi^T \mathbf{p}$$

If the jth link contains a voltage source s_j, we assign a positive value if the polarity is from positive to negative along the chosen direction of the current flow. Let v_j be the voltage across the jth resistor, then

$$\mathbf{e} = \mathbf{v} + \mathbf{s}$$

Ohm's law states that the voltage across the jth resistor is given by $v_j = i_j R_j$, where i_j and R_j are the current and resistance in the jth link. In matrix form, we have

$$\mathbf{v} = R\mathbf{i} \qquad \text{where} \qquad R = \begin{pmatrix} R_1 & & 0 \\ & \ddots & \\ 0 & & R_m \end{pmatrix}$$

Let the current sources flowing out of the ith node be given by A_{ij}, whereas those flowing into the ith node are given by $A_{\ell i}$. Then the net current inflow at node i due only to current sources will be

$$b_i = \sum_{\ell} A_{\ell i} - \sum_{j} A_{ij}$$

Kirchhoff's current law states that the net flow of current at the ith node is zero. Thus we have

$$\Psi \mathbf{i} + \mathbf{b} = 0$$

In summary, for a given set of resistance, voltage sources, and current sources, we have enough information to find the potentials at each node, the voltage across each resistor, and the current flows along the links based on the chosen ground point and proposed current flows. To solve for the node potentials, we have

$$\begin{aligned}
\mathbf{e} &= \mathbf{v} + \mathbf{s} \\
-\Psi^T \mathbf{p} &= R\mathbf{i} + \mathbf{s} \\
-R^{-1}\Psi^T \mathbf{p} - R^{-1}\mathbf{s} &= \mathbf{i} \\
-\Psi R^{-1}\Psi^T \mathbf{p} - \Psi R^{-1}\mathbf{s} &= \Psi \mathbf{i} = -\mathbf{b} \\
\left(\Psi R^{-1}\Psi^T\right)\mathbf{p} &= \left(\mathbf{b} - \Psi R^{-1}\mathbf{s}\right) \\
\mathbf{p} &= \left(\Psi R^{-1}\Psi^T\right)^{-1}\left(\mathbf{b} - \Psi R^{-1}\mathbf{s}\right) \qquad (1.20)
\end{aligned}$$

Using the values of $\mathbf{p}$, we could find the voltages across the resistors,

$$\mathbf{v} = -\Psi^T \mathbf{p} - \mathbf{s} \qquad (1.21)$$

And finally, for the current flows,

$$\mathbf{i} = R^{-1}\mathbf{v} \qquad (1.22)$$

For the network shown in Figure 1.2, suppose the values for the resistors, voltage source, and current source are given by: $\{R_1, R_2, R_3, R_4, R_5, R_6\} = \{1\ \Omega,\ 2\ \Omega,\ 3\ \Omega,\ 0.5\ \Omega,\ 0.8\ \Omega,\ 10\ \Omega\}$, $S_1 = 1$ v and $A_{3,0} = 0.2$ amp. Then the solution using equations (1.20) to (1.22) yields:

$$\mathbf{p} = \begin{pmatrix} 0.6118 \\ 0.4254 \\ 0.4643 \end{pmatrix} \text{volts,} \quad \mathbf{v} = \begin{pmatrix} 0.3882 \\ 0.1864 \\ 0.4254 \\ 0.1475 \\ 0.0389 \\ 0.4643 \end{pmatrix} \text{volts, and } \mathbf{i} = \begin{pmatrix} 0.3882 \\ 0.0932 \\ 0.1418 \\ 0.2950 \\ 0.0486 \\ 0.0464 \end{pmatrix} \text{amps}$$

Remarks:

1. R^{-1} is just a diagonal matrix containing the reciprocals of the diagonal elements of R.
2. $\left(\Psi R^{-1}\Psi^T\right)$ is an $n \times n$ symmetric matrix, and its inverse is needed in equation (1.20). If n is large, it is often more efficient to approach the same

problem using the numerical techniques that are covered in the next chapter, such as the conjugate gradient method.

The last group of properties given in Table 1.3 involves the relationship between vectorization, matrix products, and Kronecker products. These properties are very useful in reformulating matrix equations in which the unknown matrices X do not exclusively appear on the right or left position of the products in the equation. For example, a form known as **Sylvester matrix equation**, which often results from control theory as well as in finite difference solutions, is given by

$$QX + XR = C \tag{1.23}$$

where $Q[=]N \times N$, $R[=]M \times M$ are $C[=]N \times M$ are constant matrices, whereas $X[=]N \times M$ is the unknown matrix. After inserting appropriate identity matrices, the properties can be used to obtain the following result:

$$\begin{aligned}
\text{vec}\left(QXI_M + I_N XR\right) &= \text{vec}\,(C) \\
\text{vec}\left(QXI_M\right) + \text{vec}\left(I_N XR\right) &= \\
\left(I_M^T \otimes Q\right)\text{vec}\,(X) + \left(R^T \otimes I_N\right)\text{vec}\,(X) &= \\
\left(I_M^T \otimes Q + R^T \otimes I_N\right)\text{vec}\,(X) &= \text{vec}\,(C)
\end{aligned}$$

By setting $A = I_M^T \otimes Q + R^T \otimes I_N$, $\mathbf{x} = \text{vec}\,(X)$, and $\mathbf{b} = \text{vec}\,(C)$, the problem can be recast as $A\mathbf{x} = \mathbf{b}$.

EXAMPLE 1.7. In example 1.3, the finite difference equation resulted in the matrix equation given by

$$T\,(k+1) \;=\; AT(k) \;+\; T(k)B \;+\; C$$

where $A[=]N \times N$, $B[=]M \times M$, $C[=]N \times M$, and $T[=]N \times M$. At equilibrium, $T(k+1) = T(k) = T_{\text{eq}}$, a constant matrix. Thus the matrix equation becomes

$$T_{\text{eq}} = AT_{\text{eq}} + T_{\text{eq}}B + C$$

Using the vectorization properties in Table 1.4, we obtain

$$\text{vec}\left(T_{\text{eq}}\right) = \left(I_{[M]} \otimes A\right)\text{vec}\left(T_{\text{eq}}\right) + \left(B^T \otimes I_{[N]}\right)\text{vec}\left(T_{\text{eq}}\right) + \text{vec}\,(C)$$

or

$$K\mathbf{x} = \mathbf{b} \qquad \longrightarrow \qquad \mathbf{x} = K^{-1}\mathbf{b}$$

where

$$\begin{aligned}
K &= I_{[NM]} - \left(I_{[M]} \otimes A\right) - \left(B^T \otimes I_{[N]}\right) \\
\mathbf{x} &= \text{vec}\left(T_{\text{eq}}\right) \\
\mathbf{b} &= \text{vec}\,(C)
\end{aligned}$$

After solving for $\mathbf{x}$, T_{eq} can be recovered by using the reshape operator, that is, $T_{\text{eq}} = \text{reshape}\,(\mathbf{x}, N, M)$.

Table 1.6. *Properties of determinants*

1	Determinant of Products	$\mathbf{det}\left(AB\right) = \mathbf{det}\left(A\right)\mathbf{det}\left(B\right)$
2	Determinant of Triangular Matrices	$\mathbf{det}\left(A\right) = \prod_{i=1}^{N} a_{ii}$
3	Determinant of Transpose	$\mathbf{det}\left(A^T\right) = \mathbf{det}\left(A\right)$
4	Determinant of Inverses	$\mathbf{det}\left(A^{-1}\right) = \mathbf{det}\left(A\right)^{-1}$
5[†]	Let B contain permuted columns of A based on sequence K	$\mathbf{det}\left(B\right) = \epsilon(K)\mathbf{det}\left(A\right)$ where $\epsilon(K)$ is the permutation sign function
6[†]	Scaled Columns: $B = \left(\begin{array}{c\|c\|c} \beta_1\, a_{11} & & \beta_N\, a_{1N} \\ \vdots & \cdots & \vdots \\ \beta_1\, a_{N1} & & \beta_N\, a_{NN} \end{array}\right)$	$\mathbf{det}\left(B\right) = \left(\prod_{j=1}^{N} \beta_j\right)\mathbf{det}\left(A\right)$
7[†]	Multilinearity	$\left(\begin{array}{c\|c\|c\|c\|c} a_{11} & & x_1 + y_1 & & a_{1N} \\ \vdots & \cdots & \vdots & \cdots & \vdots \\ a_{N1} & & x_N + y_n & & a_{NN} \end{array}\right) = \left(\begin{array}{c\|c\|c\|c\|c} a_{11} & & x_1 & & a_{1N} \\ \vdots & \cdots & \vdots & \cdots & \vdots \\ a_{N1} & & x_N & & a_{NN} \end{array}\right) + \left(\begin{array}{c\|c\|c\|c\|c} a_{11} & & y_1 & & a_{1N} \\ \vdots & \cdots & \vdots & \cdots & \vdots \\ a_{N1} & & y_n & & a_{NN} \end{array}\right)$
8[†]	Linearly Dependent Columns	$\mathbf{det}\left(A\right) = 0$ if for some $\gamma_k \neq 0$, $\sum_{j=1}^{N} \gamma_i A_{\bullet,j} = 0$

[†] Using item 3 (i.e., that the transpose operation does not alter the determinant), a dual set of properties exists for items 5 to 8, in which the columns are replaced by rows.

1.3.2 Properties of Determinants

Because the determinant is a very important matrix operation, we devote a separate table for the properties of determinants. A summary of the properties of determinants is given in Table 1.6. The proofs for these properties are given in Section A.4.2 as an appendix.

Note that even though A and B may not commute, the determinants of both AB and BA are the same, that is,

$$\mathbf{det}\left(AB\right) = \mathbf{det}\left(A\right)\mathbf{det}\left(B\right) = \mathbf{det}\left(B\right)\mathbf{det}\left(A\right) = \mathbf{det}\left(BA\right)$$

Several properties of determinants help to improve computational efficiency. For instance, the fact that the determinant of a triangular or diagonal matrix is just the product of the diagonal means that there is tremendous advantage to finding multiplicative factors that could "diagonalize" or "triagularize" the original matrix. Later, in Chapter 3, we try to find such a nonsingular T whose effect would be to make $C = T^{-1}AT$ diagonal or triangular. Yet C and A will have the same determinant, that is,

$$\mathbf{det}\left(T^{-1}AT\right) = \mathbf{det}\left(T^{-1}\right)\mathbf{det}\left(A\right)\mathbf{det}\left(T\right) = \frac{1}{\mathbf{det}\left(T\right)}\mathbf{det}\left(A\right)\mathbf{det}\left(T\right) = \mathbf{det}\left(A\right)$$

The last property in the list is one of the key application of determinants in linear algebra. It states that if the columns of a matrix are **linearly dependent** (defined next), then the determinant is zero.

Definition 1.7. *Vectors* $\{\mathbf{v}_1, \mathbf{v}_2, \ldots, \mathbf{v}_N\}$ *are* **linearly dependent** *if*

$$\sum_{j=1}^{N} \gamma_i \mathbf{v}_i = 0 \tag{1.24}$$

for some $\gamma_k \neq 0$

This means that if $\{\mathbf{v}_1, \ldots, \mathbf{v}_N\}$ is a linearly dependent set of vectors, then any of the vectors in the set can be represented as a linear combination of the other $(N-1)$ vectors. For instance, let

$$\mathbf{v}_1 = \begin{pmatrix} 1 \\ 1 \\ 1 \end{pmatrix} \quad \mathbf{v}_2 = \begin{pmatrix} 1 \\ 2 \\ 1 \end{pmatrix} \quad \mathbf{v}_3 = \begin{pmatrix} 0 \\ 1 \\ 0 \end{pmatrix}$$

We can compute the determinant of $V = \left(\mathbf{v}_1 \mid \mathbf{v}_2 \mid \mathbf{v}_3 \right) = 0$ and conclude immediately that the columns are dependent. In fact, we check easily that $\mathbf{v}_1 = \mathbf{v}_2 - \mathbf{v}_3$, $\mathbf{v}_2 = \mathbf{v}_1 + \mathbf{v}_3$, or $\mathbf{v}_3 = \mathbf{v}_2 - \mathbf{v}_1$.

EXAMPLE 1.8. Let a tetrahedron be described by four vertices in 3D space given by $\mathbf{p}_1$, $\mathbf{p}_2$, $\mathbf{p}_3$, and $\mathbf{p}_4$, as shown in Figure 1.3. Let $\mathbf{v}_1 = \mathbf{p}_2 - \mathbf{p}_1$, $\mathbf{v}_2 = \mathbf{p}_3 - \mathbf{p}_1$, and $\mathbf{v}_3 = \mathbf{p}_4 - \mathbf{p}_1$ form a 3×3 matrix

$$V = \left(\mathbf{v}_1 \;\middle|\; \mathbf{v}_2 \;\middle|\; \mathbf{v}_3 \right)$$

It can be shown using the techniques given in Section 4.1, together with Section 4.2, that the volume of the tetrahedron can be found by the determinant formula:

$$\text{Volume} = \frac{1}{6}\,\mathbf{abs}\left(\mathbf{det}\left(V\right)\right)$$

For instance, let

$$\mathbf{p}_1 = \begin{pmatrix} 1 \\ 0 \\ 0 \end{pmatrix} \quad \mathbf{p}_2 = \begin{pmatrix} 1 \\ 1 \\ 0 \end{pmatrix} \quad \mathbf{p}_3 = \begin{pmatrix} 1 \\ 1 \\ 1 \end{pmatrix} \quad \mathbf{p}_4 = \begin{pmatrix} 0 \\ 1 \\ 1 \end{pmatrix}$$

Figure 1.3. A tetrahedron described by four points: $\mathbf{p}_1$, $\mathbf{p}_2$, $\mathbf{p}_3$, and $\mathbf{p}_4$.

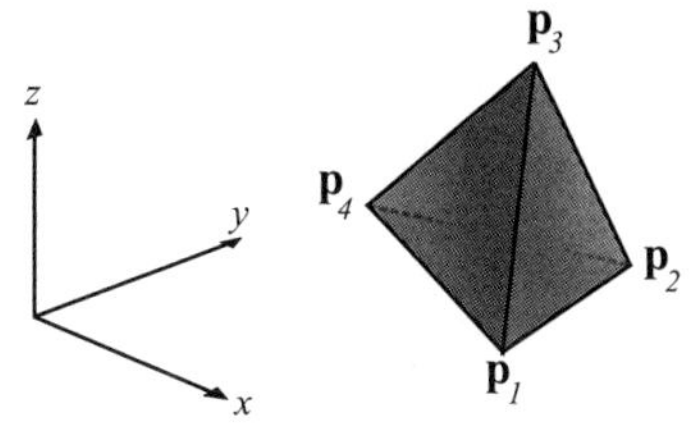

then the tetrahedron formed by vertices $\mathbf{p}_1$, $\mathbf{p}_2$, $\mathbf{p}_3$, and $\mathbf{p}_4$ yields

$$V = \left(\begin{array}{c|c|c} 0 & 0 & -1 \\ 1 & 1 & 1 \\ 0 & 1 & 1 \end{array} \right) \quad \rightarrow \quad \text{Volume} = \frac{1}{6}$$

If instead of $\mathbf{p}_4$, we have by $\widehat{\mathbf{p}}_4 = \begin{pmatrix} 1 & 0 & 1 \end{pmatrix}^T$,

$$V = \left(\begin{array}{c|c|c} 0 & 0 & 0 \\ 1 & 1 & 0 \\ 0 & 1 & 1 \end{array} \right) \quad \rightarrow \quad \text{Volume} = 0$$

which means $\mathbf{p}_1$, $\mathbf{p}_2$, $\mathbf{p}_3$, and $\widehat{\mathbf{p}}_4$ are coplanar, with $\mathbf{v}_1 = \mathbf{v}_2 - \mathbf{v}_3$.

1.3.3 Matrix Inverse Formulas

In this section, we include some of the formulas for the inverses of matrices and two important results: the matrix inversion lemma (known as the Woodbury formula) and Cramer's rule.

We start with the inverse of a diagonal matrix. The inverse of a diagonal matrix D is another diagonal matrix containing the reciprocals of the corresponding diagonal elements d_i, that is,

$$D^{-1} = \begin{pmatrix} d_1 & & 0 \\ & \ddots & \\ 0 & & d_N \end{pmatrix}^{-1} = \begin{pmatrix} d_1^{-1} & & 0 \\ & \ddots & \\ 0 & & d_N^{-1} \end{pmatrix} \tag{1.25}$$

Direct calculations can be used to show that $DD^{-1} = D^{-1}D = I$.

Next, we have a formula for the inverse of a triangular matrix T of size N.

LEMMA 1.3. *For a triangular matrix $T[=]N \times N$, let D be a diagonal matrix such that $d_{ii} = t_{ii}$ and matrix $K = D - T$. Then,*

$$T^{-1} = D^{-1}\left(I + \sum_{\ell=1}^{N-1} \left(KD^{-1}\right)^{\ell}\right) \tag{1.26}$$

PROOF. Multiply (1.26) by $T = D - K$ and expand,

$$TT^{-1} = (D - K)\, D^{-1}\left(I + \sum_{\ell=1}^{N-1} \left(KD^{-1}\right)\right) = I - \left(KD^{-1}\right)^{N}$$

but KD^{-1} is a strictly triangular matrix that is **nilpotent** matrix of degree $(N-1)$ (see exercise **E1.9**), that is, $\left(KD^{-1}\right)^{N} = 0$.

Next, we discuss an important result in matrix theory known as the **matrix inversion lemma**, also known as the **Woodbury matrix formula**.

LEMMA 1.4. *Let A, C, and $M = C^{-1} + DA^{-1}B$ be nonsingular, then*

$$(A + BCD)^{-1} = A^{-1} - A^{-1}B\left[C^{-1} + DA^{-1}B\right]^{-1} DA^{-1} \tag{1.27}$$

PROOF. With $M = C^{-1} + DA^{-1}B$, let Q be the right hand side of (1.27), that is,

$$Q = A^{-1} - A^{-1}BM^{-1}DA^{-1} \tag{1.28}$$

Then,

$$\begin{aligned}
(A + BCD)\,Q &= (AQ) + (BCDQ) \\
&= \left(AA^{-1} - AA^{-1}BM^{-1}DA^{-1}\right) \\
&\qquad + \left(BCDA^{-1} - BCDA^{-1}BM^{-1}DA^{-1}\right) \\
&= I + BCDA^{-1} - B\left(I + CDA^{-1}B\right)M^{-1}DA^{-1} \\
&= I + BCDA^{-1} - B\left(CC^{-1} + CDA^{-1}B\right)M^{-1}DA^{-1} \\
&= I + BCDA^{-1} - BC\left(C^{-1} + DA^{-1}B\right)M^{-1}DA^{-1} \\
&= I + BCDA^{-1} - BCMM^{-1}DA^{-1} \\
&= I + BCDA^{-1} - BCDA^{-1} \\
&= I
\end{aligned}$$

In a similar fashion, one can also show that $Q(A + BCD) = I$.

Remark: The matrix inversion lemma given by (1.27) is usually applied in cases in which the inverse of A is already known and the size of C is significantly smaller than A.

EXAMPLE 1.9. Let

$$G = \begin{pmatrix} 1 & 0 & 2 \\ 2 & 2 & 0 \\ -1 & -1 & 3 \end{pmatrix} = T + \mathbf{w}\mathbf{v}^T$$

where

$$T = \begin{pmatrix} 1 & 0 & 0 \\ 2 & 2 & 0 \\ -1 & -1 & 3 \end{pmatrix} \quad ; \quad \mathbf{w} = \begin{pmatrix} 1 \\ 0 \\ 0 \end{pmatrix} \quad \text{and} \quad \mathbf{v}^T = \begin{pmatrix} 0 & 0 & 2 \end{pmatrix}$$

this means we split G into a triangular matrix T and a product of a column vector $\mathbf{w}$ and row vector $\mathbf{v}^T$. We can use lemma 1.3 to find

$$T^{-1} = \begin{pmatrix} 1 & 0 & 0 \\ -1 & 1/2 & 0 \\ 0 & 1/6 & 1/3 \end{pmatrix}$$

Then with (1.27),

$$G^{-1} = \left(T^{-1} + \mathbf{w}\,[1]\,\mathbf{v}^T\right)^{-1} = T^{-1} - T^{-1}\mathbf{w}\left(1 + \mathbf{v}^T T^{-1}\mathbf{w}\right)^{-1}\mathbf{v}^T T^{-1}$$

$$= \begin{pmatrix} 1 & -1/3 & -2/3 \\ -1 & 5/6 & 2/3 \\ 0 & 1/6 & 1/3 \end{pmatrix}$$

where we took advantage of the fact that $\left(1 + \mathbf{v}^T T^{-1}\mathbf{w}\right)[=]1 \times 1$.

We complete this subsection with the discussion of a technique used in solving a subset of $A\mathbf{x} = \mathbf{b}$. Suppose we want to solve for only one of the unknowns, for example, the kth element of $\mathbf{x}$, for a given linear equation $A\mathbf{x} = \mathbf{b}$. One could extract the kth element of $\mathbf{x} = A^{-1}\mathbf{b}$, but this involves the evaluation of A^{-1}, which can be computationally expensive. As it turns out, finding the inverse is unnecessary if only one unknown is needed, by using **Cramer's rule**, as given by the following lemma:

LEMMA 1.5. *Let $A[=]N \times N$ be nonsingular, then*

$$\mathbf{x}_k = \frac{\mathbf{det}\left(A_{[k,\mathbf{b}]}\right)}{\mathbf{det}\,(A)} \tag{1.29}$$

where $A_{[k,\mathbf{b}]}$ is obtained A by replacing the kth column with $\mathbf{b}$.

PROOF. Using (1.16), $\mathbf{x} = A^{-1}\mathbf{b}$ can then be written as

$$\begin{pmatrix} x_1 \\ \vdots \\ x_N \end{pmatrix} = \frac{1}{\mathbf{det}(A)} \begin{pmatrix} \mathbf{cof}(a_{11}) & \cdots & \mathbf{cof}(a_{N1}) \\ \vdots & \ddots & \vdots \\ \mathbf{cof}(a_{1N}) & \cdots & \mathbf{cof}(a_{NN}) \end{pmatrix} \begin{pmatrix} b_1 \\ b_2 \\ \vdots \\ b_N \end{pmatrix}$$

or for the kth element,

$$x_k = \frac{\sum_{j=1}^{n} b_j\,\mathbf{cof}(a_{kj})}{\mathbf{det}(A)}$$

The numerator is just the determinant of a matrix, $A_{[k,\mathbf{b}]}$, which is obtained from A, with the kth column replaced by $\mathbf{b}$.

EXAMPLE 1.10. Let

$$A = \begin{pmatrix} 1 & 0 & 2 \\ 2 & 2 & 0 \\ -1 & -1 & 3 \end{pmatrix} \quad \text{and} \quad \mathbf{b} = \begin{pmatrix} 2 \\ 3 \\ 2 \end{pmatrix}$$

Then for $A\mathbf{x} = \mathbf{b}$, the value of x_2 can be found immediately using Cramer's rule,

$$x_2 = \frac{\mathbf{det}\left(A_{[2,\mathbf{b}]}\right)}{\mathbf{det}\,(A)} = \frac{\begin{vmatrix} 1 & 2 & 2 \\ 2 & 3 & 0 \\ -1 & 2 & 3 \end{vmatrix}}{\begin{vmatrix} 1 & 0 & 2 \\ 2 & 2 & 0 \\ -1 & -1 & 3 \end{vmatrix}} = \frac{11}{6}$$

1.4 Block Matrix Operations

A set of operations called **block matrix operations** (also known as **partitioned matrix operations**) takes advantage of special submatrix structures. The block operations are given as follows:

$$\left(\begin{array}{c|c} A & B \\ \hline C & D \end{array}\right)\left(\begin{array}{c|c} E & F \\ \hline G & H \end{array}\right) = \left(\begin{array}{c|c} AE+BG & AF+BH \\ \hline CE+DG & CF+DH \end{array}\right) \tag{1.30}$$

$$\textbf{det}\left(\begin{array}{c|c} A & 0 \\ \hline C & D \end{array}\right) = \textbf{det}\left(A\right)\textbf{det}\left(D\right) \tag{1.31}$$

$$\textbf{det}\left(\begin{array}{c|c} A & B \\ \hline C & D \end{array}\right) = \textbf{det}(A)\,\textbf{det}\left(D - CA^{-1}B\right) \quad \text{; if } A^{-1} \text{ exists} \tag{1.32}$$

$$= \textbf{det}\left(D\right)\textbf{det}\left(A - BD^{-1}C\right) \quad \text{; if } D^{-1} \text{ exists} \tag{1.33}$$

$$\left(\begin{array}{c|c} A & B \\ \hline C & D \end{array}\right)^{-1} = \left(\begin{array}{c|c} W & X \\ \hline Y & Z \end{array}\right) \tag{1.34}$$

where, W, X, Y, and Z depend on the two possible cases:

Case 1: A and $\Gamma = D - CA^{-1}B$ are nonsingular, then

$$\begin{aligned} Z &= \Gamma^{-1} \\ Y &= -\Gamma^{-1}CA^{-1} = -ZCA^{-1} \\ X &= -A^{-1}B\Gamma^{-1} = -A^{-1}BZ \\ W &= A^{-1}(I + B\Gamma^{-1}CA^{-1}) = A^{-1}(I - BY) = (I - XC)A^{-1} \end{aligned} \tag{1.35}$$

Case 2: D and $\widehat{\Gamma} = A - BD^{-1}C$ are nonsingular, then

$$\begin{aligned} W &= \widehat{\Gamma}^{-1} \\ X &= -\widehat{\Gamma}^{-1}BD^{-1} = -WBD^{-1} \\ Y &= -D^{-1}C\widehat{\Gamma}^{-1} = -D^{-1}CW \\ Z &= D^{-1}(I + C\widehat{\Gamma}^{-1}BD^{-1}) = D^{-1}(I - CX) = (I - YB)D^{-1} \end{aligned} \tag{1.36}$$

The proofs of (1.30) through (1.36) are given in Section A.4.5. The matrices $\Gamma = D - CA^{-1}B$ and $\widehat{\Gamma} = A - BD^{-1}C$ are known as the **Schur complements** of A and D, respectively.

EXAMPLE 1.11. Consider the open-loop process structure consisting of R process units as shown in Figure 1.4. The local state vector for process unit i is given by $\mathbf{x}_i = \left(x_{1i}, \quad \cdots, \quad x_{Ni}\right)^T$. For instance, x_{ki} could stand for the kth species in process unit i. The interaction among the process units is given by

$$\begin{aligned} A_1\mathbf{x}_1 + B_1\mathbf{x}_2 &= \mathbf{p}_1 \\ C_{i-1}\mathbf{x}_{i-1} + A_i\mathbf{x}_i + B_i\mathbf{x}_{i+1} &= \mathbf{p}_i \qquad \text{if } 1 < i < R \\ C_{R-1}\mathbf{x}_{R-1} + A_R\mathbf{x}_R &= \mathbf{p}_R \end{aligned}$$

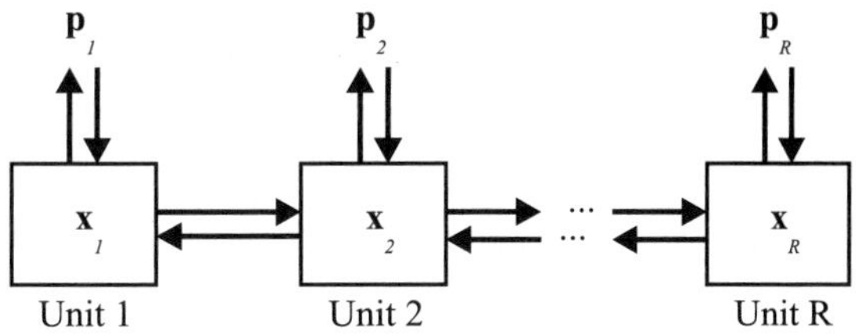

Figure 1.4. An open-loop system of R process units.

where A_i, B_i, $C_i[=]N \times N$ and $\mathbf{p}_i[=]N \times 1$. A block matrix description is

$$\mathbf{F}\mathbf{x} = \mathbf{p}$$

where

$$\mathbf{F} = \left(\begin{array}{c|c|c|c} A_1 & B_1 & & 0 \\ \hline C_1 & \ddots & \ddots & \\ \hline & \ddots & \ddots & B_{R-1} \\ \hline 0 & & C_{R-1} & A_R \end{array}\right) \qquad \mathbf{x} = \left(\begin{array}{c} \mathbf{x}_1 \\ \hline \vdots \\ \hline \mathbf{x}_R \end{array}\right) \qquad \mathbf{p} = \left(\begin{array}{c} \mathbf{p}_1 \\ \hline \vdots \\ \hline \mathbf{p}_R \end{array}\right)$$

As a numerical illustration, let $N = 2$ and $R = 2$ and

$$\mathbf{F} = \left(\begin{array}{cc|cc} -1 & 2 & 0 & 1 \\ 1 & -2 & 1 & 2 \\ \hline 0 & 1 & -2 & 1 \\ 1 & 1 & 0 & -1 \end{array}\right) \quad \text{and} \quad \mathbf{p} = \left(\begin{array}{c} 0.4 \\ 0.6 \\ \hline 0.4 \\ 0.5 \end{array}\right)$$

Using (1.36) to find $\mathbf{F}^{-1}$, we have

$$D^{-1} = \begin{pmatrix} -0.5 & -0.5 \\ 0 & -1.0 \end{pmatrix} \qquad \widehat{\Gamma}^{-1} = \left(A - BD^{-1}C\right)^{-1} = \begin{pmatrix} -0.0952 & 0.2857 \\ 0.3333 & 0 \end{pmatrix}$$

from which

$$W = \begin{pmatrix} -0.0952 & 0.2857 \\ 0.3333 & 0 \end{pmatrix} \qquad X = \begin{pmatrix} -0.0952 & 0.2857 \\ 0.3333 & 0 \end{pmatrix}$$

$$Y = \begin{pmatrix} 0.2857 & 0.1429 \\ 0.2381 & 0.2857 \end{pmatrix} \qquad Z = \begin{pmatrix} -0.4286 & 0.1429 \\ 0.1429 & -0.0476 \end{pmatrix}$$

and

$$\mathbf{x} = \left(\begin{array}{c} \mathbf{x}_1 \\ \hline \mathbf{x}_2 \end{array}\right) = \mathbf{F}^{-1}\mathbf{p} = \left(\begin{array}{c|c} W & X \\ \hline Y & Z \end{array}\right) \left(\begin{array}{c} 0.4 \\ 0.5 \\ \hline 0.4 \\ 0.6 \end{array}\right) = \left(\begin{array}{c} 0.5333 \\ 0.3333 \\ \hline 0.1 \\ 0.2667 \end{array}\right)$$

Remark: Suppose we are interested in $\mathbf{x}_1$ (or $\mathbf{x}_2$); then one needs only the values of W and X (or Y and Z, respectively).

1.5 Matrix Calculus

In this section, we establish the conventions and notations to be used in this book for derivatives and integrals of systems of multivariable functions and equations. For simplicity, we assume that the functions are sufficiently differentiable. The main advantage of matrix calculus is also to allow for compact notation and thus improve the tractability of calculating large systems of differential equations. This means that matrix algebra and matrix analysis tools can be used to study the solution and behavior of systems of differential equations, the numerical solution of systems of nonlinear algebraic equations, and the numerical optimization of multivariable functions.

1.5.1 Matrix of Univariable Functions

Let $A(t)$ be a matrix of univariable functions; then the derivative of $A(t)$ with respect to t is defined as

$$\frac{d}{dt}A(t) = \lim_{\Delta t \to 0} \frac{1}{\Delta t}\Big(A(t+\Delta t) - A(t) \Big) = \begin{pmatrix} \dfrac{d(a_{11})}{dt} & \cdots & \dfrac{d(a_{1M})}{dt} \\ \vdots & \ddots & \vdots \\ \dfrac{d(a_{N1})}{dt} & \cdots & \dfrac{d(a_{NM})}{dt} \end{pmatrix} \tag{1.37}$$

Based on (1.37), we can obtain the various properties given in Table 1.7. For the derivative of determinants, the proof is given in Section A.4.6.

Likewise, the integral of matrices of univariable functions is defined as follows:

$$\int_{t_0}^{t_f} A(t)\, dt = \lim_{\Delta t \to 0} \sum_{k=0}^{T-1} A(k\Delta t)\; \Delta t = \begin{pmatrix} \displaystyle\int_{t_0}^{t_f} a_{11}(t)dt & \cdots & \displaystyle\int_{t_0}^{t_f} a_{1N}(t)dt \\ \vdots & \ddots & \vdots \\ \displaystyle\int_{t_0}^{t_f} a_{N1}(t)dt & \cdots & \displaystyle\int_{t_0}^{t_f} a_{NN}(t)dt \end{pmatrix} \tag{1.38}$$

where $T = (t_f - t_0)/\Delta t$. Based on the linearity property of the integrals, we have the properties shown in Table 1.8.

EXAMPLE 1.12. Define the following function as the **matrix exponential**,

$$\mathbf{exp}\left(A(t)\right) = I + A(t) + \frac{1}{2!}A(t)^2 + \frac{1}{2!}A(t)^2 + \frac{1}{3!}A(t)^3 + \cdots \tag{1.39}$$

then the derivative of $\mathbf{exp}\left(A(t)\right)$ is given by

$$\frac{d}{dt}\mathbf{exp}\left(A(t)\right) = 0 + \frac{d}{dt}A(t) + \frac{1}{2!}\left(A(t)\left(\frac{d}{dt}A(t)\right) + \left(\frac{d}{dt}A(t)\right)A(t)\right) + \cdots$$

In general, $A(t)$ and its derivative are not commutative. However, for the special case in which A and its derivative commute, the matrix exponential simplifies to

$$\begin{aligned} \frac{d}{dt}\mathbf{exp}\left(A(t)\right) &= \frac{d}{dt}A(t) + A(t)\left(\frac{d}{dt}A(t)\right) + \frac{1}{2!}A(t)^2\left(\frac{d}{dt}A(t)\right) + \cdots \\ &= \mathbf{exp}\left(A(t)\right)\left(\frac{d}{dt}A(t)\right) = \left(\frac{d}{dt}A(t)\right)\mathbf{exp}\left(A(t)\right) \end{aligned}$$

One such case is when $A(t)$ is diagonal. Another case is when $A(t) = \alpha(t)M$, where M is a constant square matrix.

Table 1.7. *Properties of derivatives of matrices of univariable functions*

1	Sum of Matrices	$\frac{d}{dt}\left(M(t)+N(t)\right)=\left(\frac{d}{dt}M\right)+\left(\frac{d}{dt}N\right)$
2	Scalar Products	$\frac{d}{dt}\left(\alpha(t)M(t)\right)=\frac{d\alpha}{dt}M+\alpha\left(\frac{d}{dt}M\right)$
3	Matrix Products	$\frac{d}{dt}\left(M(t)N(t)\right)=\left(\frac{d}{dt}M\right)N+M\left(\frac{d}{dt}N\right)$
4	Hadamard Products	$\frac{d}{dt}\left(M(t)\circ N(t)\right)=\left(\frac{d}{dt}M\right)\circ N+M\circ\left(\frac{d}{dt}N\right)$
5	Kronecker Products	$\frac{d}{dt}\left(M(t)\otimes N(t)\right)=\left(\frac{d}{dt}M\right)\otimes N+M\otimes\left(\frac{d}{dt}N\right)$
6	Partitioned Matrices	$\frac{d}{dt}\left(\begin{array}{c\|c} A(t) & B(t) \\ \hline C(t) & D(t)\end{array}\right)=\left(\begin{array}{c\|c} \frac{d}{dt}A & \frac{d}{dt}B \\ \hline \frac{d}{dt}C & \frac{d}{dt}D\end{array}\right)$
7	Matrix Transpose	$\frac{d}{dt}\left(A(t)^T\right)=\left(\frac{d}{dt}A\right)^T$
8	Matrix Inverse	$\frac{d}{dt}\left(A(t)^{-1}\right)=-A^{-1}\left(\frac{d}{dt}A\right)A^{-1}$
9	Determinants	$\frac{d}{dt}\left(\mathbf{det}\left(A(t)\right)\right)=\sum_{k=1}^{N}\mathbf{det}\left(\widehat{A}_{\langle k\rangle}(t)\right)$ where $\widehat{A}_{\langle k\rangle}=\begin{pmatrix} a_{11} & \cdots & a_{1N} \\ & \vdots & \\ \frac{da_{k1}}{dt} & \cdots & \frac{da_{kN}}{dt} \\ & \vdots & \\ a_{n1} & \cdots & a_{NN}\end{pmatrix}\leftarrow k^{\text{th}}\text{row}$

EXAMPLE 1.13. Three vertices of a tetrahedron are stationary, namely $\mathbf{p}_1$, $\mathbf{p}_2$, and $\mathbf{p}_3$. The last vertex $\mathbf{p}_4(t)$ moves as a function of t. As described in Example 1.8, the volume of the tetrahedron (after applying a transpose operation) is given by

$$\text{Vol}=\frac{1}{6}\,\mathbf{det}\left(\begin{array}{c} \left(\mathbf{p}_2-\mathbf{p}_1\right)^T \\ \hline \left(\mathbf{p}_3-\mathbf{p}_1\right)^T \\ \hline \left(\mathbf{p}_4-\mathbf{p}_1\right)^T \end{array}\right)$$

Using the formula for the derivative of determinants (cf. property 9 in Table 1.7), the rate of change of Vol per change in t is given by

$$\frac{d}{dt}\text{Vol} = \frac{1}{6}\left(0 + 0 + \mathbf{det}\begin{pmatrix} \left(\mathbf{p}_2 - \mathbf{p}_1\right)^T \\ \hline \left(\mathbf{p}_3 - \mathbf{p}_1\right)^T \\ \hline \frac{d}{dt}\left(\mathbf{p}_4^T\right) \end{pmatrix} \right)$$

For instance, let the points be given by

$$\mathbf{p}_1 = \begin{pmatrix} 1 \\ 2 \\ 0 \end{pmatrix} \quad \mathbf{p}_2 = \begin{pmatrix} 0.5 \\ 1 \\ 0 \end{pmatrix} \quad \mathbf{p}_3 = \begin{pmatrix} 1 \\ 1 \\ 0 \end{pmatrix} \quad \mathbf{p}_4 = \begin{pmatrix} 2t+3 \\ -t+1 \\ t+5 \end{pmatrix}$$

then

$$\frac{d}{dt}\text{Vol} = 0 + 0 + \frac{1}{6}\,\mathbf{det}\begin{pmatrix} -0.5 & -1 & 0 \\ 0 & -1 & 0 \\ 2 & -1 & 1 \end{pmatrix} = \frac{1}{12}$$

EXAMPLE 1.14. Let $f(\theta) = \mathbf{p}^T Q(\theta)\mathbf{p}$, where $\mathbf{p}$ is constant and $Q(\theta)$ is a square matrix. Then the integral of $f(\theta)$ can be evaluated as

$$\begin{aligned} \int_0^\pi f(\theta)d\theta &= \int_0^\pi \mathbf{p}^T Q(\theta)\mathbf{p}\; d\theta \\ &= \mathbf{p}^T \left(\int_0^\pi Q(\theta)\; d\theta \right) \mathbf{p} \end{aligned}$$

For a specific example,

$$\int_0^\pi \begin{pmatrix} p_1 & p_2 \end{pmatrix} \begin{pmatrix} \cos(\theta) & \sin(\theta) \\ -\sin(\theta) & \cos(\theta) \end{pmatrix} \begin{pmatrix} p_1 \\ p_2 \end{pmatrix} d\theta = \begin{pmatrix} p_1 & p_2 \end{pmatrix} \begin{pmatrix} 0 & -2 \\ 2 & 0 \end{pmatrix} \begin{pmatrix} p_1 \\ p_2 \end{pmatrix} = 0$$

1.5.2 Derivatives of Multivariable Functions

Let x_i, $i = 1, \ldots, N$, be independent variables collected in a column vector as

$$\mathbf{x} = \begin{pmatrix} x_1 \\ \vdots \\ x_N \end{pmatrix}$$

then a multivariable, scalar function f of these variables is denoted by

$$f\left(\mathbf{x}\right) = f\left(x_1, x_2, \ldots, x_M\right)$$

Table 1.8. *Properties of integrals of matrices of univariable functions*

1	Sum of Matrices	$\int \left(M(t)+N(t)\right) dt = \int M dt + \int N dt$
2	Scalar Products	$\int \left(\alpha M\right) dt = \begin{cases} \alpha \int M dt & \text{if } \alpha \text{ is constant} \\ \left(\int \alpha dt\right) M & \text{if } M \text{ is constant} \end{cases}$
3	Matrix Products	$\int \left(MN\right) dt = \begin{cases} M \int N dt & \text{if } M \text{ is constant} \\ \left(\int M dt\right) N & \text{if } N \text{ is constant} \end{cases}$
4	Hadamard Products	$\int \left(M(t) \circ N(t)\right) dt = \begin{cases} M \circ \int N dt & \text{if } M \text{ is constant} \\ \left(\int M dt\right) \circ N & \text{if } N \text{ is constant} \end{cases}$
5	Kronecker Products	$\int \left(M(t) \otimes N(t)\right) dt = \begin{cases} M \otimes \int N dt & \text{if } M \text{ is constant} \\ \left(\int M dt\right) \otimes N & \text{if } N \text{ is constant} \end{cases}$
6	Partitioned Matrices	$\int \left(\begin{array}{c\|c} A(t) & B(t) \\ \hline C(t) & D(t) \end{array}\right) dt = \left(\begin{array}{c\|c} \int A dt & \int B dt \\ \hline \int C dt & \int D dt \end{array}\right)$
7	Matrix Transpose	$\int \left(A(t)^T\right) dt = \left(\int A\, dt\right)^T$

whereas a vector of multivariable scalar functions $\mathbf{f}(\mathbf{x})$ is also arranged in a column vector as

$$\mathbf{f}(\mathbf{x}) = \begin{pmatrix} f_1(\mathbf{x}) \\ \vdots \\ f_N(\mathbf{x}) \end{pmatrix} = \begin{pmatrix} f_1(x_1, x_2, \ldots, x_M) \\ \vdots \\ f_N(x_1, x_2, \ldots, x_M) \end{pmatrix}$$

We denote a row vector of length M known as the **gradient vector**, which is the partial derivatives of $f(\mathbf{x})$ by

$$\frac{d}{d\mathbf{x}} f(\mathbf{x}) = \left(\frac{\partial f}{\partial x_1}, \ \ldots, \ \frac{\partial f}{\partial x_M} \right) \tag{1.40}$$

When applied to each function in a vector of N functions, we obtain an $N \times M$ matrix known as the **Jacobian matrix**,

$$\frac{d}{d\mathbf{x}}\mathbf{f}(\mathbf{x}) = \frac{d}{d\mathbf{x}}\begin{pmatrix} f_1(\mathbf{x}) \\ \vdots \\ f_M(\mathbf{x}) \end{pmatrix} = \begin{pmatrix} \frac{\partial f_1}{\partial x_1} & \cdots & \frac{\partial f_1}{\partial x_M} \\ \vdots & \ddots & \vdots \\ \frac{\partial f_N}{\partial x_1} & \cdots & \frac{\partial f_N}{\partial x_M} \end{pmatrix} \tag{1.41}$$

EXAMPLE 1.15. A set of first-order consecutive chemical reactions $\mathbf{A} \to \mathbf{B} \to \mathbf{C}$ occur in an isothermal continuously stirred tank reactor under constant volume operation. The dynamic equations describing the component balance of each compound are given by

$$\begin{aligned} V\frac{dx_A}{dt} &= F_{\text{in}}\left(x_{A,\text{in}} - x_A\right) - k_A V x_A \\ V\frac{dx_B}{dt} &= F_{\text{in}}\left(x_{B,\text{in}} - x_B\right) - k_B V x_B + k_A V x_A \\ V\frac{dx_C}{dt} &= F_{\text{in}}\left(x_{C,\text{in}} - x_C\right) - k_C V x_C + k_B V x_B \end{aligned}$$

where x_A, x_B, and x_C are the concentrations of A, B, and C, respectively. The other variables in the model are treated as constants. We can collect these concentrations in a column vector $\mathbf{x}$ as

$$\mathbf{x} = \begin{pmatrix} x_A \\ x_B \\ x_C \end{pmatrix} = \begin{pmatrix} x_1 \\ x_2 \\ x_3 \end{pmatrix}$$

then the differential equations can be represented in matrix form as

$$\frac{d}{dt}\left(\mathbf{x}\right) = \mathbf{f}\left(\mathbf{x}\right)$$

where

$$\mathbf{f}\left(\mathbf{x}\right) = \begin{pmatrix} f_1\left(\mathbf{x}\right) \\ f_2\left(\mathbf{x}\right) \\ f_3\left(\mathbf{x}\right) \end{pmatrix} = \begin{pmatrix} \frac{F_{\text{in}}}{V}\left(x_{A,\text{in}} - x_1\right) - k_A x_1 \\ \frac{F_{\text{in}}}{V}\left(x_{B,\text{in}} - x_2\right) - k_B x_2 + k_A x_1 \\ \frac{F_{\text{in}}}{V}\left(x_{C,\text{in}} - x_3\right) - k_C x_3 + k_B x_2 \end{pmatrix}$$

which can be further recast in matrix product form as:

$$\mathbf{f}\left(\mathbf{x}\right) = K\mathbf{x} + \mathbf{b}$$

where K and $\mathbf{b}$ are constant matrices given by

$$K = \begin{pmatrix} -k_A - \frac{F_{\text{in}}}{V} & 0 & 0 \\ k_A & -k_B - \frac{F_{\text{in}}}{V} & 0 \\ 0 & k_B & -k_C - \frac{F_{\text{in}}}{V} \end{pmatrix} \quad ; \quad \mathbf{b} = \frac{F_{\text{in}}}{V}\begin{pmatrix} x_{A,\text{in}} \\ x_{B,\text{in}} \\ x_{C,\text{in}} \end{pmatrix}$$

The term $K\mathbf{x}$, which is just a collection of N linear combination of the elements of $\mathbf{x}$, is said to have the **linear form**. Furthermore, note that K can also be obtained by taking the Jacobian matrix of $\mathbf{f}\left(\mathbf{x}\right)$, that is,

$$\frac{d}{d\mathbf{x}}\mathbf{f}\left(\mathbf{x}\right) = \begin{pmatrix} \dfrac{\partial f_1}{\partial x_1} & \dfrac{\partial f_1}{\partial x_2} & \dfrac{\partial f_1}{\partial x_3} \\ \dfrac{\partial f_2}{\partial x_1} & \dfrac{\partial f_2}{\partial x_2} & \dfrac{\partial f_2}{\partial x_3} \\ \dfrac{\partial f_3}{\partial x_1} & \dfrac{\partial f_3}{\partial x_2} & \dfrac{\partial f_3}{\partial x_3} \end{pmatrix} = \begin{pmatrix} -k_A - \dfrac{F_{\text{in}}}{V} & 0 & 0 \\ k_A & -k_B - \dfrac{F_{\text{in}}}{V} & 0 \\ 0 & k_B & -k_C - \dfrac{F_{\text{in}}}{V} \end{pmatrix}$$

We show later that when $\mathbf{f}(\mathbf{x})$ is expanded through a Taylor series, there will always be a term that has the linear form, precisely due to the Jacobian matrix being evaluated at a prescribed point.

We define the operator $d^2/d\mathbf{x}^2$ on $f(\mathbf{x})$ as

$$\begin{aligned} \frac{d^2}{d\mathbf{x}^2}\left(f(\mathbf{x})\right) &= \frac{d}{d\mathbf{x}}\left(\left(\frac{d}{d\mathbf{x}}f(\mathbf{x})\right)^T\right) = \frac{d}{d\mathbf{x}}\begin{pmatrix} \partial f/\partial x_1 \\ \vdots \\ \partial f/\partial x_N \end{pmatrix} \\ &= \begin{pmatrix} \dfrac{\partial^2 f}{\partial x_1{}^2} & \cdots & \dfrac{\partial^2 f}{\partial x_1 \partial x_N} \\ \vdots & \ddots & \vdots \\ \dfrac{\partial^2 f}{\partial x_N \partial x_1} & \cdots & \dfrac{\partial^2 f}{\partial x_N{}^2} \end{pmatrix} \end{aligned} \tag{1.42}$$

This matrix of second-order derivatives is known as the **Hessian matrix** of $f(\mathbf{x})$. Note that because $\dfrac{\partial^2 f}{\partial x_i \partial x_j} = \dfrac{\partial^2 f}{\partial x_j \partial x_i}$, Hessian matrices are symmetric.

Both the gradients and the Hessians are useful in evaluating local maxima of multivariable functions. Specifically, we have the following theorem that gives the sufficient condition for a local minimum or local maximum.

THEOREM 1.1. *Let $f(\mathbf{x})$ be a multivariable function that is twice differentiable in $\mathbf{x}$. A point $\mathbf{x}^*$ yields a local minimum value for $f(\mathbf{x})$ if*

$$\left.\frac{d}{d\mathbf{x}}f(\mathbf{x})\right|_{\mathbf{x}=\mathbf{x}^*} = \mathbf{0}^T \tag{1.43}$$

and

$$\mathbf{v}^T\left(\left.\frac{d^2}{d\mathbf{x}^2}f(\mathbf{x})\right|_{\mathbf{x}=\mathbf{x}^*}\right)\mathbf{v} > 0 \qquad \textit{for all } \mathbf{v} \neq \mathbf{0} \tag{1.44}$$

Likewise, A point $\mathbf{x}^*$ yields a local maximum value for $f(\mathbf{x})$ if

$$\left.\frac{d}{d\mathbf{x}}f(\mathbf{x})\right|_{\mathbf{x}=\mathbf{x}^*} = \mathbf{0}^T \tag{1.45}$$

and

$$\mathbf{v}^T \left(\frac{d^2}{d\mathbf{x}^2} f(\mathbf{x}) \bigg|_{\mathbf{x}=\mathbf{x}^*} \right) \mathbf{v} < 0 \qquad \textit{for all } \mathbf{v} \neq \mathbf{0} \tag{1.46}$$

PROOF. (See A.4.9)

The conditions given in (1.44) and (1.46) are also known as **positive definiteness condition** and **negative definiteness conditions**, respectively.

Definition 1.8. *An $N \times N$ matrix A is* **positive definite**, *denoted $(A > 0)$, if*

$$\mathbf{x}^* A\mathbf{x} > 0 \qquad \textit{for all } \mathbf{x} \neq 0 \tag{1.47}$$

and A is **positive semi-definite** *if*

$$\mathbf{x}^* A\mathbf{x} \geq 0 \qquad \textit{for all } \mathbf{x} \tag{1.48}$$

For negative definite matrices, we need to just note from (1.46) that a matrix M is negative definite if $-M$ is positive definite. We include a brief discussion on some tests for positive definiteness in Section A.5 as an appendix.

EXAMPLE 1.16. Consider the function

$$f\left(x_1, x_2\right) = e^{-(x_1+1)^2} e^{-(x_2-1)^2}$$

A plot of $f(x_1, x_2)$ is shown in Figure 1.5. The gradient is given by

$$\frac{d}{d\mathbf{x}} f(\mathbf{x}) = \begin{pmatrix} \frac{\partial f}{\partial x_1} & \frac{\partial f}{\partial x_2} \end{pmatrix} = -2e^{-(x_1+1)^2} e^{-(x_2-1)^2} \begin{pmatrix} (x_1+1) & (x_2-1) \end{pmatrix}$$

The gradient is zero at the point $(x_1, x_2) = (-1, 1)$, which makes that point a candidate local maximum or minimum. Furthermore, the Hessian of $f(\mathbf{x})$ is given by

$$\frac{d^2 f}{d\mathbf{x}^2} = \begin{pmatrix} (4x_1^2 + 8x_1 + 2) & 4(x_1+1)(x_2-1) \\ 4(x_1+1)(x_2-1) & (4x_2^2 - 8x_2 + 2) \end{pmatrix} f(x_1, x_2)$$

At $(x_1, x_2) = (-1, 1)$, the Hessian becomes

$$\frac{d^2 f}{d\mathbf{x}^2} = \begin{pmatrix} -2 & 0 \\ 0 & -2 \end{pmatrix}$$

and for $\mathbf{v} \neq \mathbf{0}$,

$$\begin{pmatrix} v_1 & v_2 \end{pmatrix} \begin{pmatrix} -2 & 0 \\ 0 & -2 \end{pmatrix} \begin{pmatrix} v_1 \\ v_2 \end{pmatrix} = -2\left(v_1^2 + v_2^2\right) < 0$$

which satisfies the negative definiteness condition. Thus the point $(x_1, x_2) = (-1, 1)$ is a local maximum. This can be seen to be the case from Figure 1.5.

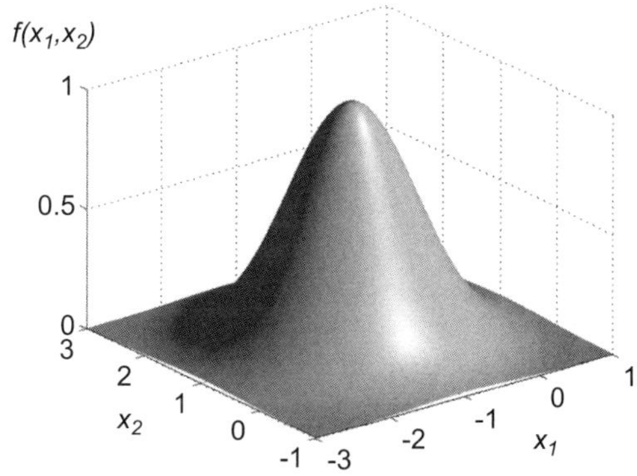

Figure 1.5. The surface plot for $f(x_1, x_2) = e^{-(x_1+1)^2} e^{-(x_2-1)^2}$.

For the special cases of linear and quadratic forms, the derivatives, gradients, and Hessians have special formulas as given by the following lemma:

LEMMA 1.6. *Let $A[=]M \times N$ and $\mathbf{x}[=]N \times 1$, then*

$$\frac{d}{d\mathbf{x}}\left(A\mathbf{x}\right) = A \tag{1.49}$$

$$\text{For } A[=]N \times N, \quad \frac{d}{d\mathbf{x}}\left(\mathbf{x}^T A\mathbf{x}\right) = \mathbf{x}^T\left(A + A^T\right) \tag{1.50}$$

$$\frac{d^2}{d\mathbf{x}^2}\left(\mathbf{x}^T A\mathbf{x}\right) = A + A^T \tag{1.51}$$

PROOF. (See section A.4.7)

Remark: Equations (1.49) to (1.51) are used in the next chapter during the solution to the least squares problem.

1.6 Sparse Matrices

We end this chapter with a brief discussion on sparse matrices because several real-world problems involve large matrices that contain a very high percentage of zero entries. Computer programs such as MATLAB include special commands to handle the calculation of sparse matrices. For instance, consider the finite difference approximation of d^2u/dx^2,

$$\frac{d^2u}{dx^2} \approx \frac{u_{i+1} - 2u_i + u_{i-1}}{\Delta x^2}$$

This leads to a matrix representation $A\mathbf{u}$, where

$$A = \frac{1}{\Delta x^2}\begin{pmatrix} -2 & 1 & & 0 \\ 1 & -2 & \ddots & \\ & \ddots & \ddots & 1 \\ 0 & & 1 & -2 \end{pmatrix}$$

Table 1.9. *Some MATLAB commands for sparse matrix operation*

MATLAB Command	Description
`A=sparse(rowA,colA,nzA,N,M)`	Creates a sparse matrix of size `N`×`M` with `rowA`, `colA`, and `nzA` as the vectors for row index, column index, and nonzero elements, respectively
`A=sparse(S)`	Converts a full formatted matrix `S` to sparse matrix format
`[rowA,colA]=find(A)`	Returns the row and column indices of nonzero elements of matrix `A`
`spy(A)`	Visualization of the sparsity pattern of matrix `A`
`speye(N)`	Creates a sparse formatted identity matrix of size `N`
`A=spdiags(V,d,M,N)`	Creates an `M`×`N` sparse matrix by placing columns of `V` along the diagonals specified by `d`
Operations: `+, -, *,\,^`	Performs sparse matrix operations and leaves the result in sparse matrix format
Functions: `exp,sin,cos, tan,log,kron,inv,`	Evaluates the functions of all elements (zero and nonzero) but leaves the results in sparse matrix format

The fraction of nonzero elements is $(3N-2)/N^2$. For instance, with $N = 100$, this fraction is, approximately 1%.

Two features of sparse matrices can immediately lead to computational advantages. The first feature is that the storage of only nonzero elements of sparse matrices will result in very significant savings by avoiding the storage of zero elements. The second feature is that when performing matrix-vector products, it should not be necessary to multiply the zero elements of the matrix with that of the vector. For instance,

$$\beta = \mathbf{v}^T\mathbf{w} = \begin{pmatrix} 2 & 1 & 0 & 0 & 0 & 3 \end{pmatrix}\begin{pmatrix} a \\ b \\ c \\ d \\ e \\ f \end{pmatrix} = \begin{pmatrix} 2 & 1 & 3 \end{pmatrix}\begin{pmatrix} a \\ b \\ f \end{pmatrix}$$

We discuss only the **coordinate format** because it is the most flexible and the simplest approach, although other schemes can have further significant storage savings. In this format, the nonzero elements can be collected in a vector along with two vectors of the same length: one vector for the row indices and the other vector for the column indices. For example, let matrix A be given as

$$A = \begin{pmatrix} 0 & a & 0 & 0 & b \\ 0 & 0 & 0 & c & 0 \\ d & 0 & 0 & e & f \\ 0 & 0 & 0 & g & 0 \end{pmatrix}$$

then the three vectors nz_A, row_A, and col_A, indicating nonzero elements, row indices, and column indices, respectively, are given by

$$nz_A = \begin{pmatrix} a \\ b \\ c \\ d \\ e \\ f \\ g \end{pmatrix} \qquad row_A = \begin{pmatrix} 1 \\ 1 \\ 2 \\ 3 \\ 3 \\ 3 \\ 4 \end{pmatrix} \qquad col_A = \begin{pmatrix} 2 \\ 5 \\ 4 \\ 1 \\ 4 \\ 5 \\ 4 \end{pmatrix}$$

This implies that the storage requirements will be thrice the length of nz_A. As a guideline, for the storage under the coordinate formats to be worthwhile, the fraction of nonzero elements must be less than $\frac{1}{3}$.

Some MATLAB commands for sparse matrices are listed on Table 1.9. In the next chapter, we often note the solution methods for the linear equation $A\mathbf{x} = \mathbf{b}$ that have been developed to preserve as much of the sparsity of matrices, because the amount of computations can be significantly reduced if sparsity of A is present.

1.7 EXERCISES

E1.1. Show that A^TA, AA^T, and $A + A^T$ are symmetric, and $A - A^T$ is skew symmetric.

E1.2. Let $A[=]10 \times 50$, $B[=]50 \times 20$, and α a scalar. From a computational view, is it better to compute $(\alpha A)\,B$, $A\,(\alpha B)$, or $\alpha\,(AB)$? Explain your conclusions, including any possible exceptions to your conclusion.

E1.3. Prove or disprove the following properties for matrix traces:

1. $\mathbf{tr}\,(A + B) = \mathbf{tr}\,(A) + \mathbf{tr}\,(B)$
2. $\mathbf{tr}\,(\alpha A) = \alpha \mathbf{tr}\,(A)$ for scalar α.
3. $\mathbf{tr}\,(A^T) = \mathbf{tr}\,(A)$.
4. $\mathbf{tr}\,(ABC) = \mathbf{tr}\,(CAB) = \mathbf{tr}\,(BCA)$, assuming $A[=]N \times M$, $B[=]M \times P$, $C[=]P \times N$. If, in addition, A, B, and C are symmetric, then $\mathbf{tr}\,(ABC) = \mathbf{tr}\,(ACB)$.
5. If A is symmetric and B is skew symmetric, then $\mathbf{tr}\,(AB) = 0$.
6. $\mathbf{tr}\,(A \otimes B) = \mathbf{tr}\,(A)\,\mathbf{tr}\,(B)$.

E1.4. Let $A[=]N \times M$ and $B[=]M \times N$; show that

$$\mathbf{det}\left(I_{[N]} - AB \right) = \mathbf{det}\left(I_{[M]} - BA \right)$$

(Hint: use (1.32) and (1.33))

E1.5. For a nonsingular matrix G, define the relative gain array of G as

$$R = G \circ G^{-T}$$

where $G^{-T} = (G^{-1})^T$. Prove the following properties for R:

1. The sum of all elements along any row or along any column is 1, that is,

$$\sum_{i=1}^{n} r_{ij} \quad = \quad 1 \qquad \text{for all } i$$

$$\sum_{j=1}^{n} r_{ij} \quad = \quad 1 \qquad \text{for all } j$$

2. if G is triangular, then $R = I$.

Note: The relative gain array is used in process control theory to determine the best pairing between manipulated and controlled variables that would reduce the interaction among the control loops.

E1.6. If $A^2 = A$ then A is said to be **idempotent**.

1. Show that if A is idempotent, then $I - A$ and A^k, $k > 1$, are also idempotent.
2. Show that if A is idempotent, then either $\mathbf{det}\left(A\right) = 1$ or $\mathbf{det}\left(A\right) = 0$. Furthermore, show that the only nonsingular idempotent matrices are the identity matrices.
3. **Projection matrices** are idempotent matrices that are also Hermitian. Verify that the following matrix is a projection matrix:

$$A = \frac{1}{6}\begin{pmatrix} 2 & -2 & -2 \\ -2 & 5 & -1 \\ -2 & -1 & 5 \end{pmatrix}$$

and evaluate its determinant.

E1.7. Let $\mathbf{p}$, $\mathbf{s}$, and $\mathbf{t}$ be three points in the (x, y)-plane, that is,

$$\mathbf{p} = \begin{pmatrix} p_x \\ p_y \end{pmatrix} \qquad \mathbf{s} = \begin{pmatrix} s_x \\ s_y \end{pmatrix} \qquad \mathbf{t} = \begin{pmatrix} t_x \\ t_y \end{pmatrix}$$

A circumcircle for $\mathbf{p}$, $\mathbf{s}$, and $\mathbf{t}$ is a circle that would pass through through all three points, as shown in Figure 1.6.

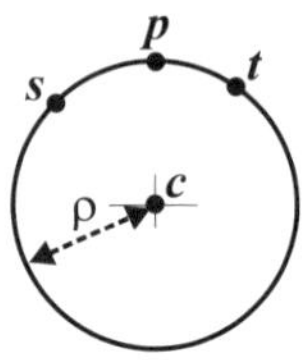

Figure 1.6. Circumcircle that passes through $\mathbf{p}$, $\mathbf{s}$, and $\mathbf{t}$.

Let $\rho_{\{\mathbf{p},\mathbf{s},\mathbf{t}\}}$ and $\mathbf{c}_{\{\mathbf{p},\mathbf{s},\mathbf{t}\}} = \begin{pmatrix} c_x & c_y \end{pmatrix}^T$ be the radius and center point for the circumcircle.

1. Show that the radius of the circumcircle has to satisfy

$$\rho^2_{\{\mathbf{p},\mathbf{s},\mathbf{t}\}} = (\mathbf{p}-\mathbf{c})^T(\mathbf{p}-\mathbf{c}) = (\mathbf{t}-\mathbf{c})^T(\mathbf{t}-\mathbf{c}) = (\mathbf{s}-\mathbf{c})^T(\mathbf{s}-\mathbf{c}) \qquad (1.52)$$

2. Show that (1.52) can be combined to yield the following matrix equation:

$$\begin{pmatrix} (\mathbf{p}-\mathbf{s})^T \\ (\mathbf{s}-\mathbf{t})^T \end{pmatrix}\mathbf{c} = \frac{1}{2}\begin{pmatrix} \mathbf{p}^T\mathbf{p} - \mathbf{s}^T\mathbf{s} \\ \mathbf{s}^T\mathbf{s} - \mathbf{t}^T\mathbf{t} \end{pmatrix}$$

3. Using the preceding results, find the radius and center of the circumcircle for

$$\mathbf{p} = \begin{pmatrix} 0 \\ 0 \end{pmatrix} \qquad \mathbf{s} = \begin{pmatrix} 1 \\ 0 \end{pmatrix} \qquad \mathbf{t} = \begin{pmatrix} 1 \\ 1 \end{pmatrix}$$

Plot the points and the circumcircle to verify your solution.

4. Find the circumcircle for

$$\mathbf{p} = \begin{pmatrix} 1 \\ -1 \end{pmatrix} \quad \mathbf{s} = \begin{pmatrix} 2 \\ 2 \end{pmatrix} \quad \mathbf{t} = \begin{pmatrix} 3 \\ 5 \end{pmatrix}$$

What do you conclude about this example ?

Note: Circumcircles are used in mesh-triangulation methods for finite elements solution of partial differential equations. One specific method is called the Delaunay triangulation method. For more details, see Chapter 14.

E1.8. Using Q defined in (1.28), complete the proof of the matrix inversion lemma given in (1.27) by showing that indeed $Q(A + BCD) = I$.

E1.9. If $A^r = 0$, with $A^{r-1} \neq 0$ for some positive integer r, then A is said to be **nilpotent of index** r. A **strictly triangular matrix** is a triangular matrix whose diagonal elements are all zero. Show that strictly triangular matrices are nilpotent.

E1.10. Matrix A is said to be **orthogonal** if $A^T A = AA^T = I$.

1. Show that permutation matrices, P, are orthogonal where

$$P = \begin{pmatrix} \mathbf{e}_{k_1}^T \\ \hline \vdots \\ \hline \mathbf{e}_{k_N}^T \end{pmatrix} \quad ; \quad k_1 \neq \cdots \neq k_N$$

and

$$\mathbf{e}_i = \begin{pmatrix} 0 \\ \vdots \\ 0 \\ 1 \\ 0 \\ \vdots \\ 0 \end{pmatrix} \leftarrow i^{\text{th}} \text{ element}$$

2. Show that the products of orthogonal matrices are also orthogonal.

E1.11. Matrix A is said to be **unitary** if $A^* A = AA^* = I$. Show that the normalized Fourier matrix, given by

$$F = \frac{1}{\sqrt{N}} \begin{pmatrix} f_{11} & \cdots & f_{1N} \\ \vdots & \ddots & \vdots \\ f_{N1} & \cdots & f_{NN} \end{pmatrix} \quad ; \quad f_{k\ell} = e^{-2\pi i(k-1)(\ell-1)/N}$$

is unitary.

E1.12. Determine whether the following statement is true or false, and explain: "Let Q be skew-symmetric, then $\mathbf{x}^T Q\mathbf{x} = 0$ for all vectors $\mathbf{x}$."

E1.13. A matrix that is useful for the method of Gaussian elimination is the matrix obtained by replacing the kth column of an identity matrix by vector $\mathbf{w}$ as follows:

$$\widehat{E}_L(k,\mathbf{w}) = \begin{pmatrix} 1 & & & w_1 & & & 0 \\ & \ddots & & \vdots & & & \\ & & 1 & w_{k-1} & & & \\ & & & w_k & & & \\ & & & w_{k+1} & 1 & & \\ & & & \vdots & & \ddots & \\ 0 & & & w_N & & & 1 \end{pmatrix}$$

1. Evaluate the determinant of $\widehat{E}_L(k,\mathbf{w})$.
2. Assuming $w_k \neq 0$, show that the inverse of $\widehat{E}_L(k,\mathbf{w})$ is given by

$$\widehat{E}_L(k,\mathbf{w})^{-1} = \begin{pmatrix} 1 & & & -(w_1/w_k) & & & 0 \\ & \ddots & & \vdots & & & \\ & & 1 & -(w_{k-1}/w_k) & & & \\ & & & (1/w_k) & & & \\ & & & -(w_{k+1}/w_k) & 1 & & \\ & & & \vdots & & \ddots & \\ 0 & & & -(w_N/w_k) & & & 1 \end{pmatrix}$$

3. Consider a matrix $A[=]N \times M$ with a nonzero element in the kth row and ℓ^{th} column, that is, $a_{k\ell} \neq 0$. Based on $a_{k\ell}$, define the elements of vector $\mathbf{g}(a_{k\ell})$ as follows:

$$g_i = \begin{cases} -\dfrac{a_{i\ell}}{a_{k\ell}} & \text{if } i \neq k \\[2ex] \dfrac{1}{a_{k\ell}} & \text{if } i = k \end{cases}$$

Discuss the effects of premultiplying A by $\widehat{E}_L\left(k, \mathbf{g}(a_{k\ell})\right)$.
(Hint: Use the following specific example,

$$A = \begin{pmatrix} 4 & -1 & 3 & 2 \\ 2 & 1 & 5 & 2 \\ 1 & 0 & 2 & 1 \end{pmatrix}$$

Construct $\widehat{E}_L\left(3, \mathbf{g}(3,1)\right)$ and $\widehat{E}_L\left(1, \mathbf{g}(1,4)\right)$, then obtain the products $\widehat{E}_L A$ and observe the effects. Using these results, infer as well as prove the general results.)

E1.14. The Vandermonde matrix is a matrix having a special structure given by

$$V = \begin{pmatrix} \lambda_1^{n-1} & \lambda_1^{n-2} & \cdots & \lambda_1 & 1 \\ \lambda_2^{n-1} & \lambda_2^{n-2} & \cdots & \lambda_2 & 1 \\ \vdots & \vdots & \ddots & \vdots & \vdots \\ \lambda_n^{n-1} & \lambda_n^{n-2} & \cdots & \lambda_n & 1 \end{pmatrix}$$

and a determinant given by

$$\mathbf{det}\left(V\right) = \prod_{i<j} (\lambda_i - \lambda_j) \tag{1.53}$$

1. Verify the formula (1.53) for the case $\lambda = \{1, -1, 2, -3\}$. What happens to the determinant if any of the λ's are repeated?
2. An $(n-1)^{\text{th}}$ degree polynomial

$$y = a_{n-1}x^{n-1} + \cdots + a_1 x + a_0$$

is needed to pass through n given points (x_i, y_i). Find a matrix G such that solving $G\mathbf{v} = \mathbf{y}$ yields the vector of coefficients $\mathbf{v} = (a_0, \ldots, a_{n-1})^T$ where $\mathbf{y} = (y_1, \ldots, y_n)^T$. What is the conditions on the points (x_i, y_i) so that a unique solution is guaranteed to exist?

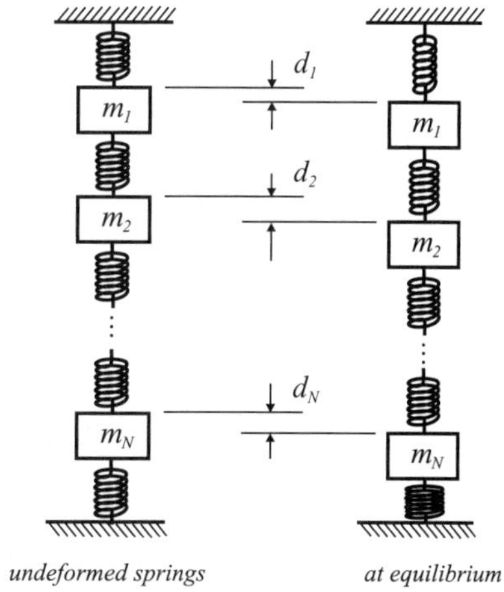

Figure 1.7. Various masses connected in series by Hookean springs. The left figure shows the position when the springs have not yet been deformed, whereas the right figure shows the new equilibrium positions due to gravitational effects.

E1.15. Consider a set of N masses m_i arranged in a series configuration as shown in Figure 1.7 connected by $N+1$ springs that satisfy Hooke's law, $f_{s,j} = k_j x_j$ where $f_{s,j}$, k_j and x_j are the spring force, spring constant, and spring elongation, respectively.[7] One end each of springs 1 and $(N+1)$ are attached to immovable points. Let d_i be the displacement of the masses at equilibrium (relative to the positions when the springs were undeformed). Then the spring elongations x_j are related to mass displacements d_i by the following equations:

$$x_j = \begin{cases} d_1 & \text{for } j = 1 \\ d_j - d_{j-1} & \text{for } 2 \le j \le N \\ -d_N & \text{for } j = N+1 \end{cases}$$

while the force balances for each mass become

$$m_i g = f_{s,i} - f_{s,i+1} = k_i x_i - k_{i+1} x_{i+1}$$

where g is the gravitational acceleration.

1. Show that the main equations relating the vector of displacements $\mathbf{d} = (d_1, \ldots, d_N)^T$ and the vector of masses $\mathbf{m} = (m_1, \ldots, m_N)^T$ can be put in matrix form as

$$\left(H \Phi H^T\right) \mathbf{d} = \mathbf{m} \qquad (1.54)$$

[7] Adapted from the discussion in G. Strang, *Introduction to Applied Mathematics*, Wellesley-Cambridge Press, Wellesley, MA, 1986, pp. 40–44.

where $H[=]N \times (N+1)$ and $\Phi[=](N+1) \times (N+1)$ are given by

$$H = \begin{pmatrix} 1 & -1 & & & 0 \\ & 1 & -1 & & \\ & & \ddots & \ddots & \\ 0 & & & 1 & -1 \end{pmatrix} \qquad \Phi = \begin{pmatrix} \frac{k_1}{g} & & 0 \\ & \ddots & \\ 0 & & \frac{k_{N+1}}{g} \end{pmatrix}$$

2. We can partition matrices H and Φ further as

$$H = \left(\widehat{H} \mid \mathbf{q} \right) \qquad \Phi = \left(\begin{array}{c|c} \widehat{\Phi} & 0 \\ \hline 0 & \alpha \end{array} \right)$$

with $\mathbf{q} = (0, \ldots, 0, -1)^T$, $\alpha = k_{N+1}/g$ and

$$\widehat{H} = \begin{pmatrix} 1 & -1 & & 0 \\ & \ddots & \ddots & \\ & & \ddots & -1 \\ 0 & & & 1 \end{pmatrix} \qquad \widehat{\Phi} = \begin{pmatrix} \frac{k_1}{g} & & 0 \\ & \ddots & \\ & & \frac{k_N}{g} \end{pmatrix}$$

Show that the inverses of $\widehat{H}$ and $\widehat{\Phi}$ are given by

$$\widehat{H}^{-1} = \begin{pmatrix} 1 & \cdots & 1 \\ & \ddots & \vdots \\ 0 & & 1 \end{pmatrix} \quad \text{and} \quad \widehat{\Phi}^{-1} = \begin{pmatrix} r_1 & & 0 \\ & \ddots & \\ 0 & & r_N \end{pmatrix}$$

where $r_j = g/k_j$.

3. Using this partitioning, we can reformulate (1.54) as

$$\left(\widehat{H}\widehat{\Phi}\widehat{H}^T + \alpha \mathbf{q}\mathbf{q}^T \right) \mathbf{d} = \mathbf{m} \tag{1.55}$$

Show that after applying the matrix inverse formula of (1.27) to (1.55), we have

$$\mathbf{d} = (AB)\mathbf{m}$$

where

$$A = \begin{pmatrix} 1 & & 0 \\ \vdots & \ddots & \\ 1 & \cdots & 1 \end{pmatrix} \begin{pmatrix} r_1 & & 0 \\ & \ddots & \\ 0 & & r_N \end{pmatrix} \begin{pmatrix} 1 & \cdots & 1 \\ & \ddots & \vdots \\ 0 & & 1 \end{pmatrix}$$

and

$$B = I - \frac{1}{R_{N+1}} \left(\begin{array}{c|c|c} 0 & & 0 \\ \vdots & \cdots & \vdots \\ 0 & & 0 \\ R_1 & & R_N \end{array} \right)$$

where $R_n = \sum_{j=1}^{n} r_j = \sum_{j=1}^{n} \frac{g}{k_j}$.

E1.16. An N-multistage evaporator system is shown in Figure 1.8. Brine having a mass fraction $x_{L_{N+1}}$, salt is fed to stage N at a rate $n_{L_{N+1}}$. Condensation of steam from stage $(i-1)$ is used to evaporate some water from the brine fed to stage i. The pressure at stage i is fixed. Furthermore, assume that the enthalpy of the brine solution and pure water are the same and that liquid enthalpy H_{L_i} and H_{V_i} are constant.

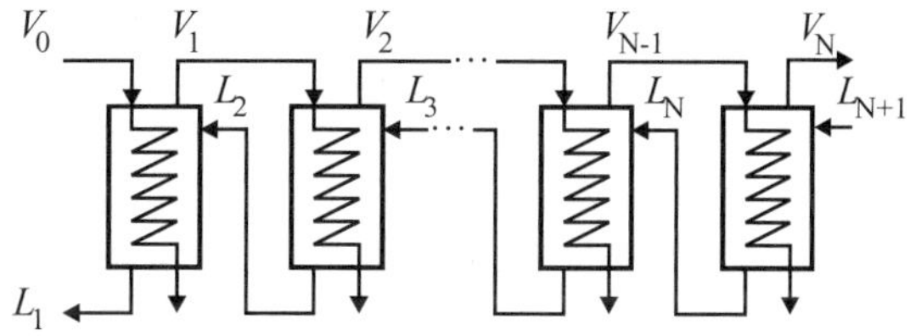

Figure 1.8. Flowsheet of an N-stage evaporator system.

For steady-state operations, the equations are given by

Enthalpy balances:

$$\left(H_{V_{i-1}} - H_{L_{i-1}}\right) n_{V_{i-1}} - n_{V_i} H_{V_i} - n_{L_i} H_{L_i} + n_{L_{i+1}} H_{L_{i+1}} = 0 \tag{1.56}$$

Mass balance:

$$n_{L_{i+1}} - n_{L_i} - n_{V_i} = 0 \tag{1.57}$$

Component balance:

$$n_{L_{i+1}} x_{L_{i+1}} - n_{L_i} x_{L_i} = 0 \tag{1.58}$$

1. Show that (1.58) yields,

$$x_{L_1} n_{L_1} = x_{L_{N+1}} n_{L_{N+1}} \tag{1.59}$$

2. Define vectors $\mathbf{n}_V$ and $\mathbf{n}_L$ as

$$\mathbf{n}_V = \begin{pmatrix} n_{V_1} & \cdots & n_{V_N} \end{pmatrix}^T \quad ; \quad \mathbf{n}_L = \begin{pmatrix} n_{L_1} & \cdots & n_{L_N} \end{pmatrix}^T$$

   Obtain matrices A, B, and C and vectors $\mathbf{q}$ and $\mathbf{g}$ such that (1.56), (1.57), and (1.59) can be represented in the following matrix equations:

$$\begin{aligned} A\mathbf{n}_V + B\mathbf{n}_L + \mathbf{q} n_{V_0} &= \mathbf{h} \\ \mathbf{n}_V + C\mathbf{n}_L &= \mathbf{f} \\ \mathbf{g}^T \mathbf{n}_L &= x_{L_{N+1}} n_{L_{N+1}} \end{aligned} \tag{1.60}$$

   where

$$\mathbf{h} = \begin{pmatrix} 0 \\ \vdots \\ 0 \\ -H_{L_{N+1}} n_{L_{N+1}} \end{pmatrix} \quad ; \quad \mathbf{f} = \begin{pmatrix} 0 \\ \vdots \\ 0 \\ n_{L_{N+1}} \end{pmatrix}$$

3. Assuming the inverses exist, show that the value of n_{V_0} can be obtained from (1.60) as

$$n_{V_0} = \frac{x_{L_{N+1}} n_{L_{N+1}} - \mathbf{g}^T (AC - B)^{-1} (A\mathbf{f} - \mathbf{h})}{\mathbf{g}^T (AC - B)^{-1} \mathbf{q}} \tag{1.61}$$

4. One could also combine (1.60) in compact form as

$$G\,\mathbf{n} = \mathbf{v} \tag{1.62}$$

   where

$$G = \left(\begin{array}{c|c|c} A & B & \mathbf{q} \\ \hline I & C & 0 \\ \hline 0 & \mathbf{g}^T & 0 \end{array}\right) \quad \mathbf{n} = \left(\begin{array}{c} \mathbf{n}_V \\ \hline \mathbf{n}_L \\ \hline n_{V_0} \end{array}\right) \quad \mathbf{v} = \left(\begin{array}{c} \mathbf{h} \\ \hline \mathbf{f} \\ \hline x_{L_{N+1}} n_{L_{N+1}} \end{array}\right)$$

Show that one can obtain (1.61) by using Cramer's rule to solve n_{V_0} from (1.62). (Hint: use (1.32) and (1.35).)[8]

E1.17. Prove that the inverse of an upper triangular matrix is also an upper triangular matrix.

E1.18. Verify the formula (1.26) for the inverse of the following matrices:

$$L = \begin{pmatrix} 1 & 0 & 0 \\ 2 & 1 & 0 \\ -2 & 3 & 2 \end{pmatrix} \qquad U = \begin{pmatrix} 1 & 2 & 2 & 0 \\ 0 & -1 & 1 & 2 \\ 0 & 0 & 2 & 1 \\ 0 & 0 & 0 & 1 \end{pmatrix}$$

Extend the inverse formula (1.26) for the triangular matrices to apply to block-triangular matrices. Verify your block triangular matrix inverse formula to find the inverse of the following matrix:

$$H = \left(\begin{array}{ccc|cc|cc} 1 & -1 & 1 & 0 & 0 & 0 & 0 \\ 2 & 0 & 1 & 0 & 0 & 0 & 0 \\ 1 & -1 & 3 & 0 & 0 & 0 & 0 \\ \hline 2 & 0 & 1 & 2 & 2 & 0 & 0 \\ 1 & 1 & 1 & 1 & 2 & 0 & 0 \\ \hline 0 & 3 & 2 & 1 & 0 & 1 & 1 \\ 1 & -1 & 1 & 0 & 2 & 2 & 3 \end{array}\right)$$

E1.19. Often in recursive data regression, a matrix A is updated as follows: $A_{\text{new}} = A + \mathbf{u}\mathbf{v}^T$, where $\mathbf{u}$ and $\mathbf{v}$ are column vectors.

1. Prove the following identity:

$$\mathbf{det}\left(A + \mathbf{u}\mathbf{v}^T\right) = \left(1 + \mathbf{v}^T A^{-1}\mathbf{u}\right)\mathbf{det}(A) \tag{1.63}$$

2. Equation (1.63) implies that the new matrix $A_{\text{new}} = A + \mathbf{u}\mathbf{v}^T$ will become singular if $\mathbf{v}^T A^{-1}\mathbf{u} = -1$. For the following case:

$$A = \begin{pmatrix} 1 & 2 & 3 \\ 0 & -1 & -1 \\ 1 & 4 & 1 \end{pmatrix}, \quad \mathbf{u} = \begin{pmatrix} 1 \\ 0 \\ 0 \end{pmatrix}, \quad \mathbf{v} = \begin{pmatrix} 0 \\ 0 \\ \alpha \end{pmatrix}$$

find the value of α that would make A_{new} singular. (This means that any nonsingular matrix can be made singular by changing the value of just a single element.)

E1.20. Let $A[=]N \times N$ and $D[=]M \times M$ be nonsingular matrices, $N \neq M$. Prove (or disprove) the following equation:

$$\left(A - BD^{-1}C\right)^{-1} BD^{-1} = A^{-1}B\left(D - CA^{-1}B\right)^{-1} \tag{1.64}$$

E1.21. The unsteady-state heat conduction in a flat rectangular plate of dimension $L \times W$ is given by

$$\frac{\partial T}{\partial t} = \alpha\left(\frac{\partial^2 T}{\partial x^2} + \frac{\partial^2 T}{\partial y^2}\right)$$

subject to the boundary conditions

$$\begin{aligned} T(0, y, t) &= 50 + \frac{W - y}{W}50 \\ T(L, y, t) &= 50 - \frac{W - y}{W}50 \end{aligned}$$

[8] In practice, one would simply apply Cramer's rule directly to determine n_{V_0} because (1.61) still contains an inverse calculation. This subproblem is just an academic exercise in applying the various properties of block matrix operations.

$$T(x,0,t) = 50$$
$$T(x,W,t) = 100$$

and initial condition

$$T(x,y,0) = 50$$

Following example 1.3, set $L = 1$, $W = 2$, $N = 20$, $M = 40$, $\alpha = 0.05$, and obtain a finite difference approximation to the differential equation, that is, find A, B, and C such that

$$T(k+1) = AT(k) + T(k)B + C$$

Then, using a computer, obtain time plots of T for $t = 0.02, 0.04, \ldots, 0.18$. (Note: The choice of Δt will be crucial to a stable simulation of the process. A smaller value will be more stable but will require more k values. For this problem, use $\Delta t = 0.0001$.)

E1.22. Consider the rectification section of a distillation column shown in Figure 1.9.[9]

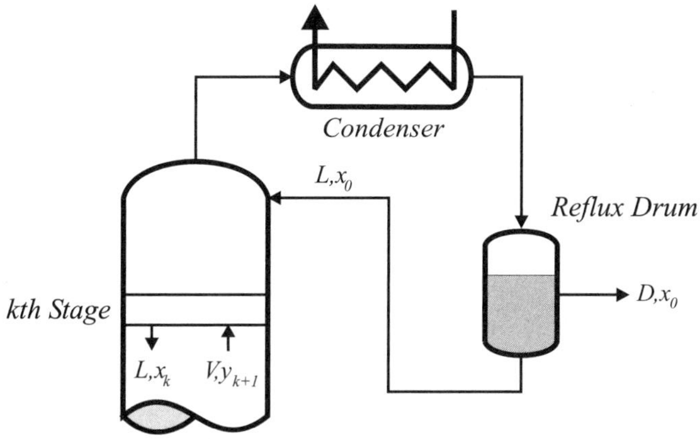

Figure 1.9. Rectification section of a distillation column under total condensation.

Under the assumptions of constant molal overflow and constant volatility, the mole fraction of liquid components exiting stage k is given by

$$\mathbf{x}_k = \left(\frac{1}{\mathbf{q}_k^T \mathbf{x}_0}\right) \mathbf{q}_k \circ \mathbf{x}_0 \tag{1.65}$$

where

$$\mathbf{q}_k = \begin{cases} \mathbf{p} & \text{for } k = 1 \\ R\left(\mathbf{q}_{k-1} \circ \mathbf{p}\right) + \left(\mathbf{x}_0^T \mathbf{q}_{k-1}\right) \mathbf{p} & \text{for } k > 1 \end{cases} \tag{1.66}$$

$\mathbf{x}_0$ contains the mole fractions in the distillate stream, $\mathbf{p}$ contains the reciprocal relative volatilities $1/\alpha_i$, and R is the (scalar) reflux ratio.

1. Evaluate the composition of the liquid out of the third stage, $\mathbf{x}_3$, using the following values:

$$\mathbf{p} = \begin{pmatrix} 7 \\ 4 \\ 1 \end{pmatrix} \quad , \quad \mathbf{x}_0 = \begin{pmatrix} 0.2 \\ 0.1 \\ 0.7 \end{pmatrix} \quad , \quad R = 0.6$$

[9] This problem is adapted from the example given in N. Amundson, *Mathematical Methods in Chemical Engineering, Volume 1, Matrices and Their Applications*, Prentice Hall, Englewood Cliffs, NJ, 1966, pp. 149–157.

2. Show that for $k > 1$, the iterative formula (1.66) can be recast as

$$\mathbf{q}_k = A\mathbf{q}_{k-1}$$

where

$$A = D + \mathbf{p}\mathbf{x}_0^T \tag{1.67}$$

and

$$D = \begin{pmatrix} Rp_1 & & 0 \\ & \ddots & \\ 0 & & Rp_n \end{pmatrix}$$

Thus show that

$$\mathbf{q}_k = A^{k-1}\mathbf{p}$$

3. For matrix A defined in (1.67), show that the determinant and the adjugate of $(A - \lambda I)$, where λ is a scalar variable, are given by

$$\mathbf{det}\,(A - \lambda I) = \psi_\lambda \omega_\lambda \tag{1.68}$$

$$\mathbf{adj}\,(A - \lambda I) = \Gamma_\lambda \left(\psi_\lambda I - \mathbf{p}\mathbf{x}_0^T \Gamma_\lambda\right) \omega_\lambda \tag{1.69}$$

where

$$\omega_\lambda = \prod_{i=1}^{n} (Rp_i - \lambda) \quad ; \quad \psi_\lambda = \left(1 + \sum_{i=1}^{n} \frac{x_{0i}p_i}{Rp_i - \lambda}\right)$$

$$\Gamma_\lambda = \begin{pmatrix} (Rp_1 - \lambda)^{-1} & & 0 \\ & \ddots & \\ 0 & & (Rp_n - \lambda)^{-1} \end{pmatrix}$$

(Hint: Use (1.63) for the determinant and (1.27) for the adjugate formula.)[10]

E1.23. Consider the tridiagonal matrix given by

$$T_N = \begin{pmatrix} a_1 & b_1 & & 0 \\ c_1 & \ddots & \ddots & \\ & \ddots & \ddots & b_{N-1} \\ 0 & & c_{N-1} & a_N \end{pmatrix} \tag{1.70}$$

Define $T_0 = [1]$ and $T_1 = [a_1]$.

1. Show that the determinant of T_k, for $k > 1$ is given by the following recursive formula (also known as the **continuant equation**):

$$\mathbf{det}\left(T_k\right) = a_k\,\mathbf{det}\left(T_{k-1}\right) - b_{k-1}c_{k-1}\mathbf{det}\left(T_{k-2}\right) \tag{1.71}$$

2. Verify this formula by finding the determinant of

$$A = \begin{pmatrix} 1 & -1 & 0 \\ -2 & 2 & 1 \\ 0 & 2 & -2 \end{pmatrix}$$

[10] This formula for the adjugate can be used further to obtain a closed formula for $\mathbf{x}_n$ based on Sylvester's formula.

3. For the special case that $a_1 = a_2 = \cdots = a_k = a$, $b_1 = b_2 = \cdots = b_k = b$ and $c_1 = c_2 = \cdots = c_k = c$. Let $D_k = \mathbf{det}\left(T_k \right)$, then (1.71) becomes a difference equation given by

$$D_{k+2} - aD_{k+1} + bcD_k = 0 \tag{1.72}$$

subject to $D_0 = 1$ and $D_1 = a$. Let $\Delta = \sqrt{a^2 - 4bc}$. Show that the solution of (1.72) is given by

$$D_k = \begin{cases} \dfrac{(a+\Delta)^{k+1} - (a-\Delta)^{k+1}}{2^{k+1}\Delta} & \text{if } \Delta \neq 0 \\ (1+k)\left(\dfrac{a}{2}\right)^k & \text{if } \Delta = 0 \end{cases} \tag{1.73}$$

(Hint: The solution given in (1.73) can be derived from (1.72) by treating it as a difference equation subject to the two initial conditions. These methods are given in Section 7.4. For this problem, all that is needed is to check whether (1.73) will satisfy (1.72).)

E1.24. The general equation of an ellipse is given by

$$\left(\frac{x}{a_1}\right)^2 + \left(\frac{y}{a_2}\right)^2 - 2\frac{xy}{a_1 a_2}\cos\delta = \sin^2\delta \tag{1.74}$$

where $a_1,\ a_2,\ \sin(\delta) \neq 0$.
Let $v = (x, y)^T$, find matrix A such that equation (1.74) can be written as

$$v^T A v = 1$$

(Note: $v^T A v = 1$ is the general equation of a conic, and if A is *positive definite*, then the conic is an ellipse.)

E1.25. Show that $\mathbf{tr}\,(AB) = \mathbf{tr}\,(BA)$ (assuming conformability conditions are met).

E1.26. Let $f(x_1, x_2)$ be given by

$$f(\mathbf{x}) = \exp\left(-3\left((x_1-1)^2 + (x_2+1)^2\right)\right) - \exp\left(-3\left((x_1+1)^2 + (x_2-1)^2\right)\right)$$

Find the gradient $df/d\mathbf{x}$ at $\mathbf{x} = (0,0)^T$, $\mathbf{x} = (1,-1)^T$ and $\mathbf{x} = (-1,1)^T$. Also, find the Hessian at $\mathbf{x} = (1,-1)^T$ and $\mathbf{x} = (-1,1)^T$ and determine whether they are positive or negative definite.

E1.27. Determine which of the following matrices are positive definite:

$$A = \begin{pmatrix} 3 & 4 \\ 1 & 2 \end{pmatrix} \qquad B = \begin{pmatrix} 3 & 4 \\ 1 & 0 \end{pmatrix} \qquad C = \begin{pmatrix} 0 & 2 \\ -2 & 0 \end{pmatrix}$$

E1.28. Let A be a square matrix containing a zero in the main diagonal. Can A be positive definite? Why or why not?

E1.29. Prove the following equality

$$\frac{d}{dt}(AB) = \left(\frac{d}{dt}A\right)B + A\left(\frac{d}{dt}B\right)$$

E1.30. If A is nonsingular, prove that

$$\frac{d}{dt}A^{-1} = -A^{-1}\frac{dA}{dt}A^{-1} \tag{1.75}$$

E1.31. A Marquardt vector update with a scalar parameter μ is defined by

$$\mathbf{p}(\mu) = -\left(J^T J + \mu I\right)^{-1} J^T \mathbf{F}$$

where $J[=]m \times n$ and $\mathbf{F}[=]m \times 1$ are constant matrices with $m \geq n$. This vector is useful in solving unconstrained minimization algorithms. Another scalar function $\phi(\mu)$ is used to indicate the proximity of $\|\mathbf{p}\| = \sqrt{\mathbf{p}^T\mathbf{p}}$ to a fixed value δ,

$$\phi(\mu) = \sqrt{\mathbf{p}^T\mathbf{p}} - \delta$$

Show that

$$\frac{d\phi}{d\mu} = -\frac{\mathbf{p}^T\left(J^TJ + \mu I\right)^{-1}\mathbf{p}}{\sqrt{\mathbf{p}^T\mathbf{p}}}$$

E1.32. Let $\mathcal{P}$ be a simply connected polygon (i.e., containing no holes) in a 2D plane described by points (x_i, y_i) indexed in a counterclockwise manner. Then the area of polygon $\mathcal{P}$, $\text{area}_{\mathcal{P}}$, can be obtained by the following calculations:[11]

$$\text{area}_{\mathcal{P}} = \frac{1}{2}\,\text{abs}\left(\mathbf{det}\begin{pmatrix} x_1 & x_2 \\ y_1 & y_2 \end{pmatrix} + \mathbf{det}\begin{pmatrix} x_2 & x_3 \\ y_2 & y_3 \end{pmatrix} + \cdots + \mathbf{det}\begin{pmatrix} x_N & x_1 \\ y_N & y_1 \end{pmatrix}\right) \tag{1.76}$$

1. Show that (1.76) can be recast as follows:

$$\text{area}_{\mathcal{P}} = \frac{1}{2}\,\text{abs}\left(\mathbf{x}^T\left(S_h - S_h^T\right)\mathbf{y}\right) \tag{1.77}$$

where S_h is the shift matrix given by

$$S_h = \left(\begin{array}{c|ccc} 0 & 1 & & 0 \\ \vdots & & \ddots & \\ 0 & 0 & & 1 \\ \hline 1 & 0 & \cdots & 0 \end{array}\right)$$

2. Verify (1.77) for the simple case of the triangle determined by $(x, y)_1 = (1, 1)$, $(x, y)_2 = (2, 1)$, and $(x, y)_3 = (1.8, 2)$. Furthermore, notice that if we replace the third point by $(x, y)_3 = (a, 2)$ for arbitrary real values for a, we will obtain the same area. Explain why this is the case, both for the figure and the formula.
3. The points covering the area of crystal growth was obtained from a scanned image and given in Table 1.10. Plot the polygon and then obtain the area using (1.77).

E1.33. Consider the generalized Poisson equation in a rectangular domain $L \times W$ given by

$$\frac{\partial^2 u}{\partial x^2} + \frac{\partial^2 u}{\partial y^2} = \Phi(x, y)u + \Gamma(x, y)$$

subject to the boundary conditions,

$$u(0, y) = f_0(y) \qquad u(x, 0) = g_0(x)$$
$$u(L, y) = f_L(y) \qquad u(x, W) = g_W(x)$$

Using the central difference approximation,

$$\frac{\partial^2 u}{\partial x^2} \approx \frac{u_{n+1,m} - 2u_{n,m} + u_{n-1,m}}{\Delta x^2} \quad \text{and} \quad \frac{\partial^2 u}{\partial y^2} \approx \frac{u_{n,m+1} - 2u_{n,m} + u_{n,m-1}}{\Delta y^2}$$

[11] M. G. Stone, *A mnemonic for areas of polygons*, The American Mathematical Monthly, vol. 93, no. 6, (1986), pp. 479–480.

Table 1.10. *Envelope points for crystal growth*

x	y	x	y	x	y
0.2293	0.3991	0.6970	0.5307	0.4343	0.1447
0.2984	0.4488	0.7431	0.5015	0.4067	0.1681
0.3583	0.4781	0.7385	0.4693	0.3813	0.2208
0.3813	0.5541	0.6901	0.4459	0.3629	0.2675
0.3813	0.6155	0.6210	0.3991	0.3445	0.2939
0.3975	0.6915	0.5634	0.4079	0.3353	0.3114
0.4459	0.7325	0.5173	0.3670	0.3168	0.3260
0.4988	0.6827	0.5081	0.2675	0.2800	0.3319
0.5311	0.5892	0.4965	0.1944	0.2569	0.3406
0.5749	0.5687	0.4666	0.1535	0.2339	0.3582
0.6555	0.5482				

where $n = 1, 2, \cdots, N$, $m = 1, 2, \cdots, M$, $\Delta x = L/(N+1)$, $\Delta y = W/(M+1)$ and $u_{n,m} = u(n\Delta x, m\Delta y)$. Show that the finite difference approximation will yield the following matrix equation:

$$AU + UB = Q \circ U + H \tag{1.78}$$

where $A = (1/\Delta x^2)T_{[N\times N]}$, $B = (1/\Delta y^2)T_{[M\times M]}$, $Q = [q_{nm}]$,

$$T_{[K\times K]} = \begin{pmatrix} -2 & 1 & & 0 \\ 1 & -2 & \ddots & \\ & \ddots & \ddots & 1 \\ 0 & & 1 & -2 \end{pmatrix} [=] K \times K$$

$$H = \begin{pmatrix} \gamma_{11} & \cdots & \gamma_{1M} \\ \vdots & \ddots & \vdots \\ \gamma_{N1} & \cdots & \gamma_{NM} \end{pmatrix} - \frac{1}{\Delta x^2} \begin{pmatrix} f_{0,1} & \cdots & f_{0,M} \\ 0 & \cdots & 0 \\ & \vdots & \\ 0 & \cdots & 0 \\ f_{L,1} & \cdots & f_{L,M} \end{pmatrix}$$

$$- \frac{1}{\Delta y^2} \begin{pmatrix} g_{0,1} & 0 & & 0 & g_{W,1} \\ \vdots & \vdots & \cdots & \vdots & \vdots \\ g_{0,N} & 0 & & 0 & g_{W,N} \end{pmatrix}$$

with $q_{nm} = \Phi(n\Delta x, m\Delta y)$, $\gamma_{nm} = \Gamma(n\Delta x, m\Delta y)$, $f_{0,m} = f_0(m\Delta y)$, $f_{L,m} = f_L(m\Delta y)$, $g_{0,n} = g_0(n\Delta x)$, and $g_{W,n} = g_W(n\Delta x)$. Furthermore, show that (1.78) can be transformed to a linear equation given by

$$Z\, \mathbf{vec}(U) = \mathbf{vec}(H)$$

where $Z = I_N \otimes A + B \otimes I_M - \mathbf{diag}\Big(\mathbf{vec}(Q)\Big)$.

E1.34. Prove that all the alternative quadratic forms given in example 1.4 are equivalent, that is, $\mathbf{x}^T Q\mathbf{x} = \mathbf{x}^T L\mathbf{x} = \mathbf{x}^T R\mathbf{x}$.

2 Solution of Multiple Equations

One of the most basic applications of matrices is the solution of multiple equations. Generally, problems involving multiple equations can be categorized as either linear or nonlinear types. If the problems involve only linear equations, then they can be readily formulated as $A\mathbf{x} = \mathbf{b}$, and different matrix approaches can be used to find the vector of unknowns given by $\mathbf{x}$. When the problem is nonlinear, more complex approaches are needed. Numerical approaches to the solution of nonlinear equations, such as the Newton method and its variants, also take advantage of matrix equations.

In this chapter, we first discuss the solution of the linear equation $A\mathbf{x} = \mathbf{b}$. This includes direct and indirect methods. The indirect methods are also known as iterative methods. The distinguishing feature between these two types of approaches is that direct methods (or noniterative) methods obtain the solution using various techniques such as reduction by elimination, factorization, forward or backward substitution, matrix splitting, or direct inversion. Conversely, the indirect (iterative) methods require an initial guess for the solution, and the solution is improved using iterative algorithms until the solution meets some specified criterion of maximum number of iterations or minimum tolerance on the errors.

The most direct approach is to simply apply the inverse formula given by

$$\mathbf{x} = A^{-1}\mathbf{b}$$

This is a good approach as long as the inverse can be found easily, for example, when the matrix is orthogonal or unitary. Also, if the matrix is diagonal or triangular, Section 1.3.3 gives some direct formulas for their inverse. However, in general, the computation of the matrix inverse using the adjoint formula given in (1.16) is not the method of choice, specially for large systems.

In Section 2.1, we discuss the first direct method known as the **Gauss-Jordan elimination method**. This simple procedure focuses on finding matrices Q and W such that QAW will yield a block matrix with an identity matrix in the upper left corner and zero everywhere else. In Section 2.2, a similar approach known as the **LU decomposition method** is discussed. Here, matrix A is factored as $A = LU$, with L and U being upper triangular and lower triangular matrices, respectively. The triangular structures of L and U allow for quick computation of the unknowns via forward and backward substitutions. Other methods such as matrix splitting techniques, which take advantage of special structures of A, are also discussed, with details given the appendix as Section B.6.

In Section 2.4, we switch to indirect (iterative) methods, which include: Jacobi, Gauss-Seidel, and the succesive over-relaxation (SOR). Other iterative methods, such as conjugate-gradient (CG) and generalized minimal residual (GMRES) methods, are also discussed briefly, with details given in the appendices as Sections B.11 and B.12.2.

In Section 2.5, we obtain the least-squares solution. These are useful in parameter estimation of models based on data. We also include a method for handling least-squares problems that involve linear equality constraints.

Having explored the various methods for solving linear equations, we turn our attention to the solution of multiple nonlinear equations in Section 2.9. We limit our discussion to numerical solutions based on Newton's methods, because they involve the application of matrix equations.

2.1 Gauss-Jordan Elimination

Assuming A is nonsingular, the **Gauss-Jordan Elimination** method basically involves the determination of two nonsingular matrices Q and W such that $QAW = I$. Assuming both matrices have been found, the linear equation can be solved as follows:

$$\begin{aligned} A\mathbf{x} &= \mathbf{b} \\ QA\mathbf{x} &= Q\mathbf{b} \\ (QAW)\, W^{-1}\mathbf{x} &= Q\mathbf{b} \\ W^{-1}\mathbf{x} &= Q\mathbf{b} \\ \mathbf{x} &= WQ\mathbf{b} \end{aligned} \tag{2.1}$$

This method can also be used to find the inverse,

$$A^{-1} = WQ \tag{2.2}$$

If A is nonsingular, there are several values of Q and W that will satisfy $QAW = I$.[1]

EXAMPLE 2.1. Consider the following:

$$\begin{pmatrix} 2 & -1 & 1 \\ 4 & 5 & -1 \\ -1 & 3 & 2 \end{pmatrix} \begin{pmatrix} x_1 \\ x_2 \\ x_3 \end{pmatrix} = \begin{pmatrix} -1 \\ 15 \\ 3 \end{pmatrix}$$

Leaving the details for obtaining Q and W for now, note that the following matrices will satisfy $QAW = I$:

$$Q = \begin{pmatrix} 0 & 1/5 & 0 \\ 0 & 3/17 & -5/17 \\ 17/50 & -1/10 & 7/25 \end{pmatrix} \qquad W = \begin{pmatrix} 0 & 1 & 13/17 \\ 1 & -4/5 & -7/17 \\ 0 & 0 & 1 \end{pmatrix}$$

[1] Suppose $QAW = I$, then $\widehat{Q} = TQ$ and $\widehat{W} = WT^{-1}$ for any nonsingular T will also yield $\widehat{Q}A\widehat{W} = I$ and $A^{-1} = WQ = \widehat{W}\widehat{Q}$.

then (2.1) yields the solution

$$\mathbf{x} = WQ\begin{pmatrix} -1 \\ 15 \\ 3 \end{pmatrix} = \begin{pmatrix} 1 \\ 2 \\ -1 \end{pmatrix}$$

Even when A is singular or A is a non-square $N \times M$ matrix, the same approach can be generalized to include these cases. That is, there exist nonsingular matrices Q and W such that QAW yields a partitioned matrix consisting of an $r \times r$ identity matrix on the upper left portion and zero matrices on the remaining partitions,

$$QAW = \left(\begin{array}{c|c} I_r & 0_{[r,M-r]} \\ \hline 0_{[N-r,r]} & 0_{[N-r,M-r]} \end{array}\right) \tag{2.3}$$

When A is nonsingular, $r = N = M$; otherwise, $r \leq \min(N, M)$. The value of r, which is the size of the identity in the upper left corner, is known as the **rank** of matrix A. The rank indicates how many columns or rows of A are linearly independent.

2.1.1 Evaluation of Q and W

The classical approach of the Gauss-Jordan elimination method is based on finding a group of elementary column and row matrix operations (cf. Section 1.2.2.1) that would sequentially replace the kth row and k^{th} column with zeros except for the diagonal entry, which is replaced by 1 (hence the name "elimination"). For instance, let A be given by

$$A = \begin{pmatrix} 1 & 1 & 1 \\ -1 & 2 & 3 \\ 2 & 4 & 3 \end{pmatrix}$$

With

$$E_L = \begin{pmatrix} 0 & 0 & 1/4 \\ 0 & 1 & -1/2 \\ 1 & 0 & -1/4 \end{pmatrix} \quad \text{and} \quad E_R = \begin{pmatrix} 0 & 1 & 0 \\ 1 & -1/2 & -3/4 \\ 0 & 0 & 1 \end{pmatrix}$$

we have

$$E_L A E_R = \left(\begin{array}{c|cc} 1 & 0 & 0 \\ \hline 0 & -2 & 3/2 \\ 0 & 1/2 & 1/4 \end{array}\right)$$

Matrix E_L and E_R can be easily formulated using the formulas given in (B.3) to (B.6). The next stage is to extract the lower right block matrix and then apply the elimination process once more. The process stops when the lower right blocks are all zeros. The complete details of the classic Gauss-Jordan elimination are given in Section B.1 as an appendix. In addition, a MATLAB program `gauss_jordan.m` is available on the book's webpage.

Another alternative to finding matrices Q and W is based on the Singular Value Decomposition (SVD) method. This approach has been found to be more stable than the Gauss-Jordan elimination approach. Details on how to use SVD for this purpose are given in Section B.2 as an appendix.

2.1.2 Singular A and Partial Rank Cases

When $r < \min(N, M)$, two outcomes are possible: either there are an infinite number of solutions or there are no solutions. Starting with $A\mathbf{x} = \mathbf{b}$, using Q and W that satisfy (2.3),

$$(QAW)\left(W^{-1}\mathbf{x}\right) = Q\mathbf{b}$$

$$\left(\begin{array}{c|c} I_r & 0_{[r,M-r]} \\ \hline 0_{[N-r,r]} & 0_{[N-r,M-r]} \end{array}\right)\left(\begin{array}{c} \mathbf{y}_{\text{upper}} \\ \hline \mathbf{y}_{\text{lower}} \end{array}\right) = \left(\begin{array}{c} Q_{\text{upper}} \\ \hline Q_{\text{lower}} \end{array}\right)\mathbf{b} \tag{2.4}$$

where $\mathbf{y} = W^{-1}\mathbf{x}$. The first r equations are then given by

$$\mathbf{y}_{\text{upper}} = Q_{\text{upper}}b \tag{2.5}$$

whereas the last $(N - r)$ equations of (2.4) are given by

$$0\ \mathbf{y}_{\text{upper}} + 0\ \mathbf{y}_{\text{lower}} = Q_{\text{lower}}\mathbf{b} \tag{2.6}$$

This means that if $(Q_{\text{lower}}\mathbf{b}) \neq 0$, no exact solution exists. However, if $(Q_{\text{lower}}\mathbf{b}) = 0$, $\mathbf{y}_{\text{lower}}$ can be any arbitrary vector, implying an infinite set of solutions. Equivalently, the existence of solutions can be determined by checking whether the rank r of A is the same as the rank of the enlarged matrix formed by appending A with $\mathbf{b}$, that is, solutions exist if and only if

$$\text{rank}\left(\begin{array}{c|c} A & \mathbf{b} \end{array}\right) = \text{rank}\left(\begin{array}{c} A \end{array}\right) \tag{2.7}$$

Note that when $\mathbf{b} = 0$, both (2.6) and (2.7) are trivially satisfied.

Suppose $Q_{\text{lower}}\mathbf{b} = 0$ in (2.6). Using $\mathbf{y}_{\text{upper}}$ from (2.5) and appending it with $(M - r)$ arbitrary constants c_i for $\mathbf{y}_{\text{lower}}$, we have

$$W^{-1}\mathbf{x} = \mathbf{y} = \left(\begin{array}{c} Q_{\text{upper}}\mathbf{b} \\ \hline c_1 \\ \vdots \\ c_{M-r} \end{array}\right) \quad \Longrightarrow \quad \mathbf{x} = W\left(\begin{array}{c} Q_{\text{upper}}\mathbf{b} \\ \hline c_1 \\ \vdots \\ c_{M-r} \end{array}\right)$$

Let W_L and W_R be matrices formed by the first r columns and the last $(M-r)$ columns of W, respectively; then the solutions are given by

$$\mathbf{x} = \left(\begin{array}{c|c} W_L & W_R \end{array} \right) \left(\begin{array}{c} Q_{\text{upper}}\mathbf{b} \\ \hline c_1 \\ \vdots \\ c_{M-r} \end{array} \right) = W_L Q_{\text{upper}}\mathbf{b} + W_R \begin{pmatrix} c_1 \\ \vdots \\ c_{N-r} \end{pmatrix} \tag{2.8}$$

EXAMPLE 2.2. Consider the following equation for $A\mathbf{x} = \mathbf{b}$,

$$\begin{pmatrix} 1 & 2 & 1 \\ 3 & 2 & 4 \\ 2 & 4 & 2 \end{pmatrix} \begin{pmatrix} x_1 \\ x_2 \\ x_3 \end{pmatrix} = \begin{pmatrix} 0 \\ 1 \\ 0 \end{pmatrix}$$

The values of r, Q, and W can be found to be $r = 2$ and

$$Q = \begin{pmatrix} 0 & 0 & 1/4 \\ 0 & 1/3 & -1/6 \\ 1 & 0 & -1/2 \end{pmatrix} \qquad W = \begin{pmatrix} 0 & 0 & 1 \\ 1 & -1/2 & -1/6 \\ 0 & 1 & -2/3 \end{pmatrix}$$

and

$$W_L = \begin{pmatrix} 0 & 0 \\ 1 & -1/2 \\ 0 & 1 \end{pmatrix} \quad \text{and} \quad W_R = \begin{pmatrix} 1 \\ -1/6 \\ -2/3 \end{pmatrix}$$

Because $Q_{\text{lower}}\mathbf{b} = 0$, we have infinite solutions. Using (2.8),

$$\mathbf{x} = \begin{pmatrix} 0 \\ -1/6 \\ 1/3 \end{pmatrix} + \begin{pmatrix} 1 \\ -1/6 \\ -2/3 \end{pmatrix} c_1$$

Remarks:

1. The difference DOF $= (M - r)$ is also known as the **degree of freedom**, and it determines the number of arbitrary constants available (assuming that solution is possible).
2. Full-rank non-square matrices can be further classified as either **full-column-rank** (when $r = M < N$), also known as **full-tall matrices**, or **full-row-rank** (when $r = N < M$), also known as **full-wide matrices**.

   For the full-tall case, assuming that $Q_{\text{lower}}\mathbf{b} = 0$, (2.8) implies that only one solution is possible, because DOF $= 0$. However, for the full wide-case, the condition given in (2.6) is not necessary to check because Q_{lower} is no longer available. This means that for full-wide matrices, infinite solutions are guaranteed with DOF $= M - r$.
3. In Section 2.5, the linear parameter estimation problem involves full-tall matrices A, in which the rows are associated with data points whereas the columns are associated with the number of unknown parameters $x_1, \ldots, x_N$. It is very likely

that the $A\mathbf{x} = \mathbf{b}$ will not satisfy (2.6), that is, there will be no exact solutions. The problem will have to be relaxed and modified to the search for $\mathbf{x}$ that would minimize the difference $(\mathbf{b} - A\mathbf{x})$ based on the Euclidean norm. The modified problem is called the **least-squares problem**, which is discussed later in Section 2.5.

2.2 LU Decomposition

Instead of using the matrices Q and W resulting from the Gauss-Jordan elimination method, one can use a factorization of a square matrix A known as the **LU decomposition** in which L and U are lower and upper triangular matrices, respectively, such that

$$A = LU \tag{2.9}$$

The special structures of L and U allow for a two-phase approach to solving the linear problem $A\mathbf{x} = \mathbf{b}$: a **forward-substitution phase** followed by a **backward-substitution phase**. Let $\mathbf{y} = U\mathbf{x}$, then the LU decomposition in (2.9) results in

$$\begin{aligned} A\mathbf{x} &= \mathbf{b} \\ L(U\mathbf{x}) &= \mathbf{b} \\ L\mathbf{y} &= \mathbf{b} \end{aligned}$$

$$\begin{pmatrix} \ell_{11} & & & 0 \\ \ell_{21} & \ell_{22} & & \\ \vdots & \vdots & \ddots & \\ \ell_{N1} & \ell_{N2} & \cdots & \ell_{NN} \end{pmatrix} \begin{pmatrix} y_1 \\ y_2 \\ \vdots \\ y_N \end{pmatrix} = \begin{pmatrix} b_1 \\ b_2 \\ \vdots \\ b_N \end{pmatrix}$$

Assuming L is nonsingular, the forward-substitution phase is the sequential evaluation of $y_1, y_2, \ldots, y_N$,

$$y_1 = \frac{b_1}{\ell_{11}} \quad\rightarrow\quad y_2 = \frac{b_2 - \ell_{21}y_1}{\ell_{11}} \quad\rightarrow\quad \cdots \quad\rightarrow\quad y_N = \frac{b_N - \sum_{i=1}^{N-1} \ell_{Ni}y_i}{\ell_{NN}} \tag{2.10}$$

Once $\mathbf{y}$ has been found, the backward-substitution phase works similarly to sequentially obtain $x_N, x_{N-1}, \ldots, x_1$, that is, with

$$U\mathbf{x} = \mathbf{y}$$

$$\begin{pmatrix} u_{11} & \cdots & u_{1,N-1} & u_{1N} \\ & \ddots & \vdots & \vdots \\ & & u_{N-1,N-1} & u_{N-1,N} \\ 0 & & & u_{NN} \end{pmatrix} \begin{pmatrix} x_1 \\ \vdots \\ x_{N-1} \\ x_N \end{pmatrix} = \begin{pmatrix} y_1 \\ \vdots \\ y_{N-1} \\ y_N \end{pmatrix}$$

then

$$x_N = \frac{y_N}{u_{NN}} \quad\rightarrow\quad x_{N-1} = \frac{y_{N-1} - u_{N-1,N}\,x_N}{u_{N-1,N-1}} \quad\rightarrow\quad \cdots \quad\rightarrow\quad x_1 = \frac{y_1 - \sum_{i=2}^{N} u_{1i}\,x_i}{u_{11}} \tag{2.11}$$

In some textbooks, LU decomposition is synonymous with **Gaussian elimination**.

EXAMPLE 2.3. Consider the linear equation

$$\begin{pmatrix} 2 & 1 & -1 \\ 4 & 0 & 2 \\ 6 & 2 & 2 \end{pmatrix} \mathbf{x} = \begin{pmatrix} -1 \\ 8 \\ 8 \end{pmatrix}$$

One LU decomposition (using methods discussed in the next section) is given by

$$L = \begin{pmatrix} 1 & 0 & 0 \\ 2 & 2 & 0 \\ 3 & 1 & 1 \end{pmatrix} \quad \text{and} \quad U = \begin{pmatrix} 2 & 1 & -1 \\ 0 & -1 & 2 \\ 0 & 0 & 3 \end{pmatrix}$$

The forward substitution yields the solution of $L\mathbf{y} = \mathbf{b}$,

$$y_1 = \frac{-1}{1} = -1 \quad \rightarrow \quad y_2 = \frac{8 - 2(-1)}{2} = 5 \quad \rightarrow \quad y_3 = \frac{8 - 3(-1) - 1(5)}{1} = 6$$

and the backward-substitution gives the solution of $U\mathbf{x} = \mathbf{y}$,

$$x_3 = \frac{6}{3} = 2 \quad \rightarrow \quad x_2 = \frac{5 - 2(2)}{-1} = -1 \quad \rightarrow \quad x_1 = \frac{-1 - (-1)(2) - 1(-1)}{2} = 1$$

Remarks: In MATLAB, one can use the backslash (\) operation for dealing with either forward or backward substitution.[2] Thus, assuming we found lower and upper matrices L and U such that $A = LU$, then for solving the equation $A\mathbf{x} = \mathbf{b}$, the forward substitution to find $\mathbf{y} = L^{-1}\mathbf{b}$ can be implemented in MATLAB as: `y=L\b;`. This is then followed by $\mathbf{x} = U^{-1}\mathbf{y}$, the backward substitution that can be implemented in MATLAB as: `x = U\y`.

2.2.1 Crout, Doolittle, and Choleski Methods for LU Decomposition

To find an LU decomposition of A, we can use the equations for a_{ij} resulting from the special structure of L and U,

$$a_{ij} = \sum_{k=1}^{N} \ell_{ik} u_{kj} = \begin{cases} \ell_{ii} u_{ij} + \sum_{k=1}^{i-1} \ell_{ik} u_{kj} & \text{if } i < j \\ \ell_{ii} u_{jj} + \sum_{k=1}^{i-1} \ell_{ik} u_{kj} & \text{if } i = j \\ \ell_{ij} u_{jj} + \sum_{k=1}^{j-1} \ell_{ik} u_{kj} & \text{if } i > j \end{cases} \tag{2.12}$$

Various LU decomposition algorithms are possible by imposing some additional conditions. **Crout's method** uses the condition $u_{jj} = 1$, whereas **Doolittle's method** uses the condition $\ell_{ii} = 1$. Another LU decomposition known as **Choleski's method** is possible if A is symmetric and positive definite, by setting $L = U^T$, that is, $LL^T = A$. These three methods are summarized in Table 2.1.

[2] When the "backslash" operation is used in MATLAB, different cases are assessed first by MATLAB, that is, the function determines first whether it is sparse, banded, tridiagonal, triangular, full, partial-ranked, and so forth, and then it chooses the appropriate algorithms. More details are available from the help file, mldivide.

Table 2.1. *Methods for LU decomposition*

Name	Algorithm (For $p = 1, \ldots, N$)			
Crout's Method	u_{pp}	$=$	1	
	ℓ_{ip}	$=$	$\left(a_{ip} - \sum_{k=1}^{p-1} \ell_{ik} u_{kp}\right)$	for $i = p, \ldots, N$
	u_{pj}	$=$	$\left(a_{pj} - \sum_{k=1}^{p-1} \ell_{pk} u_{kj}\right) / \ell_{pp}$	for $j = p+1, \ldots, N$
Doolittle's Method	ℓ_{pp}	$=$	1	
	u_{pj}	$=$	$\left(a_{pj} - \sum_{k=1}^{p-1} \ell_{pk} u_{kj}\right)$	for $j = p, \ldots, N$
	ℓ_{ip}	$=$	$\left(a_{ip} - \sum_{k=1}^{p-1} \ell_{ik} u_{kp}\right) / u_{pp}$	for $i = p+1, \ldots, N$
Choleski's Method	ℓ_{pp}	$=$	$\sqrt{a_{pp} - \sum_{k=1}^{p-1} \ell_{pk}^2}$	
	ℓ_{ip}	$=$	$\left(a_{ip} - \sum_{k=1}^{p-1} \ell_{ik} \ell_{pk}\right) / \ell_{pp}$	for $i = p+1, \ldots, N$

Remarks:

1. The nonzero elements of L and U are evaluated column-wise and row-wise, respectively. For example, in Crout's method, at the pth stage, we first set $u_{pp} = 1$ and then evaluate ℓ_{pp}. Thereafter, the pth column of L and the pth row of U are filled in.
2. The advantage of Choleski's method over the other two methods is that the required storage is reduced by half. However, Choleski's method requires the square root operation. Other factorizations are available for symmetric positive definite matrices that avoid the square root operation, for example, $A = LDL^T$, where D is a diagonal matrix.
3. Because the methods use the reciprocal of either ℓ_{ii} or u_{jj}, pivoting, that is, permutations of the rows and columns of A, are sometimes needed, as discussed next.

For nonsingular matrices, pivoting is often needed unless A is positive definite or A is diagonally dominant, that is, if

$$|a_{ii}| \geq \sum_{j \neq i} [a_{ij}] \tag{2.13}$$

A simple rule for pivoting at the pth stage is to maximize $|\ell_{pp}|$ (for Crout's method) or $|u_{pp}|$ (for Doolittle's method) by permuting the last $(N - p)$ columns or rows. In MATLAB, the command for LU factorization is given by: `[L,U,P]=lu(A)`, where $LU = PA$ and P is a permutation matrix. (A MATLAB file `crout_rowpiv.m` is available on the book's webpage that implements Crout's method with row-wise pivoting.)

If A is singular, permutation of both rows and columns of A are needed, that is, one must find permutation matrices P_L and P_R such that

$$P_L A P_R = \left(\begin{array}{c|c} L_1 & 0 \\ \hline B & 0 \end{array}\right) \left(\begin{array}{c|c} U_1 & C \\ \hline 0 & I_{(N-r)} \end{array}\right) \tag{2.14}$$

where L_1 and U_1 are lower and upper triangular matrices, respectively.

A major concern occurs when using LU factorization of large sparse matrices. During the factorization, the zero entries of A may be replaced by nonzero entries in L or U. This situation is known as "**fill-ins**" by L or U. This means that LU factorization could potentially lose the storage and efficiency advantages gained by the sparsity of A. For example, let A be given by

$$A = \begin{pmatrix} 1 & 1 & 1 & 1 & 1 & 1 & 1 \\ 1 & 0 & 1 & 0 & 0 & 0 & 0 \\ 1 & 1 & 0 & 0 & 0 & 0 & 0 \\ 1 & 0 & 0 & 1 & 0 & 0 & 0 \\ 1 & 0 & 0 & 0 & 1 & 0 & 0 \\ 1 & 0 & 0 & 0 & 0 & 1 & 0 \\ 1 & 0 & 0 & 0 & 0 & 0 & 1 \end{pmatrix}$$

then Crout's method will yield the following factors:

$$L = \begin{pmatrix} 1 & 0 & 0 & 0 & 0 & 0 & 0 \\ 1 & -1 & 0 & 0 & 0 & 0 & 0 \\ 1 & 0 & -1 & 0 & 0 & 0 & 0 \\ 1 & -1 & -1 & 2 & 0 & 0 & 0 \\ 1 & -1 & -1 & 1 & \frac{3}{2} & 0 & 0 \\ 1 & -1 & -1 & 1 & \frac{1}{2} & \frac{4}{3} & 0 \\ 1 & -1 & -1 & 1 & \frac{1}{2} & \frac{1}{3} & \frac{5}{4} \end{pmatrix} \quad \text{and} \quad U = \begin{pmatrix} 1 & 1 & 1 & 1 & 1 & 1 & 1 \\ 0 & 1 & 0 & 1 & 1 & 1 & 1 \\ 0 & 0 & 1 & 1 & 1 & 1 & 1 \\ 0 & 0 & 0 & 1 & \frac{1}{2} & \frac{1}{2} & \frac{1}{2} \\ 0 & 0 & 0 & 0 & 1 & \frac{1}{3} & \frac{1}{3} \\ 0 & 0 & 0 & 0 & 0 & 1 & \frac{1}{4} \\ 0 & 0 & 0 & 0 & 0 & 0 & 1 \end{pmatrix}$$

This example shows a significant number of fill-ins compared with the original A. Thus, if one wants to use LU factors to solve the linear equation $A\mathbf{x} = \mathbf{b}$, where A is sparse, some additional preconditioning is often performed before solution. For instance, one could find a permutation matrix P that transforms the problem into

$$\left(PAP^T\right)(P\mathbf{x}) = (P\mathbf{b}) \quad \rightarrow \quad \widehat{A}\widehat{\mathbf{x}} = \widehat{\mathbf{b}}$$

such that $\widehat{A} = \left(PAP^T\right)$ attains a new structure that would allow minimal fill-ins in L and U. One popular approach is to find $\widehat{A}$ that has a small **bandwidth**.

Definition 2.1. *For a square matrix A of size N, the* **left matrix bandwidth of** *A is the value*

$$\mathsf{BW}_{\text{left}}(A) = \max\left(k = j - i \;\middle|\; i \leq j, a_{ij} \neq 0\right)$$

Likewise, the **right matrix bandwidth of** *A is the value*

$$\mathsf{BW}_{\text{right}}(A) = \max\left(k = i - j \;\middle|\; i \geq j, a_{ij} \neq 0\right)$$

The maximum of left or right bandwidths

$$\mathsf{BW}(A) = \max\left(\mathsf{BW}_{\text{left}}(A), \mathsf{BW}_{\text{right}}(A)\right)$$

is known as the **matrix bandwidth of** *A.*

In short, the left bandwidth is the lowest level of the subdiagonal (diagonals below the main diagonal) that is nonzero, whereas the right bandwidth is the highest level of super-diagonal (diagonals above the main diagonal) that is nonzero. For instance, let

$$A = \begin{pmatrix} 1 & 0 & 0 & 0 & 0 & 0 \\ 0 & 2 & 0 & 0 & 0 & 0 \\ 0 & 0 & 0 & 1 & 0 & 0 \\ 0 & 0 & 0 & -1 & 0 & 0 \\ 0 & -1 & 0 & 0 & 3 & 0 \\ 0 & 0 & 0 & 0 & 0 & 1 \end{pmatrix}$$

then $\mathsf{BW}_{\text{left}}(A) = 3$ and $\mathsf{BW}_{\text{right}}(A) = 1$, so $\mathsf{BW}(A) = 3$. Note also that $\mathsf{BW}_{\text{left}}(A^T) = 1$ and $\mathsf{BW}_{\text{right}}(A^T) = 3$, so $\mathsf{BW}(A^T) = 3$. Thus, for symmetric matrices, the bandwidth, the left bandwidth, and the right bandwidth are all the same.

From the LU factorization calculations used in either Crout's or Doolittle's algorithm, it can be shown that the L will have the same left bandwidth as A, and U will have the same right bandwidth of A. This means that the fill-ins by L and U factors can be controlled by reducing the bandwidth of A. One algorithm for obtaining the permutation P such that $\widehat{A} = PAP^T$ has a smaller bandwidth is the **Reverse Cuthill-McKee Reordering** algorithm. Details of this algorithm are given in Section B.4 as an appendix. In MATLAB, the command is given by `p=symrcm(A)`, where `p` contains the sequence of the desired permutation.

2.2.2 Thomas Algorithm

Let A have a tri-diagonal structure given by

$$A = \begin{pmatrix} a_1 & b_1 & & 0 \\ c_1 & \ddots & \ddots & \\ & \ddots & \ddots & b_{N-1} \\ 0 & & c_{N-1} & a_N \end{pmatrix} \tag{2.15}$$

Assuming that no pivoting is needed, for example, when A is diagonally dominant, then the LU decomposition via Crout's method will yield bi-diagonal matrices L and U given by

$$L = \begin{pmatrix} z_1 & & & 0 \\ c_1 & \ddots & & \\ & \ddots & \ddots & \\ 0 & & c_{N-1} & z_N \end{pmatrix} \quad \text{and} \quad U = \begin{pmatrix} 1 & (b_1/z_1) & & 0 \\ & \ddots & \ddots & \\ & & \ddots & b_{N-1}/z_{N-1} \\ 0 & & & 1 \end{pmatrix}$$

where

$$z_1 = a_1 \qquad \text{and} \qquad z_k = a_k - \frac{b_{k-1}c_{k-1}}{z_{k-1}} \quad , \qquad k = 2, \ldots, N \tag{2.16}$$

The forward-substitution phase for solving $L\mathbf{y} = \mathbf{v}$ is then given by

$$y_1 = \frac{v_1}{z_1} \qquad \text{and} \qquad y_k = \frac{1}{z_k}\left(v_k - c_{k-1}y_{k-1}\right) \quad , \qquad k = 2, \ldots, N \tag{2.17}$$

whereas the backward-substitution phase is given by

$$x_N = y_N \qquad \text{and} \qquad x_k = y_k - \frac{b_k x_{k+1}}{z_k} \quad , \qquad k = N-1, \ldots, 1 \tag{2.18}$$

One implication of (2.16), (2.17), and (2.18) is that there is no need to form matrices L and U explicitly. Furthermore, from (2.16), we note that the storage space used for a_k can also be relieved for use by z_k. The method just described is known as the **Thomas algorithm** and it is often used in solving linear equations resulting from one-dimensional finite difference methods. (A MATLAB code `thomas.m` is available on the book's webpage, which implements the Thomas algorithm and takes advantage of storage savings.)

EXAMPLE 2.4. Consider the following two-point boundary value problem:

$$h_2(x)\frac{d^2u}{dx^2} + h_1(x)\frac{du}{dx} + h_0(x)u = f(x)$$

subject to $u(0) = u_0$ and $u(L) = u_L$. We can use the finite difference approximations of the derivatives given by

$$\frac{d^2u}{dx^2} \approx \frac{u_{k+1} - 2u_k + u_{k-1}}{\Delta x^2} \quad \text{and} \quad \frac{du}{dx} \approx \frac{u_{k+1} - u_{k-1}}{2\Delta x}$$

Let $h_{2,k} = h_2(k\Delta x)$, $h_{1,k} = h_1(k\Delta x)$, $h_{0,k} = h_0(k\Delta x)$, $f_k = f(k\Delta x)$ and $u_k = u(k\Delta x)$, where $\Delta x = L/(N+1)$. This results in the following linear equation:

$$\begin{pmatrix} a_1 & b_1 & & 0 \\ c_1 & \ddots & \ddots & \\ & \ddots & \ddots & b_{N-1} \\ 0 & & c_{N-1} & a_N \end{pmatrix} \begin{pmatrix} u_1 \\ u_2 \\ \vdots \\ u_N \end{pmatrix} = \begin{pmatrix} w_1 \\ w_2 \\ \vdots \\ w_N \end{pmatrix}$$

where,

$$a_k = h_{0,k} - \frac{2h_{2,k}}{\Delta x^2} \quad ; \quad b_k = \frac{h_{2,k}}{\Delta x^2} + \frac{h_{1,k}}{2\Delta x} \quad ; \quad c_k = \frac{h_{2,k+1}}{\Delta x^2} - \frac{h_{1,k+1}}{2\Delta x}$$

$$w_k = \begin{cases} f_1 - c_0 u_0 & \text{if } k = 1 \\ f_k & \text{if } 2 \le k \le N-1 \\ f_N - b_N u_L & \text{if } k = N \end{cases}$$

For a specific example, let $L = 4$, $h_2(x) = x^2$, $h_1(x) = x$, $h_0(x) = 7$ and

$$f(x) = \left(-9x^2 + 3x - 7\right)e^{-3x} + \left(5x^3 - 15x^2 + 40x\right)e^{-x}$$

with boundary values $u_0 = -1$ and $u_L = 0.3663$. Using $N = 100$, we apply the Thomas algorithm and obtain

$$\begin{aligned} \mathbf{z}^T &= (5.000, -1.900, 8.737, \ldots, -9.268, -9.434, -9.603) \\ \mathbf{y}^T &= (-0.824, 0.032, -0.108, \ldots, -0.0047, -0.0054, 0.3773) \\ \mathbf{u}^T &= (-1.000, -0.698, -0.423, \ldots, 0.389, 0.377, 0.366) \end{aligned}$$

The exact solution is $u(x) = 5xe^{-x} - e^{-3x}$. Plots of the numerical solution and the exact solution are shown in Figure 2.1.

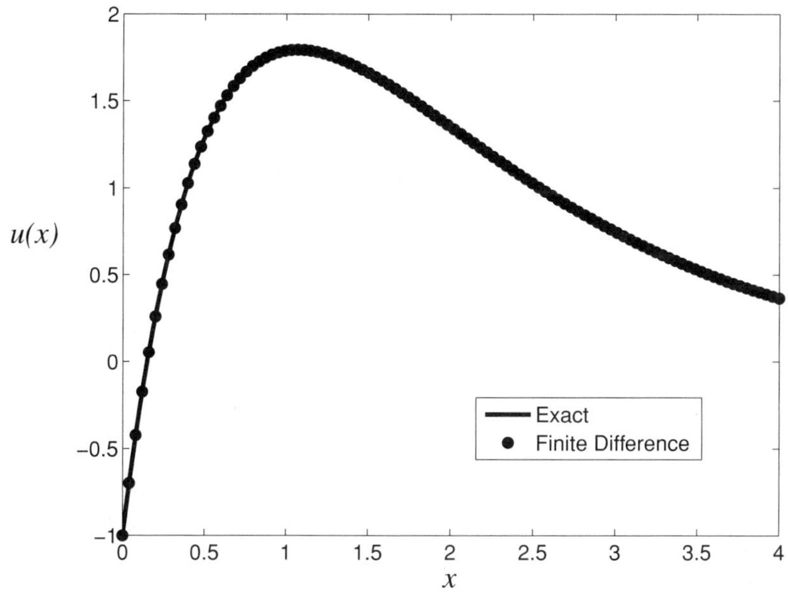

Figure 2.1. A plot of the numerical solution using finite difference approximation and the exact solution for Example 2.4.

A block matrix LU decomposition is also possible. The details for this are given in Section B.5. Likewise, a block matrix version of the Thomas algorithm is available and is left as an exercise (cf. exercise **E2.16**).

2.3 Direct Matrix Splitting

In some cases, a matrix A can be split into two matrices M and S, that is, $A = M + S$, such that M is easy to invert (e.g., diagonal, triangular, tri-diagonal, block diagonal, block triangular, or block tri-diagonal).[3] One possibility is when the rows and columns of $S = A - M$ can be permuted to contain large numbers of zero rows or zero columns; then an efficient method known as **diakoptic method** can be used to solve the linear equation.

EXAMPLE 2.5. Consider the problem $A\mathbf{x} = \mathbf{b}$, where

$$A = \begin{pmatrix} -2 & 1 & 0 & 0 & 0 \\ 1 & -2 & 1 & 0 & 0 \\ -2 & 1 & -2 & 1 & 0 \\ 3 & 0 & 1 & -2 & 1 \\ -2 & 0 & 0 & 1 & -2 \end{pmatrix} \quad \text{and} \quad \mathbf{b} = \begin{pmatrix} 1 \\ 3 \\ 0 \\ 1 \\ 2 \end{pmatrix}$$

We can split matrix A as $A = M + S$ with

$$M = \begin{pmatrix} -2 & 1 & 0 & 0 & 0 \\ 1 & -2 & 1 & 0 & 0 \\ 0 & 1 & -2 & 1 & 0 \\ 0 & 0 & 1 & -2 & 1 \\ 0 & 0 & 0 & 1 & -2 \end{pmatrix} \quad \text{and} \quad S = \begin{pmatrix} 0 & 0 & 0 & 0 & 0 \\ 0 & 0 & 0 & 0 & 0 \\ -2 & 0 & 0 & 0 & 0 \\ 3 & 0 & 0 & 0 & 0 \\ -2 & 0 & 0 & 0 & 0 \end{pmatrix}$$

[3] Formulas for finding the inverse of triangular matrices are given in (1.26) and can be extended to block triangular matrices as in Exercise E1.18.

Then we have

$$(M+S)\mathbf{x}=\mathbf{b} \quad \rightarrow \quad (I+M^{-1}S)\mathbf{x}=M^{-1}\mathbf{b}$$

Since M is tridiagonal and only the first column of S contains nonzero terms, M^{-1} need to be applied only on the first column of S and on $\mathbf{b}$, using the Thomas algorithm. This results in a lower triangular problem which can be solved using forward substitution,

$$\begin{pmatrix} 1.333 & 0 & 0 & 0 & 0 \\ 0.667 & 1 & 0 & 0 & 0 \\ 1.000 & 0 & 1 & 0 & 0 \\ -0.667 & 0 & 0 & 1 & 0 \\ 0.667 & 0 & 0 & 0 & 1 \end{pmatrix} \mathbf{x} = \begin{pmatrix} -3.5 \\ -6.0 \\ -5.5 \\ 5.0 \\ -3.5 \end{pmatrix} \quad \rightarrow \quad \mathbf{x} = \begin{pmatrix} -2.625 \\ -4.250 \\ -2.875 \\ -6.750 \\ -1.750 \end{pmatrix}$$

Another possibility is when A has a special "block-arrow" structure. These problems occur often when domain decomposition methods are applied during the finite difference solutions of partial differential equations. In those cases, the **method of Schur complements** can be used. Details for both general diakoptic methods and Schur complement methods are included in Section B.6 as an appendix.

2.4 Iterative Solution Methods

In this section, we briefly discuss a group of methods known as **stationary iterative methods**. These methods are given by the form

$$\mathbf{x}^{(k+1)} = H\mathbf{x}^{(k)} + \mathbf{c} \tag{2.19}$$

where H and $\mathbf{c}$ are constant, and $\mathbf{x}^{(k)}$ is the kth iteration for the solution $\mathbf{x}$. A solution is accepted when the iterations have converged to a vector $\mathbf{x}^*$, for example, when the norm given by

$$\|\mathbf{x}^{(k+1)} - \mathbf{x}^{(k)}\| = \sqrt{\sum_{i=1}^{N}\left(x_i^{(k+1)} - x_i^{(k)}\right)^2} \tag{2.20}$$

is less than some tolerance ϵ.

Based on the solution of difference equation (cf. section 7.4), one can show that stationary iterative methods will converge if another index based on the matrix H, called the **spectral radius**, is less than one. The spectral radius of H, denoted by $\widehat{\rho}(H)$, is defined as

$$\widehat{\rho}(H) = \max_{i=1}^{N} |\lambda_i(H)| \tag{2.21}$$

where λ_i is the ith **eigenvalue** of H. We discuss eigenvalues and their role in stability in more detail in Section 3.8, Example 3.10. For now, we simply state the the eigenvalues of a matrix $H[=]N \times N$ are the N roots of the polynomial equation,

$$\mathbf{det}(\lambda I - H) = 0 \tag{2.22}$$

Examples of the stationary iterative methods include the Jacobi method, Gauss-Seidel method, and successive over-relaxation (SOR) methods. All three methods result from a matrix splitting

$$A = A_1 + A_2 \tag{2.23}$$

such that $A_1\mathbf{x} = \mathbf{v}$ or A_1^{-1} is easy to solve. In Section 2.3, we showed how matrix splitting can be used for direct solution of linear equations if A possesses special structures such as those given in Sections B.6.1 and B.6.2. However, when the matrices are sparse, yet lacking in special structures, the inverse of $\left(I + A_1^{-1}A_2\right)$ may not be as easy to calculate directly and is likely to significantly increase the number of "fill-ins." In this case, an iterative approach is a viable alternative.

2.4.1 Jacobi Method

The Jacobi method chooses the split by extracting the diagonal of A, that is,

$$A_1 = \begin{pmatrix} a_{11} & & 0 \\ & \ddots & \\ 0 & & a_{N,N} \end{pmatrix} \qquad A_2 = \begin{pmatrix} 0 & a_{12} & \cdots & a_{1N} \\ a_{21} & 0 & \cdots & a_{2N} \\ \vdots & \ddots & \ddots & \vdots \\ a_{N1} & a_{N2} & \cdots & 0 \end{pmatrix}$$

The matrix equation can then be rearranged as follows:

$$\begin{aligned} A\mathbf{x} = (A_1 + A_2)\,\mathbf{x} &= \mathbf{b} \\ A_1\mathbf{x} &= \mathbf{b} - A_2\mathbf{x} \\ \mathbf{x} &= A_1^{-1}\mathbf{b} - A_1^{-1}A_2\mathbf{x} \end{aligned}$$

Starting with an initial guess of $\mathbf{x}^{(0)}$, the Jacobi iteration is given by

$$\mathbf{x}^{(k)} = A_1^{-1}\mathbf{b} - A_1^{-1}A_2\mathbf{x}^{(k-1)} \tag{2.24}$$

where $\mathbf{x}^{(k)}$ is the kth iterate. One often writes (2.24) in the indexed notation as

$$x_i^k = v_i - \sum_{j \neq i} s_{ij} x_j^{k-1} \tag{2.25}$$

where

$$v_i = \frac{b_i}{a_{ii}} \qquad s_{ij} = \frac{a_{ij}}{a_{ii}} \qquad j \neq i$$

Some important issues need to be addressed:

1. Pivoting is needed to make sure that $a_{ii} \neq 0$ for all i.
2. Depending on A, the Jacobi iterations may still not converge.

The first issue is straightforward, and usually pivoting is used to move the elements of maximum absolute values to the main diagonal. The second issue is more challenging. Because the Jacobi method is a stationary iterative method, convergence will depend on the spectral radius of $\left(A_1^{-1}A_2\right)$. If the spectral radius is not less than 1, additional techniques are needed.

2.4.2 Gauss-Seidel Method

The Gauss-Seidel method can be viewed as a simple improvement over the Jacobi method. Recall (2.25) and note that the formula for the kth iteration of the ith element, $\mathbf{x}_i^{(k)}$, means that the elements, $\mathbf{x}_j^{(k)}$ for $j = 1, \ldots, (i-1)$, can already be used on the right-hand side of (2.25). Thus the Gauss-Seidel method is given by

$$x_i^k = v_i - \sum_{j<i} s_{ij} x_j^k - \sum_{n \geq i} s_{ij} x_n^{k-1} \tag{2.26}$$

where

$$v_i = \frac{b_i}{a_{ii}} \qquad s_{ij} = \frac{a_{ij}}{a_{ii}} \qquad j \neq i$$

Alternatively, one can view Gauss-Seidel as using a different splitting, that is $A = \hat{A}_1 + \hat{A}_2$ where

$$\hat{A}_1 = \begin{pmatrix} a_{1,1} & & & 0 \\ a_{2,1} & \ddots & & \\ \vdots & \ddots & \ddots & \\ a_{N,1} & \cdots & a_{N,N-1} & a_{N,N} \end{pmatrix} \quad \hat{A}_2 = \begin{pmatrix} 0 & a_{1,2} & \cdots & a_{1,N} \\ & \ddots & \ddots & \vdots \\ & & \ddots & a_{N-1,N} \\ 0 & & & 0 \end{pmatrix} \tag{2.27}$$

The Gauss-Seidel matrix formulation is then given by

$$\hat{A}_1 \mathbf{x}^{(k)} = \mathbf{b} - \hat{A}_2 \mathbf{x}^{(k-1)} \tag{2.28}$$

Because $\hat{A}_1$ is lower triangular, the Gauss-Seidel equation (2.26) is simply the solution of (2.28) by using a forward-substitution procedure. The convergence of the Gauss-Seidel method is determined by the value of the spectral radius of $\left(\hat{A}_1^{-1}\hat{A}_2\right)$.

2.4.3 Successive Over-Relaxation Methods (SOR)

A further improvement for the ith element can be obtained by a linear combination of the kth Gauss-Seidel approximate and the $(k-1)^{\text{th}}$ approximate. This means that if we let $\bar{x}_i^{(k)}$ be the estimate of the ith element using Gauss-Seidel, it has been found that the convergence toward the solution can be improved by adding an extrapolation step given by

$$x_i^{(k)} = \omega \bar{x}_i^{(k)} + (1-\omega) x_i^{(k-1)} \tag{2.29}$$

Equation (2.29) is known as the **successive over-relaxation (SOR) formula**, and $0 < \omega < 2$ is called the **extrapolation factor**.

The matrix splitting used by the SOR formula is

$$A = L + D + U$$

where

$$L = \begin{pmatrix} 0 & & & 0 \\ a_{2,1} & \ddots & & \\ \vdots & \ddots & \ddots & \\ a_{N,1} & \cdots & a_{N,N-1} & 0 \end{pmatrix} \quad U = \begin{pmatrix} 0 & a_{1,2} & \cdots & a_{1,N} \\ & \ddots & \ddots & \vdots \\ & & \ddots & a_{N-1,N} \\ 0 & & & 0 \end{pmatrix}$$

$$D = \begin{pmatrix} a_{1,1} & & 0 \\ & \ddots & \\ 0 & & a_{N,N} \end{pmatrix}$$

Multiplying A by (ω/ω) and then adding $(D - D)/\omega$,

$$A = \frac{1}{\omega}\left(\omega\left(L + D + U\right) + D - D\right) = \frac{1}{\omega}\left(\left[D + \omega L\right] + \left[(\omega - 1)\, D + \omega U\right]\right)$$

Substituting this into the linear equation and then multiplying both sides by ω,

$$\begin{aligned} \omega\, A\mathbf{x} = \left(\left[D + \omega L\right] + \left[(\omega - 1)\, D + \omega U\right]\right)\mathbf{x} &= \omega\,\mathbf{b} \\ \left[D + \omega L\right]\mathbf{x} &= \left[(1 - \omega)\, D - \omega U\right]\mathbf{x} + \omega\,\mathbf{b} \end{aligned}$$

Letting $\widetilde{A}_1 = D + \omega L$ and $\widetilde{A}_2 = (\omega - 1)\, D + \omega U$, the *SOR* matrix formulation is then given by

$$\widetilde{A}_1\mathbf{x}^{(k)} = \omega\,\mathbf{b} - \widetilde{A}_2\mathbf{x}^{(k-1)} \tag{2.30}$$

Because $\widetilde{A}_1$ is also a lower triangular matrix, the solution of (2.30) will require only a forward substitution. The convergence of the SOR method is determined by the value of the spectral radius of $\left(\tilde{A}_1^{-1}\tilde{A}_2\right)$.

For positive definite, symmetric matrices, for example, $L = U^T$, $0 < \omega < 2$ should be a sufficient condition for convergence. Otherwise, the upper bound on ω may be less than 2. Also, the rate of convergence depends on the relationship between ω and the spectral radius of $\left(\tilde{A}_1^{-1}\tilde{A}_2\right)$.

EXAMPLE 2.6. For the linear equation $A\mathbf{x} = \mathbf{b}$, let

$$A = \begin{pmatrix} -4.2 & 0 & 1 & 1 & 0 & 0 & 1 & 1 \\ 1 & -4.2 & 0 & 1 & 1 & 0 & 0 & 1 \\ 1 & 1 & -4.2 & 0 & 1 & 1 & 0 & 0 \\ 0 & 1 & 1 & -4.2 & 0 & 1 & 1 & 0 \\ 0 & 0 & 1 & 1 & -4.2 & 0 & 1 & 1 \\ 1 & 0 & 0 & 1 & 1 & -4.2 & 0 & 1 \\ 1 & 1 & 0 & 0 & 1 & 1 & -4.2 & 0 \\ 0 & 1 & 1 & 0 & 0 & 1 & 1 & -4.2 \end{pmatrix}; \quad \mathbf{b} = \begin{pmatrix} -6.2 \\ -5.4 \\ 9.2 \\ 0 \\ -6.2 \\ -1.2 \\ 13.4 \\ -4.2 \end{pmatrix}$$

Using an initial guess $\mathbf{x}^{(0)} = (1, 0, 0, 0, 0, 0, 0, 0)^T$, we can compare the performance of Jacobi, Gauss-Seidel, and SOR by tracing the convergence toward the exact solution, which is given by $\mathbf{x}_{\text{exact}} = (1, 2, -1, 0, 1, 1, -2, 1)$. Let k^{th} error be defined as

$$\text{Err}^{(k)} = \sqrt{\sum_{i=1}^{8}\left(x_i^{(k)} - x_{\text{exact},i}\right)^2}$$

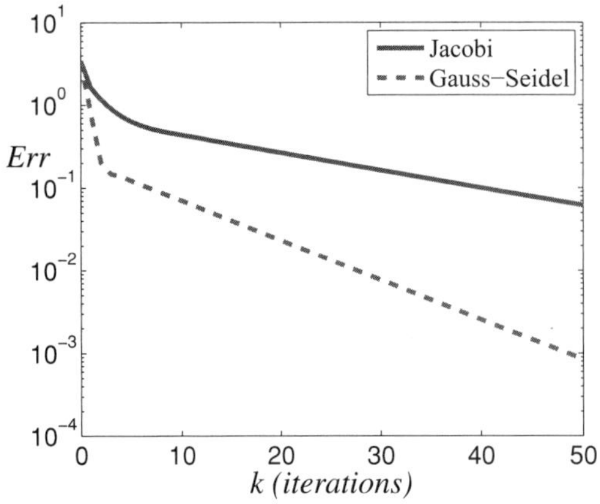

Figure 2.2. Convergence for both the Jacobian method and Gauss-Seidel method applied to the equation in Example 2.5.

then Figure 2.2 shows the convergence between the Jacobi and Gauss-Seidel methods. We see that the Gauss-Seidel method indeed improves on the convergence rate.

When $\omega = 1$, the SOR method reduces to the Gauss-Seidel method. Figure 2.3 shows the convergence performance for $\omega = \{0.8, 1, 1.3, 1.5, 1.7\}$. We see that $\omega > 1$ is needed to gain an improvement over Gauss-Seidel. The optimal choice among the values given was for $\omega = 1.3$. However, the SOR method was divergent for $\omega = 1.7$. Note that A is not a symmetric positive definite matrix.

In Section 2.7 and Section 2.8, we show two more iterative methods: the conjugate gradient method and the generalized minimal residual (GMRES) method, respectively. These methods have good convergence properties and do not sacrifice the sparsity of the matrices. The iterations involved in the conjugate gradient method and GMRES are based on optimization problems that are framed under the linear-algebra perspective of the linear equation $A\mathbf{x} = \mathbf{b}$. Thus, in the next section, we include a brief review of linear algebraic concepts, which also leads to the least-squares solution.

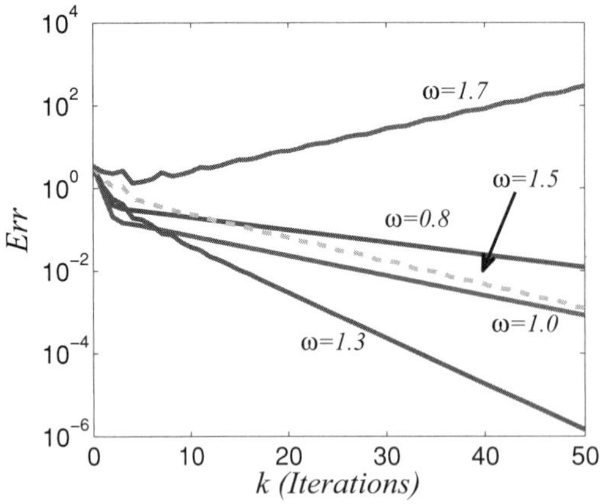

Figure 2.3. Convergence for the SOR method for $\omega = \{0.8, 1, 1.3, 1.5, 1.7\}$ applied to the equation in Example 2.5.

2.5 Least-Squares Solution

In this section, we are changing to the perspective of solving $A\mathbf{x} = \mathbf{b}$ using a linear algebra interpretation. That is, we are now looking for $\mathbf{x}$ as the weights under which the column vectors of A will be combined to match $\mathbf{b}$ as closely as possible. We have included a brief review of some basic terms and notation of linear vector algebra in Section B.7 as an appendix.

For a fixed A and $\mathbf{b}$, we define the **residual vector**, $\mathbf{r}$, as

$$\mathbf{r} = \mathbf{b} - A\mathbf{x} \tag{2.31}$$

The Euclidean norm of the residual vector will be used as a measure of closeness between $\mathbf{b}$ and $A\mathbf{x}$, where the norm of a vector $\mathbf{v}$ is defined by

$$\|\mathbf{v}\| = \sqrt{\mathbf{v}^*\mathbf{v}} \tag{2.32}$$

As noted in Section B.7, one can find $\mathbf{x}$ such that $\mathbf{r} = 0$ only if $\mathbf{b}$ resides in the span of the columns of A. Otherwise, we have to settle for the search of $\widehat{\mathbf{x}}$ that would minimize the norm of $\mathbf{r}$.

Recall from Theorem 1.1 that for $\widehat{\mathbf{x}}$ to yield a local optimum value of $f(\mathbf{x})$, it is sufficient to have have the gradient of f be zero and the Hessian of f be positive definite at $\mathbf{x} = \widehat{\mathbf{x}}$. Furthermore, for the special case in which $f(\mathbf{x})$ is given by

$$f(\mathbf{x}) = \mathbf{x}^*Q\mathbf{x} + \mathbf{w}^*\mathbf{x} + c \tag{2.33}$$

where Q is Hermitian and positive definite, the minimum will be global. To see this, first set the gradient to zero,

$$\left.\frac{d}{d\mathbf{x}}f\right|_{\mathbf{x}=\widehat{\mathbf{x}}} = \widehat{\mathbf{x}}^*Q + \mathbf{w}^* = 0_{[1\times M]} \qquad \rightarrow \qquad Q\widehat{\mathbf{x}} = -\mathbf{w}$$

Because Q is nonsingular, the value for $\widehat{\mathbf{x}}$ will be unique.

Let us now apply this result to minimize the norm of the residual vector. The value of $\widehat{\mathbf{x}}$ that minimizes $\|\mathbf{r}\|$ will also minimize $\|\mathbf{r}\|^2$. Thus,

$$\begin{aligned} \|\mathbf{r}\|^2 &= \mathbf{r}^*\mathbf{r} = (\mathbf{b} - A\mathbf{x})^*(\mathbf{b} - A\mathbf{x}) \\ &= (\mathbf{b}^*\mathbf{b}) + \mathbf{x}^*A^*A\mathbf{x} - 2\mathbf{b}^*A\mathbf{x} \end{aligned}$$

where we used the fact that $\mathbf{b}^*A\mathbf{x}$ is symmetric. Next, evaluating the derivative, $d/d\mathbf{x}\left(\|\mathbf{r}\|^2\right)$, and setting this to zero when $\mathbf{x} = \widehat{\mathbf{x}}$, we have

$$2\widehat{\mathbf{x}}^*A^*A - 2\mathbf{b}^*A = 0$$

or

$$(A^*A)\widehat{\mathbf{x}} = A^*\mathbf{b} \tag{2.34}$$

Equation (2.34) is called the **normal equation**. The matrix A^*A is also known as the **Grammian** (or **Gram matrix**) of A.[4] The solution of (2.34) for $\widehat{\mathbf{x}}$ is unique if and only if A^*A is nonsingular. This condition is equivalent to having the columns of A be an independent set, as given in the following theorem:

[4] Some textbooks refer to A^*A as the **normal matrix**. However, the term **normal matrix** is used in other textbooks (as we do in this text) to refer more generally to matrices K that satisfy $KK^* = K^*K$.

THEOREM 2.1. *The columns of $A[=]N \times M$ are linearly independent if and only if A^*A is nonsingular*

PROOF. The columns of A can be combined linearly as

$$\sum_{j=1}^{M} \alpha_j A_{\bullet,j} = 0 \quad \rightarrow \quad A \begin{pmatrix} \alpha_1 \\ \vdots \\ \alpha_M \end{pmatrix} = 0$$

Multiplying by A^*,

$$A^*A \begin{pmatrix} \alpha_1 \\ \vdots \\ \alpha_M \end{pmatrix} = 0$$

Thus, if A^*A is nonsingular, the unique solution is given by $\alpha_j = 0$ for all j, that is, the columns of A are linearly independent if and only if A^*A is nonsingular.

Returning to the optimization problem, we need to check if the minimum is achieved. The Hessian of $\|\mathbf{r}\|^2$ is given by

$$\frac{d^2}{d\mathbf{x}^2}\left(\|\mathbf{r}\|^2\right) = 2\frac{d}{d\mathbf{x}}\left((A^*A)\,\mathbf{x} - A^*\mathbf{b}\right) = 2A^*A$$

Assuming that the columns of A are linearly independent, then A^*A is positive definite.[5] In summary, suppose the columns of A are linearly independent, then the **least-squares solution** of $A\mathbf{x} = \mathbf{b}$ is given by

$$\widehat{\mathbf{x}} = (A^*A)^{-1} A^*\mathbf{b} = A^\dagger \mathbf{b} \tag{2.35}$$

where $A^\dagger$ is the **pseudo-inverse** of A.

Definition 2.2. *Let $A[=]N \times M$ whose columns are linearly independent. Then the* **pseudo-inverse** *of A, denoted by $A^\dagger$, is defined as*

$$A^\dagger = (A^*A)^{-1} A^* \tag{2.36}$$

If A is nonsingular, the pseudo-inverse is the same as the inverse of A, that is,

$$A^\dagger = (A^*A)^{-1} A^* = A^{-1}(A^*)^{-1} A^* = A^{-1}$$

Other properties of $A^\dagger$ are:

1. $AA^\dagger$ and $A^\dagger A$ are symmetric
2. $AA^\dagger A = A$
3. $A^\dagger A A^\dagger = A^\dagger$

[5] With $\mathbf{y} = A\mathbf{x}$,

$$\mathbf{x}^* A^* A\mathbf{x} = \mathbf{y}^*\mathbf{y} = \sum_{i=1}^{N} \bar{y}_i y_i \geq 0$$

Table 2.2. *Sample experimental data*

$T(^oF)$	P(mm Hg)	C(mole fraction)	$T(^oF)$	P(mm Hg)	C(mole fraction)
60.2	600	0.660	60.7	720	0.721
70.1	602	0.672	69.3	720	0.731
79.9	610	0.686	78.7	725	0.742
90.0	590	0.684	92.6	700	0.742
62.3	680	0.702	60.5	800	0.760
75.5	672	0.711	70.6	800	0.771
77.2	670	0.713	81.7	790	0.777
85.3	670	0.720	91.8	795	0.790

Remark: In MATLAB, the pseudoinverse is implemented more efficiently by the backslash (\ ) operator, thus instead of `x=inv(A'*A)*(A'*b)`, one can use the command: `x=A\b`.

One application of the least-squares solution is to obtain a linear regression model for a multivariable function, that is,

$$y = \alpha_1 v_1 + \cdots + \alpha_M v_M \tag{2.37}$$

where y is a dependent variable and v_j is the jth independent variable. Note that for the special case in which the elements of the matrices are all real, we can replace the conjugate transpose by simple transpose.

EXAMPLE 2.7. Suppose want to relate the effects of temperature and pressure on concentration by a linear model given by

$$C = \alpha T + \beta P + \gamma \tag{2.38}$$

that would fit the set of experimental data given in Table 2.2. Based on the model given in (2.38), we can formulate the problem as $A\mathbf{x} = \mathbf{b}$, with

$$A = \begin{pmatrix} 60.2 & 600 & 1 \\ 70.1 & 602 & 1 \\ \vdots & \vdots & \vdots \\ 91.8 & 795 & 1 \end{pmatrix} \qquad \mathbf{x} = \begin{pmatrix} \alpha \\ \beta \\ \gamma \end{pmatrix} \qquad \mathbf{b} = \begin{pmatrix} 0.660 \\ 0.672 \\ \vdots \\ 0.790 \end{pmatrix}$$

Because exact equality is not possible, we try for a least-squares solution instead, that is,

$$\mathbf{x} = A^{\dagger}\mathbf{b} = \begin{pmatrix} 0.0010 \\ 0.0005 \\ 0.2998 \end{pmatrix}$$

The linear model is then given by

$$C = 0.0010T + 0.0005P + 0.2998 \tag{2.39}$$

A plot of the plane described by (2.39) together with the data points given in Table 2.2 is shown in Figure 2.4.

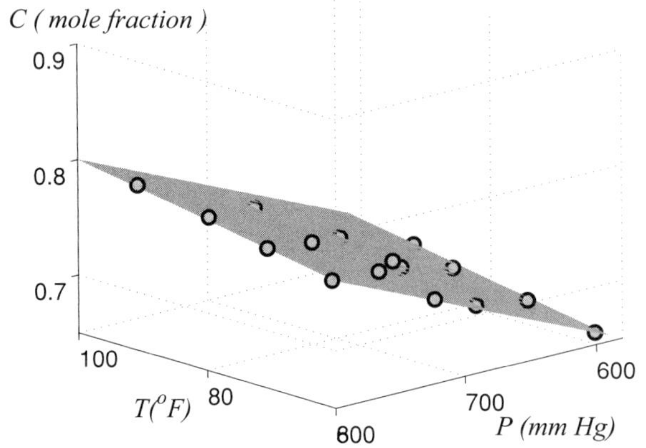

Figure 2.4. A plot of the least-squares model together with data points for Example 2.6.

The least-squares method can also be applied to a more general class of models called the **linear-in-parameter models**. This is given by

$$y(\mathbf{v}) = \sum_{j=1}^{M} \alpha_i f_j(\mathbf{v}) \tag{2.40}$$

where $v_1, \ldots, v_K$ are the independent variables and $y(\mathbf{v}), f_1(\mathbf{v}), \ldots, f_M(\mathbf{v})$ are linearly independent functions. Methods for determining whether the functions are linearly independent are given in Section B.8.

EXAMPLE 2.8. Consider the laboratory data given Table 2.3. The data are to be used to relate vapor pressure to temperature using the Antoine Equation given by:

$$\log_{10}(P_{\text{vap}}) = A - \frac{B}{T + C} \tag{2.41}$$

where P_{vap} is the vapor pressure in mm Hg and T is the temperature in °C. First, we can rearrange (2.41) as

$$T \log_{10}(P_{\text{vap}}) = -C \log_{10}(P_{\text{vap}}) + AT + (AC - B)$$

We can then relate this equation with (2.40) by setting $y = T\log_{10}(P_{\text{vap}})$, $f_1 = \log_{10}(P_{\text{vap}})$, $f_2 = T$ and $f_3(T, P) = 1$, $\alpha_1 = -C$, $\alpha_2 = A$ and $\alpha_3 = AC - B$. After applying the data from Table 2.3,

$$\begin{pmatrix} 1.291 & 29.0 & 1 \\ 1.327 & 30.5 & 1 \\ \vdots & \vdots & \vdots \\ 3.063 & 132.0 & 1 \end{pmatrix} \begin{pmatrix} \alpha_1 \\ \alpha_2 \\ \alpha_3 \end{pmatrix} = \begin{pmatrix} 37.43 \\ 40.47 \\ \vdots \\ 404.29 \end{pmatrix}$$

The normal equation yields $\alpha_1 = -222.5$, $\alpha_2 = 7.39$, and $\alpha_3 = 110.28$, or in terms of the original parameters, $A = 7.39$, $C = 222.5$, and $B = 1534$. Figure 2.5 shows the data points together with the model (2.41).

Table 2.3. *Raw data from vapor pressure experiment*

T (°C)	P_{vap} (mm Hg)	T (°C)	P_{vap} (mm Hg)
29.0	20	83.5	238
30.5	21	90.2	305
40.0	35	105.2	512
45.3	46	110.5	607
53.6	68	123.2	897
60.1	92	130.0	1092
72.0	152	132.0	1156
79.7	206		

Another common situation for some least-squares problem $A\mathbf{x} =_{\text{lsq}} \mathbf{b}$ is that they might be accompanied by linear equality constraints. Let the constraints be given by

$$C\mathbf{x} = \mathbf{z} \tag{2.42}$$

where $A[=]N \times M$, $C[=]K \times M$ with $K < M < N$, and both A and C are full rank. Using Gauss-Jordan elimination on C, we can find Q and W such that

$$QCW = \left(\begin{array}{c|c} I_{[K]} & 0_{[K\times(M-K)]} \end{array} \right)$$

then based on (2.8), the solution to (2.42) is given by

$$\mathbf{x} = W \left(\begin{array}{c} Q\mathbf{z} \\ \hline \mathbf{v} \end{array} \right) \tag{2.43}$$

where $\mathbf{v}$ is a vector containing $(M - K)$ unknown constants. Let $W = \left(W_L \;\middle|\; W_R \right)$ with $W_L[=]M \times K$ and $W_R[=]M \times (M - K)$. Then applying (2.43) to the least-squares problem,

$$\begin{aligned} A\mathbf{x} = A \left(\begin{array}{c|c} W_L & W_R \end{array} \right) \left(\begin{array}{c} Q\mathbf{z} \\ \hline \mathbf{v} \end{array} \right) &= \mathbf{b} \qquad \rightarrow \qquad \mathbf{v} = (AW_R)^{\dagger} \left(\mathbf{b} - AW_LQ\mathbf{z} \right) \\ AW_LQ\mathbf{z} + AW_R\mathbf{v} &= \mathbf{b} \end{aligned}$$

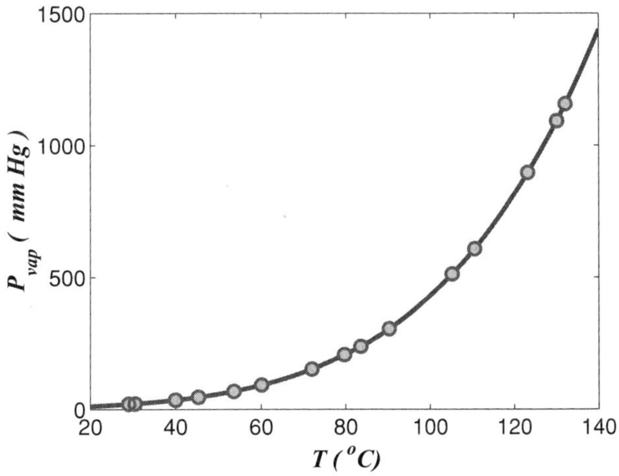

Figure 2.5. Comparison of the Antoine model and raw data.

Table 2.4. *Vapor-liquid equilibrium data*

f_L	f_V	f_L	f_V	f_L	f_V
0	0	0.2931	0.4506	0.5690	0.7549
0.0718	0.1700	0.3190	0.5257	0.6236	0.8103
0.1121	0.2332	0.3362	0.5217	0.6753	0.8142
0.1322	0.1937	0.3937	0.4032	0.7443	0.8300
0.1753	0.2530	0.4052	0.5968	0.7902	0.8972
0.1983	0.3636	0.4483	0.6522	0.9080	0.9289
0.2500	0.3478	0.5172	0.6759	0.9167	0.9802
				1.0000	1.0000

where $(AW_R)^{\dagger}$ is the pseudo-inverse of (AW_R). This result is then used to form the desired least-squares solution,

$$\mathbf{x} = W\left(\begin{array}{c} Q\mathbf{z} \\ \hline \mathbf{v} \end{array}\right) \quad \text{where} \quad \mathbf{v} = (AW_R)^{\dagger}\left(\mathbf{b} - AW_LQ\mathbf{z}\right) \tag{2.44}$$

EXAMPLE 2.9. Suppose we want to fit a second-order polynomial model to relate liquid mole fraction f_L to the vapor mole fraction f_V,

$$f_V = \alpha + \beta f_L + \gamma f_L^2 \tag{2.45}$$

using data given in Table 2.4. The least-squares problem is then given by $A\mathbf{x} = \mathbf{b}$, where

$$A = \begin{pmatrix} 1 & 0 & 0 \\ \vdots & \vdots & \vdots \\ 1 & (f_L)_n & (f_L)_n^2 \\ \vdots & \vdots & \vdots \\ 1 & 1 & 1 \end{pmatrix} \qquad \mathbf{b} = \begin{pmatrix} 0 \\ \vdots \\ (f_V)_n \\ \vdots \\ 1 \end{pmatrix} \qquad \mathbf{x} = \begin{pmatrix} \alpha \\ \beta \\ \gamma \end{pmatrix}$$

The physical constraints due to pure substances require that $f_L = 0$ at $f_V = 0$, and $f_L = 1$ at $f_V = 1$.[6] The constraints are $C\mathbf{x} = \mathbf{z}$, where

$$C = \begin{pmatrix} 1 & 0 & 0 \\ 1 & 1 & 1 \end{pmatrix} \quad \text{and} \quad \mathbf{z} = \begin{pmatrix} 0 \\ 1 \end{pmatrix}$$

Using (2.44), we obtain

$$\mathbf{x} = \begin{pmatrix} 0 \\ 1.6868 \\ -0.6868 \end{pmatrix}$$

Thus the model that satisfies the constraints is given by

$$f_V = -0.6868\, f_L^2 + 1.6868\, f_L$$

However, if the equality constraints were neglected, a different model is obtained and given by

$$f_V = -0.6348\, f_L^2 + 1.5997\, f_L + 0.0263$$

[6] For azeotropic systems, more constraints may have to be included.

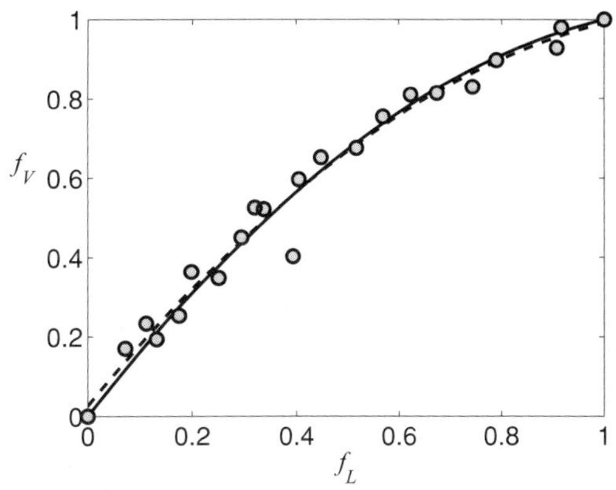

Figure 2.6. Comparison of models using constraints (solid line), models without using constraints (dashed line), and the data points (open circles).

The plots in Figure 2.6 compare both models. Although they appear close to each other, the violation of the constraints by the second model may present some complications, especially when they are applied to a process simulator.

2.6 QR Decomposition

In this section, we introduce another factorization of A known as the **QR decomposition**. This allows for an efficient solution of the least-squares problem.[7]

The QR decomposition of A is given by $A = QR$, such that the columns of Q are orthogonal to each other, that is, $Q^*Q = I$, and R is an upper triangular matrix.[8] Details of the QR algorithm are included in Section C.2.1 as an appendix. We could then apply this factorization to solve the normal equation as follows,

$$\begin{aligned} A^*A\mathbf{x} &= A^*\mathbf{b} \\ (R^*Q^*)(QR)\mathbf{x} &= \\ R^*R\mathbf{x} &= A^*\mathbf{b} \end{aligned} \tag{2.46}$$

Because R is upper triangular, R^*R is already a Choleski LU factorization of A^*A, and a forward and backward substitution can be used to find the least-squares solution $\mathbf{x}$.

Remarks:

1. In MATLAB, QR decomposition can be obtained using the command `[Q,R]=qr(A,0)`, where the option 0 will yield an "economy version" such that $Q[=]N \times M$ and $R[=]M \times M$ if $A[=]N \times M$ and $N > M$. However, if Q is not needed, as in (2.46), the command `R=qr(A)` will yield R if A is stored as a sparse matrix; otherwise, one needs to extract the upper triangular portion, that is, `R=triu(qr(A))`.
2. The QR factorization exists for all A regardless of size and rank. Thus it presents one method for the least-squares solution of $A\mathbf{x} =_{\text{lsq}} \mathbf{b}$ in case A^*A is singular.[9] A MATLAB code that implements the QR algorithm is available on the book's webpage as `QR_house.m`, where `[R,P,Q]=QR_house(A)` yields an

[7] The QR algorithm is also an efficient method for calculating eigenvalues, as discussed in Section C.2.
[8] Another classic method for finding a set of orthogonal vectors that has the same span of a given set of vectors is the Gram-Schmidt method. Details of this method are given in Section B.9.
[9] Another method is to use SVD to find the Moore-Penrose inverse, (cf. Section 3.9.1).

additional permutation matrix P such that $QR = AP$, $Q^*Q = I$ and R is an upper triangular matrix whose diagonal is arranged in decreasing magnitude. This means that if A is not full rank, R can be partitioned to be

$$R = \left(\begin{array}{c|c} \widehat{R} & 0 \\ \hline 0 & 0 \end{array}\right)$$

(see exercise **E2.13**, part b). In this case, due to the permutation P, the normal equation of (2.46) will have to be modified to become

$$R^* R\mathbf{x} = P^T A^* \mathbf{b}$$

If Q is not needed, then use instead `[R,P]=QR_house(A)`.

2.7 Conjugate Gradient Method

The conjugate gradient (CG) method is an iterative method for the solution of $A\mathbf{x} = \mathbf{b}$, where A is a real symmetric positive definite matrix. Let $\mathbf{x}^{(i)}$ be the ith update of $\mathbf{x}$.[10] The method is based on updating the value of $\mathbf{x}^{(i)}$ such that the scalar function $f(\mathbf{x})$ given by

$$f(\mathbf{x}) = \mathbf{x}^T A\mathbf{x} - \mathbf{x}^T \mathbf{b} \tag{2.47}$$

is minimized. Let $\widehat{\mathbf{x}}$ be the exact solution, then $f(\widehat{\mathbf{x}}) = 0$.

Because the conjugate gradient method is iterative, it can take advantage of the sparsity in matrix A when evaluating matrix products. However, unlike the Jacobi, Gauss-Seidel, or SOR methods, the conjugate gradient method is guaranteed to reach the solution within a maximum of N moves assuming there are no round-off errors, where N is the size of $\mathbf{x}$. If small roundoff errors are present, the Nth iteration should still be very close to $\widehat{\mathbf{x}}$.

The conjugate gradient method is currently one of the more practical methods to solve linear equations resulting from finite-element methods. The resulting matrices from finite-element methods are usually very large, sparse, and, in some cases, symmetric and positive definite.

In its simplest formulation, the method involves only a few calculations in each step. The update equations are given by

$$\mathbf{x}^{(i+1)} = \mathbf{x}^{(i)} + \alpha^{(i)} \mathbf{d}^{(i)} \tag{2.48}$$

where $\alpha^{(i)}$ and $\mathbf{d}^{(i)}$ are the ith weight and i^{th} update vector, respectively. The new value will have a residual error given by

$$\mathbf{r}^{(i+1)} = \mathbf{b} - A\mathbf{x}^{(i+1)} \tag{2.49}$$

The conjugate gradient method simply chooses the weight $\alpha^{(i)}$ and update vectors $\mathbf{d}^{(i)}$ such that

$$\left(\mathbf{d}^{(i)}\right)^T A\mathbf{r}^{(j)} = 0 \qquad \text{for } j < i \tag{2.50}$$

[10] We will use the un-bold letter x_i for the ith element of $\mathbf{x}$.

that is, $\mathbf{d}^{(i)}$ will be A-orthogonal, or conjugate, to the past residual vectors $\mathbf{r}^{(j)}$, $j < 1$ (which also happens to be the negative gradient of $f(\mathbf{x})$). This criteria will be satisfied by choosing the following:

$$\mathbf{d}^{(i)} = \begin{cases} \mathbf{r}^{(0)} & \text{for } i = 0 \\ \mathbf{r}^{(i)} - \dfrac{(\mathbf{r}^{(i)})^T A\mathbf{d}^{(i-1)}}{(\mathbf{d}^{(i-1)})^T A\mathbf{d}^{(i-1)}}\mathbf{d}^{(i-1)} & \text{for } i > 0 \end{cases} \quad \text{and} \quad \alpha^{(i)} = \frac{(\mathbf{r}^{(i)})^T \mathbf{r}^{(i)}}{(\mathbf{d}^{(i)})^T A\mathbf{d}^{(i)}} \tag{2.51}$$

The function $f(\mathbf{x})$ will in general have ellipsoidal contours surrounding the origin. The criteria given in (2.50) is based on A-orthogonality. This just means that with $A = S^T S$, the function f will have spherical contours under the new space $\mathbf{y} = S\mathbf{x}$. In this new space, each iteration is orthogonal to past moves, yet are optimally directed to the exact solution from its current subspace. This makes the convergence quite fast. More importantly, all these moves happen without having to evaluate the factor S.

The details of the conjugate gradient method are given in Section B.11 as an appendix. In that section, we include the theorems that prove the claim that criteria (2.50) is attained by using (2.51) and that the method converges in N moves (assuming no roundoff errors).

Remarks: In MATLAB, the command for conjugate gradient method is given by `x=pcg(A,b)` to solve $A\mathbf{x} = \mathbf{b}$. Also, a MATLAB function is available on the book's webpage for the conjugate gradient method `x=conj_grad(A,b,x0)`, where $x0$ is the initial guess.

2.8 GMRES

The conjugate gradient method was developed to solve the linear equation $A\mathbf{x} = \mathbf{b}$, where A is Hermitian and positive definite. For the general case, where A is nonsingular, one could transform the problem to achieve the requirements of the conjugate gradient method in several ways, including $A^*A\mathbf{x} = A^*\mathbf{b}$ or

$$\left(\begin{array}{c|c} I & A \\ \hline A^* & 0 \end{array}\right)\left(\begin{array}{c} r \\ \hline x \end{array}\right) = \left(\begin{array}{c} b \\ \hline 0 \end{array}\right)$$

Another approach, known as the **Generalized Minimal Residual (GMRES) Method**, introduces an iterative approach to solve $A\mathbf{x} = \mathbf{b}$ that updates the solution $\mathbf{x}^{(k)}$ by reducing the norms of residual errors at each iteration. Unlike the conjugate gradient method, GMRES is well suited for cases with non-Hermitian matrix A.

Briefly, in GMRES, a set of orthogonal vectors $\mathbf{u}^{(k)}$ is built sequentially using another method known as **Arnoldi's algorithm**, starting with $\mathbf{u}^{(0)} = \mathbf{r}^{(0)}$. At each introduction of $\mathbf{u}^{(k+1)}$, a matrix U_{k+1} is formed, which is then used to solve an associated least-squares problem

$$\left(U_{k+1}^* A U_k\right) \mathbf{y}_k =_{\text{lsq}} \begin{pmatrix} \|\mathbf{r}^{(0)}\| \\ 0 \\ \vdots \\ 0 \end{pmatrix}$$

to obtain vector $\mathbf{y}_k$. This result is then used to solve for the kth estimate,

$$\mathbf{x}^{(k)} = \mathbf{x}^{(0)} + U_k \mathbf{y}_k \tag{2.52}$$

The process stops when the residual $\mathbf{r}^{(k)} = \mathbf{b} - A\mathbf{x}^{(k)}$ has a norm less than a specified tolerance.

Although it looks complicated, the underlying process can be shown to minimize the residual at every iteration. Thus the rate of convergence is quite accelerated. In some cases, the solution may even be reached in much fewer iterations than N. However, if round-off errors are present or if A is not well conditioned, the algorithm may be slower, especially when U_k starts to grow too large. A practical solution to control the growth of U_k is to invoke some restarts (with the current estimate as the initial guess at the restart). This would degrade the efficiency of the algorithm because the old information would be lost at each restart. Fortunately, the new initial guesses will always be better than the initial guesses of previous restarts.

The details of GMRES, including some enhancements to accelerate the convergence further, can be found in Section B.12 as an appendix.

Remarks: In MATLAB, the command for the GMRES method is given by `x=GMRES(A,b,m)` to solve $A\mathbf{x} = \mathbf{b}$ for $\mathbf{x}$ with restarts after every m iterations. Also, a MATLAB function `gmres_method.m` is available on the book's webpage to allow readers to explore the algorithm directly without implementing restarts.

2.9 Newton's Method

Let us apply some of the methods of the previous sections to the solution of nonlinear equations. Newton's method takes the linearization of nonlinear equations and converts the problem back to a linear one around a local iteration. Consider a set of n nonlinear functions, f_i, of n variables: $x_1, \ldots, x_n$:

$$\mathbf{F}(\mathbf{x}) = \begin{pmatrix} f_1(x_1, \ldots, x_n) \\ \vdots \\ f_n(x_1, \ldots, x_n) \end{pmatrix} = 0$$

Newton's method is an iterative search for the values of $\mathbf{x}$ such that $\mathbf{F}(\mathbf{x})$ is as close to zero as the method allows. Using an initial guess, $\mathbf{x}^{(0)}$, the values of $\mathbf{x}^{(k+1)}$ is updated from $\mathbf{x}^{(k)}$ by adding a correction term $\Delta_k \mathbf{x}$,

$$\mathbf{x}^{(k+1)} = \mathbf{x}^{(k)} + \Delta_k \mathbf{x}$$

To determine the correction term, we can use the Taylor series expansion of $\mathbf{F}(\mathbf{x}^{(k+1)})$ around $\mathbf{x}^{(k)}$,

$$\mathbf{F}\left(\mathbf{x}^{(k+1)}\right) = \mathbf{F}\left(\mathbf{x}^{(k)}\right) + \left.\frac{d\mathbf{F}}{d\mathbf{x}}\right|_{\mathbf{x}=\mathbf{x}^{(k)}} \left(\mathbf{x}^{(k+1)} - \mathbf{x}^{(k)}\right) + \ldots$$

By forcing the condition that $\mathbf{F}\left(\mathbf{x}^{(k+1)}\right) = 0$ while truncating the Taylor series expansion after the second term, we obtain the update equation for Newton's method,

$$\Delta_k \mathbf{x} = \left(\mathbf{x}^{(k+1)} - \mathbf{x}^{(k)}\right) = -J_k^{-1} \mathbf{F}\left(\mathbf{x}^{(k)}\right) \tag{2.53}$$

where J_k is the Jacobian matrix of $\mathbf{F}$ evaluated at $\mathbf{x} = \mathbf{x}^{(k)}$,

$$J_k = \left.\frac{d\mathbf{F}}{d\mathbf{x}}\right|_{\mathbf{x}=\mathbf{x}^{(k)}} = \left(\begin{array}{ccc} \dfrac{\partial f_1}{\partial x_1} & \cdots & \dfrac{\partial f_1}{\partial x_n} \\ \vdots & \ddots & \vdots \\ \dfrac{\partial f_n}{\partial x_1} & \cdots & \dfrac{\partial f_n}{\partial x_n} \end{array}\right)_{\mathbf{x}=\mathbf{x}^{(k)}}$$

The updates are then obtained in an iterative manner until the norm of $\mathbf{F}\left(\mathbf{x}^{(k)}\right)$ is below a set tolerance, ϵ.

In summary, the Newton's method is given by the following procedure:

Algorithm of Newton's Method.

1. **Initialize.** Choose an initial guess: $\mathbf{x}^{(0)}$
2. **Update.** Repeat the following steps until either $\left\|\mathbf{F}\left(\mathbf{x}^{(k)}\right)\right\| \le \epsilon$ or the number of iterations have been exceeded
   (a) Calculate J_k.
   (If J_k is singular, then stop the method and declare "Singular Jacobian.")
   (b) Calculate the correction term: $\Delta_k\mathbf{x} = -J_k^{-1}\mathbf{F}\left(\mathbf{x}^{(k)}\right)$
   (c) Update $\mathbf{x}$: $\mathbf{x}^{(k+1)} = \mathbf{x}^{(k)} + \Delta_k\mathbf{x}$

Remarks:

1. In general, the formulation of the exact Jacobian may be difficult to evaluate. Instead, approximations are often substituted, including the simplest approach, called the **secant method**, which uses finite difference to approximate the partial derivatives, that is,

$$J_k = \left.\left(\begin{array}{ccc} \dfrac{\partial f_1}{\partial x_1} & \cdots & \dfrac{\partial f_1}{\partial x_N} \\ \vdots & \ddots & \vdots \\ \dfrac{\partial f_N}{\partial x_1} & \cdots & \dfrac{\partial f_N}{\partial x_N} \end{array}\right)\right|_{\mathbf{x}=\mathbf{x}^{(k)}} \approx \left(\begin{array}{ccc} s_{11}\left(\mathbf{x}^{(k)}\right) & \cdots & s_{1N}\left(\mathbf{x}^{(k)}\right) \\ \vdots & \ddots & \vdots \\ s_{N1}\left(\mathbf{x}^{(k)}\right) & \cdots & s_{NN}\left(\mathbf{x}^{(k)}\right) \end{array}\right) \tag{2.54}$$

where

$$s_{ij}\left(\mathbf{x}\right) = \frac{f_i\left(x_1, \ldots, x_j + \Delta x_j, \ldots, x_N\right) - f_i\left(x_1, \ldots, x_N\right)}{\Delta x_j}$$

Because the approximation in (2.54) may often be computationally expensive, other approaches include a group of methods known as **Quasi-Newton methods**. One of the most popular is the **Broyden method**, in which the Jacobian J_k is approximated by a matrix $B^{(k)}$ that is updated by the following rule

$$B^{(k+1)} = B^{(k)} + \frac{1}{\left(\Delta_k\mathbf{x}\right)^T \Delta_k\mathbf{x}}\left[\Delta_k\mathbf{F} - B^{(k)}\Delta_k\mathbf{x}\right]\left(\Delta_k\mathbf{x}\right)^T \tag{2.55}$$

where

$$\Delta_k\mathbf{F} = \mathbf{F}\left(\mathbf{x}^{(k+1)}\right) - \mathbf{F}\left(\mathbf{x}^{(k)}\right)$$

and $B^{(0)}$ can be initiated using (2.54). Moreover, because the Newton update in fact needs the inverse of the Jacobian, that is, $\Delta_k\mathbf{x} = -J_k^{-1}\mathbf{F}\left(\mathbf{x}^{(k)}\right) \approx -\left(B^{(k)}\right)^{-1}\mathbf{F}\left(\mathbf{x}^{(k)}\right)$, an update of the inverse of $B^{(k)}$ is more desirable. Thus, using (1.27), (2.55) can be inverted to yield

$$\left(B^{(k+1)}\right)^{-1} = \left(B^{(k)}\right)^{-1} - \frac{1}{\alpha}\left[\left(B^{(k)}\right)^{-1}\Delta_k\mathbf{F} - \Delta_k\mathbf{x}\right]\Delta_k\mathbf{x}^T\left(B^{(k)}\right)^{-1} \qquad (2.56)$$

where

$$\alpha = \Delta_k\mathbf{x}^T\left(B^{(k)}\right)^{-1}\Delta_k\mathbf{F}$$

2. One can also extend Newton's method (and its variants) to the solution of unconstrained minimization

$$\min_{\mathbf{x}} f(\mathbf{x}) \qquad (2.57)$$

by setting

$$\mathbf{F}(\mathbf{x}) = \left(\frac{d}{d\mathbf{x}}f\right)^T \qquad \text{and} \qquad J(\mathbf{x}) = \left(\frac{d^2}{d\mathbf{x}^2}f\right) \qquad (2.58)$$

where $J(\mathbf{x})$ is the Hessian matrix of $f(\mathbf{x})$. The point $\mathbf{x}^*$ becomes a minimimum of $f(\mathbf{x})$ if $(df/d\mathbf{x})^T(\mathbf{x}^*) = 0$ and $d^2f/d\mathbf{x}^2(\mathbf{x}^*) > 0$.

One practical concern with Newton methods is that the convergence to the solutions are strongly dependent on the initial guesses. There are several ways to improve convergence. One approach is known as the **line search method**, also known as the **back-tracking method**, which we discuss next. Another approach is the **double-dogleg method**, and the details of this approach are given in Section B.13 as an appendix.

Remarks: In MATLAB, the command to find the solution of a set of nonlinear equation is `fsolve`. Also, another function is available for nonlinear least squares given by `lsqnonlin`.

2.10 Enhanced Newton Methods via Line Search

Consider the scalar function $f(x) = \tanh(x)$ whose root is given by $x = 0$. Starting with an initial guess $x_0 = 0.9$, the left plot in Figure 2.7 shows that Newton's method converges very close to the root after three iterations. However, if we had chosen $x_0 = 1.1$ as the initial guess, the right plot in Figure 2.7 shows that the method diverges away from the root. This shows that the success of Newton's method is highly dependent on the initial guess. However, it also shows that if only a fraction of the correction step $\Delta_k\mathbf{x}$ had been taken, the next few iterations may move the value of $\mathbf{x}$ into a region in which the Newton method is convergent.

Instead of (2.53), let the update be

$$\Delta_k\mathbf{x} = \lambda\delta_k \qquad (2.59)$$

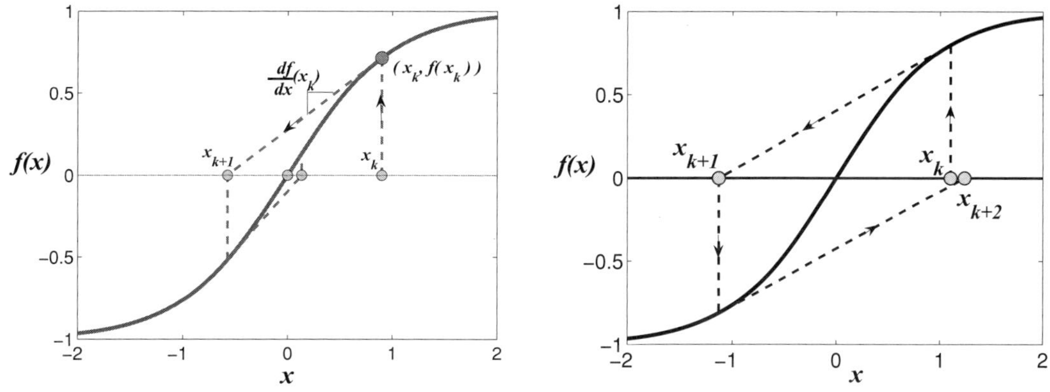

Figure 2.7. The performances of the Newton's method to finding the solution of $\tanh(x) = 0$. The left plot used $x_0 = 0.9$ as the initial guess, whereas the right plot used $x_0 = 1.1$ as the initial guess.

where

$$\delta_k = -J_k^{-1}\mathbf{F}\left(\mathbf{x}^{(k)}\right) \tag{2.60}$$

If λ is chosen too small, the number of iterations may become unnecessarily large. On the other hand, with λ close to 1, we end up with the same Newton update. To determine a good value of λ, we first define the scalar criterion function

$$\phi_k(\lambda) = \frac{1}{2}\left\|\mathbf{F}\left(\mathbf{x}^{(k)} + \lambda\delta_k\right)\right\|^2 \tag{2.61}$$

and search for the value of λ that would minimize $\phi_k(\lambda)$. However, we will use polynomial curve fits instead of using Newton's method. We will denote the line search procedure by searching for a sequence $\{\lambda_0, \lambda_1, \ldots, \lambda_m, \ldots, \lambda_\ell\}$, which terminates when $\mathbf{F}(\mathbf{x}_k + \lambda_\ell\delta_k)$ is acceptable. The conditions for acceptance of λ_ℓ is given later in (2.67). We first discuss the calculations for λ_m, $m = 0, 1, \ldots, \ell$.

The initial value of the line search, λ_0, can be found by using three terms:

$$\begin{aligned}
\phi_k(0) &= \frac{1}{2}\left\|\mathbf{F}\left(\mathbf{x}^{(k)}\right)\right\|^2 \\
\phi_k'(0) &= \left.\frac{d}{d\lambda}\phi_k\right|_{\lambda=0} = \mathbf{F}^T\left(\mathbf{x}^{(k)}\right)J_k\delta_k \\
\phi_k(1) &= \frac{1}{2}\left\|\mathbf{F}\left(\mathbf{x}^{(k)} + \delta_k\right)\right\|^2
\end{aligned}$$

These terms determine a polynomial approximation of $\phi_k(\lambda)$ given by

$$\mathcal{P}_0(\lambda) = \left(\phi_k(1) - \phi_k(0) - \phi_k'(0)\right)\lambda^2 + \left(\phi_k'(0)\right)\lambda + \left(\phi_k(0)\right) \tag{2.62}$$

Setting the derivative of (2.62) equal to zero, we obtain the value λ_0 that would minimize $\mathcal{P}_0$,

$$\lambda_0 = -\frac{\phi_k'(0)}{2\left(\phi_k(1) - \phi_k(0) - \phi_k'(0)\right)} \tag{2.63}$$

If the λ_0 is not acceptable, we continue with an iteration scheme for λ_m, $m = 1, 2, \ldots$, and use the values of ϕ_k at two previous values of λ, that is, $\phi_k(\lambda_{m-1})$ and $\phi_k(\lambda_{m-2})$, together with $\phi_k(0)$ and $\phi_k{}'(0)$. These four terms now determine a unique cubic polynomial given by

$$\mathcal{P}_m(\lambda) = a\lambda^3 + b\lambda^2 + \left(\phi_k{}'(0)\right)\lambda + \left(\phi_k(0)\right) \tag{2.64}$$

where a and b can be obtained by solving

$$\begin{pmatrix} \lambda_{m-1}^3 & \lambda_{m-1}^2 \\ \lambda_{m-2}^3 & \lambda_{m-2}^2 \end{pmatrix} \begin{pmatrix} a \\ b \end{pmatrix} = \begin{pmatrix} \phi_k(\lambda_{m-1}) - \phi_k{}'(0)\lambda_{m-1} - \phi_k(0) \\ \phi_k(\lambda_{m-2}) - \phi_k{}'(0)\lambda_{m-2} - \phi_k(0) \end{pmatrix}$$

The minimum of $\mathcal{P}_m(\lambda)$ can be found to be

$$\lambda_m = \frac{-b + \sqrt{b^2 - 3a\phi_k{}'(0)}}{3a} \tag{2.65}$$

(For $m = 1$, we set $\lambda_{m-2} = \lambda_{-1} = 1$). Although (2.63) and (2.65) generate the minima based on their corresponding polynomials, the fractional reduction needs to be controlled within a range, which is usually set as

$$\frac{1}{10} \le \left(\frac{\lambda_m}{\lambda_{m-1}}\right) \le \frac{1}{2} \tag{2.66}$$

This is to avoid producing a λ that is too small, which could mean very slow convergence of the line search updates or even a premature termination of the solution. On the other hand, if the decrease of λ is too small, the line search updates are more likely to miss the regions that accept regular Newton updates.

The remaining issue is the acceptability condition for λ. A simple criteria is that an acceptable $\lambda = \lambda_\ell$ occurs when the average rate of change of ϕ_k is at least a fraction of the initial rate of change $\phi_k{}'(0)$, for example, for $\alpha \in (0, 1]$,

$$\frac{\phi_k(\lambda_\ell) - \phi_k(0)}{\lambda_\ell} \le \alpha\phi_k{}'(0)$$

or equivalently,

$$\phi_k(\lambda_\ell) \le \phi_k(0) + \alpha\,\lambda_\ell\,\phi_k{}'(0) \tag{2.67}$$

It can be shown that with $\alpha \le 0.25$, (2.65) can be guaranteed to have real roots. However, a usual choice is to set α as low as 10^{-4}. Because $\lambda \le 1$, this parameter is often referred to as the **damping coefficient** of the line method.

To summarize, we have the following enhanced Newton line search procedure:

Algorithm of Enhanced Newton's Method with Line Search.

1. **Initialize.** Choose an initial guess: $\mathbf{x}^{(0)}$
2. **Update.** Repeat the following steps until either $\left\|\mathbf{F}\left(\mathbf{x}^{(k)}\right)\right\| \le \epsilon$ or the number of iterations have been exceeded.
   (a) Calculate J_k.
   (If J_k is singular, then stop the method and declare "Singular Jacobian.")
   (b) Calculate the correction term: $\delta_k = -J_k^{-1}\mathbf{F}\left(\mathbf{x}^{(k)}\right)$

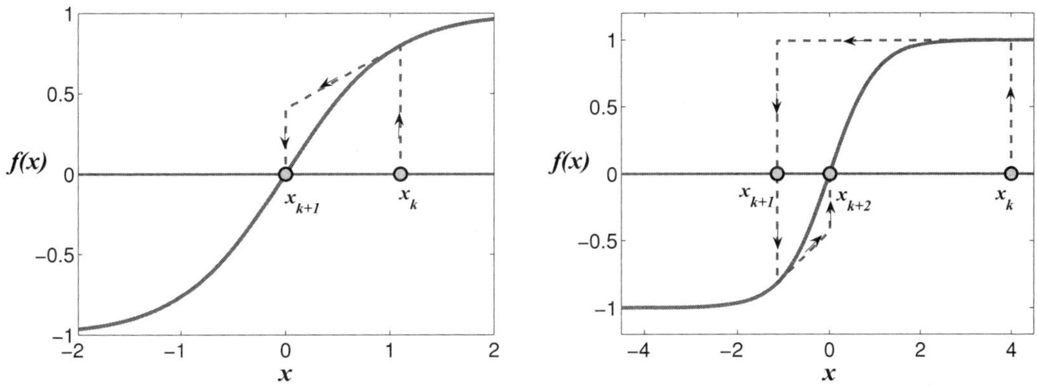

Figure 2.8. The performances of the line search method to finding the solution of $\tanh(x) = 0$. The left plot used $x_0 = 1.1$ as the initial guess, whereas the right plot used $x_0 = 4.0$ as the initial guess.

(c) Solve for damping coefficient:
Evaluate λ_m, $(m = 0, 1, 2, \ldots)$ using (2.63) for $m = 0$ and (2.65) for $m > 0$, but clipped according to the range of ratios given in (2.66), until $m = \ell$, where λ_ℓ satisfies (2.67)

(d) Update $\mathbf{x}^{(k)}$: $\mathbf{x}^{(k+1)} = \mathbf{x}^{(k)} + \lambda_\ell \delta_k$

Remarks: A MATLAB code for the enhanced Newton method is available on the book's webpage as `nsolve.m`, and the line search method is implemented when the parameter "`type`" is set to 1. Also, a MATLAB file `NewtonMin.m` is available on the book's webpage that uses the ehanced Newton method for minimization of a scalar function, where the line search method is implemented when the parameter `type` is set to 1.

In Figure 2.8, we see how the line search approach improves Newton's method. On the left plot, the initial guess of $x_0 = 1.1$ using Newton's method had yielded a divergent sequence (cf. Figure 2.7), but with the inclusion of line search, it took about one iteration to yield a value that is very close to the final solution. The right plot shows that even using an initial guess of $x_0 = 4.0$ where the slope is close to flat, the line search method reached a value close to the solution in about two iterations.

There are other alternative search directions to (2.60). These directions can be used when the Jacobian $J_k = d\mathbf{F}/d\mathbf{x}_k$ is singular, or is close to being singular. These search directions are:

1. **Gradient search:** $\delta_k = -J_k^T \mathbf{F}$
2. **Marquardt search:** $\delta_k = -\left(J_k^T J_k + \mu I\right)^{-1} J_k^T \mathbf{F}$, where μ is chosen such that $|\delta_k| \leq \beta$, $\beta > 0$.
3. **Pseudoinverse Search:** $\delta_k = -J_k^\dagger \mathbf{F}$ where $J_k^\dagger$ is the pseudo-inverse of the Jacobian.

The last two alternatives are often used when the dimension of $\mathbf{F}$ is much higher than the dimension of $\mathbf{x}$, yielding a nonlinear least-squares problem. The Levenberg-Marquardt is a method for nonlinear least squares that combines the gradient search and the Newton search directions. Details of this method are included in Section B.14 as an appendix.

2.11 EXERCISES

E2.1. The reaction for a multiple reversible first-order batch reaction shown in Figure 2.9,

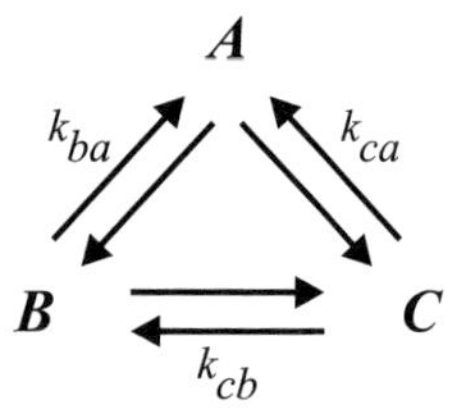

Figure 2.9. A three-component reversible reaction.

is given by

$$\frac{d\mathbf{x}}{dt} = A\mathbf{x}$$

where

$$A = \begin{pmatrix} -(k_{ab}+k_{ac}) & k_{ba} & k_{ca} \\ k_{ab} & -(k_{ba}+k_{bc}) & k_{cb} \\ k_{ac} & k_{bc} & -(k_{ca}+k_{cb}) \end{pmatrix} \quad ; \quad \mathbf{x} = \begin{pmatrix} x_a \\ x_b \\ x_c \end{pmatrix}$$

x_i is the mass fraction of component i and k_{ij} is the specific rate constant for the formation of component j from component i.
The equilibrium is obtained by setting $d\mathbf{x}/dt = 0$, i.e. $A\,\mathbf{x}_{\text{eq}} = \mathbf{0}$.

1. Show that A is singular.
2. Because A is singular, the linear equation should yield multiple equilibrium values, which is not realistic. Is there a missing equation? If so, determine the missing equation.
3. Is it possible, even with the additional equation, that a case of non-unique solution or a case with no solution may occur? Explain.

E2.2. Let $A[=]N \times N$, $B[=]M \times M$, $G[=]N \times N$, $H[=]M \times M$ and $C[=]N \times M$ by given matrices. Let $X[=]N \times M$ be the unknown matrix that must satisfy

$$AX + XB + GXH = C$$

Obtain the conditions for which the solution X will be unique, non-unique, or nonexisting.

E2.3. Find X (possibly infinite solution) that satisfies the following equation:

$$\begin{pmatrix} 3 & 2 & 1 \\ 4 & 5 & 6 \\ 1 & 1 & 1 \end{pmatrix} X + X \begin{pmatrix} 1 & 2 & 3 \\ 0 & 4 & 5 \\ 1 & 6 & 8 \end{pmatrix} = \begin{pmatrix} 9 & 16 & 23 \\ 23 & 18 & 32 \\ 8 & 22 & 31 \end{pmatrix}$$

using the matrices Q and W found based on:

1. Gauss-Jordan method
2. Singular Value Decomposition method

E2.4. Consider the network given in Figure 2.10, given $S_1 = S_3 = 15$ v, $A_{4,5} = 10$ mA, and all the resistors have 1 KΩ except for $R_6 = 100\ \Omega$. Based on the notations and equations given in Example 1.6,

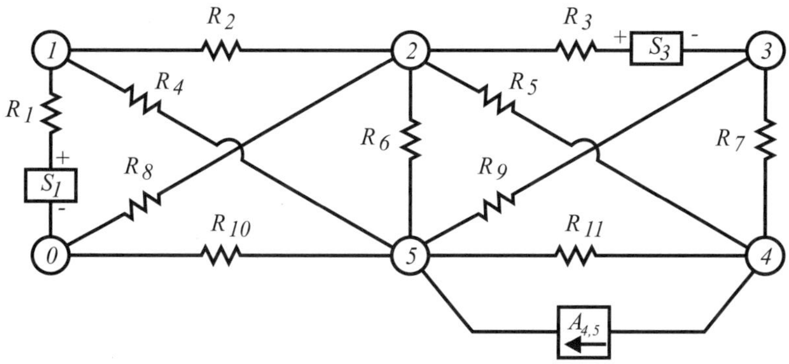

Figure 2.10. Resistive network containing two voltage sources and one current source.

we have

$$\left(\Psi R^{-1}\Psi^T\right)\mathbf{p} = \left(\mathbf{b} - \Psi R^{-1}\mathbf{s}\right)$$

where $\mathbf{p}$ is the vector of node potentials, Ψ is the node-link incidence matrix, $\mathbf{p}$ is the vector of current sources, $\mathbf{s}$ is the vector of voltage sources, and R is a diagonal matrix consisting of the resistance of each link. Use the LU factorization approach to solve for $\mathbf{p}$, and then solve for the voltages $\mathbf{v}$ across each resistor, where

$$\mathbf{v} = -\Psi^T\mathbf{p} - \mathbf{s}$$

E2.5. Redo the boundary value problem given in Example 2.4 but using the following functions:

$$h_2(x) = 2 + x \quad , \quad h_1(x) = 5x \quad , \quad h_0(x) = 2 + x$$

$$f(x) = \left(-5x^2 + 11x - 8\right)e^{-2x+6} + \left(8x^2 + 57x + 2\right)$$

and determine $u(x)$ for $0 \le x \le 10$ with the boundary conditions $u(0) = 1$ and $u(10) = 81$. Use the finite interval $\Delta x = 0.01$ and compare with the exact solution,

$$u(x) = (1 + 8x) + xe^{-2x+6}$$

(Note: Finding the exact solution for a high-order linear differential equation containing variable coefficients is often very difficult. We have an exact solution in this case because we actually obtained the differential equation and the boundary conditions using $u(x)$.)

E2.6. Let matrix M be a tri-diagonal matrix plus two additional nonzero elements $m_{1,3} = s$ and $m_{N,N-2} = t$,

$$M = \begin{pmatrix} a_1 & b_1 & s & & 0 \\ c_1 & \ddots & \ddots & & \\ & \ddots & \ddots & \ddots & \\ & & \ddots & \ddots & b_{N-1} \\ & & t & c_{N-1} & a_N \end{pmatrix} \tag{2.68}$$

1. Using Crout's method (without pivoting), M can be decomposed as $M = LU$, where

$$L = \left(\begin{array}{cccccc} z_1 & & & & & 0 \\ c_1 & z_2 & & & & \\ & \ddots & \ddots & & & \\ 0 & & c_{N-2} & z_{N-1} & & \\ \hline 0 & & t & \ell_{N,N-1} & \ell_{N,N} & \end{array}\right)$$

$$U = \left(\begin{array}{cccccc} 1 & f_1 & \frac{s}{a_1} & & 0 \\ \hline & 1 & f_2 & & 0 \\ & & \ddots & \ddots & \\ & & & 1 & f_{N-1} \\ 0 & & & & 1 \end{array}\right) \qquad (2.69)$$

with

$$z_1 = a_1 \quad ; \quad z_k = a_k - c_{k-1} f_{k-1}\ ,\ k = 2, \ldots, N$$

$$f_k = \frac{b_k}{z_k}\ ,\ \text{for } i = 1 \ldots, N-1, i \neq 2 \quad ; \quad f_2 = \frac{b_2}{z_2} - \frac{s\, c_1}{z_2 a_1}$$

and

$$\ell_{N,N-1} = c_{N-1} - t f_{N-2} \quad ; \quad \ell_{N,N} = z_N + t f_{N-1} f_{N-2}$$

Verify that $LU = M$. (Note that L and U are almost bi-diagonal, i.e., except for the additional nonzero terms: $\ell_{N,N-2} = t$ and $u_{1,3} = s/a_1$.)

2. Develop a modification to the Thomas algorithm that implements the forward and backward substitution using the L and U defined in (2.69) to solve for $M\mathbf{x} = \mathbf{v}$. (Check to see that your equations reduces to the original Thomas algorithm if $s = t = 0$.)

E2.7. Consider the equation

$$A\mathbf{x} = \begin{pmatrix} 4 & -1 & -1 & -1 & -1 & \cdots & -1 \\ -1 & 4 & -1 & 0 & 0 & \cdots & 0 \\ -1 & -1 & 4 & -1 & 0 & \cdots & 0 \\ -1 & 0 & -1 & 4 & \ddots & \ddots & \vdots \\ \vdots & \vdots & \ddots & \ddots & \ddots & \ddots & 0 \\ \vdots & \vdots & & \ddots & \ddots & \ddots & -1 \\ -1 & 0 & \cdots & \cdots & 0 & -1 & 4 \end{pmatrix} \mathbf{x} = \begin{pmatrix} 1 \\ 2 \\ \vdots \\ 2 \\ 1 \end{pmatrix}$$

where $A[=]50 \times 50$.

1. Use LU factorization of A to solve for $\mathbf{x}$. (Note that the L matrix has several fill-ins.)
2. If we permute the rows and columns of A using row permutation P with the sequence [2, . . . , 50, 1] such that $\widehat{A} = PAP^T$, and then obtain the LU factorization of $\widehat{A}$, what happens to the amount of fill-ins? Describe how one could use this fact to solve for $\mathbf{x}$.

E2.8. Use the direct matrix splitting method given in Sections 2.3, B.6 to solve the following problem:

$$\begin{pmatrix} 1 & 1 & -1 & 0 & 1 & 3 \\ -2 & 2 & 2 & 5 & 1 & 3 \\ 1 & 0 & 2 & -2 & 1 & -2 \\ 3 & 0 & 0 & -3 & -1 & 1 \\ 4 & 0 & 0 & 0 & 1 & 2 \\ 3 & 0 & 0 & 0 & 0 & 1 \end{pmatrix} \mathbf{x} = \begin{pmatrix} 6 \\ 8 \\ 2 \\ 3 \\ 7 \\ 4 \end{pmatrix}$$

by using a split $A = M + S$, with M containing the upper triangular portion of A. (Note: Based on the equations given in Section 2.3, note that one can use backward substitution when evaluating $M^{-1}\widehat{S}$ or $M^{-1}\mathbf{b}$, i.e., there is no need to explicitly find the inverse of M. This is automatically done in MATLAB when using the backslash (' \ ') or `mldivide` operator.)

E2.9. For a given data table $\{(x_1, y_1), \ldots, (x_N, y_N)\}$, with $x_{k+1} > x_k$, **cubic spline interpolation** generates a set of piecewise-continuous functions that passes through each data point using cubic polynomials. In addition, the interpolating function can be made to have continuous first- and second-order derivatives at each data point.

For $x \in [x_k, x_{k+1}]$, a polynomial defined as

$$p_k(x) = \begin{pmatrix} y_{k+1} & y_k & \gamma_{k+1} & \gamma_k \end{pmatrix} \begin{pmatrix} x_{k+1}^3 & x_k^3 & 6x_{k+1} & 6x_k \\ x_{k+1}^2 & x_k^2 & 2 & 2 \\ x_{k+1} & x_k & 0 & 0 \\ 1 & 1 & 0 & 0 \end{pmatrix}^{-1} \begin{pmatrix} x^3 \\ x^2 \\ x \\ 1 \end{pmatrix} \tag{2.70}$$

will satisfy the following conditions:

$$p_k(x_k) = y_k \; ; \; p_k(x_{k+1}) = y_{k+1} \; ; \; \left.\frac{d^2 p_k}{dx^2}\right|_{x=x_k} = \gamma_k \; ; \; \left.\frac{d^2 p_k}{dx^2}\right|_{x=x_{k+1}} = \gamma_{k+1}$$

By specifying continuity plus continuous first- and second-order derivatives between pairs of connecting polynomials, it can be shown that the following conditions result:

$$\Delta x_k \gamma_{k-1} + 2\left(\Delta x_k + \Delta x_{k+1}\right)\gamma_k + \Delta x_{k+1}\gamma_{k+1} = 6\left(\frac{\Delta y_{k+1}}{\Delta x_{k+1}} - \frac{\Delta y_k}{\Delta x_k}\right) \tag{2.71}$$

where $\Delta x_k = x_k - x_{k-1}$ and $\Delta y_k = y_k - y_{k-1}$. Equation (2.71) applies only to $k = 2, \ldots, N-1$. Two additional specifications are needed and can be obtained by either setting (i) $\gamma_1 = 0$ and $\gamma_N = 0$ (also known as the natural condition), or (ii) $\gamma_1 = \gamma_2$ and $\gamma_N = \gamma_{N-1}$, or (iii) setting γ_1, γ_2, and γ_3 to be collinear and γ_{N-2}, γ_{N-1}, and γ_N to be collinear, that is,

$$\frac{\gamma_3 - \gamma_1}{x_3 - x_1} = \frac{\gamma_3 - \gamma_2}{x_3 - x_2} \quad \text{and} \quad \frac{\gamma_{N-2} - \gamma_N}{x_{N-2} - x_N} = \frac{\gamma_{N-2} - \gamma_{N-1}}{x_{N-2} - x_{N-1}} \tag{2.72}$$

Show that (2.71) can be combined with the various types of end conditions to give the following matrix equation:

$$M \begin{pmatrix} \gamma_1 \\ \vdots \\ \gamma_N \end{pmatrix} = \begin{pmatrix} 0 \\ 6\left(\frac{\Delta y_3}{\Delta x_3} - \frac{\Delta y_2}{\Delta x_2}\right) \\ \vdots \\ 6\left(\frac{\Delta y_N}{\Delta x_N} - \frac{\Delta y_{N-1}}{\Delta x_{N-1}}\right) \\ 0 \end{pmatrix} \tag{2.73}$$

where

$$M = \begin{pmatrix} a & b & c & & & 0 \\ \Delta x_2 & 2\left(\Delta x_2 + \Delta x_3\right) & \Delta x_3 & & & \\ & \ddots & \ddots & & \ddots & \\ & & \Delta x_{N-1} & & 2\left(\Delta x_{N-1} + \Delta x_N\right) & \Delta x_N \\ 0 & & d & & e & f \end{pmatrix}$$

and

End conditions	Parameters
$\gamma_1 = 0$	$(a, b, c) = (1, 0, 0)$
$\gamma_N = 0$	$(d, e, f) = (0, 0, 1)$
$\gamma_1 = \gamma_2$	$(a, b, c) = (1, -1, 0)$
$\gamma_{N-1} = \gamma_N$	$(d, e, f) = (0, -1, 1)$
$\gamma_1, \gamma_2, \gamma_3$ Collinear	$(a, b, c) = \left(\Delta x_3, -\left(\Delta x_2 + \Delta x_3\right), \Delta x_2\right)$
$\gamma_{N-2}, \gamma_{N-1}, \gamma_N$ Collinear	$(d, e, f) = \left(\Delta x_N, -\left(\Delta x_{N-1} + \Delta x_N\right), \Delta x_{N-1}\right)$

Note that matrix M is tri-diagonal for the first two types of end conditions, but the third type (i.e., with collinearity) has the form given in (2.68). Thus (2.73) can be used to generate the values of second derivatives γ_k, $k = 1, \ldots, N$, which can then be substituted into (2.70) in the appropriate interval $[x_k, x_{k+1}]$.

Using these results, obtain a cubic spline curve that satisfies (2.72) and passes through the data points given in Table 2.5.

Table 2.5. *Data for spline curve interpolation*

x	y
−0.9931	0.9971
−0.7028	1.4942
−0.4908	1.3012
−0.3433	0.9094
−0.1406	1.0380
−0.0161	1.3070
0.2143	1.4415
0.4355	1.1023
0.6244	0.8509
1.0023	0.6930

E2.10. The 2D simple linear regression model is given by

$$z = mw + c$$

where z and w are the independent and dependent variables, respectively, m is the slope of the line, and c is the z-intercept.

Show that the well-known regression formulas

$$m = \frac{N\sum_{i=1}^{N} w_i z_i - \sum_{i=1}^{N} w_i \sum_{i=1}^{N} z_i}{N\sum_{i=1}^{N} w_i^2 - (\sum_{i=1}^{N} w_i)^2}$$

$$c = \frac{\sum_{i=1}^{N} z_i \sum_{i=1}^{N} w_i^2 - \sum_{i=1}^{N} w_i \sum_{i=1}^{N} w_i z_i}{N\sum_{i=1}^{N} w_i^2 - (\sum_{i=1}^{N} w_i)^2}$$

can be obtained from the solution of the normal equation:

$$A^T A\mathbf{x} = A^T \mathbf{b}$$

where

$$A = \begin{pmatrix} w_1 & 1 \\ \vdots & \vdots \\ w_N & 1 \end{pmatrix} \qquad \mathbf{x} = \begin{pmatrix} m \\ c \end{pmatrix} \qquad \mathbf{b} = \begin{pmatrix} z_1 \\ \vdots \\ z_N \end{pmatrix}$$

E2.11. Using data given in Table 2.6, obtain the parameters α_1, α_2, and α_3 that would yield the least-squares fit of the following model:

$$z = (\alpha_1 w^2 + \alpha_2 w + \alpha_3)\sin(2\pi w) \tag{2.74}$$

Table 2.6. *Data to be fit by (2.74)*

w	z	w	z	w	z	w	z	w	z
0.1	0.65	1.1	1.82	2.1	7.12	3.1	13.99	4.1	23.33
0.2	−0.01	1.2	3.15	2.2	11.25	3.2	24.27	4.2	36.52
0.3	0.87	1.3	4.01	2.3	14.27	3.3	23.53	4.3	40.41
0.4	−0.55	1.4	2.51	2.4	9.11	3.4	15.45	4.4	24.56
0.5	−1.02	1.5	−0.21	2.5	−1.81	3.5	0.62	4.5	0.63
0.6	−0.46	1.6	−3.39	2.6	−9.44	3.6	−16.37	4.6	−26.76
0.7	−0.08	1.7	−8.18	2.7	−18.24	3.7	−30.67	4.7	−46.66
0.8	−1.23	1.8	−7.52	2.8	−16.55	3.8	−31.63	4.8	−47.83
0.9	−1.99	1.9	−4.23	2.9	−11.20	3.9	−20.13	4.9	−31.23
1.0	0.89	2.0	0.15	3.0	0.64	4.0	−2.48	5.0	0.55

E2.12. Using the vapor liquid equilibrium data shown in Table 2.7, obtain a 5th order polynomial fit of f_V as a function of f_L that satisfies the following constraints:

$$\begin{aligned} f_V &= 0 && \text{when } f_L = 0 \\ f_V &= 1 && \text{when } f_L = 1 \\ f_V &= 0.65 && \text{when } f_L = 0.65 \end{aligned}$$

Table 2.7. *Vapor liquid equilibrium data*

f_L	f_V	f_L	f_V	f_L	f_V	f_L	f_V
0.02	0.07	0.29	0.43	0.85	0.79	0.38	0.45
0.06	0.16	0.37	0.50	0.89	0.86	0.54	0.58
0.11	0.24	0.50	0.56	0.95	0.92	0.81	0.73
0.18	0.33	0.71	0.68	0.99	0.97	0.94	0.88
0.22	0.35	0.78	0.74	0.25	0.37	0.97	0.93

E2.13. Let $A = QR$ be the QR factorization of $A[=]N \times M$ described in Section 2.6. It was mentioned that R could immediately be used as the Choleski factor in solving $R^T R\mathbf{x} = A^T\mathbf{b}$, that is, forward and backward substitution can be used to find $\mathbf{x}$ using R. Of course when N is small, for example, 2 or 3, it appears that this approach is unnecessarily complicated because the inverse of $A^T A$ may be determined just as fast. However, if $A^T A$ is large and singular, the QR factorization may offer a viable alternative.

1. Solve to following least-squares problem:

$$\begin{pmatrix} 3 & 1 \\ 4 & 1 \\ 5 & 1 \\ 6 & 1 \\ 7 & 1 \\ 8 & 1 \end{pmatrix} \mathbf{x} =_{\text{lsq}} \begin{pmatrix} 5 \\ 6 \\ 7 \\ 8 \\ 9 \\ 10 \end{pmatrix}$$

use R obtained from the QR factorization of A and solve for $\mathbf{x}$. Compare this to using $\widehat{R}$ obtained from the Choleski factorization of $A^T A$, that is, $\widehat{R}^T\widehat{R} = A^T A$.

2. Try the same approach on the following problem:

$$\begin{pmatrix} 3 & 1 & 2 \\ 4 & 1 & 3 \\ 5 & 1 & 4 \\ 6 & 1 & 5 \\ 7 & 1 & 6 \\ 8 & 1 & 7 \end{pmatrix} \mathbf{x} =_{\text{lsq}} \begin{pmatrix} 5 \\ 6 \\ 7 \\ 8 \\ 9 \\ 10 \end{pmatrix} \tag{2.75}$$

Is a situation possible in which there are no least-squares solution? If solutions exist, how would one find the infinite least-squares solution $\mathbf{x}$? Test your approach on (2.75).

E2.14. A generalization of the Euclidean norm for a vector $\mathbf{v}$ is the **Q-weighted norm** (also known as the **Riemanian norm**), denoted by $\|\mathbf{v}\|_Q$, where Q is a Hermitian positive-definite matrix, defined as

$$\|\mathbf{v}\|_Q = \sqrt{\mathbf{v}^* Q \mathbf{v}} \tag{2.76}$$

The Euclidean norm results by setting $Q = I$.

1. Show that the function $\|\bullet\|_Q$ satisfies the conditions of norms as given in Table B.4. (Hint: With Choleski factorization, there exists a nonsingular S such that $Q = S^* S$. Then use $\|\mathbf{v}\|_Q = \|S\mathbf{v}\|$ and the fact that Euclidean norms satisfy the conditions for norms.)
2. Using $\|\mathbf{r}\|_Q^2$, where $(\mathbf{r} = \mathbf{b} - A\mathbf{x})$ is the residual, show that the **weighted least-squares** solution is the solution to the **weighted normal equation**

$$A^T Q A\mathbf{x} = A^T Q\mathbf{b} \tag{2.77}$$

3. One application for the inclusion of variable weights is to weaken the influence of older data. Suppose r_k is k^{th} element of the residual vector, and r_k is a more recent data than r_{k-1}. Next, choose Q to be

$$Q = \begin{pmatrix} \lambda^{N-1} & & & 0 \\ & \ddots & & \\ & & \lambda & \\ 0 & & & 1 \end{pmatrix} \tag{2.78}$$

where $0 < \lambda < 1$ is known as the **forgetting factor**. Then $Q\mathbf{r}$ attaches heavier weights on more recent data.

Using data given in Table 2.8 and (2.77), obtain a weighted linear regression model

$$z = mw + c$$

for $\lambda = 1, 0.5, 0.1$, and plot the three cases together with the data. Explain the effects of different λ choices.

Table 2.8. *Data for weighted linear regression*

z	w	z	w	z	w
0.0104	0.0219	0.1187	0.5219	0.4459	0.7792
0.0173	0.0921	0.1671	0.5892	0.4896	0.8465
0.0196	0.1652	0.1717	0.5278	0.5012	0.8319
0.0403	0.1827	0.1970	0.5892	0.5472	0.8728
0.0472	0.3085	0.2293	0.6184	0.6048	0.8728
0.0518	0.2588	0.2477	0.6652	0.7039	0.9167
0.0588	0.1915	0.3145	0.7032	0.7730	0.9313
0.0726	0.3933	0.3353	0.7529	0.8306	0.9459
0.0841	0.3289	0.3744	0.7208	0.9528	0.9722
0.1094	0.4430	0.3952	0.7880	0.9781	0.9956

E2.15. The least-squares solution for $A\mathbf{x} = \mathbf{b}$ can also be implemented recursively; that is, as more data come in, the normal equation is resolved. Instead of having to invert A^TA at each arrival of a new set of data, we can implement a **recursive least-squares** approach.

After using N data, suppose we have stored the values of $Z_N = \left(A_N^TA_N\right)^{-1}$ and $\mathbf{w}_N = A_N^T\mathbf{b}_N$. When the new data come in, the data matrix and target vector become

$$A_{N+1} = \left(\frac{A_N}{\mathbf{v}_N^T}\right) \quad . \quad \mathbf{b}_{N+1} = \left(\frac{\mathbf{b}_N}{\beta_N}\right)$$

1. Show that

$$\mathbf{w}_{N+1} = A_{N+1}^T\mathbf{b}_{N+1} = \mathbf{w}_N + \beta_N\mathbf{v}_N \tag{2.79}$$

2. Using the Woodbury formula (cf. (1.27))[11], show that

$$Z_{N+1} = \left(A_{N+1}^TA_{N+1}\right)^{-1} = Z_N - \frac{1}{1 + \mathbf{v}_N^TZ_N\mathbf{v}_N}Z_N\mathbf{v}_N\mathbf{v}_N^TZ_N \tag{2.80}$$

3. Use (2.79) and (2.80) to obtain the correction $\mathbf{f}_N$ in the following recursive formula:

$$\mathbf{x}_{N+1} = \mathbf{x}_N + \mathbf{f}_N \tag{2.81}$$

where $\mathbf{x}_k$ is the kth least-squares estimate of $\mathbf{x}$.

4. Using Q_N as given in (2.78) with a fixed forgetting factor λ, determine how the recursive formulas can be modified to obtain a recursive weighted least-squares method.

[11] For the inverse of $(A + BCD)$, the Woodbury formula applied to the special case where $B[=]N \times 1$ and $C[=]1 \times N$ is more well known as the **Sherman-Morrison-Woodbury formula**.

Note: The recursive least-squares method, with or without forgetting factors, is used for adaptive estimation of processes with variable parameters, for example, fouling, catalyst poisoning, or wear-and-tear.

E2.16. In some cases, the linear equations will result in block tri-diagonal structures, for example, the open-loop process units in Example 1.11. Let the complete equation be given by

$$G\mathbf{x} = \mathbf{v}$$

where G is **block tri-diagonal**

$$G = \left(\begin{array}{c|c|c|c|c} A_1 & B_1 & & & 0 \\ \hline C_1 & A_2 & B_2 & & \\ \hline & \ddots & \ddots & \ddots & \\ \hline & & C_{N-2} & A_{N-1} & B_{N-1} \\ \hline 0 & & & C_{N-1} & A_N \end{array}\right)$$

where A_n, B_n, and C_n are all $K \times K$ block matrices.

1. Show that the **block**-LU **decomposition** of G is given by

$$G = LU$$

where

$$L = \left(\begin{array}{c|c|c|c} Q_1 & & & 0 \\ \hline C_1 & Q_2 & & \\ \hline & \ddots & \ddots & \\ \hline 0 & & C_{N-1} & Q_N \end{array}\right) \qquad U = \left(\begin{array}{c|c|c|c} I & W_1 & & 0 \\ \hline & \ddots & \ddots & \\ \hline & & I & W_{N-1} \\ \hline 0 & & & I \end{array}\right)$$

and, with $C_0 = 0$, $W_0 = 0$,

$$\begin{aligned} Q_n &= A_n - C_{n-1}W_{n-1} & [=]\, K \times K \\ W_n &= Q_n^{-1}B_n & [=]\, K \times K \end{aligned}$$

2. Instead of evaluating Q_n, the inverse, $Z_n = Q_n^{-1}$, will be of importance. Show that

$$\begin{aligned} Z_1 &= A_1^{-1} \\ Z_n &= \left(A_n - C_{n-1}Z_{n-1}B_{n-1}\right)^{-1} \qquad n = 2, 3, \ldots, N \end{aligned}$$

3. Returning to the main problem of solving $G\mathbf{x} = \mathbf{v}$ and assuming all Z_n exist, both $\mathbf{x}$ and $\mathbf{v}$ will have to be partitioned accordingly, that is,

$$\mathbf{x} = \begin{pmatrix} \mathbf{x}_1 \\ \vdots \\ \mathbf{x}_N \end{pmatrix} \qquad \mathbf{v} = \begin{pmatrix} \mathbf{v}_1 \\ \vdots \\ \mathbf{v}_N \end{pmatrix}$$

where $\mathbf{x}_n$ and $\mathbf{v}_n$ are vectors of length K.

Having both L and U be block-triangular matrices, show that the forward substitution becomes

$$\begin{aligned} \mathbf{y}_1 &= Z_1\mathbf{v}_1 \\ \mathbf{y}_n &= Z_n\left(\mathbf{v}_n - C_{n-1}\mathbf{y}_{n-1}\right) \quad \text{for } n = 2, \ldots, N \end{aligned}$$

whereas the backward substitution becomes

$$\begin{aligned} \mathbf{x}_N &= \mathbf{y}_N \\ \mathbf{x}_n &= \mathbf{y}_n - Z_nB_n\mathbf{x}_{n+1} \quad \text{for } n = (N-1), \ldots, 1 \end{aligned}$$

Note: This the block-matrix version of the Thomas algorithm (cf. Section 2.2.2). If storage is a significant issue, one could use the storage of A_n and replace it with Z_n, except for the case of patterned A_n's, as in the exercise that follows. Furthermore, from the backward substitutions given previously, there is no need to store W_n either.

4. For $n = 1, \ldots, 10$, let

$$A_n = \begin{pmatrix} -4 & 1 \\ 1 & -4 \end{pmatrix} \quad B_n = \begin{pmatrix} -1 & 0 \\ 0 & 1 \end{pmatrix} \quad C_n = \begin{pmatrix} 1 & 0 \\ 0 & -1 \end{pmatrix}$$

and

$$\mathbf{v}_1 = \begin{pmatrix} 2 \\ 2 \end{pmatrix} \quad \mathbf{v}_{10} = \begin{pmatrix} 3 \\ 1 \end{pmatrix} \quad \mathbf{v}_k = \begin{pmatrix} 0 \\ 0 \end{pmatrix} \text{ for } k = 2, 3, \ldots, 9$$

Use the block version of the Thomas algorithm to solve for $\mathbf{x}$.

E2.17. When solving for the steady-state solution of an elliptic partial differential equation, the resulting linear equations will involve large sparse matrices. As an example, one could have the following:

$$A = \left(\begin{array}{c|c|c|c|c} R & -I & & & 0 \\ \hline -I & R & -I & & \\ \hline & \ddots & \ddots & \ddots & \\ \hline & & -I & R & -I \\ \hline 0 & & & -I & R \end{array}\right)$$

with

$$R = \begin{pmatrix} 4 & -1 & & & 0 \\ -1 & 4 & -1 & & \\ & \ddots & \ddots & \ddots & \\ & & -1 & 4 & -1 \\ 0 & & & -1 & 4 \end{pmatrix}$$

Let $R[=]25 \times 25$ and $A[=]625 \times 625$. Solve the linear equation $A\mathbf{x} = \mathbf{b}$ for $\mathbf{x}$, when $b_k = 10$ when $k = 1, \ldots, 5$ and $b_j = 0$ for $j = 6, \ldots, 625$ using:

1. LU decomposition followed by forward and backward substitution.
2. Conjugate gradient method.
3. Block-Thomas algorithm (cf. see Exercise E2.16)

E2.18. A hot liquid stream entering the annulus of a double-pipe heat exchanger at a temperature $T_{\text{hot,in}} = 300°C$ is being cooled by cold water counter-flowing through the inner pipe and entering at $T_{\text{cold,in}} = 25°C$ from the other side of the pipe. The working equations relating the temperatures of all the entering and exiting flows are given by

$$\alpha \frac{(T_{\text{hot,in}} - T_{\text{cold,out}}) - (T_{\text{hot,out}} - T_{\text{cold,in}})}{\ln\left(\dfrac{T_{\text{hot,in}} - T_{\text{cold,out}}}{T_{\text{hot,out}} - T_{\text{cold,in}}}\right)} = T_{\text{cold,out}} - T_{\text{cold,in}}$$

$$\beta (T_{\text{hot,in}} - T_{\text{hot,out}}) = (T_{\text{cold,out}} - T_{\text{cold,in}})$$

where the parameters are

$$\alpha = \frac{U_{\text{inner}}(\pi D_{\text{inner}} L)}{\dot{m}_{\text{cold}} C_{p,\text{cold}}} = 0.3 \quad \text{and} \quad \beta = \frac{\dot{m}_{\text{hot}} C_{p,\text{hot}}}{\dot{m}_{\text{cold}} C_{p,\text{cold}}} = 0.25$$

with U as heat transfer coefficient, D is the diameter, L is the length of the pipes, C_p are the heat capacities, and $\dot{m}$ are the mass flow rates.

Let $x = T_{\text{hot,out}}$ and $y = T_{\text{cold,out}}$

1. Solve for x and y by finding the roots of the following functions:

$$\begin{aligned} f_1(x, y) &= \alpha \frac{(T_{\text{hot,in}} - y) - (x - T_{\text{cold,in}})}{\ln\left(\dfrac{T_{\text{hot,in}} - y}{x - T_{\text{cold,in}}}\right)} - (y - T_{\text{cold,in}}) \\ f_2(x, y) &= \beta(T_{\text{hot,in}} - x) - (y - T_{\text{cold,in}}) \end{aligned}$$

Apply Newton's formula using the secant approximation for the Jacobian by choosing $\delta x = 0.1$ and $\delta y = 0.1$,

$$J \approx \begin{pmatrix} \frac{1}{\delta x} & 0 \\ 0 & \dfrac{1}{\delta y} \end{pmatrix} \begin{pmatrix} f_1(x+\delta x, y) - f_1(x, y) & f_1(x, y+\delta y) - f_1(x, y) \\ f_2(x+\delta x, y) - f_2(x, y) & f_2(x, y+\delta y) - f_2(x, y) \end{pmatrix}$$

and initial guess $x = y = 325/2$.

2. If one or more equation can easily be solved explicitly, this can improve computational efficiency by reducing the dimension of the search space. For instance, one could first solve for y in terms of x in $f_2(x, y) = 0$ to obtain

$$y = T_{\text{cold,in}} + \beta(T_{\text{hot,in}} - x)$$

which can then be substituted into $f_1(x, y)$ to yield

$$f_3(x) = f_1\left(x, T_{\text{cold,in}} + \beta(T_{\text{hot,in}} - x)\right)$$

Use Newton's method using the secant approximation for the Jacobian like before, with the same initial guess $x = 325/2$. (Note: In this case, $f_3(x)$ is a scalar function, and one can plot the function to easily determine whether there are multiple roots.)

E2.19. Given R equilibrium reactions that involves $M > R$ components. Let v_{ij} be the stoichiometric coefficients of component j in reaction i and stored in a stoichiometric matrix $V[=]R \times M$ where

$$v_{ij} = \begin{cases} > 0 & \text{if component } j \text{ is a product in (forward) reaction } i \\ < 0 & \text{if component } j \text{ is a reactant in (forward) reaction } i \\ = 0 & \text{if component } j \text{ is not part of reaction } i \end{cases}$$

After a sufficient time and under constant operating temperature, all the reactions will attain equilibrium such that the mole fractions will satisfy

$$K_i = \prod_{j=1}^{M} x_j^{v_{ij}}$$

where K_i is the equilibrium constant for the ith reaction. A typical problem is to determine the mole fractions x_j based on a given set of initial moles n_{0j} for each component and equilibrium constants K_i. One effective approach to solve this problem is to use the concept of the **extent of reaction** ξ_i, which is the number of units that the ith reaction had undergone in the forward direction. For instance, let $\xi_2 = 2.5$ and stoichiometric coefficient $v_{23} = 2$;

then component 3 will gain $\xi_2 v_{23} = 5$ moles due to reaction 2. Thus if we let n_j be the number of moles of component j at equilibrium, we have

$$n_j = n_{0j} + \sum_{k=1}^{R} v_{kj}\xi_k$$

which allows the mole fractions at equilibrium to be defined in terms of ξ_k, that is,

$$x_j = \frac{n_j}{\sum_{j=1}^{M} n_j} = \frac{n_{0j} + \sum_{k=1}^{R} v_{kj}\xi_k}{\sum_{j=1}^{M} n_{0j} + \sum_{k=1}^{R} \beta_k \xi_k}$$

where

$$\beta_k = \sum_{j=1}^{M} v_{kj}$$

The problem then becomes finding the roots ξ_k, $k = 1, \ldots, R$, of a vector of functions, $\mathbf{f} = (f_1, \ldots, f_R)^T = 0$ given by

$$f_i(\xi_1, \ldots, \xi_R) = \left(\prod_{j=1}^{M} x_j^{v_{ij}}\right) - K_i = \left(\prod_{j=1}^{M}\left(\frac{n_{0j} + \sum_{k=1}^{R} v_{kj}\xi_k}{\sum_{j=1}^{M} n_{0j} + \sum_{k=1}^{R} \beta_k\xi_k}\right)^{v_{ij}}\right) - K_i = 0$$

1. Show that the $(i, k)^{\text{th}}$ element of the Jacobian of $\mathbf{f}$ is given by

$$\frac{\partial f_i}{\partial \xi_k} = (f_i + K_i)\left(\sum_{j=1}^{M} \frac{v_{ij}v_{kj}}{n_j} - \frac{\beta_i\beta_k}{n_T}\right)$$

where $n_j = n_{0j} + \sum_{k=1}^{R} v_{kj}\xi_k$ and $n_T = \sum_{j=1}^{M} n_j$. (Hint: Use the chain rule on f_i, i.e.,

$$\frac{\partial f_i}{\partial \xi_k} = \sum_{j=1}^{M} \frac{\partial f_i}{\partial x_j}\frac{\partial x_j}{\partial \xi_k}$$

and simplify further by using the definition of f_i.) Next, show that the Jacobian can also be put in matrix form as

$$J = \frac{d}{d\xi}\mathbf{f} = DS \tag{2.82}$$

where D and S are a diagonal matrix and a symmetric matrix, respectively, given by

$$D = \begin{pmatrix} f_1 + K_1 & & 0 \\ & \ddots & \\ 0 & & f_R + K_R \end{pmatrix}$$

$$S = V\left(\begin{pmatrix} n_1^{-1} & & 0 \\ & \ddots & \\ 0 & & n_M^{-1} \end{pmatrix} - \frac{1}{n_T}\begin{pmatrix} 1 & \cdots & 1 \\ \vdots & \ddots & \vdots \\ 1 & \cdots & 1 \end{pmatrix}\right)V^T$$

which means that the Newton update is given by $\Delta_k\xi = -J^{-1}\mathbf{f} = -S^{-1}D^{-1}\mathbf{f}\big|_{\xi(k)}$. (Note: Because S involves n_j^{-1}, one has to avoid choosing an initial value guess that would make any of the n_j equal to zero.)

2. Consider the following equilibrium reactions:

$$\text{Reaction 1:} \quad A + B \quad \Leftrightarrow \quad C + D$$

$$\text{Reaction 2:} \quad A + C \quad \Leftrightarrow \quad 2E$$

with equilibrium constants, $K_1 = 0.4$ and $K_2 = 4/15$. Assume initial moles to be $n_{0,A} = 100$, $n_{0,B} = 100$, $n_{0,C} = 0$, $n_{0,D} = 0$ and $n_{0,E} = 0$. Use Newton's method (i.e., write your own MATLAB m-file that implements the Jacobian formulas obtained in the previous question) to find the extents of reactions, ξ_1 and ξ_2 using as initial guesses:

(a) Case 1: $\xi_1^{(0)} = 50$ and $\xi_2^{(0)} = 20$

(b) Case 2: $\xi_1^{(0)} = 90$ and $\xi_2^{(0)} = 2$

(c) Case 3: $\xi_1^{(0)} = 90$ and $\xi_2^{(0)} = 11$

If the solutions converge, evaluate the equilibrium composition.

3. Use the nonlinear solver `nsolve.m` that is available on the book's webpage to find the solution with the initial guesses of preceding Cases 2 and 3. Implement both the line search (type 1) and the double-dogleg search (type 2). If the MATLAB Optimization toolbox is available, test these cases also using the built-in MATLAB function `fsolve`.

E2.20. Use the Levenberg-Marquardt method (cf. B.14) to find the parameters $\alpha_1, \ldots, \alpha_7$ of the model

$$y = \alpha_1 \tanh(\alpha_2 x + \alpha_3) + \alpha_4 \tanh(\alpha_5 x + \alpha_6) + \alpha_7$$

Table 2.9. *Data for problem E2.20*

x	y	x	y	x	y
0.7488	2.3816	15.2650	3.0132	31.8548	1.4167
1.7857	2.3816	19.4124	3.0044	34.1590	0.9956
3.0530	2.3728	21.2558	2.9956	35.6567	0.8377
4.4355	2.4518	25.4032	2.7061	37.6152	0.7412
5.0115	2.5658	27.1313	2.4868	40.3802	0.6711
5.9332	2.7763	27.8226	2.2939	43.7212	0.6360
7.5461	2.9956	29.3203	1.9430	47.6382	0.6360
10.5415	3.0395	31.0484	1.5658	49.4816	0.6447
13.4217	3.0570				

that would yield a least-squares fit of the data given in Table 2.9. Also, a MATLAB file `Levenmarq.m` that implements the `Levenberg-Marquardt` algorithm is available (Note: Use a random Gaussian number generator to obtain the initial guess).

3 Matrix Analysis

In Chapter 1, we started with the study of matrices based on their composition, structure, and basic mathematical operations such as addition, multiplication, inverses, determinants, and so forth, including matrix calculus operations. Then, in Chapter 2, we focused on the use of matrices to solve simultaneous equations of the form $A\mathbf{x} = \mathbf{b}$, including their applications toward the solution of nonlinear equations via Newton algorithms. Based on linear algebra, we saw that $A\mathbf{x}$ can also be taken to be a linear combination of the columns A with the elements of $\mathbf{x}$ acting as the weights. Under this perspective, the least-squares problem shifted the objective to be that of finding $\mathbf{x}$ that would minimize the residual error given by $\mathbf{r} = \mathbf{b} - A\mathbf{x}$.

In this chapter, we return to the equation $A\mathbf{x} = \mathbf{b}$ with a third perspective. Here, we consider matrix A to be an operator that will "transform" (or "map") an input vector $\mathbf{x}$ to yield an output vector $\mathbf{b}$, as shown schematically in Figure 3.1. We call this the matrix operator perspective of the linear equation. The main focus is now on A as a "machine" that needs to be analyzed, constructed, or modified to achieve some desired operational characteristics. For instance, we may want to construct a matrix A that rotates, stretches, or flips various points $\mathbf{x}_i$ described by vectors. As another example, a stress tensor (to be discussed in Chapter 4) can be represented by a matrix T, which can then be used to find the stress vector $\mathbf{s}$ pointing in the direction of a unit vector $\mathbf{n}$ by the operation $\mathbf{s} = T\mathbf{n}$.

We begin with some general matrix operators in Section 3.1. These include unitary or orthogonal operators, projection operators, reflector operators, and so forth. We also include an affine extension of the various operators that would allow translation. Then, in Section 3.2, we introduce the important properties called **eigenvalues** and objects called **eigenvectors**, which characterize the behavior of the matrix as an operator. We also outline the properties of eigenvalues and eigenvectors of specific classes of matrices. As we show in later chapters, the eigenvalues are very important tools for determining stability of differential equations or iterative processes. We also include an alternative numerical approach to the evaluation of eigenvalues and eigenvectors, known as the **QR method**. This method is an extension of another method known as the **power method**. Details of the QR method and the power method are included in Section C.2 as an appendix.

Another application of eigenvalue analysis is also known as **spectral analysis**. It provides useful factorizations such as diagonalization, Jordan-block forms, singular value decompositions, and polar decompositions. In Section 3.7, we discuss

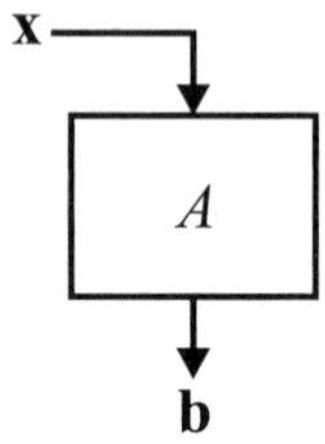

Figure 3.1. Matrix as an operator.

an important result known as the Cayley-Hamilton theorem, which is very useful in evaluating analytical matrix functions. These tools provide valuable insights into functions of a matrix operator. As one application of spectral analysis, we briefly discuss the **principal component analysis** in Section 3.9.2. This type of analysis is used to reduce the dimensions of linear models.

Finally, in Section 3.11, we provide a brief discussion on matrix norms and condition numbers. These norms are important to sensitivity and error analysis of matrix operators. For instance, the condition numbers can help determine whether the solution of $A\mathbf{x} = \mathbf{b}$ will have a large error or not when small perturbations are introduced in A or $\mathbf{b}$.

3.1 Matrix Operators

Let $A[=]N \times M$ be a matrix that transforms an M-dimensional vector $\mathbf{x}$ to an N-dimensional vector $\mathbf{b}$. This is represented by the same linear equation $A\mathbf{x} = \mathbf{b}$. Generally, $\mathbf{x}$ does not reside in the same space as $\mathbf{b}$, even when $N = M$. For instance, for the stress tensor equation $T\mathbf{n} = \mathbf{s}$, although $\mathbf{n}$ and $\mathbf{s}$ are both vectors of the same length, $\mathbf{s}$ has units of stress, that is, force per unit area, whereas $\mathbf{n}$ does not.

Matrix operators are also known as **linear mappings** because they obey the following property:

$$A\left(\alpha\mathbf{x} + \beta\mathbf{y}\right) = \alpha A\mathbf{x} + \beta A\mathbf{y} \tag{3.1}$$

where α and β are scalars. The associative properties of matrices can then be viewed as compositions of operator sequences as follows:

$$A\left(B\mathbf{x}\right) = \mathbf{b} \quad \rightarrow \quad \left(AB\right)\mathbf{x} = \mathbf{b}$$

This is shown in Figure 3.2. As we saw in Chapter 1, matrix products are not commutative in general. This means that when $C = AB$ operates on $\mathbf{x}$, operator B is applied first to yield $\mathbf{y} = B\mathbf{x}$. Then $\mathbf{y}$ is operated on by A to yield $\mathbf{b}$.

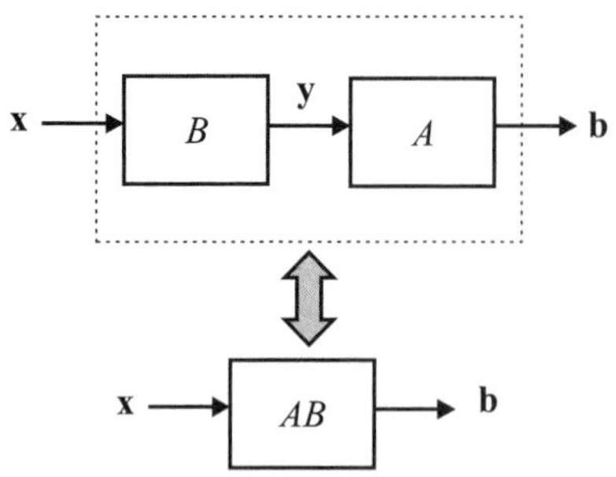

Figure 3.2. Composition of operators.

When using the same operator on several input vectors, one could also collect the input vectors into a matrix, say X, and obtain a corresponding matrix of output vectors B.

$$AX = A\left(\mathbf{x}_1 \mid \mathbf{x}_2 \mid \cdots \mid \mathbf{x}_P \right) = \left(\mathbf{b}_1 \mid \mathbf{b}_2 \mid \cdots \mid \mathbf{b}_P \right) = B$$

Note that although X and B are matrices, they are not treated as operators. This is to emphasize that, as always, the applications will dictate whether a matrix is viewed as an operator or not.

3.1.1 Orthogonal and Unitary Matrix Operators

Definition 3.1. *A square matrix A is an* **unitary** *matrix if*

$$A^*A = AA^* = I \tag{3.2}$$

A square matrix A is an **orthogonal** *matrix if*

$$A^TA = AA^T = I \tag{3.3}$$

Unitary matrix operators are operators that preserve Euclidean norms defined as

$$\|\mathbf{x}\| = \sqrt{\sum_{i=1}^{n} \overline{x}_i x_i} = \sqrt{\mathbf{x}^*\mathbf{x}}$$

To see this,

$$\|\mathbf{b}\|^2 = \|A\mathbf{x}\|^2 = (A\mathbf{x})^* \, (A\mathbf{x}) = \mathbf{x}^* \, (A^*A)\mathbf{x} = \mathbf{x}^*\mathbf{x} = \|x\|^2$$

When the matrices and vectors are real, unitary operators are synonymous with orthogonal operators. Thus we also get for real vectors and matrices,

$$\|\mathbf{b}\|^2 = \|A\mathbf{x}\|^2 = (A\mathbf{x})^T \, (A\mathbf{x}) = \mathbf{x}^T \, (A^TA)\mathbf{x} = \mathbf{x}^T\mathbf{x} = \|x\|^2$$

Note that if A is a unitary (or orthogonal) operator, the inverse operator is found by simply taking the conjugate transpose (or transpose), that is, $A^{-1} = A^*$ (or $A^{-1} = A^T$).

In the following examples, unless explicitly stated, the matrix operators and vectors are understood to have real elements. Examples of orthogonal matrices include the following:

1. **Permutation operators.** These matrices are obtained by permuting the rows of an identity matrix. When operating on an input vector, the result is a vector with the coordinates switched according to sequence in the permutation P. For instance,

$$P\begin{pmatrix} x \\ y \\ z \end{pmatrix} = \begin{pmatrix} 0 & 0 & 1 \\ 1 & 0 & 0 \\ 0 & 1 & 0 \end{pmatrix}\begin{pmatrix} x \\ y \\ z \end{pmatrix} = \begin{pmatrix} z \\ x \\ y \end{pmatrix}$$

2. **Rotation operators (also known as Givens operator).** The canonical rotation operators R are obtained from an identity matrix in which two elements in the diagonal, say, at $r_{k,k}$ and $r_{\ell,\ell}$, are replaced by $\cos(\theta)$, and two other off-diagonal

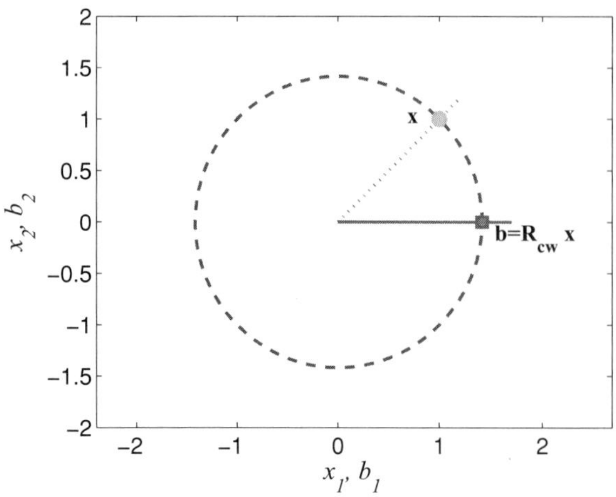

Figure 3.3. $R_{\text{cw}}(\theta)$ rotates x clockwise by θ radians.

elements at $r_{k,\ell}$ and $r_{\ell,k}$ are replaced by $\sin(\theta)$ and $-\sin(\theta)$, respectively. For a two-dimensional rotation, we have

$$R_{\text{cw}(\theta)} = \begin{pmatrix} \cos(\theta) & \sin(\theta) \\ -\sin(\theta) & \cos(\theta) \end{pmatrix} \tag{3.4}$$

as a clockwise rotation of input vectors. To see that R_{cw} is indeed an orthogonal matrix, we have

$$\begin{aligned} R_{\text{cw}}^T R_{\text{cw}} &= \begin{pmatrix} \cos(\theta) & -\sin(\theta) \\ \sin(\theta) & \cos(\theta) \end{pmatrix} \begin{pmatrix} \cos(\theta) & \sin(\theta) \\ -\sin(\theta) & \cos(\theta) \end{pmatrix} \\ &= \begin{pmatrix} \cos^2(\theta) + \sin^2(\theta) & 0 \\ 0 & \cos^2(\theta) + \sin^2(\theta) \end{pmatrix} = I \end{aligned}$$

Similarly, we can show that $R_{\text{cw}} R_{\text{cw}}^T = I$.

To illustrate the effect of clockwise rotation, consider

$$\mathbf{x} = \begin{pmatrix} 1 \\ 1 \end{pmatrix} \quad \rightarrow \quad \mathbf{b} = R_{\text{cw}(\pi/4)}\mathbf{x} = \begin{pmatrix} \sqrt{2} \\ 0 \end{pmatrix}$$

The original point was rotated $\pi/4$ radians clockwise, as shown in Figure 3.3.

Because R_{cw} is orthogonal, the inverse operator is the counterclockwise rotation obtained by simply taking the transpose of R_{cw}, that is,

$$R_{\text{ccw}(\theta)} = R_{\text{cw}}(\theta)^{-1} = \begin{pmatrix} \cos(\theta) & -\sin(\theta) \\ \sin(\theta) & \cos(\theta) \end{pmatrix} \tag{3.5}$$

For three dimensions, the clockwise rotation operators around x, y, and z, assuming the vectors are arranged in the classical order of $(x, y, z)^T$, are given by $R_{\text{cw},x}$, $R_{\text{cw},y}$, and $R_{\text{cw},z}$, respectively,

$$R_{\text{cw},x} = \begin{pmatrix} 1 & 0 & 0 \\ 0 & \cos(\theta) & \sin(\theta) \\ 0 & -\sin(\theta) & \cos(\theta) \end{pmatrix} \qquad R_{\text{cw},y} = \begin{pmatrix} \cos(\theta) & 0 & -\sin(\theta) \\ 0 & 1 & 0 \\ \sin(\theta) & 0 & \cos(\theta) \end{pmatrix}$$

$$R_{\text{cw},z} = \begin{pmatrix} \cos(\theta) & \sin(\theta) & 0 \\ -\sin(\theta) & \cos(\theta) & 0 \\ 0 & 0 & 1 \end{pmatrix}$$

and shown in Figure 3.4.

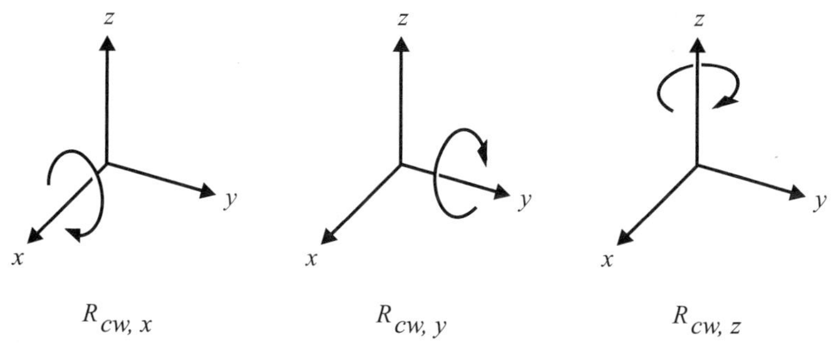

Figure 3.4. 3D clockwise rotation operators.

3. **Reflection operators** (also known as **Household operators**). These operators have the effect of reflecting a given point symmetrically along a given hyperplane. The hyperplane is determined by the direction of a nonzero vector $\mathbf{w}$ that is normal to the plane. The **Householder transformation operator**, $H_{\mathbf{w}}$, based on $\mathbf{w}$, is defined as

$$H_{\mathbf{w}} = I - \frac{2}{\mathbf{w}^*\mathbf{w}}\mathbf{w}\mathbf{w}^* \tag{3.6}$$

Note that $H_{\mathbf{w}}$ is also Hermitian, that is, $H_{\mathbf{w}}^* = H_{\mathbf{w}}$. The action of $H_{\mathbf{w}}$ on a vector $\mathbf{x}$ is to reflect it with respect to a hyperplane that is perpendicular to $\mathbf{w}$. This is shown in Figure 3.5.

To show that $H_{\mathbf{w}}$ is unitary, we have

$$\begin{aligned} H_{\mathbf{w}}^*H_{\mathbf{w}} = H_{\mathbf{w}}H_{\mathbf{w}}^* = H_{\mathbf{w}}^2 &= (I - \frac{2}{\mathbf{w}^*\mathbf{w}}\mathbf{w}\mathbf{w}^*)(I - \frac{2}{\mathbf{w}^*\mathbf{w}}\mathbf{w}\mathbf{w}^*) \\ &= I - \frac{4}{\mathbf{w}^*\mathbf{w}}\mathbf{w}\mathbf{w}^* + \frac{4}{(\mathbf{w}^*\mathbf{w})^2}\mathbf{w}\mathbf{w}^*\mathbf{w}\mathbf{w}^* \\ &= I - \frac{4}{\mathbf{w}^*\mathbf{w}}\mathbf{w}\mathbf{w}^* + \frac{4}{(\mathbf{w}^*\mathbf{w})}\mathbf{w}\mathbf{w}^* = I \end{aligned}$$

The Householder operator could also be used as an alternative way to move a particular vector into another direction specified by a vector (instead of specifying angles). For instance, let $\mathbf{x}$ be the input and $\mathbf{y}$ the desired output, where both have the same norm but pointing in different directions. By setting $\mathbf{w} = \mathbf{y} - \mathbf{x}$, $H_{(\mathbf{y}-\mathbf{x})}$ will move $\mathbf{x}$ to have the same direction as $\mathbf{y}$. To show this, use the fact that

Figure 3.5. Householder transformation as a reflector operator.

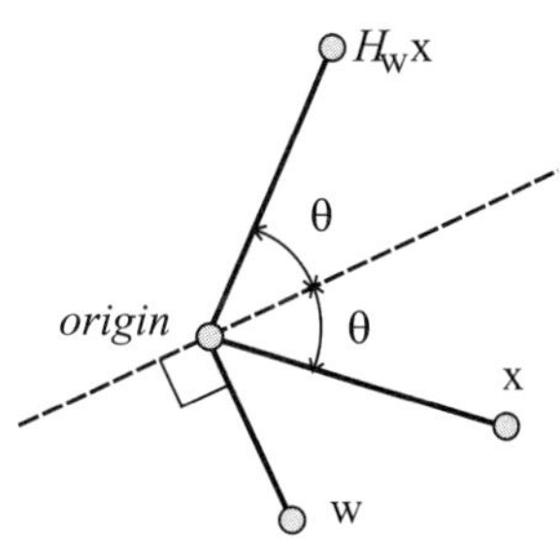

$\mathbf{x}^*\mathbf{x} = \mathbf{y}^*\mathbf{y}$ and $\mathbf{x}^*\mathbf{y} = \mathbf{y}^*\mathbf{x}$. Then,

$$\begin{aligned} H_{(\mathbf{x}-\mathbf{y})}\mathbf{x} &= \left(I - \frac{2}{(\mathbf{x}-\mathbf{y})^*(\mathbf{x}-\mathbf{y})}(\mathbf{x}-\mathbf{y})(\mathbf{x}-\mathbf{y})^*\right)\mathbf{x} \\ &= \mathbf{x} - \frac{1}{\mathbf{x}^*\mathbf{x} - \mathbf{y}^*\mathbf{x}}(\mathbf{x}\mathbf{x}^*\mathbf{x} - \mathbf{y}\mathbf{x}^*\mathbf{x} - \mathbf{x}\mathbf{y}^*\mathbf{x} + \mathbf{y}\mathbf{y}^*\mathbf{x}) \\ &= \frac{\mathbf{x}\mathbf{x}^*\mathbf{x} - \mathbf{x}\mathbf{y}^*\mathbf{x} - \mathbf{x}\mathbf{x}^*\mathbf{x} + \mathbf{y}\mathbf{x}^*\mathbf{x} + \mathbf{x}\mathbf{y}^*\mathbf{x} - \mathbf{y}\mathbf{y}^*\mathbf{x}}{\mathbf{x}^*\mathbf{x} - \mathbf{y}^*\mathbf{x}} \\ &= \mathbf{y} \end{aligned} \tag{3.7}$$

3.1.2 Projection Operators

Projection operators are another important class of operators. Generally, they separate a given space into two complementary subspaces: one subspace is the region where the outputs of P will reside, and the other subspace is specified by its complement, $(I - P)$.

> **Definition 3.2.** *A matrix operator P is* **idempotent** *if and only if $P^2 = P$. If in addition, P is Hermitian, then it is known as a* **projection operator**. *If P is a projection operator, then $(I - P)$ is known as the* **complementary projection operator**.

We can see that if P is idempotent, then so is its complement, $(I - P)$, that is,

$$\begin{aligned} (I-P)^2 &= (I-P)(I-P) \\ &= I - 2P + P^2 \\ &= I - P \end{aligned}$$

Let $\mathbf{v}$ reside in the linear space $\mathcal{L}$, and let $\mathcal{S}$ be a subspace in $\mathcal{L}$. Then a projection operation, $P_{\mathcal{S}}$, can be used to decompose a vector into two complementary vectors,

$$\begin{aligned} \mathbf{v} &= (P_{\mathcal{S}} + I - P_{\mathcal{S}})\mathbf{v} \\ &= P\mathbf{v} + (I-P)\mathbf{v} \\ &= \mathbf{v}_{\mathcal{S}} + \mathbf{v}_{\mathcal{L}-\mathcal{S}} \end{aligned}$$

where $\mathbf{v}_{\mathcal{S}}$ will reside in the subspace $\mathcal{S}$, whereas $\mathbf{v}_{\mathcal{L}-\mathcal{S}}$ will reside in the complementary subspace $\mathcal{L} - \mathcal{S}$.

Let $\{\mathbf{q}_1, \ldots, \mathbf{q}_M\}$ be a set of **orthonormal basis vectors** for $\mathcal{S}$, that is,

$$\mathbf{q}_i^*\mathbf{q}_j = \begin{cases} 0 & \text{if } i \neq j \\ 1 & \text{if } i = j \end{cases} \qquad \text{and} \qquad \mathcal{S} = \textbf{Span}\{\mathbf{q}_1, \ldots, \mathbf{q}_M\}$$

Then a projection operator onto $\mathcal{S}$ is given by

$$P_{\mathcal{S}} = \sum_{i=1}^{M} \mathbf{q}_i\mathbf{q}_i^* \tag{3.8}$$

This can be generalized to a set of linearly independent vectors $\{\mathbf{a}_1, \ldots, \mathbf{a}_M\}$, with $\mathcal{S} = \mathbf{Span}\{\mathbf{a}_1, \ldots, \mathbf{a}_M\}$. One can use the QR algorithm (cf. Section 2.6) to obtain matrices $\widehat{Q}$ and $\widehat{R}$ where

$$\widehat{Q}\widehat{R} = A \qquad \text{and} \qquad \widehat{Q}^*\widehat{Q} = I$$

Let Q be the first M columns of $\widehat{Q}$; then these columns will be a set of orthonormal basis vectors for $\mathcal{S}$. Also, let R be the first M rows of $\widehat{R}$. One can check that $QR = A$. The projection operator is then given by

$$P_{\mathcal{S}} = QQ^* \tag{3.9}$$

with the following properties:

1. $P_{\mathcal{S}}$ is Hermitian.
2. For any vector $\mathbf{v}$, $P_{\mathcal{S}}\mathbf{v}$ will be in the span $\mathcal{S}$. To show this,

$$QQ^*\mathbf{v} = \sum_{i=1}^{M} \mathbf{q}_i \left(\mathbf{q}_i^*\mathbf{v}\right) = \sum_{i=1}^{M} \alpha_i \mathbf{q}_i \in \mathcal{S}$$

   where $\alpha_i = \mathbf{q}_i^*\mathbf{v}$.
3. For any vector $\mathbf{v}$ and $\mathbf{b}$, $\mathbf{y} = P_{\mathcal{S}}\mathbf{v}$ is orthogonal to $\mathbf{z} = P_{\mathcal{L}-\mathcal{S}}\mathbf{b}$, where $P_{\mathcal{L}-\mathcal{S}} = I - P_{\mathcal{S}}$. To show this,

$$\mathbf{y}^*\mathbf{z} = \mathbf{v}^* P_{\mathcal{S}}^* P_{\mathcal{L}-\mathcal{S}}\mathbf{b} = \mathbf{v}^* QQ^* \left(I - QQ^*\right)\mathbf{b} = \mathbf{v}^*\left(QQ^* - QQ^*QQ^*\right)\mathbf{b} = \mathbf{0}$$

There is also a strong relationship between the projection operator and the least-squares solution of $A\mathbf{x} = \mathbf{b}$ given by (2.35). Using the QR factorization of A,

$$\begin{aligned}
\widehat{\mathbf{x}} &= \left(A^*A\right)^{-1} A^*\mathbf{b} = \left(R^*Q^*QR\right)^{-1} R^*Q^*\mathbf{b} \\
&= R^{-1}Q^*\mathbf{b} \\
R\widehat{\mathbf{x}} &= Q^*\mathbf{b} \\
QR\widehat{\mathbf{x}} &= QQ^*\mathbf{b} \\
A\widehat{\mathbf{x}} &= P_{\mathcal{S}}\mathbf{b}
\end{aligned}$$

Thus $A\widehat{\mathbf{x}}$ is the orthogonal projection of $\mathbf{b}$ onto the subspace spanned by the columns of A.

EXAMPLE 3.1. Suppose we want to obtain the projection operator of a three-dimensional vector $\mathbf{b}$ onto the subspace spanned by $\mathbf{a}_1$ and $\mathbf{a}_2$, where

$$\mathbf{b} = \begin{pmatrix} 0.5 \\ 0.5 \\ -1 \end{pmatrix} \quad \mathbf{a}_1 = \begin{pmatrix} 1 \\ 0 \\ -1 \end{pmatrix} \quad \mathbf{a}_2 = \begin{pmatrix} 0 \\ 1 \\ -2 \end{pmatrix}$$

Let $A = \left(\mathbf{a}_1 \quad \mathbf{a}_2\right)$. Using the QR-algorithm on A, we obtain the orthonormal basis,

$$Q = \left(\ \mathbf{q}_1 \mid \mathbf{q}_2\ \right) = \begin{pmatrix} \sqrt{2}/2 & -\sqrt{3}/3 \\ 0 & \sqrt{3}/3 \\ -\sqrt{2}/2 & -\sqrt{3}/3 \end{pmatrix}$$

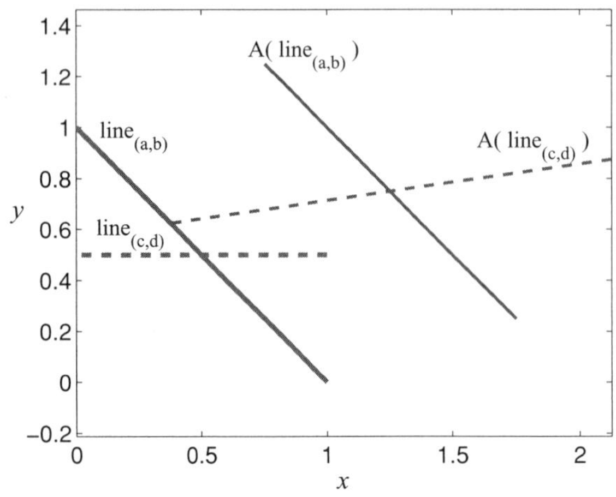

Figure 3.6. Using a linear operator A to translate $\text{line}_{(\mathbf{a},\mathbf{b})}$ may not translate another line $\text{line}_{(\mathbf{c},\mathbf{d})}$.

then the projection operators are given by

$$P_{\mathcal{S}} = QQ^* = \frac{1}{6}\begin{pmatrix} 5 & -2 & -1 \\ -2 & 2 & -2 \\ -1 & -2 & 5 \end{pmatrix} \qquad P_{(\mathcal{L}-\mathcal{S})} = I - P_{\mathcal{S}} = \frac{1}{6}\begin{pmatrix} 1 & 2 & 1 \\ 2 & 4 & 2 \\ 1 & 2 & 1 \end{pmatrix}$$

Solving for the least-squares solution,

$$\begin{pmatrix} 1 & 0 \\ 0 & 1 \\ -1 & -2 \end{pmatrix}\mathbf{x} = \begin{pmatrix} 0.5 \\ 0.5 \\ -1 \end{pmatrix} \quad \rightarrow \quad \widehat{\mathbf{x}} = R^{-1}Q^*\mathbf{b} = \frac{1}{12}\begin{pmatrix} 5 \\ 4 \end{pmatrix}$$

We can check that $A\widehat{\mathbf{x}} = P_{\mathcal{S}}\mathbf{b}$.

3.1.3 Affine Operators

A linear operator can often be constructed to translate one line at a time. However, for a given set of nonparallel lines, it is not possible to translate all these lines in the same fashion. To illustrate, consider the line segments $\text{line}_{(\mathbf{a},\mathbf{b})}$ and $\text{line}_{(\mathbf{c},\mathbf{d})}$, where

$$\mathbf{a} = \begin{pmatrix} 1 \\ 0 \end{pmatrix} \quad \mathbf{b} = \begin{pmatrix} 0 \\ 1 \end{pmatrix} \quad \mathbf{c} = \begin{pmatrix} 0 \\ 0.5 \end{pmatrix} \quad \mathbf{d} = \begin{pmatrix} 1 \\ 0.5 \end{pmatrix}$$

A matrix operator A given by

$$A = \begin{pmatrix} 1.75 & 0.75 \\ 0.25 & 1.25 \end{pmatrix}$$

will translate $\text{line}_{(\mathbf{a},\mathbf{b})}$ but not $\text{line}_{(\mathbf{c},\mathbf{d})}$, as shown in Figure 3.6.

A generalization of linear operators that include translations is called **affine operators**.

Definition 3.3. *An* **affine operator** *acting on vector* $\mathbf{v}$ *, denoted by* $\mathbf{Aff}_{(A,\mathbf{t})}(\mathbf{v})$*, is a linear operator* A *acting on* $\mathbf{v}$ *followed by a translation* $\mathbf{t}$*, that is,*

$$\mathbf{Aff}_{(A,\mathbf{t})}(\mathbf{v}) = A\mathbf{v} + \mathbf{t} \tag{3.10}$$

Strictly speaking, affine operators are not linear operators. However, we can transform the affine operation and make it a linear matrix operation. One approach is to expand both the operator A and the input vector $\mathbf{v}$ as follows:

$$\widehat{A} = \left(\begin{array}{c|c} A & \mathbf{t} \\ \hline 0 & 1 \end{array}\right) \qquad \widehat{\mathbf{v}} = \left(\begin{array}{c} \mathbf{v} \\ \hline 1 \end{array}\right) \tag{3.11}$$

where vector $\mathbf{t}$ is the desired translation. After the affine operation $\widehat{A}$ has finished operation on $\widehat{\mathbf{v}}$, one can then obtain the desired results by simply removing the last element of $\widehat{\mathbf{v}}$, that is,

$$\left(\begin{array}{c|c} I & 0 \end{array}\right)\widehat{A}\widehat{\mathbf{v}} = \left(\begin{array}{c|c} I & 0 \end{array}\right)\left(\begin{array}{c|c} A & \mathbf{t} \\ \hline 0 & 1 \end{array}\right)\left(\begin{array}{c} \mathbf{v} \\ \hline 1 \end{array}\right) = \left(\begin{array}{c|c} I & 0 \end{array}\right)\left(\begin{array}{c} A\mathbf{v}+\mathbf{t} \\ \hline 1 \end{array}\right) = A\mathbf{v}+\mathbf{t}$$

EXAMPLE 3.2. Given a line segment defined by its endpoints, $\mathbf{a} = \left(a_x, a_y\right)^T$ and $\mathbf{b} = \left(b_x, b_y\right)^T$. Suppose we want to rotate this line segment counterclockwise by θ radians at its midpoint, $\mathbf{m} = \left(m_x, m_y\right)^T$, where $m_x = \left(a_x + b_x\right)/2$ and $m_y = \left(a_y + b_y\right)/2$. This operation can be achieved by a sequence of affine operations. First, we can translate the midpoint of the line to the origin. Second, we rotate the line. Lastly, we translate the origin back to $\mathbf{m}$. In the 2-D case, this is given by

$$\begin{aligned}
\mathbf{Aff} &= \left(\begin{array}{cc|c} 1 & 0 & m_x \\ 0 & 1 & m_y \\ \hline 0 & 0 & 1 \end{array}\right)\left(\begin{array}{cc|c} \cos(\theta) & -\sin(\theta) & 0 \\ \sin(\theta) & \cos(\theta) & 0 \\ \hline 0 & 0 & 1 \end{array}\right)\left(\begin{array}{cc|c} 1 & 0 & -m_x \\ 0 & 1 & -m_y \\ \hline 0 & 0 & 1 \end{array}\right) \\
&= \left(\begin{array}{cc|c} \cos(\theta) & -\sin(\theta) & \alpha \\ \sin(\theta) & \cos(\theta) & \beta \\ \hline 0 & 0 & 1 \end{array}\right)
\end{aligned}$$

where

$$\begin{aligned}
\alpha &= \left(1 - \cos(\theta)\right) m_x + \sin(\theta)\, m_y \\
\beta &= -\sin(\theta)\, m_x + \left(1 - \cos(\theta)\right) m_y
\end{aligned}$$

To illustrate, for $\mathbf{a} = \begin{pmatrix}0.5 & 0.5\end{pmatrix}^T$, $\mathbf{b} = \begin{pmatrix}0.7 & 0.5\end{pmatrix}^T$ and $\theta = \pi/4$, the affine operation is shown in Figure 3.7.

3.2 Eigenvalues and Eigenvectors

To study the characteristics of a particular matrix operator A, one can collect several pairs of $\mathbf{x}$ and $\mathbf{b} = A\mathbf{x}$. Some of these pairs will behave more distinctively than others, and they yield more information about the operator. A group of vectors known as **eigenvectors** have the distinct property such that the only effect that A has on them is a scaling operation.

Definition 3.4. *Let $A[=]N \times N$, then $\mathbf{v} \neq 0$ is an* **eigenvector** *of A if*

$$A\mathbf{v} = \lambda\mathbf{v} \tag{3.12}$$

where λ is a scalar called the **eigenvalue**.

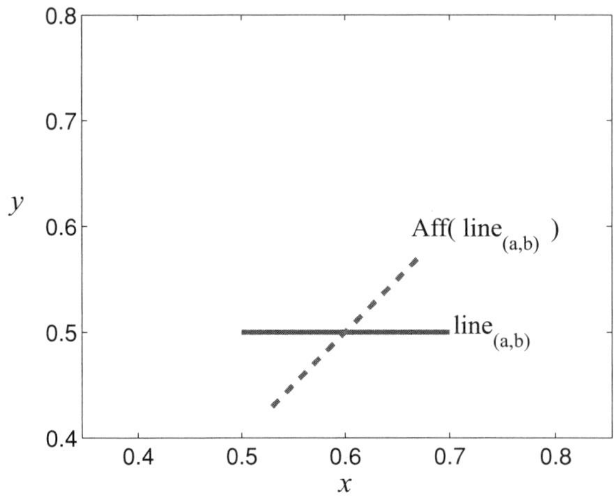

Figure 3.7. Applying affine operation **Aff** that rotates a line segment $\text{line}_{(\mathbf{a},\mathbf{b})}$ counterclockwise by $\theta = \pi/4$ radians around its midpoint.

Because $\mathbf{v} = 0$ is always a solution to (3.12), it does not give any particular information about an operator. It is known as the **trivial vector**. This is why we only consider nonzero vectors to be eigenvectors.

To evaluate the eigenvectors of A, we can use the condition given by (3.12).

$$A\mathbf{v} = \lambda\mathbf{v} \quad \rightarrow \quad (A - \lambda I)\,\mathbf{v} = 0 \tag{3.13}$$

For $\mathbf{v} \neq 0$, we need $(\lambda I - A)$ to be singular, that is,

$$\mathbf{det}\,(\,\lambda I - A\,) = 0 \tag{3.14}$$

Equation (3.14) is known as the **characteristic equation** of A, and this equation can be expanded into a polynomial of order N, where N is the size of A. Using the formula for determinants given by (1.10), we have the following lemma:

LEMMA 3.1. *Let $A[=]N \times N$. Then the characteristic equation (3.14) will yield a polynomial equation given by*

$$\text{charpoly}(\lambda) = \lambda^N + \beta_{N-1}\lambda^{N-1} + \;\cdots\; + \beta_1\lambda + \beta_0 = 0 \tag{3.15}$$

where

$$\beta_{N-k} = (-1)^k \sum_{\ell_1 < \cdots < \ell_k} \mathbf{det}\left(A_{[\ell_1,\cdots,\ell_k]}^{[\ell_1,\cdots,\ell_k]} \right)$$

and $A_{[\ell_1,\cdots,\ell_k]}^{[\ell_1,\cdots,\ell_k]}$ is the matrix obtained by extracting the rows and columns indexed by $\ell_1, \ldots, \ell_k$.

The polynomial, charpoly (A), *is known as the* **characteristic polynomial of** A. *Specifically, two important coefficients are $\beta_0 = (-1)^N \mathbf{det}\,(A)$ and $\beta_{N-1} = (-1)\,$***trace***$\,(A)$.*

Remarks: In MATLAB, the command to obtain the coefficients of the characteristic polynomial is given by `c = poly(A)`, where **c** will be a vector of the coefficients of the polynomial of descending powers of λ. Because the eigenvalues will depend on the solution of this polynomial, small errors in the coefficients can cause a large error in the values of the roots. Thus for large matrices, other methods are used to find the eigenvalues. Nonetheless, the characteristic polynomials themselves are important in other applications such as the Cayley-Hamilton theorem (which is discussed later), Other methods exist for evaluating the characteristic polynomials with increased precision. One of these methods is known as the Danilevskii method, which is included in Section C.6 as an appendix.

The characteristic polynomials will yield N roots, and the eigenvalues may either be real or complex numbers. The collection of all the eigenvalues of A, including multiplicities, are also known as the **spectrum** of A, which we denote by **Spec**(A), that is,

$$\mathbf{Spec}\,(A) = \{\lambda_1, \cdots, \lambda_N\} \tag{3.16}$$

EXAMPLE 3.3. Let A be given by

$$A = \begin{pmatrix} 1 & 0 & 7 \\ 0 & 5 & 6 \\ 7 & 0 & 1 \end{pmatrix}$$

Then expanding $\mathbf{det}\,(\lambda I - A)$, we obtain

$$\mathbf{det}\,(\lambda I - A) = \mathbf{det} \begin{pmatrix} (\lambda - 1) & 0 & -7 \\ 0 & (\lambda - 5) & -6 \\ -7 & 0 & (\lambda - 1) \end{pmatrix} = \lambda^3 - 7\lambda^2 - 38\lambda + 240$$

which is the characteristic polynomial of A. By equating the polynomial to zero, we obtain the characteristic equation

$$\lambda^3 - 7\lambda^2 - 38\lambda + 240 = 0$$

whose roots are $-6,\ 5,\ 8$. Thus, the eigenvalues of A are given by the set **Spec** $(A) = \{-6, 5, 8\}$.

Once the eigenvalues are obtained, each distinct value of λ_* can be used to generate the corresponding eigenvector, $\mathbf{v}_{\lambda_*}$, by substituting each eigenvalue one at a time into equation (3.13),

$$(\lambda_* I - A)\,\mathbf{v}_{\lambda_*} = 0 \tag{3.17}$$

Each of the equations (3.17) will yield infinite solutions, but only the directions of these vectors will be important, that is, if $\mathbf{v}$ is an eigenvector, then so is $\alpha\mathbf{v}$ for any scalar $\alpha \neq 0$. As a standard, the eigenvectors are often normalized to have a norm equal to 1, that is, with $\alpha = \|\mathbf{v}\|^{-1}$.

We can generate a procedure of finding the eigenvectors for each distinct eigenvalue λ_* of a matrix A. Let $W_{(\lambda_*,\ k)}$ be a nonsingular submatrix of $(\lambda_* I - A)$ obtained by removing the kth row and k^{th} column, that is,

$$W_{(\lambda_*,k)} = (\lambda_* I - A)_{(k,k)\downarrow} \quad \text{and} \quad \mathbf{det}\left(W_{(\lambda_*,k)}\right) \neq 0$$

Next, let vector $\mathbf{c}_{(k)}$ be the kth column of matrix A with the kth element removed, i.e.

$$\mathbf{c}_{(k)} = \begin{pmatrix} a_{1,k} \\ \vdots \\ a_{k-1,k} \\ \hline a_{k+1,k} \\ \vdots \\ a_{N,k} \end{pmatrix}$$

Now evaluate the vector $\widehat{\mathbf{v}}$ as

$$\widehat{\mathbf{v}} = W^{-1}_{(\lambda_*,k)}\mathbf{c}_{(k)} \tag{3.18}$$

Finally, the eigenvector can be obtained by inserting a 1 in the kth position of $\widehat{\mathbf{v}}$

$$\mathbf{v}_{\lambda_*} = \begin{pmatrix} \widehat{v}_1 \\ \vdots \\ \widehat{v}_{k-1} \\ \hline 1 \\ \hline \widehat{v}_k \\ \vdots \\ \widehat{v}_N \end{pmatrix} \tag{3.19}$$

To obtain a normalized vector, $\mathbf{v}_{\lambda*}$ can be scaled by $\alpha = \left\| \mathbf{v}_{\lambda_*} \right\|^{-1}$. This matrix reduction approach is not valid if $W_{(\lambda_*,k)}$ is singular for all k. In those cases, one needs to solve the equation $(\lambda_* I - A)\,\mathbf{v} = 0$ using techniques discussed in Section 2.1.

EXAMPLE 3.4. Let A be given by

$$A = \begin{pmatrix} 1 & 2 & 3 \\ 4 & 5 & 6 \\ 7 & 8 & 1 \end{pmatrix}$$

then the characteristic equation is given by

$$\textbf{det}\,(\lambda I - A) = \lambda^3 - 7\lambda^2 - 66\lambda - 24 = 0$$

whose roots are the eigenvalues: $\lambda = 12.4542, -5.0744, -0.3798$. Using (3.18) and (3.19), we obtain by using $k = 1$,

$$\mathbf{v}_{(12.4542)} = \begin{pmatrix} 1.0000 \\ 2.3492 \\ 2.2519 \end{pmatrix}, \quad \mathbf{v}_{(-5.0744)} = \begin{pmatrix} 1.0000 \\ 1.3415 \\ -2.9191 \end{pmatrix}, \quad \mathbf{v}_{(-0.3798)} = \begin{pmatrix} 1.0000 \\ -0.8991 \\ 0.1394 \end{pmatrix}$$

and their normalized versions are

$$\mathbf{v}_{(12.4542)} = \begin{pmatrix} 0.2937 \\ 0.6901 \\ 0.6615 \end{pmatrix}, \quad \mathbf{v}_{(-5.0744)} = \begin{pmatrix} 0.2972 \\ 0.3987 \\ -0.8676 \end{pmatrix}, \quad \mathbf{v}_{(-0.3798)} = \begin{pmatrix} 0.7397 \\ -0.6650 \\ 0.1031 \end{pmatrix}$$

We now list some eigenvectors corresponding to λ_* for the following special cases:

1. 2×2 **Matrices**.

$$\mathbf{det}\,(\lambda I - A) = \lambda^2 - \mathbf{tr}\,(A) + \mathbf{det}\,(A) \tag{3.20}$$

and the eigenvalues can then be obtained as

$$\lambda = \frac{\mathbf{tr}\,(A) \pm \sqrt{\mathbf{tr}\,(A)^2 - 4\,\mathbf{det}\,(A)}}{2} \tag{3.21}$$

To obtain the corresponding eigenvectors, (3.19) becomes

$$\mathbf{v}_{\lambda_*} = \begin{cases} \begin{pmatrix} \dfrac{a_{12}}{\lambda_* - a_{11}} \\ 1 \end{pmatrix} & \text{if } a_{11} \neq \lambda_* \\ \begin{pmatrix} 1 \\ \dfrac{a_{21}}{\lambda_* - a_{22}} \end{pmatrix} & \text{if } a_{22} \neq \lambda_* \end{cases} \tag{3.22}$$

In case $\lambda_* = a_{11} = a_{22}$, then l'Hospital's rule can still be used where $0/0$ occurs.

2. **Diagonal matrices.** Let $D[=]N \times N$ be diagonal, then the eigenvalues are simply the diagonal elements,

$$\mathbf{det}\,(\lambda I - D) = \mathbf{det} \begin{pmatrix} (\lambda - d_1) & & 0 \\ & \ddots & \\ 0 & & (\lambda - d_N) \end{pmatrix} = \prod_{i=1}^{N} (\lambda - d_i)$$

whose roots are $\lambda = d_1, d_2, \ldots, d_N$. The eigenvector corresponding to $\lambda = d_k$ is the kth unit vector, $\mathbf{e}_k$.

3. **Triangular matrices**. Let $L[=]N \times N$ be lower triangular. Then the eigenvalues of L are also given by the diagonal elements, that is,

$$\mathbf{det}\,(\lambda I - L) = \mathbf{det} \begin{pmatrix} (\lambda - \ell_{11}) & & 0 \\ \vdots & \ddots & \\ -\ell_{N1} & & (\lambda - \ell_{NN}) \end{pmatrix} = \prod_{i=1}^{N} (\lambda - \ell_{ii})$$

To obtain the eigenvector corresponding to $\lambda_* = \ell_{kk}$ (if λ_* is not unique, choose the largest value of k), we can form the eigenvector as $k-1$ number of zeros appended to the result of (3.19) applied to the lower $(k-1) \times (k-1)$ submatrix of L as follows:

$$\mathbf{v}_{(\ell_{kk})} = \left(\begin{array}{c} 0_{k-1} \\ \hline 1 \\ \hline \mathbf{q} \end{array} \right)$$

where

$$\mathbf{q} = \begin{pmatrix} (\ell_{k,k} - \ell_{k+1,k+1}) & & 0 \\ \vdots & \ddots & \\ -\ell_{N,k+1} & \cdots & (\ell_{k,k} - \ell_{NN}) \end{pmatrix}^{-1} \begin{pmatrix} \ell_{k+1,k} \\ \vdots \\ \ell_{N,k} \end{pmatrix}$$

If $k = N$, then $\mathbf{v}_{(\ell_{NN})} = \mathbf{e}_N$. Note that $\mathbf{q}$ can be obtained using forward substitution.

4. **Companion matrices**. These matrices have a special structure in which an $(N-1) \times (N-1)$ identity matrix is appended by a zero column on the right and a row vector of N constants at the bottom. It is used often in building matrices that would yield a characteristic polynomial with specified coefficients.

   An $N \times N$ **companion matrix** (also known as a **Frobenius matrix**) has the following form:

$$C = \begin{pmatrix} 0 & 1 & 0 & \cdots & 0 \\ 0 & 0 & 1 & \cdots & 0 \\ \vdots & \vdots & \vdots & \ddots & \vdots \\ 0 & 0 & 0 & \cdots & 1 \\ -\alpha_0 & -\alpha_1 & -\alpha_2 & \cdots & -\alpha_{N-1} \end{pmatrix} \tag{3.23}$$

   The characteristic polynomial of C is given by:

$$\lambda^n + \sum_{i=0}^{n-1} \alpha_i \lambda^i = 0$$

   and the eigenvectors corresponding to each distinct eigenvalue λ_* of C is given by

$$\mathbf{v}_{\lambda_*} = \begin{pmatrix} 1 \\ \lambda_* \\ \vdots \\ \lambda_*^{N-1} \end{pmatrix}$$

   (We leave the derivation as an exercise in **E3.8**),

Remarks: In MATLAB, the eigenvalues and normalized eigenvectors are obtained using the statement: `[V,D]=eig(A)`, which yields a diagonal matrix D whose diagonal contains the eigenvalues of A, and the kth column of matrix V is the normalized eigenvector corresponding to the kth eigenvalue. Alternatively, in case only a subset of the eigenvalues of A is needed, one could use: `eigs`.

There are other numerical methods for obtaining the eigenvalues and eigenvectors of a matrix. These methods avoid having to first obtain the characteristic polynomial, thus also avoiding the problems of accuracy associated with finding the roots of the polynomials. One such method is the QR **method**, which is based on the result of another method known as the **power method**. Details for this approach of evaluating eigenvales and eigenvectors are given in Section C.2.

3.3 Properties of Eigenvalues and Eigenvectors

We now list some useful properties and identities of eigenvalues and eigenvectors. Some of these provide a quick way for obtaining the eigenvalues or eigenvectors, whereas the other properties describe the characteristics of the eigenvalues and eigenvectors.

Property 1. The eigenvalues of diagonal and triangular matrices are the diagonal entries of these matrices. The eigenvector of a diagonal matrix D corresponding to eigenvalue d_{jj} is the unit vector $\mathbf{e}_j$.

Property 2. The eigenvalue of block diagonal, upper block triangular, or lower block triangular matrices is the collection of all the eigenvalues of each of the block matrices in the diagonal.

Property 3. Let λ be an eigenvalue of A and let α be a scalar, then $\alpha\lambda$ is an eigenvalue of αA. For $\alpha \neq 0$, the eigenvectors of A and αA are the same.

Property 4. A and A^* have the same eigenvalues. In general, they do not have the same set of eigenvectors. The eigenvectors of A^* are known as the **left eigenvectors** of A.

Property 5. Let λ be an eigenvalue of A. Then λ^k is an eigenvalue of A^k, assuming A^k exists, $k = \ldots, -2, -1, 0, 1, 2, \ldots$. For $k \neq 0$, the eigenvectors of A and A^k are the same for the corresponding eigenvalues λ and λ^k, respectively.

Property 6. Let $B = T^{-1}AT$, where T is nonsingular. The eigenvalues of A and B will be the same. (Note: $T^{-1}AT$ is called a **similarity transformation of** A.) If $\mathbf{v}$ is an eigenvector of A corresponding to the eigenvalue λ, then $T^{-1}\mathbf{v}$ is an eigenvector of B corresponding to the same eigenvalue λ.

Property 7. $\prod_{i=1}^{n} \lambda_i = \mathbf{det}A$.

Property 8. $\sum_{i=1}^{n} \lambda_i = \mathbf{tr}\,(A)$.

Property 9. The eigenvalues of Hermitian matrices are all real-valued. The eigenvalues of skew-Hermitian matrices are all pure imaginary.

Property 10. The eigenvalues of positive-definite Hermitian matrices are all positive.

Property 11. The eigenvectors of a Hermitian matrix are orthogonal.

Property 12. Eigenvectors corresponding to distinct eigenvalues are linearly independent.

Property 13. The eigenvalues of A are inside the union of **Gershgorin circles** given by

$$\left|\lambda - a_{kk}\right| \leq \sum_{i=1, i\neq k}^{n} \left|a_{ki}\right| \qquad k = 1, 2, \ldots, N \tag{3.24}$$

The proofs of these properties are given in Section C.1.1 as an appendix, with the exception of Property 13, which is left as an exercise (see Exercise **E3.20**).

Properties 1 through 6 give formulas for the calculation of eigenvalues. Specifically, we have already used property 1 in the previous section for the case of diagonal and triangular matrices. Property 2 is an extension of property 1 to block-triangular matrices.

Assuming that the eigenvalues of A are known, properties 3 to 6 give methods to determine eigenvalues for: (i) scalar products, αA; (ii) transposes, A^T; (iii) powers, A^k; and (iv) similarity transformations, $T^{-1}AT$. Property 6 states that similarity transformations will not change the set of eigenvalues. To illustrate, suppose we have A and T given by

$$A = \begin{pmatrix} 3 & 0 \\ 2 & 5 \end{pmatrix} \quad \text{and} \quad T = \begin{pmatrix} 1 & 2 \\ 2 & 1 \end{pmatrix} \quad \rightarrow \quad H = T^{-1}AT = \begin{pmatrix} 7 & 4 \\ -2 & 1 \end{pmatrix}$$

One can show that the eigenvalues of A and H are both given by $\lambda(A) = 3,\ 5$. Similarity transformations are used extensively in matrix analysis because they allow one to change the elements of a matrix and still maintain the same set of eigenvalues. This type of transformations will be used in several applications such as diagonalization and triangularization and are discussed in later sections.

Properties 7 and 8 give two interesting relationships between the elements of A and the eigenvalues of A. First, the determinant of A is equal to the product of all eigenvalues. This means that when A is singular, at least one of the eigenvalues must be zero. Second, the sum of the eigenvalues is equal to the trace of A. Thus, for real matrices, because the trace will be real, the eigenvalues will have to contain either real values and/or complex conjugate pairs.

Properties 9 through 11 apply to Hermitian matrices and symmetric real matrices. These properties have important implications in real-world applications. For instance, real-symmetric matrix operators, such as stress tensors and strain tensors, are guaranteed to have real eigenvalues and eigenvectors. In addition, if the operator is symmetric positive definite, then its action as an operator will only stretch or contract input vectors without any rotation. Property 10 gives another method for determining whether a given Hermitian matrix is positive definite. It states that if the eigenvalues of Hermitian A are all positive, then A is immediately positive definite, that is, $\mathbf{v}^*A\mathbf{v} > 0$ for any $\mathbf{v} \neq 0$. Furthermore, the eigenvectors of the Hermitian or real-symmetric matrices are also guaranteed to be orthogonal.

For instance, let A be given by

$$A = \frac{1}{6}\begin{pmatrix} 8.5 & -3.5 & 1 \\ -3.5 & 8.5 & 1 \\ 1 & 1 & 4 \end{pmatrix}$$

The eigenvalues are $\lambda = 0.5, 1, 2$ with eigenvectors

$$\mathbf{v}_{\lambda=0.5} = \begin{pmatrix} 1 \\ 1 \\ -2 \end{pmatrix} \;;\; \mathbf{v}_{\lambda=1} = \begin{pmatrix} 1 \\ 1 \\ 1 \end{pmatrix} \;;\; \mathbf{v}_{\lambda=2} = \begin{pmatrix} 1 \\ -1 \\ 0 \end{pmatrix}$$

One can show that the eigenvectors form an orthogonal set. The ellipsoid shown in Figure 3.8 is the result of A acting on the points of a unit sphere. The eigenvectors are alo shown in the figure, and they form three orthogonal directions.

Property 12 guarantees that if the eigenvalues of A are distinct, then the set of eigenvectors will be linearly independent. This means that the eigenvectors can be used as a basis for the N-dimensional space. This also means that a matrix V, whose columns are formed by the eigenvectors, will be nonsingular. This fact is used later during the diagonalization of matrices.

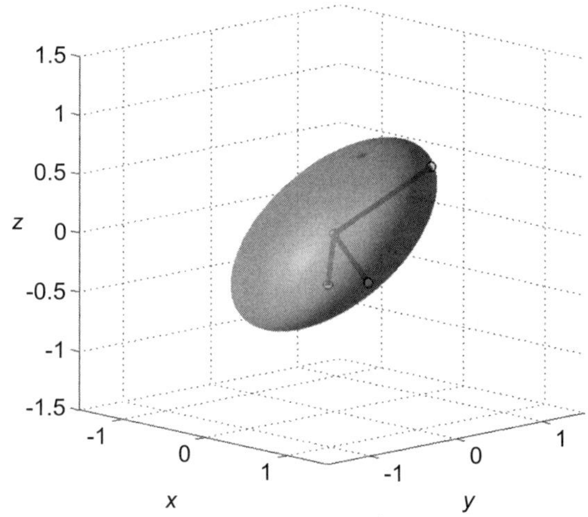

Figure 3.8. The result of Hermitian A operating on a unit sphere. Also marked are the three eigenvectors.

Finally, property 13 gives us a simple way to estimate the location of the eigenvalues without having to evaluate them exactly. To illustrate this property, let A be given by

$$A = \begin{pmatrix} 10 & 1 & 0 & 0 & 0 & 0 \\ 1 & -10 & 1 & 0 & 0 & 0 \\ 0 & 1 & 30 & 1 & 0 & 0 \\ 0 & 0 & 1 & 5 & 1 & 0 \\ 0 & 0 & 0 & 1 & 1 & 1 \\ 0 & 0 & 0 & 0 & 1 & -20 \end{pmatrix}.$$

Then eigenvalues can be found to be $\lambda = -20.0476, -10.0748, 0.8075, 5.2001, 10.0498, 30.0649$. These are plotted in Figure 3.9 together with the Gershgorin circles of A as defined in (3.24). (A MATLAB file `gershgorin.m` to plot Gershgorin circles is available on the book's webpage.)

This property is very useful, especially for large matrices. In addition, they can be used to determine initial guesses when using numerical methods to find the eigenvalues.

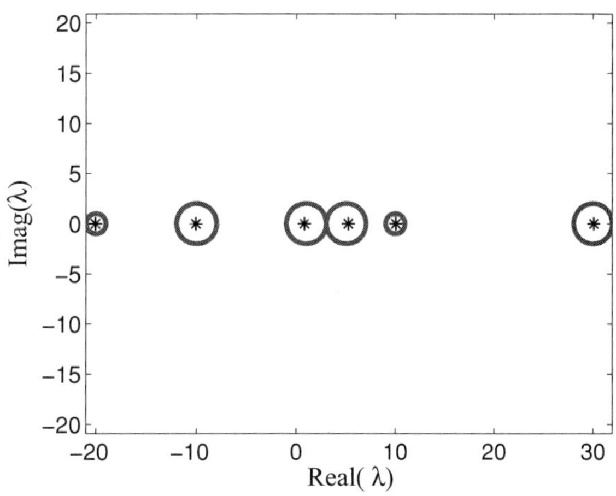

Figure 3.9. Gershgorin circles together with the eigenvalues ('$*$').

3.4 Schur Triangularization and Normal Matrices

We begin with one of the basic factorizations of a matrix operator known as **Schur triangularization**. For any square matrix A, one can find a unitary matrix operator U such that U^*AU will be an upper triangular matrix, say, T. (Equivalently, we have $A = UTU^*$.) In addition, the eigenvalues will appear in the diagonals of T. Details of the algorithm are given in Section C.4.1 as an appendix.

Remarks: In MATLAB, the command `[U,T]=schur(A,'complex')` will yield the unitary matrix U and triangular matrix T.

Schur triangularization is very useful for obtaining several properties of eigenvalues and eigenvectors. Unlike the other factorizations, such as diagonalization (discussed in the next section), Schur triangularization applies to all square matrices. One important application is for special matrices known as **normal matrices**.

Definition 3.5. *A is a* **normal matrix** *if $AA^* = A^*A$.*

Examples of normal matrices include Hermitian, real symmetric, skew-Hermitian, real skew-symmetric, unitary, orthogonal, and circulant matrices. **Circulant matrices** are matrices that have the form

$$A = \begin{pmatrix} a_0 & a_1 & a_2 & \cdots & a_n \\ a_n & a_0 & a_1 & \cdots & a_{n-1} \\ a_{n-1} & a_n & a_0 & \cdots & a_{n-2} \\ \vdots & \vdots & \vdots & \ddots & \vdots \\ a_1 & a_2 & a_3 & \cdots & a_0 \end{pmatrix} \tag{3.25}$$

They are used in Fourier analysis, and they appear during finite difference solutions of problems with periodic boundary conditions. For instance, the following is a circulant matrix:

$$A = \begin{pmatrix} -1 & 2 & 1 \\ 1 & -1 & 2 \\ 2 & 1 & -1 \end{pmatrix} \tag{3.26}$$

It is a normal matrix because

$$A^*A = \begin{pmatrix} 6 & -1 & -1 \\ -1 & 6 & -1 \\ -1 & -1 & 6 \end{pmatrix} = AA^*$$

THEOREM 3.1. *If A is a normal matrix, then there exists a unitary matrix U such that*

$$U^*AU = \text{diag}\left(\lambda_1, \ldots, \lambda_N\right)$$

where $\lambda_1, \ldots, \lambda_N$ are the eigenvalues of A and the columns U are the corresponding orthonormal eigenvectors of A. Furthermore, a matrix will have orthonormal eigvectors if and only if it is normal.

PROOF. (See Section C.1.2 for proof)

For instance, for the circulant matrix A given in (3.26), the Schur triangularization yields a unitary operator U given by

$$U = \begin{pmatrix} 0.5774 & 0.2540 + 0.5185i & 0.5185 + 0.2540i \\ 0.5774 & -0.5760 - 0.0393i & -0.0393 - 0.5760i \\ 0.5774 & 0.3220 - 0.4792i & -0.4792 + 0.3220i \end{pmatrix}$$

$$\rightarrow \quad U^*AU = \begin{pmatrix} 2 & 0 & 0 \\ 0 & -2.5 + 0.866i & 0 \\ 0 & 0 & -2.5 - 0.866i \end{pmatrix}$$

3.5 Diagonalization

For some square matrices A, there exists a nonsingular matrix T such that the similarity transformation $T^{-1}AT$ will be a diagonal matrix Λ.

$$T^{-1}AT = \begin{pmatrix} \lambda_1 & & 0 \\ & \ddots & \\ 0 & & \lambda_N \end{pmatrix} = \Lambda$$

In addition, the diagonal elements of Λ will be the eigenvalues of A. If such a matrix T exists, matrix A is called as a **diagonalizable matrix** (also known as a **semisimple** or a **nondefective** matrix). Equivalently, it yields a factorization of A as $A = T\Lambda T^{-1}$. As we show later, diagonalization of matrices, when possible, are very useful for the evaluation and analysis of matrix functions and matrix differential equations.

Three cases (not necessarily disjoint) are guaranteed to be diagonalizable. The first case is when all eigenvalues of A are distinct. The second case is when the matrix is a normal matrix. The third case applies to matrices that contain repeated eigenvalues but satisfy some rank conditions.

1. **All eigenvalues are distinct.** Let

$$V = \left(\mathbf{v}_1 \;\middle|\; \cdots \;\middle|\; \mathbf{v}_N \right) \qquad \text{and} \qquad \Lambda = \text{diag}(\lambda_1, \ldots, \lambda_N)$$

The eigenvalue equations can then combined as

$$\begin{array}{rcl} A\mathbf{v}_1 & = & \lambda_1\mathbf{v}_1 \\ & \vdots & \\ A\mathbf{v}_N & = & \lambda_N\mathbf{v}_N \end{array} \qquad \rightarrow \qquad AV = V\Lambda$$

Using Property 12 of eigenvectors (cf. Section 3.3), V will be nonsingular. Thus

$$V^{-1}AV = \Lambda \tag{3.27}$$

2. **Normal matrices.** This class includes matrices that can have repeated roots. Based on Theorem 3.1, one can find a unitary matrix U, such that $U^*AU = \Lambda$, where U can be found using Schur triangularization.

3. **Repeated eigenvalues that satisfy rank conditions.**

THEOREM 3.2. *Let λ_i be the eigenvalues of A with a multiplicity of k_i, $i = 1, 2, \ldots, p$, where p is the number of distinct eigenvalues. If*

$$\mathrm{rank}(\lambda_i I - A) = N - k_i \tag{3.28}$$

then there exists linearly independent eigenvectors $\mathbf{v}_1, \ldots, \mathbf{v}_N$

PROOF. (See Section C.1.3 for proof.)

With (3.28), one can again obtain a nonsingular V, such that $V^{-1}AV = \Lambda$.

EXAMPLE 3.5. Let

$$A = \begin{pmatrix} -1 & 0 & 0 \\ -1 & -2 & 0 \\ 0 & 0 & -2 \end{pmatrix}$$

The eigenvalues are $\{-2, -2, -1\}$, that is, $\lambda_1 = -2, k_1 = 2$ and $\lambda_2 = -1, k_2 = 1$.

$$\lambda_1 I - A = \begin{pmatrix} -1 & 0 & 0 \\ 1 & 0 & 0 \\ 0 & 0 & 0 \end{pmatrix} \qquad \lambda_2 I - A = \begin{pmatrix} 0 & 0 & 0 \\ 1 & 1 & 0 \\ 0 & 0 & 1 \end{pmatrix}$$

Because $(\mathrm{rank}(\lambda_1 I - A) = 1 = N - 2)$ and $(\mathrm{rank}(\lambda_2 I - A) = 2 = N - 1)$, the conditions of Theorem 3.2 are satisfied. This implies that A is diagonalizable.

For $\lambda_1 = -2$, we have two linearly independent eigenvectors,

$$\mathbf{v}_1 = \begin{pmatrix} 0 \\ 1 \\ 0 \end{pmatrix} \quad \text{and} \quad \mathbf{v}_2 = \begin{pmatrix} 0 \\ 0 \\ 1 \end{pmatrix}$$

and for $\lambda_2 = -1$, we have the eigenvector,

$$\mathbf{v}_3 = \begin{pmatrix} 1 \\ -1 \\ 0 \end{pmatrix}$$

Thus

$$V = \begin{pmatrix} 0 & 0 & 1 \\ 1 & 0 & -1 \\ 0 & 1 & 0 \end{pmatrix} \quad \rightarrow \quad V^{-1}AV = \begin{pmatrix} -2 & 0 & 0 \\ 0 & -2 & 0 \\ 0 & 0 & -1 \end{pmatrix}$$

3.6 Jordan Canonical Form

For nondiagonalizable matrices, the closest form to a diagonal matrix obtainable by similarity transformations is the **Jordan canonical form** given by

$$J = \left(\begin{array}{c|c|c} J_1 & & 0 \\ \hline & \ddots & \\ \hline 0 & & J_m \end{array}\right) \tag{3.29}$$

where J_i is called the **Jordan block**, which is either a scalar or a matrix block of the form

$$J_i = \begin{pmatrix} \lambda & 1 & & 0 \\ & \ddots & \ddots & \\ & & \ddots & 1 \\ 0 & & & \lambda \end{pmatrix} \tag{3.30}$$

The nonsingular matrix T such that $T^{-1}AT = J$ is known as the **modal matrix**, and the columns of the modal matrix are known as the **canonical basis** of A. Thus the **Jordan decomposition of** A, or the **Jordan canonical form of** A, is given by

$$A = TJT^{-1} \tag{3.31}$$

Note that if all the Jordan blocks are 1×1, (3.31) reduces to the diagonalization of A.

EXAMPLE 3.6. Consider the matrix A,

$$A = \begin{pmatrix} 3 & 0 & 0 & 0 & 1 \\ 0 & 3 & 0 & 1 & 1 \\ 1 & 0 & 3 & 0 & 0 \\ 0 & 0 & 0 & 2 & 0 \\ 0 & 0 & 0 & 0 & 3 \end{pmatrix}$$

Using T given by

$$T = \begin{pmatrix} 0 & 0 & 0 & -1.2992 & 0.8892 \\ -0.7071 & -0.5843 & 0 & -1.2992 & 1.1826 \\ 0 & -1.0107 & -1.2992 & 0.8892 & 1.8175 \\ 0.7071 & 0 & 0 & 0 & 0 \\ 0 & 0 & 0 & 0 & -1.2992 \end{pmatrix}$$

we have

$$J = T^{-1}AT = \left(\begin{array}{c|c|ccc} 2 & 0 & 0 & 0 & 0 \\ \hline 0 & 3 & 0 & 0 & 0 \\ \hline 0 & 0 & 3 & 1 & 0 \\ 0 & 0 & 0 & 3 & 1 \\ 0 & 0 & 0 & 0 & 3 \end{array}\right)$$

Note that there are three Jordan blocks even though there are only two distinct eigenvalues.

Details for finding the modal matrix T for a Jordan decomposition is given in Section C.3. However, existing methods for calculating modal matrices are not very reliable for large matrices. One approach is to first reduce the problem by obtaining the eigenvectors to diagonalize part of A, for example, those corresponding to unique eigenvalues, leaving the rest for a Jordan decomposition.

Remarks: In MATLAB, the command for finding Jordan block is `[T,J]= jordan(A)`, which will yield the modal matrix T and Jordan block matrix $J = T^{-1}AT$. We have also included a MATLAB function `jordan_decomp.m`, which is available on the book's webpage. The attached code allows the user to specify a tolerance, because the Jordan decomposition is known to be sensitive to round-off errors for large systems (e.g., the eigenvalues may not be precise and may possibly miss the occurrence of repeated eigenvalues).

3.7 Functions of Square Matrices

Scalar functions can often be extended to their matrix versions, for example, we can define functions of square matrices such as sin(A), cos(A) and exp(A). However, because the commutativity property for product of scalar does not extend to matrices, this and other complications imply that there are important differences between scalar functions and matrix functions. We begin with the definition of well-defined functions.

Definition 3.6. *Let $f(x)$ be a function having a power series expansion*

$$f(x) = \sum_{i=0}^{\infty} \alpha_i \, x^i \tag{3.32}$$

which is convergent for $|x| < R$. Then the function of a square matrix A defined by

$$f(A) = \sum_{i=0}^{\infty} \alpha_i \, A^i \tag{3.33}$$

is called a **well-defined function** *if each eigenvalue has an absolute value less than the radius of convergence R.*

For instance, one of the most important functions in the solution of initial value problems in ordinary differential equations is the exponential function exp (A) where A is a square matrix. Then it is defined as the matrix version of exp (λ), that is,

$$\exp(A) = I + A + \frac{1}{2}A^2 + \frac{1}{3!}A^3 + \cdots \tag{3.34}$$

Sometimes, it is not advisable to calculate the power series of a square matrix directly from the definition, especially when convergence is slow and the matrices are large. An alternative approach is to use diagonalization or Jordan canonical forms. Another alternative is to use the Cayley-Hamilton theorem to obtain an equivalent finite series. We now discuss each of these cases:

1. **Case 1: A is diagonalizable**. In this case, one can find T such that $T^{-1}AT = \Lambda$, where Λ is diagonal and contains the eigenvalues. We can then substitute the factorization $A = T\Lambda T^{-1}$ into (3.33) to yield

$$\begin{aligned} f(A) &= \alpha_0 TT^{-1} + \alpha_1 T\Lambda T^{-1} + \alpha_2 (T\Lambda T^{-1})(T\Lambda T^{-1}) + \ldots \\ &= T(\alpha_0 I + \alpha_1 \Lambda + \alpha_2 \Lambda^2 + \ldots)T^{-1} \\ &= T \begin{pmatrix} (\alpha_0 + \alpha_1\lambda_1 + \ldots) & & 0 \\ & \ddots & \\ 0 & & (\alpha_0 + \alpha_1\lambda_N + \ldots) \end{pmatrix} T^{-1} \\ &= T \begin{pmatrix} f(\lambda_1) & & 0 \\ & \ddots & \\ 0 & & f(\lambda_N) \end{pmatrix} T^{-1} \end{aligned} \tag{3.35}$$

For instance, if A is diagonalizable, then

$$\exp(A) = T \begin{pmatrix} \exp(\lambda_1) & 0 & \cdots & 0 \\ 0 & \exp(\lambda_2) & \cdots & 0 \\ \vdots & \vdots & \ddots & \vdots \\ 0 & 0 & \cdots & \exp(\lambda_n) \end{pmatrix} T^{-1}$$

An equivalent formulation of (3.35) in terms of right and left eigenvectors is given by a method based on **Sylvester's matrix theorem.** Details on this approach are given in Section C.5.

2. **Case 2: A is not diagonalizable.** If A is not diagonalizable, we need to implement the Jordan canonical form. The kth power of a Jordan block $J_i[=]n \times n$ is given by

$$J_i^k = \begin{pmatrix} \lambda_i & 1 & & 0 \\ & \ddots & \ddots & \\ & & \ddots & 1 \\ 0 & & & \lambda_i \end{pmatrix}^k = \begin{pmatrix} \lambda_i^k & \beta_{[k,k-1]}\lambda_i^{k-1} & \cdots & \beta_{[k,k-n+1]}\lambda_i^{k-n+1} \\ 0 & \lambda_i^k & \cdots & \beta_{[k,k-n+2]}\lambda_i^{k-n+2} \\ \vdots & \vdots & \ddots & \vdots \\ 0 & 0 & \cdots & \lambda_i^k \end{pmatrix} \tag{3.36}$$

where,

$$\beta_{k,j} = \begin{cases} \dfrac{k!}{(k-j)!j!} & \text{if } j \geq 0 \\ 0 & \text{otherwise} \end{cases}$$

The matrix function of the Jordan block then becomes

$$f(J_i) = \alpha_0 I + \alpha_1 J_i + \alpha_2 J_i^2 + \cdots = \begin{pmatrix} \gamma_{[i,0]} & \gamma_{[i,1]} & \cdots & \gamma_{[i,N-1]} \\ 0 & \gamma_{[i,0]} & \cdots & \gamma_{[i,N-2]} \\ \vdots & \vdots & \ddots & \vdots \\ 0 & 0 & \cdots & \gamma_{[i,0]} \end{pmatrix} \tag{3.37}$$

where,

$$\gamma_{[i,j]} = \sum_{k=0}^{\infty} \alpha_{k+j}\beta_{[k+j,k]}\lambda_i^k = \frac{1}{j!}\sum_{k=0}^{\infty} \alpha_{k+j}\frac{(k+j)!}{k!}\lambda_i^k = \frac{1}{j!}\left(\frac{d^j f(\lambda_i)}{d\lambda_i^j}\right)$$

or

$$f(J_i) = \begin{pmatrix} f(\lambda_i) & f^{(1)}(\lambda_i) & \cdots & (1/(N-1)!)f^{(N-1)}(\lambda_i) \\ 0 & f(\lambda_i) & \cdots & (1/(N-2)!)f^{(N-2)}(\lambda_i) \\ \vdots & \vdots & \ddots & \vdots \\ 0 & 0 & \cdots & f(\lambda_i) \end{pmatrix} \tag{3.38}$$

where

$$f^{(k)}(\lambda) = \frac{d^k f(\lambda)}{d\lambda^k}$$

Returning to the main issue of evaluating the function of a nondiagonalizable matrix, we have

$$f(A) = T \begin{pmatrix} f(J_1) & 0 & \cdots & 0 \\ 0 & f(J_2) & \cdots & 0 \\ \vdots & \vdots & \ddots & \vdots \\ 0 & 0 & \cdots & f(J_m) \end{pmatrix} T^{-1} \tag{3.39}$$

EXAMPLE 3.7. An example of a non-diagonalizable matrix is

$$A = \begin{pmatrix} -2 & 0 & 0 \\ 1 & -2 & 0 \\ 4 & 2 & -2 \end{pmatrix}$$

The modal matrix T and Jordan matrix J of A can be obtained to be

$$T = \begin{pmatrix} 0 & 0 & 1 \\ 0 & 1 & 0 \\ 2 & 4 & 0 \end{pmatrix} \quad \text{and} \quad J = \begin{pmatrix} -2 & 1 & 0 \\ 0 & -2 & 1 \\ 0 & 0 & -2 \end{pmatrix}$$

The function $\exp(A)$ can then be evaluated as follows:

$$\exp(A) = T \begin{pmatrix} e^{-2} & e^{-2} & \frac{1}{2}e^{-2} \\ 0 & e^{-2} & e^{-2} \\ 0 & 0 & e^{-2} \end{pmatrix} T^{-1} = e^{-2} \begin{pmatrix} 1 & 0 & 0 \\ 1 & 1 & 0 \\ 5 & 2 & 1 \end{pmatrix}$$

3. **Case 3: Using finite sums to evaluate matrix functions**. We first need to state an important theorem known as the **Cayley-Hamilton theorem**:

THEOREM 3.3. *For any square matrix $A[=]N \times N$, whose characteristic polynomial is given by*

$$\text{charpoly}(\lambda) = a_0 + a_1\lambda + \cdots + a_N\lambda^N = 0 \tag{3.40}$$

then matrix A will also satisfy the characteristic polynomial, that is,

$$\text{charpoly}(A) = a_0 I + a_1 A + \cdots + a_N A^N = 0 \tag{3.41}$$

PROOF. (See Section C.1.4 for proof)

Using the Cayley-Hamilton theorem, we can see that A^N can be written as a linear combination of A^i, $i = 0, 1, \ldots, (N-1)$.

$$A^N = \frac{-1}{a_N}(a_0 I + \cdots + a_{n-1}A^{N-1}) \tag{3.42}$$

The same is true with A^{N+1},

$$\begin{aligned}
A^{N+1} &= \frac{-1}{a_N}(a_0 A + \cdots + a_{n-1}A^N) \\
&= \frac{-1}{a_N}\left(a_0 A + \cdots + a_{N-1}\left[\frac{-1}{a_N}(a_0 I + \cdots + a_{n-1}A^{N-1})\right]\right) \\
&= \beta_0 I + \beta_1 A + \cdots + \beta_{n-1}A^{N-1} \qquad (3.43)
\end{aligned}$$

We can continue this process and conclude that A^{N+j}, $j \geq 0$, can always be recast as a linear combination of I, A, ..., A^{N-1}. Applying this fact to 3.33, we can conclude that

$$f(A) = c_0 I + c_1 A + \cdots + c_{N-1}A^{N-1} \qquad (3.44)$$

for some coefficients $c_0, \ldots, c_{N-1}$. What remains is to determine N linearly independent equations that would yield the values of these coefficients.

Because the suggested derivation of (3.44) was based on the characteristic polynomial, this equation should also hold if A is replaced by λ_i, an eigenvalue of A. Thus we can get m linearly independent equations from the m distinct eigenvalues:

$$\begin{aligned}
f(\lambda_1) &= c_0 + c_1\lambda_1 + \cdots + c_{n-1}\lambda_1^{N-1} \\
&\vdots \\
f(\lambda_m) &= c_0 + c_1\lambda_m + \cdots + c_{n-1}\lambda_m^{N-1} \qquad (3.45)
\end{aligned}$$

For the remaining equations, we can use $[d^q f(\lambda)/d\lambda^q]_{\lambda=\lambda_i}$, $q = 1, ..., r_i$, where r_i is the multiplicity of λ_i in the spectrum of A, that is,

$$\left.\frac{d^q f(\lambda)}{d\lambda^q}\right|_{\lambda=\lambda_i} = \left.\frac{d^q}{d\lambda^q}(c_0 + c_1\lambda + \cdots + c_{n-1}\lambda^{N-1})\right|_{\lambda=\lambda_i} \qquad (3.46)$$

After obtaining the required independent linear equations, $c_0, \ldots, c_{n-1}$ can be calculated and used in (3.44).[1]

EXAMPLE 3.8. Let

$$A = \begin{pmatrix} -2 & 0 & 0 \\ 0 & -3 & 1 \\ 1 & 0 & -3 \end{pmatrix}$$

The eigenvalues are $\lambda_1 = -2$ and $\lambda_2 = \lambda_3 = -3$. To find $\exp(A)$, we can first apply (3.45) and (3.46) to obtain three equations,

$$\begin{aligned}
\left.e^{\lambda}\right|_{\lambda=-2} = e^{-2} &= c_0 + c_1(-2) + c_2(-2)^2 \\
\left.e^{\lambda}\right|_{\lambda=-3} = e^{-3} &= c_0 + c_1(-3) + c_2(-3)^2 \\
\left.\frac{d\left(e^{\lambda}\right)}{d\lambda}\right|_{\lambda=-3} = e^{-3} &= c_1 + 2c_2(-3)
\end{aligned}$$

[1] Note that in cases in which the degree of degeneracy of A introduces multiple Jordan blocks corresponding to the same eigenvalue, this method may not yield n linearly independent equations. In those cases, however, there exists a polynomial of lower order than the characteristic polynomial (called minimal polynomials) such that the required number of coefficients will be equal to the number of linear equations obtained from the method just described.

to determine $c_0 = 0.5210$, $c_1 = 0.2644$, $c_2 = 0.0358$. Then

$$\exp(A) = c_0 I + c_1 A + c_2 A^2 = \begin{pmatrix} 0.1353 & 0 & 0 \\ 0.0358 & 0.0498 & 0.0498 \\ 0.0855 & 0 & 0.0498 \end{pmatrix}$$

3.8 Stability of Matrix Operators

One of the most important applications of matrix analysis is to determine the stability of matrix operators in the following sense:

Definition 3.7. *A matrix operator A is* **stable** *if, when it is applied repeatedly on a nonzero vector $\mathbf{v}$, where $\|\mathbf{v}\| < \infty$,*

$$\lim_{k\to\infty} \left\| A^k \mathbf{v} \right\| < \infty$$

Furthermore, it is **asymptotically stable** *if the limit exists, that is,*

$$\lim_{k\to\infty} \left\| A^k \mathbf{v} \right\| = \mathbf{w}$$

where $\|\mathbf{w}\| \leq \infty$.

EXAMPLE 3.9. Consider a real triangular matrix given by

$$A = \begin{pmatrix} a & 0 \\ b & c \end{pmatrix}$$

It can be shown, using the techniques given in Section 7.4, that

$$A^k = \begin{pmatrix} a^k & 0 \\ b\left(\dfrac{a^k - c^k}{a - c}\right) & c^k \end{pmatrix} \qquad \text{if } a \neq c$$

or

$$A^k = \begin{pmatrix} a^k & 0 \\ kbc^{k-1} & c^k \end{pmatrix} \qquad \text{if } a = c$$

Thus, for any nonzero vector $\mathbf{v} = \begin{pmatrix} v_1 & v_2 \end{pmatrix}^T$, we have $\mathbf{w}(k) = A^k\mathbf{v} = \begin{pmatrix} w_1(k) & w_2(k) \end{pmatrix}^T$, where

$$w_1(k) = a^k v_1 \quad \text{and} \quad w_2(k) = \begin{cases} b\left(\dfrac{a^k - c^k}{a - c}\right) v_1 + c^k v_2 & \text{if } a \neq c \\ kba^{k-1} v_1 + a^k v_2 & \text{if } a = c \end{cases}$$

If both $|a| < 1$ and $|c| < 1$, we see that $w_1(\infty) = w_2(\infty) = 0$ and conclude that A is an **asymptotically stable** operator. However, if either $|a| > 1$ or $|c| > 1$, then $w_1(k)$ or $w_2(k)$, or both, will become unbounded as $k \to \infty$, thus A will be an unstable operator.

However, if either $a = -1$, $|c| \leq 1$ or $c = -1$, $|a| \leq 1$, A will still be stable, but not asymptotically stable, because the limit may not exist. If $a = c$ with $|a| = 1$ and $b \neq 0$ then $w_2(k)$ will be unbounded.

As Example 3.9 shows, there are several cases that can lead to either asymptotically stable, stable but not asymptotically stable, or unstable operators. However, a sufficient condition for asymptotically stable matrix operator was simple to check, that is, if $|a| < 1$ and $|c| < 1$. Likewise, a sufficient condition for instability is also simple to check, that is, if $|a| > 1$ or $|c| > 1$, we have instability. For the general case, we can also find these sufficient conditions for stability by testing whether the spectral radius is less than or equal to 1.

Definition 3.8. *Given a matrix operator $A[=]N \times N$ having eigenvalues $\lambda_1, \ldots, \lambda_N$, the* **spectral radius of** *A, denoted by $\rho(A)$, is defined as*

$$\rho(A) = \max_{i=1,\ldots,N} \left| \lambda_i \right| \tag{3.47}$$

We then have the sufficient conditions for stability and instability of matrix operators:

THEOREM 3.4. *An operator $A[=]N \times N$ is asymptotically stable if $\rho(A) < 1$.*

THEOREM 3.5. *An operator $A[=]N \times N$ is unstable if $\rho(A) > 1$.*

To prove these theorems, we can use the Jordan block decomposition of A, that is,

$$A = TJT^{-1} \quad \text{where} \quad J = \left(\begin{array}{c|c|c} J_1 & & 0 \\ \hline & \ddots & \\ \hline & & J_M \end{array}\right) \quad ; \quad J_i = \begin{pmatrix} \lambda_i & 1 & & 0 \\ & \lambda_i & \ddots & \\ & & \ddots & 1 \\ 0 & & & \lambda_i \end{pmatrix}$$

Using (3.38) and (3.39), with $f(A) = A^k$, we have

$$A^k = T \left(\begin{array}{c|c|c} J_1^k & & 0 \\ \hline & \ddots & \\ \hline & & J_M^k \end{array}\right) T^{-1}$$

where

$$J_i^k = \begin{pmatrix} \lambda_i^k & \times & \cdots & \times \\ & \lambda_i^k & \ddots & \vdots \\ & & \ddots & \times \\ 0 & & & \lambda_i^k \end{pmatrix}$$

with $'\times'$ representing elements having the form $\alpha\lambda^\ell$ with some constant α and $\ell < k$. Thus if any $|\lambda_i| > 1$ for some i, that is, $\rho(A) > 1$, then A^k will be unstable. However, if $|\lambda_i| < 1$ for all i, that is, $\rho(A) < 1$, then A^k will converge to a zero matrix.

EXAMPLE 3.10. In Section 2.4, the stationary iterative methods for solving linear equations $A\mathbf{x} = \mathbf{b}$ required splitting the matrix A as $A = A_1 + A_2$ to obtain an iterative method given by,

$$\mathbf{x}^{(k+1)} = H\mathbf{x}^{(k)} + \mathbf{c} \tag{3.48}$$

where $H = -A_1^{-1}A_2$ and $\mathbf{c} = A_1^{-1}\mathbf{b}$. (For the Jacobi method, we chose A_1 as the diagonal of A, whereas for the Gauss-Siedel method, we chose A_1 to be the lower triangular portion of A.)

At the kth iteration of (3.48),

$$\mathbf{x}^{(k+1)} = H^{k+1}\mathbf{x}^{(0)} + \left(I + H + \cdots + H^k\right)\mathbf{c}$$

With $\rho(H) < 1$, we have $\lim_{k\to\infty} H^k = 0$, and $\lim_{k\to\infty} \mathbf{x}^{(k+1)} = \widehat{\mathbf{x}}$, where

$$\widehat{\mathbf{x}} = \left(I + \sum_{i=1}^{\infty} H^i\right)\mathbf{c}$$

The infinite series $\left(I + H + H^2 + \cdots\right)$, also known as the **Von Neumann series**, is convergent if $\rho(H) < 1$. Furthermore, this series can be shown to be equal to $(I - H)^{-1}$, if the inverse exist, that is,

$$(I - H)^{-1} = \left(I + H + H^2 + \cdots\right) \tag{3.49}$$

One can show the equality in (3.49) by simply multiplying both sides by $(I - H)$. Using a Jordan canonical decomposition of $H = TJT^{-1}$,

$$I - H = T(I - J)T^{-1}$$

and because $I - J$ is triangular,

$$\mathbf{det}(I - H) = \prod_{\ell=1}^{N}(1 - \lambda_\ell)$$

which is nonzero if $\rho(H) < 1$.[2] This means that if $\rho(H) < 1$, we can use (3.49) together with $A = A_1 + A_2$, $H = -A_1^{-1}A_2$, and $\mathbf{c} = A_1^{-1}\mathbf{b}$ to show that the iterative process converges to

$$\begin{aligned}
\widehat{\mathbf{x}} &= (I - H)^{-1}\mathbf{c} \\
&= \left(I + A_1^{-1}A_2\right)^{-1} A_1^{-1}\mathbf{b} \\
&= (A_1 + A_2)^{-1} A_1 A_1^{-1}\mathbf{b} \\
&= A^{-1}\mathbf{b}
\end{aligned}$$

which is the desired solution of $A\mathbf{x} = \mathbf{b}$. (Note: The closer A_1^{-1} is to A^{-1}, i.e., the closer A_2 is to 0, the faster the convergence.)

We could also extend the sufficient stability criterion of matrix operators to matrix functions, that is,

THEOREM 3.6. *$f(A)$ is asymptotically stable if* $\rho(f(A)) = \max_i \left|f(\lambda_i)\right| < 1$.

THEOREM 3.7. *$f(A)$ is unstable if* $\rho(f(A)) = \max_i \left|f(\lambda_i)\right| > 1$.

[2] Although the determinant could get very small if several eigenvalues have magnitudes close to 1 and/or N is large.

To prove these theorems, we again use the Jordan decomposition of A with (3.38) and (3.39). Let μ be the eigenvalues of $f(A)$; then the eigenvalue equation gives

$$f(A)\mathbf{v} = T\left(\begin{array}{c|c|c} f(J_1) & & 0 \\ \hline & \ddots & \\ \hline 0 & & f(J_M) \end{array}\right)T^{-1}\mathbf{v} = \mu\mathbf{v}$$

$$\left(\begin{array}{c|c|c} f(J_1) & & 0 \\ \hline & \ddots & \\ \hline 0 & & f(J_M) \end{array}\right)(T^{-1}\mathbf{v}) = \mu(T^{-1}\mathbf{v})$$

Because $f(J_i)$ are all triangular, we have the diagonal elements as the eigenvalues, that is,

$$\mu = \{f(\lambda_1)^{k_1}, \cdots, f(\lambda_N)^{k_N}\}$$

Thus we can apply Theorems 3.4 and 3.5 to these eigenvalues to obtain Theorems 3.6 and 3.7, respectively.

EXAMPLE 3.11. Let $f(A) = \exp(A)$. A sufficient condition for stability is given by $\max_i \left|\left(e^{\lambda_i}\right)\right| < 1$, which is equivalent to having the real part of λ_i be negative, for all i. On the other hand, if any λ_i has a positive real part, then $\max_i \left|\left(e^{\lambda_i}\right)\right| > 1$.

Remarks: Note that in this example, we have still defined stability based on integer powers of matrix operators, that is, the boundedness of $\mathbf{w}(k) = (\exp A)^k\, \mathbf{v}$ for some $\|\mathbf{v}\| < \infty$. We will attach an equivalent criterion for $\exp(tA)$ in section 6.6, with t being a continuous parameter. Fortunately, the sufficient stability and instability criterion ends up being the same.

3.9 Singular Value Decomposition

Thus far, discussion has mostly focused on square matrices. As we saw in the previous chapter, there are plenty of situations in which the matrix A will not be square. One of the methods for analyzing these types of matrices as operators is to use a decomposition called the **singular value decomposition**. As we did with least-squares solutions, instead of solving $A\mathbf{x} = \mathbf{b}$, we converted the problem to $A^*A\mathbf{x} = A^*\mathbf{b}$. The matrix A^*A is now square, but more importantly, it is positive semidefinite and Hermitian. Being positive semidefinite and Hermitian, the eigenvalues are all real and non-negative, and the eigenvectors are also orthogonal (cf. Section 3.3). As we see later, one of the very important applications of singular value decomposition is the principal component analysis for data analysis.

Definition 3.9. *For $A[=]N \times M$, the* **singular values** *of A are non-negative real numbers given by*

$$\sigma_i = \sqrt{\lambda_i} \tag{3.50}$$

where λ_i is the ith eigenvalue of A^*A.

For any matrix $A[=]N \times M$, there exists a decomposition called the **singular value decomposition**,

$$A = U \Sigma V^* \tag{3.51}$$

where $U[=]N \times N$ and $V[=]M \times M$ are unitary, and $\Sigma[=]N \times M$ is a diagonal matrix

$$\Sigma = \left(\begin{array}{ccc|c} \sigma_1 & & 0 & \\ & \ddots & & 0_{[r,M-r]} \\ 0 & & \sigma_r & \\ \hline & 0_{[N-r,r]} & & 0_{[N-r,M-r]} \end{array}\right)$$

$\sigma_1 \geq \sigma_2 \geq \sigma_r$ are the nonzero singular values of A, and r will be the rank of A. Details of the SVD algorithm can be found in Section C.4.2 as an appendix.

Remarks: In MATLAB, the command `[U,S,V]=svd(A)` can be used to obtain the diagonal matrix S containing the singular values in descending order and unitary matrices U and V such that $A = USV^*$. We have also included a MATLAB code that is available on the book's webpage as `svd_alg.m` and applies the algorithm described in Section C.4.2.

3.9.1 Moore-Penrose Generalized Inverse

Let us begin with the definition of a generalized inverse called the **Moore-Penrose inverse**. These "inverses" are quite versatile because they can be used to find "solutions" that are applicable to nonsingular or singular matrices, and square or non-square matrices.

Definition 3.10. *A matrix $G[=]M \times N$ is the* **Moore-Penrose inverse** *of $A[=]N \times M$ if*

1. $AGA = A$
2. $GAG = G$
3. $(AG)^* = AG$
4. $(GA)^* = GA$

The pseudo-inverse defined in (2.36), that is, $A^\dagger = (A^*A)^{-1} A^*$, is one method for evaluating the generalized inverse.[3] However, this formula is limited to matrices in which A^*A is nonsingular. A more reliable approach is to use the singular value decomposition.

Let $\Sigma^{\langle -1 \rangle}$ be constructed from Σ^T by replacing the nonzero entries by their reciprocals, that is,

$$\Sigma^{\langle -1 \rangle} = \left(\begin{array}{ccc|c} \sigma_1^{-1} & & 0 & \\ & \ddots & & 0_{[r,N-r]} \\ 0 & & \sigma_r^{-1} & \\ \hline & 0_{[M-r,r]} & & 0_{[M-r,N-r]} \end{array}\right) \tag{3.52}$$

[3] The terms "generalized inverse" and "pseudo-inverse" are synonymous. Nonetheless, we just prefer to set the term pseudo-inverse aside for $A^\dagger$ to invoke the use of the formula $(A^*A)^{-1} A^*$.

then it can be shown that

$$A^{\langle \mathcal{M} \rangle} = V\Sigma^{\langle -1 \rangle}U^* \tag{3.53}$$

is a generalized inverse of A. Thus, for $A\mathbf{x} = \mathbf{b}$, the Moore-Penrose solution is given by

$$\mathbf{x} = A^{\langle \mathcal{M} \rangle}\mathbf{b} = \left(V\Sigma^{\langle -1 \rangle}U^*\right)\mathbf{b}$$

EXAMPLE 3.12. Let

$$A = \begin{pmatrix} 1 & 2 & 3 \\ 4 & 5 & 6 \\ 7 & 8 & 9 \end{pmatrix} \qquad \mathbf{b} = \begin{pmatrix} 10 \\ 25 \\ 40 \end{pmatrix}$$

To solve for $\mathbf{x}$ such that $A\mathbf{x} = \mathbf{b}$, we first obtain the SVD decomposition of A,

$$U = \begin{pmatrix} 0.2148 & 0.8872 & -0.4082 \\ 0.5206 & 0.2496 & 0.8165 \\ 0.8263 & -0.3879 & -0.4082 \end{pmatrix} \quad V = \begin{pmatrix} 0.4797 & -0.7767 & 0.4082 \\ 0.5724 & -0.0757 & -0.8165 \\ 0.6651 & 0.6253 & 0.4082 \end{pmatrix}$$

$$\Sigma = \begin{pmatrix} 16.8481 & 0 & 0 \\ 0 & 1.0684 & 0 \\ 0 & 0 & 0 \end{pmatrix} \quad \rightarrow \quad \Sigma^{\langle -1 \rangle} = \begin{pmatrix} 0.0594 & 0 & 0 \\ 0 & 0.93600 & \\ 0 & 0 & 0 \end{pmatrix}$$

Then,

$$A^{\langle \mathcal{M} \rangle} = \begin{pmatrix} -0.6389 & -0.1667 & 0.3056 \\ -0.0556 & 0.0000 & 0.0556 \\ 0.5278 & 0.1667 & -0.1944 \end{pmatrix}$$

The Moore-Penrose solution is

$$\mathbf{x} = A^{\langle \mathcal{M} \rangle}\mathbf{b} = \frac{5}{3}\begin{pmatrix} 1 \\ 1 \\ 1 \end{pmatrix}$$

To check, we can see that, indeed, we have $A\mathbf{x} = \mathbf{b}$. Recall that this equality is possible because it still fulfils the rank condition, that is,

$$\text{rank}\left(A \mid \mathbf{b}\right) = \text{rank}\left(A\right) = 2$$

If the rank conditions are not met, equality of $A\mathbf{x} = \mathbf{b}$ will not exist. However, the Moore-Penrose solution will yield a least-squares solution.[4]

EXAMPLE 3.13. Let

$$A = \begin{pmatrix} 1 & 2 \\ 3 & 4 \\ 5 & 6 \end{pmatrix}$$

Then the SVD decomposition is given by

$$U = \begin{pmatrix} -0.2298 & 0.8835 & 0.4082 \\ -0.5247 & 0.2408 & -0.8165 \\ -0.8196 & -0.4019 & 0.4082 \end{pmatrix} \quad V = \begin{pmatrix} -0.6196 & -0.7849 \\ -0.7849 & 0.6196 \end{pmatrix}$$

[4] If A is not full rank, the least-squares solution is not unique.

$$\Sigma = \begin{pmatrix} 9.5255 & 0 \\ 0 & 0.5143 \\ 0 & 0 \end{pmatrix} \quad \rightarrow \quad \Sigma^{\langle -1 \rangle} = \begin{pmatrix} 0.1050 & 0 & 0 \\ 0 & 1.9444 & 0 \end{pmatrix}$$

and the Moore-Penrose inverse is given by

$$A^{\langle \mathcal{M} \rangle} = V\Sigma^{\langle -1 \rangle}U^* = \begin{pmatrix} -1.3333 & -0.3333 & 0.6667 \\ 1.0833 & 0.3333 & -0.4167 \end{pmatrix}$$

One can check that, because A is full rank,

$$A^{\langle \mathcal{M} \rangle} = A^{\dagger} = (A^*A)^{-1} A^*$$

When using the Moore-Penrose solution of $A\mathbf{x} = \mathbf{b}$ where $A[=]N \times M$ and $N > M$ (as is the case for least-squares problems), there appears to be redundant computations for the calculation of $A^{\langle \mathcal{M} \rangle}$. Instead, one can use the **reduced SVD decomposition** (or economical SVD) of A. This is obtained by extracting only the first M columns of U and the first M rows of Σ,

$$U^e = \left(U_{\bullet,1} \mid \cdots \mid U_{\bullet,M} \right) \tag{3.54}$$

$$\Sigma^e = \begin{pmatrix} \Sigma_{1,\bullet} \\ \hline \vdots \\ \hline \Sigma_{M,\bullet} \end{pmatrix} \tag{3.55}$$

$$V^e = V \tag{3.56}$$

Note that Σ^e will now be a square matrix. In this case, $(U^e)^* U^e = I_{[M]}$ but $(U^e)(U^e)^* \neq I_{[M]}$. This loss of full unitary property is not important for the least-squares solution. What is gained is a significant reduction in storage, because in most least-squares problems, the amount of data can be large, whereas the dimension of $\mathbf{x}$ may be much smaller, that is, $N \gg M$. More importantly, for $N > M$, the generalized inverse can be obtained as

$$A^{\langle \mathcal{M} \rangle} = V\Sigma^{\langle -1 \rangle}U^* = V^e \left(\Sigma^e\right)^{\langle -1 \rangle} \left(U^e\right)^* \tag{3.57}$$

EXAMPLE 3.14. Recall Example 3.13. Instead of the standard SVD decomposition, we can obtain the reduced SVD version. For

$$A = \begin{pmatrix} 1 & 2 \\ 3 & 4 \\ 5 & 6 \end{pmatrix}$$

$$U^e = \begin{pmatrix} -0.2298 & 0.8835 \\ -0.5247 & 0.2408 \\ -0.8196 & -0.4019 \end{pmatrix} \qquad V^e = \begin{pmatrix} -0.6196 & -0.7849 \\ -0.7849 & 0.6196 \end{pmatrix}$$

$$\Sigma^e = \begin{pmatrix} 9.5255 & 0 \\ 0 & 0.5143 \end{pmatrix} \quad \rightarrow \quad \left(\Sigma^e\right)^{\langle -1 \rangle} = \begin{pmatrix} 0.1050 & 0 \\ 0 & 1.9444 \end{pmatrix}$$

and the Moore-Penrose inverse is given by

$$A^{\langle \mathcal{M} \rangle} = V^e \left(\Sigma^e\right)^{\langle -1 \rangle} \left(U^e\right)^* = \begin{pmatrix} -1.3333 & -0.3333 & 0.6667 \\ 1.0833 & 0.3333 & -0.4167 \end{pmatrix}$$

which is exactly the same as the one obtained using the standard SVD.

Remark: In MATLAB, one can use the command `[U,S,V]=svd(A,0)` to obtain the reduced SVD decomposition.

3.9.2 Principal Component Analysis

Principal component analysis is a linear analysis technique for reducing the dimensions of data sets. Let $A[=]N \times M$, with $N \gg M$, be the data set, where a_{ij} is j^{th} variable for the ith data, that is, the N data points are assumed to reside in an M-dimensional variable space. Suppose the variance is observed to be significantly larger in a subspace of $K < M$ dimensions compared with the subspace in the remaining dimensions. Then one may wish to determine where this K-dimensional hyperspace resides in original space, that is, let $\mathbf{x} = \{x_1, x_2, \ldots, x_M\}$ be the original space; then we want to find a linear transformation S such that in the new space $\widehat{\mathbf{x}} = S\mathbf{x} = \{\widehat{x}_1, \widehat{x}_2, \ldots, \widehat{x}_M\}$, the last $M - K$ dimensions of whcih will have very low variations. The first K dimensions will then be referred to as the **principal components**, and the last $M - K$ dimensions can be removed. By removal, we mean projecting the data only to the first K dimensions.

Principal component analysis can be performed using the singular value decompositions. Given the data matrix A, each data column is first translated such that the mean is zero

$$\widehat{A} = \left(\widehat{A}_{\bullet,1} \mid \cdots \mid \widehat{A}_{\bullet,M} \right)$$

where

$$\widehat{A}_{\bullet,k} = A_{\bullet,k} - \alpha_k \begin{pmatrix} 1 \\ \vdots \\ 1 \end{pmatrix} \qquad \alpha_k = \frac{1}{N} \sum_{i=1}^{N} a_{i,k}$$

Matrix $\widehat{A}$ is known as the **mean-adjusted data**. This is because we are looking only for rotations of coordinates centered at the mean of the data cluster.

Afterward, the reduced singular-value decomposition is applied to $\widehat{A}$,

$$\widehat{A} = U^e \Sigma^e \left(V^e\right)^*$$

The columns of V^e immediately represent the unit vectors of the new coordinates, that is, the desired transformation is given by $\widehat{\mathbf{x}} = V^e \mathbf{x}$. One can then use the values of the singular values in Σ^e to decide the level of dimensional reduction depending on the ratio of singular values, for example, if $(\sigma_\ell / \sigma_1) < \epsilon$, then set $K = \ell - 1$ and reduce the dimensions by $M - K$.

Having found K, the data can now be projected to the reduced spaces. To do so, we extract the first K columns of V^e and obtain our projection operator (cf. (3.9)) as

$$P_{V^e} = \widehat{V}\widehat{V}^* \tag{3.58}$$

where

$$\widehat{V} = (V^e)^{[1,\ldots,K]}$$

This will project the data onto the K subspace. If we wish to rotate the data to the new coordinate system, we need V^e one more time. In summary, the projected data under the new coordinates can be found by

$$\widehat{A}_{\text{new}} = \widehat{A}\,(P_{V^e} V^e) \tag{3.59}$$

EXAMPLE 3.15. For a set of 300 data points, the singular value decomposition of the mean adjusted data resulted in

$$\Sigma = \begin{pmatrix} 5.2447 & 0 & 0 \\ 0 & 2.4887 & 0 \\ 0 & 0 & 0.1476 \end{pmatrix} \quad \text{and} \quad V = \begin{pmatrix} 0.6446 & 0.7632 & 0.0444 \\ 0.4448 & -0.4217 & 0.7901 \\ 0.6218 & -0.4896 & -0.6113 \end{pmatrix}$$

Based on the singular values, it appears that the last dimension can be removed because it is significantly smaller than the other singular values. Thus we can take the first two columns of V and obtain the projection operator P_V as

$$P_V = \begin{pmatrix} 0.9980 & -0.0351 & 0.0271 \\ -0.0351 & 0.3757 & 0.4830 \\ 0.0271 & 0.4830 & 0.6263 \end{pmatrix}$$

Plotting the new data using (3.59), we see the data projected and rotated in Figure 3.10.

3.10 Polar Decomposition

Given a matrix $A[=]N \times N$, the **polar decomposition** of A is given by

$$A = RS \tag{3.60}$$

where R is unitary and S is a Hermitian positive semidefinite matrix.[5] The decomposition (3.60) separates the action of A into two sequences: the first is a stretching operation by S, and the second is a unitary operation by R, which is usually a rotation or reflection.

The classic approach to polar decomposition is to first find S such that $S^2 = A^*A$, that is, S is the positive square root of A^*A.

Definition 3.11. *A matrix $B[=]N \times N$ is a* **square root** *of $A[=]N \times N$ if $B^2 = A$. If in addition, B is positive semidefinite, then B is the* **positive square root** *of A, that is, the eigenvalues of B are all non-negative.*

To obtain the square root of A^*A, we note that A^*A is Hermitian, and thus the normalized eigenvectors should form an orthonormal set. Using (3.35),

$$S = V \begin{pmatrix} \sqrt{\lambda_1} & & 0 \\ & \ddots & \\ 0 & & \sqrt{\lambda_N} \end{pmatrix} V^* \tag{3.61}$$

[5] The factorization is called the polar decomposition to imply a generalization of the scalar case $z = re^{i\theta}$, where r is a positive real number and $\exp(i\theta)$ describes a rotation.

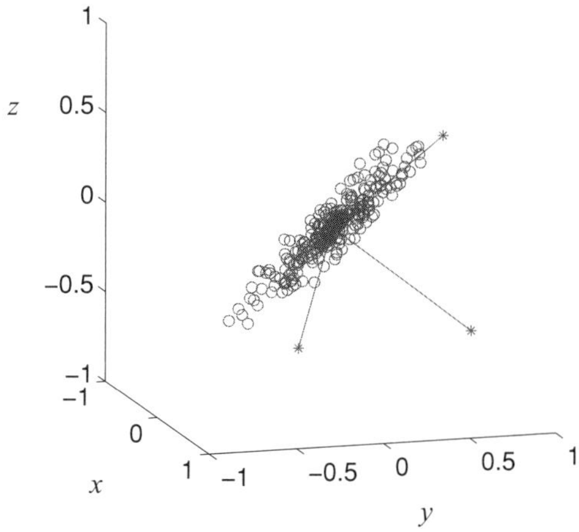

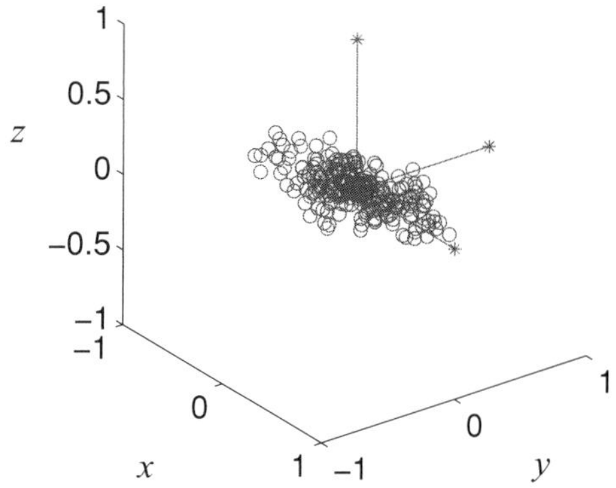

Figure 3.10. Using SVD to locate the principal components in the original mean-adjusted data (top plot). The projected data were then rotated (bottom plot).

where, λ_i is the ith eigenvalue of A^*A and $\mathbf{v}_i$ is its corresponding eigenvector of A^*A.

$$V = \left(\ \mathbf{v}_1 \ \middle| \ \cdots \ \middle| \ \mathbf{v}_N \ \right)$$

If S is nonsingular, then $R = AS^{-1}$. To show that R is unitary, we can use the fact that S is Hermitian and then show that

$$R^*R = S^{-1}A^*AS^{-1} = S^{-1}S^2S^{-1} = I$$

and

$$RR^* = AS^{-1}S^{-1}A^* = A\left(A^*A\right)^{-1}A^* = I$$

For larger matrices including singular matrices, a more efficient approach to obtain the polar decomposition is to use the singular value decomposition. Using U, Σ, and V found using the singular value decomposition, we can let

$$R = UV^* \qquad \text{and} \qquad S = V\Sigma V^* \tag{3.62}$$

Then,

$$RS = \left(UV^*\right)\left(V\Sigma V^*\right) = U\Sigma V^* = A$$

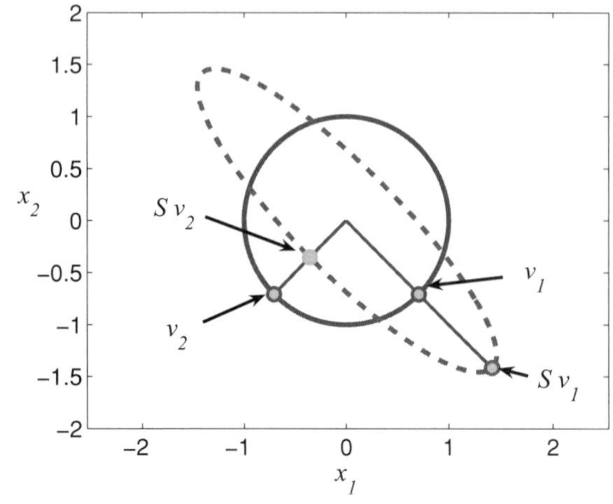

Figure 3.11. The effects of S on the unit circle.

EXAMPLE 3.16. The polar decomposition factor of

$$A = \begin{pmatrix} 1.4575 & -1.2745 \\ -0.0245 & 0.7075 \end{pmatrix}$$

can be found to be

$$S = \begin{pmatrix} 1.25 & -0.75 \\ -0.75 & 1.25 \end{pmatrix} \qquad \text{and} \qquad R = \begin{pmatrix} 0.866 & -0.500 \\ 0.500 & 0.866 \end{pmatrix}$$

Figure 3.11 shows how the points of a unit circle are stretched by S. The eigenvalues of S are $\lambda_1 = 2$ and $\lambda_2 = 0.5$ with the corresponding eigenvectors

$$\mathbf{v}_1 = \begin{pmatrix} 0.7071 \\ -0.7071 \end{pmatrix} \qquad \mathbf{v}_2 = \begin{pmatrix} -0.7071 \\ -0.7071 \end{pmatrix}$$

Along $\mathbf{v}_1$, the points are stretched to twice it original size, whereas along $\mathbf{v}_2$, the points are compressed to half its original size. After S has deformed the unit circle into an ellipse, R will rotate it by an angle $\theta = \tan^{-1}(0.5/0.866) = 30^o$ counterclockwise, as shown in Figure 3.12.

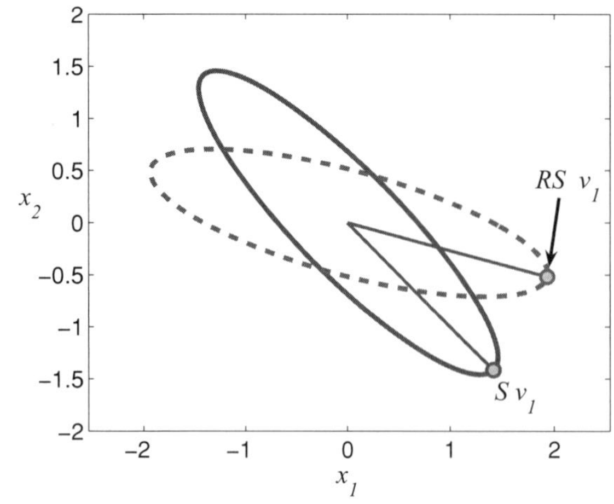

Figure 3.12. The effects of R on the ellipse resulting from the transformation of the unit circle by S.

Table 3.1. *Properties of induced matrix norms*

1	Positivity	$\|A\| \geq 0$
2	Scaling	$\|\alpha A\| = \|\alpha\| \, \|A\|$
3	Triangle Inequality	$\|A + B\| \leq \|A\| + \|B\|$
4	Unique Zero	$\|A\| = 0$ only if $A = 0$
5	Norm of Identity	$\|I\| = 1$
6	Product Inequality 1	$\|A\mathbf{x}\| \leq \|A\| \, \|\mathbf{x}\|$
7	Product Inequality 2	$\|AB\| \leq \|A\| \, \|B\|$

There is also a dual polar decomposition form given by

$$A = \widehat{S}\widehat{R} \tag{3.63}$$

where $\widehat{S}$ is a Hermitian positive definite matrix, in which $\widehat{S}$ is the square root of AA^*, and $\widehat{R}$ is unitary. In this case, the operation will involve a unitary operation followed by a stretching operation. Note that in general, $\widehat{R} \neq R$ and $\widehat{S} \neq S$. The procedure for finding the dual polar decomposition is included as an exercise in **E3.19**.

3.11 Matrix Norms

Most matrix norms are based on the norms specified for the vectors. We have two such norms defined as follows:

Definition 3.12. *For $A[=]N \times M$, the* **Frobenius norm**, *denoted by $\|A\|_F$, is defined by*

$$\|A\|_F = \left\|\text{vec}\,(A)\right\| \tag{3.64}$$

whereas the **induced matrix norm**, *denoted by $\|A\|$, is defined by*

$$\|A\| = \sup_{\mathbf{x} \neq 0} \frac{\|A\mathbf{x}\|}{\|\mathbf{x}\|} = \max_{\|\mathbf{x}\|=1} \|A\mathbf{x}\| \tag{3.65}$$

The Frobenius matrix norm can be used to measure whether a matrix is close to being a zero matrix. As an extension, it can also be used to measure whether a matrix is close to being a diagonal matrix. For instance, let $\mathcal{E} = (A - \text{diag}\,(A))$; then $\|\mathcal{E}\|_F = 0$ only if A is diagonal.[6]

For the most part, however, we focus on induced matrix norms. The properties of the induced norm are given in Table 3.1, most of which carry over from the properties of vector norms. Specifically, using the Euclidean norm $\|\mathbf{x}\|_2$, we have the induced Euclidean norm of matrix A, denoted as $\|A\|_2$, and evaluated as

$$\|A\|_2 = \max_{\|\mathbf{x}\|=1} \sqrt{\mathbf{x}^* A^* A \mathbf{x}} = \max_i \sqrt{\lambda_i\,(A^*A)} = \sigma_1\,(A) \tag{3.66}$$

where $\lambda\,(A^*A)$ is the eigenvalue of A^*A and $\sigma_1\,(A)$ is the largest singular value of A. Unless indicated, we will default to the induced Euclidean norm and drop the subscript '2.' Note, however, that generally the matrix norms are greater than or equal to the spectral radius, that is,

$$\|A\| \geq \rho\,(A) \tag{3.67}$$

[6] This norm can be used to determine whether the iterated QR method has achieved diagonalization of a positive definite matrix as contained in the SVD algorithm discussed in Section 3.9.

To see this, we use the eigenvalue equation,

$$A\mathbf{v} = \lambda\mathbf{v} \quad \rightarrow \quad \|A\| \, \|\mathbf{v}\| \geq |\lambda| \, \|\mathbf{v}\|$$

or

$$\|A\| \geq \max |\lambda_i| = \rho(A)$$

For the special case in which A is Hermitian, we have $\|A\| = \rho(A)$.

Based on matrix norms, another useful characterization of matrix A is the condition number,

Definition 3.13. *The* **condition number** *of A is defined as*

$$\gamma(A) = \max_{\|\mathbf{v}\|,\|\mathbf{w}\|=1} \frac{\|A\mathbf{v}\|}{\|A\mathbf{w}\|} \geq 1 \tag{3.68}$$

Based on Euclidean vector norms, we have

$$\gamma(A) = \max_{\|\mathbf{v}\|,\|\mathbf{w}\|=1} \frac{\sqrt{\mathbf{v}^* A^* A\mathbf{v}}}{\sqrt{\mathbf{w}^* A^* A\mathbf{w}}} = \frac{\max_i \left(|\lambda_i|\right)}{\min_j \left(|\lambda_j|\right)} = \frac{\sigma_{\max}}{\sigma_{\min}} \tag{3.69}$$

where λ_i is the ith eigenvalue of A^*A, whereas $\sigma_{\max}$ and $\sigma_{\min}$ are the maximum and minimum singular values of A, respectively.

EXAMPLE 3.17. Let us recall A given in Example 3.16,

$$A = \begin{pmatrix} 1.4575 & -1.2745 \\ -0.0245 & 0.7075 \end{pmatrix}$$

As shown in Example 3.16, A will transform the unit circle into a rotated ellipse. The points of the ellipse are bounded by two circles: an enclosing circle $C_{\max}$ with radius $\rho(C_{\max})$ and an enclosed circle $C_{\min}$ with radius $\rho(C_{\min})$, that is,

$$\mathbf{y} = A\mathbf{x} \quad \rightarrow \rho(C_{\min}) \leq \|\mathbf{y}\| \leq \rho(C_{\max})$$

These circles are shown together with the ellipse in Figure 3.13. In our case, $\rho(C_{\min}) = 0.5$ and $\rho(C_{\max}) = 2$, which turn out to be the smallest and largest singular values of A, respectively. The condition number, $\gamma(A)$, is then given by the ratio of $\rho_{C_{\max}}$ to $\rho_{C_{\min}}$, or $\gamma(A) = 4$.

From Example 3.17, we see that the condition number is an indication of the shape of the resulting ellipse (or hyper-ellipsoids). A large condition number would indicate a flatter ellipse. In the extreme case, an infinite condition number means that an N-dimensional vector will lose at least one dimension after the operation by A. For nonsingular matrices,[7] the condition number can be shown to be

$$\gamma(A) = \left\|A^{-1}\right\| \, \|A\| \tag{3.70}$$

[7] The requirement for nonsingularity of A is needed because A^{-1} does not exist for singular A. However, if we set $\left\|A^{-1}\right\| = \infty$ for nonsingular A, then (3.70) can be applied to singular matrices.

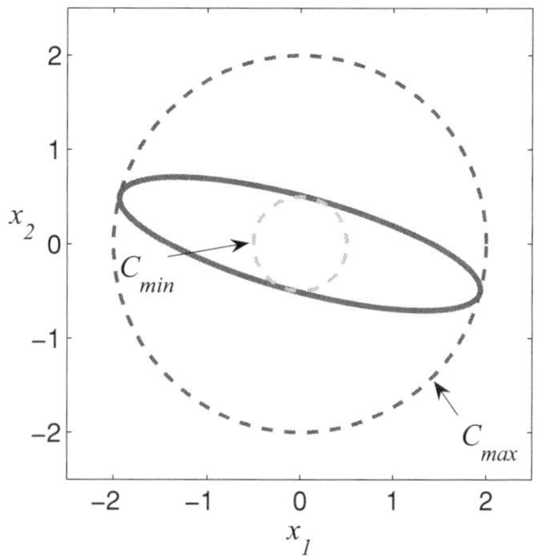

Figure 3.13. The enclosing and enclosed circles for the ellipse generated by transforming a unit circle by A.

The condition number can be used to predict how roundoff errors can affect the solution of linear equations, $A\mathbf{x} = \mathbf{b}$. When the conditions numbers of A are too large, the equation is classified as **ill-conditioned**. In these cases, the solutions become unreliable. To see this, suppose $\mathbf{b}$ is perturbed by $\Delta\mathbf{b}$; then the solution for $\mathbf{x}$ will also be perturbed by $\Delta\mathbf{x}$. Assuming the unperturbed solution is given by $\mathbf{x} = A^{-1}\mathbf{b}$, then applying the product inequality properties given in Table 3.1,

$$\begin{aligned} \mathbf{x} + \Delta\mathbf{x} &= A^{-1}(\mathbf{b} + \Delta\mathbf{b}) \\ \Delta\mathbf{x} &= A^{-1}\Delta\mathbf{b} \\ \|\Delta\mathbf{x}\| &\leq \left\|A^{-1}\right\| \|\Delta\mathbf{b}\| \end{aligned}$$

and

$$\begin{aligned} \|\mathbf{b}\| &= \|A\mathbf{x}\| \leq \|A\| \|\mathbf{x}\| \\ \frac{\|\mathbf{b}\|}{\|A\|} &\leq \|\mathbf{x}\| \end{aligned}$$

Combining these inequalities,

$$\frac{\|\Delta\mathbf{x}\|}{\|\mathbf{x}\|} \leq \|A\| \left\|A^{-1}\right\| \frac{\|\Delta\mathbf{b}\|}{\|\mathbf{b}\|} = \gamma(A) \frac{\|\Delta\mathbf{b}\|}{\|\mathbf{b}\|} \tag{3.71}$$

Equation (3.71) gives an upper bound of the relative error in $\mathbf{x}$ for a given relative error in $\mathbf{b}$. Thus the solution $\mathbf{x} + \Delta\mathbf{x}$ is unreliable if the condition number is very large.

EXAMPLE 3.18. Consider

$$A = \begin{pmatrix} 46 & 8261 & 2454 \\ -8326 & -43 & -4927 \\ -2529 & 5076 & -2 \end{pmatrix} \qquad \mathbf{b} = \begin{pmatrix} -5761 \\ -13210 \\ -7607 \end{pmatrix}$$

The solution is given by $\mathbf{x} = \begin{pmatrix} 1 & -1 & 1 \end{pmatrix}^T$. If $\mathbf{b}$ is perturbed slightly by

$$\Delta\mathbf{b} = \begin{pmatrix} -0.6 \\ 0 \\ 0.8 \end{pmatrix} \rightarrow \frac{\|\Delta\mathbf{b}\|}{\|\mathbf{b}\|} = 6.1365 \times 10^{-5}$$

Solving for the new $\mathbf{x}$,

$$\mathbf{x} + \Delta\mathbf{x} = A^{-1}\left(\mathbf{b} + \Delta\mathbf{b}\right) = \begin{pmatrix} 2.5018 \\ -0.2526 \\ -1.5444 \end{pmatrix}$$

whose relative error norm is

$$\frac{\|\Delta\mathbf{x}\|}{\|\mathbf{x}\|} = 1.7596$$

For this case, the condition number of A is $\gamma(A) = 3.1639 \times 10^4$.

One source of ill-conditioning is when the spread of singular values are so wide. In some cases, rescaling the unknown variables could reduce the condition number, but only within some limits. Another source for ill-conditioning is the proximity of A to a singularity condition, that is, when the minimum singular value is close to zero. In this case, one could put a threshold on singular values by replacing σ_k with zeros if $\sigma_k < \epsilon$. Afterward, the Moore-Penrose generalized inverse can be applied to solve the linear equation.

In summary, because the linear equations obtained from real-world applications may often contain data corrupted with measurement errors or roundoff errors, one should generally consider using the condition number to check whether the solutions are reliable.

3.12 EXERCISES

E3.1. Determine whether the following matrices are orthogonal, unitary, or normal:

$$A = \begin{pmatrix} \cos(\theta) & -\sin(\theta) & 0 \\ \sin(\theta) & \cos(\theta) & 0 \\ 0 & 0 & \cos(\theta) \end{pmatrix} \qquad B = \begin{pmatrix} 1 & 2 & 3 & 4 \\ 4 & 1 & 2 & 3 \\ 3 & 4 & 1 & 2 \\ 2 & 3 & 4 & 1 \end{pmatrix}$$

$$C = \left(I - \frac{2}{\mathbf{w}^*\mathbf{w}}\mathbf{w}\mathbf{w}^*\right)\begin{pmatrix} \cos(\theta) & -\sin(\theta) \\ \sin(\theta) & \cos(\theta) \end{pmatrix}$$

E3.2. Determine which of the following statements are always true (Show examples or counterexamples):

1. Products of unitary matrices are unitary.
2. Unitary matrices are commutative.
3. Sums of normal matrices are normal.
4. Products of normal matrices are normal.

E3.3. Find a single operator that would first rotate a point 30° counterclockwise around the z-axis then 30° clockwise around the x-axis. Verify the operators found by using sample vectors and plot the vectors before and after the operations.

E3.4. Let $\mathbf{v}$ and $\mathbf{w}$ be real vectors of length N and M, respectively.

1. By treating $\mathbf{w}^T$ as an operator, what class of vectors are allowed as inputs to $\mathbf{w}^T$? Describe the outputs of this operator.

2. Define $A = \mathbf{v}\mathbf{w}^T$ as the **matrix dyadic operator**. What class of vectors are allowed as inputs to A ? Describe the outputs of the operator.

E3.5. The curve shown in Figure 3.14 is generated using the data in Table 3.2.
1. Find an affine operator that would rotate the curve by θ radians counterclockwise around the point $(x, y) = (a, b)$. Test this operator on the data given in Table 3.2 with $(a, b) = (4, -1.5)$ and $\theta = \pi/2$.
2. Find an affine operator that would reflect the curve along a line that contains points (x_1, y_1) and (x_2, y_2). Test this operator on the data given in Table 3.2 with $(x_1, y_1) = (0, -1)$ and $(x_2, y_2) = (10, -3)$.

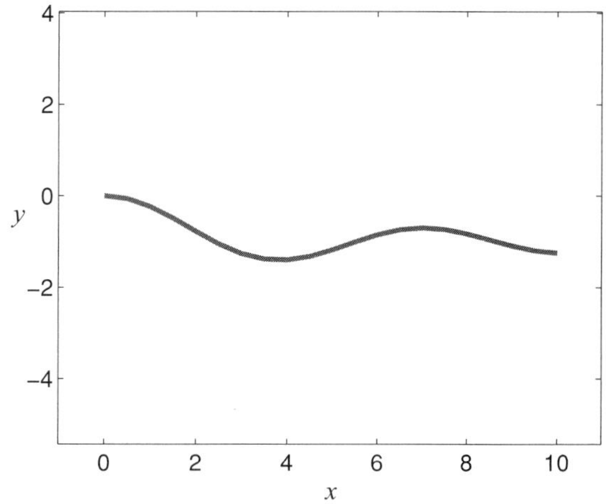

Figure 3.14. The curve used for exercise.

Table 3.2. *Data for curve shown in Figure 3.14*

x	y	x	y	x	y
0.0	0.0000	3.5	−1.3801	7.0	−0.6999
0.5	−0.0615	4.0	−1.3971	7.5	−0.7337
1.0	−0.2348	4.5	−1.3205	8.0	−0.8283
1.5	−0.4882	5.0	−1.1776	8.5	−0.9581
2.0	−0.7761	5.5	−1.0068	9.0	−1.0903
2.5	−1.0484	6.0	−0.8494	9.5	−1.1939
3.0	−1.2601	6.5	−0.7399	10.0	−1.2459

E3.6. Let P be a projection operator. What determinant values are possible for P?

E3.7. Find the projection operator for the space spanned by

$$\mathbf{v}_1 = \begin{pmatrix} 1 \\ 0 \\ 3 \\ 2 \end{pmatrix} \quad \text{and} \quad \mathbf{v}_2 = \begin{pmatrix} -2 \\ 1 \\ 3 \\ 1 \end{pmatrix}$$

E3.8. An $N \times N$ **companion matrix** (also known as a **Frobenius matrix**) has the following form:

$$C = \begin{pmatrix} 0 & 1 & 0 & \cdots & 0 \\ 0 & 0 & 1 & \cdots & 0 \\ \vdots & \vdots & \vdots & \ddots & \vdots \\ 0 & 0 & 0 & \cdots & 1 \\ -\alpha_0 & -\alpha_1 & -\alpha_2 & \cdots & -\alpha_{N-1} \end{pmatrix}$$

Show that the characteristic polynomial of C is given by:

$$\lambda^n + \sum_{i=0}^{n-1} \alpha_i \lambda^i = 0$$

Furthermore, show that the eigenvectors corresponding to each distinct eigenvalue λ_i of C is given by

$$\mathbf{v}_i = \begin{pmatrix} 1 \\ \lambda_i \\ \vdots \\ \lambda_i^{N-1} \end{pmatrix}$$

E3.9. Consider the $N \times N$ tri-diagonal matrix given by

$$T_N = \begin{pmatrix} a & b & & 0 \\ c & \ddots & \ddots & \\ & \ddots & \ddots & b \\ 0 & & c & a \end{pmatrix}$$

Using the results of exercise **E1.23** item 3, the determinant of T_N when $\Delta = \sqrt{a^2 - 4bc} \neq 0$ is given by

$$\mathbf{det}\left(T_N\right) = \frac{(a+\Delta)^{N+1} - (a-\Delta)^{N+1}}{2^{N+1}\Delta} \tag{3.72}$$

1. Let $a = 2q\sqrt{bc}$. Show that if

$$q = \cos\left(\frac{k\pi}{N+1}\right) \qquad k = 1, 2, \ldots, N \tag{3.73}$$

then $\mathbf{det}\left(T_N\right) = 0$. (Hint: First show that for $\left| T_N \right| = 0$, an equivalent condition becomes

$$\left(q + \sqrt{q^2 - 1}\right)^{2(N+1)} = 1$$

Then use de Moivre's formula for the roots of unity to find the value of q, i.e., given $z^M = 1$, the Mth roots are given by

$$z = \exp\left(\frac{2ik\pi}{M}\right) \qquad k = 1, 2, \ldots, M-1$$

where $i = \sqrt{-1}$.)
2. Using the previous result, show that the eigenvalues of the tri-diagonal matrix T_N are given by

$$\lambda_k = a + 2\left(\sqrt{bc}\right)\cos\left(\frac{k\pi}{N+1}\right) \qquad k = 1, 2, \ldots, N$$

Verify this formula for $N = 5$, with $a = 3$, $b = 1$ and $c = 2$.

E3.10. Let λ_* be an eigenvalue of A. Show that any nonzero column of $\mathbf{adj}\left(\lambda_* I - A\right)$ is an eigenvector of A corresponding to λ_*. Verify this for the case in which A is given by

$$A = \begin{pmatrix} 2 & 0 & 0 \\ 2 & 3 & 0 \\ 2 & 3 & 4 \end{pmatrix}$$

E3.11. For a given 3D real vector $\mathbf{v} = (a, b, c)^T$, the **matrix cross-product operator**, often denoted by $\mathbf{v}_\times$ is defined as

$$\mathbf{v}_\times = \begin{pmatrix} 0 & -c & b \\ c & 0 & -a \\ -b & a & 0 \end{pmatrix} \tag{3.74}$$

1. Let $a = b = c = 1$. Now collect M points of a closed flat circle as a matrix X, where

$$X = \left(\mathbf{x}_1 \;\middle|\; \cdots \;\middle|\; \mathbf{x}_M \right) \quad \text{and} \quad \mathbf{x}_k = \begin{pmatrix} \cos\left(\frac{2\pi(k-1)}{M-1}\right) \\ \sin\left(\frac{2\pi(k-1)}{M-1}\right) \\ 0 \end{pmatrix}$$

   Then obtain $B = \mathbf{v}_\times X$ and plot each column of B as points in a 3D graph. Observe that the resulting ellipse will be perpendicular to $\mathbf{v}$.
2. Show that the operator $\mathbf{v}_\times$ is skew-Hermitian. Obtain the eigenvalues and thus verify that Property 9 of eigenvalues given in Section 3.3 holds, that is, that the eigenvalues are either zero or pure imaginary. Obtain the eigenvalues and their corresponding eigenvectors for the case where $a = b = c = 1$.
3. Let $a = 0.5$, $b = 0.5$ and $c = 1$. Find the polar decomposition of $\mathbf{v}_\times = RS$ where R is orthogonal and S is the square root of $\left(\mathbf{v}_\times^T \mathbf{v}_\times\right)$. Describe the behavior of the operator S on any input $\mathbf{x}$ to $\mathbf{v}_\times$. Also describe how R will affect vectors $(S\mathbf{x})$.

E3.12. Determine which matrices below are diagonalizable and which are not. If diagonalizable, obtain matrix T that diagonalizes the matrix; otherwise, determine the modal matrix T that produces a similarity transformation to the Jordan canonical form.

$$\text{a)} \quad A = \begin{pmatrix} 1 & -2 & 0 & 2 \\ 2 & 1 & -2 & 0 \\ 0 & 2 & 1 & -2 \\ -2 & 0 & 2 & 1 \end{pmatrix}$$

$$\text{b)} \quad B = \begin{pmatrix} 4 & 1 & 2 \\ 1 & 3 & 1 \\ -1 & 0 & 2 \end{pmatrix}$$

$$\text{c)} \quad C = \begin{pmatrix} 1 & 1 & -1 \\ 0 & 2 & 0 \\ 0 & 0 & 2 \end{pmatrix}$$

E3.13. Let matrix $C_i[=]N_i \times N_i$ have the companion form (cf. (3.23)),

$$C_i = \begin{pmatrix} 0 & 1 & 0 & \cdots & 0 \\ 0 & 0 & 1 & \cdots & 0 \\ \vdots & \vdots & \vdots & \ddots & \vdots \\ 0 & 0 & 0 & \cdots & 1 \\ -\alpha_{0,i} & -\alpha_{1,i} & -\alpha_{2,i} & \cdots & -\alpha_{(N_i-1),i} \end{pmatrix}$$

Now let A be a block triangular matrix having C_i as the ith block diagonal, that is,

$$A = \left(\begin{array}{c|c|c} C_1 & & 0 \\ \hline \times & \ddots & \\ \hline \times & \times & C_K \end{array}\right)$$

Write down the characteristic equation for A having this structure.

E3.14. Let matrix A be given by

$$A = \begin{pmatrix} 26 & 0 & 0 \\ 68 & 45 & 0 \\ 12 & 20 & 10 \end{pmatrix}$$

Evaluate the functions below using: (a) series expansion (3.33), (b) diagonalization, and (c) finite sums.

1. $\cos(A)$
2. $B = A^{1/2}$, that is, such that $B^2 = A$

E3.15. Use Sylvester's formula (C.10) to evaluate A^5 where

$$A = \begin{pmatrix} 1 & 0 & 0 \\ 1 & \frac{1}{2} & 0 \\ 1 & 1 & 2 \end{pmatrix}$$

E3.16. Let A be given by

$$A = \begin{pmatrix} -2 & 0 & 0 & 0 & 0 \\ 1 & -2 & 0 & 0 & 0 \\ 5 & 2 & -2 & 0 & 0 \\ -6 & -2 & 0 & -2 & 0 \\ 1 & 1 & 1 & 1 & -2 \end{pmatrix}$$

Evaluate the functions below using: (a) Jordan decomposition and (b) finite sums.

1. $\cos(A)$
2. $\sin(1.5A)$
3. $\exp\big(\exp(A)\big)$

E3.17. Prove or disprove the following claim: "For any square matrix A, $\mathbf{det}(\exp A) = e^{\mathbf{trace}(A)}$." (Hint: Use the Jordan decomposition of A and properties of determinants of triangular matrices.)

E3.18. Determine whether the following matrix operators and matrix function operators are stable or unstable: αA, $(A + \alpha I)$, $\exp(\alpha A)$, and $\exp(A + \alpha I)$ where

$$A = \begin{pmatrix} -2 & 0 & 0 \\ 1 & -2.5 & 0 \\ 1 & 1 & -2 \end{pmatrix}$$

for the following cases: $\alpha = 0.1, 0.5$, and 2.8.

E3.19. Find the dual polar decomposition, $A = \widehat{S}\widehat{R}$, for

$$A = \begin{pmatrix} 1 & 3 & 4 \\ 4 & 5 & 5 \\ 0 & 2 & 0 \end{pmatrix}$$

where $\widehat{S}$ is Hermitian positive semidefinite matrix square root of AA^*, and $\widehat{R}$ is unitary.

E3.20. Show that each of the eigenvalues of matrix A of size N must lie in at least one of N circles known as the **Gershgorin circles** centered at a_{kk} and given by

$$k^{\text{th}} \text{ circle}: \qquad |\lambda - a_{kk}| \leq \sum_{i=1, i\neq k}^{N} |a_{ki}|$$

(Hint: Let $\mathbf{v} \neq \mathbf{0}$ be the eigenvector corresponding to eigenvalue λ. Find k such that $|v_k| = \max_{j=1,\ldots,n} |v_j|$, then extract the kth equation from $A\mathbf{v} = \lambda\mathbf{v}$. Next, use the fact that $\left|\sum_i f_i\right| \leq \sum_i |f_i|$ and $\left|\prod_i f_i\right| = \prod_i |f_i|$.)

E3.21. A square matrix A of size n is **diagonally dominant** if for all $k = 1, \ldots, n$

$$|a_{kk}| > \sum_{i=1, i\neq k}^{n} |a_{ki}|$$

Using the result about Gershogorin circles given in the previous exercise, show that a diagonally dominant matrix is nonsingular.

E3.22. Let $\mathbf{v}$ be an eigenvector matrix A. Show that the corresponding eigenvalue for $\mathbf{v}$ can be obtained by using the **Rayleigh quotient** defined as

$$\lambda = \frac{\mathbf{v}^* A\mathbf{v}}{\mathbf{v}^*\mathbf{v}} \tag{3.75}$$

E3.23. Use the power method (cf. Section C.2.2) to find the dominant eigenvalue of

$$A = \begin{pmatrix} 4 & 2 & 3 \\ -3 & 0 & 2 \\ 1 & 1 & 4 \end{pmatrix}$$

E3.24. Let A have distinct eigenvalues; any vector $\mathbf{w}$ can be represented as a linear combination of the eigenvectors, that is,

$$\mathbf{w} = \sum_{i=1}^{N} \alpha_i \mathbf{v}_i$$

where $\mathbf{v}_i$ is the ith eigenvector of A. Assume that $|\lambda_1| \geq |\lambda_k|$, $k > 1$ Then,

$$\begin{aligned} A\mathbf{w} &= \sum_{i=1}^{N} \alpha_i \lambda_i \mathbf{v}_i \\ &= \alpha_1 \lambda_1 \left(\mathbf{v}_1 + \sum_{j=2}^{N} \alpha_j \frac{\lambda_j}{\lambda_1} \mathbf{v}_j \right) \end{aligned}$$

Multiply this equation by A^k and use Property 5 of Section 3.3 to show that the power method will approach $\alpha_1 \lambda^{k+1} \mathbf{v}_1$ (or $\mathbf{v}_1$ if normalization is performed) as $k \to \infty$. This is the basis for the power method.

E3.25. Consider the data given in Table 3.3 and plotted in Figure 3.15. We want to use principal component analysis to obtain a 2D curve in 3D space.

1. Obtain the mean-adjusted data matrix $\widehat{A}$ and take the reduced singular value decomposition of $\widehat{A}$, that is, $\widehat{A} = U\Sigma V^*$
2. Obtain the projection matrix P_V that would project the mean-adjusted data onto the space spanned by the first two columns of V. (This assumes that the last column is attached to the smallest singular value).

Table 3.3. *Raw data set for the principal component analysis*

x	y	z	x	y	z	x	y	z
0.0104	0.0541	1.3846	0.1302	0.5161	0.2402	0.6302	0.5395	1.1170
0.0288	0.1944	1.0956	0.2131	0.5482	0.2274	0.7523	0.5395	1.4258
0.0311	0.2558	0.7672	0.2615	0.5629	0.2011	0.7915	0.5892	1.2890
0.0449	0.3085	0.6511	0.3145	0.6096	0.4074	0.9021	0.6213	1.2737
0.0657	0.3816	0.3595	0.4090	0.6096	0.4178	0.9389	0.7295	1.0864
0.0703	0.4371	0.1628	0.4666	0.5629	0.7435	0.9597	0.7646	1.1498
0.1048	0.4898	0.1700	0.5288	0.5512	0.9040	0.9873	0.9342	0.5464

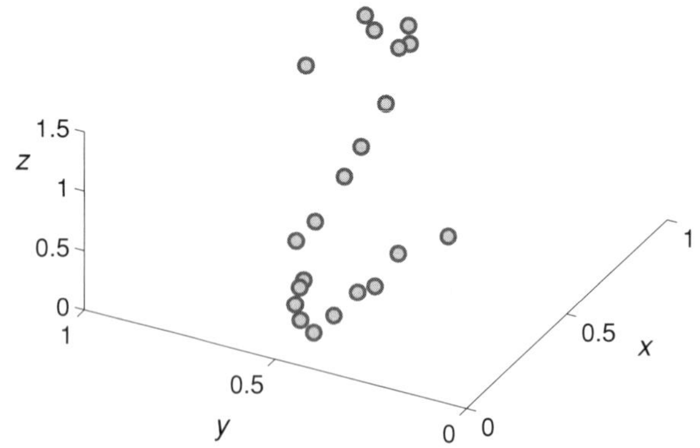

Figure 3.15. Plot of raw data for the principal component analysis.

3. Using both P and V, we end up with a 2D surface in which we could perform additional regression analysis. Thus let B be the transformed data matrix

$$B = \left(\widehat{A}\right)(PV)$$

The last column of B should be all zeros. Let x_t and y_t represent the first and second columns of B, respectively, of the transformed data.

4. A plot of x_t vs. y_t is given in Figure 3.16. Note that the curve can not be described by $y_t = y_t(x_t)$. However, we can rotate the curve 60°

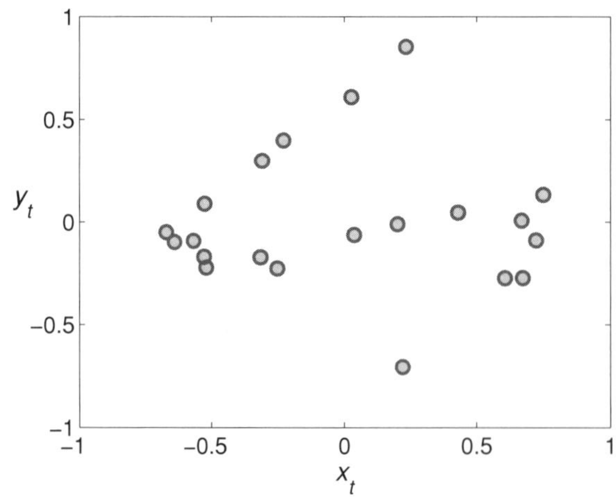

Figure 3.16. Plot of data transformed into 2D space.

counter-clockwise using $R_{ccw(60^\circ)}$. This will then make it possible for the nonlinear regression to be applied to the rotated curve. Let

$$C = \left(\; B_{\bullet,1} \; \middle| \; B_{\bullet,2} \; \right) R^T_{ccw(60^\circ)}$$

and let x_f and y_f represent the first and second columns of C. Obtain a fifth-order polynomial regression model for the rotated curve, that is,

$$y_f = a_0 + a_1 x_f + a_2 x_f^2 + a_3 x_f^3 + a_4 x_f^4 + a_5 x_f^5$$

5. By reversing the operations, we can transform the regression curve $y_f = y_f(x_f)$ back to be in terms of the original variables. Thus perform these operations on the regression curve to the curve shown in Figure 3.17.

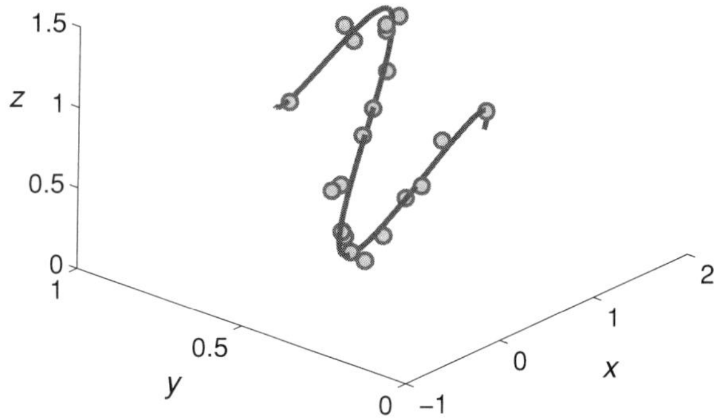

Figure 3.17. Plot of the data together with the regression curve in the original space.

E3.26. Find the least-squares solution of the following equation using the SVD decomposition using a tolerance level of 10^{-4} (i.e., set $\sigma_k = 0$ if $\sigma_k \leq 10^{-4}$) and the reduced SVD decomposition:

$$\begin{pmatrix} 0.7922 & 0.6555 & 0.1367 \\ 0.9595 & 0.1712 & 0.7883 \\ 0.6557 & 0.7060 & -0.0503 \\ 0.0357 & 0.0318 & 0.0039 \\ 0.8491 & 0.2769 & 0.5722 \\ 0.9340 & 0.0462 & 0.8878 \\ 0.6787 & 0.0971 & 0.5816 \\ 0.7577 & 0.8235 & -0.0657 \\ 0.7431 & 0.6948 & 0.0483 \\ 0.3922 & 0.3171 & 0.0751 \end{pmatrix} \mathbf{x} = \begin{pmatrix} 2.5583 \\ 3.7049 \\ 1.9351 \\ 0.1488 \\ 3.1600 \\ 3.7059 \\ 2.6256 \\ 2.2373 \\ 2.3301 \\ 1.2753 \end{pmatrix}$$

E3.27. With the Frobenius norm of $A[=]N \times M$ defined as

$$\|A\|_F = \sqrt{\sum_{i=1}^{N} \sum_{j=1}^{M} |a_{ij}|^2}$$

Show that another method can be used to evaluate $\|A\|_F$, for example,

$$\|A\|_F = \sqrt{\textbf{trace}\,(A^* A)}$$

PART II

VECTORS AND TENSORS

The next two chapters contain a detailed discussion of vector and tensor analysis.

Chapter 4 contains the basic concepts of vectors and tensors, including vector and tensor algebra. We begin with a description of vectors as an abstract object having a magnitude and direction, whereas tensors are then defined as operators on vectors. Several algebraic operations are summarized together with their matrix representations. Differential calculus of vector and tensors are then introduced with the aid of gradient operators, resulting in operations such as gradients, divergences, and curls. Next, we discuss the transformations of rectangular coordinates to curvilinear coordinates, such as cylindrical, spherical, and other general orthogonal coordinate systems.

Chapter 5 then focuses on the integral calculus of vectors. Detailed discussions of line, surface, and volume integrations are included in the appendix, including the mechanics of calculations. Instead, the chapter discusses various important integral theorems such as the divergence theorem, the *Stokes*' theorem, and the general Liebnitz formula. An application section is included to show how several physical models, especially those based on conservation laws, can be cast in terms of tensor calculus, which is independent of coordinate systems. The models generated are generally in the form of partial differential equations that are applicable to problems in mechanics, fluid dynamics, general physico-chemical processes, and electromagnetics. The solutions of these models are the subject of Part III and Part IV of the book.

4 Vector and Tensor Algebra and Calculus

In this chapter, we work with objects that possess a magnitude and a direction. These objects are known as **physical vectors** or simply **vectors**.[1] There are two types of vectors: **bound vectors**, which are fixed to a specified point in the space, and **free vectors**, which are allowed to move around in the space. Ironically, free vectors are often used when working in rigid domains, whereas bound vectors are often used when working in flowing or flexible domains. We mostly deal with bound vectors.

We denote vectors with underlined bold letters, such as $\underline{\mathbf{v}}$, and we denote scalars with nonunderlined letters, such as α, unless otherwise noted. Familiar examples of vectors are velocity, acceleration, and forces. For these vectors, the concept of direction and magnitude are natural and easy to grasp. However, it is important to note that vectors can be built depending on the user's interpretation and objectives. As long as a magnitude and direction can be attached to a physical property, then vector analysis can be used. For instance, for angular velocities of a rigid body, one needs to describe how fast the rotation is, whether the rotation is counterclockwise or clockwise, and where the axis of rotation is. By attaching an arrow whose direction is along the axis of rotation, whose length determines how fast the rotation is, and pointing in the direction consistent with a counterclockwise or clockwise convention, the angular velocity becomes a vector. In our case, we adapt the right-hand screw convention to represent the counterclockwise direction as a positive direction (see Figure 4.1).

We begin in Section 4.1 with the description of fundamental vector operations. The definitions of operations such as vector sums and different type of products, including scalar, dot, cross, and triple products, are done in a geometric sense, that is, based only on measurements of distance and angles. Later, we introduce unit basis vectors such as $\underline{\boldsymbol{\delta}}_x$, $\underline{\boldsymbol{\delta}}_y$, and $\underline{\boldsymbol{\delta}}_z$ in the rectangular coordinates pointing in the x, y, and z directions, respectively. When vectors are represented as linear combinations of the basis unit vectors, an alternative set of efficient calculations can be achieved. These basis vectors are also used to define an operator known as a **tensor**. Moreover, vectors and tensors can also be represented by matrices. In doing so, we can take advantage of various matrix properties and apply matrix analysis and computation

[1] Physical vectors can also be represented by matrix vectors. We defer this matrix representation until Section 4.3.

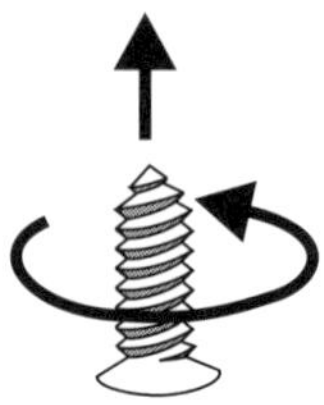

Figure 4.1. The right-hand screw convention.

to handle vector operations; that is, the concepts of eigenvalues, eigenvectors, polar decomposition, diagonalization, and so forth can be applied to the physical vectors and tensors.

In Section 4.5, we discuss the derivatives of vectors that are dependent on one variable. The derivative of such a vector is different from that of a scalar function because vectors have the additional property of directions. Furthermore, the results of the derivatives are also vectors. The derivative of sums and different types of products can also be obtained. Due to its ubiquitous importance, we briefly apply the derivatives to position vectors $\underline{\vec{\mathbf{r}}}$ and its various derivatives, for example, the velocity vector and acceleration vector, including other related items such as tangent vectors, normal vectors, binormal vectors, curvature, and torsion. Then, in Section 4.7, we discuss the differential calculus of vector fields, where a distribution of vectors are specified at different locations of the 3D space. This includes differential operations such as the gradient, divergence, curl, and Laplacian, among others.

Finally, in Sections 4.8 and 4.9, we discuss alternative coordinate systems, namely cylindrical, spherical, and general orthogonal coordinate systems. These are important coordinate systems to consider because in several real-world applications, the boundaries are cylindrical (e.g., pipe flow) or spherical (e.g., heat transfer from a spherical surface). Starting with the basic transformation rules between the coordinates, we can generate relationships with those in the rectangular coordinate systems. Unfortunately, for differential operators, changing the representations based on the other coordinate systems tends to be more complicated. Nonetheless, the formulas for gradient, divergence, curl, and Laplacian can still be generated in a straightforward manner.

4.1 Notations and Fundamental Operations

In this section, we describe the fundamental vector operations. These are summarized in Table 4.1. The geometric interpretations are shown in these tables. However, one should be careful when applying these interpretations on vectors that have physical units. We briefly explore some of these concerns.

Because most vectors have physical units, we attach all the units with the "magnitude" component of the vector. Thus, by defining the norm, $\|\underline{\mathbf{v}}\|$, as the magnitude of the $\underline{\mathbf{v}}$, we can represent $\underline{\mathbf{v}}$ as

$$\underline{\mathbf{v}} = \|\underline{\mathbf{v}}\| \; \underline{\mathbf{n}}_{\mathrm{v}} \quad \text{where} \quad \underline{\mathbf{n}}_{\mathrm{v}} = \frac{1}{\|\underline{\mathbf{v}}\|} \; \underline{\mathbf{v}} \tag{4.1}$$

Here, $\underline{\mathbf{n}}_{\mathrm{v}}$ is the "normalized" unit vector of $\underline{\mathbf{v}}$, and it will not have any physical units attached to it.

Table 4.1. *Fundamental vector operations*

	Operation	Notation	Procedure
1	Addition	$\underline{\mathbf{c}} = \underline{\mathbf{a}} + \underline{\mathbf{b}}$	Let $\underline{\mathbf{a}}$ and $\underline{\mathbf{b}}$ be attached to point O. Construct a parallelogram with $\underline{\mathbf{a}}$ and $\underline{\mathbf{b}}$ as two adjacent sides. Then $\underline{\mathbf{c}}$ is the vector at point O with its arrowhead located opposite point O in the parallelogram. 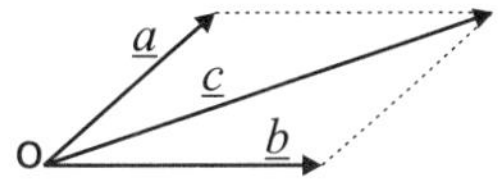
2	Norm	$\alpha = \lVert \underline{\mathbf{v}} \rVert$	α is the length of $\underline{\mathbf{v}}$, from the tail to the arrowhead. 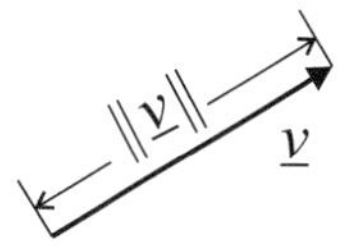
3	Scalar Product	$\underline{\mathbf{w}} = \alpha \underline{\mathbf{v}}$	Direction of $\underline{\mathbf{w}}$ = direction of $\underline{\mathbf{v}}$; $\lVert \underline{\mathbf{w}} \rVert = \alpha \lVert \underline{\mathbf{v}} \rVert$ 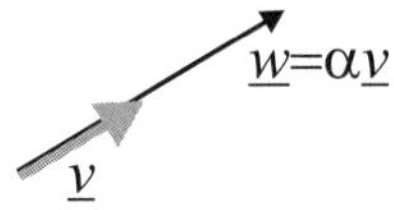
4	Dot Product	$\alpha = \underline{\mathbf{u}} \cdot \underline{\mathbf{v}}$	$\alpha = \lVert u \rVert \, \lVert v \rVert \cos(\theta)$ where, θ = smaller angle between $\underline{\mathbf{u}}$ and $\underline{\mathbf{v}}$. 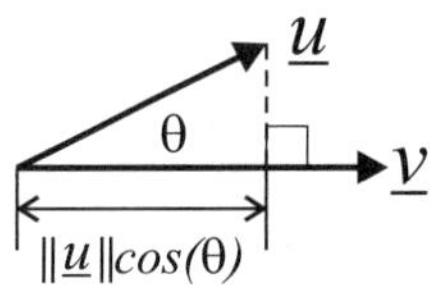
5	Cross Product	$\underline{\mathbf{c}} = \underline{\mathbf{a}} \times \underline{\mathbf{b}}$	Direction of $\underline{\mathbf{c}}$ = perpendicular to plane containing $\underline{\mathbf{a}}$ and $\underline{\mathbf{b}}$, based on the right-hand screw convention. $\lVert \underline{\mathbf{c}} \rVert = \lVert \underline{\mathbf{a}} \rVert \, \lVert \underline{\mathbf{b}} \rVert \sin\theta$ = area of parallelogram formed by $\underline{\mathbf{a}}$ and $\underline{\mathbf{b}}$. where, θ = angle from $\underline{\mathbf{a}}$ to $\underline{\mathbf{b}}$ based on the right-hand screw convention 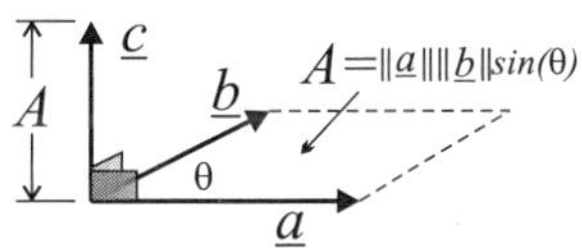
6	Triple Product	$\alpha = \underline{\mathbf{c}} \cdot (\underline{\mathbf{a}} \times \underline{\mathbf{b}})$ $= [\underline{\mathbf{c}}\ \underline{\mathbf{a}}\ \underline{\mathbf{b}}]$	α = volume of parallelepiped formed by $\underline{\mathbf{a}}$, $\underline{\mathbf{b}}$, and $\underline{\mathbf{c}}$. $\alpha = \lVert \underline{\mathbf{a}} \rVert \, \lVert \underline{\mathbf{b}} \rVert \, \lVert \underline{\mathbf{c}} \rVert \sin\theta \cos\phi$ where, θ = angle from $\underline{\mathbf{a}}$ to $\underline{\mathbf{b}}$ based on right-hand screw convention, and ϕ = angle between $\underline{\mathbf{c}}$ and a line that is perpendicular to the plane containing $\underline{\mathbf{a}}$ and $\underline{\mathbf{b}}$. 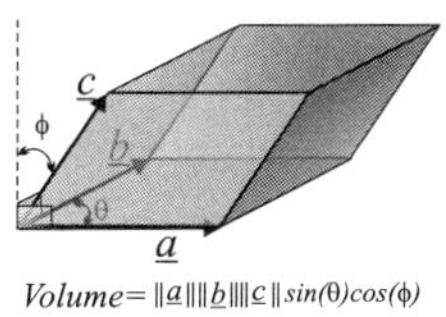

Let $\underline{\mathbf{c}}$ be the sum of vectors $\underline{\mathbf{a}}$ and $\underline{\mathbf{b}}$; then $\underline{\mathbf{c}}$ is also called the **resultant**, and $\underline{\mathbf{c}}$ should be in the plane containing both $\underline{\mathbf{a}}$ and $\underline{\mathbf{b}}$. Furthermore, $\underline{\mathbf{a}}$, $\underline{\mathbf{b}}$, and $\underline{\mathbf{c}}$ should all have the same physical units. This is necessary for the following cosine law to apply,

$$\left\|\underline{\mathbf{c}}\right\|^2 = \left\|\underline{\mathbf{a}}\right\|^2 + \left\|\underline{\mathbf{b}}\right\|^2 - 2\left\|\underline{\mathbf{a}}\right\|\left\|\underline{\mathbf{b}}\right\|\cos\left(\pi - \theta\right) \tag{4.2}$$

where θ is the angle between $\underline{\mathbf{a}}$ and $\underline{\mathbf{b}}$. Also, the angle between $\underline{\mathbf{b}}$ and $\underline{\mathbf{c}}$ is given by

$$\gamma = \sin^{-1}\left(\frac{\left\|\underline{\mathbf{a}}\right\|}{\left\|\underline{\mathbf{c}}\right\|}\sin(\theta)\right) \tag{4.3}$$

There are four types of vector products: scalar, dot, cross, and triple products. The resulting products will most likely not be the same as factored vectors, both in units or in meaning. Thus care must be taken when plotting different type of vectors in the same 3D space.

The properties for the vector operations are given in Table 4.2. The table is grouped based on the operations involved (i.e., sums, scalar products, dot products, cross products, triple products, and norms). Most of the properties for the sums, scalar products, and dot products are similar to the vectors of matrix theory. On the other hand, the properties of cross products are quite different. First, we see that cross products are anti-commutative; that is, the sign is reversed if the order of the vectors are interchanged. Next, the parallelism property states that two nonzero vectors are parallel to each other if their cross product is the zero vector.[2] Finally, cross products are not associative. This means that the use of parentheses is imperative; otherwise, the operation will be ambiguous.

The properties of both the sums and scalar products show that the space of physical vectors satisfies the conditions given in Table B.3 for a linear vector space. This means that the properties and definitions attached with linear vector spaces are also applicable to the space of physical vectors. These include the definitions of linear combination, linear independence, span, and dimension, as given in Table B.2.[3] Thus

1. A set of vectors, $\left\{\underline{\mathbf{v}}_1, \ldots, \underline{\mathbf{v}}_n\right\}$ are **linearly independent** if the only possible linear combination that results in a zero vector,

$$\alpha_1\underline{\mathbf{v}}_1 + \cdots + \alpha_n\underline{\mathbf{v}}_n = \underline{\mathbf{0}}$$

is when $\alpha_1 = \cdots = \alpha_n = 0$.
2. A set of linearly independent vectors, $V = \left\{\underline{\mathbf{v}}_1, \ldots, \underline{\mathbf{v}}_n\right\}$ is a **basis** for a space $\mathcal{S}$, if $\mathcal{S}$ is the span of the vectors of V.
3. The **dimension** of a subspace $\mathcal{S}$ is the number of linearly independent vectors that would span $\mathcal{S}$.

[2] This is a dual property to orthogonality, in which two nonzero vectors are orthogonal if their dot product is zero. However, one should note that the parallel property results in a vector, whereas the orthogonality property results in a scalar.

[3] In fact, the abstract linear space is the generalization of the space of physical vectors.

Table 4.2. *Linear properties of physical vectors*

	Vector Sums	
1	Associative	$\underline{\mathbf{v}}+\left(\underline{\mathbf{w}}+\underline{\mathbf{y}}\right)=(\underline{\mathbf{v}}+\underline{\mathbf{w}})+\underline{\mathbf{y}}$
2	Commutative	$\underline{\mathbf{v}}+\underline{\mathbf{w}}=\underline{\mathbf{w}}+\underline{\mathbf{v}}$
3	Identity is $\underline{\mathbf{0}}$	$\underline{\mathbf{0}}+\underline{\mathbf{v}}=\underline{\mathbf{v}}$
4	Inverse exist and unique	$\underline{\mathbf{v}}+(-\underline{\mathbf{v}})=\underline{\mathbf{0}}$
	Scalar Products	
1	Associative	$\alpha\,(\beta\underline{\mathbf{v}})=(\alpha\beta)\,\underline{\mathbf{v}}$
2	Identity is 1	$1\underline{\mathbf{v}}=\underline{\mathbf{v}}$
3	Vector is distributive over scalar sums	$(\alpha+\beta)\,\underline{\mathbf{v}}=\alpha\underline{\mathbf{v}}+\beta\underline{\mathbf{v}}$
4	Scalar is distributive over vector sums	$\alpha\,(\underline{\mathbf{v}}+\underline{\mathbf{w}})=\alpha\underline{\mathbf{v}}+\alpha\underline{\mathbf{w}}$
	Dot Products	
1	Squared norm	$\left\Vert\underline{\mathbf{v}}\right\Vert^2=\underline{\mathbf{v}}\cdot\underline{\mathbf{v}}$
2	Commutative	$\underline{\mathbf{v}}\cdot\underline{\mathbf{w}}=\underline{\mathbf{w}}\cdot\underline{\mathbf{v}}$
3	Distributive	$\underline{\mathbf{u}}\cdot(\underline{\mathbf{v}}+\underline{\mathbf{w}})=\underline{\mathbf{u}}\cdot\underline{\mathbf{v}}+\underline{\mathbf{u}}\cdot\underline{\mathbf{w}}$
4	Scalar Product	$\alpha\,(\underline{\mathbf{v}}\cdot\underline{\mathbf{w}})=(\alpha\underline{\mathbf{v}})\cdot\underline{\mathbf{w}}=\underline{\mathbf{v}}\cdot(\alpha\underline{\mathbf{w}})$
5	Orthogonality	$\underline{\mathbf{u}}\cdot\underline{\mathbf{v}}=0$ if $\underline{\mathbf{u}}=0,\ \underline{\mathbf{v}}=0$ or $\underline{\mathbf{u}}\perp\underline{\mathbf{v}}$
	Cross Products	
1	Anti-commutative	$\underline{\mathbf{v}}\times\underline{\mathbf{w}}=-\underline{\mathbf{w}}\times\underline{\mathbf{v}}$
2	Distributive	$\underline{\mathbf{u}}\times(\underline{\mathbf{v}}+\underline{\mathbf{w}})=\underline{\mathbf{u}}\times\underline{\mathbf{v}}+\underline{\mathbf{u}}\times\underline{\mathbf{w}}$
3	Scalar Product	$\alpha\,(\underline{\mathbf{v}}\times\underline{\mathbf{w}})=(\alpha\underline{\mathbf{v}})\times\underline{\mathbf{w}}=\underline{\mathbf{v}}\times(\alpha\underline{\mathbf{w}})$
4	Parallelism	$\underline{\mathbf{u}}\times\underline{\mathbf{v}}=0$ if $\underline{\mathbf{u}}=0,\ \underline{\mathbf{v}}=0$ or $\underline{\mathbf{u}}\parallel\underline{\mathbf{v}}$
5	Non-associativity	$\underline{\mathbf{u}}\times\left(\underline{\mathbf{v}}\times\underline{\mathbf{y}}\right)=\left(\underline{\mathbf{u}}\cdot\underline{\mathbf{y}}\right)\underline{\mathbf{v}}-(\underline{\mathbf{u}}\cdot\underline{\mathbf{v}})\,\underline{\mathbf{y}}$ $(\underline{\mathbf{u}}\times\underline{\mathbf{v}})\times\underline{\mathbf{y}}=\left(\underline{\mathbf{y}}\cdot\underline{\mathbf{u}}\right)\underline{\mathbf{v}}-\left(\underline{\mathbf{y}}\cdot\underline{\mathbf{v}}\right)\underline{\mathbf{u}}$
	Triple Products	
1	Cyclic Permutation	$\underline{\mathbf{u}}\cdot(\underline{\mathbf{v}}\times\underline{\mathbf{w}})=\underline{\mathbf{w}}\cdot(\underline{\mathbf{u}}\times\underline{\mathbf{v}})=\underline{\mathbf{v}}\cdot(\underline{\mathbf{w}}\times\underline{\mathbf{u}})$
	Norms	
1	Positivity	$\Vert\underline{\mathbf{v}}\Vert\geq 0$
2	Scaling	$\left\Vert\alpha\underline{\mathbf{v}}\right\Vert=\lvert\alpha\rvert\ \left\Vert\underline{\mathbf{v}}\right\Vert$
3	Triangle Inequality	$\left\Vert\underline{\mathbf{v}}+\underline{\mathbf{w}}\right\Vert\leq\left\Vert\underline{\mathbf{v}}\right\Vert+\left\Vert\underline{\mathbf{w}}\right\Vert$
4	Unique Zero	$\left\Vert\underline{\mathbf{v}}\right\Vert=0$ only if $\underline{\mathbf{v}}=0$

EXAMPLE 4.1. Consider the set of 3D vectors $\underline{\mathbf{a}}$, $\underline{\mathbf{b}}$, $\underline{\mathbf{c}}$, and $\underline{\mathbf{d}}$ and a 2D plane $\mathcal{S}$ as shown in Figure 4.2. Suppose further that $\underline{\mathbf{b}}$, $\underline{\mathbf{c}}$, and $\underline{\mathbf{d}}$ lie in the plane $\mathcal{S}$. Then,

- The only sets of three linearly independent vectors are: $\left\{\underline{\mathbf{a}},\underline{\mathbf{b}},\underline{\mathbf{c}}\right\}$, $\left\{\underline{\mathbf{a}},\underline{\mathbf{b}},\underline{\mathbf{d}}\right\}$, $\left\{\underline{\mathbf{a}},\underline{\mathbf{c}},\underline{\mathbf{d}}\right\}$.
- The set $\left\{\underline{\mathbf{b}},\underline{\mathbf{c}},\underline{\mathbf{d}}\right\}$ is linearly dependent.
- All the sets of any two different vectors are linearly independent, for example, $\left\{\underline{\mathbf{a}},\underline{\mathbf{b}}\right\}$, $\left\{\underline{\mathbf{a}},\underline{\mathbf{c}}\right\}$, and so forth.
- The span of $\left\{\underline{\mathbf{b}},\underline{\mathbf{c}}\right\}$ is $\mathcal{S}$, and thus the dimension of $\mathcal{S}$ is 2. Also, the spans of $\left\{\underline{\mathbf{b}},\underline{\mathbf{c}}\right\}$, $\left\{\underline{\mathbf{c}},\underline{\mathbf{d}}\right\}$, and $\left\{\underline{\mathbf{b}},\underline{\mathbf{d}}\right\}$ are all the same.

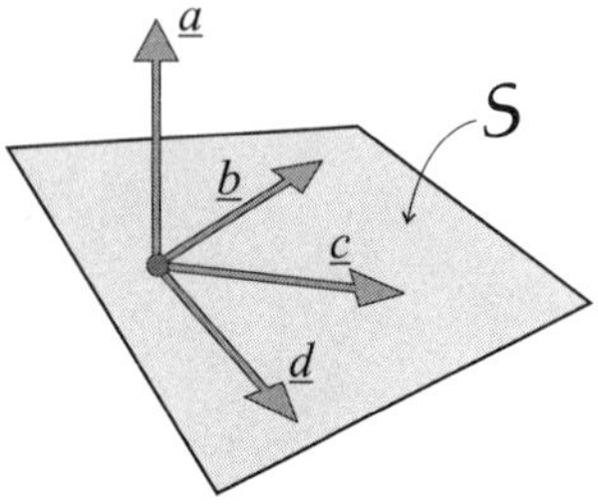

Figure 4.2. A set of 3D vectors for Example 4.1.

4.2 Vector Algebra Based on Orthonormal Basis Vectors

In most applications, vectors can be represented as a linear combination of basis vectors. For instance, if $\underline{\mathbf{b}}_1$, $\underline{\mathbf{b}}_2$, and $\underline{\mathbf{b}}_3$ are three linearly independent 3D vectors, then any 3D vector $\underline{\mathbf{v}}$ can be represented as

$$\underline{\mathbf{v}} = \alpha\underline{\mathbf{b}}_1 + \beta\underline{\mathbf{b}}_2 + \gamma\underline{\mathbf{b}}_3 \tag{4.4}$$

The scalars α, β, and γ are known as the **components** along $\mathbf{b}_1$, $\mathbf{b}_2$, and $\mathbf{b}_3$ vectors, respectively.

In particular, if the basis vectors are normalized (magnitude 1 and no physical units) and orthogonal to each other, the set of basis unit vectors are known as **orthonormal basis vectors**. We start with orthonormal basis vectors based on the **Cartesian coordinate system**, also known as the **rectangular coordinate systems**, described by (x, y, z), using the convention shown in Figure 4.3.[4]

The unit vectors based on the Cartesian coordinates are denoted by $\underline{\boldsymbol{\delta}}_x$, $\underline{\boldsymbol{\delta}}_y$, and $\underline{\boldsymbol{\delta}}_z$, each pointing in the positive x, y, and z direction, respectively. Thus

$$\underline{\mathbf{v}} = v_x\underline{\boldsymbol{\delta}}_x + v_y\underline{\boldsymbol{\delta}}_y + v_z\underline{\boldsymbol{\delta}}_z \tag{4.5}$$

and the scalars v_x, v_y, and v_z will be the x- , y-, and z- components of $\underline{\mathbf{v}}$, respectively (see Figure 4.4).

The dot products and cross products of Cartesian unit vectors can be summarized as follows:

$$\underline{\boldsymbol{\delta}}_i \cdot \underline{\boldsymbol{\delta}}_j = \delta_{ij} \tag{4.6}$$

$$\underline{\boldsymbol{\delta}}_i \times \underline{\boldsymbol{\delta}}_j = \epsilon_{ijk}\underline{\boldsymbol{\delta}}_k \tag{4.7}$$

where δ_{ij} is known as the **Kronecker delta** defined by

$$\delta_{ij} = \begin{cases} 0 & \text{if } i \neq j \\ 1 & \text{if } i = j \end{cases} \tag{4.8}$$

[4] The convention is that when looking down against the positive z-axis, the (x, y) plane, when rotated, should have the positive y-axis pointing vertically upward and the positive x-axis pointing horizontally to the right.

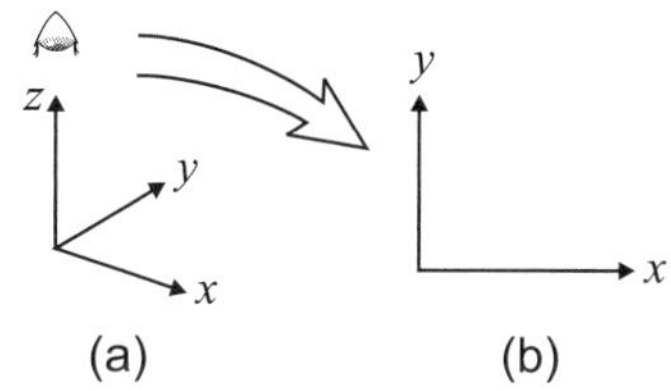

Figure 4.3. The Cartesian coordinate system with three axes perpendicular to each other. The figure on the right (a) is the relationship among x, y, and z. The figure on the left (b) is the convention by looking directly down into the positive z-direction.

and ϵ_{ijk} is the permutation symbol (also known as **Levi-Civita symbol**) defined by

$$\epsilon_{ijk}\begin{cases} 0 & \text{if } i=j \text{ or } j=k \text{ or } i=k \\ 1 & \text{if } (i,j,k)=(x,y,z) \text{ or } (i,j,k)=(z,x,y) \text{ or } (i,j,k)=(y,z,x) \\ -1 & \text{if } (i,j,k)=(x,z,y) \text{ or } (i,j,k)=(z,y,x) \text{ or } (i,j,k)=(y,x,z) \end{cases} \tag{4.9}$$

Using (4.6) and the distributive property of dot products, the x-component of $\underline{\mathbf{v}}$ can be found as follows:

$$\begin{aligned} \underline{\mathbf{v}}\cdot\underline{\boldsymbol{\delta}}_x &= \left(v_x\underline{\boldsymbol{\delta}}_x + v_y\underline{\boldsymbol{\delta}}_y + v_z\underline{\boldsymbol{\delta}}_z\right)\cdot\underline{\boldsymbol{\delta}}_x \\ &= v_x\underline{\boldsymbol{\delta}}_x\cdot\underline{\boldsymbol{\delta}}_x + v_y\underline{\boldsymbol{\delta}}_y\cdot\underline{\boldsymbol{\delta}}_x + v_z\underline{\boldsymbol{\delta}}_z\cdot\underline{\boldsymbol{\delta}}_x = v_x \end{aligned}$$

Similarly, $v_y = \underline{\mathbf{v}}\cdot\underline{\boldsymbol{\delta}}_y$ and $v_z = \underline{\mathbf{v}}\cdot\underline{\boldsymbol{\delta}}_z$.

The following identities between the permutation symbols and Kronecker deltas will also be useful during the derivation of vector operations and properties (see exercise **E4.31**):

$$\begin{aligned} \epsilon_{ijk}\epsilon_{\ell mn} &= \mathbf{det}\begin{pmatrix} \delta_{i\ell} & \delta_{im} & \delta_{in} \\ \delta_{j\ell} & \delta_{jm} & \delta_{jn} \\ \delta_{k\ell} & \delta_{km} & \delta_{kn} \end{pmatrix} \\ \sum_i \epsilon_{ijk}\epsilon_{imn} &= \delta_{jm}\delta_{kn} - \delta_{jn}\delta_{km} \end{aligned} \tag{4.10}$$

Equations (4.6) and (4.7), together with the properties given in Table 4.2 yield alternative approaches to the operations defined in Table 4.1. These are given in Table 4.3.

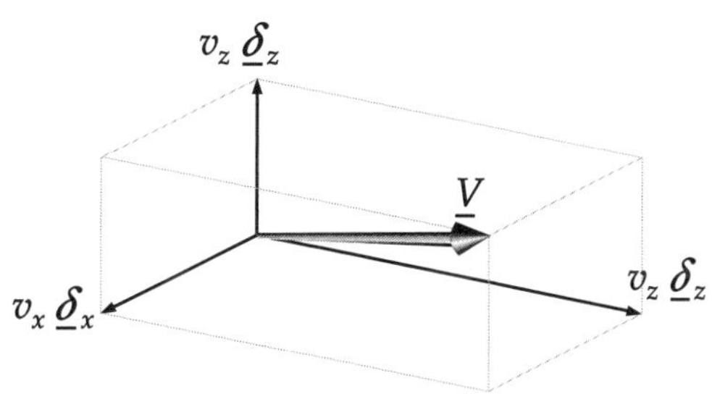

Figure 4.4. Vector $\underline{V}$ as a linear combination of unit vectors.

Table 4.3. *Vector operations based on unit vectors*

Vector Addition	$\underline{\mathbf{v}}+\underline{\mathbf{w}}$	$=$	$(v_x+w_x)\,\underline{\boldsymbol{\delta}}_x+(v_y+w_y)\,\underline{\boldsymbol{\delta}}_y+(v_z+w_z)\,\underline{\boldsymbol{\delta}}_z$
		$=$	$\sum_{i=x,y,z}(v_i+w_i)\,\underline{\boldsymbol{\delta}}_i$
Norm	$\lVert\underline{\mathbf{v}}\rVert$	$=$	$\sqrt{v_x^2+v_y^2+v_z^2}$
		$=$	$\sqrt{\sum_{i=x,y,z}v_i^2}$
Scalar Product	$\alpha\,\underline{\mathbf{v}}$	$=$	$(\alpha v_x)\,\underline{\boldsymbol{\delta}}_x+(\alpha v_y)\,\underline{\boldsymbol{\delta}}_y+(\alpha v_z)\,\underline{\boldsymbol{\delta}}_z$
		$=$	$\sum_{i=x,y,z}\alpha v_i\underline{\boldsymbol{\delta}}_i$
Dot Product	$\underline{\mathbf{v}}\cdot\underline{\mathbf{w}}$	$=$	$(v_xw_x)+(v_yw_y)+(v_zw_z)$
		$=$	$\sum_{i,j=x,y,z}\delta_{ij}\;v_i\;w_j=\sum_{i=x,y,z}v_i\;w_i$
Cross Product	$\underline{\mathbf{v}}\times\underline{\mathbf{w}}$	$=$	$(v_yw_z-v_zw_y)\,\underline{\boldsymbol{\delta}}_x+(v_zw_x-v_xw_z)\,\underline{\boldsymbol{\delta}}_y+(v_xw_y-v_yw_x)\,\underline{\boldsymbol{\delta}}_z$
		$=$	$\sum_{i,j,k=x,y,z}\epsilon_{ijk}\;v_i\;w_j\;\underline{\boldsymbol{\delta}}_k$
Triple Product	$\underline{\mathbf{u}}\cdot(\underline{\mathbf{v}}\times\underline{\mathbf{w}})$	$=$	$(v_yw_z-v_zw_y)\,u_x+(v_zw_x-v_xw_z)\,u_y+(v_xw_y-v_yw_x)\,u_z$
		$=$	$\sum_{i,j,k=x,y,z}\epsilon_{ijk}\;u_i\;v_j\;w_k$

EXAMPLE 4.2. Let us prove the identity for $\underline{\mathbf{u}}\times\left(\underline{\mathbf{v}}\times\underline{\mathbf{y}}\right)$ given in Table 4.2. First, we can expand the cross products in terms of the permutation symbols:

$$\begin{aligned}\underline{\mathbf{u}}\times\left(\underline{\mathbf{v}}\times\underline{\mathbf{y}}\right) &= \left(\sum_{i=x,y,z}u_i\underline{\boldsymbol{\delta}}_i\right)\times\left(\sum_{j,k,\ell=x,y,z}\epsilon_{jk\ell}v_jy_k\underline{\boldsymbol{\delta}}_\ell\right)\\ &= \sum_{i,j,k,\ell,m=x,y,z}\epsilon_{i\ell m}\epsilon_{jk\ell}u_iv_jy_k\;\underline{\boldsymbol{\delta}}_m\end{aligned}$$

Due to the cyclic nature of the symbols, we see that $\epsilon_{i\ell m}=\epsilon_{\ell mi}$ and $\epsilon_{jk\ell}=\epsilon_{\ell jk}$. Thus, using (4.10),

$$\begin{aligned}\underline{\mathbf{u}}\times\left(\underline{\mathbf{v}}\times\underline{\mathbf{y}}\right) &= \sum_{i,j,k,\ell,m=x,y,z}\epsilon_{\ell mi}\epsilon_{\ell jk}u_iv_jy_k\;\underline{\boldsymbol{\delta}}_m\\ &= \sum_{i,j,k,m=x,y,z}\left(\delta_{mj}\delta_{ik}-\delta_{mk}\delta_{ij}\right)u_iv_jy_k\;\underline{\boldsymbol{\delta}}_m\\ &= \sum_{i,j=x,y,z}(u_iy_i)\,v_j\underline{\boldsymbol{\delta}}_j-\sum_{i,k=x,y,z}(u_iv_i)\,y_k\underline{\boldsymbol{\delta}}_k\\ &= \left(\underline{\mathbf{u}}\cdot\underline{\mathbf{y}}\right)\underline{\mathbf{v}}-(\underline{\mathbf{u}}\cdot\underline{\mathbf{v}})\,\underline{\mathbf{y}}\end{aligned}$$

Some useful mnemonics exist for the cross product and the triple product. These are given by

$$\underline{\mathbf{v}} \times \underline{\mathbf{w}} = \mathbf{det}\begin{pmatrix} \underline{\boldsymbol{\delta}}_x & \underline{\boldsymbol{\delta}}_y & \underline{\boldsymbol{\delta}}_z \\ v_x & v_y & v_z \\ w_x & w_y & w_z \end{pmatrix} \tag{4.11}$$

$$\underline{\mathbf{u}} \cdot (\underline{\mathbf{v}} \times \underline{\mathbf{w}}) = \mathbf{det}\begin{pmatrix} u_x & u_y & u_z \\ v_x & v_y & v_z \\ w_x & w_y & w_z \end{pmatrix} \tag{4.12}$$

Note, however, that (4.11) and (4.12) are just memory tools and should not be treated as definitions.

The unit basis vectors are normalized vectors, and thus they will have no physical units. The physical units are instead attached to the x-, y-, and z-components of the vector. Moreover, because scalars can move freely across the various types of products, the products of unit vectors can be computed in pure geometric terms. For instance, to compute for work we now have

$$\begin{aligned} W = \underline{\mathbf{F}} \cdot \underline{\Delta s} &= \left(f_x\underline{\boldsymbol{\delta}}_x + f_y\underline{\boldsymbol{\delta}}_y + f_z\underline{\boldsymbol{\delta}}_z\right) \cdot \left(\Delta s_x\underline{\boldsymbol{\delta}}_x + \Delta s_y\underline{\boldsymbol{\delta}}_y + \Delta s_z\underline{\boldsymbol{\delta}}_z\right) \\ &= (f_x\underline{\boldsymbol{\delta}}_x) \cdot (\Delta s_x\underline{\boldsymbol{\delta}}_x) + \cdots \\ &= (f_x\Delta s_x)\,(\underline{\boldsymbol{\delta}}_x \cdot \underline{\boldsymbol{\delta}}_x) + \cdots \end{aligned}$$

where the products $f_x\Delta s_x$, and so forth are scalars with physical units of force times length, whereas $\underline{\boldsymbol{\delta}}_x \cdot \underline{\boldsymbol{\delta}}_x$, and so forth are either 1 or 0 and do not have any units.

4.3 Tensor Algebra

In this section, we discuss vector operators known as tensors. We begin with the definition of the building blocks for tensors, which are operators on vectors known as dyads.

Definition 4.1. *The* **dyad** *based on vectors* $\underline{\mathbf{v}}$ *and* $\underline{\mathbf{w}}$*, denoted by* $(\underline{\mathbf{vw}})$*, is an operator that transforms an input vector* $\underline{\mathbf{x}}$ *into output vector* $\underline{\mathbf{y}}$ *as follows:*

$$\begin{aligned} (\underline{\mathbf{vw}})\left(\underline{\mathbf{x}}\right) &= \underline{\mathbf{v}}\,(\underline{\mathbf{w}} \cdot \underline{\mathbf{x}}) \\ &= \alpha\underline{\mathbf{v}} = \underline{\mathbf{y}} \end{aligned} \tag{4.13}$$

where $\alpha = \underline{\mathbf{w}} \cdot \underline{\mathbf{x}}$.

EXAMPLE 4.3. One important concept in transport of properties such as material, energy, momentum, and so forth is the idea of flux, $\underline{\mathbf{f}}$. By **flux**, we mean the amount of property passing perpendicularly through a specified region per unit area of that region per unit time. This is a vector because the definition has both a magnitude and direction, although one has to be cautious about which

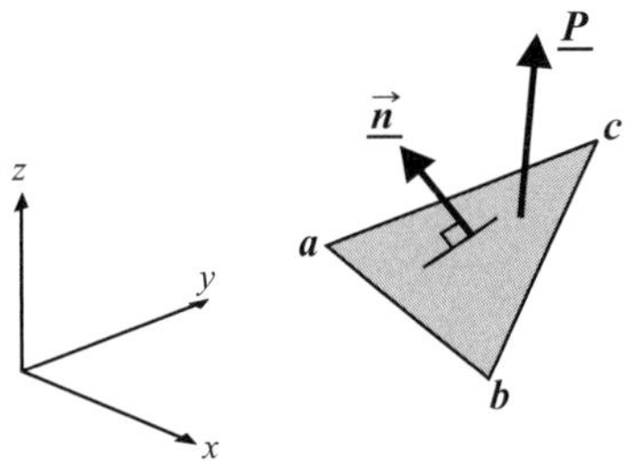

Figure 4.5. The triangle Δ_{abc} together with its normal unit vector.

perpendicular direction is of interest; for example, for a closed surface, the unit normal vector to the surface can be chosen to be the outward normal.[5]

Consider the triangle Δ_{abc} determined by three non-collinear points a, b, and c, each defined by the position vectors $\underline{\mathbf{a}}$, $\underline{\mathbf{b}}$, and $\underline{\mathbf{c}}$, respectively, and shown in Figure 4.5.[6] Assume that the sequence $\{\mathbf{a}, \mathbf{b}, \mathbf{c}\}$ yields a counterclockwise turn to yield our desired direction. Then the desired unit normal vector can be found to be

$$\underline{\vec{\mathbf{n}}} = \frac{(\underline{\mathbf{b}} - \underline{\mathbf{a}}) \times (\underline{\mathbf{c}} - \underline{\mathbf{a}})}{\left\| (\underline{\mathbf{b}} - \underline{\mathbf{a}}) \times (\underline{\mathbf{c}} - \underline{\mathbf{a}}) \right\|}$$

Assume a constant property flow, $\underline{\mathbf{P}}$ (e.g., kg of water per ft^2 of region perpendicular to its flow), which does not necessarily flow in the same direction as $\underline{\vec{\mathbf{n}}}$. The magnitude of the flux through Δ_{abc} will need to take the projection of $\underline{\mathbf{P}}$ along $\underline{\vec{\mathbf{n}}}$ using the inner product. Thus the flux of $\underline{\mathbf{P}}$ through triangle Δ_{abc} is given by

$$\underline{\mathbf{f}} = \left(\underline{\mathbf{P}} \cdot \underline{\vec{\mathbf{n}}} \right) \underline{\vec{\mathbf{n}}}$$

or, in terms of the dyad notation,

$$\underline{\mathbf{f}} = \underline{\vec{\mathbf{n}}}\, \underline{\vec{\mathbf{n}}} \left(\underline{\mathbf{P}} \right)$$

Let $\dot{P}_{\Delta abc}$ be the total rate of property flowing through the triangle. Then $\dot{P}_{\Delta abc}$ can be obtained by taking the normal component of the flux and multiplying it by the area. The area of the triangle Δ_{abc} is given by

$$A = \frac{1}{2} \left\| (\underline{\mathbf{b}} - \underline{\mathbf{a}}) \times (\underline{\mathbf{c}} - \underline{\mathbf{a}}) \right\|$$

Thus

$$\begin{aligned} \dot{P}_{\Delta abc} &= (\underline{\mathbf{f}} \cdot \underline{\vec{\mathbf{n}}}) A = \left(\underline{\mathbf{P}} \cdot \underline{\vec{\mathbf{n}}} \right) A = \frac{1}{2} \underline{\mathbf{P}} \cdot \left((\underline{\mathbf{b}} - \underline{\mathbf{a}}) \times (\underline{\mathbf{c}} - \underline{\mathbf{a}}) \right) \\ &= \underline{\mathbf{P}} \cdot (\underline{\mathbf{a}} \times \underline{\mathbf{b}}) + \underline{\mathbf{P}} \cdot (\underline{\mathbf{b}} \times \underline{\mathbf{c}}) + \underline{\mathbf{P}} \cdot (\underline{\mathbf{c}} \times \underline{\mathbf{a}}) \end{aligned}$$

[5] In this book, we avoid nonorientable surfaces such as Möbius strips.

[6] A position vector $\underline{\vec{\mathbf{r}}} = r_x \underline{\boldsymbol{\delta}}_x + r_y \underline{\boldsymbol{\delta}}_y + r_z \underline{\boldsymbol{\delta}}_z$ is a vector that starts at the origin and ends at the point defined by $(x, y, z) = (r_x, r_y, r_z)$. See Section 4.6 for more applications of position vectors.

Remarks: If we generalize the property flow vector $\underline{\mathbf{P}}$ instead to a vector field $\underline{\mathbf{P}}(x, y, z)$ and call it the **flux vector field**, then the other usage of the term **flux**, say, $\widehat{f}$, for example, as used in electromagnetics, is the integral of the dot product of the vector field with the unit vectors normal to the surface at the different points, that is,

$$\widehat{f} = \int_S \underline{\mathbf{P}}(x, y, z) \cdot \underline{\vec{\mathbf{n}}}(x, y, z) dS$$

This results in a scalar quantity with units of the property per unit time – no longer per unit area. Hopefully, the context of the application will be sufficient to distinguish the two meanings of flux.

Among the different possible dyads, an important set of nine dyads based on the Cartesian unit vectors is the set of **unit dyads**, given by

$$\begin{matrix} \left(\underline{\delta}_x\underline{\delta}_x\right), & \left(\underline{\delta}_x\underline{\delta}_y\right), & \left(\underline{\delta}_x\underline{\delta}_z\right) \\ \left(\underline{\delta}_y\underline{\delta}_x\right), & \left(\underline{\delta}_y\underline{\delta}_y\right), & \left(\underline{\delta}_y\underline{\delta}_z\right) \\ \left(\underline{\delta}_z\underline{\delta}_x\right), & \left(\underline{\delta}_z\underline{\delta}_y\right), & \left(\underline{\delta}_z\underline{\delta}_z\right) \end{matrix}$$

The concept of dyads can be extended to triads, and so forth, collectively known as **polyads**.

Definition 4.2. *An* **nth-order polyad** *based on vectors* $\underline{\mathbf{v}}_1, \ldots, \underline{\mathbf{v}}_n$ *denoted by* $(\underline{\mathbf{v}}_n \cdots \underline{\mathbf{v}}_1)$ *is a multilinear functional operator*[7] *(based on dots products) acting on a given sequence of n-input vectors* $\underline{\mathbf{x}}_1, \cdots, \underline{\mathbf{x}}_n$ *to yield a scalar output, as follows:*

$$(\underline{\mathbf{v}}_n \cdots \underline{\mathbf{v}}_1)\left(\underline{\mathbf{x}}_1, \quad \cdots, \quad \underline{\mathbf{x}}_n\right) = \prod_{i=1}^{n} (\underline{\mathbf{v}}_i \cdot \underline{\mathbf{x}}_i) \tag{4.14}$$

Note that dyads are second-order polyads. With respect to (4.13), $(\underline{\mathbf{w}} \cdot \underline{\mathbf{x}})\,\underline{\mathbf{v}}$ can continue to operate on another input vector, say, $\underline{\mathbf{a}}$, such that

$$(\underline{\mathbf{v}}\underline{\mathbf{w}})\left(\underline{\mathbf{x}} \cdot \underline{\mathbf{a}}\right) = (\underline{\mathbf{w}} \cdot \underline{\mathbf{x}})\,(\underline{\mathbf{v}} \cdot \underline{\mathbf{a}})$$

Based on the three Cartesian unit vectors, there will be 3^n number of nth-order polyads, which we refer to as the unit n^{th}-order polyads. For example, there will be 3^3 unit triads such as $\left(\underline{\delta}_x\underline{\delta}_x\underline{\delta}_x\right)$, $\left(\underline{\delta}_x\underline{\delta}_x\underline{\delta}_y\right)$, and so forth.

An nth-order tensor is defined as:

Definition 4.3. *Let* $B = \left\{\underline{\delta}_1, \cdots, \underline{\delta}_m\right\}$ *be a set of orthonormal basis vectors of an m dimensional space* $\mathcal{S}$. *The* **nth- order tensor** *under B is a linear combination of* n^{th} *order unit polyads formed by the unit basis vectors in B.*

[7] For f to be a multilinear functional operator, we mean that

$$f(x, \ldots, y, \alpha u + \beta v, w, \ldots, z) = \alpha f(x, \ldots, y, u, w, \ldots, z) + \beta f(x, \ldots, y, v, w, \ldots, z)$$

where α and β are scalars.

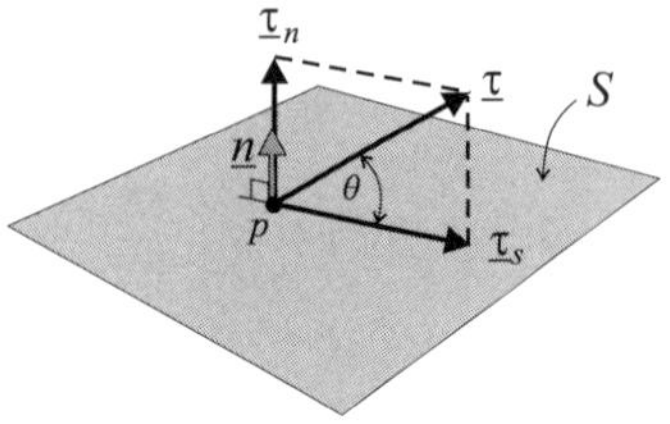

Figure 4.6. Decomposition of stress $\underline{\tau}$ into a normal stress, $\underline{\tau}_n$, and shear stress, $\underline{\textbf{\textit{tau}}}_s$, with respect to plane S at point p.

The zeroth order tensors are scalars, and first-order tensors are vectors.[8] By convention, the term "**tensors**" refers to second-order tensors. Based on this convention, we denote tensors by a letter with double underlines. We also limit our discussion to spaces with a maximum of three dimensions. Using the three Cartesian unit vectors, we have

$$\underline{\underline{T}} = T_{xx}\left(\underline{\delta}_x\underline{\delta}_x\right) + T_{xy}\left(\underline{\delta}_x\underline{\delta}_y\right) + T_{xz}\left(\underline{\delta}_x\underline{\delta}_z\right) + \cdots + T_{zz}\left(\underline{\delta}_z\underline{\delta}_z\right) \tag{4.15}$$

where the scalar T_{ij} is called the $(i, j)-$**component** of tensor $\underline{\underline{T}}$. A special tensor, called the **unit tensor**, is defined by

$$\underline{\underline{\delta}} = \underline{\delta}_x\underline{\delta}_x + \underline{\delta}_y\underline{\delta}_y + \underline{\delta}_z\underline{\delta}_z \tag{4.16}$$

When this tensor operates on any vector, the result is the same vector.

EXAMPLE 4.4. Stress Tensors. Consider a material contained in a 3D domain. Stress is defined as the amount of force F, applied at a point p, per unit area of a fixed plane S that includes the point p. In general, the force F may be on an oblique angle θ with the plane S. We can identify the plane S by a unit vector $\underline{\mathbf{n}}(S)$ that is perpendicular to S, known as its **unit normal vector**. Then the stress $\underline{\tau}(p, S)$ (i.e., at p with respect to plane S) can be decomposed into two additive vectors,

$$\underline{\tau}(p, S) = \underline{\tau}_n(p, S) + \underline{\tau}_s(p, S)$$

where $\underline{\tau}_n(p, S)$, called the **normal stress**, is pointed along the direction of the unit normal vector, and $\underline{\tau}_s(p, S)$, called the **shear stress**, is pointed along a direction perpendicular to the unit normal vector (see Figure 4.6). If the material is non-uniform in terms of properties and applied forces, then the stress vector needs to be localized by using infinitesimal planes dS instead of S. The vector $\underline{\tau}$ is also known as the **traction vector**.

[8] A vector $\underline{\mathbf{v}}$ can be considered a tensor only when it acts as an operator on another vector, say, $\underline{\mathbf{a}}$ (which in our case is via dot products),

$$\underline{\mathbf{v}}\left(\underline{\mathbf{a}}\right) = \underline{\mathbf{v}} \cdot \underline{\mathbf{a}}$$

Otherwise, a vector is mostly just an object having a magnitude and direction. Also, the nth-order tensor described in Definition 4.3 is still limited to the space of physical vectors, and the functional operations are limited to dot products. A generalization to abstract linear vectors is possible, but requires an additional vector space called the **dual vector space**.

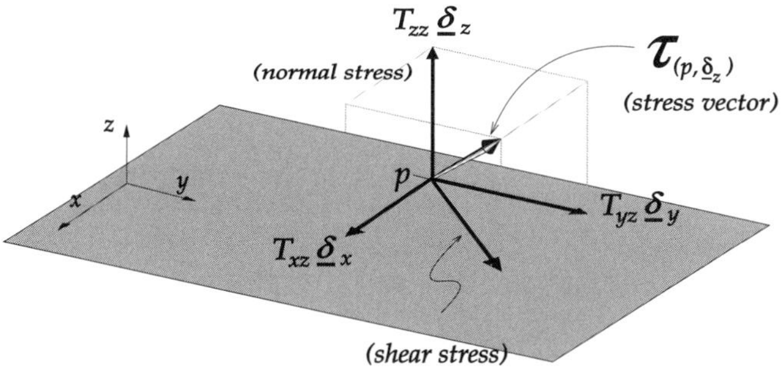

Figure 4.7. The stress vector with respect to $\underline{\delta}_z$ as a sum of the normal stress and shear stress.

By representing vectors in the Cartesian coordinate system, we can define a **stress tensor** as a mapping that contains the state of the stress of a material at a point p. Let $\underline{\underline{T}}$ be the stress tensor given by

$$\begin{aligned}\underline{\underline{T}} &= T_{xx}\underline{\delta}_x\underline{\delta}_x + T_{xy}\underline{\delta}_x\underline{\delta}_y + T_{xz}\underline{\delta}_x\underline{\delta}_z \\ &\quad + T_{yx}\underline{\delta}_y\underline{\delta}_x + T_{yy}\underline{\delta}_y\underline{\delta}_y + T_{yz}\underline{\delta}_y\underline{\delta}_z \\ &\quad + T_{zx}\underline{\delta}_z\underline{\delta}_x + T_{zy}\underline{\delta}_z\underline{\delta}_y + T_{zz}\underline{\delta}_z\underline{\delta}_z\end{aligned}$$

Let us explore the coefficients of $\underline{\underline{T}}$. Consider the (x, y) plane at point p (see Figure 4.7). The unit normal vector is $\underline{\delta}_z$. When tensor $\underline{\underline{T}}$ operates on $\underline{\delta}_z$,

$$\underline{\tau}_{(p,\underline{\delta}_z)} = \underline{\underline{T}}\,\underline{\delta}_z = T_{xz}\underline{\delta}_x + T_{yz}\underline{\delta}_y + T_{zz}\underline{\delta}_z$$

T_{zz} is the term along $\underline{\delta}_z$, and it yields the normal stress with respect to the (x, y) plane. Conversely, the other two terms, T_{xz} and T_{yz}, are the x and y components of the shear stress.

For the general case, we can represent the unit normal vector $\underline{\mathbf{n}}$ as

$$\underline{\mathbf{n}} = n_x\underline{\delta}_x + n_y\underline{\delta}_y + n_z\underline{\delta}_z$$

The stress (or traction) vector can then be obtained by letting $\underline{\underline{T}}$ operate on $\underline{\mathbf{n}}$,

$$\begin{aligned}\underline{\tau}_{(p,\underline{\mathbf{n}})} &= \underline{\underline{T}}\left(\underline{\mathbf{n}}\right) \qquad (4.17) \\ &= \left(T_{xx}\underline{\delta}_x\underline{\delta}_x + T_{xy}\underline{\delta}_x\underline{\delta}_y + \ldots + T_{yz}\underline{\delta}_y\underline{\delta}_z + T_{zz}\underline{\delta}_z\underline{\delta}_z\right)\left(n_x\underline{\delta}_x + n_y\underline{\delta}_y + n_z\underline{\delta}_z\right) \\ &= \left(T_{xx}n_x + T_{xy}n_y + T_{xz}n_z\right)\underline{\delta}_x + \left(T_{yx}n_x + T_{yy}n_y + T_{yz}n_z\right)\underline{\delta}_y \\ &\quad + \left(T_{zx}n_x + T_{zy}n_y + T_{zz}n_z\right)\underline{\delta}_z\end{aligned}$$

Note that to determine the normal stress vector we need to project this vector along the direction of $\underline{\mathbf{n}}$. (see Example **E4.7).**

Equation (4.17) is a statement of the relationship between the stress tensor and the stress vector and is known as **Cauchy's fundamental theorem for stress.**[9]

9 In some books on transport phenomenon or continuum mechanics, the position of the tensor elements are switched (i.e., the transpose of $\underline{\underline{T}}$ described previously) yielding the traction vector as $\underline{\tau} = \underline{\mathbf{n}} \cdot \underline{\underline{T}}$.

Using the properties of dot products and Definition 4.1, we have the following tensor operations:

$$
\begin{aligned}
&\textbf{Tensor Addition:} & \underline{\underline{T}} + \underline{\underline{S}} &= \sum_{i=x,y,z} \sum_{j=x,y,z} \left(T_{ij} + S_{ij}\right)\left(\underline{\delta}_i\underline{\delta}_j\right) \\
&\textbf{Scalar Multiplication of a Tensor:} & \alpha\underline{\underline{T}} &= \sum_{i=x,y,z} \sum_{j=x,y,z} \left(\alpha T_{ij}\right)\left(\underline{\delta}_i\underline{\delta}_j\right) \\
&\textbf{Inner Product of Two Tensors:} & \underline{\underline{T}} \cdot \underline{\underline{S}} &= \sum_{i=x,y,z} \sum_{j=x,y,z} \left(\sum_{k=x,y,z} T_{ik}S_{kj}\right)\left(\underline{\delta}_i\underline{\delta}_j\right) \\
&\textbf{Inner Product of a Tensor with a Vector:} & \underline{\underline{T}} \cdot \underline{\mathbf{v}} &= \sum_{i=x,y,z} \left(\sum_{j=x,y,z} T_{ij}v_j\right)\underline{\delta}_i \\
&\textbf{Double Dot Product:} & \underline{\underline{T}} : \underline{\underline{S}} &= \sum_{i=x,y,z} \sum_{j=x,y,z} \left(T_{ij}S_{ji}\right)
\end{aligned}
\tag{4.18}
$$

4.4 Matrix Representation of Vectors and Tensors

In using the Cartesian coordinates, the tensor and vector operation can also be represented by matrices. First, the unit vectors are represented by:

$$
\underline{\delta}_x = \begin{pmatrix} 1 \\ 0 \\ 0 \end{pmatrix} \qquad \underline{\delta}_y = \begin{pmatrix} 0 \\ 1 \\ 0 \end{pmatrix} \qquad \underline{\delta}_z = \begin{pmatrix} 0 \\ 0 \\ 1 \end{pmatrix} \tag{4.19}
$$

With (4.19), correspondences can be established between tensor operations and matrix operations. These are given in Table 4.4.

Once the vectors and tensors have been translated to matrix equations, the various properties of matrices apply. For instance, the eigenvectors of a symmetric tensor will be orthogonal to each other, associated with three real eigenvalues. Moreover, if the symmetric tensor is positive semi-definite, then eigenvalues are all non-negative real.

EXAMPLE 4.5. Consider the stress tensor given by

$$
\underline{\underline{T}} \;=\; 2\underline{\delta}_x\underline{\delta}_x + 2\underline{\delta}_x\underline{\delta}_y + \underline{\delta}_x\underline{\delta}_z + 2\underline{\delta}_y\underline{\delta}_x + \underline{\delta}_y\underline{\delta}_y + \underline{\delta}_y\underline{\delta}_z + \underline{\delta}_z\underline{\delta}_x + \underline{\delta}_z\underline{\delta}_y + 4\underline{\delta}_z\underline{\delta}_z
$$

It can be represented in matrix form by

$$
T = \begin{pmatrix} 2 & 2 & 1 \\ 2 & 1 & 1 \\ 1 & 1 & 4 \end{pmatrix}
$$

whose eigenvalues are -0.5688, 2.3644, and 5.2044 and eigenvectors

$$
\mathbf{v}_{\lambda=-0.5688} = \begin{pmatrix} 0.6035 \\ -0.7962 \\ 0.0422 \end{pmatrix} \;:\; \mathbf{v}_{\lambda=2.3644} = \begin{pmatrix} 0.6228 \\ 0.4377 \\ -0.6484 \end{pmatrix} \;;\; \mathbf{v}_{\lambda=5.2044} = \begin{pmatrix} 0.4479 \\ 0.4176 \\ 0.7601 \end{pmatrix}
$$

Table 4.4. *Correspondence between tensors and matrices*

Tensor notations	Matrix representation
$\underline{\mathbf{a}} = a_x\underline{\boldsymbol{\delta}}_x + a_y\underline{\boldsymbol{\delta}}_y + a_z\underline{\boldsymbol{\delta}}_z$	$\mathbf{a} = \left(a_x, \ a_y, \ a_z \right)^T$
$\underline{\mathbf{a}} + \underline{\mathbf{b}}$	$\mathbf{a} + \mathbf{b} = \left(a_x + b_x, \ a_y + b_y, \ a_z + b_z \right)^T$
$\alpha \, \underline{\mathbf{v}}$	$\alpha \, \mathbf{v} = \left(\alpha v_x, \ \alpha v_y, \ \alpha v_z \right)^T$
$\underline{\mathbf{a}} \cdot \underline{\mathbf{b}}$	$\mathbf{a}^T \mathbf{b}$
$\underline{\mathbf{a}} \, \underline{\mathbf{b}}$	$\mathbf{a}\mathbf{b}^T$
$\underline{\mathbf{a}} \times \underline{\mathbf{b}}$	$H^{[\mathbf{a}]}\mathbf{b}$ where $H^{[\mathbf{a}]} = \left(h_{ij}^{[\mathbf{a}]} \right); \ h_{ij}^{[\mathbf{a}]} = \sum_{k=1}^{3} \epsilon_{i,j,k} a_k$
$\underline{\underline{T}} = \sum_{i=x,y,z} \sum_{j=x,y,z} t_{ij} \left(\underline{\boldsymbol{\delta}}_i \underline{\boldsymbol{\delta}}_j \right)$	$T = \left(t_{ij} \right)$
$\underline{\underline{T}} + \underline{\underline{S}}$	$T + S = \left(t_{ij} + s_{ij} \right)$
$\alpha \underline{\underline{T}}$	$\alpha \, T = (\alpha t_{ij})$
$\underline{\underline{T}} \cdot \underline{\underline{S}}$	$T \, S = \left(\sum_{k=x,y,z} t_{ik} s_{kj} \right)$
$\underline{\underline{T}} \cdot \underline{\mathbf{v}}$	$T \, \mathbf{v} = \left(\sum_{j=x,y,z} t_{xj} v_j, \ \sum_{j=x,y,z} t_{yj} v_j, \ \sum_{j=x,y,z} t_{zj} v_j \right)^T$
$\underline{\underline{T}} : \underline{\underline{S}}$	$\mathrm{trace}(T \, S)$

If desired, these orthonormal eigenvectors can be put back in vector notation, for example,

$$\underline{\mathbf{v}}_{\lambda=-0.5688} = 0.6035\underline{\boldsymbol{\delta}}_x - 0.7962\underline{\boldsymbol{\delta}}_y + 0.0422\underline{\boldsymbol{\delta}}_z$$

These eigenvectors determine the **principal axes** of the tensor, which can be represented by an ellipsoid, as shown in Figure 4.8. Along the lines of the eigenvectors, there are no shear stresses, just normal stress. The eigenvector corresponding to the eigenvalue with the largest magnitude determines where the maximum normal stress is directed and is shown in the figure as line va, whereas the minimum normal stress occurs along the line shown as line vb. Along these axes, if the eigenvalue is positive, we can consider it as tension; otherwise, if the corresponding eigenvalue is negative, we have compression.[10] Thus along va we have tension, and along vb we have compression.

Remarks: The items in the right column of Table 4.4 can be translated directly to matrix operation commands in MATLAB. In addition, MATLAB provides some

[10] As usual, the sign convention for tension and compression can be switched depending on which field of study defines the tensors.

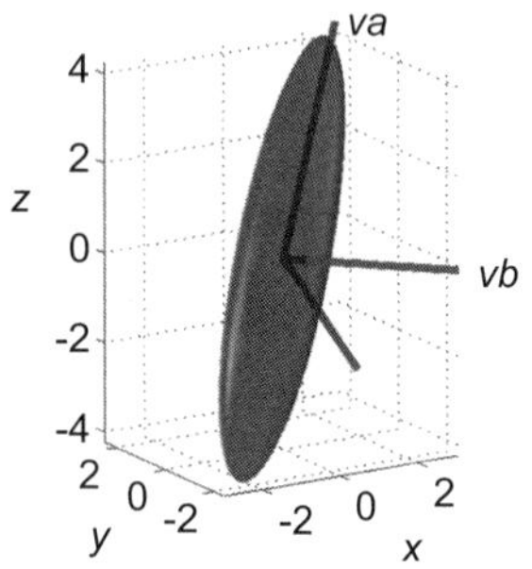

Figure 4.8. Stress tensor ellipsoid. The line along va has the maximum normal stress, whereas the line along vb has the minimum normal stress.

vector operation commands such as `q=dot(A,B)` and `C=cross(A,B)` to evaluate dot products and cross products of matrix A and B, respectively, where each column of $A[=]3 \times M$ and $B[=]3 \times M$ contains the vectors being operated on.

4.5 Differential Operations for Vector Functions of One Variable

Consider a vector fixed at a point, say, **p**, but is changing its direction and magnitude as a function of variable t. Then we can define a vector derivative as

Definition 4.4. *The* **derivative of $\underline{\mathbf{v}}(t)$ with respect to** *t, denoted by $d\underline{\mathbf{v}}/dt$, is given by*

$$\frac{d}{dt}\underline{\mathbf{v}} = \lim_{\Delta t \to 0} \frac{1}{\Delta t}\left(\underline{\mathbf{v}}\,(t+\Delta t) - \underline{\mathbf{v}}\,(t) \right) \tag{4.20}$$

In Cartesian coordinates, the vector $\underline{\mathbf{v}}(t) = v_x(t)\underline{\boldsymbol{\delta}}_x + v_y(t)\underline{\boldsymbol{\delta}}_y + v_z(t)\underline{\boldsymbol{\delta}}_z$ will have a derivative given by

$$\begin{aligned}\frac{d}{dt}\underline{\mathbf{v}}(t) &= \lim_{\Delta t \to 0} \frac{1}{\Delta t}\Bigg(\left(v_x(t+\Delta t)\underline{\boldsymbol{\delta}}_x + v_y(t+\Delta t)\underline{\boldsymbol{\delta}}_y + v_z(t+\Delta t)\underline{\boldsymbol{\delta}}_z\right) \\ &\qquad - \left(v_x(t)\underline{\boldsymbol{\delta}}_x + v_y(t)\underline{\boldsymbol{\delta}}_y + v_z(t)\underline{\boldsymbol{\delta}}_z\right)\Bigg) \\ &= \frac{dv_x}{dt}\underline{\boldsymbol{\delta}}_x + \frac{dv_y}{dt}\underline{\boldsymbol{\delta}}_y + \frac{dv_z}{dt}\underline{\boldsymbol{\delta}}_z \end{aligned} \tag{4.21}$$

A geometric interpretation of (4.21) is shown in Figure 4.9. Note that the vector $d\underline{\mathbf{v}}/dt$ does not necessarily point in the same direction as $\underline{\mathbf{v}}$, and neither does it have the same physical units as $\underline{\mathbf{v}}$ (unless t is unitless). Using (4.21) together with the properties of vector operations given in Table 4.2, the following identities can be obtained:

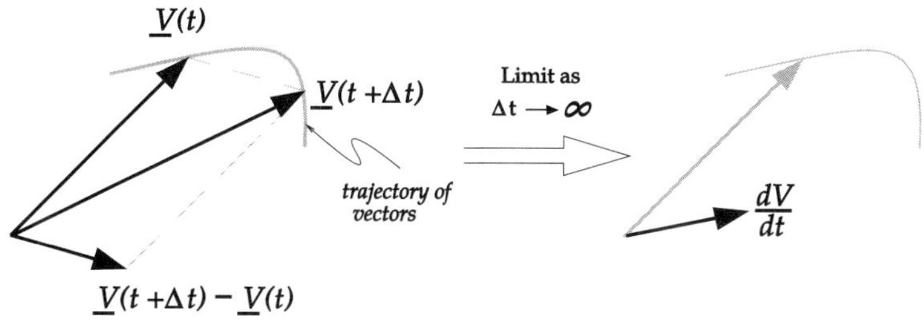

Figure 4.9. Definition of dV/dt.

Sums: $$\frac{d}{dt}\Big(\underline{\mathbf{u}}(t)+\underline{\mathbf{v}}(t)\Big) = \frac{d\underline{\mathbf{u}}}{dt}+\frac{d\underline{\mathbf{v}}}{dt}$$

Scalar Products: $$\frac{d}{dt}\Big(\psi(t)\underline{\mathbf{v}}(t)\Big) = \frac{d\psi}{dt}\underline{\mathbf{v}}+\psi\frac{d\underline{\mathbf{v}}}{dt}$$

Dot Products: $$\frac{d}{dt}\Big(\underline{\mathbf{u}}(t)\cdot\underline{\mathbf{v}}(t)\Big) = \frac{d\underline{\mathbf{u}}}{dt}\cdot\underline{\mathbf{v}}+\underline{\mathbf{u}}\cdot\frac{d\underline{\mathbf{v}}}{dt}$$

Cross Products: $$\frac{d}{dt}\Big(\underline{\mathbf{u}}(t)\times\underline{\mathbf{v}}(t)\Big) = \frac{d\underline{\mathbf{u}}}{dt}\times\underline{\mathbf{v}}+\underline{\mathbf{u}}\times\frac{d\underline{\mathbf{v}}}{dt} \tag{4.22}$$

Note that the order is important for the derivatives of cross product because cross products are anti-commutative.

4.6 Application to Position Vectors

One of the most useful vectors is the **position vector**, which we denote by $\underline{\vec{\mathbf{r}}}$

$$\underline{\vec{\mathbf{r}}} = x\underline{\boldsymbol{\delta}}_x + y\underline{\boldsymbol{\delta}}_y + z\underline{\boldsymbol{\delta}}_z$$

If $x(t)$, $y(t)$ and $z(t)$ are functions of a real parameter t, then a path is traced as t increases from a starting value t_0 to a final value t_F. If the path begins and ends at the same point, that is, $x(t_0) = x(t_F)$, $y(t_0) = y(t_F)$, and $z(t_0) = z(t_F)$, then we say that the path is a **closed path**. If the path does not intersect itself before the final point, then we call it a **simple path**. The parameter t is often chosen to be time or elapse time.

If we now apply the derivative operation to the position vector, we have the **velocity vector**, $\underline{\vec{\mathbf{v}}}(t)$,[11]

$$\underline{\vec{\mathbf{v}}}(t) = \vec{v}_x\underline{\boldsymbol{\delta}}_x + \vec{v}_y\underline{\boldsymbol{\delta}}_y + \vec{v}_z\underline{\boldsymbol{\delta}}_z = \frac{d}{dt}\underline{\vec{\mathbf{r}}} = \frac{dx}{dt}\underline{\boldsymbol{\delta}}_x + \frac{dy}{dt}\underline{\boldsymbol{\delta}}_y + \frac{dz}{dt}\underline{\boldsymbol{\delta}}_z \tag{4.23}$$

where the components are $\vec{v}_x = dx/dt$, $\vec{v}_y = dy/dt$ and $\vec{v}_z = dz/dt$.

The norm of $\underline{\vec{\mathbf{v}}}(t)$ is the **speed**, denoted by $\dot{s}$,

$$\dot{s}(t) = \left\|\underline{\vec{\mathbf{v}}}(t)\right\| = \sqrt{\vec{v}_x^2 + \vec{v}_y^2 + \vec{v}_z^2} \tag{4.24}$$

[11] We reserve the arrow sign above the variable to signify the derivatives of position vectors, including velocity, acceleration, and so forth, and their components.

In as much as the position (x, y, z) changes with the parameter t, the distance traveled along the path will also increase. We call this distance the **arc length** at t, denoted by $s(t)$ and defined as

$$s(t) = \int_{t_0}^{t} \left\| \underline{\vec{\mathbf{v}}}(\tau) \right\| d\tau \tag{4.25}$$

Note that the integrand is monotonic and nondecreasing, which means $s(t) \geq 0$. Differentiation will yield back the speed, that is,

$$\frac{ds}{dt} = \left\| \underline{\vec{\mathbf{v}}}(t) \right\| = \dot{s}(t) = \sqrt{\vec{v}_x^2 + \vec{v}_y^2 + \vec{v}_z^2}$$

Vector $\underline{\vec{\mathbf{v}}}(t)$ will be oriented tangent to the path at t. Thus to find the **unit tangent vector at** s, denoted by $\underline{\vec{\mathbf{t}}}(t)$, a normalization of $\underline{\vec{\mathbf{v}}}(s)$ is sufficient,

$$\underline{\vec{\mathbf{t}}}(t) = \frac{\underline{\vec{\mathbf{v}}}(t)}{\left\| \underline{\vec{\mathbf{v}}}(t) \right\|} \tag{4.26}$$

A derivative of $\underline{\vec{\mathbf{t}}}(t)$ with respect to t is also possible. Because $\underline{\vec{\mathbf{t}}}(t + dt)$ and $\underline{\vec{\mathbf{t}}}(t)$ both have magnitude equal to 1, the only change will be in the direction. Using this fact, we can apply the formula for the derivative of dot products given in (4.22),

$$\begin{aligned} \left\| \underline{\vec{\mathbf{t}}}(t) \right\|^2 = \underline{\vec{\mathbf{t}}} \cdot \underline{\vec{\mathbf{t}}} = 1 \quad \rightarrow \quad \frac{d}{dt}\left(\underline{\vec{\mathbf{t}}} \cdot \underline{\vec{\mathbf{t}}}\right) &= 0 \\ 2\left(\underline{\vec{\mathbf{t}}} \cdot \frac{d\underline{\vec{\mathbf{t}}}}{dt}\right) &= 0 \end{aligned}$$

Thus $d\underline{\vec{\mathbf{t}}}/dt$ is perpendicular to $\underline{\vec{\mathbf{t}}}$. If we normalize this new vector to have a magnitude of 1, we obtain the **unit normal to the curve at** t, denoted by $\underline{\vec{\mathbf{n}}}$, that is,

$$\underline{\vec{\mathbf{n}}} = \frac{d\underline{\vec{\mathbf{t}}}/dt}{\left\| d\underline{\vec{\mathbf{t}}}/dt \right\|} \tag{4.27}$$

Once we have $\underline{\vec{\mathbf{t}}}(t)$ and $\underline{\vec{\mathbf{n}}}(t)$, we can find a third unit vector, called the **binormal unit vector**, denoted by $\underline{\vec{\mathbf{b}}}$, that is perpendicular to both $\underline{\vec{\mathbf{t}}}$ and $\underline{\vec{\mathbf{n}}}$ defined by

$$\underline{\vec{\mathbf{b}}} = \underline{\vec{\mathbf{t}}} \times \underline{\vec{\mathbf{n}}} \tag{4.28}$$

The vectors $d\underline{\vec{\mathbf{t}}}/dt$ and $d\underline{\vec{\mathbf{b}}}/dt$ should yield information about how fast the shape of the path is changing such as curvature and torsion, but they are incomplete because both vectors are rates with respect to t, which is not ideal if only the shape is needed. For the purpose of measuring the curvature of the path, one choice is instead to take the ratio of the change in $\underline{\vec{\mathbf{t}}}(t)$ per change in arclength $s(t)$,

$$\frac{d\underline{\vec{\mathbf{t}}}}{ds} = \frac{d\underline{\vec{\mathbf{t}}}/dt}{ds/dt}$$

The magnitude of this ratio is a measure known as the **path curvature** at t and denoted by $\kappa(t)$,

$$\kappa(t) = \left\| \frac{d\underline{\vec{\mathbf{t}}}/dt}{ds/dt} \right\| = \frac{\left\| d\underline{\vec{\mathbf{t}}}/dt \right\|}{\dot{s}(t)}$$

The **radius of curvature**, denoted by r_{curve}, is defined as the reciprocal of the curvature,

$$r_{\text{curve}}(t) = \frac{1}{\kappa(t)} = \frac{\dot{s}(t)}{\left\| d\vec{\underline{\mathbf{t}}}/dt \right\|}$$

This is the radius of a circle that is tangent to the curve at t and two differentially close neighboring points.

For the torsion of the path, we have a similar ratio. This time we define **torsion of the path at** t, denoted by $\tau(t)$, as the norm of the change in the binormal unit vector $\vec{\underline{\mathbf{b}}}(t)$ per change in arclength $s(t)$, that is,

$$\tau(t) = \left\| \frac{d\vec{\underline{\mathbf{b}}}/dt(t)}{ds/dt} \right\| = \frac{\left\| d\vec{\underline{\mathbf{b}}}/dt \right\|}{\dot{s}(t)}$$

The **radius of torsion**, denoted by r_{torsion}, is the reciprocal of torsion,

$$r_{\text{torsion}} = \frac{1}{\tau} = \frac{\dot{s}}{\left\| d\vec{\underline{\mathbf{b}}}/dt \right\|}$$

Finally, we can take the derivative of velocity $\vec{\underline{\mathbf{v}}}(t)$ to obtain the **acceleration vector**, denoted by $\vec{\underline{\mathbf{a}}}$,

$$\vec{\underline{\mathbf{a}}}(t) = \vec{a}_x \underline{\boldsymbol{\delta}}_x + \vec{a}_y \underline{\boldsymbol{\delta}}_y + \vec{a}_z \underline{\boldsymbol{\delta}}_z = \frac{d\vec{\underline{\mathbf{v}}}}{dt} = \frac{d^2x}{dt^2}\underline{\boldsymbol{\delta}}_x + \frac{d^2y}{dt^2}\underline{\boldsymbol{\delta}}_y + \frac{d^2z}{dt^2}\underline{\boldsymbol{\delta}}_z \tag{4.29}$$

where the components are given by $\vec{a}_x = d^2x/dt^2$, $\vec{a}_y = d^2y/dt^2$ and $\vec{a}_z = d^2z/dt^2$. Alternatively, we could represent acceleration in terms of $\vec{\underline{\mathbf{t}}}$ and $\vec{\underline{\mathbf{n}}}$ as follows,

$$\begin{aligned}
\vec{\underline{\mathbf{a}}}(t) &= \frac{d\vec{\underline{\mathbf{v}}}}{dt} = \frac{d\left(\dot{s}\vec{\underline{\mathbf{t}}}\right)}{dt} \\
&= \frac{d\dot{s}}{dt}\vec{\underline{\mathbf{t}}} + \dot{s}\frac{d\vec{\underline{\mathbf{t}}}}{dt} = \frac{d\dot{s}}{dt}\vec{\underline{\mathbf{t}}} + \dot{s}\left\| \frac{d\vec{\underline{\mathbf{t}}}}{dt} \right\| \vec{\underline{\mathbf{n}}} \\
a_T\vec{\underline{\mathbf{t}}} + a_N\vec{\underline{\mathbf{n}}} &= \frac{d\dot{s}}{dt}\vec{\underline{\mathbf{t}}} + \dot{s}^2\kappa\vec{\underline{\mathbf{n}}}
\end{aligned}$$

Thus the tangential and normal components of the acceleration vector are given by $a_T = d\dot{s}/dt$ and $a_N = \dot{s}^2\kappa$, respectively.

EXAMPLE 4.6. Consider the helical path described by $x(t) = cos(t)$, $y(t) = sin(t)$ and $z(t) = t$. Then we have the following:

Position Vector:	$\vec{\underline{\mathbf{r}}}$	$=$	$\cos(t)\underline{\boldsymbol{\delta}}_x + \sin(t)\underline{\boldsymbol{\delta}}_y + t\underline{\boldsymbol{\delta}}_z$
Velocity:	$\vec{\underline{\mathbf{v}}}$	$=$	$-\sin(t)\underline{\boldsymbol{\delta}}_x + \cos(t)\underline{\boldsymbol{\delta}}_y + \underline{\boldsymbol{\delta}}_z$
Speed:	$\dot{s}$	$=$	$\sqrt{2}$
Unit Tangent:	$\vec{\underline{\mathbf{t}}}$	$=$	$-\frac{\sin(t)}{\sqrt{2}}\underline{\boldsymbol{\delta}}_x + \frac{\cos(t)}{\sqrt{2}}\underline{\boldsymbol{\delta}}_y + \frac{1}{\sqrt{2}}\underline{\boldsymbol{\delta}}_z$
Unit Normal:	$\vec{\underline{\mathbf{n}}}$	$=$	$-\cos(t)\underline{\boldsymbol{\delta}}_x - \sin(t)\underline{\boldsymbol{\delta}}_y$
Unit Binormal:	$\vec{\underline{\mathbf{b}}}$	$=$	$\frac{\sin(t)}{\sqrt{2}}\underline{\boldsymbol{\delta}}_x - \frac{\cos(t)}{\sqrt{2}}\underline{\boldsymbol{\delta}}_y + \frac{1}{\sqrt{2}}\underline{\boldsymbol{\delta}}_z$
Curvature:	κ	$=$	$1/2$
Torsion:	τ	$=$	$1/2$
Acceleration:	$\vec{\underline{\mathbf{a}}}$	$=$	$-\cos(t)\underline{\boldsymbol{\delta}}_x - \sin(t)\underline{\boldsymbol{\delta}}_y$ or $\vec{\underline{\mathbf{n}}}$

Let a 3D surface $S(x, y, z) = 0$ be parameterized by ξ and η,

$$x = x(\xi, \eta);\ y = y(\xi, \eta);\ z = z(\xi, \eta)$$

Then to find the normal to the surface S at a point $a = \{a_x, a_y, a_z\}$, we could use the previous results to first find a vector that is tangent to the curve in S at point a along constant ξ and another vector that is tangent to the curve in S at point a along constant η. Thus

$$\vec{\underline{\mathbf{t}}}_\xi = \frac{\partial \vec{\underline{\mathbf{r}}}}{\partial \xi} = \frac{\partial x}{\partial \xi}\underline{\boldsymbol{\delta}}_x + \frac{\partial y}{\partial \xi}\underline{\boldsymbol{\delta}}_y + \frac{\partial z}{\partial \xi}\underline{\boldsymbol{\delta}}_z$$

$$\vec{\underline{\mathbf{t}}}_\eta = \frac{\partial \vec{\underline{\mathbf{r}}}}{\partial \eta} = \frac{\partial x}{\partial \eta}\underline{\boldsymbol{\delta}}_x + \frac{\partial y}{\partial \eta}\underline{\boldsymbol{\delta}}_y + \frac{\partial z}{\partial \eta}\underline{\boldsymbol{\delta}}_z$$

Then we take the cross product of both tangent vectors to find a vector that is perpendicular to the plane that contains both tangent vectors and then normalize the result to find a unit normal vector,

$$\vec{\underline{\mathbf{n}}} = \frac{\vec{\underline{\mathbf{t}}}_\xi \times \vec{\underline{\mathbf{t}}}_\eta}{\left\| \vec{\underline{\mathbf{t}}}_\xi \times \vec{\underline{\mathbf{t}}}_\eta \right\|} \tag{4.30}$$

(See Exercise **E4.10** as a sample problem.)

Another approach to finding the unit normal vector to the surface is to use the gradient function. This is discussed in the next section.

4.7 Differential Operations for Vector Fields

Let $\psi(\mathbf{p})$ be scalar field that is a function of the spatial position $\mathbf{p}$. We assume that the scalar fields will be sufficiently differentiable. Likewise, let $\underline{\mathbf{v}}(\mathbf{p})$ be a vector field that is a function of position $\mathbf{p}$. We assume that the components of $\underline{\mathbf{v}}$ are sufficiently differentiable. Under the Cartesian coordinate system, this means

$$\underline{\mathbf{v}}(x, y, z) = v_x(x, y, z)\,\underline{\boldsymbol{\delta}}_x + v_y(x, y, z)\,\underline{\boldsymbol{\delta}}_y + v_z(x, y, z)\,\underline{\boldsymbol{\delta}}_z$$

where $v_x(x, y, z)$, $v_y(x, y, z)$, and $v_z(x, y, z)$ are scalar fields that are assumed to be sufficiently differentiable.

In the following sections, we first discuss the various differential operations of vector fields based on the Cartesian coordinate systems. Later, we show how these differential operations can be defined in other coordinate systems, such as cylindrical and spherical coordinates systems.

Remarks: In MATLAB, the commands `quiver` and `quiver3` plot arrows depicting the vector field at the specified mesh positions.

EXAMPLE 4.7. Let $\mathbf{v}(x, y, z)$ be a given velocity field of a fluid domain. A set of curves known as **streamlines** can be found such that at any point, say, $\mathbf{p} = (x^*, y^*, z^*)$, in a streamline, $\mathbf{v}(\mathbf{p})$ will be tangent to that curve at $\mathbf{p}$. If the fluid flow is changing with time, the snapshots of the collection of streamlines will also change. However, under steady state, the streamlines will show the path lines of fluid particles. Thus streamlines offer another view of fluid motion; that is, instead of a distribution of arrows, we have a portrait containing several curves.

Let the position of a streamline curve be parameterized by s, that is, $\underline{\vec{\mathbf{r}}} = \underline{\vec{\mathbf{r}}}(s)$; then we could use the parallelism property of cross products to obtain the condition

$$\frac{d\underline{\vec{\mathbf{r}}}}{ds} \times \mathbf{v}(\underline{\vec{\mathbf{r}}}(s)) = \underline{\mathbf{0}} \tag{4.31}$$

This condition is satisfied by the following equalities:

$$\frac{dx}{ds} = v_x; \qquad \frac{dy}{ds} = v_y; \qquad \frac{dz}{ds} = v_z \tag{4.32}$$

subject to initial conditions $\underline{\vec{\mathbf{r}}}(0) = \underline{\vec{\mathbf{r}}}_0$. However, in general, these simultaneous differential equations may not be that easy to evaluate. One could use numerical ODE solvers, discussed in Chapter 7. In MATLAB, the command `streamline` uses finite difference approximations to obtain the streamlines by specifying the starting points. Another MATLAB command `streamtube` adds tube-width information to show the magnitude of the velocity instead of a simple curve.

To illustrate, consider a 2D velocity field described by

$$\mathbf{v} = \left(\frac{5}{3}x^2 - y + \frac{17}{6}\right)\underline{\boldsymbol{\delta}}_x + \left(-\frac{4}{3}x^2 + (y - x) - \frac{29}{30}\right)\underline{\boldsymbol{\delta}}_y$$

We can plot the velocity field and streamlines as shown in Figure 4.10. This figure was generated using a MATLAB file that are available in the book's webpage, `streamline_test([0 0 0.04 0.1 0.18 0.3],[2.05 2.2 2 2 2 2],0,1,2,3)`

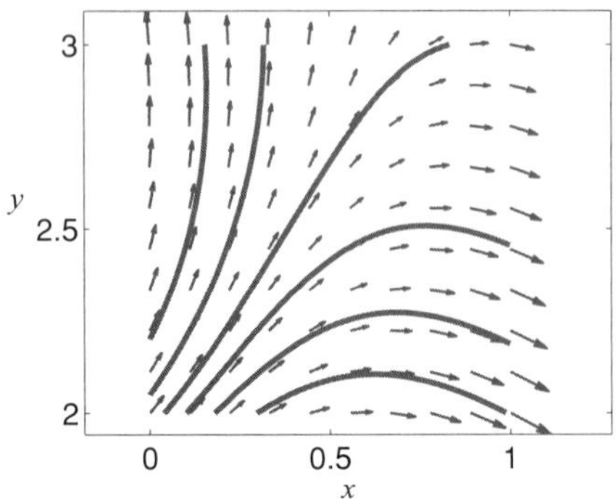

Figure 4.10. A velocity field together with some streamlines.

4.7.1 Gradients

Consider a 3D scalar field $\psi(x, y, z)$. For example, let $u(x, y, z)$ be a scalar temperature field, that is, a temperature distribution in the 3D space. Usually, the gradation of scalar fields will determine the driving forces in mass, momentum, or energy transport. Then it is important to determine a vector field based on the gradations of the scalar field that can be used to determine the directions (and magnitude) of driving forces. The desired vector field is known as the **gradient vector field**. The **gradient** of $\psi(x, y, z)$ is then defined as

$$\textbf{gradient}(\psi(x, y, z)) = \frac{\partial \psi}{\partial x}\underline{\boldsymbol{\delta}}_x + \frac{\partial \psi}{\partial y}\underline{\boldsymbol{\delta}}_y + \frac{\partial \psi}{\partial z}\underline{\boldsymbol{\delta}}_z \tag{4.33}$$

This assumes that the scalar field is continuous and smooth (i.e., differentiable in each independent variable). The gradient definition can be simplified by introducing a differential operator called the **gradient operator**, or **grad** for short.[12] The grad operator is denoted by ∇, and it is defined as

$$\nabla = \underline{\boldsymbol{\delta}}_x \frac{\partial}{\partial x} + \underline{\boldsymbol{\delta}}_y \frac{\partial}{\partial y} + \underline{\boldsymbol{\delta}}_z \frac{\partial}{\partial z} \tag{4.34}$$

Thus (4.33) can be rewritten as

$$\textbf{gradient}(\psi(x, y, z)) = \nabla \psi(x, y, z)$$

EXAMPLE 4.8. Consider the following scalar field, which is independent of z, given by

$$\psi(x, y) = e^{-\left(x^2+y^2\right)}$$

then the gradient is given by

$$\nabla \psi\,(x, y) = \left(-2xe^{-\left(x^2+y^2\right)}\right)\underline{\boldsymbol{\delta}}_x - \left(2ye^{-\left(x^2+y^2\right)}\right)\underline{\boldsymbol{\delta}}_y$$

A plot of $\psi(x, y)$ is given in Figure 4.11. Let us look at the gradient at position $(x, y) = (-0.5, -1)$. The x-component of the gradient is given by the slope of a curve at $y = -1$, as shown in the figure. The result is $\partial\psi/\partial x = 0.287$. The y-component of the gradient is given by the slope of a curve at $x = -0.5$, also shown in the figure. The resulting value of the y-component of the gradient is

[12] In some texts, it is also called the "del" operator.

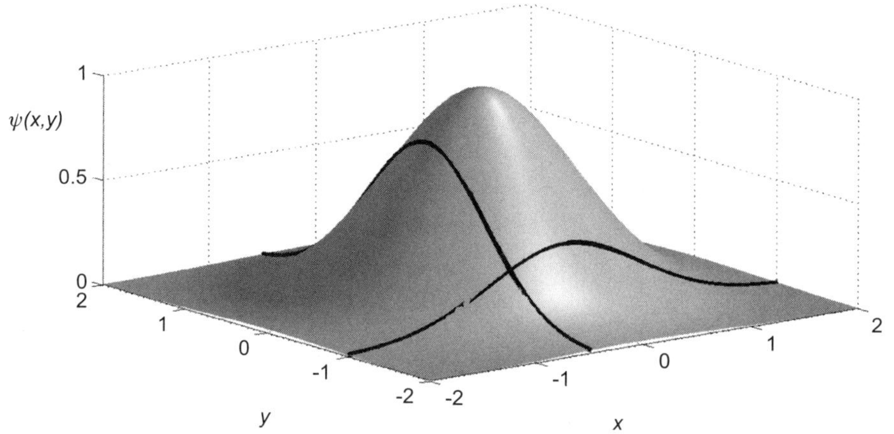

Figure 4.11. The surface plot of $\psi(x, y)$, together one curve at constant $y = -1$ and another curve at constant $x = -0.5$.

$\partial\psi/\partial y = 0.573$ (about twice as steep as the other slope). The gradient is given by

$$\nabla\psi\,(-0.5, -1) = 0.287\underline{\boldsymbol{\delta}}_x + 0.573\underline{\boldsymbol{\delta}}_y$$

whose magnitude is $\left\|\nabla\psi\,(-0.5, -1)\right\| = 0.641$.

If we had chosen the point $(x, y) = (-1.5, -1)$, the resulting gradient would be

$$\nabla\psi\,(0, -1) = 0.116\underline{\boldsymbol{\delta}}_x + 0.078\underline{\boldsymbol{\delta}}_y$$

where the x-component is about twice as large as the y-component, and the magnitude of the gradient vector is $\left\|\nabla\psi\,(-1.5, -1)\right\| = 0.14$. One could look at Figure 4.11 and see that indeed the gradient at the point $(x, y) = (-0.5, -1)$ is much steeper than $(x, y) = (-1.5, -1)$. Furthermore, one can see that the gradient is pointing at an ascending direction.

The gradient vector could also be used to obtain a directional derivative of a 3D scalar field.

Definition 4.5. *For a multivariable function $f(x_1, \ldots, x_n)$ and normalized vector $\mathbf{v} = (v_1, \ldots, v_n)$, the* **directional derivative of f along unit vector v**, *denoted by $D_{\mathbf{v}}f\,(\mathbf{x})$, is defined as*

$$D_{\mathbf{v}}f\,(\mathbf{x}) = \lim_{\alpha\to 0^+} \frac{f\,(\mathbf{x} + \alpha\mathbf{v}) - f\,(\mathbf{x})}{\alpha} \tag{4.35}$$

For $\psi\,(x, y, z)$, the numerator of (4.35) sets $\Delta x = \alpha v_x$, $\Delta y = \alpha v_y$, and $\Delta z = \alpha v_z$. Then for small changes,

$$\Delta\psi = \frac{\partial\psi}{\partial x}\Delta x + \frac{\partial\psi}{\partial y}\Delta y + \frac{\partial\psi}{\partial z}\Delta z = \frac{\partial\psi}{\partial x}\alpha v_x + \frac{\partial\psi}{\partial y}\alpha v_y + \frac{\partial\psi}{\partial z}\alpha v_z$$

After substitution to (4.35) and taking the limit, we have

$$D_{\mathbf{v}}\psi\,(x, y, z) = \frac{\partial\psi}{\partial x}v_x + \frac{\partial\psi}{\partial y}v_y + \frac{\partial\psi}{\partial z}v_z = \nabla\psi \cdot \mathbf{v} \tag{4.36}$$

The directional derivative is the scalar projection of the gradient along the direction of vector $\underline{\mathbf{v}}$. Based on the definition (4.35), the value of the directional derivative is zero when ψ is a constant along $\mathbf{v}$. For a 2D case, for example, $\psi = \psi(x, y) = \kappa$, this defines a curve (possibly closed) called the **contour curves** or **contours** of $\psi\,(x, y)$ at different levels determined by the parameter κ. However, for a 3D scalar field, that is, $\psi = \psi\,(x, y, z) = \kappa$, these surfaces are called the **contour surfaces** or **isosurfaces** determined by the parameter κ. Because the directional derivative is zero along the contour lines or contour surfaces, we see from (4.36), that the gradient $\nabla\psi$ is perpendicular to either the contour lines for the 2D case or the contour surfaces for the 3D case.

Remarks: In MATLAB, the commands `contour` and `contour3` gives the contour plots of a surface, for example, of a 2D scalar field. For a 3D scalar field, one can instead use `contourf` to obtain contour slice plots or use the command `isosurface` to find the contour surfaces at fixed values (one can then use the command `hold` to allow several surface plots while introducing NaN's to produce cutouts of the surfaces). For the gradients, there is the MATLAB command `gradient`, which calculates the gradient based on finite differences; but the accuracy depends on the resolution of the mesh. The result of the gradient can then be plotted using `quiver` or `quiver3`.

EXAMPLE 4.9. Suppose the dimensionless temperature distribution $u = (T - T_{\text{center}})/T_{\text{center}}$ is given by the following scalar field:

$$u = \left(\frac{x}{3}\right)^2 + y^2 + z^2$$

The contour surfaces, that is, $u = \kappa$=constant, are ellipsoids centered at the origin. The surfaces corresponding to $\kappa = 0.5, 1.0, 1.5$ are shown in Figure 4.12.

Consider two points at the surface defined by $\kappa = 1$: $a = (2.9, 0, 0.2560)$ and $b = (0, 0, 1)$. At point a, we find that the gradient is given by $\nabla u|_a = 0.644\underline{\delta}_x + 0.5121\underline{\delta}_z$ and $\|\nabla u|_a\| = 0.823$. At point b, we have the gradient $\nabla u|_b = 2\underline{\delta}_z$ and $\|\nabla u|_b\| = 2$. Both gradients can be seen to be normal to the contour surfaces. Note that the distance between the surface at $\kappa = 1$ and $\kappa = 1.5$ around point a is farther than the distance between surfaces around point b, and yet the magnitude of the gradient at point a is less than half the magnitude of the gradient at point b. Thus one should be cautious when reading contour maps or contour surfaces: the closer the adjacent contours are to each other, the greater the magnitude of the gradient.

The gradient vector can also be used to determine the rate of change in ψ along a path C. The differential of $\psi(x, y, z)$ is given by

$$d\psi(x, y, z) = \frac{\partial\psi}{\partial x}dx + \frac{\partial\psi}{\partial y}dy + \frac{\partial\psi}{\partial z}dz$$

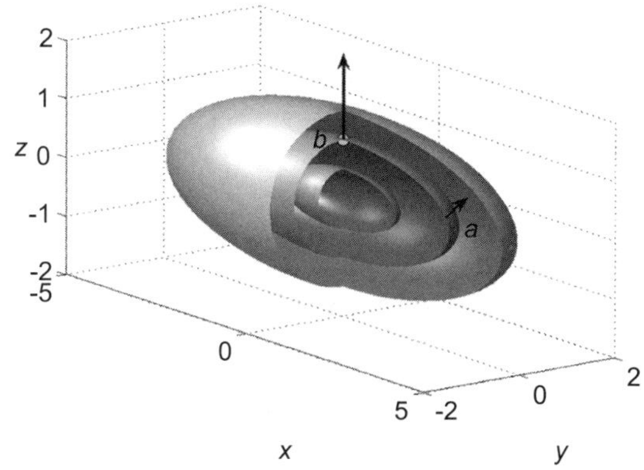

Figure 4.12. The contour surfaces of u at $\kappa = 0.5$ (innermost ellipsoid), $\kappa = 1.0$ (middle ellipsoid), and $\kappa = 1.5$ (outermost ellipsoid). Also, the gradients at points a and b are shown to be perpendicular to the surface corresponding to $\kappa = 1$.

Along the path, the rate of change in ψ per change in arclength s is given by

$$\frac{d\psi(x, y, z)}{ds} = \frac{\partial\psi}{\partial x}\frac{dx}{ds} + \frac{\partial\psi}{\partial y}\frac{dy}{ds} + \frac{\partial\psi}{\partial z}\frac{dz}{ds} \tag{4.37}$$

The right-hand side of the equation can now be factored into a dot product,

$$\begin{aligned}\frac{d\psi(x, y, z)}{ds} &= (\nabla\psi)\cdot\left(\frac{dx}{ds}\underline{\boldsymbol{\delta}}_x + \frac{dy}{ds}\underline{\boldsymbol{\delta}}_y + \frac{dz}{ds}\underline{\boldsymbol{\delta}}_z\right)\\ &= (\nabla\psi)\cdot\vec{\underline{\mathbf{t}}}\end{aligned} \tag{4.38}$$

where $\vec{\underline{\mathbf{t}}}$ is the unit tangent vector (cf. (4.26)). Thus the rate of change of ψ along the path C is the directional derivative along the unit tangent vectors of the path at the desired point.

Let θ be the angle between $\nabla\psi$ and $\vec{\underline{\mathbf{t}}}$. Then (4.38) becomes

$$\frac{d\psi}{ds} = \|\nabla\psi\|\cos\theta \tag{4.39}$$

This means that at a point, the maximum value for increase of ψ occurs when $\cos\theta = 1$ or $\theta = 0$, whereas the maximum rate of decrease of ψ occurs when $\cos\theta = -1$ or $\theta = \pi$. This generates one of the methods used for the search for local optima of $\psi\,(x, y, z)$, that is, to choose an update path C directed along the gradient $\nabla\psi$ to find the local maximum, called the **gradient ascent method**. In other cases, we choose an update path negative to the direction of $\nabla\psi$ to find a local mimimum, which is called the **gradient descent method**. If the rate of change, $d\psi/ds$, at a point is zero for all θ, that is, for all paths passing through this point, then the gradient must be zero at this point. This indicates an **extreme point** or **critical point**, which is either an optimum or a saddle point. Thus a necessary condition for optimality is $\nabla\phi = 0$.

Finally, the gradient operation can be used to find the unit normal of any smooth surface. Let the given surface be given as $f(x, y, z) = C$. As an alternative to methods used in Section 4.6, for example, (4.30), the unit normal to $f(x, y, z) = C$ at the point $a = (a_x, a_y, a_z)$ is given by

$$\underline{\mathbf{n}} = \frac{\nabla f}{\left\|\nabla f\right\|} \tag{4.40}$$

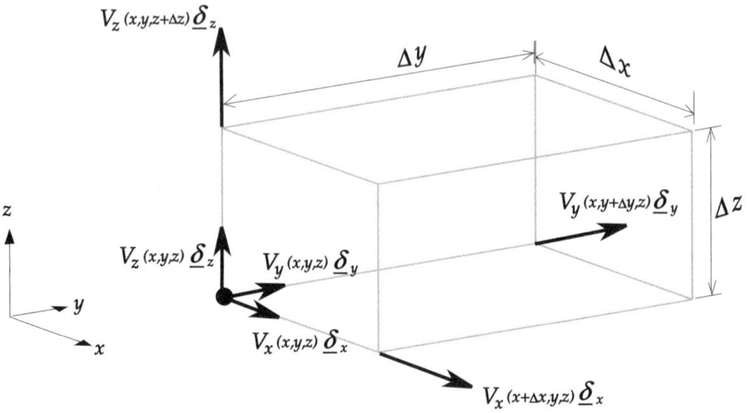

$$\nabla \bullet \underline{V} = \lim_{\Delta \to 0} \left(\frac{V_x(x+\Delta x, y, z) - V_x(x,y,z)}{\Delta x} + \frac{V_y(x, y+\Delta y, z) - V_y(x,y,z)}{\Delta y} + \frac{V_z(x,y,z+\Delta z) - V_z(x,y,z)}{\Delta z} \right)$$

Figure 4.13. Evaluation of the divergence of a vector field at a point.

4.7.2 Divergence

Consider the vector field $\underline{\mathbf{v}}\,(x, y, z)$,

$$\underline{\mathbf{v}}(x, y, z) = v_x(x, y, z)\underline{\boldsymbol{\delta}}_x + v_y(x, y, z)\underline{\boldsymbol{\delta}}_y + v_z(x, y, z)\underline{\boldsymbol{\delta}}_z$$

The **divergence** of a vector field at a point is defined as

$$\textbf{divergence}\left(\underline{\mathbf{v}}\,(x, y, z)\right) = \frac{\partial v_x}{\partial x} + \frac{\partial v_y}{\partial y} + \frac{\partial v_z}{\partial z} \tag{4.41}$$

In terms of the ∇ operator, the divergence is written as $\nabla \cdot \underline{\mathbf{v}}$,

$$\begin{aligned} \nabla \cdot \underline{\mathbf{v}} &= \left(\sum_{i=x,y,z} \underline{\boldsymbol{\delta}}_i \frac{\partial}{\partial i} \right) \cdot \left(\sum_{j=x,y,z} v_j \underline{\boldsymbol{\delta}}_j \right) = \sum_{i,j=x,y,z} \frac{\partial v_j}{\partial i} \underline{\boldsymbol{\delta}}_i \cdot \underline{\boldsymbol{\delta}}_j \\ &= \sum_{i=x,y,z} \frac{\partial v_i}{\partial i} = \frac{\partial v_x}{\partial x} + \frac{\partial v_y}{\partial y} + \frac{\partial v_z}{\partial z} \end{aligned}$$

Based on this definition, the evaluation of divergence is shown in Figure 4.13. The divergence of a vector field yields scalar fields. This is in contrast to the gradient of a scalar field, which yields a vector field. For instance, suppose the vector field is a flux field, that is, the rate of transport of a quantity (momentum, mass or energy) per unit area across the flow. The divergence of the flux field at a point will then measure how much the transported quantity will "diverge" at that point due to the field. Specifically, for mass flux, we can write the **continuity equation** (or differential mass balance) as

$$\nabla \cdot \Big(\rho(x, y, z)\underline{\mathbf{v}}(x, y, z) \Big) = 0$$

where ρ is the density and $\underline{\mathbf{v}}$ is the velocity field. Note that in the most general case, both ρ and $\underline{\mathbf{v}}$ will depend on the position. A vector field $\underline{\mathbf{v}}$ is called **solenoidal** if $\nabla \cdot \underline{\mathbf{v}} = 0$.

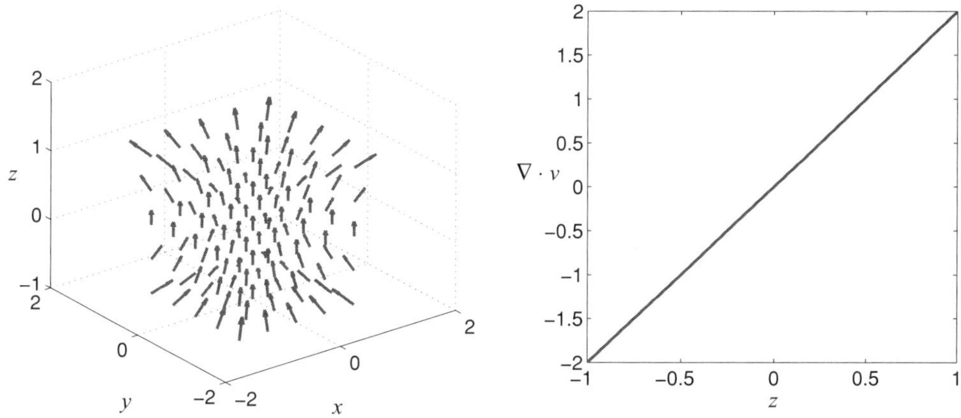

Figure 4.14. The plot of the vector field $\underline{\mathbf{v}}$ with $f(z) = z$ and the divergence $\nabla \cdot \underline{\mathbf{v}} = 2z$.

Remarks: In MATLAB, the command `divergence` calculates the divergence field based on finite difference. The result can then be plotted using `surf` or `surfl` for 2D fields. For 3D fields, one can use the command `slice` to obtain a set of color/shade graded slices.

EXAMPLE 4.10. Consider the vector field, $\underline{\mathbf{v}} = (xf(z))\,\underline{\boldsymbol{\delta}}_x + (yf(z))\,\underline{\boldsymbol{\delta}}_y + \underline{\boldsymbol{\delta}}_z$. The divergence is then given by $\nabla \cdot \underline{\mathbf{v}} = 2f(z)$. Thus, for this example, the divergence yields a scalar field that is dependent only in z.

Let us now look at two cases. First, consider $f(z) = z$. The vector field is shown in Figure 4.14. The divergence varies linearly with z, and one can see that the divergence is positive when $z > 0$ and the field appears to be diverging, whereas the divergence is negative when $z < 0$ and the vector field appears to be converging.

For the second case, consider $f(z) = \sqrt{z^2}$. The vector field is shown in Figure 4.15. As the vector field shows, there appears to be no region where the vector field is converging; thus the divergence is always positive, except at $z = 0$.

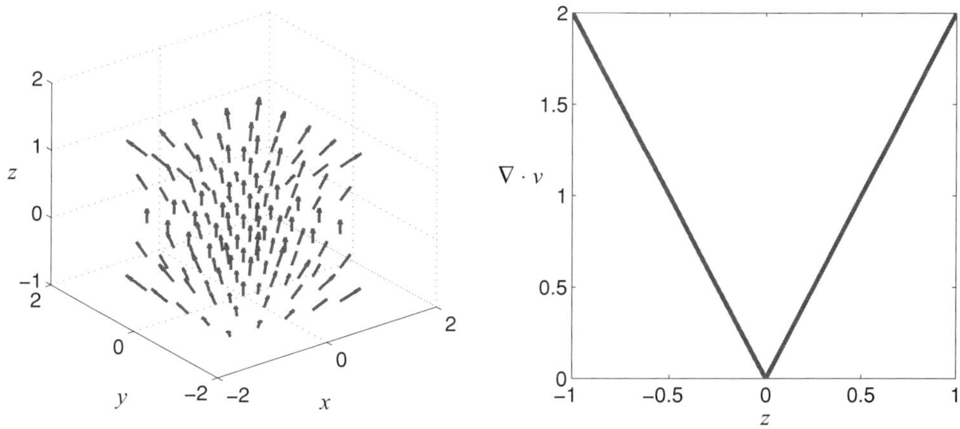

Figure 4.15. The plot of the vector field $\underline{\mathbf{v}}$ with $f(z) = \sqrt{z^2}$ and the divergence $\nabla \cdot \underline{\mathbf{v}} = 2\sqrt{z^2}$.

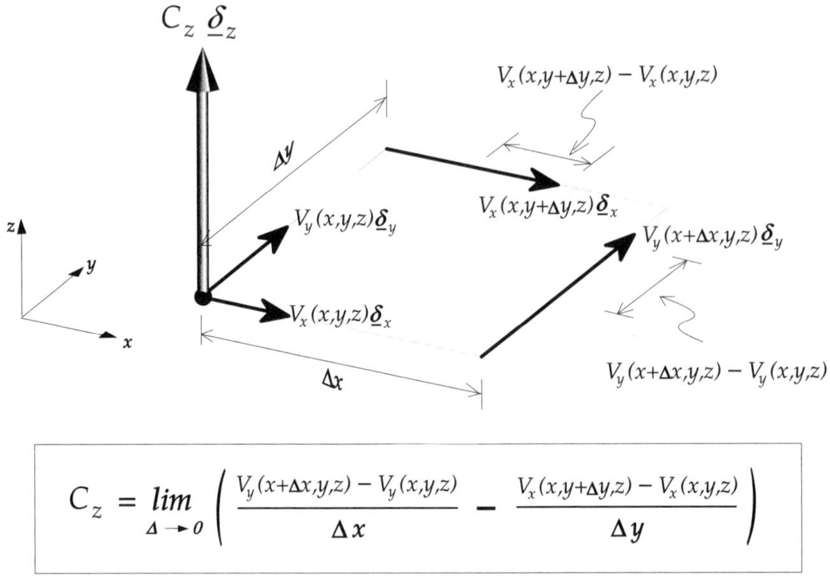

Figure 4.16. Evaluation of the curl of a vector field at a point across the (x, y) plane.

4.7.3 Curl

Due to the distribution in a vector field, neighboring vectors will affect the differential volume at a point to twirl or rotate. In Figure 4.16, for a particle (i.e., a differential volume) at point $P = (P_x, P_y, P_z)$, we consider the projection of the vector fields in the (x, y) plane at P_z level. Let $C_z\underline{\delta}_z$ be a vector that describes the tendency of the particle at P to rotate based on the relative vector distributions of $v_x(x, y, P_z)$ along y-direction and the relative vector distribution of $v_y(x, y, P_z)$ along the x-direction. Using the right-hand screw convention, the total effect is given by

$$\begin{aligned} C_z &= \lim_{\Delta \to 0} \left(\frac{v_y(x+\Delta x, y, z) - v_y(x, y, z)}{\Delta x} - \frac{v_x(x, y+\Delta y, z) - v_x(x, y, z)}{\Delta y} \right) \\ &= \frac{\partial v_y}{\partial x} - \frac{\partial v_x}{\partial y} \end{aligned}$$

Following the same analysis in the (x, z) plane, we get $C_y\underline{\delta}_y$, where

$$C_y = \frac{\partial v_x}{\partial z} - \frac{\partial v_z}{\partial x}$$

and in the (y, z) plane, we have $C_x\underline{\delta}_x$, where

$$C_x = \frac{\partial v_z}{\partial y} - \frac{\partial v_y}{\partial z}$$

Adding the three vectors yields a vector known as the **Curl** of the vector field at the point (x, y, z),

$$\mathbf{Curl}(\underline{\mathbf{v}}(x, y, z)) = \left(\frac{\partial v_z}{\partial y} - \frac{\partial v_y}{\partial z} \right) \underline{\delta}_x + \left(\frac{\partial v_x}{\partial z} - \frac{\partial v_z}{\partial x} \right) \underline{\delta}_y + \left(\frac{\partial v_y}{\partial x} - \frac{\partial v_x}{\partial y} \right) \underline{\delta}_z \tag{4.42}$$

Using the ∇ operator, the curl is written as the cross product of ∇ with $\underline{\mathbf{v}}$,

$$
\begin{aligned}
\nabla \times \underline{\mathbf{v}} &= \left(\sum_{i=x,y,z} \underline{\boldsymbol{\delta}}_i \frac{\partial}{\partial i}\right) \times \left(\sum_{j=x,y,z} v_j \underline{\boldsymbol{\delta}}_j\right) = \sum_{ij} \frac{\partial v_j}{\partial i} \underline{\boldsymbol{\delta}}_i \times \underline{\boldsymbol{\delta}}_j = \sum_{ijk} \epsilon_{ijk} \frac{\partial v_j}{\partial i} \underline{\boldsymbol{\delta}}_k \\
&= \left(\frac{\partial v_z}{\partial y} - \frac{\partial v_y}{\partial z}\right) \underline{\boldsymbol{\delta}}_x + \left(\frac{\partial v_x}{\partial z} - \frac{\partial v_z}{\partial x}\right) \underline{\boldsymbol{\delta}}_y + \left(\frac{\partial v_y}{\partial x} - \frac{\partial v_x}{\partial y}\right) \underline{\boldsymbol{\delta}}_z
\end{aligned} \tag{4.43}
$$

With the rectangular coordinates, each unit vector will have the same direction and magnitude at any point in space. Under the Cartesian coordinate system, the following mnemonic is valid:

$$
\nabla \times \underline{\mathbf{v}} = \mathbf{det} \begin{pmatrix} \underline{\boldsymbol{\delta}}_x & \underline{\boldsymbol{\delta}}_y & \underline{\boldsymbol{\delta}}_z \\ \frac{\partial}{\partial x} & \frac{\partial}{\partial y} & \frac{\partial}{\partial z} \\ v_x & v_y & v_z \end{pmatrix}
$$

For the special case in which $\underline{\mathbf{v}}$ is a velocity field, $\nabla \times \underline{\mathbf{v}}$ is the vector field known as the **vorticity** of $\underline{\mathbf{v}}$. The angular velocity of a rigid body is related to the curl by

$$
\underline{\Omega} = \frac{1}{2} \nabla \times \underline{\mathbf{v}} \tag{4.44}
$$

A vector field $\underline{\mathbf{v}}$ is called **irrotational** if $\nabla \times \underline{\mathbf{v}} = 0$.

Remarks: In MATLAB, the command `curl` can be used to generate the curl velocity field. Another MATLAB command `streamribbon` can be used to attach the curl information by the twisting of a ribbon attached to the streamline in which the width of the ribbon represents the magnitude of the curl at that point.

EXAMPLE 4.11. Consider the following vector field:

$$
\underline{\mathbf{v}} = \left(-\frac{y}{0.1 + \sqrt{x^2 + y^2}}\right) \underline{\boldsymbol{\delta}}_x + \left(\frac{x}{0.1 + \sqrt{x^2 + y^2}}\right) \underline{\boldsymbol{\delta}}_y + \underline{\boldsymbol{\delta}}_z
$$

then the curl is given by

$$
\nabla \times \underline{\mathbf{v}} = \left(20 \frac{1 + 5\sqrt{x^2 + y^2}}{\left(1 + 10\sqrt{x^2 + y^2}\right)^2}\right) \underline{\boldsymbol{\delta}}_z
$$

The vector field is shown in Figure 4.17, where the flow appears to follow a helical path. The curl is also a vector field, but for this particular example, it turns out to only have a z-component that is independent of z. Thus we can plot the curl as as function only of x and y, which is also shown in Figure 4.17. From the figure, one sees that the curl increases radially outward in any (x, y) plane. Also, note that none of the vectors located far from the z-axis are lying in a plane that is parallel to the (x, y) plane, and yet the curl at all points are directed in the positive z-direction. This shows that the curl is not necessarily perpendicular to the vector $\mathbf{v}$ at that point. This is because the curl is a differential operation on a vector field and not a cross product of two vectors.

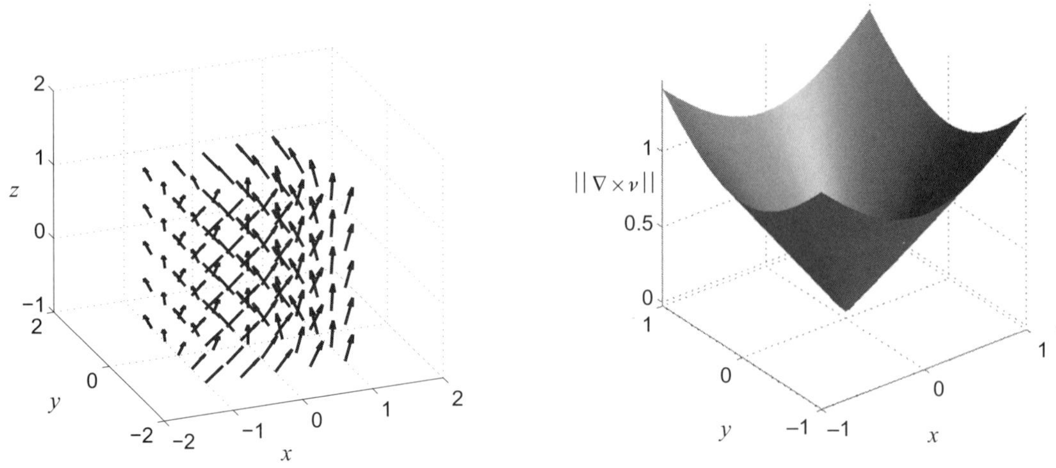

Figure 4.17. The plot of the vector field $\underline{\mathbf{v}}$ and $\|\nabla \times \underline{\mathbf{v}}\|$.

As shown in Example 4.11, the curl is not necessarily perpendicular to the vector field. However, there are situations in which the curl is perpendicular to the vector field. If $\underline{\mathbf{v}} \cdot (\nabla \times \underline{\mathbf{v}}) = 0$, then $\underline{\mathbf{v}}$ is known as a **complex lamellar vector field**. An example of a complex lamellar vector field is

$$\underline{\mathbf{v}} = v_x\,(x, y)\,\underline{\boldsymbol{\delta}}_x + v_y\,(x, y)\,\underline{\boldsymbol{\delta}}_y + 0\underline{\boldsymbol{\delta}}_z$$

Conversely, there are also vector fields whose curls are always parallel to the vector field. If $\underline{\mathbf{v}} \times (\nabla \times \underline{\mathbf{v}}) = 0$, then $\underline{\mathbf{v}}$ is known as a **Beltrami vector field**. An example of a Beltrami vector field is

$$\underline{\mathbf{v}} = \Big(\alpha \sin z + \beta \cos z\Big)\,\underline{\boldsymbol{\delta}}_x + \Big(-\beta \sin z + \alpha \cos z\Big)\,\underline{\boldsymbol{\delta}}_y + 0\underline{\boldsymbol{\delta}}_z$$

where α and β are constant.

4.7.4 Laplacian

As we discussed earlier, the gradient of a scalar field $\psi(x, y, z)$ will result in a vector field $\nabla\psi$. By taking the divergence of $\nabla\psi$, we obtain the **Laplacian** of $\psi(x, y, z)$,

$$\textbf{Laplacian}(\psi(x, y, z)) = \nabla \cdot (\nabla\psi) \tag{4.45}$$

$$= \frac{\partial^2 \psi}{\partial x^2} + \frac{\partial^2 \psi}{\partial y^2} + \frac{\partial^2 \psi}{\partial z^2} \tag{4.46}$$

The Laplacian operator is often given the shorthand notation of ∇^2 to mean

$$\nabla^2 = \nabla \cdot \nabla \tag{4.47}$$

Thus the Laplacian of ψ is commonly denoted as $\nabla^2\psi$.

As a simple example, consider the heat conduction process of a 3D solid; we can take the energy balance to be given by

Rate of Change in Energy in Control Volume

= Negative Divergence of Energy Flow

The energy flow following Fourier's law is proportional to the negative gradient of temperature, that is, energy flow = $-k\nabla T$, or (divergence of energy flow) = $\nabla \cdot (-k\nabla T)$. The rate of energy is given by $d\left(\rho C_p T\right)/dt$. With constant heat conductivity, density, and heat capacity, the energy balance reduces to

$$\frac{\partial T}{\partial t} = \alpha \nabla^2 T$$

where $\alpha = k/(\rho Cp)$, k is thermal conductivity, ρ is density, and C_p is the heat capacity. (See Section 5.4.1 for a more detailed derivation that involves the divergence theorem.)

EXAMPLE 4.12. Consider the following 2D temperature scalar field

$$T\left(x, y\right) = e^{-\left(x^2+y^2\right)}$$

then

$$\nabla T = -2T\left(x\underline{\delta}_x + y\underline{\delta}_y\right) \qquad \text{and} \qquad \nabla^2 T = 4\left(r^2 - 1\right)T$$

where $r = \sqrt{x^2 + y^2}$. Thus the Laplacian of T is a negative factor of T when $r < 1$, zero when $r = 1$, and a positive factor of T when $r > 1$.

The scalar field, gradient field, and Laplacian field are shown in Figure 4.18. One can see that the gradient field is directed toward the center and zero at the origin. Because the gradients are all pointing toward the origin around the origin, a negative divergence is expected there, and this can be seen for the the Laplacian.

Scalar functions ψ that satisfy $\nabla^2\psi = 0$ in a region S are known as **harmonic functions**. Harmonic functions in a closed region S have special direct relationships with the values of ψ at the boundaries of S, and these relationships can be derived using the divergence theorem and Green's identities, which are topics in the next chapter.

The Laplacian operator can also be applied to vectors and tensors, for instance $\nabla^2\underline{\mathbf{v}} = \nabla \cdot \nabla\underline{\mathbf{v}}$. However, we need to first define $\nabla\underline{\mathbf{v}}$, which is the gradient of a vector. This is discussed in the next section.

4.7.5 Other Differential Operators

In addition to the four major differential operators just discussed, we list some additional differential operators:

1. **Gradient-Vector Dyad**

   When the gradient operator is combined with a vector $\underline{\mathbf{v}}$, they form a tensor operator,

$$\nabla\underline{\mathbf{v}} = \sum_{k=x,y,z} \underline{\delta}_k \frac{\partial}{\partial k}\left(\sum_m v_m \underline{\delta}_m\right) = \left(\sum_{k,m=x,y,z} \frac{\partial v_m}{\partial k}\underline{\delta}_k\underline{\delta}_m\right) \tag{4.48}$$

   with scalar weights $\partial v_m/\partial k$ for the $(\underline{\delta}_k\underline{\delta}_m)$-component.

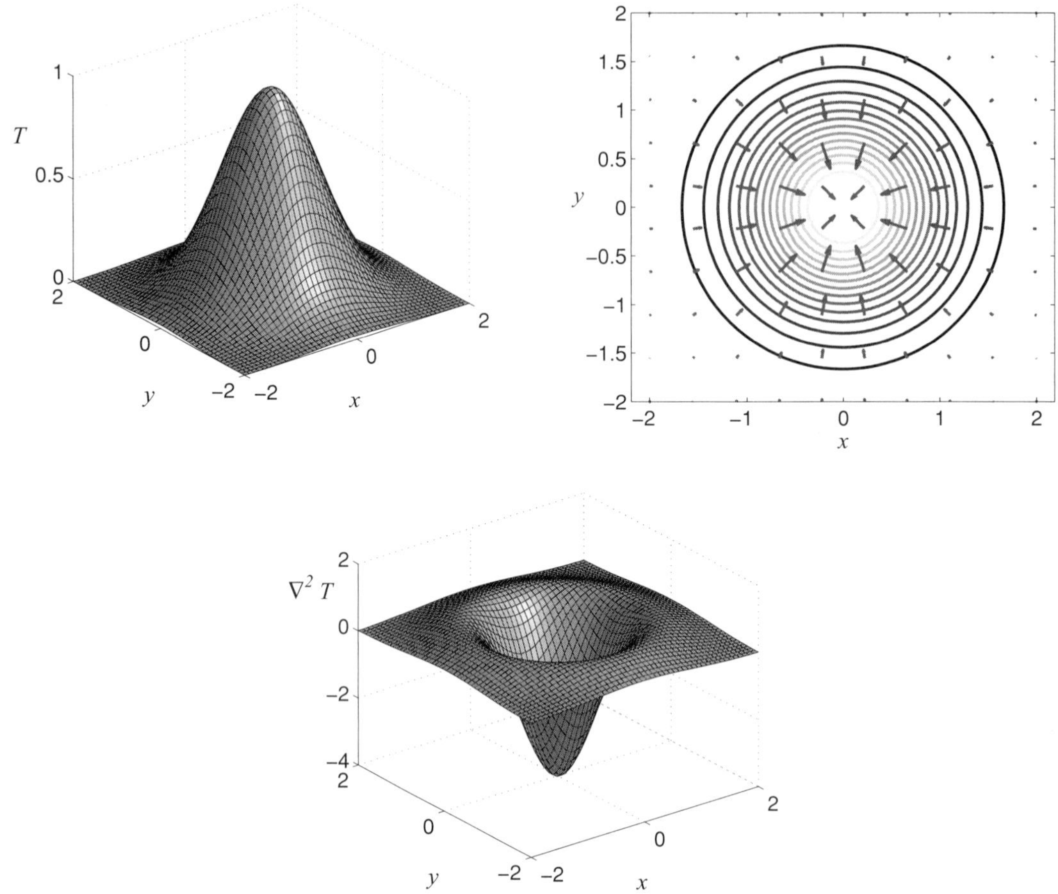

Figure 4.18. The plot of the temperature field, $T\ (x, y)$, gradient vector field ∇T, and Laplacian field $\nabla^2 T$.

If one were to write out the tensor field $\nabla\underline{\mathbf{v}}$ as a matrix, we have

$$[\nabla\underline{\mathbf{v}}] = \begin{pmatrix} \dfrac{\partial v_x}{\partial x} & \dfrac{\partial v_y}{\partial x} & \dfrac{\partial v_z}{\partial x} \\[2ex] \dfrac{\partial v_x}{\partial y} & \dfrac{\partial v_y}{\partial y} & \dfrac{\partial v_z}{\partial y} \\[2ex] \dfrac{\partial v_x}{\partial z} & \dfrac{\partial v_y}{\partial z} & \dfrac{\partial v_z}{\partial z} \end{pmatrix}$$

which is a transposed Jacobian matrix.

The inner product of another vector $\underline{\mathbf{w}}$ on $\nabla\underline{\mathbf{v}}$ will yield

$$\underline{\mathbf{w}} \cdot \nabla\underline{\mathbf{v}} = \sum_{\ell=x,y,z} w_\ell \underline{\boldsymbol{\delta}}_\ell \cdot \left(\sum_{k,m=x,y,z} \frac{\partial v_m}{\partial k} \underline{\boldsymbol{\delta}}_k \underline{\boldsymbol{\delta}}_m \right) = \sum_{m,k=x,y,z} \left(w_k \frac{\partial v_m}{\partial k} \right) \underline{\boldsymbol{\delta}}_m \quad (4.49)$$

2. **Dot-Grad**
We define the **dot-grad** operator based on $\underline{\mathbf{w}}$ as

$$\underline{\mathbf{w}} \cdot \nabla = w_x \frac{\partial}{\partial x} + w_y \frac{\partial}{\partial y} + w_z \frac{\partial}{\partial z} \tag{4.50}$$

Then for scalar field ψ and vector field $\underline{\mathbf{v}}$,

$$(\underline{\mathbf{w}} \cdot \nabla)\, \psi = w_x \frac{\partial \psi}{\partial x} + w_y \frac{\partial \psi}{\partial y} + w_z \frac{\partial \psi}{\partial z}$$

and

$$(\underline{\mathbf{w}} \cdot \nabla)\, \underline{\mathbf{v}} = w_x \frac{\partial \underline{\mathbf{v}}}{\partial x} + w_y \frac{\partial \underline{\mathbf{v}}}{\partial y} + w_z \frac{\partial \underline{\mathbf{v}}}{\partial z} = \sum_{k,m=x,y,z} \left(w_k \frac{\partial v_m}{\partial k} \right) \underline{\boldsymbol{\delta}}_m$$

where the last equation is the same as that in (4.49).

3. **Cross-Grad**
The **cross-grad** operator based on $\underline{\mathbf{w}}$ is defined as

$$\underline{\mathbf{w}} \times \nabla = \underline{\boldsymbol{\delta}}_x \left(w_y \frac{\partial}{\partial z} - w_z \frac{\partial}{\partial y} \right) + \underline{\boldsymbol{\delta}}_y \left(w_z \frac{\partial}{\partial x} - w_x \frac{\partial}{\partial z} \right) + \underline{\boldsymbol{\delta}}_z \left(w_x \frac{\partial}{\partial y} - w_y \frac{\partial}{\partial x} \right) \tag{4.51}$$

for which the mnemonic, applicable only for the rectangular coordinate system, is given by

$$\underline{\mathbf{w}} \times \nabla = \mathbf{det} \begin{pmatrix} \underline{\boldsymbol{\delta}}_x & \underline{\boldsymbol{\delta}}_y & \underline{\boldsymbol{\delta}}_z \\ w_x & w_y & w_z \\ \dfrac{\partial}{\partial x} & \dfrac{\partial}{\partial y} & \dfrac{\partial}{\partial z} \end{pmatrix}$$

Using the cross-grad operator on a scalar field ψ will yield the following identity,

$$(\underline{\mathbf{w}} \times \nabla)\, \psi = \underline{\mathbf{w}} \times (\nabla \psi)$$

Another important identity is obtained when we extend the divergence operation, based on grad operators, to the cross-grad operators,

$$(\underline{\mathbf{w}} \times \nabla) \cdot \underline{\mathbf{v}} = \underline{\mathbf{w}} \cdot (\nabla \times \underline{\mathbf{v}}) \tag{4.52}$$

4. **Laplacian of vector fields**
The Laplacian was defined earlier as the divergence of a gradient. We can apply the same operation to vectors. Thus, taking the divergence of $\nabla \underline{\mathbf{v}}$,

$$\begin{aligned} \nabla^2 \underline{\mathbf{v}} &= \nabla \cdot (\nabla \underline{\mathbf{v}}) = \sum_{k=x,y,z} \underline{\boldsymbol{\delta}}_k \frac{\partial}{\partial k} \cdot \left(\sum_{\ell,m=x,y,z} \frac{\partial v_m}{\partial \ell}\, \underline{\boldsymbol{\delta}}_\ell\, \underline{\boldsymbol{\delta}}_m \right) \\ &= \sum_{k,m=x,y,z} \frac{\partial^2 v_m}{\partial k^2} \underline{\boldsymbol{\delta}}_m \\ &= \sum_{m=x,y,z} \left(\nabla^2 v_m \right) \underline{\boldsymbol{\delta}}_m \end{aligned} \tag{4.53}$$

Note that the formula given in (4.53) is only for the rectangular coordinates. For other coordinate systems, one could start from the definition of the Laplacian as a divergence of a gradient.

4.7.6 Vector Differential Identities

A list of some important identities involving the gradient, divergence, curl, and Laplacian is given next. These identities can be proved by applying the definitions directly to both sides of the equations.

$$\nabla\left(\psi\omega\right) = \psi\nabla\omega + \omega\nabla\psi \tag{4.54}$$

$$\nabla\cdot\psi\underline{\mathbf{v}} = \psi\nabla\cdot\underline{\mathbf{v}} + \underline{\mathbf{v}}\cdot\nabla\psi \tag{4.55}$$

$$\nabla\times\psi\underline{\mathbf{v}} = \psi\nabla\times\underline{\mathbf{v}} + (\nabla\psi)\times\underline{\mathbf{v}} \tag{4.56}$$

$$\nabla\cdot(\underline{\mathbf{u}}\times\underline{\mathbf{v}}) = \underline{\mathbf{v}}\cdot\nabla\times\underline{\mathbf{u}} - \underline{\mathbf{u}}\cdot\nabla\times\underline{\mathbf{v}} \tag{4.57}$$

$$\begin{aligned}\nabla\times(\underline{\mathbf{u}}\times\underline{\mathbf{v}}) &= \underline{\mathbf{v}}\cdot\nabla\underline{\mathbf{u}} - \underline{\mathbf{u}}\cdot\nabla\underline{\mathbf{v}}\\ &\quad + \underline{\mathbf{u}}\nabla\cdot\underline{\mathbf{v}} - \underline{\mathbf{v}}\nabla\cdot\underline{\mathbf{u}}\end{aligned} \tag{4.58}$$

$$\begin{aligned}\nabla(\underline{\mathbf{u}}\cdot\underline{\mathbf{v}}) &= \underline{\mathbf{u}}\cdot\nabla\underline{\mathbf{v}} + \underline{\mathbf{v}}\cdot\nabla\underline{\mathbf{u}}\\ &\quad + \underline{\mathbf{u}}\times(\nabla\times\underline{\mathbf{v}}) + \underline{\mathbf{v}}\times(\nabla\times\underline{\mathbf{u}})\end{aligned} \tag{4.59}$$

$$\nabla\times\nabla\psi = \underline{0} \tag{4.60}$$

$$\nabla\cdot\nabla\times\underline{\mathbf{v}} = 0 \tag{4.61}$$

$$\nabla\times(\nabla\times\underline{\mathbf{v}}) = \nabla(\nabla\cdot\underline{\mathbf{v}}) - \nabla^2\underline{\mathbf{v}} \tag{4.62}$$

$$\underline{\mathbf{v}}\cdot\nabla\underline{\mathbf{v}} = \frac{1}{2}\nabla\left(\underline{\mathbf{v}}\cdot\underline{\mathbf{v}}\right) - \underline{\mathbf{v}}\times\left(\nabla\times\underline{\mathbf{v}}\right) \tag{4.63}$$

The first three identities involve operations on scalar products, including product of two scalar fields and the scalar product of a vector field. They are direct results of implementing the properties of derivatives of products. Note that in (4.56), the order of $\nabla\psi\times\underline{\mathbf{v}}$ is crucial.

EXAMPLE 4.13. Using the divergence theorem, one can derive the conservation of mass (see (5.24)) to be

$$\frac{\partial\rho}{\partial t} + \nabla\cdot(\rho\underline{\mathbf{v}}) = 0$$

where ρ and $\underline{\mathbf{v}}$ are the density field and velocity field, respectively. This can be rewritten in a more familiar form by using the identity (4.55) to replace the second term as follows:

$$\nabla\cdot(\rho\underline{\mathbf{v}}) = \rho\nabla\cdot\underline{\mathbf{v}} + \underline{\mathbf{v}}\cdot\nabla\rho$$

Then, the conservation equation becomes

$$\frac{\partial \rho}{\partial t} + \underline{\mathbf{v}} \cdot \nabla \rho + \rho \nabla \cdot \underline{\mathbf{v}} = 0 \tag{4.64}$$

or by defining another operator $D/Dt(\bullet)$ known as the **substantial rate of change operator** (or **substantial time derivative**) defined by

$$\frac{D}{Dt}(\bullet) = \left(\frac{\partial}{\partial t} + \underline{\mathbf{v}} \cdot \nabla \right)(\bullet) \tag{4.65}$$

the continuity equation (4.64) becomes

$$\frac{D\rho}{Dt} + \rho \nabla \cdot \underline{\mathbf{v}} = 0 \tag{4.66}$$

For the special case of incompressible fluid (i.e., ρ is constant), (4.66) reduces to $\nabla \cdot \underline{\mathbf{v}} = 0$.

Equation (4.57) shows that with the gradient operator, the usual cyclic-permutation properties of triple product no longer apply, except when either $\underline{\mathbf{u}}$ or $\underline{\mathbf{v}}$ is constant. Similarly, for (4.58), the identities for the usual triple vector products no longer apply as well.

The identity in (4.59) is surprisingly complicated. On the left-hand side is the gradient of a dot product. However, the right-hand side identity includes cross products of curls plus dot products of vectors with gradient-vector dyads. Note that (4.63) is a consequence of (4.59) with $\underline{\mathbf{v}} = \underline{\mathbf{u}}$.

Equation (4.60) states that gradient vector fields are irrotational. However, (4.61) states that curls have zero divergence, which means that curl fields (e.g., vorticity fields) are solenoidal. Both these identities are very useful in solving vector differential equations. For instance, in Navier-Stokes equations, a pressure gradient appears in the momentum balances. By simply taking the curl of the equation, the dependence on pressure disappears because of identity (4.60). Likewise, if one needs to remove curls in an equation, one simply needs to take the divergence of that equation.

The identity given by (4.62) relates the curl of a curl with terms involving the Laplacian of a vector field, as well as the gradient of a divergence. In some cases, this formula is used to find the Laplacian of a vector field represented in other coordinate systems.

Equation (4.63) can be used to yield an alternative definition for a Beltrami vector field; that is, with $\underline{\mathbf{v}} \times (\nabla \times \underline{\mathbf{v}}) = 0$, a vector field is a Beltrami field if and only if

$$\underline{\mathbf{v}} \cdot \nabla \underline{\mathbf{v}} = \frac{1}{2} \nabla (\underline{\mathbf{v}} \cdot \underline{\mathbf{v}})$$

Finally, equation (4.63) is also very useful because the term $\underline{\mathbf{v}} \cdot \nabla \underline{\mathbf{v}}$ appears in momentum balance equations and is known as the **inertial terms**. Thus (4.63) is often used to introduce the role of vorticity, $\nabla \times \underline{\mathbf{v}}$, in the equations of fluid dynamics.

The proofs of some of the identities can be lengthy but more or less straightforward. The following example shows the proof of the identity given in (4.59).

EXAMPLE 4.14. Let us prove the identity given in (4.59). We begin with an expansion of the left-hand side of the identity,

$$\nabla\left(\underline{\mathbf{u}}\cdot\underline{\mathbf{v}}\right) = \sum_{i=x,y,z} \underline{\boldsymbol{\delta}}_i \frac{\partial}{\partial i}\left(\sum_{k=x,y,z} u_k v_k\right) = \sum_{i,k=x,y,z} \underline{\boldsymbol{\delta}}_i \left(v_k \frac{\partial u_k}{\partial i} + u_k \frac{\partial v_k}{\partial i}\right) \tag{4.67}$$

Next, we show each term in the right-hand side of the identity,

$$\begin{aligned} \underline{\mathbf{u}} \times \left(\nabla \times \underline{\mathbf{v}}\right) &= \sum_{j=x,y,z} \underline{\boldsymbol{\delta}}_j u_j \times \left(\sum_{m,k,\ell=x,y,z} \epsilon_{\ell k m} \frac{\partial v_k}{\partial \ell} \underline{\boldsymbol{\delta}}_m\right) \\ &= \sum_{i,j,m,k,\ell=x,y,z} \underline{\boldsymbol{\delta}}_i\, \epsilon_{ijm}\epsilon_{\ell k m} u_j \frac{\partial v_k}{\partial \ell} = \sum_{i,k} \underline{\boldsymbol{\delta}}_i \left(u_k \frac{\partial v_k}{\partial i} - u_k \frac{\partial v_i}{\partial k}\right) \end{aligned} \tag{4.68}$$

$$\left(\underline{\mathbf{u}}\cdot\nabla\right)\underline{\mathbf{v}} = \left(\sum_{k=x,y,z} u_k \frac{\partial}{\partial k}\right)\left(\sum_{i=x,y,z} v_i \underline{\boldsymbol{\delta}}_i\right) = \sum_{i,k=x,y,z} \underline{\boldsymbol{\delta}}_i \left(u_k \frac{\partial v_i}{\partial k}\right) \tag{4.69}$$

Combining (4.68) and (4.69), we get

$$\underline{\mathbf{u}} \times \left(\nabla \times \underline{\mathbf{v}}\right) + \left(\underline{\mathbf{u}}\cdot\nabla\right)\underline{\mathbf{v}} = \sum_{i,k} \underline{\boldsymbol{\delta}}_i \left(u_k \frac{\partial v_k}{\partial i}\right) \tag{4.70}$$

Reversing the role of $\underline{\mathbf{u}}$ and $\underline{\mathbf{v}}$, we have

$$\underline{\mathbf{v}} \times \left(\nabla \times \underline{\mathbf{u}}\right) + \left(\underline{\mathbf{v}}\cdot\nabla\right)\underline{\mathbf{u}} = \sum_{i,k} \underline{\boldsymbol{\delta}}_i \left(v_k \frac{\partial u_k}{\partial i}\right) \tag{4.71}$$

Finally, adding (4.70) and (4.71), we arrive at a sum that is equal to (4.67).

4.8 Curvilinear Coordinate System: Cylindrical and Spherical

In this section, we discuss the cylindrical and spherical coordinate systems. The properties and differential operators are summarized in tabular form. The geometric derivations, with the aid of matrix methods, can be found in Sections D.2 and D.3 in the appendix.

4.8.1 Cylindrical Coordinate System

The **cylindrical coordinate system** is a coordinate system shown in Figure 4.19 and defined by the following invertible transformations:

$$\begin{aligned} r &= \sqrt{x^2+y^2} \\ \theta &= \tan^{-1}(y/x) \\ z &= z \end{aligned} \qquad \leftrightarrow \qquad \begin{aligned} x &= r\cos\theta \\ y &= r\sin\theta \\ z &= z \end{aligned} \tag{4.72}$$

A summary of the relationships between cylindrical and rectangular coordinates is given in Table 4.5. The derivation of the items in this table can be done via geometric arguments. Details of these derivations, aided by matrix methods, are given in Section D.2.

Figure 4.19. Cylindrical coordinate system.

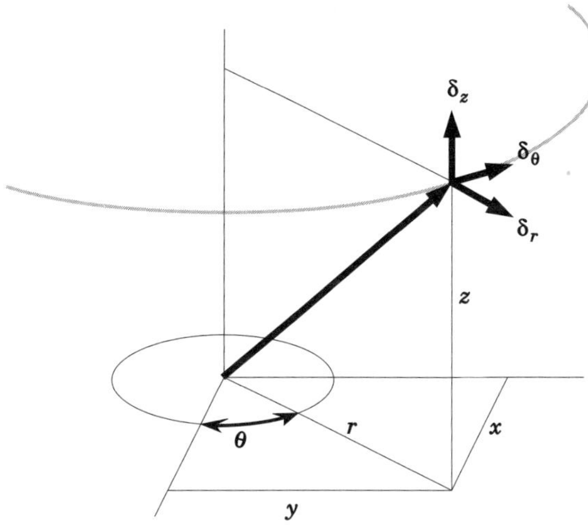

The gradients, curls, and Laplacian in cylindrical coordinates have to be evaluated using the definition of the various operators and the distributive properties of dot products and cross products. To illustrate, the divergence formula in cylindrical coordinates can be obtained as follows:

$$\begin{aligned}
\nabla\cdot\underline{\mathbf{v}} &= \left(\underline{\boldsymbol{\delta}}_r\frac{\partial}{\partial r}+\frac{1}{r}\underline{\boldsymbol{\delta}}_\theta\frac{\partial}{\partial\theta}+\underline{\boldsymbol{\delta}}_z\frac{\partial}{\partial z}\right)\cdot\left(v_r\underline{\boldsymbol{\delta}}_r+v_\theta\underline{\boldsymbol{\delta}}_\theta+v_z\underline{\boldsymbol{\delta}}_z\right)\\
&= \left(\frac{\partial v_r}{\partial r}+\frac{\partial v_z}{\partial z}\right)+\left(\frac{1}{r}\underline{\boldsymbol{\delta}}_\theta\frac{\partial}{\partial\theta}\cdot\left(v_r\underline{\boldsymbol{\delta}}_r+v_\theta\underline{\boldsymbol{\delta}}_\theta\right)\right)\\
&= \left(\frac{\partial v_r}{\partial r}+\frac{\partial v_z}{\partial z}\right)+\frac{1}{r}\underline{\boldsymbol{\delta}}_\theta\cdot\left(v_r\underline{\boldsymbol{\delta}}_\theta+\frac{\partial v_r}{\partial\theta}\underline{\boldsymbol{\delta}}_r-v_\theta\underline{\boldsymbol{\delta}}_r+\frac{\partial v_\theta}{\partial\theta}\underline{\boldsymbol{\delta}}_\theta\right)\\
&= \frac{\partial v_r}{\partial r}+\frac{1}{r}\frac{\partial v_\theta}{\partial\theta}+\frac{v_r}{r}+\frac{\partial v_z}{\partial z}
\end{aligned} \tag{4.73}$$

Likewise, for the curl,

$$\begin{aligned}
\nabla\times\underline{\mathbf{v}} &= \left(\underline{\boldsymbol{\delta}}_r\frac{\partial}{\partial r}+\frac{1}{r}\underline{\boldsymbol{\delta}}_\theta\frac{\partial}{\partial\theta}+\underline{\boldsymbol{\delta}}_z\frac{\partial}{\partial z}\right)\times\left(v_r\underline{\boldsymbol{\delta}}_r+v_\theta\underline{\boldsymbol{\delta}}_\theta+v_z\underline{\boldsymbol{\delta}}_z\right)\\
&= \left(\frac{\partial v_\theta}{\partial r}\underline{\boldsymbol{\delta}}_z-\frac{\partial v_z}{\partial r}\underline{\boldsymbol{\delta}}_\theta\right)+\left(\frac{\partial v_r}{\partial z}\underline{\boldsymbol{\delta}}_\theta-\frac{\partial v_\theta}{\partial z}\underline{\boldsymbol{\delta}}_r\right)\\
&\qquad+\frac{1}{r}\underline{\boldsymbol{\delta}}_\theta\times\left(\frac{\partial v_r}{\partial\theta}\underline{\boldsymbol{\delta}}_r+v_r\underline{\boldsymbol{\delta}}_\theta+\frac{\partial v_\theta}{\partial\theta}\underline{\boldsymbol{\delta}}_\theta-v_\theta\underline{\boldsymbol{\delta}}_r+\frac{\partial v_z}{\partial\theta}\underline{\boldsymbol{\delta}}_z\right)\\
&= \left(\frac{\partial v_\theta}{\partial r}\underline{\boldsymbol{\delta}}_z-\frac{\partial v_z}{\partial r}\underline{\boldsymbol{\delta}}_\theta\right)+\left(\frac{\partial v_r}{\partial z}\underline{\boldsymbol{\delta}}_\theta-\frac{\partial v_\theta}{\partial z}\underline{\boldsymbol{\delta}}_r\right)\\
&\qquad+\frac{1}{r}\left(-\frac{\partial v_r}{\partial\theta}\underline{\boldsymbol{\delta}}_z+v_\theta\underline{\boldsymbol{\delta}}_z+\frac{\partial v_\theta}{\partial\theta}\underline{\boldsymbol{\delta}}_r\right)\\
&= \left(\frac{1}{r}\frac{\partial v_z}{\partial\theta}-\frac{\partial v_\theta}{\partial z}\right)\underline{\boldsymbol{\delta}}_r+\left(-\frac{\partial v_z}{\partial r}+\frac{\partial v_r}{\partial z}\right)\underline{\boldsymbol{\delta}}_\theta+\frac{1}{r}\left(\frac{\partial(rv_\theta)}{\partial r}-\frac{\partial v_r}{\partial\theta}\right)\underline{\boldsymbol{\delta}}_z
\end{aligned} \tag{4.74}$$

Table 4.5. *Relationship between rectangular and cylindrical coordinates*

	Rectangular	Cylindrical
1	Unit Vectors	
	$\underline{\boldsymbol{\delta}}_x = \cos\theta\underline{\boldsymbol{\delta}}_r - \sin\theta\underline{\boldsymbol{\delta}}_\theta$ $\underline{\boldsymbol{\delta}}_y = \sin\theta\underline{\boldsymbol{\delta}}_r + \cos\theta\underline{\boldsymbol{\delta}}_\theta$ $\underline{\boldsymbol{\delta}}_z = \underline{\boldsymbol{\delta}}_z$	$\underline{\boldsymbol{\delta}}_r = \cos\theta\underline{\boldsymbol{\delta}}_x + \sin\theta\underline{\boldsymbol{\delta}}_y$ $\underline{\boldsymbol{\delta}}_\theta = -\sin\theta\underline{\boldsymbol{\delta}}_x + \cos\theta\underline{\boldsymbol{\delta}}_y$ $\underline{\boldsymbol{\delta}}_z = \underline{\boldsymbol{\delta}}_z$
2	Vector Components	
	$\underline{\mathbf{v}} = v_x\underline{\boldsymbol{\delta}}_x + v_y\underline{\boldsymbol{\delta}}_y + v_z\underline{\boldsymbol{\delta}}_z$	$\underline{\mathbf{v}} = v_r\underline{\boldsymbol{\delta}}_r + v_\theta\underline{\boldsymbol{\delta}}_\theta + v_z\underline{\boldsymbol{\delta}}_z$
	$v_x = v_r\cos\theta - v_\theta\sin\theta$ $v_y = v_r\sin\theta + v_\theta\cos\theta$ $v_z = v_z$	$v_r = v_x\cos\theta + v_y\sin\theta$ $v_\theta = -v_x\sin\theta + v_y\cos\theta$ $v_z = v_z$
3	Partial Differential Operators	
	$\frac{\partial}{\partial x} = \cos\theta\frac{\partial}{\partial r} - \left(\frac{\sin\theta}{r}\right)\frac{\partial}{\partial\theta}$ $\frac{\partial}{\partial y} = \sin\theta\frac{\partial}{\partial r} + \left(\frac{\cos\theta}{r}\right)\frac{\partial}{\partial\theta}$ $\frac{\partial}{\partial z} = \frac{\partial}{\partial z}$	$\frac{\partial}{\partial r} = \cos\theta\frac{\partial}{\partial x} + \sin\theta\frac{\partial}{\partial y}$ $\frac{\partial}{\partial\theta} = -r\sin\theta\frac{\partial}{\partial x} + r\cos\theta\frac{\partial}{\partial y}$ $\frac{\partial}{\partial z} = \frac{\partial}{\partial z}$
4	Gradient Operators	
	$\nabla = \underline{\boldsymbol{\delta}}_x\frac{\partial}{\partial x} + \underline{\boldsymbol{\delta}}_y\frac{\partial}{\partial y} + \underline{\boldsymbol{\delta}}_z\frac{\partial}{\partial z}$	$\nabla = \underline{\boldsymbol{\delta}}_r\frac{\partial}{\partial r} + \underline{\boldsymbol{\delta}}_\theta\frac{1}{r}\frac{\partial}{\partial\theta} + \underline{\boldsymbol{\delta}}_z\frac{\partial}{\partial z}$
5	Derivatives of Unit Vectors	
	$\frac{\partial\underline{\boldsymbol{\delta}}_k}{\partial m} = 0$	$\frac{\partial\underline{\boldsymbol{\delta}}_r}{\partial\theta} = \underline{\boldsymbol{\delta}}_\theta \;;\; \frac{\partial\underline{\boldsymbol{\delta}}_\theta}{\partial\theta} = -\underline{\boldsymbol{\delta}}_r$
	for $k, m = x, y, z$	zero for all other cases

EXAMPLE 4.15. A fluid flowing through a pipe of radius R attains a steady-state velocity of a Poiselle flow given by

$$\underline{\mathbf{v}}(r,\theta,z) = v_{\max}\left(1 - \frac{r^2}{R^2}\right)\underline{\boldsymbol{\delta}}_z$$

which is a paraboloid-shaped velocity profile that is symmetric along the z-axis (flow direction), as shown in Figure 4.20. The divergence of $\underline{\mathbf{v}}$ is zero; that is, it is a solenoidal field. The curl field is given by

$$\nabla \times \underline{\mathbf{v}} = 2v_{\max}\frac{r}{R^2}\underline{\boldsymbol{\delta}}_\theta$$

The curl field is also given in Figure 4.20, which shows the curl field varying linearly with increasing radius. This means that very small particles near $r = R$ would experience the maximum curl due to the velocity field around it and zero curl at the center.

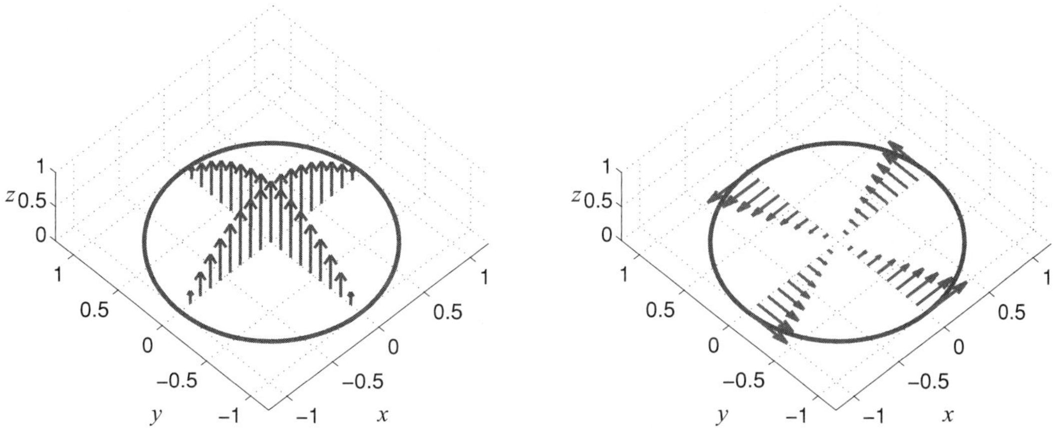

Figure 4.20. The Poiselle velocity field at $z = 0$ (left plot) and the corresponding curl field (right plot).

4.8.2 Spherical Coordinate System

The **spherical coordinate system** is a coordinate system as shown in Figure 4.21 and defined by the following invertible transformations:

$$\begin{array}{rclcrcl} r & = & \sqrt{x^2+y^2+z^2} & & x & = & r\sin\theta\cos\phi \\ \theta & = & \tan^{-1}(\sqrt{x^2+y^2}/z) & \leftrightarrow & y & = & r\sin\theta\sin\phi \\ \phi & = & \tan^{-1}(y/x) & & z & = & r\cos\theta \end{array} \tag{4.75}$$

A summary of important relationships between the rectangular and spherical coordinate systems is given in Table 4.6. In the table, as well as in several places in this chapter, we use the shorthand notation for sines and cosines (i.e., $s_\theta = \sin\theta$, $c_\theta = \cos\theta$, $s_\phi = \sin\phi$, and $c_\phi = \cos\phi$). The derivation of the items in this table can be done via geometric arguments. Details of these derivations, aided by matrix methods, are given in Section D.3.

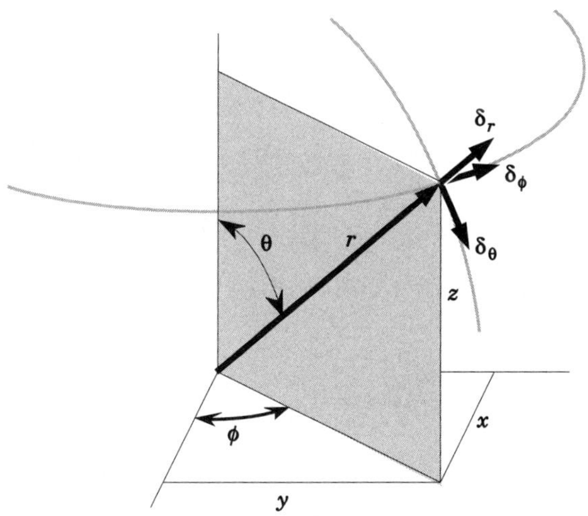

Figure 4.21. The spherical coordinate system.

Table 4.6. *Relationship between rectangular and spherical coordinates*

	Rectangular	Spherical
1	**Unit Vectors**	
	$\underline{\boldsymbol{\delta}}_x = s_\theta c_\phi \underline{\boldsymbol{\delta}}_r + c_\theta c_\phi \underline{\boldsymbol{\delta}}_\theta - s_\phi \underline{\boldsymbol{\delta}}_\phi$	$\underline{\boldsymbol{\delta}}_r = s_\theta c_\phi \underline{\boldsymbol{\delta}}_x + s_\theta s_\phi \underline{\boldsymbol{\delta}}_y + c_\theta \underline{\boldsymbol{\delta}}_z$
	$\underline{\boldsymbol{\delta}}_y = s_\theta s_\phi \underline{\boldsymbol{\delta}}_r + c_\theta s_\phi \underline{\boldsymbol{\delta}}_\theta + c_\phi \underline{\boldsymbol{\delta}}_\phi$	$\underline{\boldsymbol{\delta}}_\theta = c_\theta c_\phi \underline{\boldsymbol{\delta}}_x + c_\theta s_\phi \underline{\boldsymbol{\delta}}_y - s_\theta \underline{\boldsymbol{\delta}}_z$
	$\underline{\boldsymbol{\delta}}_z = c_\theta \underline{\boldsymbol{\delta}}_r - s_\theta \underline{\boldsymbol{\delta}}_\theta$	$\underline{\boldsymbol{\delta}}_\phi = - s_\phi \underline{\boldsymbol{\delta}}_x + c_\phi \underline{\boldsymbol{\delta}}_y$
2	**Vector Components**	
	$\underline{\mathbf{v}} = v_x \underline{\boldsymbol{\delta}}_x + v_y \underline{\boldsymbol{\delta}}_y + v_z \underline{\boldsymbol{\delta}}_z$	$\underline{\mathbf{v}} = v_r \underline{\boldsymbol{\delta}}_r + v_\theta \underline{\boldsymbol{\delta}}_\theta + v_\phi \underline{\boldsymbol{\delta}}_\phi$
	$v_x = s_\theta c_\phi v_r + c_\theta c_\phi v_\theta - s_\phi v_\phi$	$v_r = s_\theta c_\phi v_x + s_\theta s_\phi v_y + c_\theta v_z$
	$v_y = s_\theta s_\phi v_r + c_\theta s_\phi v_\theta + c_\phi v_\phi$	$v_\theta = c_\theta c_\phi v_x + c_\theta s_\phi v_y - s_\theta v_z$
	$v_z = c_\theta v_r - s_\theta v_\theta$	$v_\phi = - s_\phi v_x + c_\phi v_y$
3	**Partial Differential Operators**	
	$\frac{\partial}{\partial x} = s_\theta c_\phi \frac{\partial}{\partial r} + \left(\frac{c_\theta c_\phi}{r}\right)\frac{\partial}{\partial \theta} - \left(\frac{s_\phi}{r s_\theta}\right)\frac{\partial}{\partial \phi}$	$\frac{\partial}{\partial r} = s_\theta c_\phi \frac{\partial}{\partial x} + s_\theta s_\phi \frac{\partial}{\partial y} + c_\theta \frac{\partial}{\partial z}$
	$\frac{\partial}{\partial y} = s_\theta s_\phi \frac{\partial}{\partial r} + \left(\frac{c_\theta s_\phi}{r}\right)\frac{\partial}{\partial \theta} + \left(\frac{c_\phi}{r s_\theta}\right)\frac{\partial}{\partial \phi}$	$\frac{\partial}{\partial \theta} = r\, c_\theta c_\phi \frac{\partial}{\partial x} + r\, c_\theta s_\phi \frac{\partial}{\partial y} - r\, s_\theta \frac{\partial}{\partial z}$
	$\frac{\partial}{\partial z} = c_\theta \frac{\partial}{\partial r} - \left(\frac{s_\theta}{r}\right)\frac{\partial}{\partial \theta}$	$\frac{\partial}{\partial \phi} = -r\, s_\theta s_\phi \frac{\partial}{\partial x} + r\, s_\theta c_\phi \frac{\partial}{\partial y}$
4	**Gradient Operators**	
	$\nabla = \underline{\boldsymbol{\delta}}_x \frac{\partial}{\partial x} + \underline{\boldsymbol{\delta}}_y \frac{\partial}{\partial y} + \underline{\boldsymbol{\delta}}_z \frac{\partial}{\partial z}$	$\nabla = \underline{\boldsymbol{\delta}}_r \frac{\partial}{\partial r} + \underline{\boldsymbol{\delta}}_\theta \frac{1}{r}\frac{\partial}{\partial \theta} + \underline{\boldsymbol{\delta}}_\phi \frac{1}{r s_\theta}\frac{\partial}{\partial \phi}$
5	**Derivatives of Unit Vectors**	
	$\frac{\partial \underline{\boldsymbol{\delta}}_k}{\partial m} = 0$	$\frac{\partial \underline{\boldsymbol{\delta}}_r}{\partial \theta} = \underline{\boldsymbol{\delta}}_\theta \,;\; \frac{\partial \underline{\boldsymbol{\delta}}_\theta}{\partial \theta} = -\underline{\boldsymbol{\delta}}_r \,;\; \frac{\partial \underline{\boldsymbol{\delta}}_r}{\partial \phi} = \underline{\boldsymbol{\delta}}_\phi s_\theta$
		$\frac{\partial \underline{\boldsymbol{\delta}}_\theta}{\partial \phi} = \underline{\boldsymbol{\delta}}_\phi c_\theta \,;\; \frac{\partial \underline{\boldsymbol{\delta}}_\phi}{\partial \phi} = -\underline{\boldsymbol{\delta}}_r s_\theta - \underline{\boldsymbol{\delta}}_\theta c_\theta$
	for $k, m = x, y, z$	zero for all other cases

Using these operators, we can find the divergence and curl of a vector field to be

$$\nabla \cdot \underline{\mathbf{v}} = \left(\frac{\partial v_r}{\partial r} + \frac{1}{r}\frac{\partial v_\theta}{\partial \theta} + \frac{1}{r \sin\theta}\frac{\partial v_\phi}{\partial \phi}\right) + \left(2\frac{v_r}{r} + \frac{v_\theta \cos\theta}{r \sin\theta}\right) \tag{4.76}$$

and

$$\begin{aligned} \nabla \times \underline{\mathbf{v}} &= \left(\frac{1}{r}\frac{\partial v_\phi}{\partial \theta} - \frac{1}{r\sin\theta}\frac{\partial v_\theta}{\partial \phi} + \frac{v_\phi \cos\theta}{r \sin\theta}\right)\underline{\boldsymbol{\delta}}_r + \left(\frac{1}{r \sin\theta}\frac{\partial v_r}{\partial \phi} - \frac{\partial v_\phi}{\partial r} - \frac{v_\phi}{r}\right)\underline{\boldsymbol{\delta}}_\theta \\ &+ \left(\frac{\partial v_\theta}{\partial r} - \frac{1}{r}\frac{\partial v_r}{\partial \theta} + \frac{v_\theta}{r}\right)\underline{\boldsymbol{\delta}}_\phi \end{aligned} \tag{4.77}$$

EXAMPLE 4.16. Let the $\underline{\mathbf{v}}$ be a vector field that is a function of the position vector $\underline{\vec{\mathbf{r}}}$, away from the origin, given by

$$\underline{\mathbf{v}} = \frac{\alpha}{(\underline{\vec{\mathbf{r}}} \cdot \underline{\vec{\mathbf{r}}})^{n/2}} \underline{\vec{\mathbf{r}}}$$

Instead of using the rectangular coordinate system, where $\underline{\vec{\mathbf{r}}} = x\underline{\boldsymbol{\delta}}_x + y\underline{\boldsymbol{\delta}}_y + z\underline{\boldsymbol{\delta}}_z$, the spherical coordinates become more convenient because $\underline{\vec{\mathbf{r}}} = r\underline{\boldsymbol{\delta}}_r$, which results in

$$\underline{\mathbf{v}} = \frac{\alpha}{r^{n-1}} \underline{\boldsymbol{\delta}}_r$$

With $v_\theta = v_\phi = 0$ and $v_r = v_r(r)$, the curl can be evaluated using (4.77) to yield a zero vector

$$\nabla \times \underline{\mathbf{v}} = 0$$

that is, $\underline{\mathbf{v}}$ is irrotational. Conversely, the divergence of $\underline{\mathbf{v}}$ can be evaluated using (4.76) to yield

$$\nabla \cdot \underline{\mathbf{v}} = \frac{\partial v_r}{\partial r} + 2\frac{v_r}{r} = \alpha \frac{3-n}{r^n}$$

and $\underline{\mathbf{v}}$ becomes solenoidal for $n = 3$.

4.9 Orthogonal Curvilinear Coordinates

4.9.1 Definitions and Notations

We now generalize the results of cylindrical and spherical coordinate systems to other orthogonal curvilinear coordinates. Let the new coordinates be given by (a, b, c), that is,

$$\begin{aligned} x &= x(a,b,c) \\ y &= y(a,b,c) \\ z &= z(a,b,c) \end{aligned} \quad \leftrightarrow \quad \begin{aligned} a &= a(x,y,z) \\ b &= b(x,y,z) \\ c &= c(x,y,z) \end{aligned} \tag{4.78}$$

Then, according to the implicit value theorem, the transformation between the two coordinates will exist if the Jacobian matrix given by

$$J_{(x,y,z)\rightarrow(a,b,c)} = \begin{pmatrix} \dfrac{\partial x}{\partial a} & \dfrac{\partial x}{\partial b} & \dfrac{\partial x}{\partial c} \\ \dfrac{\partial y}{\partial a} & \dfrac{\partial y}{\partial b} & \dfrac{\partial y}{\partial c} \\ \dfrac{\partial z}{\partial a} & \dfrac{\partial z}{\partial b} & \dfrac{\partial z}{\partial c} \end{pmatrix} \tag{4.79}$$

is nonsingular.

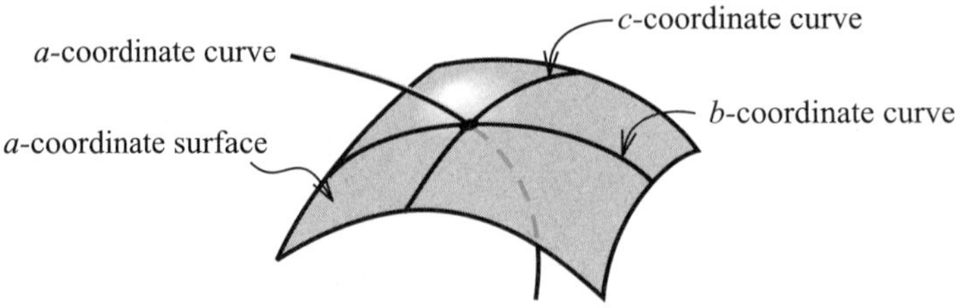

Figure 4.22. The a-coordinate curve and the a-coordinate surface.

Definition 4.6. *Let the Jacobian in (4.79) be nonsingular for the new coordinate (a, b, c) given in (4.78). The a-***coordinate curve** *(or a-curve) is the locus of points where b and c are fixed. The a-***coordinate surface** *(or a-surface) is the surface defined at a fixed a.*

Figure 4.22 shows the a-coordinate curve together with the a-coordinate surface. Using (4.78), the a-curve at $b = b_0$ and $c = c_0$ is given by a set of points described by the position vectors

$$\underline{\vec{\mathbf{r}}}\,(a, b_0, c_0) = x\,(a, b_0, c_0)\,\underline{\boldsymbol{\delta}}_x + y\,(a, b_0, c_0)\,\underline{\boldsymbol{\delta}}_y + z\,(a, b_0, c_0)\,\underline{\boldsymbol{\delta}}_z$$

whereas the a-surface at $a = a_0$ is described by the scalar function

$$a_0 = a\,(x, y, z)$$

Definition 4.7. *Let* **p** *be a point in the 3D space. A vector $\underline{\nu}_a\,(\mathbf{p})$ that is tangent to the a-coordinate curve at* **p** *is called the a-***base vector** *and defined as*

$$\underline{\nu}_a\,(\mathbf{p}) = \frac{\partial}{\partial a}\underline{\vec{\mathbf{r}}}\,\bigg|_{\mathbf{p}} = \sum_{k=x,y,z} \frac{\partial k}{\partial a}\,\underline{\boldsymbol{\delta}}_k\,\Bigg|_{\mathbf{p}} \tag{4.80}$$

A vector $\widehat{\underline{\nu}}_a(\mathbf{p})$ that is normal to the a-coordinate surface at **p** *is called the a-***reciprocal base vector**, *or a-***dual base vector**, *and defined as*

$$\widehat{\underline{\nu}}_a(\mathbf{p}) = \nabla\left(a\,(x, y, z)\right)\,\bigg|_{\mathbf{p}} = \sum_{k=x,y,z} \frac{\partial a}{\partial k}\,\underline{\boldsymbol{\delta}}_k\,\Bigg|_{\mathbf{p}} \tag{4.81}$$

The a-base vector $\underline{\nu}_a$ and a-reciprocal base vector $\widehat{\underline{\nu}}_a$ at P are shown in Figure 4.23. Note that $\underline{\nu}_a$ and $\widehat{\underline{\nu}}_a$ do not necessarily point in the same direction. Furthermore, neither $\underline{\nu}_a$ nor $\widehat{\underline{\nu}}_a$ will necessarily have unit length. The base vectors and reciprocal base vectors can be normalized easily by dividing by their respective norms. We denote the **unit base vectors** by $\underline{\boldsymbol{\delta}}_a$, and we denote the **unit reciprocal base vectors** by $\widehat{\underline{\delta}}_a$, where

$$\underline{\boldsymbol{\delta}}_a = \frac{\underline{\nu}_a}{\left\|\underline{\nu}_a\right\|} \quad \text{and} \quad \widehat{\underline{\delta}}_a = \frac{\widehat{\underline{\nu}}_a}{\left\|\widehat{\underline{\nu}}_a\right\|}$$

A similar set of definitions apply to the b-coordinate curves, b-coordinate surfaces, b-base vectors, and b-reciprocal base vectors, as well as to c-coordinate curves, c-coordinate surfaces, c-base vectors, and c-reciprocal base vectors.

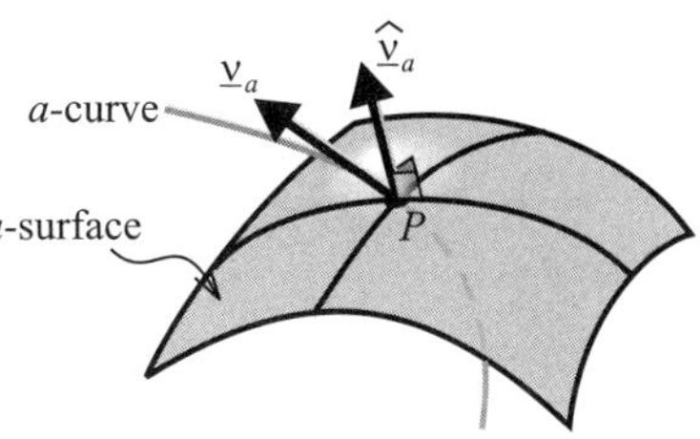

Figure 4.23. The a-base vector and the a-reciprocal base vector.

Based on Definitions 4.6 and 4.7, we have the following results and observations:

1. Vector $\widehat{\underline{\nu}}_a$ is orthogonal to $\underline{\nu}_b$ and $\underline{\nu}_c$,

$$\widehat{\underline{\nu}}_a \cdot \underline{\nu}_b = \frac{\partial a}{\partial x}\frac{\partial x}{\partial b} + \frac{\partial a}{\partial y}\frac{\partial z}{\partial b} + \frac{\partial a}{\partial z}\frac{\partial z}{\partial b} = \frac{\partial a}{\partial b} = 0$$

$$\widehat{\underline{\nu}}_a \cdot \underline{\nu}_c = \frac{\partial a}{\partial x}\frac{\partial x}{\partial c} + \frac{\partial a}{\partial y}\frac{\partial z}{\partial c} + \frac{\partial a}{\partial z}\frac{\partial z}{\partial c} = \frac{\partial a}{\partial c} = 0 \tag{4.82}$$

because a is independent of b and c. Similarly, $\widehat{\underline{\nu}}_b$ is orthogonal to $\underline{\nu}_a$ and $\underline{\nu}_c$, and $\widehat{\underline{\nu}}_c$ is orthogonal to $\underline{\nu}_a$ and $\underline{\nu}_b$.

2. The dot product of $\widehat{\underline{\nu}}_a$ and $\underline{\nu}_a$ is unity, that is,

$$\widehat{\underline{\nu}}_a \cdot \underline{\nu}_a = \frac{\partial a}{\partial x}\frac{\partial x}{\partial a} + \frac{\partial a}{\partial y}\frac{\partial z}{\partial a} + \frac{\partial a}{\partial z}\frac{\partial z}{\partial a} = \frac{\partial a}{\partial a} = 1 \tag{4.83}$$

Similarly, $\widehat{\underline{\nu}}_b \cdot \underline{\nu}_b = 1$ and $\widehat{\underline{\nu}}_c \cdot \underline{\nu}_c = 1$.

3. The set $(\underline{\nu}_a, \underline{\nu}_b, \underline{\nu}_c)$ forms a linearly independent set of vectors that span the 3D space. Thus any vector $\underline{\mathbf{v}}$ can be represented by a linear combination of the base vectors, that is,

$$\underline{\mathbf{v}} = \bar{\upsilon}_a \underline{\nu}_a + \bar{\upsilon}_b \underline{\nu}_b + \bar{\upsilon}_c \underline{\nu}_c$$

Likewise, the reciprocal base vectors $(\widehat{\underline{\nu}}_a, \widehat{\underline{\nu}}_b, \widehat{\underline{\nu}}_c)$ form another linearly independent set of basis vectors. However, they are used more as the basis for the gradient operator ∇. To see this, we start with the gradient of a scalar field in rectangular coordinates,

$$\begin{aligned}
\nabla\psi &= \sum_{k=x,y,z} \underline{\delta}_k \frac{\partial \psi}{\partial k} = \sum_{k=x,y,z} \underline{\delta}_k \sum_{m=a,b,c} \frac{\partial \psi}{\partial m}\frac{\partial m}{\partial k} \\
&= \sum_{m=a,b,c} \frac{\partial \psi}{\partial m} \sum_{k=x,y,z} \underline{\delta}_k \frac{\partial m}{\partial k} = \sum_{m=a,b,c} \frac{\partial \psi}{\partial m} \nabla m \\
&= \left(\widehat{\underline{\nu}}_a \frac{\partial}{\partial a} + \widehat{\underline{\nu}}_b \frac{\partial}{\partial b} + \widehat{\underline{\nu}}_c \frac{\partial}{\partial c} \right) \psi
\end{aligned}$$

or

$$\nabla = \widehat{\underline{\nu}}_a \frac{\partial}{\partial a} + \widehat{\underline{\nu}}_b \frac{\partial}{\partial b} + \widehat{\underline{\nu}}_c \frac{\partial}{\partial c} \tag{4.84}$$

4.9.2 Orthogonal Coordinate Systems

We now consider the special case of orthogonal coordinate systems, in which the base vectors and unit base vectors form an orthogonal set such that

$$\underline{\delta}_a = \underline{\delta}_b \times \underline{\delta}_c \qquad \underline{\delta}_b = \underline{\delta}_c \times \underline{\delta}_a \qquad \underline{\delta}_c = \underline{\delta}_a \times \underline{\delta}_b \tag{4.85}$$

Because $\widehat{\underline{\nu}}_a$ is normal to the a-surface, the orthogonality of the base vectors means that $\widehat{\underline{\nu}}_a$ and $\underline{\nu}_a$ are pointing in the same direction. After normalizing the base vectors and reciprocal base vectors, we have

$$\widehat{\underline{\delta}}_a = \underline{\delta}_a \quad ; \quad \widehat{\underline{\delta}}_b = \underline{\delta}_b \quad ; \quad \widehat{\underline{\delta}}_c = \underline{\delta}_c \tag{4.86}$$

In addition, we can use the fact that $\widehat{\underline{\nu}}_a \cdot \underline{\nu}_a = 1$, $\widehat{\underline{\nu}}_b \cdot \underline{\nu}_b = 1$, and $\widehat{\underline{\nu}}_c \cdot \underline{\nu}_c = 1$ to show that

$$\|\widehat{\underline{\nu}}_a\| = \frac{1}{\|\underline{\nu}_a\|} \quad ; \quad \|\widehat{\underline{\nu}}_b\| = \frac{1}{\|\underline{\nu}_b\|} \quad ; \quad \|\widehat{\underline{\nu}}_c\| = \frac{1}{\|\underline{\nu}_c\|} \tag{4.87}$$

We call the norms of the base vectors **scaling factors**, denoted by α_a, α_b and α_c,

$$\alpha_a = \left\| \underline{\nu}_a \right\| = \left\| \frac{\partial \vec{\underline{\mathbf{r}}}}{\partial a} \right\| \quad ; \quad \alpha_b = \left\| \underline{\nu}_b \right\| = \left\| \frac{\partial \vec{\underline{\mathbf{r}}}}{\partial b} \right\| \quad ; \quad \alpha_c = \left\| \underline{\nu}_c \right\| = \left\| \frac{\partial \vec{\underline{\mathbf{r}}}}{\partial c} \right\| \tag{4.88}$$

This means that for orthogonal base vectors, the gradient operator can also be written in terms of the base vectors or unit base vectors as follows:

$$\begin{aligned} \nabla &= \widehat{\underline{\nu}}_a \frac{\partial}{\partial a} + \widehat{\underline{\nu}}_b \frac{\partial}{\partial b} + \widehat{\underline{\nu}}_c \frac{\partial}{\partial c} \\ &= \|\widehat{\underline{\nu}}_a\| \hat{\delta}_a \frac{\partial}{\partial a} + \|\widehat{\underline{\nu}}_b\| \hat{\delta}_b \frac{\partial}{\partial b} + \|\widehat{\underline{\nu}}_c\| \hat{\delta}_c \frac{\partial}{\partial c} \\ &= \frac{1}{\alpha_a} \underline{\delta}_a \frac{\partial}{\partial a} + \frac{1}{\alpha_b} \underline{\delta}_b \frac{\partial}{\partial b} + \frac{1}{\alpha_c} \underline{\delta}_c \frac{\partial}{\partial c} \end{aligned} \tag{4.89}$$

EXAMPLE 4.17. Consider the parabolic coordinate system (a, b, θ) defined by the following equations:

$$x = ab\cos\theta \quad ; \quad y = ab\sin\theta \quad ; \quad z = \frac{1}{2}\left(a^2 - b^2\right)$$

which is a valid coordinate system in a domain where the following Jacobian is nonsingular:

$$J_{\text{rect}\rightarrow\text{parabolic}} = \begin{pmatrix} b\cos\theta & a\cos\theta & -ab\sin\theta \\ b\sin\theta & a\sin\theta & ab\cos\theta \\ a & -b & 0 \end{pmatrix}$$

Solving for a, we have

$$\begin{aligned} x^2 + y^2 &= a^2 b^2 \\ b^2 &= a^2 - 2z \end{aligned} \quad \rightarrow \quad a = \sqrt{z + r}$$

where $r = \sqrt{x^2+y^2+z^2}$. At fixed values of a, this gives the a-coordinate surface. Likewise, for the b-coordinate surface, we have

$$\begin{array}{rcl} x^2+y^2 & = & a^2b^2 \\ a^2 & = & b^2+2z \end{array} \quad \rightarrow \quad b = \sqrt{-z+r}$$

whereas for the θ-coordinate surface, we have $\theta = \mathrm{Atan}\left(\frac{y}{x}\right)$, which is the same as the cylindrical coordinate θ, that is, a plane containing the z-axis.

Let $\vec{\underline{\mathbf{r}}}$ be the position vector. Then the base vectors are given by

$$\begin{aligned} \underline{v}_a &= \frac{\partial \vec{\underline{\mathbf{r}}}}{\partial a} = b\cos\theta\underline{\boldsymbol{\delta}}_x + b\sin\theta\underline{\boldsymbol{\delta}}_y + a\underline{\boldsymbol{\delta}}_z \\ \underline{v}_b &= \frac{\partial \vec{\underline{\mathbf{r}}}}{\partial b} = a\cos\theta\underline{\boldsymbol{\delta}}_x + a\sin\theta\underline{\boldsymbol{\delta}}_y - b\underline{\boldsymbol{\delta}}_z \\ \underline{v}_\theta &= \frac{\partial \vec{\underline{\mathbf{r}}}}{\partial \theta} = -ab\sin\theta\underline{\boldsymbol{\delta}}_x + ab\cos\theta\underline{\boldsymbol{\delta}}_y \end{aligned}$$

This is an orthogonal coordinate system because

$$\underline{v}_a \cdot \underline{v}_a = a^2 + b^2 = \underline{v}_b \cdot \underline{v}_b \quad ; \quad \underline{v}_\theta \cdot \underline{v}_\theta = a^2b^2$$

and the other dot products are zero. The scaling factors are given by the square root of these dot products, that is,

$$\alpha_a = \alpha_b = \sqrt{a^2+b^2} \quad ; \quad \alpha_\theta = ab$$

which then yield the gradient operator in the parabolic coordinate system to be

$$\nabla = \frac{1}{\sqrt{a^2+b^2}}\,\underline{\boldsymbol{\delta}}_a\,\frac{\partial}{\partial a} + \frac{1}{\sqrt{a^2+b^2}}\,\underline{\boldsymbol{\delta}}_b\,\frac{\partial}{\partial b} + \frac{1}{ab}\,\underline{\boldsymbol{\delta}}_\theta\,\frac{\partial}{\partial \theta}$$

Based on (4.89), we can obtain some of the important differential identities in terms orthogonal coordinates (a, b, c) and the unit base vectors:

1. **Gradient**

$$\nabla\psi = \frac{1}{\alpha_a}\,\underline{\boldsymbol{\delta}}_a\frac{\partial\psi}{\partial a} + \frac{1}{\alpha_b}\,\underline{\boldsymbol{\delta}}_b\frac{\partial\psi}{\partial b} + \frac{1}{\alpha_c}\,\underline{\boldsymbol{\delta}}_c\frac{\partial\psi}{\partial c} \tag{4.90}$$

2. **Divergence**

$$\nabla\cdot\underline{\mathbf{w}} = \frac{1}{\alpha_a\alpha_b\alpha_c}\left(\frac{\partial}{\partial a}\alpha_b\alpha_c w_a + \frac{\partial}{\partial b}\alpha_c\alpha_a w_b + \frac{\partial}{\partial c}\alpha_a\alpha_b w_c\right) \tag{4.91}$$

3. **Curl**

$$\nabla\times\underline{\mathbf{w}} = \frac{1}{\alpha_a\alpha_b\alpha_c}\,\mathbf{det}\begin{pmatrix} \alpha_a\underline{\boldsymbol{\delta}}_a & \alpha_b\underline{\boldsymbol{\delta}}_b & \alpha_c\underline{\boldsymbol{\delta}}_c \\ \dfrac{\partial}{\partial a} & \dfrac{\partial}{\partial b} & \dfrac{\partial}{\partial c} \\ \alpha_a w_a & \alpha_b w_b & \alpha_c w_c \end{pmatrix} \tag{4.92}$$

4. **Laplacian of Scalar Fields**

$$\nabla^2\psi = \frac{1}{\alpha_a\alpha_b\alpha_c}\left[\frac{\partial}{\partial a}\left(\frac{\alpha_b\alpha_c}{\alpha_a}\frac{\partial\psi}{\partial a}\right) + \frac{\partial}{\partial b}\left(\frac{\alpha_a\alpha_c}{\alpha_b}\frac{\partial\psi}{\partial b}\right) + \frac{\partial}{\partial c}\left(\frac{\alpha_a\alpha_b}{\alpha_c}\frac{\partial\psi}{\partial c}\right)\right] \tag{4.93}$$

5. **Gradient-Vector Dyad**

$$\nabla\underline{\mathbf{w}} = \sum_{k=a,b,c}\sum_{m=a,b,c}\frac{1}{\alpha_m}\frac{\partial w_k}{\partial m}\underline{\boldsymbol{\delta}}_m\underline{\boldsymbol{\delta}}_k + \sum_{k=a,b,c}\sum_{m=a,b,c}\frac{w_k}{\alpha_m}\underline{\boldsymbol{\delta}}_m\frac{\partial\underline{\boldsymbol{\delta}}_k}{\partial m} \tag{4.94}$$

The proofs of (4.90) to (4.94) are given in the appendix as Section D.1. Applying these results to cylindrical and spherical coordinates, we obtain an alternative set of combined formulas given by the following:

Gradient of Scalar Fields:
$$\nabla\psi = \sum_k \frac{1}{\alpha_k}\frac{\partial\psi}{\partial k}\underline{\boldsymbol{\delta}}_k$$

Partial Derivative of Vector Fields:
$$\frac{\partial}{\partial m}\underline{\mathbf{v}} = \sum_k\left(\frac{\partial v_k}{\partial m}\right)\underline{\boldsymbol{\delta}}_k + \left\{\underline{\mathbf{p}}_m(\underline{\mathbf{v}})\right\}$$

Divergence of Vector Fields:
$$\nabla\cdot\underline{\mathbf{v}} = \sum_k\left(\frac{1}{\alpha_k}\frac{\partial v_k}{\partial k}\right) + \left\{D(\underline{\mathbf{v}})\right\}$$

Curl of Vector Fields:
$$\nabla\times\underline{\mathbf{v}} = \mathbf{det}\begin{pmatrix} \underline{\boldsymbol{\delta}}_a & \underline{\boldsymbol{\delta}}_b & \underline{\boldsymbol{\delta}}_c \\ \frac{1}{\alpha_a}\frac{\partial}{\partial a} & \frac{1}{\alpha_b}\frac{\partial}{\partial b} & \frac{1}{\alpha_c}\frac{\partial}{\partial c} \\ v_a & v_b & v_c \end{pmatrix} + \left\{\underline{\mathbf{q}}(\underline{\mathbf{v}})\right\}$$

(where $(a,b,c) = (x,y,z)$, (r,θ,z) or (r,θ,ϕ))

Laplacian of Scalar Fields:
$$\nabla^2\psi = \sum_k\frac{1}{\alpha_k^2}\frac{\partial^2\psi}{\partial k^2} + \left\{L(\psi)\right\}$$

Gradient-vector dyads:
$$\nabla\underline{\mathbf{v}} = \sum_m\sum_k\left(\frac{1}{\alpha_m}\frac{\partial v_k}{\partial m}\right)\underline{\boldsymbol{\delta}}_m\underline{\boldsymbol{\delta}}_k + \left\{\underline{\underline{T}}(\underline{\mathbf{v}})\right\} \tag{4.95}$$

The values of parameters α_k, $\underline{\mathbf{q}}$, D, L, $\underline{\mathbf{p}}$, and $\underline{\underline{T}}$ are given in Table 4.7 for both the cylindrical and spherical coordinates. Note that the preceding equations also apply to rectangular coordinates by setting $\alpha_x = \alpha_y = \alpha_z = 1$ and disregarding all the terms with curly brackets.

Table 4.7. *Parameters for (4.95) for cylindrical and spherical coordinates*

	Cylindrical	Spherical
α	$\alpha_r = 1, \alpha_\theta = r, \alpha_z = 1$	$\alpha_r = 1, \alpha_\theta = r, \alpha_\phi = rs_\theta$
C	$\underline{\mathbf{q}}(\underline{\mathbf{v}}) = \frac{v_\theta}{r}\underline{\boldsymbol{\delta}}_z$	$\underline{\mathbf{q}}(\underline{\mathbf{v}}) = \frac{v_\phi c_\theta}{rs_\theta}\underline{\boldsymbol{\delta}}_r - \frac{v_\phi}{r}\underline{\boldsymbol{\delta}}_\theta + \frac{v_\theta}{r}\underline{\boldsymbol{\delta}}_\phi$
D	$D(\underline{\mathbf{v}}) = \frac{v_r}{r}$	$D(\underline{\mathbf{v}}) = 2\frac{v_r}{r} + \frac{v_\theta c_\theta}{rs_\theta}$
L	$L(\psi) = \frac{1}{r}\frac{\partial\psi}{\partial r}$	$L(\psi) = \frac{2}{r}\frac{\partial\psi}{\partial r} + \frac{c_\theta}{r^2 s_\theta}\frac{\partial\psi}{\partial\theta}$
P	$\underline{\mathbf{p}}_\theta(\underline{\mathbf{v}}) = -v_\theta\underline{\boldsymbol{\delta}}_r + v_r\underline{\boldsymbol{\delta}}_\theta$	$\underline{\mathbf{p}}_\theta(\underline{\mathbf{v}}) = -v_\theta\underline{\boldsymbol{\delta}}_r + v_r\underline{\boldsymbol{\delta}}_\theta$ $\underline{\mathbf{p}}_\phi(\underline{\mathbf{v}}) = -v_\phi s_\theta\underline{\boldsymbol{\delta}}_r - v_\phi c_\theta\underline{\boldsymbol{\delta}}_\theta + (v_r s_\theta + v_\theta c_\theta)\underline{\boldsymbol{\delta}}_\phi$
T	$\underline{\underline{T}}(\underline{\mathbf{v}}) = -\frac{v_\theta}{r}\underline{\boldsymbol{\delta}}_\theta\underline{\boldsymbol{\delta}}_r + \frac{v_r}{r}\underline{\boldsymbol{\delta}}_\theta\underline{\boldsymbol{\delta}}_\theta$	$\underline{\underline{T}}(\underline{\mathbf{v}}) = -\frac{v_\theta}{r}\underline{\boldsymbol{\delta}}_\theta\underline{\boldsymbol{\delta}}_r + \frac{v_r}{r}\underline{\boldsymbol{\delta}}_\theta\underline{\boldsymbol{\delta}}_\theta - \frac{v_\phi}{r}\underline{\boldsymbol{\delta}}_\phi\underline{\boldsymbol{\delta}}_r - \frac{v_\phi c_\theta}{rs_\theta}\underline{\boldsymbol{\delta}}_\phi\underline{\boldsymbol{\delta}}_\theta + \left(\frac{v_r}{r} + \frac{v_\theta c_\theta}{rs_\theta}\right)\underline{\boldsymbol{\delta}}_\phi\underline{\boldsymbol{\delta}}_\phi$

Note: Absent items are considered zero, e.g. $\underline{\mathbf{p}}_r = \underline{0}$. (Arranged in alphabetical order.)

EXAMPLE 4.18. Let us show that the formulas given in (4.92) and (4.95) for the curl of a vector field in spherical coordinates are the same as that given by (4.77).

Applying (4.92) for the spherical coordinates, with $(a, b, c) = (r, \theta, \phi)$ and $(\alpha_a, \alpha_b, \alpha_c) = (1, r, r\sin\theta)$, gives

$$\begin{aligned}
\nabla \times \underline{\mathbf{v}} &= \frac{1}{r^2\sin\theta}\,\mathbf{det}\begin{pmatrix} \underline{\boldsymbol{\delta}}_r & r\underline{\boldsymbol{\delta}}_\theta & r\sin\theta\underline{\boldsymbol{\delta}}_\phi \\ \frac{\partial}{\partial r} & \frac{\partial}{\partial\theta} & \frac{\partial}{\partial\phi} \\ v_r & rv_\theta & r\sin\theta v_\phi \end{pmatrix} \\
&= \frac{1}{r^2\sin\theta}\underline{\boldsymbol{\delta}}_r\left(\frac{\partial(r\sin\theta v_\phi)}{\partial\theta} - \frac{\partial(rv_\theta)}{\partial\phi}\right) + \frac{1}{r\sin\theta}\underline{\boldsymbol{\delta}}_\theta\left(\frac{\partial v_r}{\partial\phi} - \frac{\partial(r\sin\theta v_\phi)}{\partial r}\right) \\
&\quad + \frac{1}{r}\underline{\boldsymbol{\delta}}_\phi\left(\frac{\partial(rv_\theta)}{\partial r} - \frac{\partial v_r}{\partial\theta}\right)
\end{aligned}$$

which can be shown to be the same as (4.77).

Alternatively, let us check (4.95) for the curl in spherical coordinates,

$$\begin{aligned}
\nabla \times \underline{\mathbf{v}} &= \mathbf{det}\begin{pmatrix} \underline{\boldsymbol{\delta}}_r & \underline{\boldsymbol{\delta}}_\theta & \underline{\boldsymbol{\delta}}_\phi \\ \frac{\partial}{\partial r} & \frac{1}{r}\frac{\partial}{\partial\theta} & \frac{1}{r\sin\theta}\frac{\partial}{\partial\phi} \\ v_r & v_\theta & v_\phi \end{pmatrix} + \left(\frac{v_\phi\cos\theta}{r\sin\theta}\right)\underline{\boldsymbol{\delta}}_r - \frac{v_\phi}{r}\underline{\boldsymbol{\delta}}_\theta + \frac{v_\theta}{r}\underline{\boldsymbol{\delta}}_\phi \\
&= \underline{\boldsymbol{\delta}}_r\left(\frac{1}{r}\frac{\partial v_\phi}{\partial\theta} - \frac{1}{r\sin\theta}\frac{\partial v_\theta}{\partial\phi}\right) + \underline{\boldsymbol{\delta}}_\theta\left(\frac{1}{r\sin\theta}\frac{\partial v_r}{\partial\phi} - \frac{\partial v_\phi}{\partial r}\right) + \underline{\boldsymbol{\delta}}_\phi\left(\frac{\partial v_\theta}{\partial r} - \frac{1}{r}\frac{\partial v_r}{\partial\theta}\right) \\
&\quad + \left(\frac{v_\phi\cos\theta}{r\sin\theta}\right)\underline{\boldsymbol{\delta}}_r - \frac{v_\phi}{r}\underline{\boldsymbol{\delta}}_\theta + \frac{v_\theta}{r}\underline{\boldsymbol{\delta}}_\phi
\end{aligned}$$

which can also be shown to yield (4.77).

4.10 EXERCISES

E4.1. Verify all the properties in Table 4.2 using the following values:

$$\alpha = 1.5 \quad ; \quad \beta = -2$$
$$\underline{\mathbf{u}} = \underline{\boldsymbol{\delta}}_x + 2\underline{\boldsymbol{\delta}}_y + 2\underline{\boldsymbol{\delta}}_z \quad ; \quad \underline{\mathbf{v}} = \underline{\boldsymbol{\delta}}_x + \underline{\boldsymbol{\delta}}_y + 2\underline{\boldsymbol{\delta}}_z$$
$$\underline{\mathbf{w}} = 2\underline{\boldsymbol{\delta}}_y + \underline{\boldsymbol{\delta}}_z \quad ; \quad \underline{\mathbf{y}} = \underline{\boldsymbol{\delta}}_x + 4\underline{\boldsymbol{\delta}}_z$$

E4.2. Prove or disprove: The dot product and cross product can be interchanged in a triple product, that is,

$$\underline{\mathbf{u}} \cdot (\underline{\mathbf{v}} \times \underline{\mathbf{w}}) = (\underline{\mathbf{u}} \times \underline{\mathbf{v}}) \cdot \underline{\mathbf{w}}$$

E4.3. Consider three non-collinear points $A = (a_x, a_y, a_z)$, $B = (b_x, b_y, b_z)$, and $C = (c_x, c_y, c_z)$. The minimum distance d from point C to the line passing through A and B is given by the following formula

$$d = \left\| \frac{\underline{\mathbf{u}} \times \underline{\mathbf{v}}}{\|\underline{\mathbf{v}}\|} \right\| \tag{4.96}$$

where

$$\underline{\mathbf{u}} = (c_x - a_x)\,\underline{\boldsymbol{\delta}}_x + (c_y - a_y)\,\underline{\boldsymbol{\delta}}_y + (c_z - a_z)\,\underline{\boldsymbol{\delta}}_z$$
$$\underline{\mathbf{v}} = (b_x - a_x)\,\underline{\boldsymbol{\delta}}_x + (b_y - a_y)\,\underline{\boldsymbol{\delta}}_y + (b_z - a_z)\,\underline{\boldsymbol{\delta}}_z$$

Show that the same value results if the roles of A and B had been interchanged.

E4.4. Prove or disprove the following identities:

1. $(\underline{\mathbf{a}} \times \underline{\mathbf{b}}) \cdot \underline{\mathbf{c}} = \underline{\mathbf{a}} \cdot (\underline{\mathbf{b}} \times \underline{\mathbf{c}})$
2. $(\underline{\mathbf{a}} \times \underline{\mathbf{b}}) \cdot (\underline{\mathbf{c}} \times \underline{\mathbf{d}}) = (\underline{\mathbf{a}} \cdot \underline{\mathbf{c}})(\underline{\mathbf{b}} \cdot \underline{\mathbf{d}}) - (\underline{\mathbf{a}} \cdot \underline{\mathbf{d}})(\underline{\mathbf{b}} \cdot \underline{\mathbf{c}})$
3. $(\underline{\mathbf{a}} \times \underline{\mathbf{b}}) \times (\underline{\mathbf{c}} \times \underline{\mathbf{d}}) = (\underline{\mathbf{a}} \cdot (\underline{\mathbf{c}} \times \underline{\mathbf{d}}))\,\underline{\mathbf{b}} - (\underline{\mathbf{b}} \cdot (\underline{\mathbf{c}} \times \underline{\mathbf{d}}))\,\underline{\mathbf{a}}$

Also, verify the proven identities using

$$\underline{\mathbf{a}} = 2\underline{\boldsymbol{\delta}}_x + 3\underline{\boldsymbol{\delta}}_y - \underline{\boldsymbol{\delta}}_z \quad ; \quad \underline{\mathbf{b}} = 3\underline{\boldsymbol{\delta}}_x + 2\underline{\boldsymbol{\delta}}_z$$
$$\underline{\mathbf{c}} = 4\underline{\boldsymbol{\delta}}_x - 2\underline{\boldsymbol{\delta}}_y + 2\underline{\boldsymbol{\delta}}_z \quad ; \quad \underline{\mathbf{d}} = -2\underline{\boldsymbol{\delta}}_y + 3\underline{\boldsymbol{\delta}}_z$$

E4.5. Determine a formula for the volume of the triangular pyramid whose vertices are (x_i, y_i, z_i), $i = 1, 2, 3, 4$. Under what conditions will the volume be zero?

E4.6. A triangle in 3D space is determined by three non-collinear vertices given by $a = (a_x, a_y, a_z)$, $b = (b_x, b_y, b_z)$, and $c = (c_x, c_y, c_z)$. Let the **oriented triangular area**, denoted by $\underline{\mathbf{A}}_\Delta(\underline{\mathbf{a}}, \underline{\mathbf{b}}, \underline{\mathbf{c}})$, be the vector defined by

$$\underline{\mathbf{A}}_\Delta(\underline{\mathbf{a}}, \underline{\mathbf{b}}, \underline{\mathbf{c}}) = \frac{1}{2}(\underline{\mathbf{b}} - \underline{\mathbf{a}}) \times (\underline{\mathbf{c}} - \underline{\mathbf{a}}) \tag{4.97}$$

where $\underline{\mathbf{a}}$, $\underline{\mathbf{b}}$, and $\underline{\mathbf{c}}$ are the position vectors based on the coordinates of a, b, and c.

1. Show that the area of the triangle Δ_{abc} is given by $\|\underline{\mathbf{A}}_\Delta(\underline{\mathbf{a}}, \underline{\mathbf{b}}, \underline{\mathbf{c}})\|$. Use this to find the area formed by $a = (1, 1, 1)$, $b = (1, 0, 1)$ and $c = (1, -1, 1)$.
2. Show that an alternative formula is given by

$$\underline{\mathbf{A}}_\Delta(\underline{\mathbf{a}}, \underline{\mathbf{b}}, \underline{\mathbf{c}}) = \frac{1}{2}(\underline{\mathbf{a}} \times \underline{\mathbf{b}} + \underline{\mathbf{b}} \times \underline{\mathbf{c}} + \underline{\mathbf{c}} \times \underline{\mathbf{a}})$$

thus also show that

$$\underline{\mathbf{A}}_\Delta(\underline{\mathbf{a}}, \underline{\mathbf{b}}, \underline{\mathbf{c}}) = \underline{\mathbf{A}}_\Delta(\underline{\mathbf{b}}, \underline{\mathbf{c}}, \underline{\mathbf{a}}) = \underline{\mathbf{A}}_\Delta(\underline{\mathbf{c}}, \underline{\mathbf{a}}, \underline{\mathbf{b}})$$

3. Consider the tetrahedron described by four vertices a, b, c, and d as shown in Figure 4.24.

Figure 4.24. A tetrahedron described by four points: a, b, c, and d.

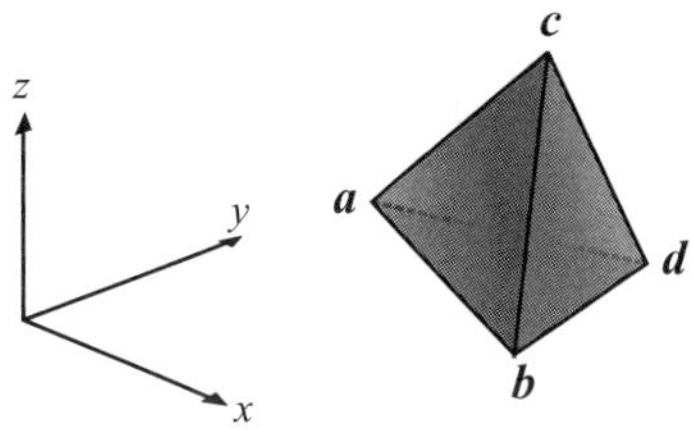

Based on the right-hand rule, the outward normal vectors will be based on the sequences (within a cyclic permutation) *abc*, *bdc*, *cda*, and *dba*. Show that the vector sum of the oriented areas will be a zero vector, that is,

$$\underline{\mathbf{A}}_{\Delta}\left(\underline{\mathbf{a}}, \underline{\mathbf{b}}, \underline{\mathbf{c}}\right) + \underline{\mathbf{A}}_{\Delta}\left(\underline{\mathbf{b}}, \underline{\mathbf{d}}, \underline{\mathbf{c}}\right) + \underline{\mathbf{A}}_{\Delta}\left(\underline{\mathbf{c}}, \underline{\mathbf{d}}, \underline{\mathbf{a}}\right) + \underline{\mathbf{A}}_{\Delta}\left(\underline{\mathbf{d}}, \underline{\mathbf{b}}, \underline{\mathbf{a}}\right) = \underline{\mathbf{0}}$$

Verify this fact using $a = (1, 2, 1)$, $b = (2, 1, 1)$, $c = (2, 2, 4)$ and $d = (3, 3, 2)$.

E4.7. Consider the following symmetric stress tensor $\underline{\underline{T}}$ and a unit normal $\underline{\mathbf{n}}$ given by

$$\underline{\underline{T}} = \begin{pmatrix} 2 & 3 & 1 \\ 3 & 1 & 1 \\ 1 & 1 & 2 \end{pmatrix} \quad ; \quad \underline{\mathbf{n}} = \frac{1}{\sqrt{3}} \begin{pmatrix} 1 \\ 1 \\ 1 \end{pmatrix}$$

1. Find the normal stress vector and shear stress vector that correspond to the surface determined by the unit normal vector $\underline{\mathbf{n}}$.
2. Show that, in general, given stress tensor $\underline{\underline{T}}$ and unit normal $\underline{\mathbf{n}}$, the normal stress vector $\underline{\tau}_{\text{normal}}$ and shear stress vector $\underline{\tau}_{\text{shear}}$ are given by

$$\underline{\tau}_{\text{normal}} = \left(\underline{\mathbf{n}}\,\underline{\mathbf{n}}\right) \cdot \left(\underline{\underline{T}} \cdot \underline{\mathbf{n}}\right)$$

$$\underline{\tau}_{\text{shear}} = \left(\underline{\underline{I}} - \underline{\mathbf{n}}\,\underline{\mathbf{n}}\right) \cdot \left(\underline{\underline{T}} \cdot \underline{\mathbf{n}}\right)$$

   where $\underline{\underline{I}}$ is the identity tensor. Verify these formulas by comparing them with the results you found previously.
3. Find the unit vectors, $\underline{\mathbf{v}}_1$, $\underline{\mathbf{v}}_2$, and $\underline{\mathbf{v}}_3$ along the principal axes of the stress tensor $\underline{\underline{T}}$. Show that the normal stress along $\underline{\mathbf{v}}_1$ is equal to $\underline{\mathbf{v}}_1$ multiplied by the corresponding eigenvalue of $\underline{\underline{T}}$ while the shear stress is zero. Show that the same is true for $\underline{\mathbf{v}}_2$ and $\underline{\mathbf{v}}_3$.

E4.8. A set of 3D vectors, $\underline{\mathbf{u}}$, $\underline{\mathbf{v}}$, and $\underline{\mathbf{w}}$, spans the whole 3D space. Another related set of three vectors is called the **reciprocal vectors**. A vector $\widehat{\underline{\mathbf{u}}}$ is the reciprocal vector of $\underline{\mathbf{u}}$ if: (1) it is orthogonal to $\underline{\mathbf{v}}$ and $\underline{\mathbf{w}}$, and (2) $\widehat{\underline{\mathbf{u}}} \cdot \underline{\mathbf{u}} = 1$. Find the formulas for the reciprocal vectors, $\widehat{\underline{\mathbf{u}}}$, $\widehat{\underline{\mathbf{v}}}$, and $\widehat{\underline{\mathbf{w}}}$. (Hint: Use cross products and triple products.)
Note: The reciprocal vectors are used to define tensors in case the basis vectors are not orthogonal.

E4.9. For the curve C given by $x(t) = 3\cos(t)$, $y = \sin(t)$, and $z = t/(1 + 2t)$, evaluate the following at $t = 5$:

1. Velocity $\vec{\underline{\mathbf{v}}}$ and acceleration $\vec{\underline{\mathbf{a}}}$
2. The tangent unit vector $\vec{\underline{\mathbf{t}}}$, normal unit vector $\vec{\underline{\mathbf{n}}}$, and binormal unit vector $\vec{\underline{\mathbf{b}}}$
3. The tangential and normal components of acceleration $\vec{\underline{\mathbf{a}}}$
4. The curvature κ and torsion τ

5. Verify the following equations known as the **Frenet formulas** at the given point:

$$\frac{d\vec{\underline{t}}}{ds} = \kappa\vec{\underline{n}}$$
$$\frac{d\vec{\underline{n}}}{ds} = -\kappa\vec{\underline{t}} + \tau\vec{\underline{b}}$$
$$\frac{d\vec{\underline{b}}}{ds} = -\tau\vec{\underline{n}}$$

Hint: For a vector $\underline{\mathbf{c}}$, use the relationship

$$\frac{d\underline{\mathbf{c}}}{ds} = \frac{d\underline{\mathbf{c}}/dt}{\dot{s}}$$

E4.10. A solid top is shown in Figure 4.25 and has its surface described by the following equations:

$$x = \sin(\theta)\cos(\phi)$$
$$y = \sin(\theta)\sin(\phi)$$
$$z = 1.5\cos(\theta/2)$$

for $0 \le \theta \le \pi$ and $0 \le \phi \le 2\pi$.

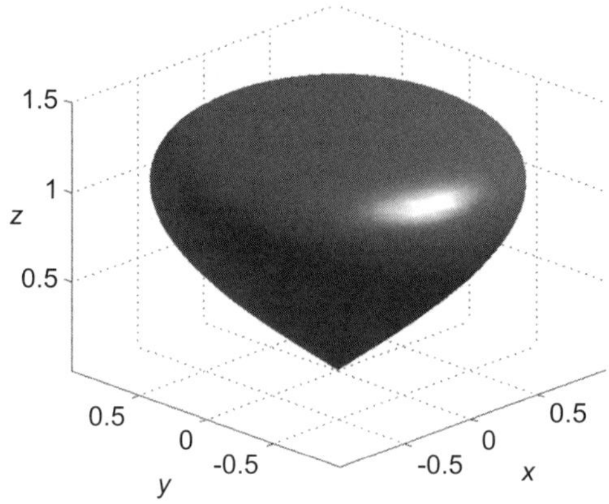

Figure 4.25. Solid surface given for problem **E4.10**.

Obtain the unit normal pointing outward at the point corresponding to $\theta = \phi = \pi/4$. Verify by plotting the unit normal along with the figure. (What about the unit normal at $\theta = \pi$?)

E4.11. Householder Tensor. We could develop the Householder operator (cf. (3.6)) for physical vectors. This tensor is supposed to reflect any given vector $\underline{\mathbf{v}}$ across a chosen plane $\mathcal{P}$. Let us denote this tensor as $\underline{\underline{H}}_{\mathbf{w}}$, where $\underline{\mathbf{w}}$ is a vector that is normal to the chosen plane. Also, define vectors $\underline{\mathbf{a}}$, $\underline{\mathbf{b}}$, and $\underline{\mathbf{u}}$ by those shown in Figure 4.26.

1. Obtain $\underline{\underline{D}}$, the dyad that would project $\underline{\mathbf{v}}$ onto $\underline{\mathbf{w}}$ and pointing along $\underline{\mathbf{w}}$, that is,

$$\underline{\mathbf{a}} = \underline{\underline{D}}\left(\underline{\mathbf{v}}\right)$$

2. From Figure 4.26, we see that the desired vector $\underline{\mathbf{u}}$ is given by

$$\underline{\mathbf{u}} = \underline{\mathbf{b}} + (-\underline{\mathbf{a}})$$

Figure 4.26. Householder operation on $\underline{\mathbf{v}}$, based on $\underline{\mathbf{w}}$.

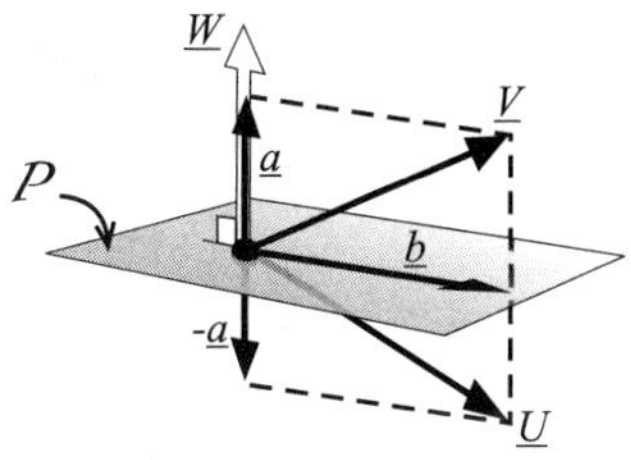

where $\underline{\mathbf{b}} = \underline{\mathbf{v}} - \underline{\mathbf{a}}$. Using these results, obtain the Householder tensor that takes vector $\underline{\mathbf{v}}$ and yields vector $\underline{\mathbf{w}}$.

E4.12. Metric Tensors. Consider the general case in which the invertible transformation between the Cartesian coordinate system and another coordinate systems (a, b, c) are given by:

$$\begin{array}{rcl} a &=& a(x,y,z) \\ b &=& b(x,y,z) \\ c &=& c(x,y,z) \end{array} \quad \Longleftrightarrow \quad \begin{array}{rcl} x &=& x(a,b,c) \\ y &=& y(a,b,c) \\ z &=& z(a,b,c) \end{array}$$

Let $\underline{\vec{\mathbf{r}}}$ be the position vector. A **Riemann space** is a space equipped with a metric given by

$$ds^2 = d\underline{\vec{\mathbf{r}}} \cdot d\underline{\vec{\mathbf{r}}}$$

$$\begin{aligned} d\underline{\vec{\mathbf{r}}} &= \frac{\partial \underline{\vec{\mathbf{r}}}}{\partial x}dx + \frac{\partial \underline{\vec{\mathbf{r}}}}{\partial y}dy + \frac{\partial \underline{\vec{\mathbf{r}}}}{\partial z}dz \\ &= \frac{\partial \underline{\vec{\mathbf{r}}}}{\partial a}da + \frac{\partial \underline{\vec{\mathbf{r}}}}{\partial b}db + \frac{\partial \underline{\vec{\mathbf{r}}}}{\partial c}dc \end{aligned}$$

1. For the Cartesian coordinates, we know that $ds^2 = dx^2 + dy^2 + dz^2$. Show that the matrix $G = J^T J$, where J is the Jacobian matrix,

$$J = \begin{pmatrix} \dfrac{\partial x}{\partial a} & \dfrac{\partial x}{\partial b} & \dfrac{\partial x}{\partial c} \\ \dfrac{\partial y}{\partial a} & \dfrac{\partial y}{\partial b} & \dfrac{\partial y}{\partial c} \\ \dfrac{\partial z}{\partial a} & \dfrac{\partial z}{\partial b} & \dfrac{\partial z}{\partial c} \end{pmatrix}$$

will satisfy the equation,

$$ds^2 = \begin{pmatrix} da & db & dc \end{pmatrix} G \begin{pmatrix} da \\ db \\ dc \end{pmatrix}$$

The elements of matrix G contains the components of $\underline{\underline{G}}$, the **metric tensor** (also known as the **fundamental tensor**).

2. A coordinate system will be orthogonal if G is diagonal. Show that the cylindrical coordinate system, that is, with $(a, b, c) = (r, \theta, z)$, and the spherical coordinate system, that is, with $(a, b, c) = (r, \theta, \phi)$, are two examples of orthogonal coordinates systems.

3. The unit vectors of the new coordinate system can be obtained as

$$\underline{\delta}_a = \frac{\partial \vec{\mathbf{r}}/\partial a}{\left\| \partial \underline{\mathbf{r}}/\partial a \right\|} \qquad \underline{\delta}_b = \frac{\partial \vec{\mathbf{r}}/\partial b}{\left\| \partial \underline{\mathbf{r}}/\partial b \right\|} \qquad \underline{\delta}_c = \frac{\partial \vec{\mathbf{r}}/\partial c}{\left\| \partial \underline{\mathbf{r}}/\partial c \right\|}$$

For the special case of orthogonal coordinate systems, the formula for ds^2 will result in the relationship between the unit vectors of the rectangular coordinate system and the unit vectors of the other orthogonal system:

$$\begin{pmatrix} \underline{\delta}_x \\ \underline{\delta}_y \\ \underline{\delta}_z \end{pmatrix} = \Phi \begin{pmatrix} \underline{\delta}_a \\ \underline{\delta}_b \\ \underline{\delta}_c \end{pmatrix}$$

where

$$\Phi = J \,\mathrm{diag}\left(\frac{1}{\sqrt{G_{11}}} \quad \frac{1}{\sqrt{G_{22}}} \quad \frac{1}{\sqrt{G_{33}}} \right)$$

because $G = J^T J$ is diagonal for orthogonal systems. Verify that the orthogonal matrices $R_{r\to c}$ and $R_{r\to s}$ given in (D.4) and (D.11), respectively, can be obtained by evaluating Φ for each case.

4. For orthogonal systems, obtain the diagonal matrix Λ such that

$$\begin{pmatrix} \dfrac{\partial}{\partial a} \\[2mm] \dfrac{\partial}{\partial b} \\[2mm] \dfrac{\partial}{\partial c} \end{pmatrix} = \Lambda \Phi^T \begin{pmatrix} \dfrac{\partial}{\partial x} \\[2mm] \dfrac{\partial}{\partial y} \\[2mm] \dfrac{\partial}{\partial z} \end{pmatrix}$$

Verify your results by applying them to both the cylindrical and spherical coordinate systems.

Note that at this point, the basic ingredients, that is, Φ and Λ, are now available to find ∇ and $\partial \underline{\delta}_k/\partial m$, for $k = a, b, c$ and $m = a, b, c$.

E4.13. Gradient Ascent Method. Consider the scalar function

$$\psi(x, y) = \frac{1}{18}\left(8x^2 + 8x - 4xy + 16y + 5y^2 + 38\right)$$

1. Obtain the gradient field of $\psi(x, y)$.
2. Let $x^{(0)} = -2$ and $y^{(0)} = 0$. Find the equation of the line that passes through $\left(x^{(0)}, y^{(0)}\right)$ and oriented along the gradient at that point. Call this line L_1.
3. Plot the magnitude of the directional derivative of ψ along the line L_1 from $\left(x^{(0)}, y^{(0)}\right)$ up to a point $\left(x^{(1)}, y^{(1)}\right)$ where the directional derivative is zero.
4. Repeat the process with $\left(x^{(1)}, y^{(1)}\right)$ instead of $\left(x^{(0)}, y^{(0)}\right)$, and iterate one more time to find $\left(x^{(2)}, y^{(2)}\right)$ and $\left(x^{(3)}, y^{(3)}\right)$.
5. Obtain a contour plot of $\psi(x, y)$ and overlay this plot with the path obtained by the **gradient ascent approach**, that is, the segments connecting, in sequence: $\left(x^{(0)}, y^{(0)}\right)$, $\left(x^{(1)}, y^{(1)}\right)$, $\left(x^{(2)}, y^{(2)}\right)$ and $\left(x^{(3)}, y^{(3)}\right)$.[13]

[13] Instead of ending each iteration at the point where the directional derivative is zero, which could mean solving nonlinear equations, one could simply take a path along the gradient line and end at

E4.14. Prove the following identity

$$\nabla \cdot (\underline{\mathbf{v}\mathbf{w}}) = (\nabla \cdot \underline{\mathbf{v}})\,\underline{\mathbf{w}} + \underline{\mathbf{v}} \cdot \nabla \underline{\mathbf{w}} \tag{4.98}$$

E4.15. Prove the following identity:

$$\nabla^2 (\nabla \cdot \underline{\mathbf{v}}) = \nabla \cdot \nabla^2 \underline{\mathbf{v}} \tag{4.99}$$

E4.16. Prove the following identity:

$$\nabla \cdot \left(\underline{\underline{T}} \cdot \underline{\mathbf{v}}\right) = \underline{\mathbf{v}} \cdot \left(\nabla \cdot \underline{\underline{T}}\right) + \underline{\underline{T}} : \nabla \underline{\mathbf{v}} \tag{4.100}$$

where

$$\nabla \cdot T = \sum_{i,j=x,y,z} \frac{\partial T_{ij}}{\partial i} \underline{\boldsymbol{\delta}}_j \qquad \text{and} \qquad \underline{\underline{T}} : \nabla \underline{\mathbf{v}} = \sum_{i,j=x,y,z} T_{ij} \frac{\partial v_j}{\partial i}$$

E4.17. Let ψ and ω be scalar fields. Using the identities given in (4.57) and (4.60), show that

$$\nabla \cdot (\nabla \psi \times \nabla \omega) = \nabla \cdot (\nabla \omega \times \nabla \psi) = 0$$

that is, $\nabla \psi \times \nabla \omega$ is solenoidal.

E4.18. Using the identity (4.98) and the continuity equation (4.64), show that following is true

$$\rho \frac{D\underline{\mathbf{v}}}{Dt} = \frac{\partial \rho \underline{\mathbf{v}}}{\partial t} + \nabla \cdot \rho \underline{\mathbf{v}}\, \underline{\mathbf{v}} \tag{4.101}$$

where D/Dt is the substantial time derivative defined in (4.65).

E4.19. The stress tensor $\underline{\underline{T}}$ of a fluid that follows Newton's law of viscosity is given by

$$\underline{\underline{T}} = -\mu \left(\nabla \underline{\mathbf{v}} + (\nabla \underline{\mathbf{v}})^T\right) + \left(\frac{2}{3}\mu - \kappa\right) (\nabla \cdot \underline{\mathbf{v}})\,\underline{\underline{I}} \tag{4.102}$$

where μ, κ, and $\underline{\underline{I}}$ are the viscosity, dilatational viscosity, and identity tensor, respectively.

1. Define the tensor $(\nabla \underline{\mathbf{v}})^T$ by

$$(\nabla \underline{\mathbf{v}})^T = \sum_{j,k=x,y,z} \frac{\partial v_j}{\partial k} \underline{\boldsymbol{\delta}}_j \underline{\boldsymbol{\delta}}_k$$

where $\underline{\mathbf{v}}$ is the velocity vector field. Then show that

$$\nabla \cdot (\nabla \underline{\mathbf{v}})^T = \nabla (\nabla \cdot \underline{\mathbf{v}})$$

2. For the case where density ρ is constant, the equation of continuity given in (4.66) becomes $\nabla \cdot \underline{\mathbf{v}} = 0$. Using this assumption, show that the divergence of the stress tensor given in (4.102) becomes

$$\nabla \cdot \underline{\underline{T}} = -\mu \nabla^2 \underline{\mathbf{v}} \tag{4.103}$$

This means that equation of motion given in (5.26) reduces to the equation known as the **Navier-Stokes equation of motion**,

$$\rho \frac{D\underline{\mathbf{v}}}{Dt} = -\nabla p + \mu \nabla^2 \underline{\mathbf{v}} + \rho \underline{\mathbf{g}}$$

where p and $\underline{\mathbf{g}}$ are the pressure and gravitational acceleration, respectively.

a fixed ratio of the length of the gradient. Using a small ratio has a better chance of not missing the optimum along the path. However, a small ratio slows down the convergence.

E4.20. A material following the Euler's equation of motion (e.g., inviscid flow) is described in (5.28), which can be rewritten to be

$$\frac{\partial \underline{\mathbf{v}}}{\partial t} + \underline{\mathbf{v}} \cdot \nabla \underline{\mathbf{v}} = -\nabla \left(Q + \Phi \right) \tag{4.104}$$

where Q is a potential body force field per unit mass and Φ is a scalar field whose gradient is defined by $\nabla \Phi = (\nabla p)/\rho$, assuming ρ is a function of pressure p only.

1. Using the identities given in Section 4.7.6, show that

$$\frac{\partial \underline{\boldsymbol{\omega}}}{\partial t} + \underline{\mathbf{v}} \cdot \nabla \underline{\boldsymbol{\omega}} = \underline{\boldsymbol{\omega}} \cdot \nabla \underline{\mathbf{v}} - \underline{\boldsymbol{\omega}} \left(\nabla \cdot \underline{\mathbf{v}} \right)$$

where $\underline{\boldsymbol{\omega}} = \nabla \times \underline{\mathbf{v}}$ is the vorticity of $\underline{\mathbf{v}}$. Or, using the substantial time derivative operator defined in (4.65),

$$\frac{D\,\underline{\boldsymbol{\omega}}}{Dt} = \underline{\boldsymbol{\omega}} \cdot \nabla \underline{\mathbf{v}} - \underline{\boldsymbol{\omega}} \left(\nabla \cdot \underline{\mathbf{v}} \right) \tag{4.105}$$

2. By dividing (4.105) by ρ and then using the equation of continuity given in (4.66), show that

$$\frac{D}{Dt}\left(\frac{1}{\rho} \underline{\boldsymbol{\omega}} \right) = \frac{1}{\rho} \underline{\boldsymbol{\omega}} \cdot \nabla \underline{\mathbf{v}} \tag{4.106}$$

This equation is known as the **Helmholtz vorticity equation** for inviscid fluids with $\rho = \rho(p)$.

E4.21. Obtain $h(\underline{\mathbf{v}}) = \underline{\mathbf{v}} \cdot (\nabla \times \underline{\mathbf{v}})$ in rectangular coordinates. When $h(\underline{\mathbf{v}}) \neq 0$, then $\underline{\mathbf{v}}$ is not perpendicular to $(\nabla \times \underline{\mathbf{v}})$. Show that this is the case for

$$\underline{\mathbf{v}}\,(x, y, z) = xyz\left(\underline{\boldsymbol{\delta}}_x + \underline{\boldsymbol{\delta}}_y \right) + \underline{\boldsymbol{\delta}}_z$$

for $x \neq y$ and $z \neq 0$. For the special case where $\underline{\mathbf{v}} = v_x\,(x, y)\,\underline{\boldsymbol{\delta}}_x + v_y\,(x, y)\,\underline{\boldsymbol{\delta}}_y + 0\underline{\boldsymbol{\delta}}_z$, what is $h\,(\underline{\mathbf{v}})$? What can one say about $h\,(\underline{\mathbf{v}})$ when $\underline{\mathbf{v}} = \nabla \psi\,(x, y, z)$, where ψ is any smooth scalar function ?

E4.22. Obtain $\nabla^2 \underline{\mathbf{v}}$ in cylindrical coordinates.

E4.23. Prove the following identity:

$$\left(\underline{\mathbf{v}} \cdot \nabla \right) \underline{\mathbf{v}} = \frac{1}{2}\left(\nabla \left\| \underline{\mathbf{v}} \right\|^2 \right) - \underline{\mathbf{v}} \times \left(\nabla \times \underline{\mathbf{v}} \right)$$

(Hint: Use identity (4.59).)

E4.24. Prove or disprove the following statement:

$$\nabla \left(\left\| \nabla \psi \right\|^2 \right) = 2\nabla \psi \cdot \nabla \left(\nabla \psi \right)$$

E4.25. The temperature distribution for a flat rectangular plate is given by

$$T\,(x, y) = 1 - e^{g(x,y)}$$

where,

$$g\,(x, y) = -4\left(\,(x - 0.5)^2 + (y + 0.5)^2\, \right)$$

for $-1 \leq x \leq 1$ and $-1 \leq y \leq 1$.

1. Calculate $\nabla T\,(x, y)$.
2. Obtain a contour plot of the temperature distribution and overlay it with the gradient vector field plot.
3. Evaluate $\nabla^2 T$ and its gradient.

4. Obtain another contour plot, this time of the Laplacian of T. Overlay it with the gradient field of the Laplacian.
5. What are the differences between the two plots?

E4.26. Find the relationship between the unit vectors of the cylindrical coordinate system with the unit vectors of the spherical coordinate system, that is, determine $R_{c\to s}$ such that

$$\begin{pmatrix} \underline{\delta}_{\widehat{r}} \\ \underline{\delta}_{\widehat{\theta}} \\ \underline{\delta}_{\phi} \end{pmatrix} = R_{c\to s} \begin{pmatrix} \underline{\delta}_{r} \\ \underline{\delta}_{\theta} \\ \underline{\delta}_{z} \end{pmatrix}$$

(Hint: Use the matrices found in Sections D.2 and D.3.)
(Note that we use $\widehat{r}$ and $\widehat{\theta}$ for the spherical coordinates to distinguish them from the r and θ variables, respectively, used for the cylindrical coordinates.)

E4.27. For a tensor $\underline{\underline{T}}$, find the relationship between the components in the rectangular coordinates and the components in the spherical coordinates, for example, $T_{rr} = T_{rr}\left(T_{xx}, \ldots, T_{zz}\right)$, and so forth.

E4.28. Show that (4.93) is equal to the formula for the Laplacian of a scalar field ψ given in (4.95) for the case of spherical coordinates.

E4.29. Given the vector field in the spherical coordinates,

$$\underline{\mathbf{v}}\left(r, \theta, \phi\right) = 2\left(\frac{1}{r^4} - \frac{1}{r^3}\right)\underline{\delta}_r$$

and a constant vector field $\underline{\mathbf{w}}$ in rectangular coordinates

$$\underline{\mathbf{w}} = 2\underline{\delta}_x + 4\underline{\delta}_y$$

Evaluate $(\underline{\mathbf{w}} \cdot \nabla\underline{\mathbf{v}})$ at $r = 1.2$ and $\theta = \phi = \pi/4$ and give the results in spherical coordinates, as well as in rectangular coordinates.

E4.30. For the parabolic coordinate system described in Example 4.17, obtain the divergence and curl of a vector field $\underline{\mathbf{v}}$ in this coordinate system. Also, obtain the reciprocal base vectors $\widehat{\underline{\nu}}_a$, $\widehat{\underline{\nu}}_b$, and $\widehat{\underline{\nu}}_\theta$ and show that $\underline{\nu}_a \cdot \widehat{\underline{\nu}}_a = 1$, $\underline{\nu}_b \cdot \widehat{\underline{\nu}}_b = 1$ and $\underline{\nu}_\theta \cdot \widehat{\underline{\nu}}_\theta = 1$.

E4.31. Prove the identities in (4.10). Hint:

$$\epsilon_{ijk} = \mathbf{det}\left(\mathbf{e}_i | \mathbf{e}_j | \mathbf{e}_k\right).$$

5 Vector Integral Theorems

In this chapter, we discuss the major integral theorems that are used to develop physical laws based on integrals of vector differential operations. The general theorems include the divergence theorem, the Stokes' theorem, and various lemmas such as the Green's lemma.

The divergence theorem is a very powerful tool in the development of several physical laws, especially those that involve conservation of physical properties. It connects volume integrals with surface integrals of fluxes of the property under consideration. In addition, the divergence theorem is also key to yielding several other integral theorems, including the Green's identities, some of which are used extensively in the development of finite element methods.

Stokes' theorem involves surface integrals and contour integrals. In particular, it relates curls of velocity fields with circulation integrals. In addition to its usefulness in developing physical laws, Stokes' theorem also offers a key criteria for path independence of line integrals inside a given region that can be determined to be **simply connected**. We discuss how to determine whether the regions are simply connected in Section 5.3.

In Section 5.5, we discuss the Leibnitz theorems involving the derivative of volume integrals in both 1D and 3D space with respect to a parameter α in which the boundaries and integrands are dependent on the same parameter α. These are important when dealing with time-dependent volume integrals.

In applying the various integral theorems, some computations may be necessary. To this end, we have included extensive discussions of the various types of integrals in Sections E.1 through E.3 (i.e., line integrals, surface integrals, and volume integrals). In these sections, we include the details for computation and some examples to appreciate the implications and evaluations of the various integral theorems useful for the integral theorems that are covered in this chapter.

In Section 5.4, we discuss two very important applications of the integral theorems. These are the development of conservation laws and the development of Maxwell equations for electricity and magnetism. Both these applications expose the power of vector differential operators for the development of the partial differential equations that model several physical systems.

We do not cover the actual solution of partial differential equations in this chapter. The solution of these partial differential equations can be very complex and is often difficult to obtain in closed analytical form except for specific cases.

Instead, their solutions are dealt with in later chapters, starting with Chapter 6 and succeeding chapters for handling problems that can be reduced to ordinary differential equations; Chapters 10 through 14 handle solution of certain classes of partial differential equations, containing both analytical and numerical approaches.

5.1 Green's Lemma

We begin with one of the basic formulas that consists of the relationship between a surface integral on surface S and its boundary C.

LEMMA 5.1. **Green's Lemma.** *Let $F(u, v)$ and $G(u, v)$ be differentiable functions in the domain $D \in \mathbb{R}^2$. Then*

$$\oint_C \left(F(u,v)du + G(u,v)dv \right) = \int_S \left(\frac{\partial G}{\partial u} - \frac{\partial F}{\partial v} \right) dudv \tag{5.1}$$

where

1. *C is a sectionally smooth boundary of the surface of integration S.*
2. *The positive direction of contour C is consistent with the definition of the positive direction of the vector*

$$\vec{\underline{\mathbf{n}}} = \frac{(\partial\vec{\underline{\mathbf{r}}}/\partial u) \times (\partial\vec{\underline{\mathbf{r}}}/\partial v)}{\left\| \partial\vec{\underline{\mathbf{r}}}/\partial u \times (\partial\vec{\underline{\mathbf{r}}}/\partial v) \right\|}$$

where $\vec{\underline{\mathbf{r}}}$ is the position vector.

PROOF. (See Section E.5.1 for proof.)

For the special case in which the surface of integration S lies in the (x, y)-plane, Green's lemma is stated in terms of x and y, that is, in (5.1) u and v are replaced by x and y, respectively. In this case, the positive direction of the contour is counterclockwise. Equivalently, the positive contour direction is chosen such that the region in S is always to the left of the contour's path.

EXAMPLE 5.1. Consider $F(x, y, z)$ and $G(x, y, z)$ given by

$$\begin{aligned} F(x,y,z) &= x^2 + y \\ G(x,y,z) &= 2x + 3y - z + 5 \end{aligned}$$

Let the surface of integration to be the top portion of the unit sphere with $0 \le \theta \le \pi/4$ and $0 \le \phi \le 2\pi$, as shown in Figure 5.1.

Using the parameterization based on the spherical coordinates, $u = \phi$ and $v = \theta$,

$$x = \sin(v)\cos(u) \quad ; \quad y = \sin(v)\sin(u) \quad ; \quad z = \cos(v)$$

and

$$\begin{aligned} F(u,v) &= (\sin(v)\cos(u))^2 + \sin(v)\sin(u) \\ G(u,v) &= 2\sin(v)\cos(u) + 3\sin(v)\sin(u) - \cos(v) + 5 \end{aligned}$$

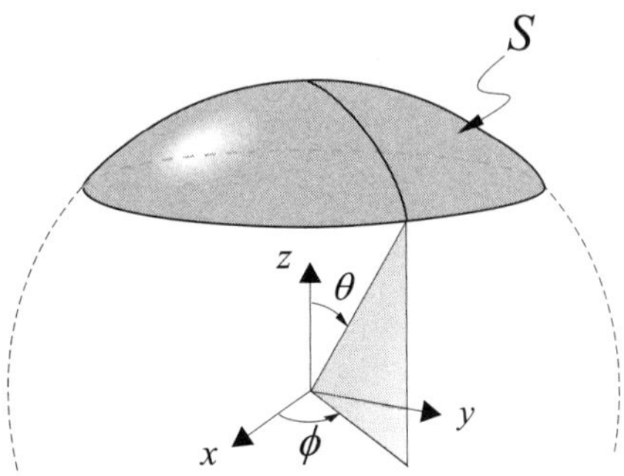

Figure 5.1. The surface of integration given by the top of a unit sphere.

which then yield

$$\begin{aligned}\frac{\partial F}{\partial v} &= \cos(v)\left(\sin(u) + 2\sin(v)\cos^2(u)\right)\\ \frac{\partial G}{\partial u} &= \sin(v)\left(3\cos(u) - 2\sin(u)\right)\end{aligned}$$

and the surface integral in the right-hand side of (5.1).

$$\int_0^{\pi/4}\int_0^{2\pi}\left(\frac{\partial G}{\partial u} - \frac{\partial F}{\partial v}\right) du dv = -\frac{\pi}{2}$$

The closed contour in the (u, v) plane is shown in Figure 5.2. The positive direction is counterclockwise, and it yields a normal vector of points in S that is outward from the center of the sphere.

Based on Figure 5.2, the line integrals can be calculated to be

$$\oint_C F(u,v)du = \int_0^{2\pi} F(u,0)\,du + \int_{2\pi}^{0} F\left(u,\frac{\pi}{4}\right) du = -\frac{\pi}{2}$$

$$\begin{aligned}\oint_C G(u,v)dv &= \int_0^{\pi/4} G(2\pi, v)\,dv + \int_{\pi/4}^{0} G(0,v)\,dv\\ &= \left(-\frac{3\sqrt{2}}{2} + 2 + 5\frac{\pi}{4}\right) + \left(\frac{3\sqrt{2}}{2} - 2 - 5\frac{\pi}{4}\right) = 0\end{aligned}$$

Combining all the results, we see that Green's lemma applies, that is,

$$\oint_C F du + \oint_C G dv = \int_S \left(\frac{\partial G}{\partial u} - \frac{\partial F}{\partial v}\right) du dv$$

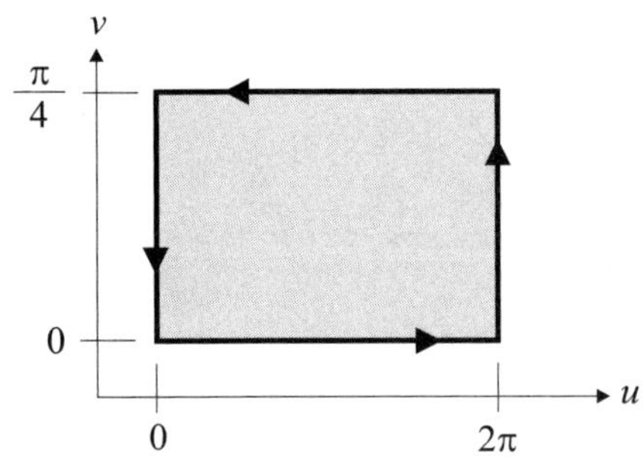

Figure 5.2. The closed path of integration in the (u, v) plane.

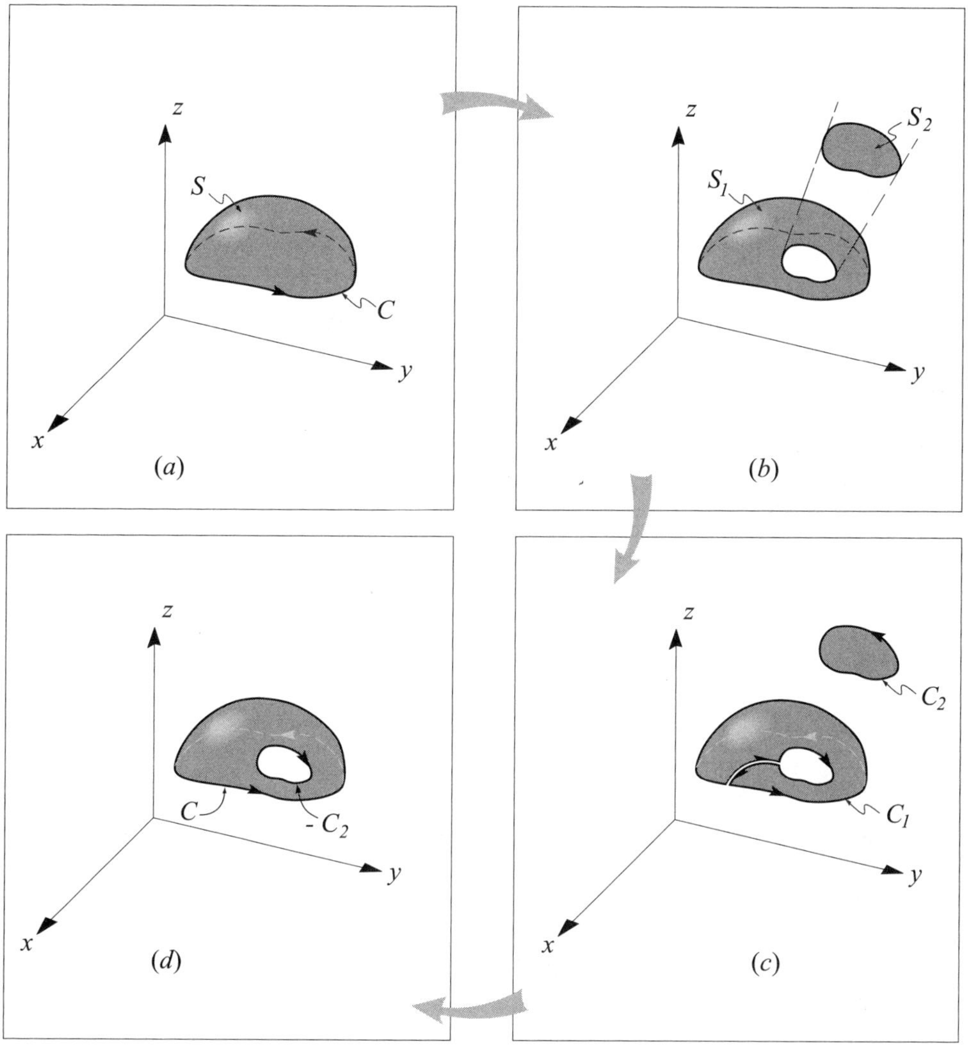

Figure 5.3. The surface of integration with holes: (a) the original surface S and contour C, (b) S_1, obtained by removing S_2, (c) the continuous contour C_1 for S_1, (d) due to the cancelation of line integrals in oppositely directed segments, $\oint_{C_1} = \oint_C - \oint_{C_2}$.

Let Γ_1 and Γ_2 be two opposite paths, that is, they travel through the same set of points but having switched the end points. Then we note that

$$\int_{\Gamma_1} F(u, v)ds = -\int_{\Gamma_2} F(u, v)ds \quad \rightarrow \quad \int_{\Gamma_1} F(u, v)ds + \int_{\Gamma_2} F(u, v)ds = 0$$

We can then use this fact to evaluate surface integrals in surfaces that contain holes. For illustration purposes, consider the surface shown in Figure 5.3b. A surface S_1 was obtained by removing S_2 from the original surface S. In Figure 5.3c, a continuous contour C_1 can be generated as the boundary of S_1. At some point along the outer boundary, contour C_1 cuts a path to reach the inner hole. The contour then traverses the boundary of this hole. After reaching the point where the contour first entered the hole, the contour C_1 retraces the same cutting path backward to continue tracking the outer path.

The two line integrals that have traversed the same path segments, but in opposite directions, will cancel each other out. Thus, as shown in Figure 5.3d, the line integral along C_1 will be equal to the line integral along C minus the line integral along C_2, which is the contour of the removed surface S_2, that is,

$$\oint_{C_1} f\, ds = \oint_{C} f\, ds - \oint_{C_2} f\, ds \tag{5.2}$$

where f is any appropriate function. From the definition of surface integrals, we know that

$$\int_{S_1} g\, dS = \int_{S} g\, dS - \int_{S_2} g\, dS \tag{5.3}$$

where g is any appropriate function.

Then applying Green's lemma and (5.2) to (5.3),

$$\begin{aligned} \int_{S_1} \left(\frac{\partial G}{\partial u} - \frac{\partial F}{\partial v}\right) du dv &= \oint_{C} (F du + G dv) - \oint_{C_2} (F du + G dv) \\ &= \oint_{C_1} (F du + G dv) \end{aligned} \tag{5.4}$$

Thus Green's lemma can be applied to surfaces with holes by using the strategy of cutting through the region to connect the outer contour with the inner contours of the holes.

5.2 Divergence Theorem

Consider a vector field $\underline{\mathbf{f}}$. The **divergence theorem**[1] states that the volume integral of the divergence $\nabla \cdot \underline{\mathbf{f}}$ in a given volume region V can also be evaluated indirectly by taking the surface integral of the flux of $\underline{\mathbf{f}}$, that is, $\underline{\mathbf{f}} \cdot \underline{\mathbf{n}}$, over the surface S that bounds the region V. In other words, the net sum of sinks and sources of a property inside a volume region can be evaluated by calculating the total flux of that property coming out of the surface of that region.

THEOREM 5.1. *Let $\underline{\mathbf{f}}(x, y, z)$ be a vector field which is differentiable in V then*

$$\int_S \underline{\mathbf{f}} \cdot \underline{\vec{\mathbf{n}}}\, dS = \int_V \nabla \cdot \underline{\mathbf{f}}\, dV \tag{5.5}$$

PROOF. (See Section E.5.2 for proof.)

The divergence theorem can be applied to (4.55) and (4.57) to yield,

$$\int_S \psi \underline{\mathbf{f}} \cdot \underline{\vec{\mathbf{n}}}\, dS = \int_V \left(\nabla \psi \cdot \underline{\mathbf{f}} + \psi \nabla \cdot \underline{\mathbf{f}}\right) dV \tag{5.6}$$

$$\int_S \left(\underline{\mathbf{f}} \times \underline{\mathbf{g}}\right) \cdot \underline{\vec{\mathbf{n}}}\, dS = \int_V \left(\underline{\mathbf{g}} \cdot (\nabla \times \underline{\mathbf{f}}) - \underline{\mathbf{f}} \cdot \left(\nabla \times \underline{\mathbf{g}}\right)\right) dV \tag{5.7}$$

[1] The theorem is also known as the **Gauss-Ostrogradski-Green divergence theorem**. However, because the same names appear in several versions of the divergence theorem, we just call the main theorem "divergence theorem" and leave the names to their specific forms (e.g., Green's theorem and Gauss' theorem).

$$\int_S \psi \vec{\underline{\mathbf{n}}}\, dS = \int_V \nabla\psi dV \tag{5.8}$$

$$\int_S \vec{\underline{\mathbf{n}}} \times \underline{\mathbf{f}}\, dS = \int_V \nabla \times \underline{\mathbf{f}}\, dV \tag{5.9}$$

Equations (5.8) can be obtained from (5.6) by setting $\underline{\mathbf{f}}$ to be a constant vector. Likewise, (5.9) can be obtained from (5.7) by setting $\underline{\mathbf{g}}$ to be a constant vector. Note, however, that the integrands in (5.6) and (5.7) are scalar fields, whereas the integrands in (5.8) and (5.9) are vector fields.

By letting $\underline{\mathbf{f}} = \nabla\phi$ in (5.6), we obtain **Green's theorem**, also known as **Green's identities**.

THEOREM 5.2. *Let ϕ and ψ be scalar functions that are twice differentiable. Then*

$$\int_S (\psi\nabla\phi)\cdot\vec{\underline{\mathbf{n}}}\, dS = \int_V (\nabla\phi\cdot\nabla\psi)\ dV + \int_V (\psi\nabla^2\phi)\ dV \tag{5.10}$$

$$\int_S (\psi\nabla\phi - \phi\nabla\psi)\cdot\vec{\underline{\mathbf{n}}}\, dS = \int_V \left(\psi\nabla^2\phi - \phi\nabla^2\psi\right)\ dV \tag{5.11}$$

PROOF. (See Section E.5.3 for proof.)

One could also view the divergence theorem as a reduction in the dimension of the integration, that is, a volume integral on one side while a surface integral on the other. Thus if the volume integration is difficult to obtain, the reduction to a surface integral would usually make it easier to evaluate, at least numerically. However, there are several instances in which volume integration would actually be easier to obtain than the surface integral. This is specially true if the divergence operation simplifies several terms in the integrand, as shown in the next example.

EXAMPLE 5.2. Consider the vector field

$$\begin{aligned}\underline{\mathbf{f}} &= \left(a_x x^2 + b_x x + f(y,z)\right)\underline{\boldsymbol{\delta}}_x + \left(a_y y^2 + b_y y + g(x,z)\right)\underline{\boldsymbol{\delta}}_y \\ &\quad + \left(a_z z^2 + b_z z + h(x,y)\right)\underline{\boldsymbol{\delta}}_z\end{aligned}$$

where a_x, a_y, a_z, b_x, b_y, and b_z are constants. Let the surface of integration S be a sphere of radius ρ. The surface integral $\int_S \underline{\mathbf{f}}\cdot\vec{\underline{\mathbf{n}}}\, dS$ could be very difficult to evaluate directly, especially when f, g, and h are complicated functions. However, the divergence of $\underline{\mathbf{f}}$ yields

$$\nabla\cdot\underline{\mathbf{f}} = 2\left(a_x x + a_y y + a_z z\right) + \left(b_x + b_y + b_z\right)$$

Using the divergence theorem,

$$\begin{aligned}\int_S \underline{\mathbf{f}}\cdot\vec{\underline{\mathbf{n}}}\, dS &= 2a_x\int_V x dV + 2a_y\int_V y dV + 2a_z\int_V z dV + \left(b_x + b_y + b_z\right)\int_V dV \\ &= \frac{4\left(b_x + b_y + b_z\right)}{3}\pi\rho^3\end{aligned}$$

Another application of the divergence theorem, via the Green's theorem given previously, is in the development of "weak solutions" needed by the finite elements methods. Briefly, the finite elements method partitions domains in which the first derivatives are satisfied. However, at the edges where the elements are patched together, the smoothness is no longer guaranteed. Green's theorem allows for volume integrals of Laplacians such as $\nabla^2\phi$ to be replaced by the surface integrals involving gradients. Details for the weak-solution formulation of the finite element methods are covered later in Section 14.1.

A very important application of the divergence theorem is the **Gauss' theorem**, which is also useful in solving partial differential equations.

THEOREM 5.3.

$$\int_S \frac{1}{r^2}\underline{\delta}_r \cdot \vec{\underline{n}}\, dS = \begin{cases} 0 & \textit{if origin is outside of } S \\ 4\pi & \textit{if origin is inside of } S \end{cases} \tag{5.12}$$

PROOF. (See Section E.5.4 for proof.)

5.3 Stokes' Theorem and Path Independence

Another important theorem is the **Stokes' theorem**. It gives a relationship between the curl of a vector field $\underline{\mathbf{f}}$ in a region S and the tangential projection of $\underline{\mathbf{f}}$ along the boundary of S. It can also be seen as a generalization of Green's lemma for a plane.[2]

THEOREM 5.4. *Let $\underline{\mathbf{f}}$ be a differentiable vector field in a surface S bounded by the closed curve C, where S is a connected and sectionally smooth surface. Then*

$$\oint_C \underline{\mathbf{f}} \cdot d\vec{\underline{\mathbf{r}}} = \int_S (\nabla \times \underline{\mathbf{f}}) \cdot \vec{\underline{n}}\, dS \tag{5.13}$$

PROOF. (See Section E.5.5 for proof.)

The line integral,

$$\oint_C \underline{\mathbf{f}} \cdot d\vec{\underline{\mathbf{r}}} = \oint_C \underline{\mathbf{f}} \cdot \vec{\underline{\mathbf{t}}}\, ds$$

is known as the **circulation** of $\underline{\mathbf{f}}$ along the closed path C. Thus (5.13) simply states that the sum of all the curls of $\underline{\mathbf{f}}$ fluxing out of the surface S bounded by C is equal to the circulation of $\underline{\mathbf{f}}$ along C (see Figure 5.4).

Stokes' theorem has several application in physics, including one approach in the development of Maxwell's equations for electric and magnetic intensity. Another application is to use it to assess path independence of line integrals.

[2] In fact, based on exterior calculus, a general Stokes' theorem can be developed that also generalizes the divergence theorem.

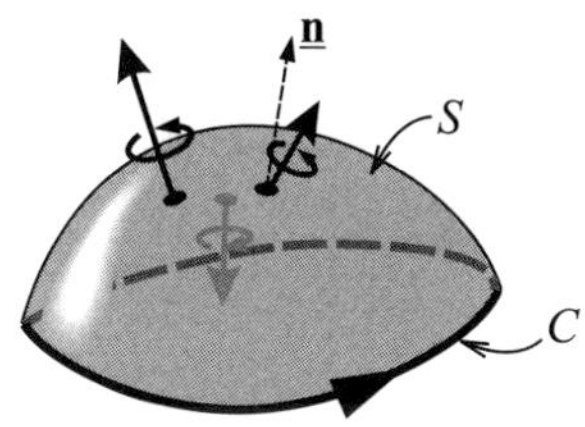

Figure 5.4. A visual description of stokes theorem.

Definition 5.1. *Let* $\underline{\mathbf{f}}$ *be a given vector field and* $\vec{\underline{\mathbf{r}}}$ *be the position vector, then the line integral*

$$I_C = \int_C \underline{\mathbf{f}} \cdot d\vec{\underline{\mathbf{r}}} \tag{5.14}$$

is said to be **path independent** *in a region* V *if, for any pair of curves,* C_1 *and* C_2 *inside the region* V,

$$I_{C_1} = I_{C_2} \tag{5.15}$$

where C_1 *and* C_2 *are continuous and sectionally smooth curves that have the same initial point* (x_i, y_i, z_i) *and the same end point* (x_f, y_f, z_f).

When V is not specified, then it is usually understood that the path independence refers to the whole 3D space.

Consider two paths $C_{1,AB}$ and $C_{2,AB}$ in region V that do not intersect each other except at the start and end points. If the line integrals are independent of path,

$$\int_{C_{1,AB}} \underline{\mathbf{f}} \cdot d\vec{\underline{\mathbf{r}}} = \int_{C_{2,AB}} \underline{\mathbf{f}} \cdot d\vec{\underline{\mathbf{r}}} \tag{5.16}$$

which could be combined into one integral,

$$\begin{aligned} 0 &= \int_{C_{1,AB}} \underline{\mathbf{f}} \cdot d\vec{\underline{\mathbf{r}}} - \int_{C_{2,AB}} \underline{\mathbf{f}} \cdot d\vec{\underline{\mathbf{r}}} \\ &= \int_{C_{1,AB}-C_{2,AB}} \underline{\mathbf{f}} \cdot d\vec{\underline{\mathbf{r}}} \\ 0 &= \oint_C \underline{\mathbf{f}} \cdot d\vec{\underline{\mathbf{r}}} \end{aligned} \tag{5.17}$$

where $C = C_{1,AB} - C_{2,AB}$ is a simple closed path, as shown in Figure 5.5. With any choice of $C_{1,AB}$ and $C_{2,AB}$ (so far assumed to be nonintersecting at midpath), by reversing the path direction of $C_{2,AB}$, a closed path C is generated. Path independence guarantees that the line integral using the closed path will have to be zero.

Alternatively, we could have rearranged (5.16) to be

$$\int_{C_{2,AB}-C_{1,AB}} \underline{\mathbf{f}} \cdot d\vec{\underline{\mathbf{r}}} = 0 = \oint_{C'} \underline{\mathbf{f}} \cdot d\vec{\underline{\mathbf{r}}} \tag{5.18}$$

where $C' = C_{2,AB} - C_{1,AB}$ is also a simple closed path but in the opposite direction of C.

Now consider the situation shown in Figure 5.6 where path $C_{1,AB}$ and $C_{2,AB}$ might intersect each other somewhere between the end points, say, at point D. Path $C_{1,AB}$ can then be partitioned to be the sum of two subpaths: $C_{1,AD}$ and $C_{1,DB}$. Likewise, path $C_{2,AB}$ can also be partitioned to be the sum of two subpaths: $C_{2,AD}$ and $C_{2,DB}$.

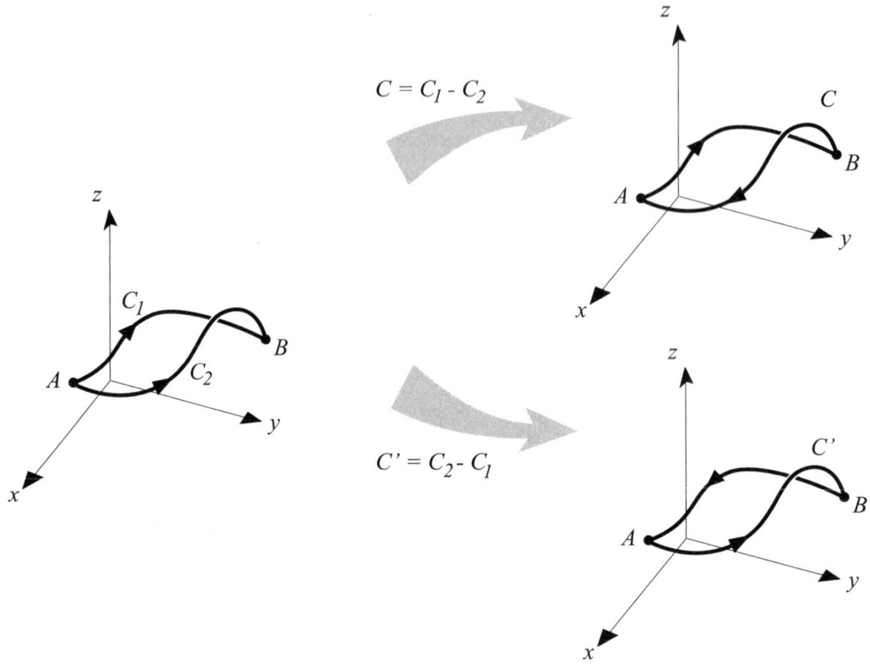

Figure 5.5. Two paths with the same start point A and end point B.

Equation (5.16) can now be expanded to be

$$
\begin{aligned}
\int_{C_{1,AB}} \underline{\mathbf{f}} \cdot d\underline{\vec{\mathbf{r}}} - \int_{C_{2,AB}} \underline{\mathbf{f}} \cdot d\underline{\vec{\mathbf{r}}} &= 0 \\
\int_{C_{1,AD}} \underline{\mathbf{f}} \cdot d\underline{\vec{\mathbf{r}}} + \int_{C_{1,DB}} \underline{\mathbf{f}} \cdot d\underline{\vec{\mathbf{r}}} - \int_{C_{2,AD}} \underline{\mathbf{f}} \cdot d\underline{\vec{\mathbf{r}}} - \int_{C_{2,DB}} \underline{\mathbf{f}} \cdot d\underline{\vec{\mathbf{r}}} &= 0 \\
\int_{C_{1,AD}} \underline{\mathbf{f}} \cdot d\underline{\vec{\mathbf{r}}} - \int_{C_{2,AD}} \underline{\mathbf{f}} \cdot d\underline{\vec{\mathbf{r}}} + \int_{C_{1,DB}} \underline{\mathbf{f}} \cdot d\underline{\vec{\mathbf{r}}} - \int_{C_{2,DB}} \underline{\mathbf{f}} \cdot d\underline{\vec{\mathbf{r}}} &= 0 \\
\int_{C_{1,AD}-C_{2,AD}} \underline{\mathbf{f}} \cdot d\underline{\vec{\mathbf{r}}} + \int_{C_{1,DB}-C_{2,DB}} \underline{\mathbf{f}} \cdot d\underline{\vec{\mathbf{r}}} &= 0 \\
\oint_{C_{AD}} \underline{\mathbf{f}} \cdot d\underline{\vec{\mathbf{r}}} - \oint_{C_{DB}} \underline{\mathbf{f}} \cdot d\underline{\vec{\mathbf{r}}} &= 0 \qquad (5.19)
\end{aligned}
$$

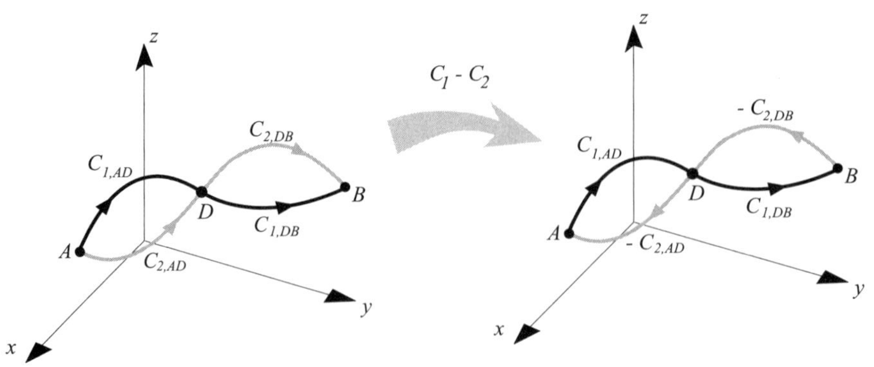

Figure 5.6. Two paths with the same start point A and end point B plus an intersection at point D.

where C_{AD} is the closed path formed by adding path $C_{1,AD}$ and reverse of path $C_{2,AD}$. Similarly, C_{DB} is the closed path formed by adding path $C_{1,DB}$ and reverse of path $C_{2,DB}$. Because the definition of path independence applies to the subpaths, the two closed paths will each generate a zero line integral also. Thus the condition for path independence is equivalent to having the line integral of any closed paths in region V all be zero, including whichever direction the closed path takes.

Having established this equivalence, we can use Stokes' theorem to determine path independence. Recall that Stokes' theorem relates the line integral in a closed path to a surface integral involving the curl of vector field $\underline{\mathbf{f}}$:

$$\oint_C \underline{\mathbf{f}} \cdot d\underline{\vec{\mathbf{r}}} = \int_S \underline{\vec{\mathbf{n}}} \cdot (\nabla \times \underline{\mathbf{f}})\, dS \tag{5.20}$$

If $\nabla \times \underline{\mathbf{f}} = 0$ for all points in region V, then Stokes' theorem suggests that this condition will also guarantee path independence, because for any surface S inside region V, a zero curl implies a zero line integral on the left-hand side of (5.20).

One more detail is still needed, however. Stokes' theorem requires that the integrand in the surface integral be bounded. This means that in the chosen region V, no singularities of the integrand can be present. In relation to the closed path integral, this kind of region is formally referred to as a **simply connected** region.

Definition 5.2. *A region V is* **simply connected** *if any simple closed path in the region can be continuously deformed to a single point. Otherwise, the region is* **multiply connected**.

A set of examples is shown in Figure 5.7. The first case in Figure 5.7(a) is a full rectangular box. In this case, we see that any closed path contained in the region can be deformed continuously into a point inside the box. The second case is shown in Figure 5.7(b). For this case, a cylindrical subregion has been removed from the center. Even though there are some closed paths that could deform to a point, the existence of at least one closed path that will not continuously deform to a point is sufficient to have this region be multiply connected. The third case is given in Figure 5.7(c). In this case, the rectangular box region has a spherical subregion removed. However, unlike case (b), any closed path inside the region of case (c) can still be continuously deformed to a point. Thus the third case is a simply connected region.

We can now summarize the preceding discussion in the following theorem:

THEOREM 5.5. *Let $\nabla \times \underline{\mathbf{f}} = \mathbf{0}$ inside a simply connected region V; then the integral $\int_C \underline{\mathbf{f}} \cdot d\underline{\vec{\mathbf{r}}}$ is independent of path inside region V.*

EXAMPLE 5.3. Consider the following vector fields:

$$\begin{aligned}
\underline{\mathbf{h}} &= y\,\underline{\delta}_x + x\,\underline{\delta}_y + 2z\,\underline{\delta}_z \\
\underline{\mathbf{g}} &= \left(x^2+y^2+z^2\right)\underline{\delta}_x + 2xy\,\underline{\delta}_y + (y-z)\,\underline{\delta}_z \\
\underline{\mathbf{f}} &= -\frac{y+2}{(x-1)^2+(y+2)^2}\,\underline{\delta}_x + \frac{x-1}{(x-1)^2+(y+2)^2}\,\underline{\delta}_y + 2\,\underline{\delta}_z
\end{aligned}$$

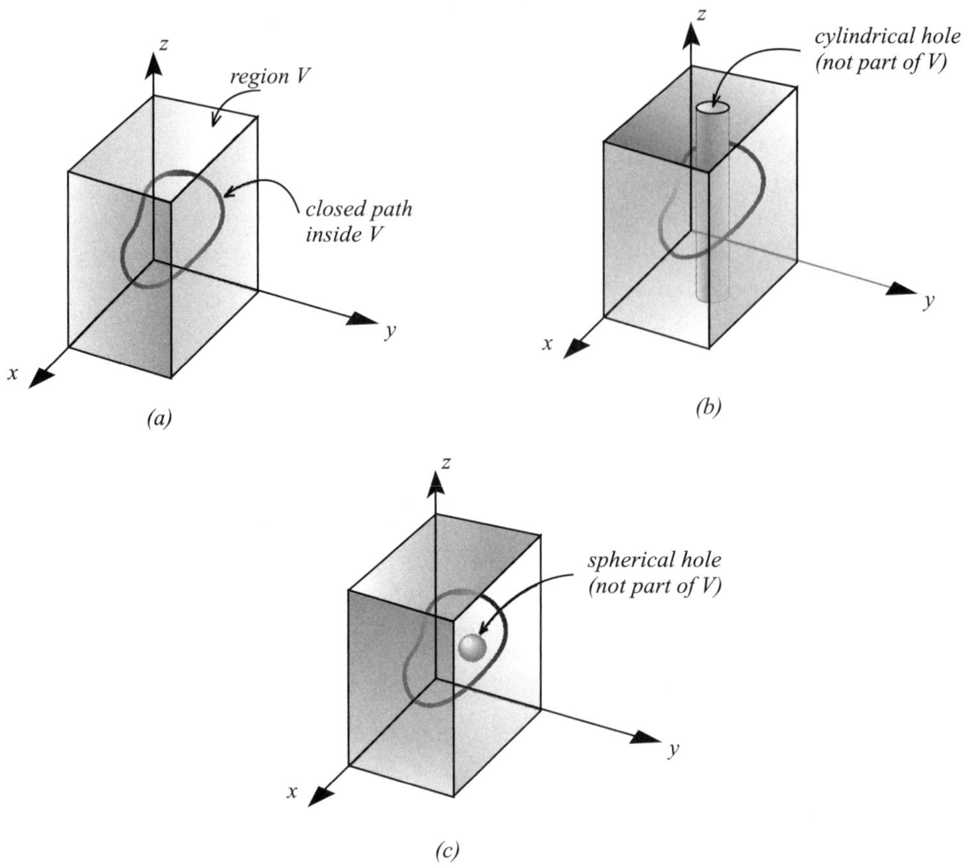

Figure 5.7. Examples of simply and multiply connected regions. (a) Solid rectangular region: simply connected; (b) solid rectangular region with a cylindrical subregion removed from the center: multiply connected; (c) solid rectangular region with a spherical subregion removed from the center: simply connected.

and the following paths with the parameter t ranging from 0 to 1:

Path C_1:
$$x = 3 + \cos(3\pi t), \quad y = 3 + \sin(3\pi t), \quad z = 2t$$

Path C_2:
$$x = -12t^2 + 10t + 4, \quad y = 3 + 4t(1-t), \quad z = 2 - 2(1-t)^2$$

Path C_3:
$$x = -6t^2 + 5t + 1, \quad y = 2t^2 - 5t, \quad z = -6t^2 + 7t$$

Path C_4:
$$x = 10t^2 - 11t + 1, \quad y = 2t^2 - 5t, \quad z = 10t^2 - 9t$$

Figure 5.8 shows paths C_1 and C_2. Note that C_1 has a helical form. Figure 5.9 shows paths C_3 and C_4 together. Included in the figure is a line described by $(x, y, z) = (1, -2, z)$. This line includes the singular point of vector field $\underline{\mathbf{f}}$. Thus the line is only relevant when considering $\underline{\mathbf{f}}$.

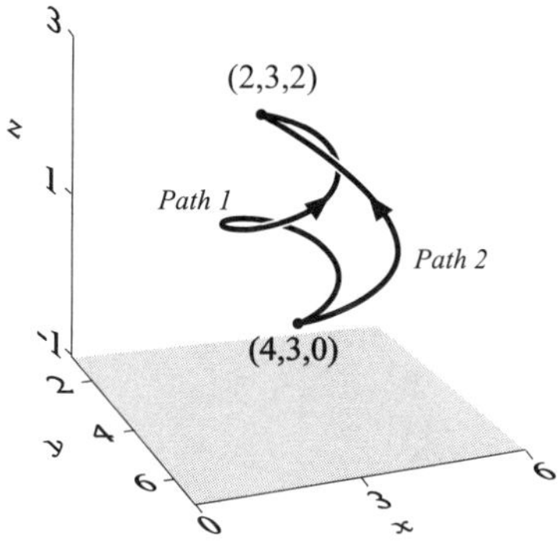

Figure 5.8. Paths C_1 and C_2.

We can tabulate the results of calculating the various type of line integrals of $\int_C \underline{\mathbf{h}} \cdot d\vec{\underline{r}}$, $\int_C \underline{\mathbf{g}} \cdot d\vec{\underline{r}}$, and $\int_C \underline{\mathbf{f}} \cdot d\vec{\underline{r}}$ for different paths. This is shown in Table 5.1, together with curl of the respective vector fields.

As expected, the line integrals for $\underline{\mathbf{h}}$ are path independent because the curl is zero and the whole 3D space is simply connected. For $\underline{\mathbf{g}}$, because the curl is not zero, the line integrals depend on the path. Finally, for $\underline{\mathbf{f}}$, because there exists a simply connected region that covers C_1 and C_2, the two line integrals are expected to be equal. However, as we increase the region to contain paths C_3 and C_4, the region (after removal of regions containing singularities) is multiply connected. Thus the theorem states that we expect that path independence is no longer guaranteed, even though the curls along paths C_3 and C_4 are zero.

5.4 Applications

In this section, we discuss two major applications of the vector integral theorems. The first application mainly uses the divergence theorem for obtaining the various conservation laws. The second application is in the field of electrodynamics. Note that the main activity here is to obtain integral and differential equations that describe the physical laws. The solution of these differential equations will not be treated here. Instead, the analytical and numerical solutions of these equations are discussed in later chapters.

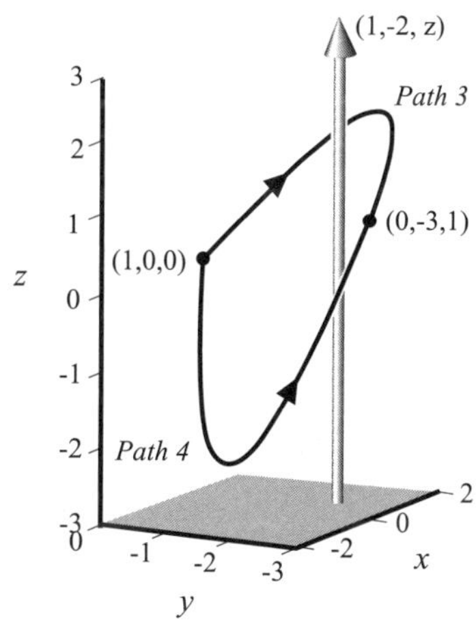

Figure 5.9. Paths C_3 and C_4. Also included here is the line $(1, -2, z)$.

Table 5.1. *Line integrals based on different vector fields and paths*

	Paths				
	Start: (4, 3, 0) End: (2, 3, 2)		Start: (1, 0, 0) End: (0, −3, 1)		
Vector fields	C_1	C_2	C_3	C_4	Curl
$\underline{\mathbf{h}}$	−2	−2	1	1	0
$\underline{\mathbf{g}}$	−36.062	−45.2	−4.4	−11.333	$-\underline{\delta}_x + 2x\underline{\delta}_y$
$\underline{\mathbf{f}}$	4.343	4.343	−1.927	4.356	0

5.4.1 Conservation Laws and Balance Equations

One of the major uses of vector analysis and the integral theorem is to use the law of conservation and balance equations within a closed region based on volume integrals (**weak forms**) to obtain differential vector laws (**strong equations**). We start by considering the rate of change of a quantity $\boldsymbol{\phi}(V)$ in a fixed volume V enclosed by a closed surface S. Often these quantities $\boldsymbol{\phi}$ can be represented instead by an associated density, which we denote by $\widehat{\boldsymbol{\phi}}_V$. For instance, if $\boldsymbol{\phi}$ is the mass, then $\widehat{\boldsymbol{\phi}}_V$ will be the density ρ. The general rate of change equations specified by the boundary of V is now

$$\begin{pmatrix}\text{Rate of Change}\\ \text{of } \boldsymbol{\phi}\\ \text{inside } V\end{pmatrix} = -\begin{pmatrix}\text{Net flux of } \boldsymbol{\phi}\\ \text{out of } V\\ \text{across surface } S\end{pmatrix} + \begin{pmatrix}\text{Internal rate of generation}\\ \text{of } \boldsymbol{\phi}\\ \text{inside } V\end{pmatrix} + \begin{pmatrix}\text{External effects}\\ \text{on } V \text{ and } S\end{pmatrix} \tag{5.21}$$

Equation (5.21) is quite general and applicable to several objects being considered. In integral form, we have

$$\frac{\partial}{\partial t}\int_V \widehat{\boldsymbol{\phi}}_V dV = -\int_S \widehat{\boldsymbol{\phi}}_V \underline{\mathbf{v}} \cdot \underline{\vec{\mathbf{n}}}\, dS - \int_S \underline{\mathbf{f}} \cdot \underline{\vec{\mathbf{n}}}\, dS + \int_V \mathcal{G}\, dV + \int_V \mathcal{E}_V\, dV + \int_S \mathcal{E}_S\, dS \tag{5.22}$$

where $\mathcal{G}$ is the generation of ϕ per unit volume, whereas $\mathcal{E}_V$ and $\mathcal{E}_S$ are external effects on ϕ per unit volume and per unit surface area, respectively. Here we have also separated the convective flux $\widehat{\boldsymbol{\phi}}_V \underline{\mathbf{v}}$ from other mechanisms of fluxes described by $\underline{\mathbf{f}}$. We treat some of the major applications, including mass, momentum, chemical component, and energy. For each of these cases, each term (e.g., flux terms and external effects) will vary and requires extra constitutive equations obtained from physical and empirical laws (often constrained under specific conditions).

1. **Mass Balance**
   Here, we have $\boldsymbol{\phi} = m$, which is mass; then the density is $\widehat{\boldsymbol{\phi}}_V = \rho$. The flux of mass out is given only by convective flow normal to the surface boundary,

that is,

$$\text{Net flow of mass out of } S = \int_S \rho \underline{\mathbf{v}} \cdot \underline{\vec{\mathbf{n}}}\, dS$$

where $\underline{\vec{\mathbf{n}}}$ is the unit normal vector in S pointing outward from dS. Because mass cannot be created nor destroyed, we have $\mathcal{G} = 0$, $\mathcal{E}_V = \mathcal{E}_S = 0$ in (5.22). Thus (5.22) becomes

$$\frac{\partial}{\partial t} \int_V \rho dV = -\int_S \rho \underline{\mathbf{v}} \cdot \underline{\vec{\mathbf{n}}}\, dS$$

Because V is assumed fixed, we can move the time derivative inside the integral.[3] Then, after applying the divergence theorem, we obtain

$$\int_V \left(\frac{\partial \rho}{\partial t} + \nabla \cdot \rho \underline{\mathbf{v}} \right) dV = 0 \tag{5.23}$$

which is the **integral form of the continuity equation**. Because V has been set arbitrarily, we can extract the integrand and obtain

$$\frac{\partial \rho}{\partial t} + \nabla \cdot \rho \underline{\mathbf{v}} = 0 \tag{5.24}$$

which is the **differential form of the continuity equation**. Using the substantial derivative operator defined in (4.65), that is, $D/Dt = \partial/\partial t + \underline{\mathbf{v}} \cdot \nabla$, (5.24) can be rewritten as

$$\frac{D\rho}{Dt} + \rho \nabla \cdot \underline{\mathbf{v}} = 0$$

2. **Momentum Balance**

Let $\phi = \underline{\mathbf{v}}$ be momentum per unit mass and $\widehat{\boldsymbol{\phi}}_V = \rho \underline{\mathbf{v}}$ becomes momentum per unit volume. We now assume that there is no internal generation of momentum (e.g., no reactions or explosions). Furthermore, we can take the external forces to be due only to pressure normal to surface S and gravitational acceleration $\underline{\mathbf{g}}$ acting on the volume V. Thus the external effects are given by

$$\text{External effects} = -\int_S p\, \underline{\vec{\mathbf{n}}}\, dS + \int_V \rho\, \underline{\mathbf{g}}\, dV$$

The momentum flux has two components: one is due to convective flow, and the other is due to the stress at the boundary. The flow of momentum out across S is given by

$$\text{Convective momentum flow out} = \int_S (\rho \underline{\mathbf{v}}) \left(\underline{\mathbf{v}} \cdot \underline{\vec{\mathbf{n}}} \right) dS$$

The other flux, also known as the **molecular momentum flux**, is due to the material stress traction vector and is given by the stress vectors pointing outward of S. Based on the discussion given in Example 4.4, we have

$$\text{Molecular momentum flow out} = \int_S \underline{\underline{T}} \cdot \underline{\vec{\mathbf{n}}}\, dS$$

[3] If V is not fixed, one needs to apply Leibnitz rules, as discussed in Section 5.5.

where $\underline{\underline{T}}$ is the **stress tensor** field. Combining the various terms, (5.22) now becomes

$$\frac{\partial}{\partial t}\int_V \rho\,\underline{\mathbf{v}}\,dV = -\int_S \left(\rho\,\underline{\mathbf{v}}\,\underline{\mathbf{v}} + \underline{\underline{T}}\right)\cdot\underline{\vec{\mathbf{n}}}\,dS - \int_S p\,\underline{\vec{\mathbf{n}}}\,dS + \int_V \rho\,\underline{\mathbf{g}}\,dV$$

After applying the divergence theorem, and consideration that V is fixed but arbitrary, we arrive at

$$\frac{\partial}{\partial t}(\rho\underline{\mathbf{v}}) = -\nabla\cdot(\rho\,\underline{\mathbf{v}}\,\underline{\mathbf{v}}) - \nabla\cdot\underline{\underline{T}} - \nabla p + \rho\underline{\mathbf{g}} \tag{5.25}$$

Equation (5.25) is known as the **Cauchy equation of motion**[4] or, in terms of the substantial time derivative operator,

$$\frac{D}{Dt} = \frac{\partial}{\partial t} + \underline{\mathbf{v}}\cdot\nabla$$

(5.25) together with (5.24) (see Exercise **E4.18**) will be reduced to

$$\rho\frac{D\underline{\mathbf{v}}}{Dt} = -\nabla\cdot\underline{\underline{T}} - \nabla p + \rho\underline{\mathbf{g}} \tag{5.26}$$

Two special cases are often used in the study of fluid dynamics. The first is when the material is incompressible and Newtonian, and we have $\nabla\cdot\underline{\underline{T}} = -\mu\nabla^2\underline{\mathbf{v}}$ (see exercise **E4.19**). This reduces (5.26) to

$$\rho\frac{D\underline{\mathbf{v}}}{Dt} = -\nabla p + \mu\nabla^2\underline{\mathbf{v}} + \rho\underline{\mathbf{g}} \tag{5.27}$$

The other special case is when $\nabla\cdot\underline{\underline{T}}$ is negligible, e.g., for an inviscid fluid, then (5.26) reduces to **Euler's equation of motion**,

$$\rho\frac{D\underline{\mathbf{v}}}{Dt} = -\nabla p + \rho\underline{\mathbf{g}} \tag{5.28}$$

3. **Energy Balance**

Let $\phi = \widehat{E}$ be the energy per unit mass; then $\widehat{\boldsymbol{\phi}}_V = \rho\widehat{E}$. The energy per unit mass is the sum of three terms: the specific internal energy $\widehat{u}$, the specific kinetic energy $\|\underline{\mathbf{v}}\|^2/2$, and the specific potential energy $\widehat{e}_p$, that is,

$$\widehat{E} = \widehat{u} + \frac{\|\underline{\mathbf{v}}\|^2}{2} + \widehat{e}_p$$

The flow of energy due to convection across surface S is given by

$$\text{Flow of energy through } S = \int_S \rho\widehat{E}\,(\underline{\mathbf{v}}\cdot\underline{\vec{\mathbf{n}}})\;dS$$

According to the first law of thermodynamics, there is no internal generation of energy.[5] For the external effects, there is the heat energy coming from the surroundings and the net work done by the surroundings. Thus we have

$$\text{External rate of heat input} = -\int_S \underline{\mathbf{q}}\cdot\underline{\vec{\mathbf{n}}}\,dS + \int_V \Theta\,dV$$

[4] As noted during the derivation, the assumptions used (e.g., no internal generation, no other body forces such as magnetic effects are present, etc.) need to hold when using (5.25); otherwise extra terms are needed.

[5] In some texts, they consider heat generated by reaction as a generation term. In our case, we consider heats of reaction and other latent heats as included in the specification of internal energy.

where $\underline{\mathbf{q}}$ is the heat flux due to heat transfer and Θ is the rate of heat input per volume due to other mechanisms such as radiation, electric field, and so forth. The net work done by the surroundings is given by

$$\text{Rate of net work done by surroundings} = -\int_S \left(p\ \underline{\mathbf{v}} + \underline{\underline{T}} \cdot \underline{\mathbf{v}}\right) \cdot \underline{\vec{\mathbf{n}}}\, dS$$

Then, after applying the divergence theorem and including the equation of continuity to put it in terms of substantial time derivative, (5.21) becomes

$$\frac{D\widehat{E}}{Dt} = -\nabla \cdot p\ \underline{\mathbf{v}} - \nabla \cdot \left(\underline{\underline{T}} \cdot \underline{\mathbf{v}}\right) - \nabla \cdot \underline{\mathbf{q}} + \Theta \tag{5.29}$$

known as the **total energy balance equation**.

For the special case of potential energy being based only on gravity, we have $\nabla \widehat{e}_p = -\underline{\mathbf{g}}$, which can be assumed to be independent of time. Then the substantial time derivative of $\widehat{e}_p$ becomes

$$\frac{D\,\widehat{e}_p}{Dt} = -\rho \underline{\mathbf{v}} \cdot \underline{\mathbf{g}} \tag{5.30}$$

We could also find the substantial rate of change of kinetic energy by taking a dot product of the equation of motion given in (5.26) by $\underline{\mathbf{v}}$. Doing so, we can combine the substantial derivative of both kinetic and potential energy to be

$$\begin{aligned} \rho \frac{D}{Dt}\left(\frac{1}{2}\left\|\underline{\mathbf{v}}\right\|^2 + \widehat{e}_p\right) &= -\underline{\mathbf{v}} \cdot \nabla p - \underline{\mathbf{v}} \cdot \left(\nabla \cdot \underline{\underline{T}}\right) \\ &= -\left(\nabla \cdot p\ \underline{\mathbf{v}} - p\left(\nabla \cdot \underline{\mathbf{v}}\right)\right) - \left(\nabla \cdot \left(\underline{\underline{T}} \cdot \underline{\mathbf{v}}\right) - \underline{\underline{T}}\mathbf{:}\nabla \underline{\mathbf{v}}\right) \end{aligned} \tag{5.31}$$

where we used the identity given in (4.100), with the last term being

$$\underline{\underline{T}}\mathbf{:}\nabla \underline{\mathbf{v}} = \sum_{i,j=x,y,z} T_{ij} \frac{\partial v_j}{\partial i}$$

Equation (5.31) is also known as the **mechanical energy balance equation**. We can then remove two terms, namely the substantial time derivative of the specific kinetic and potential energy, from the total energy balance in (5.29) by using (5.31) to obtain

$$\rho\left(\frac{D\widehat{u}}{Dt}\right) = -p\left(\nabla \cdot \underline{\mathbf{v}}\right) - \underline{\underline{T}}\mathbf{:}\nabla \underline{\mathbf{v}} - \nabla \cdot \underline{\mathbf{q}} + \Theta \tag{5.32}$$

which is known as the **thermal energy balance equation**.

As a simple application, consider the energy balance for a stationary solid in which the only mode of heat transfer from surroundings is via conduction. Applying the equation for internal energy in terms of constant heat capacity C_v with $\underline{\mathbf{v}} = 0$

$$\rho \frac{D\widehat{u}}{Dt} = \rho C_v \frac{\partial \hat{T}}{\partial t}$$

where we use $\hat{T}$ for temperature. For the heat transfer, we use **Fourier's law**, that is,

$$\underline{\mathbf{q}} = -k \nabla \hat{T}$$

where k is the **thermal conductivity coefficient**. With $\Theta = 0$, we end up with the **thermal diffusion equation** given by

$$\frac{\partial \widehat{T}}{\partial t} = \frac{k}{\rho C_p} \nabla \cdot \nabla \widehat{T} = \alpha \nabla^2 \widehat{T} \tag{5.33}$$

where $\alpha = k/(\rho C_p)$ is the **thermal diffusivity coefficient**.

4. **Component Balance**

We limit our discussion to binary component systems containing substance A and substance B. The balance will then be on $\boldsymbol{\phi}$ being the moles of A, for which the density becomes $\widehat{\boldsymbol{\phi}}_V = c_A$, which is the concentration of A in units of moles per volume of solution. The flow out due to molar flux across surface S is now given by two terms, one by convection and the other by diffusion,

$$\text{Net flow out of } A = \int_S (c_A \underline{\mathbf{v}}^\star + \underline{\mathcal{D}}) \cdot \underline{\vec{\mathbf{n}}}\, dS$$

where the velocity $\underline{\mathbf{v}}^\star$ is the molar average velocity defined by

$$\underline{\mathbf{v}}^\star = \frac{c_A \underline{\mathbf{v}}_A + c_B \underline{\mathbf{v}}_B}{c_A + c_B}$$

and $\underline{\mathcal{D}}$ is the flux due to diffusion. The internal generation term will be the net rate of production of A due to reaction.[6] This is given by

$$\text{Rate of net generation of mole } A \text{ via reaction} = \int_V R_A dV$$

Substituting all the terms to (5.21) and then applying the divergence theorem, we obtain the **binary component balance equation** given by

$$\frac{\partial c_A}{\partial t} + \nabla \cdot (c_A \underline{\mathbf{v}}^\star) = -\nabla \cdot \underline{\mathcal{D}} + R_A \tag{5.34}$$

Using **Fick's law**,

$$\underline{\mathcal{D}} = -c\, \mathcal{D}_{AB} \nabla x_A$$

where $c = c_A + c_B$ is the total concentration, x_A is the mole fraction of A and $\mathcal{D}_{AB}$ is the **diffusivity of** A in the AB solution. With the added assumption of constant density and constant diffusivity $\mathcal{D}_{AB}$, (5.34) will reduce to the **reaction-diffusion equation**,

$$\frac{\partial c_A}{\partial t} + \underline{\mathbf{v}}^* \cdot \nabla c_A = \mathcal{D}_{AB} \nabla^2 c_A + R_A \tag{5.35}$$

5.4.2 Electromagnetics

One of the important applications of vector analysis, in fact a major impetus for the development of vector calculus, is the study of electricity and magnetism. The

[6] Because we are treating only binary mixtures, we are assuming only the reaction $A \rightarrow B$.

Table 5.2. *Terms and relations in electromagnetics*

Terms	Notation	Formula
	Fields	
Electric field	$\mathbf{E}$	
Magnetic field	$\mathbf{H}$	
Charge density (field)	Q	
	Flux Densities	
Electric flux density	$\mathbf{D}$	
Magnetic flux density	$\mathbf{B}$	
Conduction current density	$\mathbf{J_c}$	$\sigma\mathbf{E}$
Displacement current density	$\mathbf{J_d}$	$\dfrac{\partial \mathbf{D}}{\partial t}$
Current density	$\mathbf{J}$	$\mathbf{J_c} + \mathbf{J_d}$
	Integrals and Fluxes	
Magnetic flux	ϕ	$\int_S \mathbf{B} \cdot \underline{\vec{\mathbf{n}}}\, dS$
Current	i	$\int_S \mathbf{J} \cdot \underline{\vec{\mathbf{n}}}\, dS$
Total charge	q	$\int_V Q\, dV$
	Parameters	
Permisivity	ϵ	
Permeability	μ	
Conductivity	σ	

different fields, flux densities, and fluxes[7] are defined and given in Table 5.2, where $\mathbf{J} = \mathbf{J}_c + \mathbf{J}_d$ is Maxwell's decomposition of the current density.[8]

Based on experimental studies, several laws were obtained that relate the different items in Table 5.2. These are

$$\int_C \mathbf{E} \cdot d\underline{\vec{\mathbf{r}}} = -\frac{\partial \phi}{\partial t} \tag{5.36}$$

$$\int_C \mathbf{H} \cdot d\underline{\vec{\mathbf{r}}} = i \tag{5.37}$$

$$\int_S \mathbf{D} \cdot \underline{\vec{\mathbf{n}}}\, dS = q \tag{5.38}$$

$$\int_S \mathbf{B} \cdot \underline{\vec{\mathbf{n}}}\, dS = 0 \tag{5.39}$$

[7] As mentioned earlier in Example 4.3, the term "flux" in electromagnetics refers to a scalar quantity that integrates the flow of a flux density vector field through a given closed surface S. The term "flux" in fluid transport equations refers to a vector field.

[8] We temporarily suspend our convention of using small bold letters for vector fields to comply with the usual conventions of electromagnetics. Likewise, please note that i is set as current and not the imaginary number.

where $\underline{\vec{r}}$ is the position vector along curve C and $\underline{\vec{n}}$ is the outward unit normal vector to the surface S. Equations (5.36) and (5.37) are known as **Faraday's law** and **Ampere's law**, respectively, whereas (5.38) and (5.39) are known as **Gauss' law for electric fields** and **Gauss' law for magnetic field**, respectively. Note that Faraday's law and Ampere's law involve the line integrals of the electric field and magnetic field projected along closed paths C and yield the work resulting from each of these fields. On the other hand, both Gauss' law for electric fields and magnetic fields are surface integral of a closed surface S, where the surface integral in (5.38) depends on the total charge inside the region bounded by S, whereas the surface integral in (5.39) is always zero.

Let us assume a simple linear relationship between **E** and **D** and between **B** and **H** given by

$$\mathbf{D} = \epsilon \mathbf{E} \tag{5.40}$$

$$\mathbf{H} = \frac{1}{\mu}\mathbf{B} \tag{5.41}$$

with ϵ and μ constant. Applying Stokes' theorem to Faraday's law and Ampere's law, and applying the divergence theorem to both of the Gauss' laws, while assuming C, S, and V to be fixed, will yield the following:

$$\begin{aligned} \int_C \mathbf{E}\cdot d\underline{\vec{r}} &= -\frac{\partial}{\partial t}\int_S \mathbf{B}\cdot\underline{\vec{n}}\,dS \\ \int_S (\nabla\times\mathbf{E})\cdot\underline{\vec{n}}\,dS &= -\int_S \frac{\partial \mathbf{B}}{\partial t}\cdot\underline{\vec{n}}\,dS \end{aligned} \tag{5.42}$$

$$\begin{aligned} \int_C \mathbf{H}\cdot d\underline{\vec{r}} &= \int_S \mathbf{J}\cdot\underline{\vec{n}}\,dS \\ \int_S (\nabla\times\mathbf{H})\cdot\underline{\vec{n}}\,dS &= \int_S \left(\sigma\mathbf{E} + \epsilon\frac{\partial \mathbf{E}}{\partial t}\right)\cdot\underline{\vec{n}}\,dS \end{aligned} \tag{5.43}$$

$$\begin{aligned} \int_S \mathbf{D}\cdot\underline{\vec{n}}\,dS &= q \\ \int_V (\nabla\cdot\mathbf{D})\,dV &= \int_V Q\,dV \end{aligned} \tag{5.44}$$

$$\begin{aligned} \int_S \mathbf{B}\cdot\underline{\vec{n}}\,dS &= 0 \\ \int_V (\nabla\cdot\mathbf{B})\,dV &= 0 \end{aligned} \tag{5.45}$$

Because the value for C, S, and V are arbitrary, equations (5.42) to (5.45) reduce to the set of equations known as **Maxwell's equations for electricity and magnetism**:

$$\nabla\times\mathbf{E} = -\frac{\partial \mathbf{B}}{\partial t} = -\mu\frac{\partial \mathbf{H}}{\partial t} \tag{5.46}$$

$$\nabla\times\mathbf{H} = \sigma\mathbf{E} + \epsilon\frac{\partial \mathbf{E}}{\partial t} \tag{5.47}$$

$$\nabla \cdot \mathbf{D} = \epsilon \nabla \cdot \mathbf{E} = Q \tag{5.48}$$

$$\nabla \cdot \mathbf{B} = \mu \nabla \cdot \mathbf{H} = 0 \tag{5.49}$$

where the rightmost equations of the first two equations show the coupled relationships between the fields **E** and **H**. These equations could be uncoupled by further taking the curl of (5.46) and (5.47), together with (5.48), (5.49), and identity (4.62),

$$\begin{aligned} \nabla \times (\nabla \times \mathbf{E}) = \nabla (\nabla \cdot \mathbf{E}) - \nabla^2 \mathbf{E} &= -\mu \frac{\partial}{\partial t} (\nabla \times \mathbf{H}) \\ \frac{1}{\epsilon} \nabla Q - \nabla^2 \mathbf{E} &= -\mu\sigma \left(\frac{\partial \mathbf{E}}{\partial t} \right) - \mu\epsilon \left(\frac{\partial^2 \mathbf{E}}{\partial t^2} \right) \end{aligned}$$

$$\begin{aligned} \nabla \times (\nabla \times \mathbf{H}) = \nabla (\nabla \cdot \mathbf{H}) - \nabla^2 \mathbf{H} &= \sigma \left(\nabla \times \mathbf{E} \right) + \epsilon \frac{\partial (\nabla \times \mathbf{E})}{\partial t} \\ -\nabla^2 \mathbf{H} &= -\mu\sigma \left(\frac{\partial \mathbf{H}}{\partial t} \right) - \mu\epsilon \left(\frac{\partial^2 \mathbf{H}}{\partial t^2} \right) \end{aligned}$$

which could be written in operator form as

$$\mathfrak{M}\left(\mathbf{E}\right) = -\frac{1}{\epsilon} \nabla Q \tag{5.50}$$

$$\mathfrak{M}\left(\mathbf{H}\right) = \mathbf{0} \tag{5.51}$$

where

$$\mathfrak{M} = \mu\epsilon \frac{\partial^2}{\partial t^2} + \mu\sigma \frac{\partial}{\partial t} - \nabla^2$$

EXAMPLE 5.4. For the special case of static electric fields, that is, $\mathbf{E} \neq \mathbf{E}(t)$, the operator $\mathfrak{M}$ reduces to the negative Laplacian, $-\nabla^2$, yielding

$$\nabla^2 \mathbf{E} = \frac{1}{\epsilon} \nabla Q$$

However, this consists of solving three differential equations because it is a vector differential equation. An interesting alternative is to observe that (5.46) under static conditions reduces to

$$\nabla \times \mathbf{E} = \mathbf{0}$$

Based on Theorem 5.5, work done by **E** along a path inside a simply connected region will be path independent or, equivalently, **E** is a **conservative field**. For conservative fields, such as **E** in this case, we can make use of the vector differential identity (4.60), that is, $\nabla \times \nabla \psi = 0$, and let $\mathbf{E} = \nabla \psi$, where ψ is now the unknown potential field. Substituting this representation into (5.48), we have

$$\nabla \cdot \mathbf{E} = \nabla^2 \psi = \frac{Q}{\epsilon}$$

which has a form known as **Poisson's equation**. For the additional condition that $Q = 0$, this reduces to

$$\nabla^2 \psi = 0$$

a form known as **Laplace's equation**. Once ψ has been solved from either the Poisson equation or Laplace equation together with the given boundary conditions, $\mathbf{E}$ can be obtaind by simply taking the gradient of ψ.

Unfortunately, from (5.47), one observes that even for static magnetic fields, $\mathbf{H}$ is generally nonconservative. Nonetheless, one can still use $\mathfrak{M} = -\nabla^2$, under static conditions, and solve

$$\nabla^2 \mathbf{H} = \mathbf{0}$$

5.5 Leibnitz Derivative Formula

The last integral formula included in this chapter is the Leibnitz formula, which gives the method for evaluating the derivative of an integral with respect to a parameter α.

THEOREM 5.6. *Given a function $F(\alpha, x)$ that is differentiable in x and α, then*

$$\begin{aligned} \frac{d}{d\alpha}\int_{g(\alpha)}^{h(\alpha)} F(\alpha, x)\, dx &= \int_{g(\alpha)}^{h(\alpha)} \frac{\partial F(\alpha, x)}{\partial \alpha} dx \\ &\quad + F(\alpha, h(\alpha))\frac{\partial h(\alpha)}{\partial \alpha} - F(\alpha, g(\alpha))\frac{\partial g(\alpha)}{\partial \alpha} \end{aligned} \tag{5.52}$$

PROOF. (See Section E.5.6, item 1, for proof.)

In three dimensions, we have

THEOREM 5.7. *Let $f(x, y, z, \alpha)$ be differentiable in an open region containing 3D region $V(\alpha)$. Then,*

$$\frac{d}{d\alpha}\int_{V(\alpha)} f(x, y, z, \alpha)\, dV = \int_{V(\alpha)} \frac{\partial f}{\partial \alpha} dV + \int_{S(\alpha)} f(x, y, z, \alpha)\, \underline{\vec{\mathbf{n}}} \cdot \frac{\partial \underline{\vec{\mathbf{r}}}}{\partial \alpha} dS \tag{5.53}$$

where $S(\alpha)$ is the boundary surface of $V(\alpha)$, $\underline{\vec{\mathbf{n}}}$ is the unit outward normal vector, and $\underline{\vec{\mathbf{r}}}$ is the position vector of the points in the boundary $S(\alpha)$.

PROOF. (See Section E.5.6, item 2, for proof.)

In several applications of the Leibnitz rule, the parameter α is taken to be the time variable t. Thus it can be used to obtain integral balance equation around dynamic regions, that is, $V = V(t)$ and $S = S(t)$.

EXAMPLE 5.5. Let us verify Liebnitz rule for the time derivative of the following volume integral

$$\int_V f(x, y, z, t)\, dV = \int_0^1 \int_0^2 \int_0^{e^{-t}} t\left(x^2 + y^2 + z^2\right)\, dx\, dy\, dz$$

The derivative of this integral can be found by direct computation to be

$$\frac{d}{dt}\int_V f(x,y,z,t)\,dV = \left(\frac{2}{3} - 2t\right)e^{-3t} + \frac{10}{3}(1-t)\,e^{-t} \tag{5.54}$$

which forms the left-hand side of (5.53).

For the first term, that is, the volume integral on the right-hand side of (5.53), we have

$$\int_V \frac{\partial f}{\partial t}\,dV = \int_0^1 \int_0^2 \int_0^{e^{-t}} \left(x^2+y^2+z^2\right)\,dx\,dy\,dz = \frac{2}{3}e^{-3t} + \frac{10}{3}e^{-t} \tag{5.55}$$

As for the surface integral term on the right-hand side of (5.53), of the six possible faces, only one of the surface integrals is nonzero, which is the face at $x = e^{-t}$ whose normal is $\underline{\delta}_x$. Thus we have

$$\int_S f\,\underline{\mathbf{n}}\cdot\frac{\partial\vec{\underline{\mathbf{r}}}}{\partial t}\,dS = \int_0^1\int_0^2 t\left(e^{-2t}+y^2+z^2\right)\left(-e^{-t}\right)\,dy\,dz = -2te^{-3t} - \frac{10}{3}te^{-t} \tag{5.56}$$

Adding the results of (5.55) and (5.56) will yield the same result given by (5.54).

5.6 EXERCISES

E5.1. Let a particle move along a path C due to an attractive force $\underline{\mathbf{f}}$ directed toward the origin, $\underline{\mathbf{f}} = k\underline{\mathbf{r}}$. The tangential unit vector along the path is given by

$$\underline{\delta}_t = \frac{d\vec{\underline{\mathbf{r}}}}{\left\|d\vec{\underline{\mathbf{r}}}\right\|}$$

1. With the projection of a force $\underline{\mathbf{f}} = k\vec{\underline{\mathbf{r}}}$ onto $\underline{\delta}_t$ given by

$$\underline{\mathbf{f}}_t = k\left(\vec{\underline{\mathbf{r}}}\cdot d\vec{\underline{\mathbf{r}}}\right)\frac{d\vec{\underline{\mathbf{r}}}}{\left\|d\vec{\underline{\mathbf{r}}}\right\|^2}$$

where k is a constant coefficient, show that the component of $\underline{\mathbf{f}}$ that is normal to the path will be given by

$$\underline{\mathbf{f}}_n = k\frac{\left\|d\vec{\underline{\mathbf{r}}}\right\|^2\vec{\underline{\mathbf{r}}} - \left(\vec{\underline{\mathbf{r}}}\cdot d\vec{\underline{\mathbf{r}}}\right)d\vec{\underline{\mathbf{r}}}}{\left\|d\vec{\underline{\mathbf{r}}}\right\|^2}$$

2. With friction proportional to the normal component, friction $= \mu\left\|\underline{\mathbf{f}}_n\right\|\underline{\delta}_t$, show that the sum of the work due to $\underline{\mathbf{f}}$ and the frictional force along path C is given by

$$W = k\int_C \left(\frac{\vec{\underline{\mathbf{r}}}\cdot d\vec{\underline{\mathbf{r}}}}{\left\|d\vec{\underline{\mathbf{r}}}\right\|} + \mu\frac{\sqrt{\left\|\vec{\underline{\mathbf{r}}}\right\|^2\left\|d\vec{\underline{\mathbf{r}}}\right\|^2 - \left(\vec{\underline{\mathbf{r}}}\cdot d\vec{\underline{\mathbf{r}}}\right)^2}}{\left\|d\vec{\underline{\mathbf{r}}}\right\|}\right)ds$$

3. Evaluate the work due to $\underline{\mathbf{f}}$ and its associated friction force for the cyclone path shown in Figure 5.10 given by

$$x = \frac{t^3}{4}\cos(16\pi t) \qquad y = \frac{t^3}{4}\sin(16\pi t) \qquad z = \frac{\sqrt{3}}{4}t^3$$

from $t = 0.1$ to $t = 1.5$ in terms of μ and k.

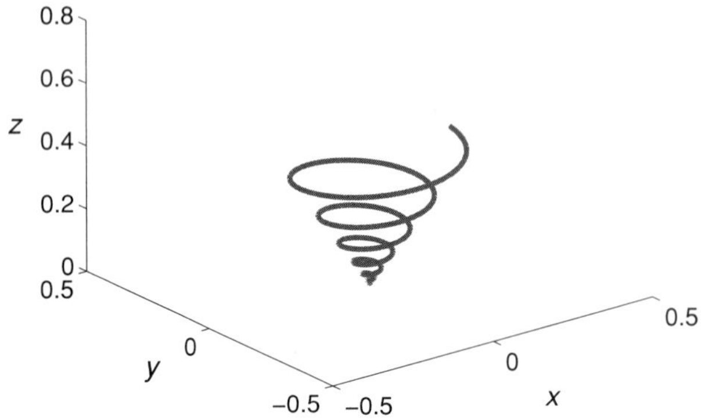

Figure 5.10. A cyclonic path.

E5.2. Let $\psi(x, y, z)$ be the function

$$\psi(x, y, z) = a_{xx}x^2 + a_{yy}y^2 + a_{zz}z^2 + a_{xy}xy + a_{yz}yz + a_{xz}xz + a_x x + a_y y + a_z z + c$$

where $a_{xx}, a_{yy}, \ldots a_{xz}, a_x, a_y, a_z, c$ are constants. Calculate the volume integral $\int_V \psi(x, y, z)\, dV$, where the volume of integration is the sphere centered at the origin with a radius ρ. Determine the volume integral if the sphere V had its center at (x_0, y_0, z_0) instead.

E5.3. Consider the top in exercise **E4.10** whose surface was described by

$$x = \sin(\theta)\cos(\phi)$$
$$y = \sin(\theta)\sin(\phi)$$
$$z = 1.5\cos(\theta/2)$$

for $0 \leq \theta \leq \pi$ and $0 \leq \phi \leq 2\pi$.

1. Find the volume of the solid.
2. Assuming a density function

$$\rho(x, y, z) = 2\left(x^2 + y^2 + z^2\right)$$

Find the center of mass, $\underline{\mathbf{c}}$, given by

$$\underline{\mathbf{c}} = \frac{\int_V \rho\, \vec{\underline{\mathbf{r}}}\, dV}{\int_V \rho\, dV}$$

where $\vec{\underline{\mathbf{r}}}$ is the position vector.

E5.4. The moment of inertia of a rigid body with respect to an axis of rotation is given by

$$I = \int \rho\, D^2\, dV$$

where ρ is the density field of the rigid body, and D is the distance of a point of the body from the axis of rotation (see Exercise **E4.3** for the formula of distance of a point from a line).

Find the moment of inertia of a sphere of radius R with the center at the origin, having a density field

$$\rho(x, y, z) = 2x^2 + y^2$$

around an axis that passes through the points $A = (1, 1, 0)$ and $B = (1, 0, 1)$.

E5.5. Prove the identities given in (5.6) through (5.9).

E5.6. A vertical cylinder A of radius R_A is parameterized by θ and s, where $0 \leq \theta \leq 2\pi$ and $-\infty < s < \infty$ via $(x, y, z) = \left(R_A \cos(\theta), R_A \sin(\theta), s\right)$. Another horizontal cylinder B is parameterized by ϕ and t, where $0 \leq \phi \leq 2\pi$ and $-\infty < t < \infty$, via $(x, y, z) = \left(t, R_B \sin(\phi), R_B \cos(\phi)\right)$. Assume $R_B < R_A$.

1. Cylinder B cuts two separate pieces out of cylinder A. Show that one of the cut pieces is a bounded surface given by $(x, y, z) = \left(R_A \cos(\theta), R_A \sin(\theta), s\right)$ with new bounds $-\theta^* \leq \theta \leq \theta^*$ where

$$\theta^* = \left|\sin^{-1}\left(\frac{R_B}{R_A}\right)\right|$$

and $|s| \leq s^*(\theta)$ where

$$s^*(\theta) = R_B \sqrt{1 - \left(\frac{R_A}{R_B}\right)^2 \sin^2(\theta)}$$

This means that the surface integral for this cut piece is given by

$$\int_S f(x, y, z) dS = \int_{-\theta^*}^{\theta^*} \int_{-s^*(\theta)}^{s^*(\theta)} g(s, \theta)\, ds d\theta$$

2. Let functions F and G be given by

$$F = 4z \quad \text{and} \quad G = x^2 + 2y$$

Verify Green's lemma (5.1) with $u = \theta$ and $v = s$ for the surface of the cut piece described Previously and $R_A = 1.5$ and $R_B = 1$. (Note: For the contour integral, the closed path can be parameterized by λ for points (θ, s) of the path, ranging from 0 to 1 with the following correspondences: $(\lambda = 0) \rightarrow (-\theta^*, 0)$, $(\lambda = 0.5) \rightarrow (\theta^*, 0)$, and $(\lambda = 1) \rightarrow (-\theta^*, 0)$).

E5.7. Let S be a closed surface, and show that the divergence theorem leads to the following identities:

$$\begin{aligned}
V &= \frac{1}{3}\int_S \underline{\mathbf{n}} \cdot \vec{\mathbf{r}}\, dS \\
0 &= \int_S \underline{\mathbf{n}} \cdot \left(\nabla \times \underline{\mathbf{v}}\right) dS \\
\int_V \nabla\psi\, dV &= \int_S \underline{\mathbf{n}}\psi\, dS \\
\int_V \nabla \times \underline{\mathbf{v}}\, dV &= \int_S \underline{\mathbf{n}} \times \underline{\mathbf{v}}\, dS
\end{aligned}$$

where V is the volume enclosed by S and $\underline{\mathbf{v}}$ is a vector field. (Hint: For the last two identities, apply the divergence theorem on $\psi\underline{\mathbf{a}}$ and $\underline{\mathbf{v}} \times \underline{\mathbf{a}}$, respectively, where $\underline{\mathbf{a}}$ is a constant vector.)

Note that the last two identities plus the divergence theorem can be combined into one mnemonic equation,

$$\int_V \nabla \diamond \left[\mathfrak{F}\right] dV = \int_S \underline{\mathbf{n}} \diamond \left[\mathfrak{F}\right] dS \tag{5.57}$$

where

	$\diamond$	$[\mathfrak{F}]$
Divergence	dot product	$\underline{\mathbf{v}}$
Gradient	scalar product	ψ
Curl	cross product	$\underline{\mathbf{v}}$

E5.8. Let C be a closed curve, and show that the following line integrals are zero:

$$L_1 = \oint_C \underline{\mathbf{a}} \cdot d\vec{\underline{\mathbf{r}}}$$

$$L_2 = \oint_C \vec{\underline{\mathbf{r}}} \cdot d\vec{\underline{\mathbf{r}}}$$

$$L_3 = \oint_C \nabla\psi \cdot d\vec{\underline{\mathbf{r}}}$$

$$L_4 = \oint_C (v\nabla w + w\nabla v) \cdot d\vec{\underline{\mathbf{r}}}$$

where $\underline{\mathbf{a}}$ is any constant vector, $\vec{\underline{\mathbf{r}}}$ is the position vector, and ψ, v, and w are scalar functions.

E5.9. Let C be a closed curve, and using Stokes' theorem, show that

$$\oint_C \psi \, d\vec{\underline{\mathbf{r}}} = \int_S \underline{\mathbf{n}} \times \nabla\psi \, dS$$

(Hint: Apply Stokes' theorem on $\psi\underline{\mathbf{a}}$, where $\underline{\mathbf{a}}$ is a constant vector.)

Note that if we also rewrite Stokes' theorem, using (4.52), as

$$\oint_C d\vec{\underline{\mathbf{r}}} \cdot \underline{\mathbf{v}} = \int_S (\underline{\mathbf{n}} \times \nabla) \cdot \underline{\mathbf{v}} \, dS$$

then the two equations can be combined into one mnemonic equation,

$$\oint_C d\vec{\underline{\mathbf{r}}} \diamond [\mathfrak{F}] = \int_S (\underline{\mathbf{n}} \times \nabla) \diamond [\mathfrak{F}] \, dS \tag{5.58}$$

where

	$\diamond$	$[\mathfrak{F}]$
Cross – Divergence	dot product	$\underline{\mathbf{v}}$
Cross – Gradient	scalar product	ψ

E5.10. Rewriting Euler's equation of motion (5.28) as

$$\frac{\partial \underline{\mathbf{v}}}{\partial t} + (\underline{\mathbf{v}} \cdot \nabla)\, \underline{\mathbf{v}} = \underline{\mathbf{g}} - \frac{1}{\rho}\nabla p$$

For the case where ρ is constant, show that

$$\frac{\partial \underline{\mathbf{v}}}{\partial t} - \underline{\mathbf{v}} \times (\nabla \times \underline{\mathbf{v}}) = -\nabla \left(\widehat{e}_p + \frac{p}{\rho} + \frac{1}{2} \|\underline{\mathbf{v}}\|^2 \right)$$

where $\underline{\mathbf{g}} = -\nabla \widehat{e}_p$, with $\widehat{e}_p$ being the potential energy per unit mass. Then for steady state and irrotational velocity field, this equation reduces to the **Bernoulli equation of motion**,

$$\widehat{e}_p + \frac{p}{\rho} + \frac{1}{2}\left\|\underline{\mathbf{v}}\right\|^2 = \text{constant}$$

along a streamline.

E5.11. One can show that the stress tensor, in the absence of couple-moments, is symmetric.

1. Let the stress tensor field of a body be given by $\underline{\underline{T}}$ and a body force per unit volume given by $\underline{\mathbf{f}}$. Show that for equilibrium, a necessary condition is given by

$$\nabla \cdot \underline{\underline{T}}^T + \underline{\mathbf{f}} = 0 \tag{5.59}$$

(Hint: The summation of forces along each coordinate should be zero. The forces include the surface integral of the force due to the stress tensor on the body surface and the volume integral of the body forces.)

2. Assuming no couple-moments, the total moments due to both body force and stress tensor on the surface should also be zero, that is,

$$\int_S \vec{\underline{\mathbf{r}}} \times \left(\underline{\underline{T}} \cdot \vec{\underline{\mathbf{n}}}\right) dS + \int_V \vec{\underline{\mathbf{r}}} \times \underline{\mathbf{f}}\, dV = 0 \tag{5.60}$$

Show that after implementation of the divergence formula and (5.59), the stress tensors indeed need to be symmetric.

E5.12. Obtain the differential volume at a point $\vec{\underline{\mathbf{r}}}$ under the paraboloid coordinate system described in Example 4.17.

E5.13. Another orthogonal coordinate system is the **torroidal coordinate system**, (τ, σ, ϕ), with $\tau \geq 0$, $-\pi \leq \sigma \leq \pi$, and $0 \leq \phi \leq 2\pi$, described by the following equations:

$$x = \frac{a\,\sinh(\tau)\,\cos(\phi)}{\cosh(\tau) - \cos(\sigma)} \quad ; \quad y = \frac{a\,\sinh(\tau)\,\sin(\phi)}{\cosh(\tau) - \cos(\sigma)} \quad ; \quad z = \frac{a\,\sin(\sigma)}{\cosh(\tau) - \cos(\sigma)} \tag{5.61}$$

1. Obtain the differential volume dV in this coordinate system.
2. A torus-surface can be described by fixing τ to be constant, with $\tau \rightarrow \infty$ yielding a circle (torus with no volume) of radius a. Let a torus be defined by parameters R and C (with $C > R$), where R is the radius of a circle located in any plane containing the z-axis and whose center is located in the (x, y)-plane at a distance C away from the z-axis. Show that the value of a and τ that correspond to parameters R and C are given by

$$\tau = \sinh^{-1}\left(\frac{C}{R}\right) \quad \text{and} \quad a = (R + C)\,\frac{\cosh(\tau) - 1}{\sinh(\tau)}$$

3. We want to determine the amount of property q flowing out of the surface of a torus in a flow field per unit area given by

$$\underline{\mathbf{v}} = 4\exp\left(\frac{-r^2}{10}\right)\underline{\delta}_r \tag{5.62}$$

where r is the spherical radius coordinate. Instead of having to find the unit normal vectors at the surface of the torus followed by the evaluation of surface integral, an alternative approach is to use the divergence and replace the problem with evaluation of a volume integral

$$\int_V \left(\nabla \cdot \underline{\mathbf{v}}\right) dV \tag{5.63}$$

Find the total flow of property q out of a torus with $R = 1$ and $C = 2$ for the flow field per unit area given in (5.62).

E5.14. A spherical solid of radius r containing substance s is decaying in time. The distribution of s is described by a time-dependent field

$$\phi(x, y, z, t) = 0.8\, e^{-t(x^2+y^2)/3}\left(\frac{z}{2r} + \frac{3}{2}\right)$$

where the center of the sphere is located at the origin. Suppose the radius of the sphere is also shrinking, symmetrically with respect to the center, due to erosion. The radius was found to be shrinking at a rate

$$\frac{dr}{dt} = -0.1r \qquad r(0) = R = 10$$

1. Find $r = r(t)$.
2. Determine the rate of change of s present in the solid at time $t = 10$.
3. Verify Leibnitz rule (5.53) for this rate of change for any time $t > 0$.

E5.15. Consider the regions $R1$ through $R3$ and singularities contained in $S1$ through $S4$ given in Table 5.3. Determine which pairs (i.e., $R1$ with $S1$ removed, $R1$ with $S2$ removed, etc.) are simply connected.

Table 5.3. *Regions R and singularities S for* **E5.15**

	Regions		Singularities
$R1$:	$-4 \le x \le 4$ $y \ge 0$ $-\infty < z < \infty$	$S1$:	$(x, y, z) = (0, 0, 0)$
$R2$:	spherical $r \ge 2$	$S2$:	Surface $2x + y - z = 0$
$R3$:	cylindical $r \ge 2$	$S3$:	Sphere with radius $r = 0.1$ and center at $(4, 0, 0)$
		$S4$:	Line passing through $(1, 1, 2)$ and $(2, 1, -1)$

E5.16. Let $\underline{\mathbf{f}}$ and $\underline{\mathbf{g}}$ be given by

$$\underline{\mathbf{f}} = \frac{2}{r-1}\underline{\delta}_r \quad \text{(under spherical coordinates)}$$

$$\underline{\mathbf{g}} = \frac{2}{\sin\left(\theta + \frac{\pi}{3}\right)}\underline{\delta}_r \quad \text{(under cylindrical coordinates)}$$

$$\underline{\mathbf{h}} = xz\underline{\delta}_x + \frac{y}{y+3}\underline{\delta}_y + z\underline{\delta}_z$$

Determine whether the following integrals are path independent or not:

1. $\int_C \underline{\mathbf{f}} \cdot d\vec{\underline{\mathbf{r}}}$, with path C restricted to the region: $r \leq 2$ (where r is the spherical radius).
2. $\int_C \underline{\mathbf{g}} \cdot d\vec{\underline{\mathbf{r}}}$, with path C restricted to the region: $0 \leq \theta \leq \pi$ (where θ is cylindrical angle).
3. $\int_C \underline{\mathbf{h}} \cdot d\vec{\underline{\mathbf{r}}}$, with path C restricted to the region: $y \geq 0$

PART III

ORDINARY DIFFERENTIAL EQUATIONS

Several models of physical systems come in the form of differential equations. The main advantage of these models lies in their flexibility through the specifications of initial and/or boundary conditions or forcing functions. Although several physical models result in partial differential equations, there are also several important cases in which the models can be reduced to ordinary differential equations. One major class involves dynamic models (i.e., time-varying systems) in which the only independent variable is the time variable, known as initial value problems. Another case is when only one of the spatial dimensions is the only independent variable. For this case, it is possible that boundary conditions are specified at different points, resulting in multiple-point boundary value problems.

There are four chapters included in this part of the book to handle the analytical solutions, numerical solutions, qualitative analysis, and series solutions of ordinary differential equations. Chapter 6 discusses the analytic approaches to solving first- and second-order differential equations, including similarity transformation methods. For higher order linear differential equations, we apply matrix methods to obtain the solutions in terms of matrix exponentials and matrizants. The chapter also includes the use of Laplace transforms for solving the high-order linear ordinary differential equations.

Numerical methods for solving ordinary differential equations are discussed in detail in Chapter 7, including implicit and explicit Runge-Kutta methods and multistep methods such as Adams-Bashforth and Adams-Moulton and the backward difference formulas (BDF) methods. We also discuss some simple error-control approaches by adaptive time-interval methods. The second part of the chapter is devoted to the solution of boundary value problems such as the shooting method for both linear and nonlinear ordinary differential equations and the Ricatti method. The last part of Chapter 7 is a relatively brief, but crucial, discussion of difference equations and stability. The analysis of difference equations is important for the application of the Dahlquist tests to determine stability regions. Using the ideas of stability, one can better descirbe "stiff" differential equations.

Chapter 8 discusses the qualitative analysis of differential equations such as phase-plane analysis; stability analysis of equilibrium points, including Lyapunov methods and linearization techniques; and limit cycles. This chapter also discusses bifurcation analysis based on one- or two-parameter systems.

Finally, in Chapter 9, we discuss the series solution methods of Frobenius for second-order and higher order linear differential equations containing variable coefficients. Among the most important applications of the series method are the solutions of the Legendre equations, the associated Legendre equations, and Bessel equations. These differential equations often result from the analytical solution of partial differential equations, which are covered in Part IV of the book.

6 Analytical Solutions of Ordinary Differential Equations

In this chapter, we discuss the major approaches to obtain analytical solutions of ordinary differential equations. We begin with the solutions of first-order differential equations. Several first-order differential equations can be transformed into two major solution approaches: the separation of variables approach and the exact differential approach. We start with a brief review of both approaches, and then we follow them with two sections on how to reduce other problems to either of these methods. First, we discuss the use of similarity transformations to reduce differential equations to become separable. We show that these transformations cover other well-known approaches, such as homogeneous-type differential equations and isobaric differential equations, as special cases. The next section continues with the search for integrating factors that would transform a given differential equation to become exact. Important special cases of this approach include first-order linear differential equations and the Bernoulli equations (after some additional variable transformation).

Next, we discuss the solution of second-order differential equations. We opted to focus first on the nonlinear types, leaving the solution of linear second-order differential equations to be included in the later sections that handle high-order linear differential equations. The approaches we consider are those that would reduce the order of the differential equations, with the expectation that once they are first-order equations, techniques of the previous sections can be used to continue the solution process. Specifically, we use a change of variables to handle the cases in which either the independent variable or dependent variable are explicitly absent in the differential equation. In addition, we also note that if the differential equation admits similarity transformations, a new pair of independent variable and dependent variable can be used to reduce the order. We also include the case of Euler-Cauchy differential equations in which we can transform the equation to one that is a linear differential equation with constant coefficients.

Before going to higher order differential equations, we include a section that discusses some important topics when handling first- or second-differential equations. One topic is the general Ricatti equation, in which we actually increase the order of the differential equation to attain linearity. Another topic is the use of Legendre transformations to handle equations in which the derivatives are present in nonlinear forms, whereas the normal variables are in linear forms. Finally, we also discuss the issue of singular solutions, which appear only in nonlinear equations. Singular solutions do not contain any arbitrary constants, and they provide an envelop that helps determine regions of the domains where solutions exist.

Next, we begin our discussion of higher order differential equations with the state-space formulation. Even when one decides to use numerical approaches, the initial step, more often than not, is to first recast the differential equations in the state-space forms. We limit our discussion of analytical solutions only to linear high order differential equations. The nonlinear cases are often handled much better using numerical methods. Nonetheless, even by limiting our problems to linear differential equations, we still have two cases: those with constant coefficient and those with nonconstant coefficients. For the linear equations with constant coefficients represented by a constant matrix A, the solutions are pretty standard, with the matrix exponentials e^{At} as the key element of the general solutions. This means we need the results of earlier chapters (i.e., Chapters 1 through 3) to help in evaluating the required matrices and functions. For instance, when the matrix A is diagonalizable (cf. Section 3.5), then $e^{At} = Ve^{\Lambda t}V^{-1}$, where Λ is a diagonal matrix of eigenvalues and V is the matrix of eigenvectors. For the case in which A is not diagonalizable, we include a brief section discussing the application of the Cayley-Hamilton theorem to provide a finite-sum approach to evaluating the solution.

For the case in which the system of linear high-order differential equations have nonconstant coefficients (i.e., $A = A(t)$), we have the generalization of the solution in the form of **fundamental matrices**, also known as **matrizants**. Although the evaluation of matrizants is difficult to find in general, some special cases do exist in which the matrizants can be found directly. For instance, when $A(t)A(\tau)$ is commutative, the matrizant will involve a simple matrix integral. More importantly, we use the concept of matrizants in the next chapter, when we solve boundary value problems of linear differential equations with nonconstant coefficients using numerical approaches.

Next, we discuss the idea of decoupling of differential equations. This is possible when the matrix A is diagonalizable. The idea that the original system can be transformed into a set of decoupled differential equations is introduced in this section. The main issue is not so much the solution of the differential equations, because they are simple restatements of the solution involving exponential matrices e^{At}. The distinct feature of the decoupled system lies in its offering of an alternative space where the decoupling condition allows the tracking of solution in a one-dimensional space. Potentially, it allows for easier design, interpretation, and analysis of physical experiments. We include one such example in the form of the Wei-Prater method for the determination of kinetic constants of multiple interacting equilibrium reactions.

Finally, we include a brief discussion of Laplace transform methods for the solution of multiple equations. Specifically, we show that this approach also yields the same results of the approach given in the earlier sections.

6.1 First-Order Ordinary Differential Equations

In this section, we limit our discussion to the solution of first-order ordinary differential equations (ODEs) that can be cast either in the **derivative form**, given by

$$\frac{dy}{dx} = f(x, y) \tag{6.1}$$

or in the **differential form**, also known as the **Pfaffian form**, given by[1]

$$M(x, y)\, dx + N(x, y)\, dy = 0 \tag{6.2}$$

The equivalence between the two forms can be obtained by setting

$$f(x, y) = -\frac{M(x, y)}{N(x, y)} \tag{6.3}$$

Although both forms are useful, (6.1) has the additional interpretation (or constraint) of fixing x and y as the independent and dependent variable, respectively, due to the definition of derivatives. Conversely, (6.2) treats both variables as independent variables of an implicit solution, i.e., $S(x, y) = C$, where C is an arbitrary constant to fit initial or boundary conditions.[2] When required, and if it is possible, an explicit solution can be attempted by rearranging $S(x, y) = C$ to obtain a form $y = y(x, C)$. For this reason, we predominantly treat the solution of (6.2), except where the given ODE can be identified more easily with standard (canonical) forms that are given in the derivative forms, such as the **linear first-order ODE** and **Bernoulli equations**.

Most approaches fall under two major categories: those that are reducible to **separation of variables approach** and those that are reducible to **exact differential approach**. In this perspective, several techniques focus simply on additional transformations, for example, additional terms, multiplicative factors, or the change of variables, to reduce the problems into one of these categories.

Suppose that after a set transformation is applied on x and y, new variables $\widehat{x}$ and $\widehat{y}$ are obtained resulting in a **separable form** given by

$$\widehat{M}(\widehat{x})\, d\widehat{x} + \widehat{N}(\widehat{y})\, d\widehat{y} = 0 \tag{6.4}$$

then the solution approach is known as the **method of separation of variables**, yielding

$$\widehat{S}(\widehat{x}, \widehat{y}) = \int \widehat{M}(\widehat{x})\, d\widehat{x} + \int \widehat{N}(\widehat{y})\, d\widehat{y} = \widehat{C} \tag{6.5}$$

with $\widehat{C}$ as arbitrary constant.

However, if after the transformation to new variables, we obtain

$$\frac{\partial \widehat{M}}{\partial \widehat{y}} = \frac{\partial \widehat{N}}{\partial \widehat{x}} \tag{6.6}$$

known as the **exactness condition**, then we say that the transformed differential equation is an **exact differential equation** and the solution is given by

$$\widehat{S} = \int_{\widehat{y}\text{ held constant}} \widehat{M}(\widehat{x}, \widehat{y})\, d\widehat{x} \;+\; g(\widehat{y}) = \int_{\widehat{x}\text{ held constant}} \widehat{N}(\widehat{x}, \widehat{y})\, d\widehat{y} \;+\; h(\widehat{x}) \tag{6.7}$$

where $g(\widehat{y})$ and $h(\widehat{x})$ are determined by matching terms in the rightmost equation in (6.7).

[1] A more general form is given by

$$F\left(x, y, \frac{dy}{dx}\right) = 0$$

which we discuss only in the context of Legendre transformations or singular solutions.

[2] There are solutions that do not allow for the inclusion of an arbitrary constant. These are known as singular solutions, which are discussed later in Section F.2.

EXAMPLE 6.1. Consider the differential equation

$$(x + y)\ dx + x\ dy = 0 \tag{6.8}$$

With $M(x, y) = x + y$ and $N(x, y) = x$, we have $\frac{\partial M}{\partial y} = 1$ and $\frac{\partial N}{\partial x} = 1$, which satisfies the exactness condition given by (6.8),and no transformations are needed. Thus the solution is given by

$$S(x, y) = \frac{x^2}{2} + xy = C \qquad \rightarrow \qquad y = \frac{1}{x}\left(C - \frac{x^2}{2}\right)$$

Alternatively, if we let $\widehat{y} = y/x$ and $\widehat{x} = x$, then (6.8) can be transformed to be

$$\frac{1}{\widehat{x}}\, d\widehat{x} + \frac{1}{2\widehat{y} + 1}\, d\widehat{y} = 0$$

which is a separable form. The solution then becomes

$$\widehat{y} = \frac{1}{2}\left(\frac{1}{\widehat{x}^2} - \widehat{C}\right) \qquad \rightarrow \qquad y = \frac{1}{x}\left(\frac{\widehat{C}}{2} - \frac{x^2}{2}\right)$$

which is the same solution with $\widehat{C} = 2C$.

Remarks:

1. In the next section, we group several techniques as similarity transformation methods. Most of these techniques should be familiar to most readers. It turns out that they are just special cases of similarity transformations.
2. Later, in the section after next, we focus on the search for an integrating factor needed to achieve exactness. Unfortunately, it is often that a systematic search may yield equations that might even be more formidable than the original problem. Thus we simply outline the guiding equations from which some heuristics will be needed to make the approach more practical.
3. It is possible that the resulting integrals are still not easily reduced to closed forms. Thus numerical integrations may still be needed to obtain the solutions.
4. Although an integrating factor is sometimes hard to find, the use of exact differentials have resulted in other benefits. For instance, in thermodynamics, Caratheodory successfully used the reciprocal of absolute temperature $1/T$ as an integrating factor to show that the **change in entropy**, Δs, defined as the integral of the ratio of differential heat transfer to temperature (i.e., $\int \delta Q/T$), becomes a path-independent function.

6.2 Separable Forms via Similarity Transformations

We begin with a general definition of **symmetry transformations**, under which **similarity transformations** are special cases.

Definition 6.1. *A pair of transformations*

$$\widehat{x} = \widehat{x}(x, y) \qquad ; \qquad \widehat{y} = \widehat{y}(x, y) \tag{6.9}$$

is called a **symmetry transformation (pair)** *for a first-order differential equation*

$$M(x, y)\, dx + N(x, y)\, dy = 0$$

if, after substituting the new variable defined by (6.9), the new differential equation is given by

$$M(\widehat{x}, \widehat{y})\, d\widehat{x} + N(\widehat{x}, \widehat{y})\, d\widehat{y} = 0$$

that is, the same functionalities for M and N remain, except that x and y are replaced by $\widehat{x}$ and $\widehat{y}$, respectively. Furthermore, we say that the new differential equation has attained **symmetry** *based on the transformations.*

There exist general approaches to obtain symmetry transformations, and the notion of symmetry transformations can also be generalized for higher order differential equations. However, even for the first-order systems, the symmetry transformations are generally difficult to obtain, and in some cases they require the solution of partial differential equations. We limit our discussion only to a special class of symmetry transformations known as **similarity transformations**, which are among the simplest symmetry transformations to try.

Definition 6.2. *A pair of transformations*

$$\widetilde{x} = \lambda^{\alpha} x \qquad ; \qquad \widetilde{y} = \lambda^{\beta} y \tag{6.10}$$

is called a **similarity transformation (pair)** *for a first-order differential equation*

$$M(x, y)\, dx + N(x, y)\, dy = 0$$

if, after substituting (6.10), the new differential equation attains symmetry, given by

$$M(\widetilde{x}, \widetilde{y})\, d\widetilde{x} + N(\widetilde{x}, \widetilde{y})\, d\widetilde{y} = 0$$

where λ is called the **similarity transformation parameter**, *and α and β are nonzero real constants.*

If a similarity transformation exists, that is, if one can find real values for α and β in (6.10) to attain symmetry, then we can combine the variables to obtain a new variable u known as the **similarity variable** or **invariant** given by

$$u = \frac{\widetilde{y}^{\alpha}}{\widetilde{x}^{\beta}} = \frac{y^{\alpha}}{x^{\beta}} \tag{6.11}$$

EXAMPLE 6.2. Consider the differential equation given by

$$\left(y + 2x^2\right)^2 dx - x^3 dy = 0$$

Substituting $x = \lambda^{-\alpha}\widetilde{x}$ and $y = \lambda^{-\beta}\widetilde{y}$, we get

$$\lambda^{-2\beta-\alpha}\left(\widetilde{y} + 2\lambda^{\beta-2\alpha}\widetilde{x}^2\right)^2 d\widetilde{x} - \lambda^{-3\alpha-\beta}\widetilde{x}^3 d\widetilde{y} = 0$$

To attain symmetry, that is, to remove the presence of λ, we need $\beta = 2\alpha$. In particular, we could set $\beta = 2$ and $\alpha = 1$. Doing so, we have the invariant[3]

$$u = \frac{y^{\alpha}}{x^{\beta}} = \frac{y}{x^2}$$

Having determined the invariant u, the next theorem guarantees that the original differential equation can be made separable.

THEOREM 6.1. *Let the differential equation*

$$M(x, y)\, dx + N(x, y)\, dy = 0$$

admit a set of similarity transformations given by $\widetilde{x} = \lambda^{\alpha}$ *and* $\widetilde{y} = \lambda^{\beta}$, *where* $\alpha \neq 0$ *and* $\beta \neq 0$. *Then, using the invariant* $u = y^{\alpha}x^{-\beta}$ *as a new variable, while maintaining* x *as the other variable, will transform the differential equation into a separable variables form given by*

$$\frac{1}{x}\, dx + \frac{1}{\beta u - \alpha u^{(\alpha-1)/\alpha} G(u)}\, du = 0 \tag{6.12}$$

where

$$G(u) = \begin{cases} -\dfrac{M(x, ux)}{N(x, ux)} & \text{if } \alpha = \beta \\[2ex] -\dfrac{M\left(x, \left(ux^{\beta}\right)^{1/\alpha}\right)}{N\left(x, \left(ux^{\beta}\right)^{1/\alpha}\right)} x^{(\alpha-\beta)/\alpha} & \text{if } \alpha \neq \beta \end{cases}$$

PROOF. (See Section F.4.1 for proof.)

EXAMPLE 6.3. In Example 6.2, we found $u = y/x^2$ to be an invariant for the differential equation

$$\left(y + 2x^2\right)^2 dx - x^3 dy = 0$$

Applying Theorem 6.1, we have

$$\frac{dx}{x} - \frac{du}{u^2 + 2u + 4} = 0$$

which is now separable, and the solution can be obtained as

$$\frac{\sqrt{3}}{3} \tan^{-1}\left(\frac{\sqrt{3}}{3}(u+1)\right) = \ln(Cx) \quad \rightarrow \quad y = x^2\left(\sqrt{3}\tan\left(\sqrt{3}\ln(Cx)\right) - 1\right)$$

[3] If we had set $\beta = 1$ and $\alpha = 1/2$, another possible invariant is given by

$$v = \frac{\sqrt{y}}{x} = \sqrt{u}$$

In fact, any function of the invariant, $\theta(u)$, is also a valid invariant for the differential equation.

A large class of differential equations that immediately admits similarity transformations are the **isobaric** first-order differential equations given by the form

$$x^{n-1} f\left(\frac{y}{x^n}\right) dx - dy = 0 \tag{6.13}$$

where $f(\eta)$ is any differentiable function of η. This admits a similarity transformation with $\beta = n\alpha$. In particular, with $\alpha = 1$, $\beta = n$, and $u = yx^{-n}$, Theorem 6.1 reduces (6.13) to the following separable form:

$$\frac{dx}{x} + \frac{du}{nu - f(u)} = 0 \tag{6.14}$$

EXAMPLE 6.4. Using the equation in Example 6.2 one more time, this differential equation can be rearranged to be

$$x\left(\frac{y}{x^2} + 2\right)^2 dx - dy = 0$$

which is an isobaric equation with $n = 2$ and $f(\eta) = (\eta + 2)^2$, where $\eta = yx^{-2}$. Based on (6.14) and $u = yx^{-2}$, we get

$$\frac{dx}{x} + \frac{du}{2u - (u+2)^2} = 0$$

which is the same separable equation obtained in Example 6.3.

One special case of isobaric equations occurs when $n = 1$. These equations are known as the **homogeneous-type first-order differential equations**, that is,

$$f\left(\frac{y}{x}\right) dx - dy = 0 \tag{6.15}$$

which can be put into separable form in terms of the variables $u = y/x$ and x,

$$\frac{dx}{x} + \frac{du}{u - f(u)} = 0 \tag{6.16}$$

Another set of equations for which similarity transformations may apply after some additional transformations is given by the form

$$f(\eta)dx - dy = 0 \tag{6.17}$$

where η is a **ratio of affine terms**, that is,

$$\eta = \frac{a_1x + a_2y + a_3}{b_1x + b_2y + b_3} \tag{6.18}$$

where a_1, a_2, a_3, b_1, b_2, and b_3 are constants, and $a_1b_2 \neq a_2b_1$.[4] Let $z = a_1x + a_2y + a_3$ and $w = b_1x + b_2y + b_3$. Then with $\eta = z/w$ and

$$\begin{pmatrix} dx \\ dy \end{pmatrix} = \frac{1}{a_1b_2 - a_2b_1}\begin{pmatrix} b_2 & -a_2 \\ -b_1 & a_1 \end{pmatrix}\begin{pmatrix} dz \\ dw \end{pmatrix}$$

[4] If $a_1b_2 = a_2b_1$, one can just set $\eta = a_1x + a_2y$ or $\eta = b_1x + b_2y$, whichever is nonconstant. This will immediately reduce the original equation to be separable.

the original differential equation can be transformed into a homogeneous-type differential equation given by

$$\left(b_2 f\left(\frac{z}{w}\right) + b_1\right) dz - \left(a_2 f\left(\frac{z}{w}\right) + a_1\right) dw = 0$$

which can be made separable under the variables w and $\zeta = z/w$, that is,

$$\frac{dw}{w} = \frac{b_2 f(\zeta) + b_1}{(a_2 - b_2\zeta)\, f(\zeta) + (a_1 - b_1\zeta)} d\zeta \tag{6.19}$$

EXAMPLE 6.5. Consider the differential equation

$$\frac{dy}{dx} = \left(\frac{5y+1}{x+2}\right)^2 + 1$$

Then applying (6.19) with $w = x+2$ and $\zeta = (5y+1)/(x+2)$ yields

$$\frac{dw}{w} = \frac{d\zeta}{5\,(\zeta^2+1) - \zeta}$$

which is a separable differential equation whose solution is given by

$$\ln(w) + \ln(C) = \frac{2}{3\sqrt{11}} \arctan\left[\frac{1}{3\sqrt{11}}(10\zeta - 1)\right]$$

or in terms of x and y,

$$y = \left[\frac{3\sqrt{11}\,\tan\left(\frac{3}{2}\sqrt{11}\,\ln\left(C[x+2]\right)\right) + 1}{50}\right](x+2) - \frac{1}{5}$$

where C is an arbitrary constant.

6.3 Exact Differential Equations via Integrating Factors

We now consider the method of the inclusion of a function $\phi(x, y)$ known as the **integrating factor** to yield an exact solution. Recall that for

$$M(x, y)dx + N(x, y)dy = 0$$

the exactness condition is given by $\dfrac{\partial M}{\partial y} = \dfrac{\partial N}{\partial x}$. If the differential equation is not exact, we need to find an integrating factor, $\phi(x, y)$, such that

$$\phi(x, y)M(x, y)dx + \phi(x, y)N(x, y)dy = 0$$

becomes exact, that is,

$$\frac{\partial(\phi M)}{\partial y} = \frac{\partial(\phi N)}{\partial x} \quad \rightarrow \quad M\frac{\partial \phi}{\partial y} - N\frac{\partial \phi}{\partial x} = \phi\left(\frac{\partial N}{\partial x} - \frac{\partial M}{\partial y}\right) \tag{6.20}$$

In general, the partial differential equation in (6.20) may be difficult to solve. A limited approach is to assume $\phi = \phi(\eta)$ where $\eta = \eta(x, y)$. This reduces (6.20) to

$$\frac{d\phi}{\phi} = \left(\frac{\left(\frac{\partial M}{\partial y}\right) - \left(\frac{\partial N}{\partial x}\right)}{\left(N\frac{\partial \eta}{\partial x}\right) - \left(M\frac{\partial \eta}{\partial y}\right)} \right) d\eta \tag{6.21}$$

If η could be found such that

$$\frac{\left(\frac{\partial M}{\partial y}\right) - \left(\frac{\partial N}{\partial x}\right)}{\left(N\frac{\partial \eta}{\partial x}\right) - \left(M\frac{\partial \eta}{\partial y}\right)} = F(\eta) \tag{6.22}$$

then the integrating factor ϕ is immediately given by

$$\phi = \exp\left(\int F(\eta) d\eta\right) \tag{6.23}$$

EXAMPLE 6.6. Given $(xy - y^2)\, dx + dy = 0$, we have

$$F = \frac{\partial M/\partial y - \partial N/\partial x}{(N)(\partial \eta/\partial x) - (M)(\partial \eta/\partial y)} = \frac{x - 2y}{\eta_x - (xy - y^2)\, \eta_y}$$

With $\eta_y = 1/y$, the denominator can be made closer in form to the numerator. Thus $\eta = \ln(y) + f(x)$. Substituting η back to F,

$$F = \frac{-2\left(-\frac{x}{2} + y\right)}{\left(\frac{df}{dx} - x\right) + y}$$

which can be made constant by setting $\frac{df}{dx} - x = -\frac{x}{2}$, or $f(x) = \frac{x^2}{4}$. Finally, we can obtain the integrating factor:

$$\eta = \ln(y) + \frac{x^2}{4} \quad \rightarrow \quad \phi(x, y) = \frac{1}{y^2}\exp\left(-\frac{x^2}{2}\right)$$

The conditions for exactness can be verified directly,

$$\frac{\partial \phi M}{\partial y} = -\frac{x}{y^2}\exp\left(-\frac{x^2}{2}\right) = \frac{\partial \phi N}{\partial x}$$

and the solution to the differential equation is

$$\frac{1}{y}\exp\left(-\frac{x^2}{2}\right) + \sqrt{\frac{\pi}{2}}\operatorname{erf}\left(\frac{x}{\sqrt{2}}\right) = C$$

where C is the arbitrary constant of integration, and $\operatorname{erf}(z)$ is the error function defined by

$$\operatorname{erf}(z) = \frac{2}{\pi}\int_0^z e^{-t^2} dt$$

Based on (6.22), two special cases are worth noting. These are:

$$\textbf{Case 1} \quad \frac{\partial M/\partial y - \partial N/\partial x}{N} = p(x) \rightarrow \left(\eta = x\right) \rightarrow \phi = \exp\left(\int p(x)dx\right) \tag{6.24}$$

$$\textbf{Case 2} \quad \frac{\partial M/\partial y - \partial N/\partial x}{M} = q(y) \rightarrow \left(\eta = y\right) \rightarrow \phi = \exp\left(-\int q(y)dy\right) \tag{6.25}$$

EXAMPLE 6.7. For the differential equation given by

$$xy\, dx - \left(x^2 + y\right)\, dy = 0$$

we have

$$\frac{\partial M/\partial y - \partial N/\partial x}{M} = \frac{3}{y}$$

Using (6.25), we get $\phi = y^{-3}$. The solution is then given by

$$\frac{x^2 + 2y}{2y^2} + C = 0$$

The **first-order linear equation** is given by the standard form

$$\frac{dy}{dx} + P(x)y = Q(x) \tag{6.26}$$

This can be rearranged to be

$$\left(P(x)y - Q(x)\right) dx + dy = 0$$

which can be shown to fall under the case satisfying (6.24). Thus the integrating factor is

$$\phi = \exp\left(\int P(x)dx\right) \tag{6.27}$$

and the solution is given by

$$F(x, y) = y \exp\left(\int P(x)dx\right) - \int \left[\exp\left(\int P(x)dx\right) Q(x)\right] dx = C$$

or

$$y = \exp\left(-\int P(x)dx\right) \left(\int \left[\exp\left(\int P(x)dx\right) Q(x)\right] dx + C\right) \tag{6.28}$$

EXAMPLE 6.8. The component balance for a liquid reactant concentration C_A in a continuously stirred reactor undergoing a first-order reaction $A \rightarrow P$ is described by

$$\frac{dC_A}{dt} = \frac{F(t)}{V}\left(C_{A,in}(t) - C_A\right) - kC_A$$

where F, V, k, and $C_{A,in}$ is the volumetric flow rate, reactor volume, specific kinetic constant, and inlet concentration of A, respectively. This can be rewritten in the standard form of a first-order differential equation,

$$\frac{dC_A}{dt} + \left(\frac{F(t)}{V} + k\right) C_A = \frac{F(t)}{V} C_{A,in}(t)$$

where we can identify $P(t) = k + (F(t)/V)$ and $Q(t) = F(t)C_{A,in}(t)/V$, and the solution via (6.28) is

$$C_A(t) = \exp\left(-\int \left(k + \frac{F(t)}{V}\right) dt\right)$$
$$\times \left(\int \left[\exp\left(\int \left(k + \frac{F(t)}{V}\right) dt\right) \frac{F(t)}{V} C_{A,in}(t)\right] dt + C_A(0)\right)$$

One nonlinear extension of the first-order linear differential equation is to introduce a factor y^n to $Q(x)$. This is known as the **Bernouli equation**,

$$\frac{dy}{dx} + P(x)y = Q(x)y^n \tag{6.29}$$

where $n \neq 1$.[5] Instead of finding another integrating factor for (6.29), another transformation is used instead. By multiplying (6.29) by $(1-n)y^{-n}$ and letting $z = y^{1-n}$, the Bernouli equation is reduced to one that is first-order linear type, that is,

$$\frac{dz}{dx} + (1-n)\,P(x)z = (1-n)\,Q(x)$$

and the solution is given by

$$z = \phi(x)^{-1}\left[\int (1-n)\,Q(x)\phi(x)\,dx + C\right] \tag{6.30}$$

where

$$\phi(x) = \exp\left(\int (1-n)\,P(x)\,dx\right)$$

6.4 Second-Order Ordinary Differential Equations

We will limit our discussion to second-order differential equations that have the form given by

$$\frac{d^2y}{dx^2} = f\left(x, y, \frac{dy}{dx}\right) \tag{6.31}$$

In this section, we focus on obtaining transformations that will reduce the order of the equation. Case 1 that follows is likely familiar to most readers, whereas Case 2 invokes similarity transformations to reduce the order.

In a later section, we also include the special case of Euler-Cauchy equation, where the transformation converts it to a linear differential equation with constant

[5] If $n = 1$, the Bernouli equation becomes separable.

coefficients. The direct solution of linear second-order or higher order equations is deferred to Section 6.5 for cases with constant coefficients or in Chapter 9 for those obtained using Frobenius series methods.

6.4.1 Case 1: Missing Explicit Dependencies on x or y

Let $p = dy/dx$ or, equivalently, $dy = p\ dx$, then the second derivative d^2y/dx^2 can be put in terms of p and x, or in terms of p and y, that is,

$$\frac{d^2y}{dx^2} = \frac{dp}{dx} \qquad \text{or} \qquad \frac{d^2y}{dx^2} = p\frac{dp}{dy}$$

This will allow the reduction of the second-order differential equation to a first-order differential equation, depending on whether $f(x, y, dy/dx)$ is missing dependencies on x or y, as follows:

$$\frac{d^2y}{dx^2} = f_1\left(x, \frac{dy}{dx}\right) \quad \rightarrow \quad \frac{dp}{dx} = f_1(x, p) \tag{6.32}$$

$$\frac{d^2y}{dx^2} = f_2\left(y, \frac{dy}{dx}\right) \quad \rightarrow \quad \frac{dp}{dy} = \frac{1}{p}f_2(y, p) \tag{6.33}$$

In either case, a first-order equation will need to be solved, yielding $p = S_1(x, C_1)$ or $p = S_2(y, C_1)$, where C_1 is an arbitrary constant. With $p = dy/dx$, the solution is completed by solving another first-order differential equation, that is,

$$\frac{dy}{dx} = S_1(x, C_1) \qquad \text{or} \qquad \frac{dy}{dx} = S_2(y, C_1)$$

which should introduce another arbitrary constant C_2.

EXAMPLE 6.9. The steady-state, one-dimensional concentration profile of substance A under a concentration-dependent diffusivity and flowing through a pipe in the axial direction is described by

$$v\frac{dC_A}{dz} = \frac{d}{dz}\left(\alpha C_A \frac{dC_A}{dz}\right)$$

where α and v are the diffusivity at unit concentration and constant flow velocity, respectively. This can be rearranged to become

$$\frac{d^2C_A}{dz^2} = \beta\frac{1}{C_A}\frac{dC_A}{dz} - \frac{1}{C_A}\left(\frac{dC_A}{dz}\right)^2$$

where $\beta = v/\alpha$. This falls under the case where the differential equation is not explicitly dependent on z. With $p = dC_A/dz$, we obtain a first-order linear differential equation given by

$$\frac{dp}{dC_A} + \frac{1}{C_A}p = \frac{\beta}{C_A}$$

and the solution is

$$p\left(=\frac{dC_A}{dz}\right)=\beta+\frac{m_1}{C_A}$$

where m_1 is an arbitrary constant. This is a separable differential equation in C_A and z, and the solution is

$$\frac{1}{\beta}C_A-\frac{m_1}{\beta^2}\ln\left(\beta C_A+m_1\right)=z+r$$

where r is another arbitrary constant. This could be simplified to be

$$q\,e^q=\xi(z)$$

where

$$q=k_1C_A-1 \quad \text{and} \quad \xi(z)=k_2e^{k_1\beta z}$$

with k_1 and k_2 as the new pair of arbitrary constants. Thus an explicit solution for $C_A(z)$ is given by

$$C_A(z)=\frac{1}{k_1}\left(W\left(\xi(z)\right)+1\right)$$

where $W(\bullet)$ is **Lambert's W-function** (also known as the **Omega function**), defined as the inverse relation of $f(w)=w\,e^w$, that is,

$$t=qe^q \leftrightarrow q=W(t) \tag{6.34}$$

6.4.2 Case 2: Order Reduction via Similarity Transformations

We can extend the definition of similarity transformations, introduced in Definition 6.2 for first-order differential equations, as follows:

Definition 6.3. *A pair of transformations*

$$\widetilde{x}=\lambda^\alpha x \qquad ; \qquad \widetilde{y}=\lambda^\beta y \tag{6.35}$$

is called a **similarity transformation (pair)** *for a second-order differential equation*

$$\frac{d^2y}{dx^2}=f\left(x,y,\frac{dy}{dx}\right) \tag{6.36}$$

if, after substituting (6.35), the new differential equation attains symmetry, given by

$$\frac{d^2\widetilde{y}}{d\widetilde{x}^2}=f\left(\widetilde{x},\widetilde{y},\frac{d\widetilde{y}}{d\widetilde{x}}\right)$$

where λ is the similarity transformation parameter, and α and β are nonzero real constants.

If $\alpha \neq 0$, then without loss of generality, we could set $\alpha = 1$, and obtain a new independent variable u and a new dependent variable v given by

$$u = \frac{y}{x^{\beta}} = \frac{\widetilde{y}}{\widetilde{x}^{\beta}} \tag{6.37}$$

$$v = \left(x^{1-\beta}\right)\frac{dy}{dx} = \left(\widetilde{x}^{1-\beta}\right)\frac{d\widetilde{y}}{d\widetilde{x}} \tag{6.38}$$

Using these new variables, the original second-order differential equation can be reduced to a first-order differential equation.[6]

THEOREM 6.2. *Let the differential equation (6.36) admit a set of similarity transformations given by (6.35). Using similarity variables u and v defined in (6.37) and (6.38), respectively, the differential equation (6.36) can be reduced to a first-order differential equation given by*

$$\frac{dv}{du} = \frac{G(u,v) + (1-\beta)\,v}{v - \beta u} \tag{6.39}$$

where

$$G(u,v) = x^{2-\beta} f\left(x, y, \frac{dy}{dx}\right)$$

PROOF. (See Section F.4.2 for proof.)

EXAMPLE 6.10. Given the second-order differential equation,

$$x^4\frac{d^2y}{dx^2} = x\frac{dy}{dx}\left(y + x^2\right) - 2y^2$$

we can determine whether a similarity transformation is possible. To do so, let

$$\widetilde{x} = \lambda^{\alpha}x\,, \qquad \widetilde{y} = \lambda^{\beta}y$$

then

$$\lambda^{-(\beta+2\alpha)}\widetilde{x}^4\frac{d^2\widetilde{y}}{d\widetilde{x}^2} = \lambda^{-2\beta}\widetilde{x}\widetilde{y}\frac{d\widetilde{y}}{d\widetilde{x}} + \lambda^{-(\beta+2\alpha)}\widetilde{x}^3\frac{d\widetilde{y}}{d\widetilde{x}} - 2\lambda^{-2\beta}\widetilde{y}$$

For symmetry, we need $\beta = 2\alpha$. Taking $\alpha = 1$ and $\beta = 2$, we have the new variables,

$$u = \frac{y}{x^2} \qquad \text{and} \qquad v = \frac{1}{x}\frac{dy}{dx}$$

Applying these to (6.39)

$$G(u,v) = uv + v - 2u^2 \quad \rightarrow \quad \frac{dv}{du} = u$$

[6] For the more general case of reducing an nth order ordinary differential equation that admits a similarity transformation, see Exercise **E10.8**.

and the solution is

$$v = \frac{u^2}{2} + C \quad \rightarrow \quad \frac{1}{x}\frac{dy}{dx} = \frac{1}{2}\left(\frac{y}{x^2}\right)^2 + C$$

where C is an arbitrary integration constant. The last equation is an isobaric type (cf. (6.13)) in which the same combination $u = y/x^2$ will transform the equation to a separable type, that is,

$$\frac{2du}{u^2 - 4u + 2C} = \frac{dx}{x}$$

Solving for u and then for y, the general solution can be simplified to be

$$y = 2x^2\left[1 + k_1 \tan\left(k_1 \ln(k_2 x)\right)\right]$$

where k_1 and k_2 are a new pair of arbitrary constants.

6.4.3 Case 3: Euler-Cauchy Differential Equation

The second-order **Euler-Cauchy equations** are differential equations having the following special form:

$$a_2 x^2 \frac{d^2y}{dx^2} + a_1 x \frac{dy}{dx} + a_0 y = f(x) \tag{6.40}$$

where a_0, a_1, and a_2 are constants. These equations can be converted to an associated second-order differential equation that is linear with constant coefficients. Unlike the other two cases, this transformation does not reduce the order of the equation. However, the method for solving linear ordinary differential equations with constant coefficients are well established, and they are discussed in Section 6.5.

The key transformation is to set $z = \ln(x)$, whose differential is $dz = dx/x$. This will transform the derivatives with respect to x to derivatives with respect to z as follows:

$$\frac{dy}{dx} = \frac{1}{x}\frac{dy}{dz} \quad \text{and} \quad \frac{d^2y}{dx^2} = \frac{1}{x^2}\left(-\frac{dy}{dz} + \frac{d^2y}{dz^2}\right)$$

Substituting these into (6.40) yields

$$a_2 \frac{d^2y}{dz^2} + (a_1 - a_2)\frac{dy}{dz} + a_0 y = f\left(e^z\right) \tag{6.41}$$

which is the desired second-order linear equation with constant coefficients.

Remarks:

1. The nth-order Euler-Cauchy equations are given by

$$\sum_{i=0}^{n} a_i x^i \frac{d^i y}{dx^i} = f(x) \tag{6.42}$$

and the same change of variable $z = \ln(x)$ will transform it into an nth-order linear differential equation with constant coefficients, involving derivatives with respect to z.

2. Euler-Cauchy equations are special cases of linear differential equations in which the coefficients are analytic; that is, they can be represented by Taylor series. The general approach for these types of differential equations is the Frobenius series solution method, which is covered in Chapter 9. However, the transformation technique described in this section will yield the same solution as the Frobenius series solution, and it has the advantage of being able to immediately determine the character of the solution based on the value of the coefficients a_0, a_1, and a_2.

We end this section with a note that there are several other techniques to solve differential equations. We include three of these in Section F.1, namely general Ricatti equations, Legendre transformations, and singular solutions.

6.5 Multiple Differential Equations

For the general analysis and solutions of multiple ordinary differential equations, we assume that the system of equations can be cast in the **state-space** formulation given by

$$\frac{d}{dt}\mathbf{x} = \mathbf{f}(t, \mathbf{x}) \tag{6.43}$$

where $\mathbf{x}$ is an $n \times 1$ column vector that we refer to as the **state vector** whose element, x_i, is known as the ith state variable,

$$\mathbf{x} = \begin{pmatrix} x_1 \\ \vdots \\ x_n \end{pmatrix} \tag{6.44}$$

and $\mathbf{f}$ is an $n \times 1$ column vector of functions, where each f_i is in general a nonlinear function of the independent variable t and the state variables,

$$\mathbf{f}(t, \mathbf{x}) = \begin{pmatrix} f_1(t, x_1, \ldots, x_n) \\ \vdots \\ f_n(t, x_1, \ldots, x_n) \end{pmatrix} \tag{6.45}$$

If $\mathbf{f} = \mathbf{f}(\mathbf{x})$, then the system of equations is known as **autonomous**. Note that for these sections, we use t as the independent variable and use $\mathbf{x}$ and/or $\mathbf{y}$ as the dependent variable.

Figure 6.1. System of three simultaneous reversible reaction.

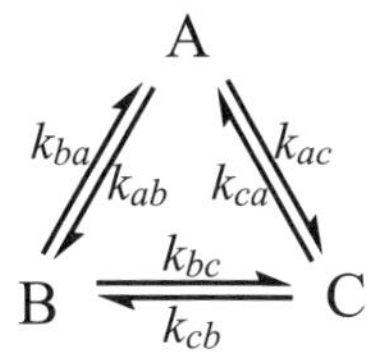

EXAMPLE 6.11. Consider a reactor in which three first-order reactions are occurring simultaneously, as shown in Figure 6.1.

The equations for the rate of change of concentrations of components A, B, and C are

$$\begin{aligned}
\frac{dC_A}{dt} &= (-k_{AB}C_A + k_{BA}C_B) + (-k_{AC}C_A + k_{CA}C_C) \\
\frac{dC_B}{dt} &= (-k_{BC}C_B + k_{CB}C_C) + (-k_{BA}C_B + k_{AB}C_A) \\
\frac{dC_C}{dt} &= (-k_{CA}C_C + k_{AC}C_A) + (-k_{CB}C_C + k_{BC}C_B)
\end{aligned}$$

These equations can now be formulated in state-space form

$$\frac{d\mathbf{C}}{dt} = \mathbf{K}\,\mathbf{C}$$

where,

$$\mathbf{C} = \begin{pmatrix} C_A \\ C_B \\ C_C \end{pmatrix}, \quad \mathbf{K} = \begin{pmatrix} -(k_{AB}+k_{AC}) & k_{BA} & k_{CA} \\ k_{AB} & -(k_{BA}+k_{BC}) & k_{CB} \\ k_{AC} & k_{BC} & -(k_{CA}+k_{CB}) \end{pmatrix}$$

Aside from a set of first-order differential equations that are already of the form given in (6.43), higher order equations can also be cast in state-space forms. Consider a high-order differential equation in which the highest order derivative can be explicitly written as follows:

$$\frac{d^n y}{dt^n} = f\left(t, y, \frac{dy}{dt}, \ldots, \frac{d^{n-1}y}{dt^{n-1}}\right) \tag{6.46}$$

By assigning states to each of y and its derivatives all the way to the $(n-1)$th derivative,

$$x_1 = y, \quad x_2 = \frac{dy}{dt}, \quad \cdots \quad x_n = \frac{d^{n-1}y}{dt^{n-1}}$$

(6.46) can then be written as

$$\frac{d}{dt}\begin{pmatrix} x_1 \\ \vdots \\ x_{n-1} \\ x_n \end{pmatrix} = \begin{pmatrix} x_2 \\ \vdots \\ x_n \\ f(t, x_1, \ldots, x_n) \end{pmatrix} \tag{6.47}$$

EXAMPLE 6.12. Consider Van der Pol's equation given by

$$\frac{d^2y}{dt^2} + (y^2 - b)\frac{dy}{dt} + y = 0$$

Let $x_1 = y$ and $x_2 = \frac{dy}{dt}$, then

$$\begin{aligned}\frac{dx_1}{dt} &= x_2 \\ \frac{dx_2}{dt} &= x_2(b - x_1^2) - x_1\end{aligned}$$

or

$$\frac{d}{dt}\mathbf{x} = \mathbf{f}(\mathbf{x}) = \begin{pmatrix} x_2 \\ bx_2 - x_1^2 x_2 - x_1 \end{pmatrix}$$

The state-space formulation is not only helpful in obtaining analytical solutions. It is in fact the standard form used in numerical solutions of high-order differential equations, as is discussed in Chapter 7.

6.5.1 System of Linear Equations

We now limit the discussion to multiple differential equations that are linear. Let the functions f_i be a linear combination of the state variables x_j, $j = 1, \ldots, n$, and a forcing functions $b_i(t)$, that is,

$$f_i(t, x_1, \ldots, x_n) = a_{i1}(t)x_i + \cdots a_{in}(t)x_n + b_i(t)$$

Then equation (6.43) can be written in matrix form as

$$\frac{d}{dt}\mathbf{x} = A(t)\mathbf{x} + \mathbf{b}(t) \tag{6.48}$$

where

$$A = \begin{pmatrix} a_{11}(t) & \cdots & a_{1n}(t) \\ \vdots & \ddots & \vdots \\ a_{n1}(t) & \cdots & a_{nn}(t) \end{pmatrix} \quad ; \quad \mathbf{b}(t) = \begin{pmatrix} b_1(t) \\ \vdots \\ b_n(t) \end{pmatrix}$$

In the next two sections, we solve (6.48) by the introduction of matrix exponentials when A is constant. The details for the analytical solution are given in Section 6.5.2. A concise formula for the explicit solution for the case with constant A is given in (F.21).

When $A = A(t)$, the matrix exponentials are generalized to matrizants. The solutions for these cases are difficult to generalize. One exception is when $A(t)$ and $A(\tau)$ commutes.

6.5.2 Matrix Exponentials

For constant A, the solution involves the use of the matrix exponential, e^{At}. We can apply (3.34) from Section 3.7

$$e^{At} = I + tA + \frac{t^2}{2!}A^2 + \frac{t^3}{3!}A^3 + \cdots \tag{6.49}$$

Some of the properties of e^{At} are given by the following theorem:

THEOREM 6.3. *Let A and W be square matrices of the same size and let t and s be scalars. Then e^{At} satisfies the following properties:*

$$\text{(i)}\quad e^{As}e^{At} = e^{A(s+t)} \tag{6.50}$$

$$\text{(ii)}\quad \left(e^{At}\right)^{-1} = e^{-At} \tag{6.51}$$

$$\text{(iii)}\quad e^{At}e^{Wt} = e^{(A+W)t} \qquad \textit{if and only if}\quad AW = WA \tag{6.52}$$

$$\text{(iv)}\quad \frac{d}{dt}e^{At} = Ae^{At} = e^{At}A \tag{6.53}$$

PROOF. (See Section F.4.3 for proof.)

Using Theorem 6.3, with $\mathbf{x} = \mathbf{x}(t)$,

$$\begin{aligned}
\frac{d}{dt}\left(e^{-At}\mathbf{x}\right) &= \left(\frac{d}{dt}e^{-At}\right)\mathbf{x} + e^{-At}\left(\frac{d}{dt}\mathbf{x}\right) = -e^{-At}A\mathbf{x} + e^{-At}\left(\frac{d}{dt}\mathbf{x}\right) \\
&= e^{-At}\left(\frac{d}{dt}\mathbf{x} - A\mathbf{x}\right)
\end{aligned} \tag{6.54}$$

Rearranging (6.48) and premultiplying by e^{-At},

$$e^{-At}\left(\frac{d}{dt}\mathbf{x} - A\mathbf{x}\right) = e^{-At}\mathbf{b}\left(t\right)$$

then with (6.54),

$$\frac{d}{dt}e^{-At}\mathbf{x} = e^{-At}\mathbf{b}\left(t\right)$$

Integrating from 0 to t, with $e^{A0} = I$,

$$\begin{aligned}
e^{-At}\mathbf{x}(t) - \mathbf{x}(0) &= \int_0^t e^{-A\tau}\mathbf{b}\left(t\right)d\tau \\
\mathbf{x}(t) &= e^{At}\mathbf{x}(0) + \int_0^t e^{A(t-\tau)}\mathbf{b}\left(\tau\right)d\tau
\end{aligned} \tag{6.55}$$

If matrix A is diagonalizable (cf. Section 3.5), with $A = V\Lambda V^{-1}$, (6.55) can be simplified to be

$$\begin{aligned}\mathbf{x}(t) &= Ve^{\Lambda t}V^{-1}\mathbf{x}(0) + Ve^{\Lambda t}V^{-1}\int_0^t Ve^{-\Lambda\tau}V^{-1}\mathbf{b}(\tau)d\tau \\ &= Ve^{\Lambda t}\left(V^{-1}\mathbf{x}(0) + \int_0^t e^{-\Lambda\tau}V^{-1}\mathbf{b}(\tau)d\tau\right)\end{aligned} \tag{6.56}$$

EXAMPLE 6.13. Given the differential equation,

$$\frac{d\mathbf{x}}{dt} = A\mathbf{x} + \mathbf{b}(t)$$

with

$$A = \begin{pmatrix} -3 & 2 & 1 \\ -2 & 1 & 2 \\ 2 & -2 & -4 \end{pmatrix} \quad ; \quad \mathbf{b}(t) = \begin{pmatrix} 4 - e^{-t} \\ 2 \\ 1 + e^{-2t} \end{pmatrix}$$

subject to the initial condition: $\mathbf{x}(0) = \begin{pmatrix} 1 \\ 0 \\ 1 \end{pmatrix}$.

The eigenvalues of A are $(-3, -2, -1)$, and A is diagonalizable. Thus $A = V\Lambda V^{-1}$, with

$$V = \begin{pmatrix} 0 & 1 & 1 \\ -1 & 0 & 1 \\ 2 & 1 & 0 \end{pmatrix} \quad ; \quad \Lambda = \begin{pmatrix} -3 & 0 & 0 \\ 0 & -2 & 0 \\ 0 & 0 & -1 \end{pmatrix}$$

Let

$$\begin{aligned}q &= \int_0^t e^{-\Lambda\tau}V^{-1}\mathbf{b}(\tau)d\tau = \begin{pmatrix} -\frac{7}{6} + e^t + \frac{1}{2}e^{2t} - \frac{1}{3}e^{3t} \\ \frac{1}{2} - t - 2e^t + \frac{3}{2}e^{2t} \\ t + e^t - e^{-t} \end{pmatrix} \\ r &= Ve^{\Lambda t}q = \begin{pmatrix} \frac{5}{2} + (t-2)\,e^{-t} + \left(-t - \frac{1}{2}\right)e^{-2t} \\ \frac{4}{3} + \left(t - \frac{1}{2}\right)e^{-t} - 2e^{-2t} + \frac{7}{6}e^{-3t} \\ \frac{5}{6} - e^{-t} + \left(-t + \frac{5}{2}\right)e^{-2t} - \frac{7}{3}e^{-3t} \end{pmatrix} \\ s &= Ve^{\Lambda t}V^{-1}\mathbf{x}(0) = \begin{pmatrix} e^{-2t} \\ 0 \\ e^{-2t} \end{pmatrix}\end{aligned}$$

The solution is then given by

$$\mathbf{x} = r + s = \begin{pmatrix} \frac{5}{2} + (t-2)\,e^{-t} + \left(-t + \frac{1}{2}\right)e^{-2t} \\ \frac{4}{3} + \left(t - \frac{1}{2}\right)e^{-t} - 2e^{-2t} + \frac{7}{6}e^{-3t} \\ \frac{5}{6} - e^{-t} + \left(-t + \frac{7}{2}\right)e^{-2t} - \frac{7}{3}e^{-3t} \end{pmatrix}$$

If A is not diagonalizable, we could use the Jordan canonical form. However, a more efficient alternative is to use the method of finite sums (cf. Section 3.7, Case 3) for evaluating e^{At}.

EXAMPLE 6.14. Given the differential equation

$$\frac{d\mathbf{x}}{dt} = A\mathbf{x}$$

with

$$A = \begin{pmatrix} -2 & 0 & 0 \\ -\frac{1}{2} & -2 & -\frac{1}{2} \\ 1 & 2 & 0 \end{pmatrix}$$

The eigenvalues of A are $\lambda = (-2, -1, -1)$. Following the method discussed in Section 3.7, the Cayley-Hamilton theorem suggests that there exist scalar functions c_2, c_1, and c_0 such that[7]

$$e^{At} = c_2 t^2 A^2 + c_1 t A + c_0 I$$

To determine these unknown coefficients, we need three independent equations.

The first two equations are obtained by applying the equation given by

$$e^{\lambda t} = c_2 t^2 \lambda^2 + c_1 t \lambda + c_0$$

to each distinct eigenvalue, $\lambda = -2$ and $\lambda = -1$, that is,

$$\begin{aligned} e^{-2t} &= 4c_2 t^2 - 2c_1 t + c_0 \\ e^{-t} &= c_2 t^2 - c_1 t + c_0 \end{aligned}$$

The third equation is obtained by taking the derivative, with respect to λ, of both sides of equation (6.14) and then setting $\lambda = -1$ (because this is the repeated root),

$$te^{-t} = -2c_2 t^2 + c_1 t$$

Combining all three equations, we solve for $c_2 t^2$, $c_1 t$, and c_0 instead of c_2, c_1, and c_0. This saves us from having to invert a nonconstant matrix, and it also adds to the efficiency in the succeeding steps. Thus

$$\begin{pmatrix} 4 & -2 & 1 \\ 1 & -1 & 1 \\ -2 & 1 & 0 \end{pmatrix} \begin{pmatrix} c_2 t^2 \\ c_1 t \\ c_0 \end{pmatrix} = \begin{pmatrix} e^{-2t} \\ e^{-t} \\ te^{-t} \end{pmatrix}$$

and

$$\begin{pmatrix} c_2 t^2 \\ c_1 t \\ c_0 \end{pmatrix} = \begin{pmatrix} (t-1)\,e^{-t} + e^{-2t} \\ (3t-2)\,e^{-t} + 2e^{-2t} \\ 2te^{-t} + e^{-2t} \end{pmatrix}$$

[7] In Section 3.7 Case 3, these scalars are constants. However, in applying those same methods here, we take the variable t as a parameter to apply Cayley-Hamilton's theorem. This will necessarily result in the coefficients c_i to be functions of t.

Next, we apply these coefficients to obtain e^{At},

$$\begin{aligned} e^{At} &= c_2 t^2 A^2 + c_1 t A + c_0 I \\ &= \begin{pmatrix} e^{-2t} & 0 & 0 \\ \frac{1}{2}\left(e^{-2t} - e^{-t}\right) & (-t+1)\,e^{-t} & -\frac{t}{2}e^{-t} \\ \left(-e^{-2t} + e^{-t}\right) & 2te^{-t} & (t+1)\,e^{-t} \end{pmatrix} \end{aligned}$$

and the solution is then given by

$$\mathbf{x} = \begin{pmatrix} e^{-2t} & 0 & 0 \\ \frac{1}{2}\left(e^{-2t} - e^{-t}\right) & (-t+1)\,e^{-t} & -\frac{t}{2}e^{-t} \\ \left(-e^{-2t} + e^{-t}\right) & 2te^{-t} & (t+1)\,e^{-t} \end{pmatrix} \mathbf{x}_0$$

where $\mathbf{x}_0 = \mathbf{x}(0)$ is the vector of initial conditions.

A general formulation of the finite series method, including details, is given in Section F.3 as an appendix.

6.5.3 Matrizants

Consider the linear matrix differential equation given by

$$\frac{d}{dt}\mathbf{x} = A(t)\mathbf{x} + \mathbf{b}(t) \tag{6.57}$$

subject to initial condition: $\mathbf{x}(0) = \mathbf{x}_0$

Let us first consider the solution of the homogenous equation, that is, with $\mathbf{b}(t) = 0$,

$$\frac{d}{dt}\mathbf{x} = A(t)\mathbf{x} \tag{6.58}$$

One solution approach is **Picard's method**. It begins by integrating the differential equation,

$$d\mathbf{x} = A(t)\mathbf{x}dt \quad \rightarrow \quad \mathbf{x}(t) = \mathbf{x}_0 + \int_0^t \left(A(\tau_1)\mathbf{x}(\tau_1)\right) d\tau_1 \tag{6.59}$$

After recursive application of (6.59),

$$\begin{aligned} \int_0^t \left(A(\tau_1)\mathbf{x}(\tau_1)\right) d\tau_1 &= \int_0^t A(\tau_1)\left(\mathbf{x}_0 + \int_0^{\tau_1} A(\tau_2)\mathbf{x}(\tau_2)d\tau_2\right) d\tau_1 \\ &= \int_0^t A(\tau_1)d\tau_1\mathbf{x}_0 + \int_0^t A(\tau_1)\int_0^{\tau_1} \left(A(\tau_2)\mathbf{x}(\tau_2)\right) d\tau_2 d\tau_1 \end{aligned}$$

Let Q_k be defined as

$$Q_1(t) = \int_0^t A(\tau_1)d\tau_1 \qquad \text{for } k = 1 \tag{6.60}$$

and

$$Q_k(t) = \int_0^t A(\tau_1) \int_0^{\tau_1} A(\tau_2) \int_0^{\tau_2} A(\tau_3) \cdots \int_0^{\tau_{k-1}} A(\tau_k) d\tau_k \cdots d\tau_3\, d\tau_2\, d\tau_1 \quad (6.61)$$

for $k > 1$. Assuming convergence, the solution becomes an infinite series given by

$$\mathbf{x}(t) = \mathbf{x}_0 + \left(\sum_{k=1}^{\infty} Q_k \right) \mathbf{x}_0 = M(t)\mathbf{x}_0 \quad (6.62)$$

where

$$M(t) = I + \sum_{k=1}^{\infty} Q_k(t) \quad (6.63)$$

$M(t)$ is known as the **matrizant** or **fundamental matrix** of the differential equation given in (6.58). It has the following properties:

$$M(0) = I \qquad \text{and} \qquad \frac{dM}{dt} = A(t)M(t) \quad (6.64)$$

THEOREM 6.4. *Let the elements of the state matrix A be bounded; then the matrizant M defined by (6.63), with Q_k defined in (6.61), is invertible.*

PROOF. (See Section F.4.4 for proof.)

Corresponding to (6.64), we have

$$M^{-1}(0) = I \qquad \text{and} \qquad dM^{-1}/dt = -M^{-1}(t)A(t) \quad (6.65)$$

To show this,

$$\begin{aligned} \frac{d}{dt}\left(M^{-1}M\right) = \frac{d}{dt}\left(M^{-1}\right)M + M^{-1}\frac{d}{dt}(M) &= 0 \\ \frac{d}{dt}\left(M^{-1}\right) + M^{-1}\left(\frac{d}{dt}(M)\right)M^{-1} &= 0 \\ \frac{d}{dt}\left(M^{-1}\right) + M^{-1}AMM^{-1} &= 0 \\ \frac{d}{dt}\left(M^{-1}\right) &= -\, M^{-1}A \end{aligned}$$

Applying (6.65), we can establish the following identity,

$$\begin{aligned} \frac{d}{dt}\left(M^{-1}\mathbf{x}\right) &= \left(\frac{d}{dt}M^{-1}\right)\mathbf{x} + M^{-1}\left(\frac{d}{dt}\mathbf{x}\right) = -M^{-1}A\mathbf{x} + M^{-1}\left(\frac{d}{dt}\mathbf{x}\right) \\ &= M^{-1}\left(\frac{d}{dt}\mathbf{x} - A\mathbf{x}\right) \end{aligned} \quad (6.66)$$

We can now solve the nonhomogeneous case, that is, $\mathbf{b}(t) \neq 0$,

$$\begin{aligned}
\frac{d}{dt}\mathbf{x} - A(t)\mathbf{x} &= \mathbf{b}(t) \\
M^{-1}\left(\frac{d}{dt}\mathbf{x} - A\mathbf{x}\right) &= M^{-1}\mathbf{b}(t) \\
\frac{d}{dt}\left(M^{-1}\mathbf{x}\right) &= M^{-1}\mathbf{b}(t) \\
d\left(M^{-1}\mathbf{x}\right) &= M^{-1}\mathbf{b}(t)dt \\
M^{-1}(t)\mathbf{x}(t) - M^{-1}(0)\mathbf{x}(0) &= \int_0^t M^{-1}(\tau)\mathbf{b}(\tau)d\tau
\end{aligned}$$

$$\mathbf{x}(t) = M(t)\mathbf{x}(0) + M(t)\int_0^t M^{-1}(\tau)\mathbf{b}(\tau)d\tau \tag{6.67}$$

where we used the fact that $M(0) = M^{-1}(0) = I$.

Equation (6.67) is the general solution of the linear nonhomogeneous equation.[8] However, matrizants M are generally difficult to evaluate. One important case that is easier to solve is when matrices $A(t)$ and $A(\tau)$ commute. If $A(t)A(\tau) = A(\tau)A(t)$ for $\tau \neq t$, then A and Q_k will also commute. Thus

$$M(t) = e^{Q_1} \qquad \text{and} \qquad M^{-1}(t) = e^{(-Q_1)} \tag{6.68}$$

where $Q_1 = \int_0^t A(\tau_1)d\tau_1$. (See Exercise **E6.19** for an example of this case.) When A is constant, the matrizant is simply given by $M = e^{At}$ as before.

We use matrizants later to help us solve linear boundary value problems via the shooting method in Section 7.5. However, in that section, we just use numerical initial value problem (IVP) solvers developed in Chapter 7 to evaluate the matrizants at the terminal point T.

6.6 Decoupled System Descriptions via Diagonalization

If the system matrix $\mathbf{A}$ is constant and diagonalizable, a change of coordinates can yield a decoupled set of differential equations. Let $A = V\Lambda V^{-1}$, where Λ is the diagonal matrix of eigenvalues, and V is the matrix containing the eigenvectors of $\mathbf{A}$. Now let $\mathbf{z}$ be the new set of state variables defined by

$$\mathbf{z} = \begin{pmatrix} z_1 \\ \vdots \\ z_n \end{pmatrix} = V^{-1}\mathbf{x} \tag{6.69}$$

where $\mathbf{x}$ is the original set of state variables, then

$$\frac{d}{dt}\mathbf{z} = V^{-1}\frac{d}{dt}\mathbf{x} \qquad \text{and} \qquad \mathbf{z}_o = \begin{pmatrix} z_{o,1} \\ \vdots \\ z_{o,n} \end{pmatrix} = V^{-1}\mathbf{x}_o$$

[8] This solution approach, where the solution of the linear homogeneous case $\mathbf{b}(t) = 0$ can be used to extend the solution to the linear nonhomogeneous case $\mathbf{b}(t) \neq 0$, is also known as **Duhamel's principle**.

The original linear equation can be transformed as follows:

$$\begin{aligned}
\frac{d}{dt}\mathbf{x} &= A\mathbf{x} + \mathbf{b}(t) = V\Lambda V^{-1}\mathbf{x} + \mathbf{b}(t) \\
V^{-1}\frac{d}{dt}\mathbf{x} &= \Lambda V^{-1}\mathbf{x} + V^{-1}\mathbf{b}(t) \\
\frac{d}{dt}\mathbf{z} &= \Lambda\mathbf{z} + Q(t)
\end{aligned} \tag{6.70}$$

where

$$Q(t) = V^{-1}\mathbf{b}(t) = \begin{pmatrix} q_1(t) \\ \vdots \\ q_n(t) \end{pmatrix}$$

Because Λ is diagonal, (6.70) can be rewritten as

$$\begin{aligned}
\frac{dz_1}{dt} &= \lambda_1 z_1 + q_1(t) \\
&\vdots \\
\frac{dz_n}{dt} &= \lambda_n z_n + q_n(t)
\end{aligned} \tag{6.71}$$

which is a set of decoupled differential equation. Each decoupled differential equation is a first-order linear differential equation whose solution is given by

$$z_k(t) = e^{\lambda_k t} z_{o,k} + \int_0^t e^{\lambda_k(t-\tau)} q(\tau) d\tau \tag{6.72}$$

The original state variables, $\mathbf{x}$, can be recovered using (6.69),

$$\mathbf{x} = V\mathbf{z} \tag{6.73}$$

which is the same solution given in (6.56).

Under the decoupled system description, the solution given in (6.72) can be used to assess the stability of the solution. If any of the eigenvalues, say, λ_k, have positive real parts, the value of z_k in (6.72) will keep increasing unboundedly with increasing values of t. Furthermore, because $\mathbf{x}$ is a linear combination of the elements of $\mathbf{z}$, the same stability conclusion is the same for $\mathbf{x}(t)$. This conclusion is also true for the non-diagonalizable case, as stated in the following theorem:

THEOREM 6.5. *For the linear system described by*

$$\frac{d}{dt}\mathbf{x} = A\mathbf{x} + \mathbf{b}(t) \tag{6.74}$$

with A constant and $\mathbf{b}(t)$ bounded, then $\mathbf{x}(t)$ is unstable if any of the eigenvalues of A has a positive real part.

PROOF. (See Section F.4.5 for proof.)

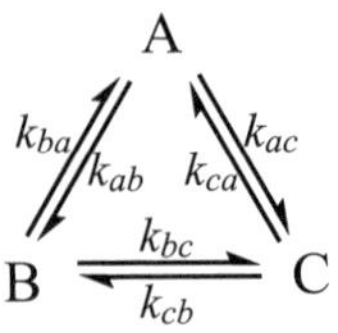

Figure 6.2. System of three simultaneous reversible reactions.

When decoupling is possible, the trajectories will move along straight lines. In the following example, this fact has been used to solve a parameter estimation problem in which strong interaction among the original set of variables is present.

EXAMPLE 6.15. Wei Prater Kinetics Wei and Prater[9] used the idea of decoupling to obtain kinetic parameters of simultaneous reversible first-order reactions of N chemical components.

For three components undergoing first-order kinetics as shown in Figure 6.2, the system is described by

$$\frac{d}{dt}\begin{pmatrix} x_A \\ x_B \\ x_C \end{pmatrix} = K\mathbf{x} = \begin{pmatrix} -(k_{AB}+k_{AC}) & k_{BA} & k_{CA} \\ k_{AB} & -(k_{BA}+k_{BC}) & k_{CB} \\ k_{AC} & k_{BC} & -(k_{CA}+k_{CB}) \end{pmatrix}\begin{pmatrix} x_A \\ x_B \\ x_C \end{pmatrix}$$

where x_i denotes mass fraction and $x_A + x_B + x_C = 1$.

In matrix form,

$$\frac{d}{dt}\mathbf{x} = K\mathbf{x}$$

where

$$\mathbf{x} = \begin{pmatrix} x_A \\ x_B \\ x_c \end{pmatrix} \qquad K = \begin{pmatrix} -(k_{AB}+k_{AC}) & k_{BA} & k_{CA} \\ k_{AB} & -(k_{BA}+k_{BC}) & k_{CB} \\ k_{AC} & k_{BC} & -(k_{CA}+k_{CB}) \end{pmatrix}$$

The objective is to estimate the six kinetic coefficients, k_{ij}, using experimental data. A typical graph is shown in Figure 6.3. Two things are worth noting in Figure 6.3: (1) all curves converge to a single point, and (2) there are two straight-line reactions. The main challenge in this parameter estimation problem is to determine the kinetic coefficients when no experiment exists that isolates dependence on only one component at a time. Wei and Prater decided to look for an alternative set of coordinates such that under new coordinates, the pseudo-concentrations are decoupled.

Thus let us define a set of pseudo-concentration variables as follows:

$$\mathbf{y} = \begin{pmatrix} y_1 \\ y_2 \\ y_3 \end{pmatrix} = V^{-1}\mathbf{x}$$

where V is the matrix of the eigenvectors of K.[10] This should result in a decoupled system

$$\frac{d\mathbf{y}}{dt} = \Lambda\mathbf{y}$$

[9] Wei, J. and Prater, C. D., The Structure and Analysis of Complex Reaction Systems, *Advances in Catalysis*, **13**, Academic Press, New York (1962).

[10] This naturally assumes that K is diagonalizable.

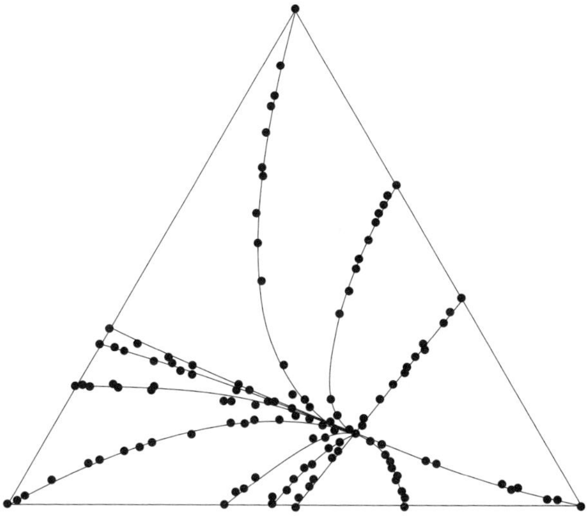

Figure 6.3. Experimentally observed compositions for butene isomerization. (Data adapted from Froment and Bischoff, *Chemical Reactor Analysis and Design*, J. Wiley and Sons, 1979, p. 21).

where Λ is the diagonal matrix of eigenvalues of K. The solution of the decoupled system is given by

$$y_1 = y_1(0)e^{\lambda_1 t} \quad , \quad y_2 = y_2(0)e^{\lambda_2 t} \quad \text{and} \quad y_3 = y_3(0)e^{\lambda_3 t}$$

In terms of the original concentration, this becomes

$$\begin{aligned} \mathbf{x} = V\mathbf{y} &= y_1\mathbf{v}_1 + y_2\mathbf{v}_2 + y_3\mathbf{v}_3 \\ &= y_1(0)e^{\lambda_1 t}\mathbf{v}_1 + y_2(0)e^{\lambda_2 t}\mathbf{v}_2 + y_3(0)e^{\lambda_3 t}\mathbf{v}_3 \end{aligned} \tag{6.75}$$

Note that $\det(K) = 0$ because the last row is the negative of the sum of the upper two rows. Recall from one of the properties of eigenvalues (cf. Property 7 of Section 3.3) that the product of the eigenvalues of matrix K is equal to the determinant of K. Because K is singular, then at least one of the eigenvalues of K must be zero. Without loss of generality, set λ_1 to zero. Then the corresponding eigenvector, $\mathbf{v}_1$, behaves according to definition of eigenvectors,

$$K\mathbf{v}_1 = \lambda_1\mathbf{v}_1 = 0$$

Thus, had the experiment started at $\mathbf{x}^* = y_1(0)e^{\lambda_1 t}\mathbf{v}_1$,

$$\frac{d}{dt}\mathbf{x}^* = K\mathbf{x}^* = 0$$

which means $\mathbf{x}^*$ is the equilibrium point of the process, $\mathbf{x}_{eq}$, that is,

$$\mathbf{x}_{eq} = y_1(0)\mathbf{v}_1$$

Let us now look at the deviations from equilibrium: $\mathbf{x} - \mathbf{x}_{eq}$. From (6.75),

$$\mathbf{x} - \mathbf{x}_{eq} = y_2(0)e^{\lambda_2 t}\mathbf{v}_2 + y_3(0)e^{\lambda_3 t}\mathbf{v}_3 \tag{6.76}$$

which is a linear combination of vectors $\mathbf{v}_2$ and $\mathbf{v}_3$.

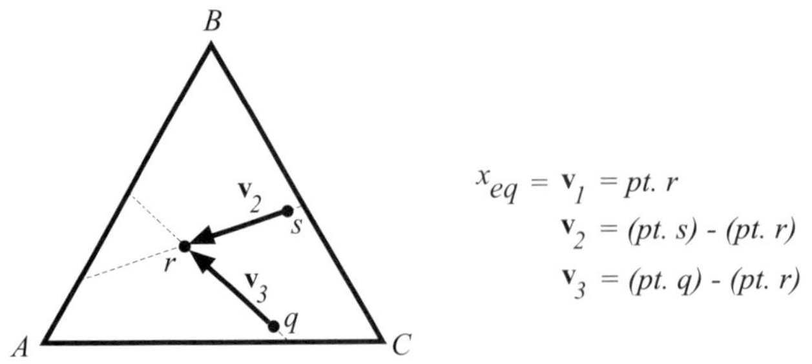

Figure 6.4. Using straight-line reactions to identify $\mathbf{v}_2$ and $\mathbf{v}_3$.

If we start the process at an initial point where $y_2(0) > 0$ and $y_3(0) = 0$, we obtain a reaction path that follows the direction of $\mathbf{v}_2$, that is, a straight-line path. Along this path, only the coefficient given by $y_2(0)e^{\lambda_2 t}$ decreases with time. The eigenvalue λ_2 can then be found using the least-squares method by estimating the slope of the linear equation

$$\ln(\|\mathbf{x} - \mathbf{x}_{eq}\|) = \ln(\|y_2(0)\mathbf{v}_2\|) + \lambda_2 t$$

as we follow the path along $\mathbf{v}_2$.

Using the other straight-line reaction, we can obtain λ_3 by starting at $y_3(0) > 0$ and $y_2(0) = 0$, which will be a path along $\mathbf{v}_3$. Thus the eigenvalue λ_3 can be found in a similar manner, using the least-squares method to estimate the slope of the linear equation

$$\ln(\|\mathbf{x} - \mathbf{x}_{eq}\|) = \ln(\|y_3(0)\mathbf{v}_3\|) + \lambda_3 t$$

Eigenvectors $\mathbf{v}_2$ and $\mathbf{v}_3$ can be obtained directly from the data, as shown in Figure 6.4, whereas the equilibrium point $\mathbf{v}_1 = \mathbf{x}_{eq}$ can be read off the plot (point r). By subtracting the mass fractions at the start of one of the straight lines (point s in the figure) and the equilibrium point (point r), the resulting vector can be designated as $\mathbf{v}_2$. Likewise, using the other straight-line reaction that is not along the previous line, one could also subtract the start point (point q) and the end point (point r) to determine $\mathbf{v}_3$.

Combining all the results so far: $\mathbf{v}_1 = \mathbf{x}_{eq}$, $\mathbf{v}_2$, $\mathbf{v}_3$, $\lambda_1 = 0$, λ_2 and λ_3, we can build matrices V and Λ,

$$V = (\mathbf{x}_{eq}|\mathbf{v}_2|\mathbf{v}_3) \qquad \Lambda = \begin{pmatrix} 0 & 0 & 0 \\ 0 & \lambda_2 & 0 \\ 0 & 0 & \lambda_3 \end{pmatrix}$$

Finally, matrix K can be reconstructed as follows:

$$K = V\Lambda V^{-1}$$

6.7 Laplace Transform Methods

For a multiple first-order linear differential equations, where the state matrix A is constant, that is,

$$\frac{d}{dt}\mathbf{x} = A\mathbf{x} + \mathbf{b}(t) \tag{6.77}$$

the method of Laplace transforms can be used to transform the set of differential equations into a set of algebraic equations.

The Laplace transform of a function $f(t)$ is defined as

$$\mathcal{L}[f(t)] = \int_0^\infty f(t)e^{-st}dt \tag{6.78}$$

where s is the Laplace transform variable, which spans the right-half complex plane. Details, including several properties of Laplace transforms, can be found in Section 12.4. In our case, we use the property of derivatives given by

$$\mathcal{L}\left[\frac{df}{dt}\right] = s\mathcal{L}[f(t)] - f(0) \tag{6.79}$$

and the convolution theorem

$$\mathcal{L}[f * g] = \mathcal{L}[f]\mathcal{L}[g] \tag{6.80}$$

where the convolution of $f(t)$ and $g(t)$ is defined by

$$f * g = \int_0^t f(t-\tau)g(\tau)\,d\tau \tag{6.81}$$

For the special case of $\mathbf{f} = e^{At}$, recall (6.53) and apply (6.79),

$$\begin{aligned} \mathcal{L}\left[\frac{d}{dt}e^{At}\right] &= \mathcal{L}\left[Ae^{At}\right] \\ s\,\mathcal{L}\left[e^{At}\right] - I &= A\mathcal{L}\left[e^{At}\right] \end{aligned}$$

or

$$\mathcal{L}\left[e^{At}\right] = \left(sI - A\right)^{-1} \tag{6.82}$$

Applying (6.79) and (6.82) to (6.77),

$$\begin{aligned} sI\,\mathcal{L}[\mathbf{x}] - \mathbf{x}(0) &= A\mathcal{L}[\mathbf{x}] + \mathcal{L}[\mathbf{b}] \\ \mathcal{L}[\mathbf{x}] &= \left(sI - A\right)^{-1}\left(\mathbf{x}(0) + \mathcal{L}[\mathbf{b}]\right) \end{aligned}$$

Next, use the inverse Laplace transform, $\mathcal{L}^{-1}[\cdot]$ defined by $\mathcal{L}^{-1}\left[\,\mathcal{L}[f]\,\right] = f$,

$$\begin{aligned} \mathbf{x} &= \mathcal{L}^{-1}\left[\,\mathcal{L}\left[e^{At}\right]\mathbf{x}(0)\,\right] + \mathcal{L}^{-1}\left[\,\mathcal{L}\left[e^{At}\right]\mathcal{L}[\mathbf{b}]\,\right] \\ &= e^{At}\mathbf{x}(0) + \left(e^{At} * (\mathbf{b})\right) \\ &= e^{At}\mathbf{x}(0) + \int_0^\infty e^{A(t-\tau)}\mathbf{b}(\tau)d\tau \end{aligned}$$

which is the same solution given by (6.55).

6.8 EXERCISES

E6.1. Consider the differential equation,

$$\left(S(y)\frac{df}{dx}\right)dx + \left(Q(y)f(x) + R(y)\right)dy = 0$$

1. Find the integrating factor using either (6.24) or (6.25).

2. Apply the result to solve the following differential equation:

$$(ydx + dy)\, e^{\alpha x} + R(y)dy = 0$$

E6.2. The axial flow along the z-direction of an incompressible fluid through a circular tube of radius R is given by the following differential equation:

$$\frac{d}{dr}\left(r\frac{dv_z}{dr}\right) = \beta r$$

where

$$\beta = \frac{P_L - P_0}{\mu L}$$

Obtain $v_z(r)$ under the conditions that v_z is finite at $r = 0$ and $v_z = 0$ at $r = R$.

E6.3. The steady-state model for the biodegradation of substance A in a spherical pellet is given by

$$\frac{d^2C_A}{dr^2} + \frac{2}{r}\frac{dC_A}{dr} = \alpha C_A^2$$

where $\alpha = k/D_{AB}$ is constant. Use the similarity transformation to reduce the order of the equation.

E6.4. Obtain the general solution for the following differential equations:

1. $\dfrac{dy}{dx} = \dfrac{4x + 5y + 2}{x + y - 3}$

2. $\dfrac{dy}{dx} = \dfrac{3x - y + 3}{y - 3x + 2}$

3. $\dfrac{dy}{dx} = (2x + y)\,(2x + y + 4)$

E6.5. Find the integrating factor for the following equation:

$$\left(\frac{a}{y}\sin\left(ax + by\right)\right) dx + \left(\frac{b}{y}\sin\left(ax + by\right) + 2\right) dy = 0$$

E6.6. Consider a thin cylindrical wire of length L and diameter D exposed to a constant surrounding temperature T_a, while the ends are fixed to be T_0 and T_L at $x = 0$ and $x = L$, respectively. The temperature is assumed to be uniform at any cross-section and the temperature profile along the length of the wire is described by the following equation resulting from the energy balance:[11]

$$\frac{d}{dx}\left[k(T)\frac{dT}{dx}\right] = \beta(T - T_a)$$

where $\beta = \pi DU/A$, with D, A, and U as diameter, cross-sectional area, and heat transfer coefficient, respectively. Suppose the material has a temperature-dependent conductivity given by

$$k(T) = k_0 - \alpha T$$

[11] This problem is adapted from an example given in Jenson and Jeffreys, *Mathematical Methods in Chemical Engineering*, 2nd Ed., Academic Press, 1977.

1. Show that the relationship between x and T is given by

$$x = \int_{T(0)}^{T(x)} \frac{k_m - \alpha (T - T_a)}{\sqrt{C + \frac{1}{3}\beta (T - T_a)^2 (-2\alpha (T - T_a) + 3k_m)}} dT$$

where C is an arbitrary constant and $k_m = k_0 - \alpha T_a$.

2. Let the parameters be given by $\alpha = 0.0067 \frac{W}{cm(^oC)^2}$, $k_0 = 1.5 \frac{W}{cm \cdot {}^oC}$, $T_a = 30^oC$, $T_0 = 60^oC$, $T_L = 100^oC$, $L = 12cm$ and $\beta = 0.012 \frac{W}{cm^3 \cdot {}^oC}$. Determine the value of C.

E6.7. The growth of microbial colony is modeled by

$$\frac{dN}{dt} = kN - \beta N^2$$

where N is the population of the colony of bacteria, $k > 0$ is the specific growth rate, and $\beta > 0$ is the death rate coefficient.

1. Let k and β be constant. Obtain the solution with $N(0) = N_0$. The solution is called the **logistic solution**.
2. Determine the equilibrium value, N_{eq}, of the logistic solution, in terms of k and β.
3. Plot the logistic solution for $0 < N_o < \dfrac{N_{eq}}{2}$ and show that a lag will become more prominent as N_o is closer to zero.
4. Solve the differential equation when $\beta = \beta(t)$, where $k < \beta < \beta_{\max}$.

E6.8. Generate a first-order differential equation that admits a similarity transformation and yet is not an isobaric differential equation.

E6.9. Consider the Nth order Euler-Cauchy differential equation:

$$\sum_{k=0}^{N} \left(a_k x^k \frac{d^k y}{dx^k} \right) = f(x)$$

1. Using the change of variable $z = \ln(x)$, show that the kth derivatives $d^k y/dx^k$ can be put in terms of the derivatives $d^k y/dz^k$ as follows:

$$\begin{aligned}
\frac{dy}{dx} &= \frac{1}{x}\frac{dy}{dz} \\
\frac{d^2y}{dx^2} &= \frac{1}{x^2}\left(-\frac{dy}{dz} + \frac{d^2y}{dx^2} \right) \\
&\vdots \\
\frac{d^ky}{dx^k} &= \frac{1}{x^k}\sum_{\ell=1}^{k} b_{k,\ell}\frac{d^\ell y}{dz^\ell} \\
&\vdots
\end{aligned}$$

where

$$b_{k,\ell} = \begin{cases} -(k-1)\, b_{k-1,1} & \text{if } \ell = 1 \\ b_{k-1,\ell-1} - (k-1)\, b_{k-1,\ell} & \text{if } 1 < \ell < k \\ 1 & \text{if } \ell = k \end{cases}$$

2. Solve the following third-order Euler-Cauchy equation:

$$x^3\frac{d^3y}{dx^3}+x^2\frac{d^2y}{dx^2}+x\frac{dy}{dx}+y=\frac{1}{2x}$$

E6.10. Let A and B given by

$$A=\begin{pmatrix}-1 & 1\\ 0 & -1\end{pmatrix}\qquad B=\begin{pmatrix}-1 & 0\\ 1 & -1\end{pmatrix}$$

Show that because A and B is not commutative, we have

$$e^{At}e^{Bt}\neq e^{(A+B)t}$$

E6.11. Let n be an integer. Find the value of k such that the following equation admits a similarity transformation:

$$\frac{dy}{dx}=y^k f\left(xy^n\right)$$

Solve the following equation using a similarity transformation:

$$\frac{1}{y^2}\frac{dy}{dx}=xy+1$$

E6.12. A guided missile is fired at a target that is moving at a constant speed s_T. Let the missile also travel at a constant speed $s_m > s_T$ while being aimed at the target (see Figure 6.5), with

$$\frac{d\underline{\mathbf{r}}}{dt}=k(t)\left(\underline{\mathbf{r}}_T-\underline{\mathbf{r}}\right)\quad\rightarrow\quad\begin{aligned}\frac{dx}{dt}&=k(t)\left(x_T-x\right)\\ \frac{dy}{dt}&=k(t)\left(y_T-y\right)\end{aligned}$$

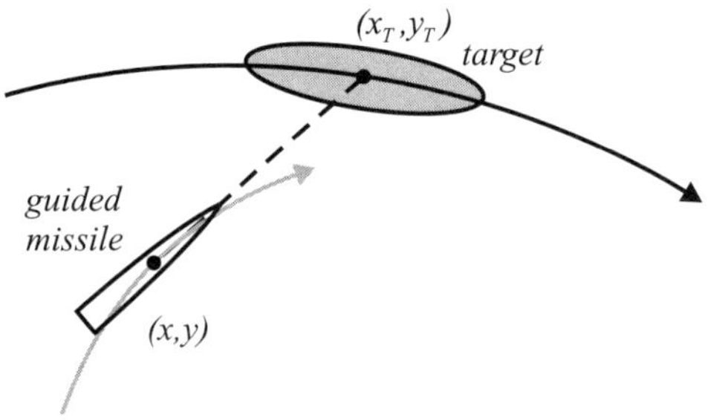

Figure 6.5. The tangent of the path of a guided missile is directed at the target at all times.

where $\underline{\mathbf{r}}$ and $\underline{\mathbf{r}}_T$ are the position vectors of the missile and target, respectively, and k is a proportionality function. After dividing dx/dt by dy/dt, and then taking another time derivative followed by one more division by dy/dt, show that we can end up with

$$\frac{d^2x}{dy^2}\left(y_T-y\right)+\frac{\sqrt{1+(dx/dy)^2}}{s_m}\left(\frac{dx}{dy}\frac{dy_T}{dt}-\frac{dx_T}{dt}\right)=0$$

where we used the fact that $s_m=\sqrt{(dx/dt)^2+(dy/dt)^2}$. Consider the simple case in which the target follows a straight path with $y_T(t)=Y_{T0}$ and $x_T(t)=X_{T0}+s_Tt$, where Y_{T0} and X_{T0} are constants, and initial conditions $x(0)=0$ and $y(0)=0$. Find the relationship between x and y, also known as the **pursuit path.**[12]

[12] Problem based on Burghes, D. N. and Borrie, M. S. *Modeling with Differential Equations*, Ellis Horword, Ltd, 1981.

E6.13. For the Ricatti equation given by

$$\frac{dy}{dx} = y^2 + \frac{\beta}{x}y + \frac{\gamma}{x^2}$$

Obtain the general solution if $(1+\beta)^2 > 4\gamma$. (See Section F.1.1 for an approach to solving Ricatti equations.)

E6.14. Given the set of differential equations,

$$\frac{d}{dt}\mathbf{x} = \begin{pmatrix} -3 & 2 & -4 & 2 \\ 0 & -2 & 1 & -1 \\ 1 & -1 & 0 & 0 \\ 1 & -1 & 2 & -2 \end{pmatrix}\mathbf{x}$$

Obtain the solution when the initial condition is given by $\mathbf{x}^T = \begin{pmatrix} 1 & 1 & 0 & 0 \end{pmatrix}$.

E6.15. Let $A[=]2 \times 2$, such that

$$\alpha = \frac{\mathbf{trace}(A)}{2} \qquad \text{and} \qquad \beta = \frac{\sqrt{\mathbf{trace}(A)^2 - 4\,\mathbf{det}(A)}}{2} \neq 0$$

Show that

$$\exp\left(At\right) = \frac{e^{\alpha t}}{\beta}\left(p\left(t\right)I + q(t)A\right)$$

where

$$p\left(t\right) = \beta\cosh\left(\beta t\right) - \alpha\sinh\left(\beta t\right) \qquad \text{and} \qquad q(t) = \sinh\left(\beta t\right)$$

Thus show that the solution of

$$\frac{d}{dt}\mathbf{x} = A\mathbf{x} + \mathbf{b}(t) \qquad \mathbf{x}(0) = \mathbf{x}_0 \tag{6.83}$$

is given by

$$\mathbf{x}(t) = \frac{e^{\alpha t}}{\beta}\left(p\left(t\right)I + q(t)A\right)\mathbf{x}_0 + \int_0^t \frac{e^{\alpha(t-\tau)}}{\beta}\left(p\left(t-\tau\right)I + q(t-\tau)A\right)\mathbf{b}(\tau)d\tau$$

Also, using l'Hospital rule, show that when $\beta = 0$,

$$\exp(At) = e^{\alpha t}\left(\left(1-\alpha t\right)I + At\right)$$

and thus the solution to (6.83) when $\beta = 0$ becomes

$$\mathbf{x}(t) = e^{\alpha t}\left((1-\alpha t)I + At\right)\mathbf{x}_0 + \int_0^t e^{\alpha(t-\tau)}\left([1-\alpha(t-\tau)]\,I + A\left[t-\tau\right]\right)\mathbf{b}(\tau)d\tau$$

E6.16. Consider an Nth-order linear differential equation,

$$\frac{d^N}{dt^N}y + \alpha_{N-1}\frac{d^{N-1}}{dt^{N-1}}y + \cdots + \alpha_1\frac{dy}{dt} + \alpha_0 y = u\left(t\right)$$

1. By defining N variables x_i as

$$x_i = \frac{d^{i-1}y}{dt^{i-1}} \qquad i = 1, \ldots, N$$

show that the Nth order differential equation can be put in state-space form

$$\frac{d}{dt}\mathbf{x} = \mathbf{A}\mathbf{x} + \mathbf{b}(t)$$

where **A** is in the companion form (cf. **E3.8**),

$$\mathbf{A} = \begin{pmatrix} 0 & 1 & 0 & \cdots & 0 \\ 0 & 0 & 1 & \cdots & 0 \\ \vdots & \vdots & \vdots & \ddots & \vdots \\ 0 & 0 & 0 & \cdots & 1 \\ -\alpha_0 & -\alpha_1 & -\alpha_2 & \cdots & -\alpha_{N-1} \end{pmatrix}$$

with $y = x_1$.

2. From **E3.8**, if the eigenvalues are distinct, the matrix of eigenvectors V, for the companion matrix A, is given by the Vandermonde matrix,

$$V = \left(\begin{array}{c|c|c} 1 & & 1 \\ \vdots & \cdots & \vdots \\ \lambda_1^{N-1} & & \lambda_N^{N-1} \end{array} \right)$$

whose inverse is given by

$$V^{-1} = \left(h_{ij} \right)$$

where

$$\begin{aligned} h_{ij} &= \frac{\omega_{i,j-1}}{\beta_i} \\ \prod_{j \neq i} (\lambda - \lambda_j) &= \omega_{i,N-1}\lambda^{N-1} + \cdots + \omega_{i,1}\lambda + \omega_{i,0} \\ \beta_i &= \prod_{j \neq i} (\lambda_i - \lambda_j) \end{aligned}$$

Using these formulas, find the solution of

$$\frac{d^3y}{dt^3} + 5\frac{d^2y}{dt^2} + 7.75\frac{dy}{dt} + 3.75y = 3 + e^{-2t} \quad y(0) = 1, \frac{dy}{dt}(0) = \frac{d^2y}{dt^2}(0) = 0$$

using matrix exponentials.

E6.17. Table 6.1 gives data for the set of reversible reactions given in Example 6.15. Find the values of kinetic rate constants k_{AB}, k_{BA}, k_{BC}, k_{CB}, k_{CA}, and k_{AC}.[13]

E6.18. Use the Legendre transformation (see Section F.1.2) to solve the following differential equation:

$$\left(\frac{dy}{dx} + a \right)^2 = ax + y$$

E6.19. Let the system matrix **A** be time-varying, given by

$$\mathbf{A}\,(t) = f(t)A_1 + g(t)A_2$$

where A_1 and A_2 are constant matrices.

1. Show that $\mathbf{A}(t)$ and $\mathbf{A}(\tau)$ commutes if A_1 and A_2 commutes. Generalize this for the case where

$$A(t) = \sum_{k=1}^{N} h_k(t)A_k$$

[13] A set of MATLAB programs are available on the book's webpage for obtaining ternary plots. See the document **plotTernaryTutorial.pdf** for instructions.

Table 6.1. *Data sets for Wei-Prater kinetics*

Data set 1			Data set 1			Data set 1		
Time	C_A	C_B	Time	C_A	C_B	Time	C_A	C_B
0.0	0.900	0.000	0.0	0.100	0.000	0.0	0.100	0.900
0.5	0.744	0.077	1.0	0.220	0.043	1.0	0.152	0.745
1.0	0.629	0.135	2.0	0.279	0.090	2.0	0.190	0.629
1.5	0.543	0.178	3.0	0.305	0.134	3.0	0.219	0.544
2.0	0.480	0.209	4.0	0.315	0.171	4.0	0.240	0.481
2.5	0.433	0.233	5.0	0.317	0.201	5.0	0.255	0.434
3.0	0.399	0.250	6.0	0.316	0.225	6.0	0.267	0.399
3.5	0.373	0.263	7.0	0.314	0.243	7.0	0.276	0.373
4.0	0.354	0.272	8.0	0.311	0.257	8.0	0.282	0.354
4.5	0.340	0.279	10.0	0.307	0.276	9.0	0.287	0.340
5.0	0.329	0.285	11.0	0.305	0.282	10.0	0.290	0.330
7.0	0.309	0.295	13.0	0.303	0.290	11.0	0.293	0.322
9.0	0.302	0.298	14.0	0.302	0.292	13.0	0.296	0.312
11.0	0.300	0.299	18.0	0.300	0.297	15.0	0.298	0.307
12.0	0.300	0.299	21.0	0.300	0.299	21.0	0.300	0.301

2. Let

$$A_1 = \begin{pmatrix} 1 & 2 \\ -1 & -2 \end{pmatrix} \qquad A_2 = \begin{pmatrix} 1 & 1 \\ -\frac{1}{2} & -\frac{1}{2} \end{pmatrix}$$

and

$$f(t) = -e^{-2t} \qquad g(t) = -2e^{-5t}$$

Obtain the solution of the following set of differential equations using matrizants:

$$\frac{d}{dt}\mathbf{x} = \Big(f(t)A_1 + g(t)A_2 \Big)\mathbf{x}$$

subject to the initial condition, $\mathbf{x}^T(0) = \begin{pmatrix} 1 & -0.6 \end{pmatrix}$.

E6.20. An Emden-Fowler equation is given by

$$\frac{d^2y}{dx^2} + \frac{a}{x}\frac{dy}{dx} + by^n = 0$$

Show that for $n \neq 1$, a similarity transformation can reduce the equation to a first-order differential equation.

E6.21. The relationships between current $i(t)$ and voltage $v(t)$ across inductors, capacitors, and resistors are given by

$$\begin{aligned} v &= R\,i && \text{Resistors} \\ \frac{dv}{dt} &= \frac{1}{C}\,i && \text{Capacitors} \\ v &= L\,\frac{di}{dt} && \text{Inductors} \end{aligned}$$

A simple third-order Butterworth filter based on the Cauer topology using two inductors, one capacitor, and one resistor is shown in Figure 6.6. Using

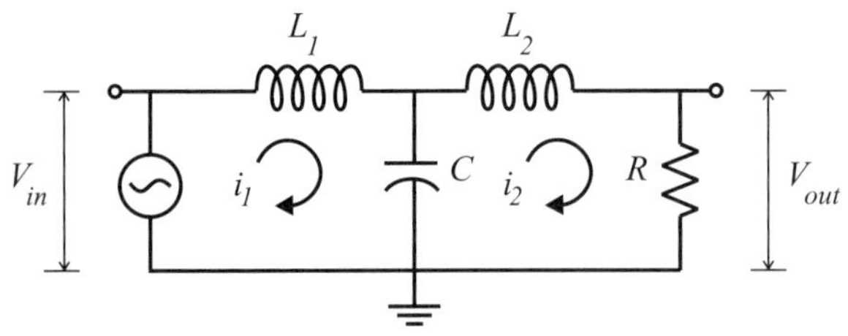

Figure 6.6. A Butterworth filter using a Cauer Topology.

Kirchhoff's laws, we end up with the following differential equations:

$$\frac{1}{C}(i_1 - i_2) + L_1\frac{d^2 i_1}{dt^2} = \frac{dv_{in}}{dt}$$

$$\frac{1}{C}(i_2 - i_1) + L_2\frac{d^2 i_2}{dt^2} + R\frac{di_2}{dt} = 0$$

where i_1 and i_2 are the currents through inductors L_1 and L_2, respectively, and $v_{in}(t)$ is a given input voltage.

1. Obtain the state-space formulation based on the state vector defined by $(x_1, x_2, x_3, x_4)^T = (i_1, di_1/dt, i_2, di_2/dt)$.
2. Determine the characteristic function for this linear system. For $R = 1\Omega$ choose values of L_1, L_2, and C so that the characteristic equation reduces to

$$s\left(s^3 + 2s^2 + 2s + 1\right) = 0$$

3. Solve for $v_{out} = i_2(t)R$ using the following parameters and initial conditions: $L_1 = 1.5$, $L_2 = 0.5$, $C = 4/3$, $R = 1$, $v_{in}(t) = \sin(0.2t) - 0.1\sin(5t)$, and $\mathbf{x} = \mathbf{0}$. Plot $v_{out}(t)$ vs. t together with $v_{in}(t)$ vs. t to see the performance of the Butterworth filter for this example.

E6.22. Show that the complete solution of

$$\left(\frac{dy}{dx} - 5\right)^2 + 5x = y + 2$$

is given by

$$y(x) = \left[\frac{1}{2}(x - 5 + C)\right]^2 + 5x - 2$$

where C is arbitrary. Find the singular solution for this differential equation and plot the singular solution together with the complete solution, that is, using different values for C, thus verify that the region where the solution exists is determined by an envelope that is obtained from the singular solution.

E6.23. A **spring and dashpot system** shown in Figure 6.7 is often used to model the dynamics of mechanical systems in motion. The force balance yields a linear second-order system given by

$$m\frac{d^2x}{dt^2} + C\frac{dx}{dt} + kx = F(t)$$

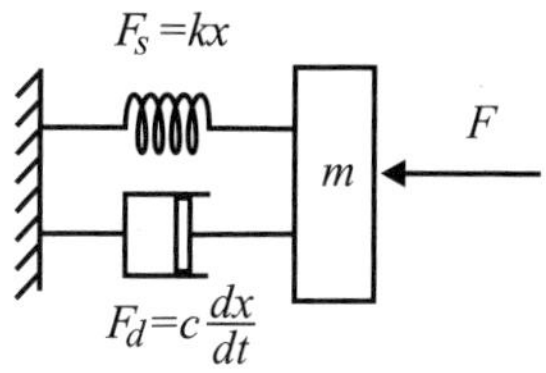

Figure 6.7. A spring and dashpot system.

1. Obtain a state-space model for a system of N spring and dashpots as shown in Figure 6.8.

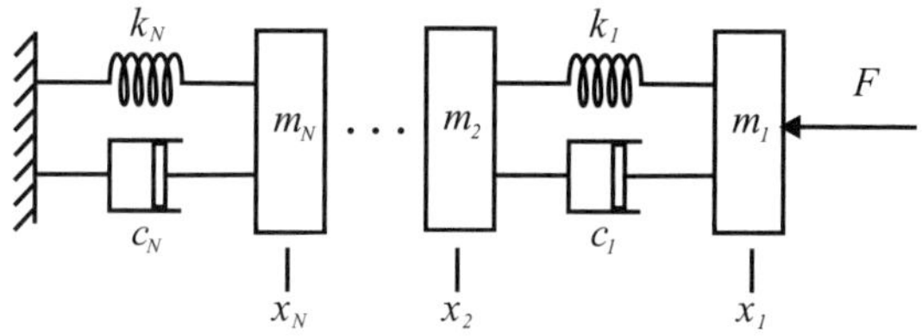

Figure 6.8. A series of springs and dashpots system.

2. Let $p_0(s) = 1$ and $p_1(s) = m_1 s^2 + C_1 s + k_1$. Then it can be shown that the characteristic equation for this system is given by $p_N(s)/(\prod_{j=1}^{N} m_j) = 0$ for $N \geq 1$, where

$$p_\ell = \left(m_\ell s^2 + C_\ell s + k_\ell\right) p_{\ell-1} - \left(C_{\ell-1} s + k_{\ell-1}\right)^2 p_{\ell-2} \quad ; \quad \ell > 1$$

Verify this to be the case for $N = 2$.

3. Let $N = 3$ and $F = 0.25\tanh(t-5)$ together with the following parameters

i	m_i	k_i	C_i
1	10	5	0.1
2	8	5	0.2
3	12	10	0.1

Will the system be stable or unstable?

E6.24. An equilibrium reaction from A to B is given by

$$2A \Leftrightarrow B$$

The kinetics are given by

$$\begin{aligned} \frac{1}{2}\frac{dC_A}{dt} &= k_b C_B - k_f C_A^2 \\ \frac{dC_B}{dt} &= k_f C_A^2 - k_b C_B \end{aligned}$$

1. Show that the equations can be combined to yield

$$\frac{d^2 C_A}{dt^2} + \left(4k_f C_A + k_b\right)\frac{dC_A}{dt} = 0$$

2. Solve for $C_A(t)$ and $C_B(t)$ with the initial conditions $C_A(0) = C_{A0}$ and $C_B(0) = C_{B0}$.

E6.25. Let C_R be the RNA concentration and C_E as the enzyme concentration. One model for protein synthesis is given by

$$\begin{aligned}\frac{dC_R}{dt} &= r(C_E) - k_1 C_R \\ \frac{dC_E}{dt} &= k_2 C_R - k_3 C_E\end{aligned}$$

where $r(C_E)$ is the rate of production of protein based solely on the enzyme concentration, and k_1, k_2 and k_3 are constants.

1. Show that this can be combined to yield a second-order reaction given by

$$\frac{d^2 C_E}{dt^2} + \alpha \frac{dC_E}{dt} = k_2 r(C_E) - \beta C_E$$

2. Assuming Michaelis-Menten kinetics for $r(C_E)$, that is,

$$r(C_E) = \frac{k_4 C_E}{k_5 + C_E}$$

Reduce the equation into a first-order differential equation with dependent variable $p = dC_E/dt$ and independent variable $z = C_E$.

E6.26. Prove that if $A(t)A(\tau)$ is commutative, then with $Q_1(t)$ and $Q_k(t)$ as defined in (6.60) and (6.61), respectively, becomes

$$Q_k(t) = \frac{1}{k!} Q_1(t)^k$$

and thus showing that the matrizant formula given in (6.68)

$$\mathbf{M}(t) = e^{Q_1(t)}$$

is valid.

7 Numerical Solution of Initial and Boundary Value Problems

In several cases, the analytical solution of ordinary differential equations, including high-order, multiple, and nonlinear types, may not be easy to obtain or evaluate. In some cases, it requires truncation of an infinite series, whereas in other cases, it may require numerical integration via quadratures.

An alternative approach is to determine the solution directly by numerical methods. This means that the solution to a differential equation will not be given as a function of the independent variables. Instead, the numerical solution is a set of points discretized over the chosen range of the independent variables. These points can then be plotted and processed further for subsequent analysis. Thus, unlike analytical solutions, numerical solutions do not yield compact formulas. Nonetheless, numerical methods are able to handle a much larger class of ordinary differential equations.

We begin the chapter with the problems in which all the fixed conditions are set at the initial point, for example, $t = 0$ or $x = x_0$, depending on which the independent variable is. These problems are known as **initial value problems**, or **IVP** for short. We discuss some of the better known methods for solving initial value problems, such as the one-step methods (e.g., Euler methods and Runge-Kutta methods) and multistep methods (e.g., the Adams-Bashforth methods and Adams-Moulton methods).

The actual development of the different methods came out of several approaches, including rigorous algebraic manipulations and collocation methods. However, we implement a simplistic approach by implementing the general numerical methods to a specific differential equation given by

$$\frac{dy}{dt} = f(y) = y$$

to generate the required parameters. The main advantage is that a Taylor series expansion of the solution turns out to be sufficiently differentiable and rich in information to produce several necessary conditions. Surprisingly, applying this technique on various numerical methods yields a majority of the necessary conditions, matching those coming from the more elegant, yet lengthy, approaches. In some cases, it yields the same complete sufficient equations, for example, for the Adam's multistep and backward difference formula (BDF) methods, including the generation of the coefficients for variable step-size BDFs.

We include simple examples to compare the performance of the IVP solvers we discuss. Primarily, we show that even though explicit methods tend to be simpler to apply in several cases, they suffer from stability problems. The implicit versions tend to be more stable, but even some implicit methods may end up being unstable.

A section on varying step sizes by error-control strategies is also provided in Section G.5 as an appendix. This includes a brief discussion of embedded formulas in explicit Runge-Kutta schemes. Varying step size is needed to allow the methods to adjust the progression along the simulation to finer steps when accuracy requires it, but the schemes also coarsen the steps in regions where errors can be tolerated by large steps.

Next, in Section 7.4, we give a brief discussion on the solution of difference equations and their stability criteria. We use these results and apply them to the standard stability analysis of linear IVP solvers; specifically, we introduce the Dahlquist test. By using this standard, we can compare the stability regions of the different methods, either implicit or explicit and either one-step or multistep.

We then turn our attention to showing how we can extend the utility of IVP solvers to solve boundary value problems (BVP). Specifically, we discuss a systematic approach of the shooting method by applying the theory of matrizants for the solution of linear two-point BVPs. It is then a simple matter to include Newton methods to extend the shooting method to nonlinear two-point BVPs.

Finally, we include a brief section on differential-algebraic equations (DAE). We limit the discussion to problems that are classified as index-1, and we outline both a BDF approach and Runge-Kutta approach.

In Section G.1, we include a supplementary section of a very brief tutorial on how the initial value problems, boundary value problems, and differential algebraic problems can be solved using the built-in solvers in MATLAB.

7.1 Euler Methods

We begin with the simplest approach, called the **Euler methods**. For the differential equation,

$$\frac{dy}{dt} = f(t, y) \tag{7.1}$$

subject to the initial condition, $y(0) = y_o$. We can replace the derivative by its finite difference approximation given by

$$\left.\frac{dy}{dt}\right|_{t=t_k} \approx \frac{\Delta_k y}{\Delta_k t} = \frac{y_{k+1} - y_k}{t_{k+1} - t_k}$$

Let $h_k = \Delta_k t = t_{k+1} - t_k$ for $k \geq 0$, then

$$y_{k+1} = y_k + h_k f(t_k, y_k) \tag{7.2}$$

Starting with $y_0 = y(0)$, then $y_1, y_2, \ldots$ can be obtained recursively by implementing (7.2). For simplicity, we now assume a fixed step size, that is, $h = h_k$ is constant. However, we do allow the step size to vary when we need to improve computational efficiency using procedures that are collectively known as **error control** techniques. These topics are included instead in Section G.5 as an appendix. Based on the definition of derivatives, accuracy should improve with smaller values of h. However, roundoff errors also impose their own limits on how small h can be, and smaller values of h will require a larger number of computations.

Equation (7.2) is known as the **explicit Euler method** or **forward Euler method**. A variation to the explicit Euler method can be obtained by using a different approximation of the derivative,

$$\frac{dy}{dt} \approx \frac{y_k - y_{k-1}}{h} \qquad \rightarrow \qquad y_k - y_{k-1} = hf(t_k, y_k)$$

or

$$y_{k+1} = y_k + hf(t_{k+1}, y_{k+1}) \tag{7.3}$$

Equation (7.3) is also known as the **implicit Euler Method** or **backward Euler Method**. It is "implicit" because y_{k+1} appears on both sides of equation (7.3), thus requiring additional steps for the evaluation of y_{k+1}.

Recall from Section 6.5 that a system of higher order differential equation can be recast in a state-space formulation given by

$$\frac{d}{dt}\mathbf{y} = \mathbf{f}(t, \mathbf{y}) \tag{7.4}$$

where $\mathbf{y}$ is the state vector and $\mathbf{f}$ is a vector of multivariable functions $f_i(t, \mathbf{y})$. The Euler methods are then given by

$$\mathbf{y}_{k+1} = \mathbf{y}_k + h\mathbf{f}(t_k, \mathbf{y}_k) \qquad : \text{Explicit Euler} \tag{7.5}$$

$$\mathbf{y}_{k+1} = \mathbf{y}_k + h\mathbf{f}(t_{k+1}, \mathbf{y}_{k+1}) \qquad : \text{Implicit Euler} \tag{7.6}$$

For the linear case, $\mathbf{f}(t, \mathbf{y}) = A(t)\mathbf{y} + \mathbf{b}(t)$, which results in

$$\mathbf{y}_{k+1} = \left(I + hA(t_k)\right)\mathbf{y}_k + h\mathbf{b}(t_k) \qquad : \text{Explicit Euler} \tag{7.7}$$

$$\mathbf{y}_{k+1} = \left(I - hA(t_{k+1})\right)^{-1}\left(\mathbf{y}_k + h\mathbf{b}(t_{k+1})\right) \qquad : \text{Implicit Euler} \tag{7.8}$$

EXAMPLE 7.1. Consider the simple first-order linear differential equation

$$\tau\frac{dy}{dt} + y = -e^{-\beta t}$$

subject to $y(0) = y_0$. The analytical solution is given by

$$y(t) = \left(y_0 + \frac{1}{1-\beta\tau}\right)e^{-t/\tau} + \frac{1}{\beta\tau - 1}e^{-\beta\tau} \tag{7.9}$$

Let $y_k^{(\mathrm{ExEu})}$ be the value for $y(t_k)$ using the explicit Euler formula, then

$$\begin{aligned} y_{k+1}^{(\mathrm{ExEu})} &= y_k^{(\mathrm{ExEu})} - \frac{h}{\tau}\left(y_k^{(\mathrm{ExEu})} + e^{-\beta t_k}\right) \\ &= \left(1 - \frac{h}{\tau}\right)y_k^{(\mathrm{ExEu})} - \frac{h}{\tau}e^{-\beta t_k} \end{aligned} \tag{7.10}$$

Let $y_k^{(\mathrm{ImEu})}$ be the value for $y(t_k)$ using the implicit Euler formula, then

$$\begin{aligned} y_{k+1}^{(\mathrm{ImEu})} &= y_k^{(\mathrm{ImEu})} - \frac{h}{\tau}\left(y_{k+1}^{(\mathrm{ImEu})} + e^{-\beta t_{k+1}}\right) \\ &= \left(\frac{\tau}{\tau+h}\right) y_k^{(\mathrm{ImEu})} - \left(\frac{h}{\tau+h}\right) e^{-\beta t_{k+1}} \end{aligned} \tag{7.11}$$

As a particular example, we can set $\tau = 0.001$, $\beta = 100$, and $y_0 = 1.0$. Then Figure 7.1 shows the performance of both the explicit and the implicit Euler method for $h = 0.0001, 0.001, 0.002$.

At the critical value of $h = \tau = 0.001$, we note that both explicit and implicit Euler methods are still stable methods, containing larger error values near the initial point. When $h = 0.002$, the explicit method became unstable, whereas the implicit method was stable. Conversely, for $h = 0.0001$, the error dropped significantly as expected, but at the cost of increasing the number of computations and amount of storage.

Generally, the explicit methods are used more often than the implicit methods because they avoid the additional steps of solving (possibly nonlinear) equations for y_{k+1}. Example 7.1 shows that as long as the increments h are sufficiently small, the explicit Euler should be reasonably satisfactory. However, very small values of h imply larger computational loads. The implicit Euler method for Example 7.1 involved only a simple inverse. However, in general, the implicit solution for y_{k+1} could be more difficult, often involving nonlinear solvers. Nonetheless, as shown in Example 7.1, the implicit methods are more stable. The issues of stability is discussed in more detail in Section 7.4.

7.2 Runge Kutta Methods

A class of methods known as the **Runge-Kutta methods** can improve the accuracy of Euler methods even under the same values of increment h. The main idea is to obtain a linear combination among several evaluations of the derivative functions $\mathbf{f}(t, \mathbf{y})$ at intermediate points between t_k and t_{k+1} to obtain a transition (or update) term δ_k, such as

$$\mathbf{y}_{k+1} = \mathbf{y}_k + \delta_k$$

As with Euler methods, there are explicit Runge-Kutta methods and implicit Runge-Kutta methods.

Let $s \geq 1$ be an integer that we refer to as the number of stages.[1] Then the s-stage Runge Kutta method for $dy/dt = f(t, y)$ is given by:

$$\delta_{kj} = hf\left[(t_k + a_j h), \left(y_k + \sum_{\ell=1}^{s} b_{j\ell}\delta_{k\ell}\right)\right] \qquad j = 1, \ldots, s \tag{7.12}$$

$$y_{k+1} = y_k + \sum_{j=1}^{s} c_j \delta_{kj} \tag{7.13}$$

[1] The stage s determines the number of parameters. However, the "order" refers to the accuracy based on the terms used in the Taylor series expansion.

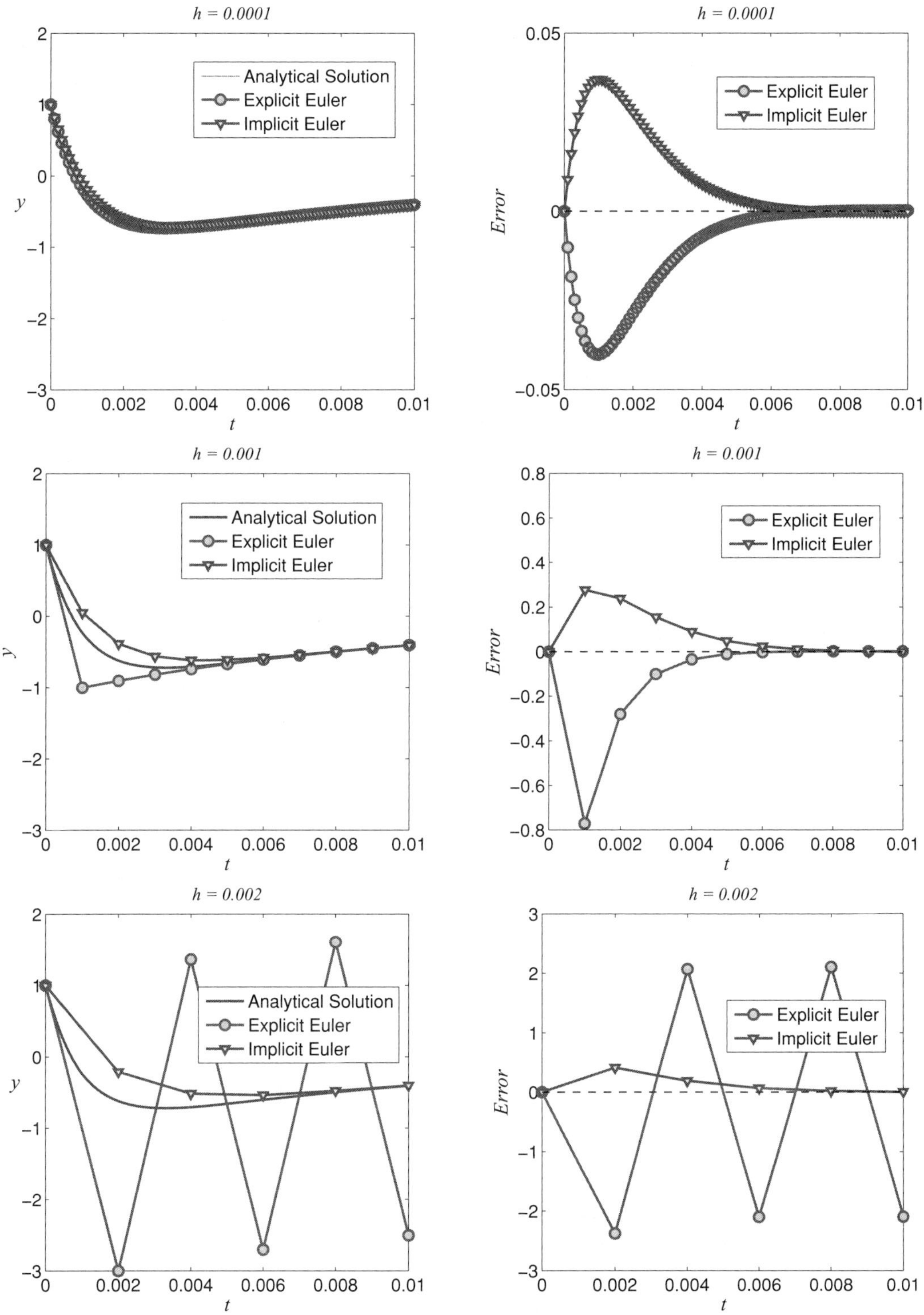

Figure 7.1. Performance comparison of the explicit and implicit Euler methods for example 7.1. The plots on the right column shows the errors of the Euler methods at different values of h.

where δ_{kj} are the intermediate update terms. Thus in (7.13), the update δ_k is a linear combination of δ_{kj} weighted by c_j. The parameters of the Runge-Kutta methods are $a_1, \ldots, a_s$, $c_1, \ldots, c_s$, and $b_{11}, b_{12} \ldots, b_{ss}$. The parameters a_j affect only t_k, whereas the parameters $b_{j,\ell}$ affect only the y_k during the evaluations of $f(t, y)$. All three sets of parameters are usually arranged in a table called the **Runge-Kutta Tableau** as follows:

$$\left(\begin{array}{c|c} \mathbf{a} & B \\ \hline & \mathbf{c}^T \end{array}\right) = \begin{array}{|c|ccc|} \hline a_1 & b_{11} & \cdots & b_{1s} \\ \vdots & \vdots & \ddots & \vdots \\ a_s & b_{s1} & \cdots & b_{ss} \\ \hline & c_1 & \cdots & c_s \\ \hline \end{array} \qquad (7.14)$$

Based on the structure of matrix B, the type of Runge-Kutta method can be classified as follows:

1. **Explicit Runge-Kutta (ERK).** If matrix B is strictly triangular, that is, $b_{ij} = 0$, $i = 1, \ldots, s$, $j \geq i$, the method is known an **explicit Runge-Kutta method**. For these methods, the intermediate updates, δ_{kj}, given in (7.12) can be evaluated sequentially.

2. **Implicit Runge-Kutta (IRK).** If matrix B is not strictly triangular, the method is known as an **implicit Runge-Kutta** (IRK) method. Some of the special cases are:

   (a) **Diagonally Implicit Runge-Kutta (DIRK).**

$$\begin{aligned} b_{ij} &= 0 \qquad j > i \\ b_{\ell\ell} &\neq 0 \qquad \text{for some } 1 \leq \ell \leq s \end{aligned} \qquad (7.15)$$

   (b) **Singly Diagonally Implicit Runge-Kutta (SDIRK).** This is a special case of DIRK in which all the diagonal elements are equal, that is, $b_{\ell\ell} = \gamma$, $\ell = 1, \ldots, s$.

Note that a system of nonautonomous equations, that is, $d\mathbf{y}/dt = \mathbf{f}(t, \mathbf{y})$, can be replaced by an autonomous system $d\mathbf{x}/dt = \mathbf{g}(\mathbf{x})$ after appending the original state $\mathbf{y}$ with t, and extending $\mathbf{f}$ by the constant 1, that is,

$$\mathbf{x} = \left(\frac{\mathbf{y}}{t}\right) \qquad \text{and} \qquad \mathbf{g} = \left(\frac{\mathbf{f}(\mathbf{x})}{1}\right) \qquad (7.16)$$

The advantage of (7.16) is that we will no longer need the parameters a_j during the Runge-Kutta calculations. Nonetheless, in some cases, there are still advantages to using the original nonautonomous system because the stability solution properties apply only to $\mathbf{y}$ (note that t is allowed to be unbounded).

If we require (7.12) to hold for the special case of $[y = t, f = 1]$ and the case $[f(y) = f(t)]$, we have the following consistency condition:

$$a_j = \sum_{\ell=1}^{s} b_{j\ell} \qquad ; j = 1, \ldots, s \qquad (7.17)$$

The parameters of the Runge-Kutta methods can be obtained by using Taylor series expansisons. The function $y(t_{k+1})$ can be expanded as a Taylor series around $y(t_k)$,

$$y(t_{k+1}) = y(t_k) + h\left.\frac{dy}{dt}\right|_{t=t_k} + \frac{h^2}{2!}\left.\frac{d^2y}{dt^2}\right|_{t=t_k} + \frac{h^3}{3!}\left.\frac{d^3y}{dt^3}\right|_{t=t_k} + \cdots$$

In terms of $f(t, y) = dy/dt$, we have

$$y_{k+1} = y_k + hf(t_k, y_k) + \frac{h^2}{2!}\left(f\left.\frac{\partial f}{\partial y}\right|_{t_k, y_k} + \left.\frac{\partial f}{\partial t}\right|_{t_k, y_k}\right) + \cdots \tag{7.18}$$

To obtain an nth-order approximation, the series will be truncated after the $(n+1)^{\text{th}}$ term. Thus the Euler method is nothing but a first-order Taylor series approximation. In the next two sections, we discuss the fourth-order explicit Runge-Kutta method and the fourth-order implicit Runge-Kutta method.

7.2.1 Fourth-Order Explicit Runge-Kutta Method

The fourth-order explicit Runge-Kutta method is one of the most used form of Runge-Kutta method. Specifically, we refer to the method given by the tableau in (7.19). The details on how these parameters were obtained are given in Section G.2.1 as an appendix.

$$\left(\begin{array}{c|c} \mathbf{a} & B \\ \hline & \mathbf{c}^T \end{array}\right) = \begin{array}{|c|cccc|} \hline 0 & 0 & 0 & 0 & 0 \\ \frac{1}{2} & \frac{1}{2} & 0 & 0 & 0 \\ \frac{1}{2} & 0 & \frac{1}{2} & 0 & 0 \\ 1 & 0 & 0 & 1 & 0 \\ \hline & \frac{1}{6} & \frac{1}{3} & \frac{1}{3} & \frac{1}{6} \\ \hline \end{array} \tag{7.19}$$

or, written explicitly,

$$\begin{aligned} \delta_{k1} &= hf\left(t_k, y_k\right) \\ \delta_{k2} &= hf\left(t_k + \frac{h}{2}, y_k + \frac{\delta_{k1}}{2}\right) \\ \delta_{k3} &= hf\left(t_k + \frac{h}{2}, y_k + \frac{\delta_{k2}}{2}\right) \\ \delta_{k4} &= hf\left(t_k + h, y_k + \delta_{k3}\right) \\ y_{k+1} &= y_k + \frac{1}{6}\left(\delta_{k1} + 2\delta_{k2} + 2\delta_{k3} + \delta_{k4}\right) \end{aligned} \tag{7.20}$$

For the multivariable case

$$\frac{d}{dt}\mathbf{y} = \mathbf{f}\left(t, \mathbf{y}\right)$$

where

$$\mathbf{y} = \begin{pmatrix} y^{[1]} \\ \vdots \\ y^{[M]} \end{pmatrix} \qquad \mathbf{f}\left(t, \mathbf{y}\right) = \begin{pmatrix} f^{[1]}\left(t, \mathbf{y}\right) \\ \vdots \\ f^{[M]}\left(t, \mathbf{y}\right) \end{pmatrix}$$

the Runge-Kutta can then be directly generalized as follows:

$$\begin{aligned}
\underline{\boldsymbol{\delta}}_{k,1} &= h\mathbf{f}\left(t_k, \mathbf{y}_k\right) \\
\underline{\boldsymbol{\delta}}_{k,2} &= h\mathbf{f}\left(t_k + \frac{h}{2}, \mathbf{y}_k + \frac{1}{2}\underline{\boldsymbol{\delta}}_{k,1}\right) \\
\underline{\boldsymbol{\delta}}_{k,3} &= h\mathbf{f}\left(t_k + \frac{h}{2}, \mathbf{y}_k + \frac{1}{2}\underline{\boldsymbol{\delta}}_{k2}\right) \\
\underline{\boldsymbol{\delta}}_{k,4} &= h\mathbf{f}\left(t_k + h, \mathbf{y}_k + \underline{\boldsymbol{\delta}}_{k3}\right) \\
\mathbf{y}_{k+1} &= \mathbf{y}_k + \frac{1}{6}\left(\underline{\boldsymbol{\delta}}_{k1} + 2\underline{\boldsymbol{\delta}}_{k2} + 2\underline{\boldsymbol{\delta}}_{k3} + \underline{\boldsymbol{\delta}}_{k4}\right)
\end{aligned} \tag{7.21}$$

7.2.2 A Fourth-Order Implicit Runge-Kutta Method

The **Gauss-Legendre method** is one type of implicit Runge-Kutta method that has good stability properties. The details for the derivation of the Runge-Kutta parameters are included in Section G.2.2 as an appendix. The Runge-Kutta tableau is given by

$$\left(\begin{array}{c|c} \mathbf{a} & B \\ \hline & \mathbf{c}^T \end{array}\right) = \begin{array}{|c|cc|} \hline \left(\frac{1}{2} - \frac{\sqrt{3}}{6}\right) & \frac{1}{4} & \left(\frac{1}{4} - \frac{\sqrt{3}}{6}\right) \\ \left(\frac{1}{2} + \frac{\sqrt{3}}{6}\right) & \left(\frac{1}{4} + \frac{\sqrt{3}}{6}\right) & \frac{1}{4} \\ \hline & \frac{1}{2} & \frac{1}{2} \\ \hline \end{array} \tag{7.22}$$

or, in actual equations,

$$\begin{aligned}
\delta_{k1} &= hf\left(t + \left(\frac{1}{2} - \frac{\sqrt{3}}{6}\right)h,\ y_k + \frac{1}{4}\delta_{k1} + \left(\frac{1}{4} - \frac{\sqrt{3}}{6}\right)\delta_{k2}\right) \\
\delta_{k2} &= hf\left(t + \left(\frac{1}{2} + \frac{\sqrt{3}}{6}\right)h,\ y_k + \left(\frac{1}{4} + \frac{\sqrt{3}}{6}\right)\delta_{k1} + \frac{1}{4}\delta_{k2}\right) \\
y_{k+1} &= y_k + \frac{1}{2}\left(\delta_{k1} + \delta_{k2}\right)
\end{aligned} \tag{7.23}$$

For the multivariable case, we have

$$\begin{aligned}
\underline{\boldsymbol{\delta}}_{k,1} &= h\mathbf{f}\left(t_k + a_1 h, \mathbf{y}_k + b_{11}\underline{\boldsymbol{\delta}}_{k,1} + b_{12}\underline{\boldsymbol{\delta}}_{k,2}\right) \\
\underline{\boldsymbol{\delta}}_{k,2} &= h\mathbf{f}\left(t_k + a_2 h, \mathbf{y}_k + b_{21}\underline{\boldsymbol{\delta}}_{k,1} + b_{22}\underline{\boldsymbol{\delta}}_{k,2}\right) \\
\mathbf{y}_{k+1} &= \mathbf{y}_k + \frac{1}{2}\left(\underline{\boldsymbol{\delta}}_{k1} + \underline{\boldsymbol{\delta}}_{k2}\right)
\end{aligned} \tag{7.24}$$

where

$$a_1 = \frac{1}{2} - \frac{\sqrt{3}}{6} \quad ; \quad a_2 = \frac{1}{2} + \frac{\sqrt{3}}{6} \quad \text{and} \quad B = \begin{pmatrix} \dfrac{1}{4} & \dfrac{1}{4} - \dfrac{\sqrt{3}}{6} \\ \dfrac{1}{4} + \dfrac{\sqrt{3}}{6} & \dfrac{1}{4} \end{pmatrix}$$

Remarks: Two MATLAB functions, `rk4.m` and `glirk.m`, are available on the book's webpage and implement the fixed-step fourth-order explicit Runge-Kutta and the fourth-order implicit (Gauss-Legedre) Runge-Kutta methods, respectively.

EXAMPLE 7.2. Let us now compare the performance of the fourth-order explicit and implicit Runge-Kutta methods applied to the same differential equation given in Example 7.1, that is,

$$\tau\frac{dy}{dt} + y = -e^{-\beta t}$$

subject to $y(0) = y_0$. Let $y_k^{(\mathrm{ExRK})}$ be the value for $y(t_k)$ using the explicit Runge-Kutta formula given in (7.20), then

$$\begin{aligned}
\delta_{k1} &= h\left(y_k^{(\mathrm{ExRK})} + e^{-\beta t_k}\right) \\
\delta_{k2} &= h\left(y_k^{(\mathrm{ExRK})} + \frac{\delta_{k1}}{2} + e^{-\beta(t_k + (h/2))}\right) \\
\delta_{k3} &= h\left(y_k^{(\mathrm{ExRK})} + \frac{\delta_{k2}}{2} + e^{-\beta(t_k + (h/2))}\right) \\
\delta_{k4} &= h\left(y_k^{(\mathrm{ExRK})} + \delta_{k3} + e^{-\beta(t_k + h)}\right) \\
y_{k+1}^{(\mathrm{ExRK})} &= y_k^{(\mathrm{ExRK})} + \frac{1}{6}\left(\delta_{k1} + 2\delta_{k2} + 2\delta_{k3} + \delta_{k4}\right)
\end{aligned}$$

Let $y_k^{(\mathrm{ImRK})}$ be the value for $y(t_k)$ using the implicit Runge-Kutta based on the Gauss-Legendre formulas given in (7.23); then, after further rearrangements of the equations,

$$y_{k+1}^{(\mathrm{ImRK})} = q_1 y_k^{(\mathrm{ImRK})} + q_2 e^{-\beta t_k} \tag{7.25}$$

where

$$q_1 = 1 - R\begin{pmatrix} 1 \\ 1 \end{pmatrix} \quad \text{and} \quad q_2 = -R\begin{pmatrix} e^{-a_1\beta h} \\ e^{-a_2\beta h} \end{pmatrix}$$

with $a_1 = \frac{1}{2} - \frac{\sqrt{3}}{6}$, $a_2 = \frac{1}{2} + \frac{\sqrt{3}}{6}$ and

$$R = \begin{pmatrix} \frac{1}{2} & \frac{1}{2} \end{pmatrix} \begin{pmatrix} \left(\frac{\tau}{h} + \frac{1}{4}\right) & \left(\frac{1}{4} - \frac{\sqrt{3}}{6}\right) \\ \left(\frac{1}{4} + \frac{\sqrt{3}}{6}\right) & \left(\frac{\tau}{h} + \frac{1}{4}\right) \end{pmatrix}^{-1}$$

Using the parameters used in Example 7.1, $\tau = 0.001$, $\beta = 100$, and $y_0 = 1.0$. Figure 7.2 shows the performance of both the explicit and the implicit Runge-Kutta methods for $h = 0.001, 0.002, 0.003$.

Compared with the Euler method, the accuracy has improved significantly, as expected. At $h = 0.001$, the Runge-Kutta methods have much smaller errors, even compared with Euler methods using an increment that is ten times smaller – at $h = 0.0001$. The explicit Euler method had become unstable at $h = 0.2$, whereas the explicit Runge-Kutta did not. The explicit Runge-Kutta did become unstable when $h = 0.03$. In all the cases, the implicit Runge-Kutta using the Gauss-Legendre formulation had the best performance. It can be shown later in Section 7.4.2 that the Gauss-Legendre implicit Runge-Kutta method will be stable for all h when $\tau > 0$, but the errors will still increase as h increases. This means that even for the implicit Runge-Kutta method, h may need to be varied to control the errors. A variable step method should be able to improve this approach further.

From the previous examples, we observe that both accuracy and stability of the solutions are often improved by using a smaller step size h. However, smaller step sizes require more computation time and storage. Moreover, different step sizes are needed at different regions of the solution to attain the desired level of accuracy while being balanced with computational efficiency. The process of continually adjusting the step sizes of the numerical methods to attain balance between accuracy and computational loads at the appropriate regions of the solution is generally known as **error control**. Two popular variations to the explicit Runge-Kutta methods that include error control are the Fehlberg 4/5 method and the Dormand-Prince 4/5 method. Details for both these approaches are included in section G.5 of the appendix.

7.3 Multistep Methods

Both the Euler methods and Runge-Kutta methods are known as **one-step methods** because they use only the current estimate y_k to predict the next value y_{k+1}. A more general approach is to include additional past values (i.e., y_{k-1}, y_{k-2}, etc.). These methods are known as **multistep methods**.

We limit our development first to the scalar autonomous differential equations, that is, $dy/dt = f(y)$. A general formulation of the m-step multistep method becomes

$$y_{k+1} = \sum_{i=0}^{m} a_i\, y_{k-i} + h \sum_{j=-1}^{m} b_j\, f(y_{k-j}) \tag{7.26}$$

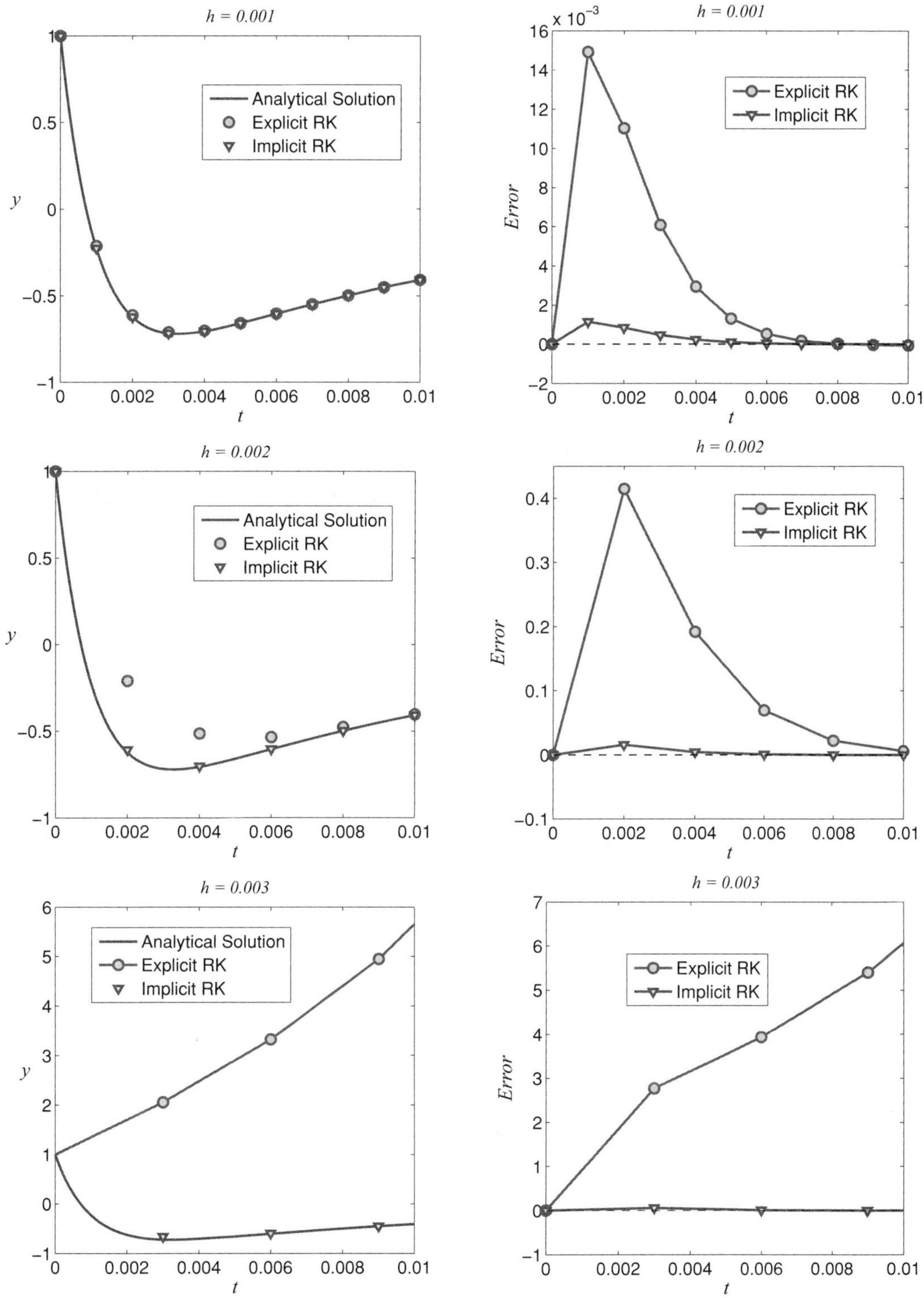

Figure 7.2. Performance comparison of the explicit and implicit Runge-Kutta methods for example 7.2

If $b_{-1} = 0$, then the method is known as an **explicit multistep method**; otherwise it is known as an **implicit multistep method**. The form given in (7.26) can be generalized further by allowing nonautonomous functions $f(t_{k-j}, y_{k-j})$. However, because our specific development will initially be applied only to $f(y) = y$, it is sufficient to limit our form to that given in (7.26). We address the nonautonomous cases later during the extension to multivariable cases.

One of the most used explicit multistep methods is the Adams-Bashforth method, whereas the most used implicit multistep method is the Adams-Moulton method. In both cases, they assume $a_0 = 1$ and $a_1 = \cdots = a_m = 0$. The order is generally not equal to the value of m. (We show later that the method is a fourth-order Adams-Bashforth method for $m = 3$ and the method is a fourth-order Adams-Moulton method for $m = 2$).

We also include in this section the idea of using *Adams-Bashforth* as a "predictor," followed by *Adams-Moulton* as a "corrector." This is one type of a class of methods known as **predictor-corrector** methods.

Finally, we also introduce a group of methods known as the **backward-difference formulas** (BDF) methods. It can be shown to be a generalization of the backward Euler method but achieves a higher order of accuracy. These methods are also implicit multistep methods.

7.3.1 Adams-Bashforth Method

For an nth-order Adams-Bashforth method, let $m = n - 1$, then the parameters for b_k can be obtained by solving the following equation:

$$\begin{pmatrix} 1 & 1 & 1 & \cdots & 1 \\ 0 & 1 & 2 & \cdots & m \\ \vdots & \vdots & \vdots & \ddots & \vdots \\ 0 & 1 & 2^m & \cdots & m^m \end{pmatrix} \begin{pmatrix} b_0 \\ b_1 \\ \vdots \\ b_m \end{pmatrix} = \begin{pmatrix} 1 \\ -\frac{1}{2} \\ \vdots \\ \frac{(-1)^m}{m+1} \end{pmatrix} \tag{7.27}$$

Details on how this equation was obtained are given in Section G.3 as an appendix. The matrix multiplying the vector of b coefficients is a Vandermonde matrix, which is nonsingular. This means that the coefficients will be unique.

For the fourth-order Adams-Bashforth method,

$$\begin{pmatrix} 1 & 1 & 1 & 1 \\ 0 & 1 & 2 & 3 \\ 0 & 1 & 4 & 9 \\ 0 & 1 & 8 & 27 \end{pmatrix} \begin{pmatrix} b_0 \\ b_1 \\ b_2 \\ b_3 \end{pmatrix} = \begin{pmatrix} 1 \\ -\frac{1}{2} \\ \frac{1}{3} \\ -\frac{1}{4} \end{pmatrix}$$

which yields

$$b_0 = \frac{55}{24} \qquad b_1 = \frac{-59}{24} \qquad b_2 = \frac{37}{24} \qquad b_3 = \frac{-9}{24}$$

Thus the fourth-order Adams-Bashforth method is given by

$$y_{k+1} = y_k + \frac{h}{24}\left(55f(y_k) - 59f(y_{k-1}) + 37f(y_{k-2}) - 9f(y_{k-3})\right) \tag{7.28}$$

or for the multivariable case,

$$\mathbf{y}_{k+1} = \mathbf{y}_k + \frac{h}{24}\left(55\,\mathbf{f}(\mathbf{y}_k) - 59\,\mathbf{f}(\mathbf{y}_{k-1}) + 37\,\mathbf{f}(\mathbf{y}_{k-2}) - 9\,\mathbf{f}(\mathbf{y}_{k-3})\right) \tag{7.29}$$

Based on the multivariable case, we could handle the nonautonomous case by including t as one of the elements in the state vector $\mathbf{y}$, that is, with $y^{[M]} = t$ and $dy^{[M]}/dt = 1$. It turns out, however, that with $f^{[M]} = 1$, (7.28) simply reduces to

$$y_{k+1}^{[M]} = y_k^{[M]} + h$$

which is just the same equation as $t_{k+1} = t_k + h$. This means that the Adams-Bashforth formulas can be extended immediately to handle the nonautonomous case, that is,

$$\mathbf{y}_{k+1} = \mathbf{y}_k + \frac{h}{24}\left(55\,\mathbf{f}(t_k, \mathbf{y}_k) - 59\,\mathbf{f}(t_{k-1}, \mathbf{y}_{k-1}) + 37\,\mathbf{f}(t_{k-2}, \mathbf{y}_{k-2}) - 9\,\mathbf{f}(t_{k-3}, \mathbf{y}_{k-3})\right) \tag{7.30}$$

7.3.2 Adams-Moulton Method

For the Adams-Moulton method, the procedure is similar except for the inclusion of b_{-1}. After this modification, the resulting matrix equation is given by

$$\begin{pmatrix} 1 & 1 & 1 & \cdots & 1 \\ -1 & 0 & 1 & \cdots & m \\ \vdots & \vdots & \vdots & \ddots & \vdots \\ (-1)^{m+1} & 0 & 1 & \cdots & m^{m+1} \end{pmatrix} \begin{pmatrix} b_{-1} \\ b_0 \\ \vdots \\ b_m \end{pmatrix} = \begin{pmatrix} 1 \\ -\dfrac{1}{2} \\ \vdots \\ \dfrac{(-1)^{m+1}}{m+2} \end{pmatrix} \tag{7.31}$$

For the second-order Adams-Moulton method, we have

$$\begin{pmatrix} 1 & 1 \\ -1 & 0 \end{pmatrix} \begin{pmatrix} b_{-1} \\ b_0 \end{pmatrix} = \begin{pmatrix} 1 \\ -\dfrac{1}{2} \end{pmatrix} \quad \longrightarrow \quad b_{-1} = b_0 = \frac{1}{2}$$

or

$$y_{k+1} = y_k + \frac{h}{2}\left(f\left(y_{k+1}\right) + f\left(y_k\right)\right) \tag{7.32}$$

This case is also known as the **trapezoidal method**.

For the fourth-order Adams-Moulton method,

$$\begin{pmatrix} 1 & 1 & 1 & 1 \\ -1 & 0 & 1 & 2 \\ 1 & 0 & 1 & 4 \\ -1 & 0 & 1 & 8 \end{pmatrix} \begin{pmatrix} b_{-1} \\ b_0 \\ b_1 \\ b_2 \end{pmatrix} = \begin{pmatrix} 1 \\ -\frac{1}{2} \\ +\frac{1}{3} \\ -\frac{1}{4} \end{pmatrix}$$

which yields $b_{-1} = 9/24$, $b_0 = 19/24$, $b_1 = -5/24$, and $b_2 = 1/24$. Thus the fourth-order Adams-Moulton method is given by

$$y_{k+1} = y_k + \frac{h}{24}\left(9f\left(y_{k+1}\right) + 19f\left(y_k\right) - 5f\left(y_{k-1}\right) + f\left(y_{k-2}\right) \right) \tag{7.33}$$

Note that because multistep methods require past values of y, one still needs to use one-step methods such as the Runge-Kutta methods for the initialization step. This means that for a fourth-order Adams-Bashforth or Adams-Moulton method, we need to use one-step methods to evaluate y_1, y_2, and y_3 (in addition to y_0) before the Adams-Bashforth could proceed. For the fourth-order Adams-Moulton method, we need the initial values of y_0, y_1, and y_2.

Notice that with $y = t$ and $f(y) = 1$, we also end up with $t_{k+1} = t_k + h$. Thus the formulas can be extended immediately to include nonautonomous cases and multivariable cases as follows:

$$\mathbf{y}_{k+1} = \mathbf{y}_k + \frac{h}{24}\left(9\,\mathbf{f}\left(t_{k+1}, y_{k+1}\right) + 19\,\mathbf{f}\left(t_k, y_k\right) - 5\,\mathbf{f}\left(t_{k-1}, y_{k-1}\right) + \mathbf{f}\left(t_{k-2}, y_{k-2}\right) \right) \tag{7.34}$$

7.3.3 Adams Predictor-Corrector Methods

Explicit formulas often have the advantage of direct evaluations, whereas implicit formulas usually have the advantages of improved stability. To merge the two methods, one could first use explicit formulas as predictors and then use the implicit formulas as correctors. This is done by applying the predicted value on one side of the implicit formulas, yielding a new value of the updates before moving to the next integration step.

To be specific, we can apply the Adams formulas as follows:

1. **Predictor:**

$$\begin{aligned} \mathbf{z}_{k+1} &= \mathbf{y}_k + \frac{h}{24}\Big(55\,\mathbf{f}\left(t_k, \mathbf{y}_k\right) - 59\,\mathbf{f}\left(t_{k-1}, \mathbf{y}_{k-1}\right) \\ &\quad + 37\,\mathbf{f}\left(t_{k-2}, \mathbf{y}_{k-2}\right) - 9\,\mathbf{f}\left(t_{k-3}, \mathbf{y}_{k-3}\right)\Big) \end{aligned} \tag{7.35}$$

2. **Corrector:**

$$\begin{aligned}\mathbf{y}_{k+1} &= \mathbf{y}_k + \frac{h}{24}\Big(9\,\mathbf{f}\left(t_{k+1}, \mathbf{z}_{k+1}\right) + 19\,\mathbf{f}\left(t_k, \mathbf{y}_k\right) \\ &\qquad - 5\,\mathbf{f}\left(t_{k-1}, \mathbf{y}_{k-1}\right) + \mathbf{f}\left(t_{k-2}, \mathbf{y}_{k-2}\right)\Big)\end{aligned} \tag{7.36}$$

One could vary the correction stage by performing additional iterations, for example, with $\mathbf{w}_{k+1}^{(0)} = \mathbf{z}_{k+1}$,

$$\mathbf{w}_{k+1}^{(j+1)} = \mathbf{y}_k + \frac{h}{24}\left(9\mathbf{f}\left(\mathbf{w}_{k+1}^{(j)}\right) + 19\mathbf{f}\left(\mathbf{y}_k\right) - 5\mathbf{f}\left(\mathbf{y}_{k-1}\right) + \mathbf{f}\left(\mathbf{y}_{k-2}\right)\right) \tag{7.37}$$

until $\left\|\mathbf{w}_{k+1}^{(j+1)} - \mathbf{w}_{k+1}^{(j)}\right\| < \epsilon$, where ϵ is a chosen tolerance, and then set $\mathbf{y}_{k+1} = \mathbf{w}_{k+1}^{(j+1)}$. Even though this may converge to the stable value of $\mathbf{w}_{k+1}^{(j+1)}$, it may still not necessarily be an accurate value. Instead of (7.37), it is more efficient to decrease the step-size h, using error-control strategies such as those discussed in Section G.5.

7.3.4 Backward-Difference Formula (BDF) Methods

Another important set of implicit multistep methods is the **backward-difference formula (BDF) method** (also known as the **Gear's method**). Based on the general form of multistep methods given in (7.26),

$$y_{k+1} = \sum_{i=0}^{m} a_i\, y_{k-i} + h \sum_{j=-1}^{m} b_j\, f\left(y_{k-j}\right)$$

the BDF method sets $b_j = 0$ for $j \geq 0$, reducing the general formula to be

$$y_{k+1} = \sum_{i=0}^{m} a_i\, y_{k-i} + hb_{-1}\, f\left(y_{k+1}\right) \tag{7.38}$$

As shown next, we need $m = p - 1$ to obtain p^{th}-order accuracy.

Following the same approach as before, we choose the simplest function for f to be $f(t, y) = y$. Doing so, we get the analytical solution

$$y_{k+\ell} = e^{\ell h} y_k \qquad \rightarrow \qquad y_{k-\ell} = e^{-(\ell+1)h} y_{k+1}$$

Substituting these values of $y_{k-\ell}$ into (7.38), with $f\left(y_{k+1}\right) = y_{k+1}$, we have

$$\begin{aligned} y_{k+1} &= \sum_{i=0}^{m} \alpha_i e^{-(i+1)h} y_{k+1} + hb_{-1} y_{k+1} \\ 1 &= hb_{-1} + \sum_{i=0}^{m} \alpha_i \left(1 + \sum_{q=1}^{\infty} \frac{\left[-(i+1)h\right]^q}{q!}\right)\end{aligned}$$

To obtain p^{th}-order accuracy, we can let $m = p - 1$ and truncate terms with h^{p+1} and higher. This yields

$$\sum_{i=0}^{p-1} \alpha_i = 1 \;;\; \sum_{i=0}^{p-1} \alpha_i\left[-(i+1)\right] + b_{-1} = 0 \;;\; \sum_{i=0}^{p-1} \frac{\left[-(i+1)\right]^q}{q!}\alpha_i = 0\,,\; q = 2, \ldots, p$$

These equations can be put in matrix form, as given here:

$$\begin{pmatrix} 0 & 1 & 1 & \cdots & 1 \\ -1 & 1 & 2 & \cdots & p \\ 0 & 1 & 2^2 & \cdots & p^2 \\ \vdots & \vdots & \vdots & \ddots & \vdots \\ 0 & 1 & 2^p & \cdots & p^p \end{pmatrix} \begin{pmatrix} b_{-1} \\ a_0 \\ a_1 \\ \vdots \\ a_{p-1} \end{pmatrix} = \begin{pmatrix} 1 \\ 0 \\ 0 \\ \vdots \\ 0 \end{pmatrix} \tag{7.39}$$

For a fourth-order BDF,

$$\begin{pmatrix} b_{-1} \\ a_0 \\ a_1 \\ a_2 \\ a_3 \end{pmatrix} = \begin{pmatrix} 0 & 1 & 1 & 1 & 1 \\ -1 & 1 & 2 & 3 & 4 \\ 0 & 1 & 2^2 & 3^2 & 4^2 \\ 0 & 1 & 2^3 & 3^3 & 4^3 \\ 0 & 1 & 2^4 & 3^4 & 4^4 \end{pmatrix}^{-1} \begin{pmatrix} 1 \\ 0 \\ 0 \\ 0 \\ 0 \end{pmatrix} = \begin{pmatrix} \frac{12}{25} \\ \frac{48}{25} \\ -\frac{36}{25} \\ \frac{16}{25} \\ -\frac{3}{25} \end{pmatrix}$$

or

$$y_{k+1} = \frac{48}{25} y_k - \frac{36}{25} y_{k-1} + \frac{16}{25} y_{k-2} - \frac{3}{25} y_{k-3} + \frac{12}{25} h f\left(y_{k+1}\right) \tag{7.40}$$

The mulitvariable case is given by

$$\mathbf{y}_{k+1} = \frac{48}{25}\,\mathbf{y}_k - \frac{36}{25}\,\mathbf{y}_{k-1} + \frac{16}{25}\,\mathbf{y}_{k-2} - \frac{3}{25}\,\mathbf{y}_{k-3} + \frac{12}{25} h\,\mathbf{f}\left(\mathbf{y}_{k+1}\right) \tag{7.41}$$

BDF methods can handle several stiff differential equations well, but stability properties do deteriorate at higher orders; that is, they are only stable for orders less than 7.

Unfortunately, care is needed when extending this result to nonautonomous differential equations because setting $y = t$ and $f(y) = 1$ will now yield

$$t_{k+1} = \frac{48}{25} t_k - \frac{36}{25} t_{k-1} + \frac{16}{25} t_{k-2} - \frac{3}{25} t_{k-3} + \frac{12}{25} h$$

which does reduce to $t_{k+1} = t_k + h$ only if the step size h is held constant. Thus for fixed h, the nonautonomous approach implies the simple replacement of $\mathbf{f}(\mathbf{y}_{k+1})$ by $\mathbf{f}\left(t_{k+1}, \mathbf{y}_{k+1}\right)$. When the step sizes are variable, the coefficient will also have to vary at each step. The generalized formulas for the BDF coefficients when the step sizes vary are discussed in Section G.4.

EXAMPLE 7.3. We now compare the performance of the four multistep methods just discussed, applied to the same linear scalar system used in Examples 7.1 and 7.2,

$$\frac{1}{\tau}\frac{dy}{dt} + y = -e^{-\beta t}$$

We have the following iteration equations:

1. Adams-Bashforth (AB)

$$y_{k+1} = y_k + \frac{h}{24\tau}\Big(55\left(-y_k - e^{-\beta t_k}\right) - 59\left(-y_{k-1} - e^{-\beta t_{k-1}}\right) + 37\left(-y_{k-2} - e^{-\beta t_{k-2}}\right) - 9\left(-y_{k-3} - e^{-\beta t_{k-3}}\right)\Big)$$

2. Adams-Moulton (AM)

$$y_{k+1} = \frac{24\tau}{24\tau + 9h}\Big(y_k + \frac{h}{24\tau}\Big[9\left(-e^{-\beta t_{k+1}}\right) + 19\left(-y_k - e^{-\beta t_k}\right) - 5\left(-y_{k-1} - e^{-\beta t_{k-1}}\right) + \left(-y_{k-2} - e^{-\beta t_{k-2}}\right)\Big]\Big)$$

3. Adams Predictor-Corrector (APC)

$$z_{k+1} = y_k + \frac{h}{24\tau}\Big(55\left(-y_k - e^{-\beta t_k}\right) - 59\left(-y_{k-1} - e^{-\beta t_{k-1}}\right) + 37\left(-y_{k-2} - e^{-\beta t_{k-2}}\right) - 9\left(-y_{k-3} - e^{-\beta t_{k-3}}\right)\Big)$$

$$y_{k+1} = y_k + \frac{h}{24\tau}\Big(9\left(-z_{k+1} - e^{-\beta t_{k+1}}\right) + 19\left(-y_k - e^{-\beta t_k}\right) - 5\left(-y_{k-1} - e^{-\beta t_{k-1}}\right) + \left(-y_{k-2} - e^{-\beta t_{k-2}}\right)\Big)$$

4. Backward-Difference Formula (BDF)

$$y_{k+1} = \frac{\tau}{25\tau + 12h}\left(48y_k - 36y_{k-1} + 16y_{k-2} - 12\frac{h}{\tau}e^{-\beta t_{k+1}}\right)$$

Figure 7.3 shows the performance of these iteration formulas of the various multistep methods for $\tau = 0.001$, $\beta = 100$, and $y_0 = 1.0$. In practical use, one would use one-step methods such as the Runge-Kutta methods to evaluate the first four or three iterations. For our example here, however, we used the exact values coming from the analytical solution to allow us to assess the performance of the multistep methods, independent of the Runge-Kutta methods.

All the methods performed well for $h = 0.0001$. However, when compared with Figure 7.3 for the Runge-Kutta methods, the accuracy and stability of the explicit Adams-Bashforth method did not appear to be as good at $h = 0.001$. The stability of the predictor-corrector formulas is better than that of the explicit Adams-Bashforth method, but it was unstable when $h = 0.002$. The two implicit methods are the Adams-Moulton and BDF methods. It appears for small h values, the Adams-Moulton performs better than the BDF method. However, as we see in Figure 7.4, the BDF maintained stability even at $h = 0.004$, but the Adams-Moulton did not. This shows that although stability is improved by using implicit methods, some methods have greater stability ranges than others. This is one reason that, among multistep methods, the BDF methods are often chosen to handle stiff[2] differential equations, and then they are coupled with other enhancements, such as step-size control, to improve accuracy.

[2] See Section 7.4 for a description of stiff differential equations.

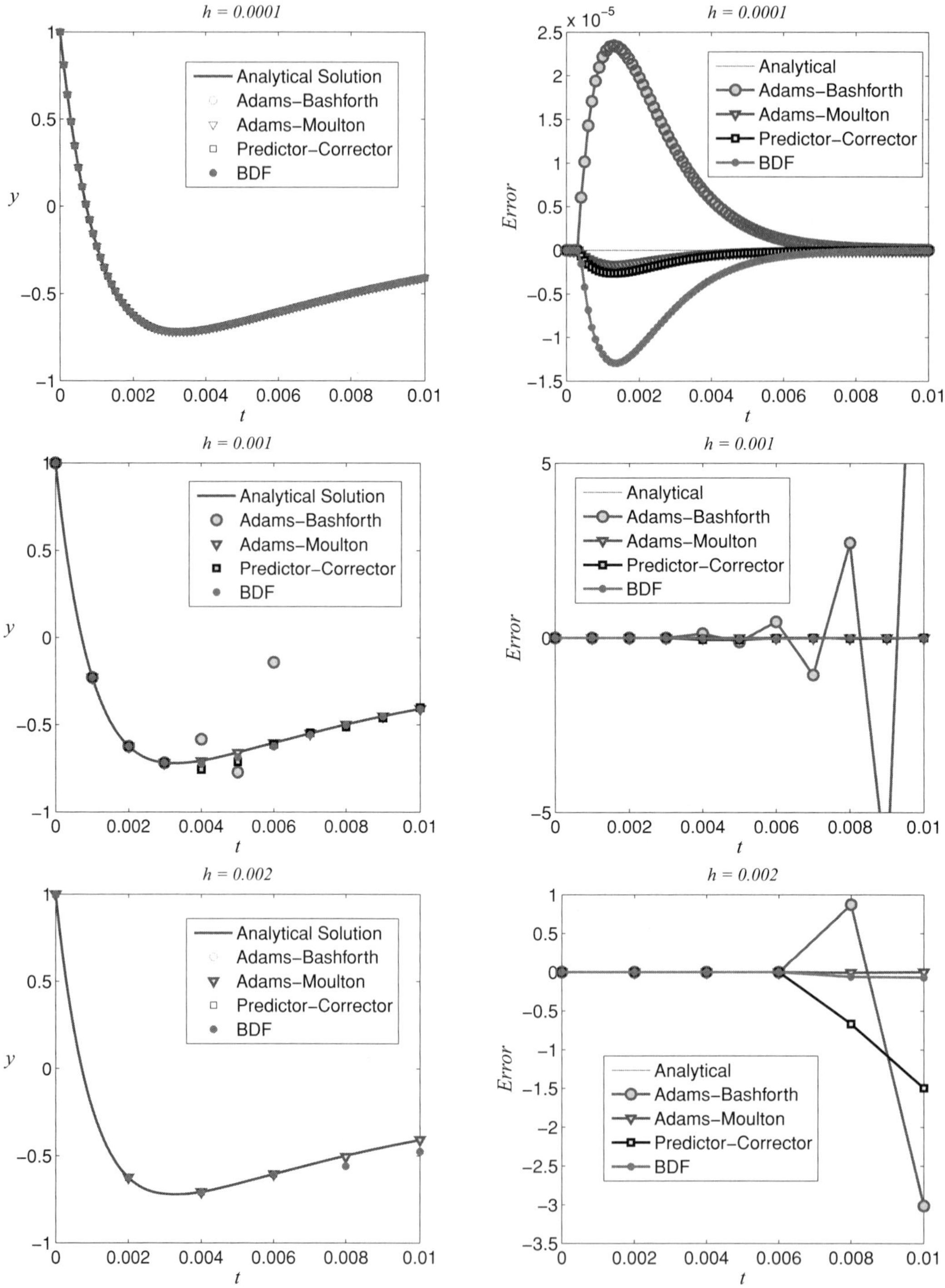

Figure 7.3. Performance comparison of the various multistep methods for example 7.3

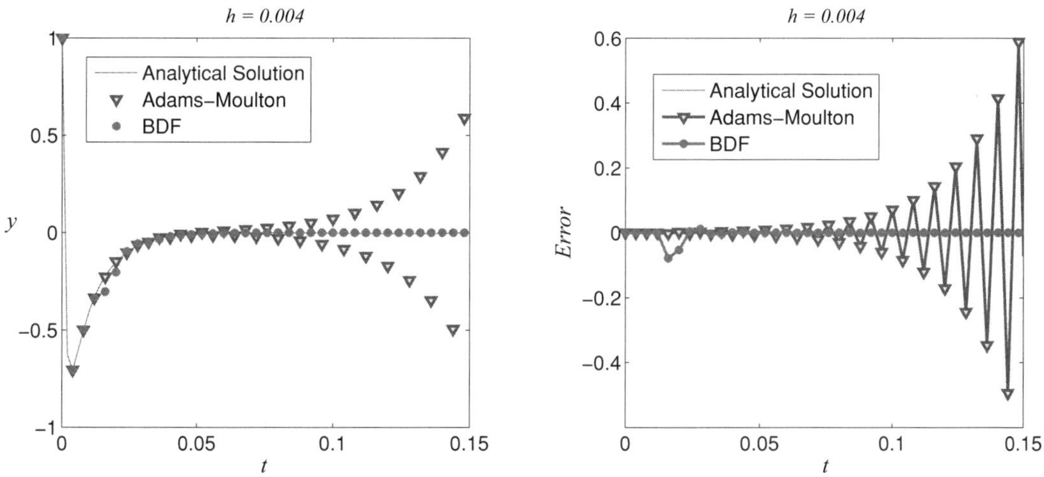

Figure 7.4. Performance comparison between Adams-Moulton and BDF at $h = 0.04$ for example 7.3

7.4 Difference Equations and Stability

The numerical methods presented in the previous sections are approximations of the exact solutions in the form of iteration schemes. These iteration equations are actually difference equations. The stability properties of the various approaches will depend on the properties of difference equations rather than on properties of differential equations. In the sections that follow, we start with a brief overview of difference equations, with particular attention to their stability properties. Once the general stability results have been established, we turn to the technique known as the Dahlquist test which involves the testing of particular numerical solution techniques to one of the simplest differential equation

$$\frac{dy}{dt} = \lambda y$$

One can then build the stability regions in the complex plane.

When looking at the stability regions, one could argue that once we have an accurate method such as the fourth-order Runge Kutta method, we should just be able to vary step size h to maintain stability. The implication of this idea is very problematic for some differential equations, because the step size needed to maintain stability for the chosen method could mean a very small step size. This is true even when error-control strategies are incorporated to numerical methods such as the $4/5^{\text{th}}$ embedded Runge-Kutta methods. The result is what appears to be a "hanging" process because the schemes might have entered a region requiring very small values of h. These systems are generally described as "stiff differential equations." The general strategy that is often prescribed is to first use explicit high-order methods because these methods are quick and accurate, and there are still a large class of non-stiff systems in several applications. However, when the appearance of "hanging" process occurs, then the system can be deemed "stiff," and implicit methods are used instead.

7.4.1 Linear Difference Equations with Constant Coefficients

We begin with the definition of the **shift operator**, $\mathcal{Q}$,

$$\mathcal{Q}\left(y_n\right) = y_{n+1} \tag{7.42}$$

where $y_n = y(nh)$ and $h = \Delta t$. A pth-order linear difference equation with constant coefficients has the following form

$$\left(\sum_{i=0}^{p} \alpha_p \mathcal{Q}^p\right) y_n = f(n) \tag{7.43}$$

where α_i, $i = 0, \ldots, p$ are constant. If $f(n) = 0$, then we say that (7.43) is **homogeneous**.

Using the constants given in the left-hand side of (7.43), we can define the **characteristic equation** as

$$\sum_{i=0}^{p} \alpha_p \, \sigma^p = 0 \tag{7.44}$$

The roots of (7.44) will be used to form the **complementary solution**, that is, the solution of the homogeneous part of (7.43), as stated by the following theorem:

THEOREM 7.1. *Let $f(n) = 0$ in the difference equation (7.43), and let the distinct (possibly repeated) roots of the pth-order characteristic equation in (7.44) be $\sigma = \left\{\sigma_1, \ldots, \sigma_M\right\}$ where σ_j is a k_j-fold root (i.e., repeated $(k_j - 1)$ times and $\sum_{j=1}^{M} k_j = p$). The solution of the homogeneous difference equation is then given by*

$$y_n = \mathcal{Q}^n y_0 = \sum_{j=1}^{M} \mathcal{S}\left(j, n\right) \tag{7.45}$$

where

$$\mathcal{S}\left(j, n\right) = \left(\sum_{\ell=0}^{k_j-1} c_{j,\ell}\, n^{\ell}\right) \left(\sigma_j\right)^n \tag{7.46}$$

and $c_{j,\ell}$ are arbitrary constants that are used to fit initial conditions.

PROOF. (See Section G.6 for proof.)

EXAMPLE 7.4. Consider the following difference equation:

$$\begin{aligned} y_{n+5} - 2.5y_{n+4} + 1.96y_{n+3} - 7.6 \times 10^{-2}y_{n+2} \\ - 5.453 \times 10^{-1}y_{n+1} + 1.7885 \times 10^{-1}y_n = 0 \end{aligned}$$

subject to the following initial conditions: $y_{-4} = y_{-3} = y_{-2} = 1$, $y_{-1} = y_0 = 0$.

The roots of the characteristic equations are: $-0.5, 0.7, 0.7, 0.8 + 0.3i, 0.8 - 0.3i$. Following the solution provided by Theorem 7.1, we have

$$y_n = C_1\left(-0.5\right)^n + \left(C_2 + C_3 n\right)\left(0.7\right)^n + C_4\left(0.8 + 0.3i\right)^n + C_5\left(0.8 - 0.3i\right)^n$$

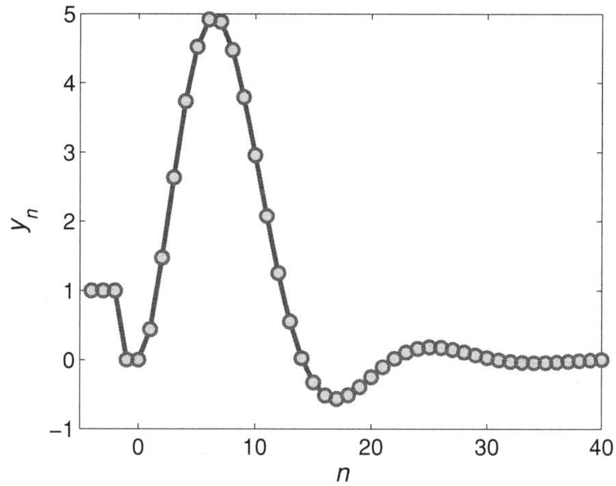

Figure 7.5. Plot of points from difference solution (open circles) and data obtained by recursion of the difference equation (solid line).

or using the polar form and Euler's identity on the complex roots,

$$y_n = C_1(-0.5)^n + (C_2 + C_3 n)(0.7)^n + r^n\left(A\cos(n\theta) + B\sin(n\theta)\right)$$

where

$$r = \sqrt{0.8^2 + 0.3^2} \qquad ; \qquad \theta = \tan^{-1}\left(\frac{0.3}{0.8}\right)$$

To match the initial conditions, we obtain: $C_1 = 4.9052 \times 10^{-2}$, $C_2 = 1.0209 \times 10^1$, $C_3 = 2.4539$, $A = -1.0258 \times 10^1$, and $B = -6.3538 \times 10^{-1}$. A plot of the solution is given in Figure 7.5, together with the curve obtained by iteration of the difference equation.

Recall that the asymptotic stability of *differential equations* requires that the real part of all eigenvalues are negative. This is not the case for difference equations. The stability of difference equations instead requires the roots of the characteristic equation to have magnitudes that are less than unity.[3] In this example, four of the eigenvalues have positive real parts, but all the roots have magnitudes less than 1. Thus the process is stable.

EXAMPLE 7.5. Let D_N be the determinant of matrix $\mathbf{A}_N$, where $\mathbf{A}_N$ is the $N \times N$ tri-diagonal matrix given by

$$\mathbf{A}_N = \begin{pmatrix} a & b & & 0 \\ c & \ddots & \ddots & \\ & \ddots & \ddots & b \\ 0 & & c & a \end{pmatrix}$$

By expanding along the first column, we can show that

$$D_N = aD_{N-1} - bcD_{N-2} \qquad \text{or} \qquad D_{N+2} - aD_{N+1} + bcD_N = 0$$

[2] See Theorem 7.2.

which is a difference equation for $N \geq 1$, with $D_1 = a$ and $D_2 = a^2 - bc$. The characteristic equation is

$$\sigma^2 - a\sigma + bc = 0$$

whose two roots can be a complex-conjugate pair, a set of repeated real numbers, or a pair of different real numbers, depending on whether the term $(a^2 - 4bc)$ is negative, zero, or positive, respectively.

Take the specific case of $a = 3$, $b = -2$, $c = -2$. Then the roots are $\sigma = 1.5 \pm i\sqrt{7}/2$ yielding

$$D_N = 2^N \left(\cos(N\theta) + \frac{3}{\sqrt{7}} \sin(N\theta) \right)$$

where $\theta = \tan^{-1}\left(\frac{\sqrt{7}}{3}\right)$.

The particular solutions of difference equations are usually found by the method of undetermined coefficients. Suppose the nonhomogeneous function of (7.43) is given by

$$f(n) = a\, n^q\, b^n\, \cos(n\theta) \qquad \text{or} \qquad f(n) = a\, n^q\, b^n\, \sin(n\theta) \tag{7.47}$$

where a, b, q, and θ are constants, with q as a positive integer. The particular solution, Y, can then be formulated as

$$Y = \left(n^K\right) b^n \left(\left(A_0 + \cdots + A_q n^q\right) \cos(n\theta) + \left(B_0 + \cdots + B_q n^q\right) \sin(n\theta) \right) \tag{7.48}$$

where $K = 0$ if $R = \left(be^{i\theta}\right)$ is not a root of the characteristic equation (7.44). If $R = be^{i\theta} = \sigma_j$, where σ_j is a k_j-fold root of the characteristic equation (7.44), then we need to set $K = k_j$. The coefficients $A_0, \ldots, A_q$ and $B_0, \ldots, B_q$ can then be determined after substituting Y into the difference equation.

7.4.2 Stability

As long as the term b in (7.47) has a magnitude less than 1, $f(n)$ is bounded as $n \to \infty$. This means that the particular solution will also be bounded as $n \to \infty$. One can then conclude that as long as the nonhomogeneous term $f(n)$ is bounded, the source of instability can only come from the complementary solution. Using Theorem 7.1, the stability of a linear difference equation will then depend on the roots of the characteristic equation.

THEOREM 7.2. *For the linear difference equation (7.43), with $f(n)$ as a linear combination of terms having the form given in (7.47), let $f(n) < \infty$ as $n \to \infty$, and let $\sigma = (\sigma_1, \ldots, \sigma_M)$ be the set of distinct roots of the characteristic equation (7.44). Then the solution of the difference equation, y_n, is stable if*[4]

$$\left| \sigma_j \right| < 1 \qquad \text{for } j = 1, \ldots, M$$

[4] The theorem gives a sufficient condition. For each σ_j that is not repeated, the stability condition could be relaxed to be $|\sigma_j| \leq 1$. However, because round-off errors are usually present, it may be more reasonable to use the strict inequality when applying it to the stability analysis of numerical solutions.

The extension to linear multivariable difference equations is straightforward. The standard form, however, is to put it back into the familiar state-space formulation, that is,

$$\mathbf{x}_{n+1} = A\mathbf{x}_n + \mathbf{b}_n \tag{7.49}$$

where

$$\mathbf{x}_n = \begin{pmatrix} x_n^1 \\ \vdots \\ x_n^M \end{pmatrix} \qquad A = \begin{pmatrix} a_{11} & \cdots & a_{1M} \\ \vdots & \ddots & \vdots \\ a_{M1} & \cdots & a_{MM} \end{pmatrix}$$

Let matrix T be the nonsingular matrix that would yield the Jordan canonical decomposition of A: $TJT^{-1} = A$. The solution of (7.49) is

$$\begin{aligned} \mathbf{x}_1 &= A\mathbf{x}_0 + \mathbf{b}_0 \\ &\vdots \\ \mathbf{x}_n &= A^n\mathbf{x}_0 + \sum_{i=0}^{n-1} A^{i-1}\mathbf{b}_{i-1} \end{aligned}$$

or

$$\mathbf{x}_n = TJ^nT^{-1}\mathbf{x}_0 + \sum_{i=1}^{n-1} TJ^{i-1}T^{-1}\mathbf{b}_{i-1} \tag{7.50}$$

When J is diagonal,

$$J = \begin{pmatrix} \sigma_1 & & 0 \\ & \ddots & \\ 0 & & \sigma_M \end{pmatrix} \rightarrow \quad \mathbf{x}_n = T\begin{pmatrix} \sigma_1^n & & 0 \\ & \ddots & \\ 0 & & \sigma_M^n \end{pmatrix} T^{-1}\mathbf{x}_0$$

$$+ \sum_{i=1}^{n-1} T \begin{pmatrix} \sigma_1^i & & 0 \\ & \ddots & \\ 0 & & \sigma_M^i \end{pmatrix} T^{-1}\mathbf{b}_{i-1}$$

When J is a full Jordan block (cf. (3.36)),

$$J = \text{diag}\begin{pmatrix} \sigma & 1 & & 0 \\ & \ddots & \ddots & \\ & & \ddots & 1 \\ 0 & & & \sigma \end{pmatrix} \rightarrow$$

$$\begin{aligned} \mathbf{x}_n &= T\begin{pmatrix} \sigma^n & \beta_{n,n-1}\sigma^{n-1} & \cdots & \beta_{n,n-M+1}\sigma^{n-M+1} \\ & \ddots & & \beta_{n,n-M+2}\sigma^{n-M+2} \\ & & \ddots & \vdots \\ & & & \sigma^n \end{pmatrix} T^{-1}\mathbf{x}_0 \\ &+ \sum_{i=1}^{n-1} T \begin{pmatrix} \sigma^i & \beta_{i,i-1}\sigma^{i-1} & \cdots & \beta_{i,i-M+1}\sigma^{i-M+1} \\ & \ddots & & \beta_{i,i-M+2}\sigma^{i-M+2} \\ & & \ddots & \vdots \\ & & & \sigma^i \end{pmatrix} T^{-1}\mathbf{b}_{i-1} \end{aligned}$$

where

$$\beta_{k,j} = \begin{cases} \dfrac{k!}{(k-j)!j!} & \text{if } j \geq 0 \\ 0 & \text{otherwise} \end{cases}$$

In either case, the stability again depends on the eigenvalues of A; that is, as long as $\mathbf{b}_i$ in (7.49) are bounded, then stability is guaranteed if $|\sigma_i| < 1$ for all $i = 1, \ldots, M$.

Based on the preceding canonical forms, for most numerical methods, including single-step or multistep, implicit or explicit, the stability is applied to the the **Dahlquist test case**. For the single-step methods, the test involves the following steps:

1. Apply the chosen numerical method on

$$\frac{dy}{dt} = \lambda y$$

   with $y(0) = 1$.
2. With $h = \Delta t$ and $z = \lambda h$, rearrange the equation to be

$$y_{n+1} = g(z)\, y_n$$

3. Find the stability region, which is the region such that

$$\text{Stab} : \left\{ z \ : \ \left|g(z)\right| < 1 \right\}$$

EXAMPLE 7.6. When the s-stage and s-order explicit Runge-Kutta methods are applied to the Dahlquist test case, the difference equation becomes

$$y_{n+1} = \left(\sum_{i=0}^{s} \frac{1}{i!} (\lambda h)^i \right) y_n$$

The stability region is then shown in Figure 7.6 for $s = 1, 2, 3, 4$. Note that, although the stability region increases with the number of stages s, the explicit Runge-Kutta methods will be conditionally stable; that is, the step size is constrained by the value of λ.

Applying the backward Euler method to the Dahlquist test, the difference equation becomes

$$y_{n+1} = \frac{1}{1 - \lambda h} y_n$$

The stability region is shown in Figure 7.7. This means that if $\text{Real}(\lambda) < 0$, the backward Euler method will be stable for any step size.

Finally, we can assess the stability region for the Gauss-Legendre method, which is an implicit fourth-order Runge-Kutta. The difference equation after applying the Gauss-Legendre method to the Dahlquist test is given by

$$y_{n+1} = \frac{12 + 6\lambda h + (\lambda h)^2}{12 - 6\lambda h + (\lambda h)^2} y_n$$

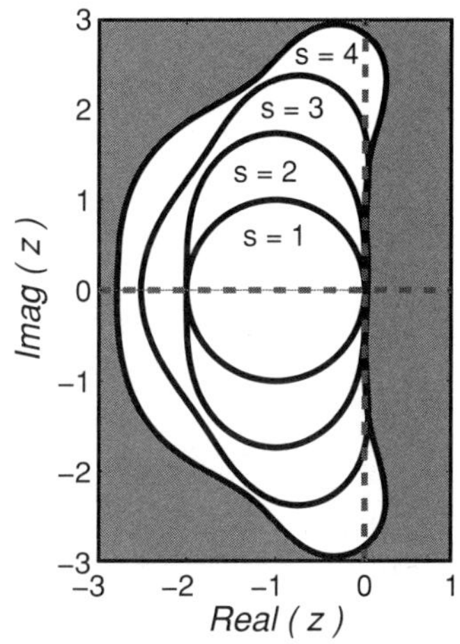

Figure 7.6. Stability region for $z = \lambda h$ using the explicit Runge-Kutta methods (unshaded regions) of order $s = 1, 2, 3, 4$.

The stability region is shown in Figure 7.8. For Real $(\lambda) < 0$, the Gauss-Legendre method will also be stable for any step size h.

For multistep methods, the procedure is similar. Instead, the characteristic equations will yield multiple roots, which are functions of $z = \lambda h$. One of the roots, usually denoted by σ_1, will have a value of 1 at $z = z^*$. This root is known as the **principal root**. The other roots are known as the **spurious roots**. For accuracy and convergence properties, the principal root is the most critical, whereas the effects of spurious roots die out eventually, often quickly, as long as the method is stable. The stability regions can again be obtained by applying the numerical method to the Dahlquist test case to obtain the characteristic equation. If any of the roots at $z = z^*$, $\sigma_k(z^*)$, have a magnitude greater than 1, then z^* belongs to the unstable region.

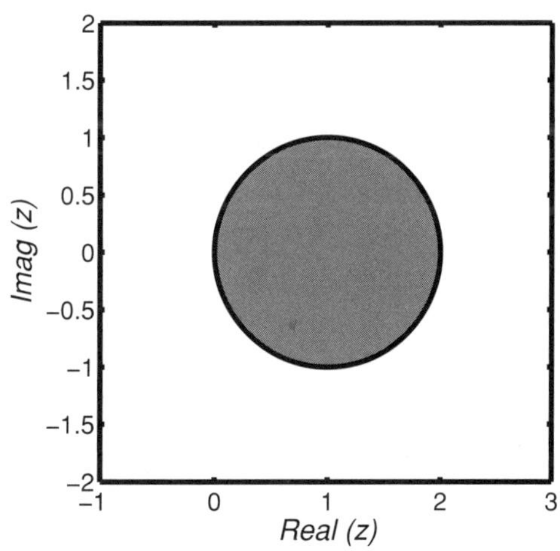

Figure 7.7. Stability region for $z = \lambda h$ using the backward Euler methods (unshaded regions).

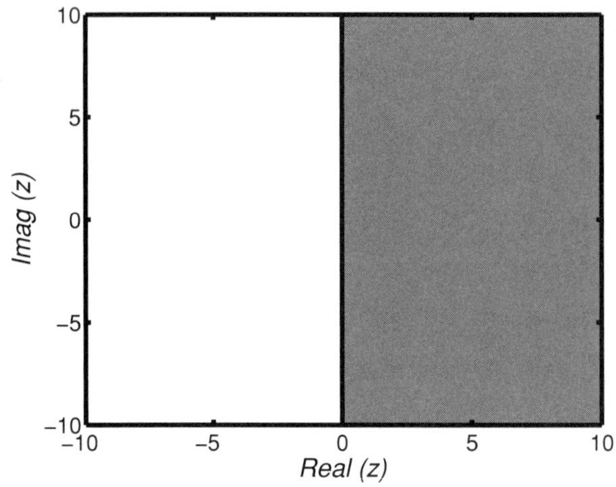

Figure 7.8. Stability region for $z = \lambda h$ using the Gauss-Legendre methods (unshaded regions).

EXAMPLE 7.7. Let us apply the Dahlquist test case, that is, $f(t, y) = \lambda y$, to the fourth-order Adams-Moulton method given in (7.33); then we obtain the following difference equation:

$$y_{k+1} = y_k + \frac{z}{24}\left(9y_{k+1} + 19y_k - 5y_{k-1} + y_{k-2}\right)$$

where $z = \lambda h$, and whose characteristic equation is then given by

$$\left(-1 + \frac{9}{24}z\right)\sigma^3 + \left(1 + \frac{19}{24}z\right)\sigma^2 - \left(\frac{5}{24}z\right)\sigma + \left(\frac{1}{24}z\right) = 0$$

Because the characteristic equation is linear in z, it is simpler to determine the boundary separating the stable region from the unstable region by finding z as a function of $\sigma = e^{i\theta}$, $-\pi \leq \theta \leq \pi$, that is,

$$z(\theta) = \frac{24\left(e^{3\theta i} - e^{2\theta i}\right)}{9e^{3\theta i} + 19e^{2\theta i} - 5e^{\theta i} + 1}$$

Thus for the Adams-Moulton implicit method, this region is shown in Figure 7.9. Note that, as suggested in Example 7.3, specifically in Figure 7.4, even though the Adams-Moulton is an implicit method, the stability range is still limited, thereby making it still not suitable for solving stiff differential equations.

From the stability regions shown in Example 7.6, we note that the explicit Runge-Kutta methods will need small values of step size as Real (λ) becomes more negative. However, both the backward Euler and the Gauss-Legendre methods are unconditionally stable for Real $(\lambda) < 0$. This type of stability is also known as ***A*-stability**. There are other types of stability, some of which are shown in Figure 7.10. Some of these alternative types of stability are needed to discriminate among different schemes, especially multistep methods.

As we have noted, for some numerical methods, especially explicit methods, stability requirements may demand smaller time steps than accuracy requires. When this situation occurs, we say that the system is a **stiff differential equation**. In several cases, differential equations are classified as stiff when the difference between the

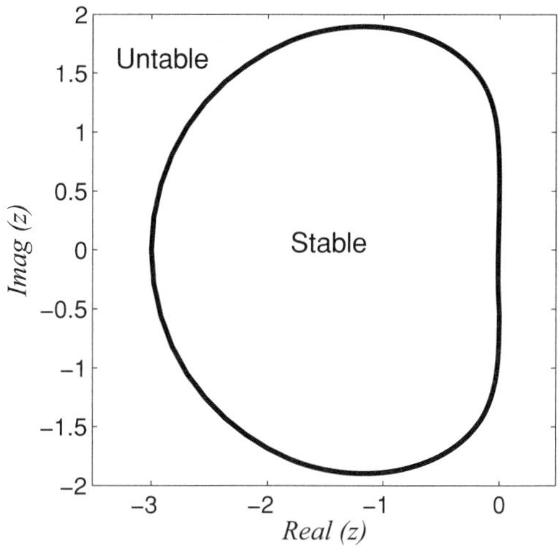

Figure 7.9. The stabliity region of the Adams-Moulton implicit method for the Dahlquist test case.

magnitudes of the largest and smallest eigenvalues is very large. For stiff differential equations, implicit methods such as the Gauss-Legendre IRK or multistep BDF schemes are usually preferred. Specifically, D-stability (cf. Figure 7.10), that is, where the stability is guaranteed inside the region whose real part is less than $D < 0$, is also known as **stiffly stable**. It can be shown that BDF schemes of lower order (e.g., order ≤ 6) are stiffly stable.

Finally, note that the stability issues discussed in this section are all based on linear systems. For nonlinear systems, linear stability analysis can be applied locally via linearization. Other sophisticated approaches are needed for global analysis and are not covered here.

7.5 Boundary Value Problems

Most of the numerical methods discussed thus far are based on the assumption that all the conditions are given at one point, for example, at $t = 0$. Those problems are known as **initial value problems** (IVP). The numerical methods could also be applied at $t = T$ and integrated backward by appropriate changes of the independent variables. For instance, with $\tau = T - t$, the problem becomes an initial value problem with respect to τ.

However, there are several applications in which some of the conditions are specified at more than one point. For example, some conditions may be set at $t = 0$, and the rest are set at $t = T$. These problems are known as **boundary value problems** (BVP). When this is the case, we cannot just use the initial value problems (IVP) solvers such as Runge-Kutta methods, Adams methods, or BDF methods directly.

We limit our discussion to two-point boundary value problems: one at $t = 0$ and the other at $t = T$. There are two types of boundary conditions:

1. **Separated boundary conditions.**

$$\begin{aligned} \mathbf{q}_0\left(\mathbf{x}(0)\right) &= 0 \\ \mathbf{q}_T\left(\mathbf{x}(T)\right) &= 0 \end{aligned} \tag{7.51}$$

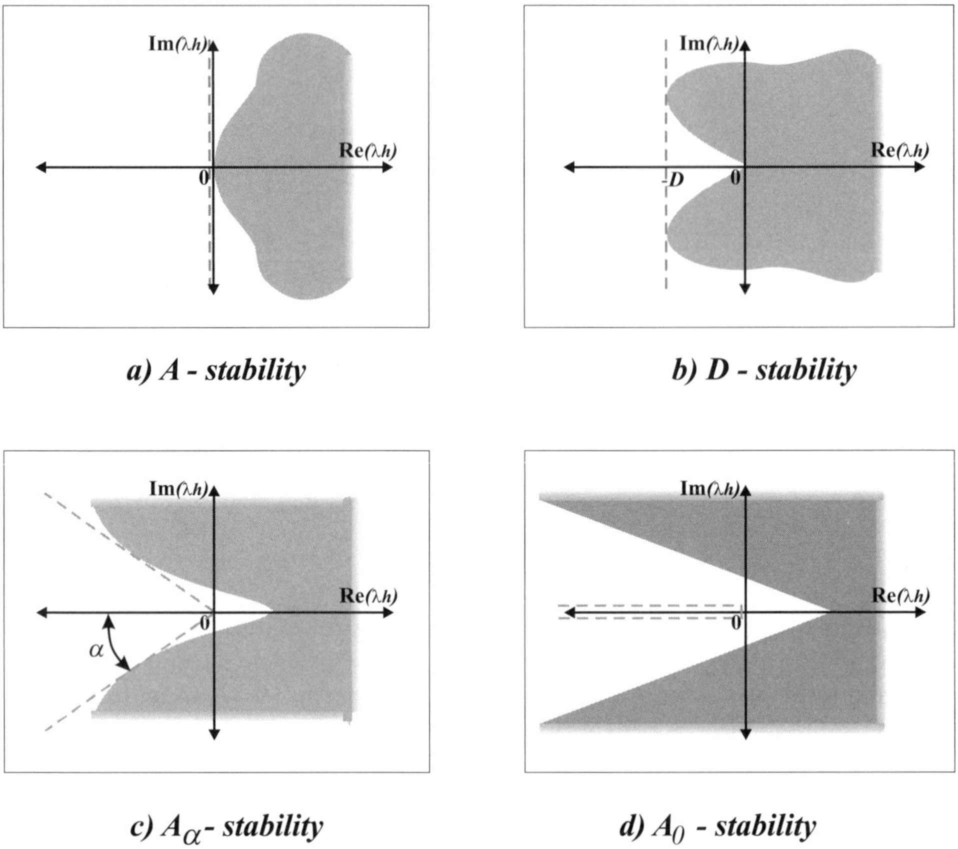

Figure 7.10. Different types of numerical stability. The unshaded regions are stable; for example, for A_α-stability, the shaded regions do not cross the dotted lines.

where $\mathbf{q}_0$ and $\mathbf{q}_T$ are general nonlinear functions. In the linear case, we have

$$\begin{aligned} Q_0\,\mathbf{x}(0) &= \beta_0 \\ Q_T\,\mathbf{x}(T) &= \beta_T \end{aligned} \tag{7.52}$$

where Q_0 and Q_T are constant matrices of sizes $(k \times n)$ and $[(n-k) \times n]$, respectively.

2. **Mixed boundary conditions.**

$$\mathbf{q}\left(\mathbf{x}(0), \mathbf{x}(T)\right) = 0 \tag{7.53}$$

where $\mathbf{q}$ is a nonlinear function. In the linear case, we have

$$Q_a\mathbf{x}(0) + Q_b\mathbf{x}(T) = \mathbf{p} \tag{7.54}$$

where Q_a and Q_b are constant square matrices of size $n \times n$.

In the next section, we discuss the shooting method for linear boundary value problems, which we later generalize to handle the nonlinear boundary value problems. Another approach, known as the **Ricatti method**, is included in Section G.8 as an appendix. Yet another approach is the finite difference method, which is discussed in Chapter 13.

7.5.1 Linear Boundary Value Problems

Consider the general linear differential equation,

$$\frac{d}{dt}\mathbf{x} = \mathbf{A}(t)\mathbf{x} + \mathbf{b}(t) \tag{7.55}$$

subject to a mixed boundary conditions at $t = 0$ and $t = T$, given by

$$Q_a\mathbf{x}(0) + Q_b\mathbf{x}(T) = \mathbf{p} \tag{7.56}$$

where $\mathbf{x}$ is the vector of n state variables and $\mathbf{p}$ is a vector of n constants.

To build upon the tools available for initial value problems, we can pose the problem as the search for initial conditions, $\mathbf{x}(0)$, such that (7.56) will be satisfied. This approach is known as the **shooting method**.

Recall from Section 6.5.3 that the solution of (7.55) is given by

$$\mathbf{x}(t) = \mathbf{M}(t)\mathbf{x}(0) + \mathbf{z}(t) \tag{7.57}$$

where $\mathbf{M}(t)$ is the matrizant of the system that satisfies the homogenous part, that is,

$$\frac{d}{dt}\mathbf{M} = \mathbf{A}(t)\mathbf{M} \quad ; \quad \mathbf{M}(0) = I \tag{7.58}$$

and $\mathbf{z}$ is the particular solution given by

$$\mathbf{z}(t) = \mathbf{M}(t)\int_0^t \mathbf{M}(\tau)^{-1}\mathbf{b}(\tau)d\tau \quad ; \quad \mathbf{z}(0) = 0 \tag{7.59}$$

Using (7.57), at $t = T$,

$$\mathbf{x}(T) = \mathbf{M}(T)\mathbf{x}(0) + \mathbf{z}(T) \tag{7.60}$$

Substituting (7.60) into the boundary conditions (7.56),

$$\begin{aligned} Q_a\mathbf{x}(0) + Q_b\left(\mathbf{M}(T)\mathbf{x}(0) + \mathbf{z}(T)\right) &= \mathbf{p} \\ \left(Q_a + Q_b\mathbf{M}(T)\right)\mathbf{x}(0) &= \mathbf{p} - Q_b\mathbf{z}(T) \\ \mathbf{x}(0) &= \left(Q_a + Q_b\mathbf{M}(T)\right)^{-1}\left(\mathbf{p} - Q_b\mathbf{z}(T)\right) \end{aligned} \tag{7.61}$$

In case $\mathbf{M}(T)$ and $\mathbf{z}(T)$ can be evaluated analytically, $\mathbf{x}(0)$ can be calculated and substituted into (7.57). This would yield an analytical solution for $\mathbf{x}(t)$. However, in other cases, one may have to rely on numerical methods to estimate $\mathbf{M}(T)$ and $\mathbf{z}(T)$.

To estimate $\mathbf{M}(T)$, apply the properties given in (7.58). Using IVP solvers such as Runge-Kutta methods, Adams methods, or BDF methods, while setting initial conditions to $\mathbf{e}_j$ (the jth column of the identity matrix), we could integrate the homogenous part of the differential equation until $t = T$. This yields $\mathbf{m}_j(T)$, that is, the jth column of $\mathbf{M}(T)$. Thus

$$\textbf{IVS}\left[\left(\frac{d}{dt}\mathbf{m}_1 = \mathbf{A}(t)\mathbf{m}_1\right),\ \mathbf{m}_1(0) = \begin{pmatrix} 1 \\ 0 \\ \vdots \\ 0 \\ 0 \end{pmatrix}\right] \quad \longrightarrow \quad \mathbf{m}_1(t = T)$$

$$\vdots$$

$$\textbf{IVS}\left[\left(\frac{d}{dt}\mathbf{m}_n = \mathbf{A}(t)\mathbf{m}_n\right),\ \mathbf{m}_n(0) = \begin{pmatrix} 0 \\ 0 \\ \vdots \\ 0 \\ 1 \end{pmatrix}\right] \quad \longrightarrow \quad \mathbf{m}_n(t = T)$$

where **IVS** denotes any initial value solver such as Runge-Kutta methods, Adams methods, or BDF methods. Afterward, we can combine the results to form the matrizant at $t = T$, that is,

$$\mathbf{M}(T) = \left(\mathbf{m}_1(T) \;\middle|\; \mathbf{m}_2(T) \;\middle|\; \cdots \;\middle|\; \mathbf{m}_n(T) \right)$$

Likewise, to estimate $\mathbf{z}(T)$, apply the properties given in (7.59). Using a zero initial condition, we could integrate the nonhomogeneous part of the differential equation until $t = T$ to obtain $\mathbf{z}(T)$, that is,

$$\textbf{IVS}\left[\left(\frac{d}{dt}\mathbf{z} = \mathbf{A}(t)\mathbf{z} + \mathbf{b}(t)\right),\ \mathbf{x}_0 = \begin{pmatrix} 0 \\ \vdots \\ 0 \end{pmatrix}\right] \quad \longrightarrow \quad \mathbf{z}(t = T)$$

Once $\mathbf{M}(T)$ and $\mathbf{z}(T)$ have been estimated, equation (7.61) can be used to determine the required initial condition, $\mathbf{x}(0)$. Finally, we can use the initial value solvers once more and integrate the (nonhomogenous) differential equation (7.55) using $\mathbf{x}(0)$.

EXAMPLE 7.8. Consider the following linear differential system of equations:

$$\frac{d}{dt}\mathbf{x} = \begin{pmatrix} -5e^{-3t} & 1 \\ -e^{-t/2} & -1 \end{pmatrix}\mathbf{x} + \begin{pmatrix} 1 \\ 0 \end{pmatrix}$$

subject to $x_1(0) = 0$ and $x_1(4) - x_2(4) = -4$. To put the boundary conditions into the form of (7.56), we have

$$Q_a = \begin{pmatrix} 1 & 0 \\ 0 & 0 \end{pmatrix} \quad ; \quad Q_b = \begin{pmatrix} 0 & 0 \\ 1 & -1 \end{pmatrix} \quad ; \quad \mathbf{p} = \begin{pmatrix} 0 \\ -4 \end{pmatrix}$$

Following the procedure, we find

$$M(4) = \begin{pmatrix} 4.4240 \times 10^{-2} & 2.9052 \times 10^{-1} \\ -2.7606 \times 10^{-3} & -6.0691 \times 10^{-2} \end{pmatrix} \quad \text{and} \quad \mathbf{z}(4) = \begin{pmatrix} 2.4966 \times 10^{0} \\ -3.8973 \times 10^{-1} \end{pmatrix}$$

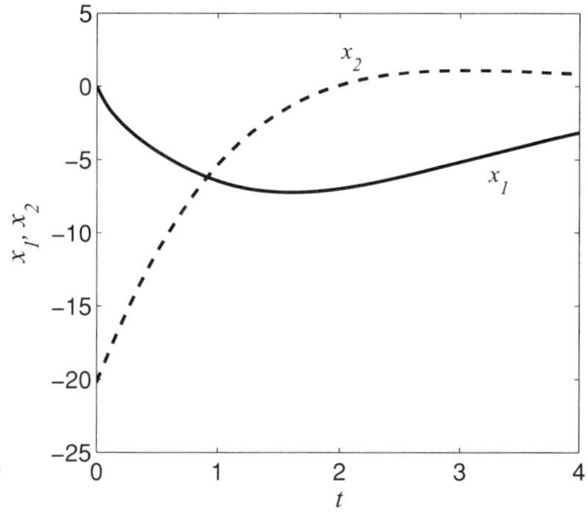

Figure 7.11. Plot of $\mathbf{x}(t)$ for the linear boundary value problem of Example 7.8.

which can now be applied to solve for the required $\mathbf{x}(0)$,

$$\mathbf{x}(0) = \left(Q_a + Q_b\mathbf{M}(T)\right)^{-1}\left(\mathbf{p} - Q_b\mathbf{z}(T)\right) = \begin{pmatrix} 0 \\ 20.182 \end{pmatrix}$$

A plot of $\mathbf{x}(t)$ is shown in Figure 7.11, which shows that the boundary conditions are indeed satisfied.

The extension to nonlinear boundary value problems can be handled by using Newton-type algorithms. Details of these approaches are given in Section G.7.

7.6 Differential Algebraic Equations

There are several physical applications that result in a set of differential equations together with a set of nonlinear algebraic equations. For instance, in the field of process control, the dynamics of the system could be described generally by

$$\mathbf{f}_{\text{sys}}\left(t, \mathbf{y}, \frac{d}{dt}\mathbf{y}, \mathbf{u}\right) = 0 \tag{7.62}$$

where t, $\mathbf{y}$, and $\mathbf{u}$ are time, state variable, and control variable, respectively, and $\mathbf{f}_{\text{sys}}$ and $\mathbf{y}$ have the same size. A feedforward control strategy specifies $\mathbf{u} = \mathbf{u}(t)$ independent of $\mathbf{y}$, and (7.62) is a nonlinear ODE system. Conversely, a feedback strategy could be implemented by specifying a control law given by

$$\mathbf{f}_{\text{ctrl}}\left(t, \mathbf{y}, \mathbf{u}\right) = 0 \tag{7.63}$$

Taken together, (7.62) and (7.63) become a set of equations of the form

$$\begin{pmatrix} \mathbf{f}_{\text{sys}}\left(t, \mathbf{y}, \frac{d}{dt}\mathbf{y}, \mathbf{u}\right) \\ \\ \mathbf{f}_{\text{ctrl}}\left(t, \mathbf{y}, \mathbf{u}\right) \end{pmatrix} = \mathbf{f}\left(t, \mathbf{z}, \frac{d}{dt}\mathbf{z}\right) = 0 \tag{7.64}$$

where $\mathbf{z} = (\mathbf{y}, \mathbf{u})^T$ is an extended state vector. The combined system in (7.64) is an example of a **differential algebraic equation** (DAE), and it is a generalized formulation of ordinary differential equations (ODE).

In some cases, as in (7.64), the DAE system takes the form known as the **semi-explicit DAE** form given by

$$\begin{aligned} \frac{d}{dt}\mathbf{y} &= \mathbf{f}_1\left(t, \mathbf{y}, \mathbf{u}\right) \\ 0 &= \mathbf{f}_2\left(t, \mathbf{y}, \mathbf{u}\right) \end{aligned} \tag{7.65}$$

Assuming we can solve for $\mathbf{u}$ explicitly from $\mathbf{f}_2$, that is,

$$0 = \mathbf{f}_2\left(t, \mathbf{y}, \mathbf{u}\right) \quad \rightarrow \quad \mathbf{u} = \mathbf{q}(t, \mathbf{y})$$

The DAE is immediately reduced to an ODE set by substitution, that is,

$$\frac{d}{dt}\mathbf{y} = \mathbf{f}_1\left(t, \mathbf{y}, \mathbf{q}\left(t, \mathbf{y}\right)\right)$$

In several instances, however, finding $\mathbf{u} = \mathbf{q}(t, \mathbf{y})$ explicitly is not easy nor possible. As an alternative approach, we could take the derivative of both equations in (7.65) to yield

$$\begin{aligned} \frac{d^2}{dt^2}\mathbf{y} &= \frac{\partial \mathbf{f}_1}{\partial t} + \frac{\partial \mathbf{f}_1}{\partial \mathbf{y}}\frac{d\mathbf{y}}{dt} + \frac{\partial \mathbf{f}_1}{\partial \mathbf{u}}\frac{d\mathbf{u}}{dt} \\ 0 &= \frac{\partial \mathbf{f}_2}{\partial t} + \frac{\partial \mathbf{f}_2}{\partial \mathbf{y}}\frac{d\mathbf{y}}{dt} + \frac{\partial \mathbf{f}_2}{\partial \mathbf{u}}\frac{d\mathbf{u}}{dt} \end{aligned}$$

which becomes an ODE set if $\partial\mathbf{f}_2/\partial\mathbf{u}$ remains nonsingular for all t, $\mathbf{u}$, and $\mathbf{y}$. This trick of taking derivatives of a strict DAE set can be done repeatedly until it becomes an ODE set. The minimum number of times a differentiation process is needed for this to occur is known as the **index of the DAE**. Thus if $\partial\mathbf{f}_2/\partial\mathbf{u}$ remains nonsingular for the system (7.65), it can be classified as an **index-1 DAE**. Likewise, an ODE set is also known as an **index-0 DAE**.

Several general-purpose DAE solvers are available to solve index-1 DAEs. Most of them are based on the BDF methods. There are also implicit Runge-Kutta methods available. There are specific features of numerical solution of DAEs that distinguish them from that of simple ODEs such as consistency of initial conditions, but we refer to other sources that discuss these issues in great detail.[5] Instead, we simply outline the general approach for either implicit Runge-Kutta and multistep BDF methods.

Assume that the following DAE set is index-1:

$$F\left(t, \mathbf{y}, \frac{d}{dt}\mathbf{y}\right) = 0 \quad \text{subject to } \mathbf{y}\left(t_0\right) = \mathbf{y}_0 \quad \text{and} \quad \frac{d}{dt}\mathbf{y}(0) = \mathbf{z}_0 \tag{7.66}$$

The BDF method (generated in Section G.4) results in the following nonlinear equation that needs to be solved for $\mathbf{y}_{k+1}$:

$$F\left(t_{k+1}, \mathbf{y}_{k+1}, \frac{1}{h_k}\sum_{i=-1}^{m} \gamma_{(i|k)}\, \mathbf{y}_{k-i} \right) = 0 \tag{7.67}$$

where the coefficients $\gamma_{(\ell|k)}$ are given in (G.22). Because the BDF method is a multistep method, one needs to have values to initiate the recursion. One approach

[5] See, for example, K. E. Brenan, S. L. Campbell, and L. R. Petzold, *Numerical Solution of Initial-Value Problems in Differential-Algebraic Equations*, North-Holland, New York, 1989.

is to use one-step BDF, followed by two-step and so on, until the maximum number of steps is reached (which has to be less than 7 for stability). This would also mean that to achieve good accuracy at the start of the BDF method, variable step sizes may be needed.

For s-stage Runge-Kutta methods, the working equation is given by

$$F\left((t_k + a_j h),\ \left(\mathbf{y}_k + \sum_{\ell=1}^{s} b_{j\ell}\underline{\boldsymbol{\delta}}_{k,\ell}\right),\ \underline{\boldsymbol{\delta}}_{k,\ell} \right) = 0 \qquad j = 1, \ldots, s \tag{7.68}$$

$$\mathbf{y}_{k+1} = \mathbf{y}_k + \sum_{j=1}^{s} c_j \underline{\boldsymbol{\delta}}_{k,\ell} \tag{7.69}$$

In most general-purpose DAE programs, such as those found in MATLAB, the form is restricted to the DAEs that can be recast in a **mass matrix form**, that is,

$$M(t, \mathbf{y}) \frac{d}{dt}\mathbf{y} = \mathbf{f}(t, \mathbf{y})$$

where M is known as the mass matrix. If the system is a strict DAE, M will be singular. An example of a DAE problem is given in Exercise **E7.14**.

7.7 EXERCISES

E7.1. Consider the following linear system

$$\frac{d}{dt}\mathbf{x} = \begin{pmatrix} -20 & -5 \\ 10 & -35 \end{pmatrix}\mathbf{x} \qquad \mathbf{x}(0) = \begin{pmatrix} 1 \\ 1 \end{pmatrix}$$

1. Obtain an analytical solution of the system, for example, using techniques in section 6.5.2 or (F.21).
2. Apply the explicit fourth-order Runge-Kutta method and compare the numerical solution with the analytic solution. Use $\Delta t = 0.01, 0.1, 0.2$.
3. Apply the Fehlberg 4/5 embedded Runge-Kutta method (see Section G.5) and compare the numerical solution with the analytic solution. Use $\Delta t = 0.01, 0.1, 0.2$ as the initial time increment, with a tolerance $\epsilon = 10^{-8}$.
4. Apply the backward Euler method and compare the numerical solution with the analytical solution. Use $\Delta t = 0.01, 0.1, 0.2$.

E7.2. Obtain fifth-order versions for Adams-Bashforth, Adams-Moulton, and BDF methods.

E7.3. Consider the following time-varying linear system:

$$\frac{d}{dt}\mathbf{x} = \begin{pmatrix} -0.8 - \cos(0.5t) & 2 \\ e^{-0.1t} & -3 \end{pmatrix}\mathbf{x} \qquad \mathbf{x}(0) = \begin{pmatrix} 1 \\ 1 \end{pmatrix}$$

1. Applying the fourth-order BDF method to this linear system, rewrite (7.40) into an explicit equation, that is, determine matrices $B_0(t)$, $B_1(t)$, $B_2(t)$ and $B_3(t)$ such that

$$\mathbf{x}_{k+1} = B_0(t)\mathbf{x}_k + B_1(t)\mathbf{x}_{k-1} + B_2(t)\mathbf{x}_{k-2} + B_3(t)\mathbf{x}_{k-3}$$

2. Implement the explicit formulas you found in the previous question using $\Delta t = 0.01$ and compare with the numerical results obtained by using Gauss-Legendre IRK method. (Note: The fourth-order BDF method is a multistep method. You can use the fourth-order explicit Runge-Kutta to initialize the first four terms needed by the BDF method.)

E7.4. One example of a stiff differential system is given by the chemical reaction system known as Robertson's kinetics:

$$\begin{aligned} A &\rightarrow B \\ B+B &\rightarrow C+B \\ B+C &\rightarrow A+C \end{aligned}$$

and the kinetic equations are given by

$$\begin{aligned} \frac{dC_A}{dt} &= -0.04C_A + 10^4 C_B C_C \\ \frac{dC_B}{dt} &= 0.04C_A - 10^4 C_B C_C - 3\times 10^7 C_B^2 \\ \frac{dC_C}{dt} &= 3\times 10^7 C_B^2 \end{aligned}$$

Let the initial conditions be: $C_A(0) = 1$, $C_B(0) = 0$ and $C_C(0) = 0$. We want the numerical solution from $t = 0$ to $t = 100$.

1. Use the Gauss-Legendre fourth-order IRK to obtain the numerical solution. (Hint: Use an initial time interval of $\Delta t = 0.01$ and a tolerance of $\epsilon = 1 \times 10^{-8}$).
2. Apply the fixed fourth-order Runge-Kutta method using $\Delta t = 0.1$, 0.01, and 0.001 by simulating from $t = 0$ to $t = 10$. Plot the results to note which time increment yields a stable result.
3. Use the MATLAB solver `ODE45` to solve the system from $t = 0$ to $t = 10$. (Note that the method will appear to "*hang*." What is occuring is that the error correction is choosing a very small time increment to avoid the instability caused by the large range of kinetic rates present in the Robertson system.)

E7.5. We can use the numerical solution (e.g., the fourth-order explicit Runge-Kutta) to estimate the parameters of the process. Because the experimental data are sometimes obtained at time instants that are different from those obtained from the numerical solution, an extra step is needed to insert these time instants. For instance, suppose the numerical solution provided the following time sequence:

$$t_{\text{num}} = [0, 0.1, 0.2, 0.3, 0.4, 0.5]$$

If the data were collected at

$$t_{\text{data}} = [0, 0.15, 0.2, 0.34, 0.48]$$

then the numerical solution would need to fill in the additional time instants to be

$$t^*_{\text{num}} = [0, 0.1, 0.15, 0.2, 0.3, 0.34, 0.4, 0.48, 0.5]$$

Let $\mathbf{x}_{\text{num}}$ and $\mathbf{x}_{\text{data}}$ be the numerical (simulated) values and the data values, respectively, at the time instants given in t_{data}. Let α be the parameters of the system. Then the parameters can be estimated by performing the following optimization:

$$\alpha^{\text{opt}} = \arg\left(\min_{\alpha} \left\| \left(\mathbf{x}_{\text{num}}(\alpha) - \mathbf{x}_{\text{data}}\right) \right\| \right)$$

Table 7.1. *Data for bioprocess*

Time	x_1	x_2	Time	x_1	x_2	Time	x_1	x_2
0.2419	1.0363	0.9146	2.2235	1.3591	0.3111	6.4171	1.5088	0.1661
0.6106	1.1018	0.8117	2.7074	1.4199	0.2456	7.4539	1.5088	0.1661
1.0023	1.1813	0.6573	3.3756	1.4480	0.1942	8.3525	1.5275	0.1661
1.3940	1.2327	0.5497	4.1590	1.4760	0.1801	9.0438	1.5135	0.1614
1.7166	1.3076	0.4515	4.9654	1.4854	0.1708	9.6198	1.5368	0.1614

For the bioprocess given by the equations,

$$\frac{dx_1}{dt} = (\mu - D)\,x_1$$

$$\frac{dx_2}{dt} = D\,(x_{2f} - x_2) - \frac{\mu x_1}{Y}$$

where

$$\mu = \frac{\mu_{\max} x_2}{k_m + x_2}$$

Fix D and x_{2f} to be constants. Use an initial guess of $D = 0.3$ and $x_{2f} = 5.0$. Also, set initial estimates of the parameters to be: $\mu_{\max} = Y = 0.5$ and $k_m = 0.1$. Using the optimization described, use the fourth-order Runge Kutta method to estimate the parameters for the least-squares fit of the data given in Table 7.1. Assume $x_1(0) = x_2(0) = 1.0$.

(Hint: Alternatively, you could use the code suggested in `oderms.m` found in the attached CD-ROM to evaluate the error criterion, and then use minimization routines.)

E7.6. Heun's method is a second-order explicit Runge-Kutta method sometimes written as follows:

$$\tilde{y}_{k+1} = y_k + h\,f\,(t_k, y_k) \tag{7.70}$$

$$y_{k+1} = y_k + \frac{h}{2}\Big(f\,(t_k, y_k) + f\,(t_{k+1}, \tilde{y}_{k+1})\Big) \tag{7.71}$$

1. Construct the Runge-Kutta tableau for Heun's method.
2. Determine the stability region based on the Dalhquist test.

E7.7. Consider the following boundary value problem:

$$\frac{d^2y}{dx^2} + \left(\frac{3+2x}{1+x}\right)\frac{dy}{dx} + \left(\frac{2+4x+x^2}{(1+x)^2}\right)y = 0$$

subject to

$$y(0) = 1 \qquad y(2) = 0.5$$

Use the shooting method and obtain the plot of $y(x)$ from $x = 0$ to $x = 2$.

E7.8. Obtain a closed formula for the determinant of the $N \times N$ matrix A_N given by

$$A_N = \begin{pmatrix} 5 & -2 & & 0 \\ -1 & 5 & \ddots & \\ & \ddots & \ddots & -2 \\ 0 & & -1 & 5 \end{pmatrix}$$

(Hint: Follow the method used in Example 7.5.)

E7.9. Obtain the solution for the two-point boundary value problem given by:

$$\begin{aligned}
\frac{dC_a}{dt} &= -k_1 C_a C_b + k_2 C_c - k_3 C_a C_c \\
\frac{dC_b}{dt} &= -k_1 C_a C_b + k_2 C_c \\
\frac{dC_c}{dt} &= k_1 C_a C_b - k_2 C_c - k_3 C_a C_c \\
\frac{dC_d}{dt} &= k_3 C_a C_c
\end{aligned}$$

subject to the boundary conditions:

$$\begin{aligned}
C_c(10) &= C_d(10) \\
C_d(0) &= 0 \\
C_a(10) &= 0.01 \\
C_b(10) &= 0.08
\end{aligned}$$

where $k_1 = 12$, $k_2 = 0.1$ and $k_3 = 2.5$. (Hint: Use the initial guess of $C_a(0) = C_b(0) = C_c(0) = 0.3$ and $C_d = 0$)

E7.10. For the fourth-order BDF method given in (7.40),

1. Find the principal root for $\lambda = 0$ and the spurious roots.
2. Determine the stability region for the Dahlquist test case.
3. Show that this method is stiffly stable by finding the value of D that would make it D-stable. Is it also A_α-stable?

E7.11. Show that the seventh-order BDF method is no longer stable for the Dahlquist test case with $\lambda = 0$. (Hint: Solve for the roots and show that some of the spurious roots are greater than 1 when $z = \lambda h = 0$.)

E7.12. Another class of an mth-order implicit multistep method is the **Milne-Simpson method** for the equation $dy/dt = f(y)$, which is given by the specific form

$$y_{n+1} = y_{n-1} + h \sum_{j=-1}^{m-1} b_j f_{n-j}$$

where $f_k = f(y_k)$. Use the same approach as in Section 7.3 (and Section G.3) of using $f(y) = y$ to generate the necessary conditions to show that the matrix equation for the parameter values of b_j that is given by

$$\begin{pmatrix}
1 & 1 & 1 & \cdots & 1 \\
-1 & 0 & 1 & \cdots & (m-1) \\
\vdots & \vdots & \vdots & \ddots & \vdots \\
(-1)^m & 0 & 1 & \cdots & (m-1)^m
\end{pmatrix}
\begin{pmatrix} b_{-1} \\ b_0 \\ \vdots \\ b_{m-1} \end{pmatrix}
=
\begin{pmatrix} \gamma_1 \\ \gamma_2 \\ \vdots \\ \gamma_{m+1} \end{pmatrix}$$

where

$$\gamma_k = \begin{cases} 0 & \text{if } k \text{ is even} \\ \dfrac{2}{k} & \text{if } k \text{ is odd} \end{cases}$$

Thus show that for $m = 4$, we have

$$y_{n+1} = y_{n-1} + \frac{h}{90}\left(29f_{n+1} + 124f_n + 24f_{n-1} + 4f_{n-2} - f_{n-3}\right)$$

E7.13. Determine the conditions under which the linear two-point boundary conditions have a unique solution. Under what conditions will the two-point boundary value problems have no solution?

E7.14. Consider the following set of DAE equations describing the equilibrium flow rates for a catalytic reaction, $A \rightarrow B + C$, in a one-dimensional flow through a packed column, assuming ideal gas behavior,[6]

$$\begin{aligned} -\frac{d\dot{n}_A}{ds} &= \frac{d\dot{n}_B}{ds} = \frac{d\dot{n}_C}{ds} = k_r \cdot \left(c_{As} - c_{Bs}\left(\frac{P}{K_s}\right)\left(\frac{\dot{n}_C}{\dot{n}_{tot}}\right)\right) \\ \frac{dP}{ds} &= -\frac{\psi}{\dot{V}_{in}}\frac{\dot{n}_{tot}RT}{P} \\ 0 &= c_{As} - K_{aA}P\left(\frac{\dot{n}_A}{\dot{n}_{tot}}\right)(c_{tot} - c_{As} - c_{Bs}) \\ 0 &= c_{Bs} - K_{aB}P\left(\frac{\dot{n}_B}{\dot{n}_{tot}}\right)(c_{tot} - c_{As} - c_{Bs}) \end{aligned}$$

where the states $\dot{n}_A$, $\dot{n}_B$, and P are the molar flow rate of A, molar flow rate of B, and pressure, respectively, at a point where the total weight of catalyst away from the inlet is given by s. The total molar flow is $\dot{n}_{tot} = \dot{n}_A + \dot{n}_B + \dot{n}_C + \dot{n}_G$, where $\dot{n}_G$ is the molar flow rate of an inert gas. The variable $\dot{V}_{in}$ is the volumetric flow rate at the inlet. The other variables c_{As} and c_{Bs} are the adsorbed concentration of A and B per unit mass of catalyst. The total number of catalytic sites is assumed to be a constant c_{tot}. The parameter ψ results from using the Ergun equation; k_r is the specific rate of reaction at the sites; and K_s, K_{aA}, and K_{aB} are equilibrium constants for the reaction, adsorption of A, and adsorption of B, respectively. Finally, $R = 8.3145\ \text{Pa}\cdot\text{m}^3\,(\text{mole K})^{-1}$ is the universal gas constant.

Assume the following parameter set and inlet conditions:

$$\psi = 11.66\,\frac{\text{Pa}}{\text{kg Cat}} \qquad c_{tot} = 10^{-3}\,\frac{\text{moles sites}}{\text{kg Cat}} \qquad k_r = 30\,\frac{\text{moles}}{\text{moles sites}}\text{sec}^{-1}$$

$$K_{aA} = 4.2 \times 10^{-5}\ \text{Pa}^{-1} \qquad K_{aB} = 2.5 \times 10^{-5}\ \text{Pa}^{-1} \qquad K_s = 9.12 \times 10^5\ \text{Pa}$$

$$T_{in} = T = 373\ \text{K} \qquad \dot{V}_{in} = 0.001\,\frac{\text{m}^3}{\text{sec}} \qquad P_{in} = 1\ \text{atm}$$

$$p_{A,in} = 0.1\ \text{atm} \qquad p_{B,in} = p_{C,in} = 0\ \text{atm}$$

[6] Based on an example from K. Beers, *Numerical Methods for Chemical Engineering*, Cambridge University Press, Cambridge, UK, 2007.

Based on ideal gas behavior, the initial conditions are:

$$P(0) = P_{in} \qquad \dot{n}_B(0) = \dot{n}_C(0) = 0$$

$$\dot{n}_A(0) = \frac{p_{A,in}\dot{V}_{in}}{RT_{in}} \qquad \dot{n}_G(0) = \frac{(P_{in} - p_{A,in})\,\dot{V}_{in}}{RT_{in}} = \dot{n}_G$$

1. Reduce the DAE to an ODE by first solving for c_{As} and c_{Bs} in terms of $\dot{n}_{tot}$ and P, and then substitute the results into the differential equations. Solve the ODE using the available ODE solvers (e.g., in MATLAB).
2. By converting it to a DAE using the mass matrix form, solve the DAE directly using available DAE solvers (e.g., in MATLAB). Compare the solution with that obtained from the previous method.

8 Qualitative Analysis of Ordinary Differential Equations

In some applications, the qualitative behavior of the solution, rather than the explicit solution, is of interest. For instance, one could be interested in the determination of whether operating at an equilibrium point is stable or not. In most cases, we may want to see how the different solutions together form a portrait of the behavior around particular neighborhoods of interest. The portraits can show how different points such as sources, sinks, or saddles are interacting to affect neighboring solutions. For most scientific applications, a better understanding of a process requires the larger portrait, including how they would change with variations in critical parameters.

We begin this chapter with a brief summary on the existence and uniqueness of solutions to differential equations. Then we define and discuss the equilibrium points of autonomous sets of differential equations, because these points determine the sinks, sources, or saddles in the solution domain. Next, we explain some of the technical terms, such as integral curves, flows, and trajectories, which are used to define different types of stability around equilibrium points. Specifically, we have Lyapunov stability, quasi-asymptotic stability, and asymptotic stability.

We then briefly investigate the various types of behavior available for a linear second-order system, $d\mathbf{x}/dt = A\mathbf{x}, A[=]2 \times 2$, for example, nodes, focus, and centers. Using the tools provided in previous chapters, we end up with a convenient map that relates the different types of behavior, stable or unstable, to the trace and determinant of A.

Afterward, we discuss the use of linearization to assess the type of stability around the equilibrium points. However, this approach only applies to equilibrium points whose linearized eigenvalues have real parts that are non-zero. For the rest of the cases, we turn to the use of Lyapunov functions. These are functions that are often related to system energy, yielding a sufficient condition for asymptotic stability. The main issue with Lyapunov functions, however powerful and general, is that there are no general guaranteed methods for finding them.

Next, we move our attention to limit cycles. These are special periodic trajectories that are isolated; that is, points nearby are ultimately trapped in the cycle. Some important oscillators such as van der Pol equations exhibit this behavior. Two theorems, namely Bendixson and Poincaré -Bendixson's theorems, are available for existence (or nonexistence) of limit cycle in a given region. Another important tool is the Poincaré map, which transforms the analysis to discrete transition maps. We explore the use of Poincaré maps together with Lyapunov analysis to show the existence and uniqueness of a limit cycle for a class of nonlinear systems known as the Liènard system. We also include a discussion of nonlinear centers, because these are also periodic trajectories but are not isolated; thus they are not limit cycles.

A brief discussion on bifurcation analysis is also included in Section H.1 as an appendix. These analyses are aimed at how the phase portraits (i.e., collective behavior of the system) are affected, as some of the parameters are varied. It could mean the addition or loss of equilibrium points or limit cycles, as well as changes in their stabilities.

Qualitative analysis of dynamical systems encompasses many other tools and topics that we do not discuss, such as nonautonomous systems and chaos.

8.1 Existence and Uniqueness

Before we characterize the solutions around neighborhoods of initial points $\mathbf{x}_0$, we need the conditions for the existence and uniqueness of solutions of a differential system. These conditions are given in the following theorem:

THEOREM 8.1. *For a system given by*

$$\frac{d\mathbf{x}}{dt} = \mathbf{f}(t, \mathbf{x})$$

If $\mathbf{f}$ *is continuous and continuously differentiable in* t *and* $\mathbf{x}$ *around* $t = t_0$ *and* $\mathbf{x} = \mathbf{x}_0$, *then there exists a unique solution* $\mathbf{x}(t)$ *in a neighborhood around* $t = t_0$ *and* $\mathbf{x} = \mathbf{x}_0$ *where the conditions of continuity are satisfied.*

This theorem is well established.[1] We can sketch the proof as follows:

1. The conditions for the continuity of the partial derivatives $\partial\mathbf{f}/\partial t$ and $\partial\mathbf{f}/\partial\mathbf{x}$, together with the fundamental theorems of calculus, imply that for every $(t, \mathbf{x}) \in R$, where R is a region where $\mathbf{f}(t, \mathbf{x})$ is continuous and continuously differentiable, then there is a constant $K > 0$ such that

$$\left\|\mathbf{f}(t, \hat{\mathbf{x}}) - \mathbf{f}(t, \mathbf{x})\right\| \le K \left\|\hat{\mathbf{x}} - \mathbf{x}\right\| \qquad \text{where } \hat{\mathbf{x}}, \mathbf{x} \in R$$

   These conditions are also known as **Lipschitz conditions**.

2. To establish the existence of a solution, Picard's iterative solution is used to find the kth approximate solution, $\mathbf{x}_k$

$$\mathbf{x}_k(t) = \mathbf{x}_0 + \int_0^t \mathbf{f}(\tau, \mathbf{x}_{k-1})\, d\tau$$

   which yields

$$\mathbf{x}_k(t) - \mathbf{x}_{k-1}(t) = \int_0^t \left[\mathbf{f}(\tau, \mathbf{x}_{k-1}) - \mathbf{f}(\tau, \mathbf{x}_{k-2})\right] d\tau$$

   The Lipschitz conditions are then used to show that Picard's iteration is convergent to the solution $\mathbf{x}(t)$, thus showing the existence of a solution.

[1] See, for example, A. C. King, J. Billingham, and S. R. Otto, *Differential Equations: Linear, Nonlinear, Ordinary and Partial*, Cambridge University Press, UK, 2003.

3. To establish uniqueness, let $\mathbf{y}(t)$ and $\mathbf{z}(t)$ be two solutions, that is,

$$\mathbf{y}(t) = \mathbf{x}_0 + \int_{t_0}^{t} \mathbf{f}(s, \mathbf{y})\, ds \quad \text{and} \quad \mathbf{z}(t) = \mathbf{x}_0 + \int_{t_0}^{t} \mathbf{f}(s, \mathbf{z})\, ds$$

or

$$\mathbf{y}(t) - \mathbf{z}(t) = \int_{t_0}^{t} [\mathbf{f}(s, \mathbf{y}) - \mathbf{f}(s, \mathbf{z})]\, ds$$

Taking the norm of the left-hand side and applying the Lipschitz condition, we get

$$\left\| \mathbf{y}(t) - \mathbf{z}(t) \right\| \leq \int_{t_0}^{t} \left\| \mathbf{f}(s, \mathbf{y}) - \mathbf{f}(s, \mathbf{z}) \right\| ds \leq K \int_{t_0}^{t} \left\| \mathbf{y}(s) - \mathbf{z}(s) \right\| ds$$

Finally, another lemma known as **Gronwall's inequality** is used. The lemma states that given $L \geq 0$ and non-negative $h(t)$ and $g(t)$ for $t \in [a, b]$, then

$$h(t) \leq L + \int_{a}^{t} h(s) g(s) ds$$

implies

$$h(t) \leq L \exp\left(\int_{a}^{t} g(s) ds \right)$$

This lemma can then be applied by setting $h(t) = \left\| \mathbf{y} - \mathbf{z} \right\|$, $g(t) = K$ and $L = 0$ to yield

$$\left\| \mathbf{y}(t) - \mathbf{z}(t) \right\| = 0$$

Thus the conditions of continuity for both $\mathbf{f}$ and its partial derivatives are also sufficient to yield a unique solution.

Although the existence and uniqueness theorem is a local result, that is, in the neighborhood of the initial point $\mathbf{x}_0$, this region can be as large as the conditions allow, for example, in a region where the continuity of both $\mathbf{f}$ and its derivatives are satisfied. A discontinuity in $\mathbf{f}(t, \mathbf{x})$ is necessary for non-uniqueness. A simple example of such a case is given in Exercise **E8.1**. This implies that when one needs to model a dynamic system by patching together different regions, it may be worthwhile to ensure smoothness in $\mathbf{f}$; for example, using cubic splines instead of linear interpolations guarantees unique solutions of the models.

8.2 Autonomous Systems and Equilibrium Points

Recall that if the function $\mathbf{f}(t, \mathbf{x}) = \mathbf{f}(\mathbf{x})$ (i.e., it is not an explicit function of t), the system

$$\frac{d\mathbf{x}}{dt} = \mathbf{f}(\mathbf{x}) \tag{8.1}$$

is **autonomous**. Otherwise it is known as **nonautonomous**. We limit our discussions in this chapter only to autonomous systems.

A point $\mathbf{x}_e$ is an **equilibrium point** (also known as **fixed point** or **stationary point**) of (8.1) if $\mathbf{f}(\mathbf{x}_e) = 0$ for $t \geq 0$. Because $\mathbf{f}(\mathbf{x})$ is generally a set of nonlinear equations, there may be more than one equilibrium point. In some cases, where $\mathbf{x}_e$ has to be real-valued and finite, there may even be no equilibrium points. The presence of multiple, isolated equilibrium points is a special feature of nonlinear systems.[2]

EXAMPLE 8.1. The equilibrium points of the following system:

$$\frac{dx_1}{dt} = x_2 \quad \text{and} \quad \frac{dx_2}{dt} = ax_1^2 + bx_1 + c$$

are given by

$$x_{1e} = \frac{-b \pm \sqrt{b^2 - 4ac}}{2a} \quad \text{and} \quad x_{2e} = 0$$

If $b^2 < 4ac$, the values for x_{1e} are complex numbers. Thus if the $\mathbf{x}$ is constrained to be real-valued, we say that no equilibrium points exist for this case. However, if $b^2 = 4ac$, we see that the only equilibrium point is for $x_{2e} = 0$ and $x_{1e} = -b/(2a)$. Finally, if $b^2 > 4ac$, we have two possible equilibrium points,

$$[\mathbf{x}_e]_1 = \begin{pmatrix} \dfrac{-b - \sqrt{b^2 - 4ac}}{2a} \\ 0 \end{pmatrix} \quad \text{and} \quad [\mathbf{x}_e]_2 = \begin{pmatrix} \dfrac{-b + \sqrt{b^2 - 4ac}}{2a} \\ 0 \end{pmatrix}$$

Note that in most cases, numerical methods such as the Newton method given in Section 2.9 may be needed to find the equilibrium points.

Remark: One of the tricks used in Chapter 7 to handle nonautonomous systems via numerical solvers of initial value problems was to extend the state $\mathbf{x}$ by adding another state $x_{n+1} = t$ to form $\hat{\mathbf{x}}$. This is accompanied by extending $\mathbf{f}$ with $f_{n+1} = 1$ to form $\hat{\mathbf{f}}$. This method might be appealing because one might argue that we get back an autonomous system. However, doing so immediately presents a problem because with $f_{n+1} = 1, \hat{\mathbf{f}} \neq 0$ for all $\hat{\mathbf{x}}$. Thus if one applies the state extension approach to analyze a nonautonomous system, the definition of equilibrium points or other solutions such as periodic solutions will have to be modified to apply strictly to the original state vector $\mathbf{x}$, that is, without the time-variable, $x_{n+1} = t$.

8.3 Integral Curves, Phase Space, Flows, and Trajectories

Assuming that (8.1) can be solved around the neighborhood of initial conditions, $\mathbf{x}_0 \in D$, where D is an open set in $\mathbb{R}^n$, we can represent the solutions of (8.1) by **integral curves**, $\mathbf{C}(\mathrm{t})$ of (8.1), which are simply a set of curves whose tangents are specified by $\mathbf{f}(\mathbf{x})$, that is,

$$\mathbf{C}(t) = \left[x_1(t), \ldots, x_n(t)\right]^T \tag{8.2}$$

and

$$\frac{d}{dt}\mathbf{C}(t) = \begin{pmatrix} f_1(x_1, \ldots, x_N) \\ \vdots \\ f_n(x_1, \ldots, x_N) \end{pmatrix} \tag{8.3}$$

[2] We need the descriptor "isolated" because linear systems can also have multiple equilibrium points but they would not be isolated.

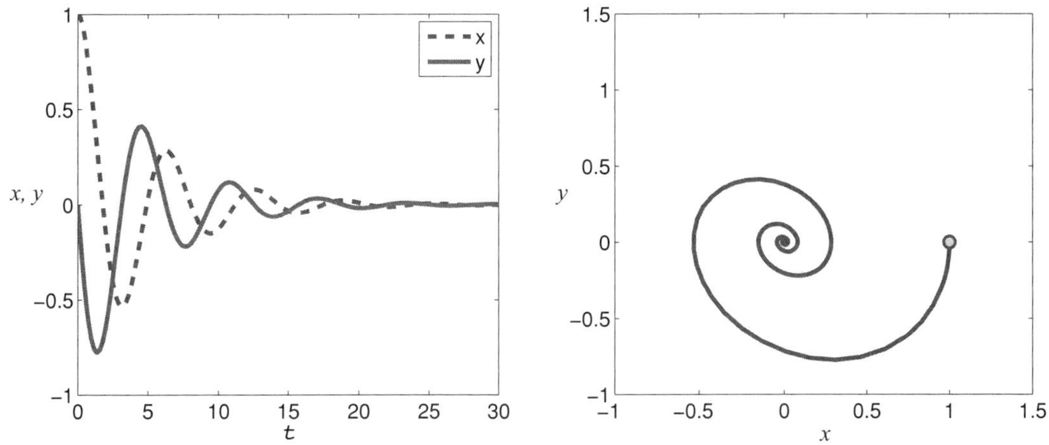

Figure 8.1. On the left is the plot of solutions x and y as functions of t. On the right is the integral curve shown in the phase plane (with the initial point shown as an open circle).

This appears to be redundant because $\mathbf{C}(t)$ is nothing but $\mathbf{x}(t)$ that satisfies (8.1). One reason for creating another descriptor such as "integral curves" is to stress the geometric character of the curves $\mathbf{C}(t)$. For instance, an immediate consequence is that because $\mathbf{C}(t)$ are simply curves parameterized by t, we can analyze and visualize the system behavior in a space involving only the components of $\mathbf{x}$, that is, without the explicit information introduced by t. We refer to the space spanned by the integral curves (i.e., spanned by $x_1, \ldots, x_n$) as the **phase space**, and the analysis of integral curves in this space is also known as **phase-space analysis**. For the special case of a two-dimensional plane, we call it a **phase-plane analysis**.

EXAMPLE 8.2. Consider the autonomous system given by

$$\frac{dx}{dt} = y \quad \text{and} \quad \frac{dy}{dt} = -1.04x - 0.4y$$

There is only a single equilibrium point, which is at the origin. For the initial condition at $\mathbf{x}_0 = (1, 0)^T$, the solutions $x(t)$ and $y(t)$ are plotted together in Figure 8.1 as functions of t. Also included in Figure 8.1 is the phase-plane plot of y versus x of the integral curve starting at $(x, y) = (1, 0)$. When exploring the solutions starting at different initial conditions, the advantage of using phase-space plot becomes clearer – various integral curves can be shown together in one figure.

Another consequence of the concept of integral curves is that the curves can have parameterizations other than t, for example, if $f_1 \neq 0$, the curves can be described by

$$\frac{dx_2}{dx_1} = \frac{f_2}{f_1}, \quad \ldots, \quad \frac{dx_n}{dx_1} = \frac{f_n}{f_1} \tag{8.4}$$

The solution of (8.4) should yield the same curves.

The equations in (8.4), or the original set given in (8.1), may not be easy to integrate analytically. In general, numerical IVP (initial value problem) solvers discussed in Chapter 7 are needed. As a supplementary approach, (8.4) suggests that the slopes (independent of t) can present visual cues to the shapes of the integral curves. This leads us to the use of **direction field** plots. A **direction field**, which we denote $\vec{\underline{\mathbf{d}}}$, is a vector field that gives the slopes of the tangents of the integral curves in phase space. It is not the same as the velocity field, because the vectors in a direction field have the same magnitudes at all points except at the equilibrium points. The components of a direction field for a given $\mathbf{f}(\mathbf{x})$ can be obtained as

$$d_i(\mathbf{x}) = \begin{cases} 0 & \text{if } f_i = 0 \\ \alpha \dfrac{f_i(\mathbf{x})}{\|\mathbf{f}(\mathbf{x})\|} & \text{otherwise} \end{cases} \tag{8.5}$$

where α is a scaling constant chosen based on visual aspects of the field. In the formulation in (8.5), the equilibrium points are associated with points rather than vectors to avoid division by zero.[3] The main advantage of direction fields is that the formulas given in (8.5) are often much easier to evaluate. The direction fields are often evaluated at points specified by rectangular, cylindrical, or spherical meshes.

Furthermore, one could collect the locus of points having the same slopes to form another set of curves known as **isoclines**. A special case of isoclines are those that collect points with slopes that are zero in one of the dimensions, and these are known as **nullclines**. For instance, for the 2D case and rectangular coordinates (x, y), the nullclines are the lines where the x components or y components are zero. Alternatively, for the 2D case under polar coordinates (r, θ), the nullclines are the lines where the slopes are radially inward or outward (i.e., no angular components) or those where the slopes are purely angular (i.e., no radial components).

EXAMPLE 8.3. Take the same system given in Example 8.2,

$$\frac{dx}{dt} = y \quad \text{and} \quad \frac{dy}{dt} = -1.04x - 0.4y$$

Then the direction field at a rectangular grid around the origin is shown in Figure 8.2. Also, there are four nullclines shown in the right plot in Figure 8.2.

In addition to integral curves, we need another concept, known as **flows**. We use flows during our definition of stability of equilibrium points.

Definition 8.1. *The* **flow** *of (8.1) is a mapping* $\varphi : \mathbb{R}^n \times \mathbb{R} \rightarrow \mathbb{R}^n$, *such that*

1. *The derivative with respect to t are given by*

$$\frac{\partial}{\partial t}\varphi(\mathbf{x}, t) = \mathbf{f}(\varphi(\mathbf{x}, t)) \tag{8.6}$$

2. $\varphi(\mathbf{x}, 0) = \mathbf{x}$
3. $\varphi(\mathbf{x}, s + t) = \varphi(\varphi(\mathbf{x}, s), t)$

Essentially, flows are the mechanism by which the path of an integral curve can be traversed. Thus flows specify the forward or backward movements at a

[3] In the terminology of direction fields, equilibrium points are called **singular points**, whereas the rest are called **regular points**.

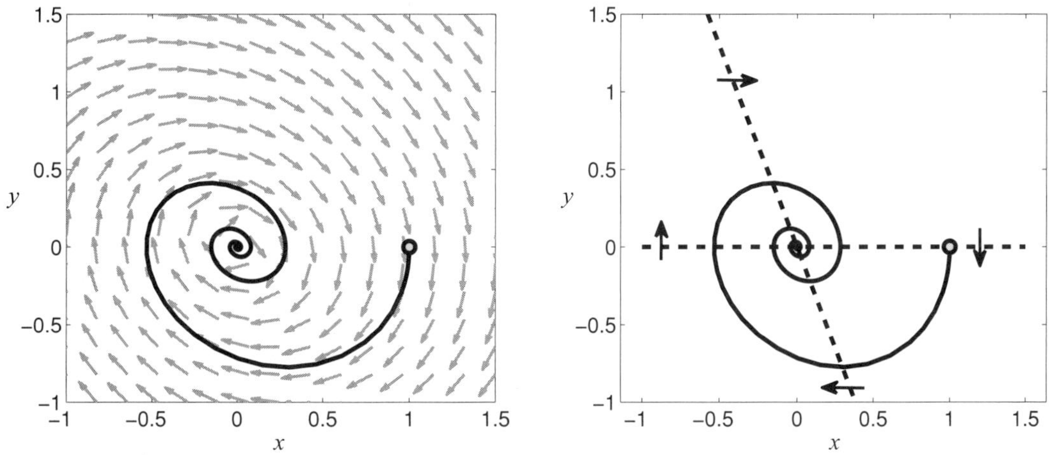

Figure 8.2. The figure on the left shows the direction field for Example 8.3, whereas the figure on the right shows the nullclines under the rectangular coordinates as dotted lines, that is, locus of purely left, purely right, purely up, and purely down slopes.

specified point in the phase space, thereby yielding a definite direction in the movement along the paths. In this respect, integral curves equipped with flows are called **trajectories**.[4]

8.4 Lyapunov and Asymptotic Stability

There are several concepts and types of stability of equilibrium points. These include Lyapunov stability, quasi-asymptotic stability, and asymptotic stability.

Definition 8.2. *An equilibrium point* $\mathbf{x}_e$ *of (8.1) is* **Lyapunov stable** *if for every* $\epsilon > 0$ *there exists a* $\delta > 0$ *such that if* $\|\mathbf{x}_e - \mathbf{y}\| < \delta$ *then* $\|\varphi(\mathbf{x}_e, t) - \varphi(\mathbf{y}, t)\| < \epsilon$ *for* $t \geq 0$*, where* $\varphi(\mathbf{x}, t)$ *is the flow of (8.1) at the point* $\mathbf{x}$.

Lyapunov stability means that if a point $\mathbf{y}$ is close to the equilibrium point $\mathbf{x}_e$, then the flow originating from both $\mathbf{x}$ and $\mathbf{y}$ will remain close, as shown in Figure 8.3. Another type of stability of an equilibrium point $\mathbf{x}_e$ is quasi-asymptotic stability (also known as **attractive property**).

Definition 8.3. *An equilibrium point* $\mathbf{x}_e$ *of (8.1) is* **quasi-asymptotically stable** *if there exists a* $\delta > 0$ *such that if* $|\mathbf{x}_e - \mathbf{y}| < \delta$ *then*

$$\lim_{t\to\infty} \left|\varphi(\mathbf{x}_e, t) - \varphi(\mathbf{y}, t)\right| = 0$$

Unlike Lyapunov stability, a quasi-asymptotic stability does not allow arbitrary specification of $\epsilon > 0$ that bounds $\|\varphi(\mathbf{y}, t) - \mathbf{x}_e\|$. Instead, all it requires is that eventually the distance will converge to zero, as shown in Figure 8.4.

[4] In most texts, there is no distinction between integral curves and trajectories. However, because we have claimed that integral curves can be considered simply as curves in the phase space, they can be reparameterized also by $-t$ for autonomous systems. Thus we suggest that the term "trajectories" is a more appropriate term when the direction of the path as t increases is important.

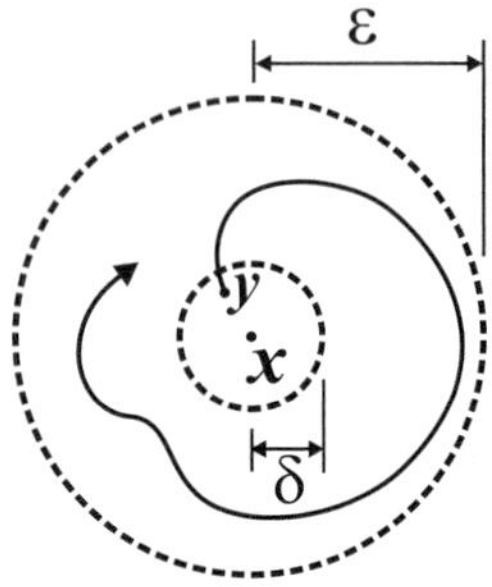

Figure 8.3. Lyapunov stability around the equilibrium point, $\mathbf{x}_e$.

Some equilibrium points can be Lyapunov stable and not quasi-asymptotic (see, e.g., Exercise E8.2), whereas others can be quasi-asymptotically stable but not Lyapunov stable. An example of a system that is quasi-asymptotically stable but not Lyapunov stable is the Vinograd system described in the following example.

EXAMPLE 8.4. A system developed by R. E. Vinograd is given by the following pair of nonlinear autonomous differential equations:

$$\begin{aligned}\frac{dx}{dt} &= f_1(x,y) = \frac{x^2(y-x)+y^5}{(x^2+y^2)\left(1+(x^2+y^2)^2\right)} \\ \frac{dy}{dt} &= f_2(x,y) = \frac{y^2(y-2x)}{(x^2+y^2)\left(1+(x^2+y^2)^2\right)}\end{aligned} \tag{8.7}$$

The only equilibrium point is the origin. A plot of the direction field for the Vinograd system (8.7) is shown in Figure 8.5.

As an alternative, we can represent the system in polar coordinates, that is,

$$\frac{dr}{dt} = \hat{f}_r(r,\theta) \quad \text{and} \quad \frac{d\theta}{dt} = \hat{f}_\theta(r,\theta) \tag{8.8}$$

where

$$\hat{f}_r = \frac{r^3 h_1(\theta) + r h_2(\theta)}{1+r^4} \quad \text{and} \quad \hat{f}_\theta = \frac{r^2 h_3(\theta) + h_4(\theta)}{1+r^4}$$

with

$$\begin{aligned}h_1(\theta) &= \sin\theta\cos\theta(\cos\theta-1)^2(\cos\theta+1)^2 \\ h_2(\theta) &= \sin\theta\cos\theta\left(3\cos^2\theta-2\right)+\left(1-2\cos^2\theta\right) \\ h_3(\theta) &= [(\cos\theta-1)(\cos\theta+1)]^3 \\ h_4(\theta) &= \cos\theta\left(\sin\theta-3\cos\theta+3\cos^3\theta\right)\end{aligned}$$

Note that $\lim_{r\to 0}(\hat{f}_r/r) = h_2(\theta)$ and $\lim_{r\to 0}\hat{f}_\theta = h_4(\theta)$.

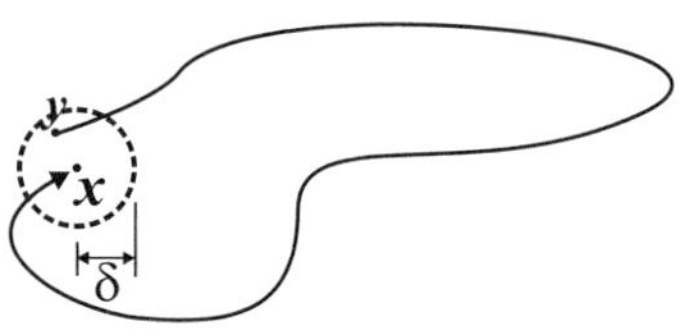

Figure 8.4. Quasi-asymptotic stability around $\mathbf{x}$, an equilibrium point.

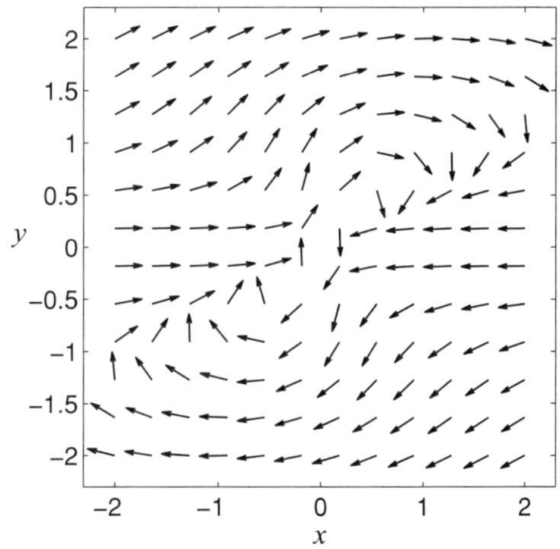

Figure 8.5. The direction field for the Vinograd system given in (8.7).

Using the nullclines of (8.8), we can find different sectors where $\hat{f}_r$ and $\hat{f}_\theta$ change signs, as shown in Figure 8.6. In the figure, we have the shaded region where $\hat{f}_r > 0$, and the unshaded region as $\hat{f}_r < 0$, with $\hat{f}_r = 0$ at the boundaries. However, inside the regions $ABDCA$, $AJKIA$, $AGHA$, and $AEFA$, we have $\hat{f}_\theta > 0$, and outside these regions, $\hat{f}_\theta < 0$. Along the curves ABD, ACD, AIK, and AJK, the tangents to the trajectories are pointing radially outward (i.e., $v_\theta = 0$), whereas along the curves AG, AH, AF, and AE, the tangent to the trajectories is pointing radially inward. This shows that local trajectories northwest of $GACD$ or southeast of $FAIK$ will be repelled from the origin, whereas the other regions will be attracted to the origin.

The trajectories for the Vinograd system starting at different initial points can also be obtained by numerical evaluation of (8.7) using the IVP solvers such as Runge-Kutta methods given in the previous chapter. These are shown in Figure 8.7.[5] The plot in Figure 8.7 is consistent with both Figures 8.5 and 8.6. It shows that initial conditions starting in some locations of the (x, y)-plane will go directly to the origin, whereas starting at other locations may initially diverge away from the origin but ultimately will converge to the origin. This is an example of a case in which the equilibrium point is quasi-asymptotically stable but not Lyapunov stable.

The third type of stability is asymptotic stability.

Definition 8.4. *An equilibrium point* $\mathbf{x}_e$ *of (8.1) is* **asymptotically stable** *if it is both Lyapunov stable and quasi-asymptotic stable.*

Figure 8.8 shows the property of asymptotic stability around $\mathbf{x}$. The equilibrium point for the system given in Example 8.2 is asymptotically stable.

[5] Because the equilibrium point is both unstable and attractive, the numerical round-off errors may produce artificial errors, possibly showing apparent chaotic behavior. A simple fix is to provide smaller error tolerance and also setting the values of the derivative functions $f_1(x, y)$ and $f_2(x, y)$ to be zero if they are within the chosen error tolerance.

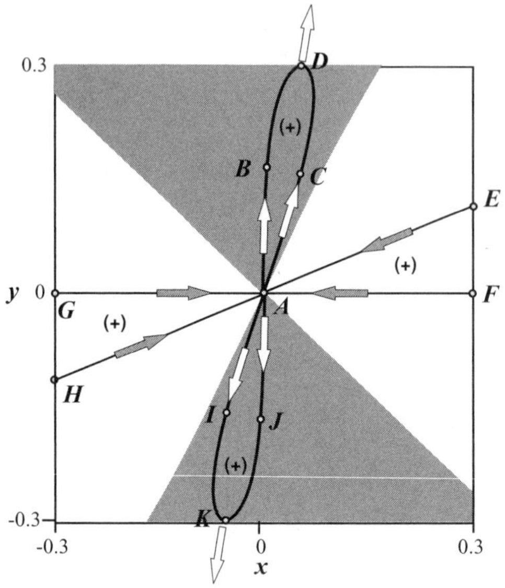

Figure 8.6. Different regions based on whether v_r and v_θ are positive or negative. Shaded region shows where $v_r > 0$, whereas regions marked by (+) show where $v_\theta > 0$.

Asymptotic stability is a much stronger type of stability than either Lyapunov stability or quasi-asymptotic stability, and therefore, it is the more desirable type of stability for several physical processes. It specifies that the flows of neighboring points around the equilibrium point can be bounded, and in addition, they will ultimately reach the equilibrium point.

Often, in engineering systems such as in a manufacturing process, a chosen steady state is often considered to be a target based on optimization between productivity and cost. Controllers are often attached to the system to ensure that these targets will be achieved and will stay put. Thus being able to maintain stability is often the key to sustained operation. Although asymptotic stability is often desired, in some cases, bounding the states to within a satisfactory neighborhood around the target is sufficient for practical purposes (i.e., Lyapunov stability is all that is needed). Indeed, stability analysis is often the main impetus for using a qualitative (or descriptive) analysis of most engineering systems. If necessary, one can

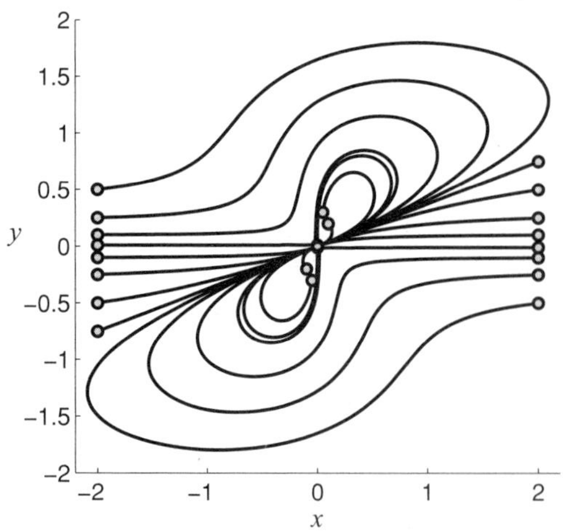

Figure 8.7. Different trajectories of the Vinograd system using different initial points shown by the open circles.

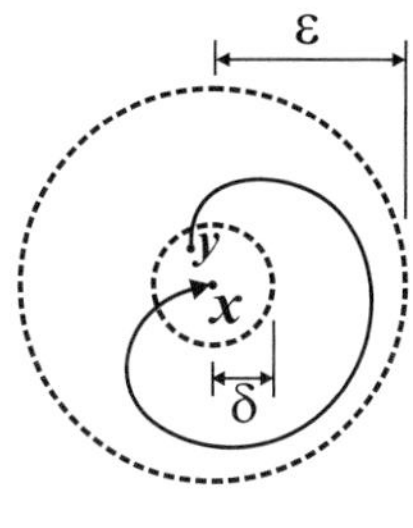

Figure 8.8. Asymptotic stability around $\mathbf{x}$, an equilibrium point.

engineer additional components and procedures such as feedback control to achieve the required behavior.

In the next two sections, we discuss the phase-plane analysis of a linear second order and the linearization approximations of nonlinear systems. The stability analysis of these approaches can be achieved using eigenvalue analysis and is often used to determine asymptotic stability. However, the linearization approach is only limited to situations in which the eigenvalues do not contain zeros or pure imaginary values. We return to Lyapunov and aysmptotic stability of nonlinear systems after these two sections, and we discuss a more general (albeit sometimes elusive) approach of using positive definite functions known as **Lyapunov functions**.

8.5 Phase-Plane Analysis of Linear Second-Order Autonomous Systems

In this section, we obtain different phase-plane portraits of linear autonomous second-order systems given by

$$\frac{d}{dt}\mathbf{x} = A\mathbf{x} = \begin{pmatrix} a_{11} & a_{12} \\ a_{21} & a_{22} \end{pmatrix} \mathbf{x} \tag{8.9}$$

Aside from having a better grasp of the different types of possible behaviors based on A, these results will remain useful for several second-order nonlinear systems because the linearized approximation of their trajectories will be sufficiently close to the original nonlinear forms, at least around the equilibrium points.

The characteristic polynomial, eigenvalues, and eigenvectors can be found in terms of the trace and determinant of A as given in (3.20), (3.21), and (3.22), which are repeated below:

Characteristic polynomial: $\lambda^2 - \mathbf{tr}(A)\lambda + \mathbf{det}(A) = 0$

Eigenvalues: $\lambda = \dfrac{\mathbf{tr}(A) \pm \sqrt{\mathbf{tr}(A)^2 - 4\mathbf{det}(A)}}{2}$

Eigenvectors:

$$\mathbf{v}_{\lambda_*} = \begin{cases} \begin{pmatrix} \dfrac{a_{12}}{\lambda_* - a_{11}} \\ 1 \end{pmatrix} & \text{if } a_{11} \neq \lambda_* \\ \\ \begin{pmatrix} 1 \\ \dfrac{a_{21}}{\lambda_* - a_{22}} \end{pmatrix} & \text{if } a_{22} \neq \lambda_* \end{cases}$$

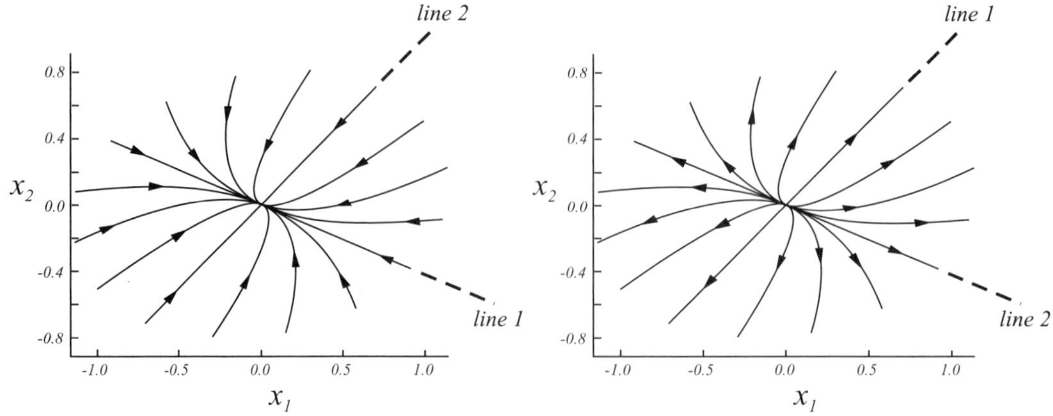

Figure 8.9. The right figure shows the trajectories around a stable node with $0 > \lambda_1 > \lambda_2$. The left figure shows the trajectories around an unstable node with $\lambda_1 > \lambda_2 > 0$.

There are only three possible cases when $\lambda_* = a_{11} = a_{22}$ – either A is strictly diagonal, upper triangular, or lower triangular, all of which have λ in the diagonals. In the strictly diagonal case, the eigenvectors are $\mathbf{e}_1$ and $\mathbf{e}_2$. For the triangular cases, there is only one linearly independent eigenvector: $\mathbf{e}_1$ for the upper triangular case, and $\mathbf{e}_2$ for the lower triangular case.

When A is nonsingular, the origin is the only equilibrium point. If A is singular and of rank 1, then the equilibrium points will lie in a line containing the eigenvector that corresponds to $\lambda = 0$. Lastly, when $A = 0$, the set of all equilibrium points is the whole space; that is, no motion occurs.

Let λ_1 and λ_2 be the eigenvalues of A. If the eigenvalues are both real, then the trajectories are classified as either nodes, stars, improper nodes, saddles, or degenerate. Otherwise, if the eigenvalues are complex-valued, the trajectories are either focuses or centers, where centers occur when the eigenvalues are pure imaginary. We discuss each case next.

1. **Nodes.** When $\mathbf{det}(A) > 0$ and $\mathbf{tr}(A)^2 > 4\,\mathbf{det}(A)$, then both λ_1 and λ_2 are real-valued and have the same sign. Both eigenvalues are negative when $\mathbf{tr}(A) < 0$, and both are positive when $\mathbf{tr}(A) > 0$. In either case, using the diagonalization procedure of Section 6.6, the solution of (8.9) is given by

$$\mathbf{x}(t) = \left(z_{10}e^{\lambda_1 t}\right)\mathbf{v}_1 + \left(z_{20}e^{\lambda_2 t}\right)\mathbf{v}_2 \tag{8.10}$$

where

$$\begin{pmatrix} z_{10} \\ z_{20} \end{pmatrix} = \left(\mathbf{v}_1 \mid \mathbf{v}_2 \right)^{-1} \mathbf{x}_0$$

If both eigenvalues are positive, the equilibrium points are classified as **unstable nodes**. Otherwise, if both eigenvalues are negative, the equilibrium points are **stable nodes**.

Based on (8.10), the trajectories $\mathbf{x}(t)$ are linear combinations of the eigenvectors $\mathbf{v}_1$ and $\mathbf{v}_2$. If the initial point $\mathbf{x}_0$ happens to be along either of the eigenvectors, then the trajectories will travel along the same line that contains the eigenvectors. Otherwise, the trajectories will be "half-U-shaped," where the center of the "U" is along the eigenvector that corresponds to the eigenvalue with the larger absolute value. Typical plots of both stable and unstable nodes are shown in Figure 8.9.

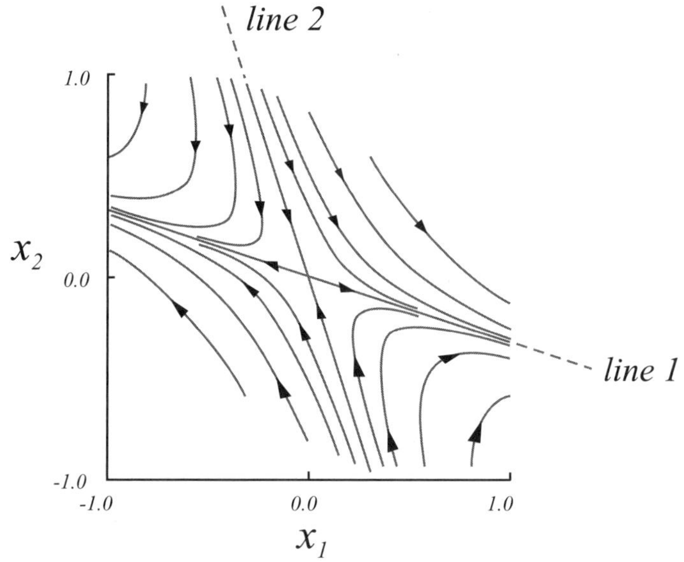

Figure 8.10. The trajectory around a saddle with $\lambda_1 > 0 > \lambda_2$ where line 1 is along $\mathbf{v}_1$ and line 2 is along $\mathbf{v}_2$.

2. **Saddles.** When $\mathbf{det}(A) < 0$, the eigenvalues are both real-valued, but one of them will be positive, whereas the other will be negative. Thus let $\lambda_1 > 0 > \lambda_2$; then, based on (8.10), $\mathbf{x}(t)$ will be a linear combination of an unstable growth along $\mathbf{v}_1$ and a stable decay along $\mathbf{v}_2$. The equilibrium points will then be classified as **saddles**. Typical plots of trajectories surrounding saddles are shown in Figure 8.10.
3. **Stars.** When $A = a\,I$, with $a \neq 0$, both eigenvalues will be equal to a, and the matrix of eigenvectors becomes $V = \left(\ \mathbf{v}_1 \ \middle|\ \mathbf{v}_2\ \right) = I$. This further implies $\mathbf{z} = V\mathbf{x}_0 = \mathbf{x}_0$. Thus (8.10) reduces to

$$\mathbf{x}(t) = e^{at}\mathbf{x}_0 \tag{8.11}$$

and the trajectories will follow along the vector $\mathbf{x}_0$. The equilibrium points are then classified as **stars**, and depending on a, they could be **stable stars** (if $a < 0$) or **unstable stars** (if $a > 0$). Typical plots of trajectories surrounding stars are shown in Figure 8.11.
4. **Improper Nodes.** Suppose the eigenvalues of A are equal to each other, that is, $\lambda_1 = \lambda_2 = \lambda \neq 0$. Using the method of finite sums (cf. 6.14) to solve this case with repeated roots, the solution is given by

$$\mathbf{x}(t) = e^{\lambda t}\left[I + t\left(A - \lambda I\right)\right]\mathbf{x}_0 \tag{8.12}$$

Note that if $A = \lambda I$, we get back the equation of the trajectories around that of a star node, as given in (8.11). Thus add another condition that A is not diagonal. In this case, there will be only one eigenvector, which is given by

$$\mathbf{v} = \begin{pmatrix} a_{12} \\ \lambda - a_{11} \end{pmatrix} \qquad \text{or} \qquad \mathbf{v} = \begin{pmatrix} \lambda - a_{22} \\ a_{21} \end{pmatrix}$$

whichever is nontrivial.

If the initial point $\mathbf{x}_0 = \alpha\mathbf{v}$ (i.e., it lies along the line containing $\mathbf{v}$), then the trajectories will travel along that line because $(A - \lambda I)\mathbf{v} = 0$ and (8.12) becomes $\mathbf{x}(t) = \alpha e^{\lambda}\mathbf{v}$. If $\mathbf{x}_0$ is outside of this line, the trajectories will be curved either

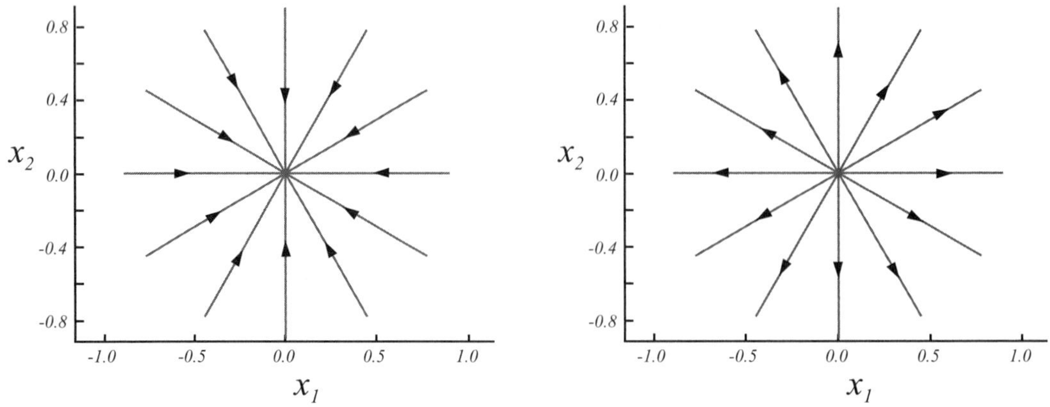

(a) Stable Star *(b) Unstable Star*

Figure 8.11. The trajectories surrounding (a) stable stars and (b) unstable stars.

with a **half-S-shape** (S-type for short) or with a **reverse-half-S-shape** (Z-type for short). In either case, the equilibrium points for this case are classified as **improper nodes**. Again, stability will depend on λ: stable if $\lambda < 0$ and unstable if $\lambda > 0$. Typical plots of the trajectories surrounding both types of stable improper nodes are shown in Figure 8.12.

To determine whether the improper node is S-type or Z-type, one can show (see, e.g., Exercise **E8.5**) that the following conditions can be used:

$$\begin{array}{llll} S\text{-type:} & \text{If} \quad a_{12}\lambda > 0 & ; & \text{as long as} \quad a_{12} \neq 0 \\ & \text{or} \quad a_{21}\lambda < 0 & ; & \text{as long as} \quad a_{21} \neq 0 \\ \\ Z\text{-type:} & \text{If} \quad a_{12}\lambda < 0 & ; & \text{as long as} \quad a_{12} \neq 0 \\ & \text{or} \quad a_{21}\lambda > 0 & ; & \text{as long as} \quad a_{21} \neq 0 \end{array} \tag{8.13}$$

5. **<u>Focus and Centers.</u>** When $\mathbf{tr}(A) < 4\,\mathbf{det}(A)$, the eigenvalues of A are a complex conjugate pair given by $\lambda_1 = \rho + i\omega$ and $\lambda_2 = \rho - i\omega$, where

$$\begin{aligned} \rho &= \frac{\mathbf{tr}(A)}{2} \\ \omega &= \frac{1}{2}\sqrt{4\,\mathbf{det}(A) - \mathbf{tr}(A)^2} \end{aligned} \tag{8.14}$$

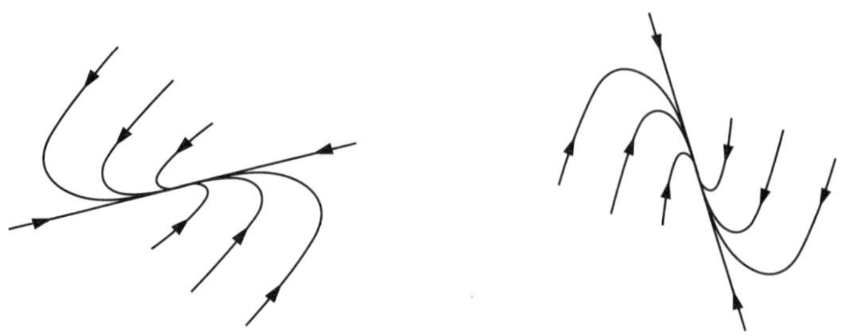

(a) S-Type Improper Node *(b) Z-Type Improper Node*

Figure 8.12. Trajectories surrounding stable (a) S-type and (b) Z-type improper nodes.

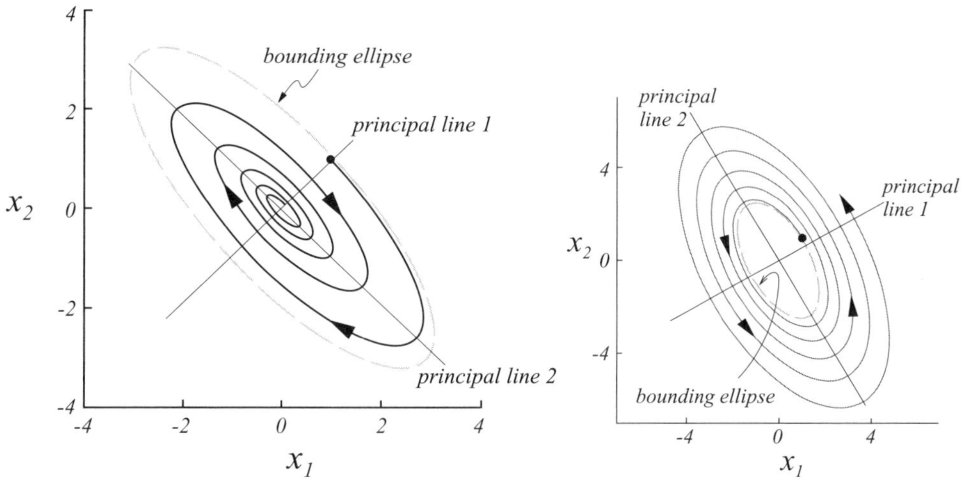

Figure 8.13. The left figure shows a stable focus and the right figure shows an unstable focus. Included in both figures are the bounding ellipses. The principal lines are the polar nullclines, where the points of these lines have no radial components.

Alternatively, a necessary and sufficient condition for eigenvalues to have nonzero imaginary parts is $\left[(a_{11} - a_{22})^2 + 4a_{12}a_{21} < 0\right]$.

Based on (8.10), the solution can be simplified in terms of trigonometric functions to be

$$\mathbf{x}(t) = e^{\rho t}\left[\cos(\omega t)\, I + \frac{\sin(\omega t)}{2\omega}\begin{pmatrix} a_{11} - a_{22} & 2a_{12} \\ 2a_{21} & a_{22} - a_{11} \end{pmatrix}\right]\mathbf{x}_0 \qquad (8.15)$$

When $\rho = 0$, the responses are periodic, and the trajectories in the phase plane will be ellipses centered at the equilibrium point, and thus the equilibrium point is known as a **center**. However, if $\rho \neq 0$, the trajectories will be elliptical spirals, and the equilibrium point is known as a **focus**. When $\rho < 0$, the spiral moves toward the equilibrium point, which is then called a **stable focus**, and when $\rho > 0$, the spiral moves away from the equilibrium point, which is then called an **unstable focus**.

Two more pieces of information can be determined from A: the direction of the rotation and the bounding ellipse. Surprisingly, the rotation depends only on the sign of a_{12}. It can be shown that the trajectories around centers and focuses will be clockwise if $a_{12} > 0$ and counterclockwise if $a_{12} < 0$ (see Exercise **E8.4**). The bounding ellipse of a stable (unstable) focus is the minimal (maximal) ellipse that contains the initial point of which the trajectory will not cross again. The points of the bounding ellipse $\bar{\mathbf{x}}$ can be found as

$$\bar{\mathbf{x}}(\psi) = \left[\cos(\psi)\; I + \sin(\psi)\frac{1}{\sqrt{\mathbf{det}(A)}}A\right]\mathbf{x}_0 \quad ; \quad 0 \leq \psi \leq 2\pi \qquad (8.16)$$

Typical plots of the trajectories surrounding stable and unstable focuses are shown in Figure 8.13, whereas the trajectories surrounding centers are shown in Figure 8.14.

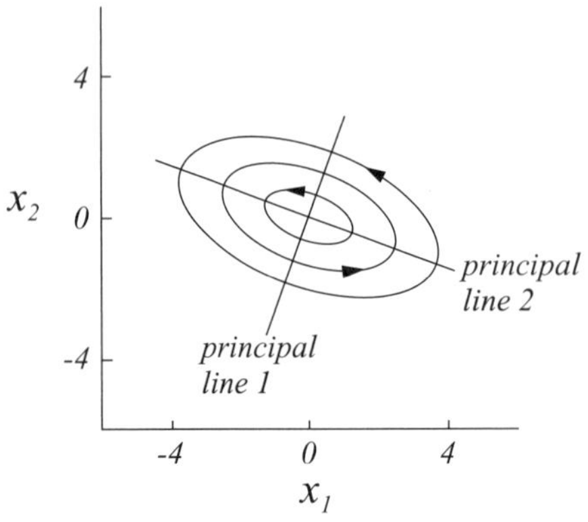

Figure 8.14. Trajectories surround a center.

6. **Degenerate Points.** When one or both eigenvalues of A are zero, there will be more than one equilibrium point. Both these cases are classified as **degenerate points** or **nonisolated equilibrium**.

   Let $\lambda_1 = 0$ and $\lambda_2 \neq 0$; then (8.10) will reduce to

$$\mathbf{x} = z_{01}\mathbf{v}_1 + z_{02}e^{\lambda_2 t}\mathbf{v}_2 \tag{8.17}$$

This means that when $\mathbf{x}_0 = \alpha\mathbf{v}_1$ then $\mathbf{x}(t) = \alpha\mathbf{v}_1$; that is, $\mathbf{v}_1$ will lie in the line that contains all the (nonisolated, non-unique) equilibrium points. Outside this line, the trajectories are parallel to $\mathbf{v}_2$, that is, an affine operation on $\mathbf{v}_2$. Again, the equilibrium points will be stable if $\lambda_2 < 0$ and unstable if $\lambda_2 > 0$. Typical plots of trajectories surrounding degenerate equilibrium points are shown in Figure 8.15.

If both eigenvalues happen to be equal to zero, then we must have both the trace and determinant of A to be zero. This implies $a_{11} = -a_{22}$ and $a_{12}a_{21} = -a_{11}^2$. From (8.12), we have

$$\mathbf{x}(t) = (I + At)\,\mathbf{x}_0$$

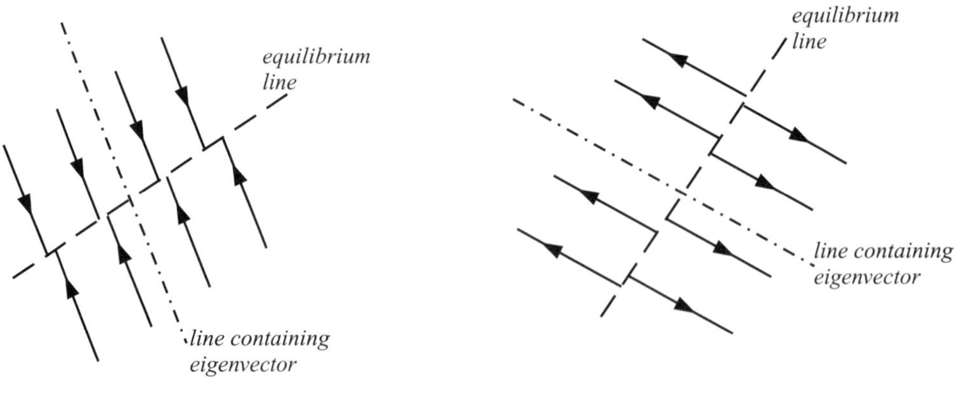

a) Stable Degenerate Case *b) Unstable Degenerate Case*

Figure 8.15. Trajectories surrounding (a) stable and (b) unstable degenerate points.

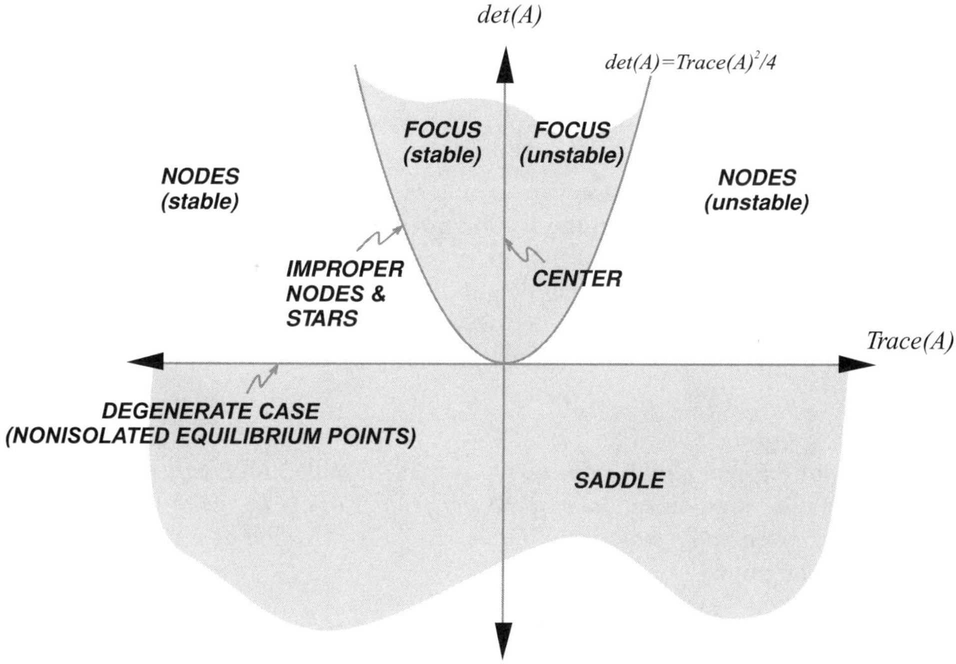

Figure 8.16. The regions containing different types of trajectories.

If $A = 0$, no motion is present. If $A \neq 0$ with $\lambda_1 = \lambda_2 = 0$, then no motion is present along the line containing the eigenvector $\mathbf{v}$, where

$$\mathbf{v} = \begin{cases} \begin{pmatrix} a_{12} \\ -a_{11} \end{pmatrix} & \text{if } a_{12} \neq 0 \\ \\ \begin{pmatrix} -a_{22} \\ a_{21} \end{pmatrix} & \text{if } a_{21} \neq 0 \end{cases}$$

Outside this line, the trajectories are along straight lines that are parallel to $\mathbf{v}$. The trajectories for $A \neq 0$ are left as Exercise **E8.6**.

The different types of equilibrium points can be summarized in terms of the trace and determinant of A, as shown in Figure 8.16.

8.6 Linearization Around Equilibrium Points

Let $\mathbf{x}_e$ be an equilibrium point of the autonomous system described by $d\mathbf{x}/dt = \mathbf{f}(\mathbf{x})$; then by expanding $\mathbf{f}$ using Taylor series around $\mathbf{x} = \mathbf{x}_e$, we obtain

$$\mathbf{f}(\mathbf{x}) = \mathbf{f}(\mathbf{x}_e) + \left.\frac{d\mathbf{f}}{d\mathbf{x}}\right|_{\mathbf{x}=\mathbf{x}_e} (\mathbf{x} - \mathbf{x}_e) + \mathcal{O}(\|\mathbf{x} - \mathbf{x}_e\|^2) \tag{8.18}$$

Also, let $J_{(\mathbf{x}_e)} = \left.\dfrac{d\mathbf{f}}{d\mathbf{x}}\right|_{\mathbf{x}=\mathbf{x}_e}$, that is, the Jacobian matrix at the equilibrium point $\mathbf{x}_e$. After truncating the higher order terms, $\mathcal{O}(\|\mathbf{x} - \mathbf{x}_e\|^2)$, together with the fact that because $\mathbf{f}(\mathbf{x}_e) = 0$, we have

$$\mathbf{f}(\mathbf{x}) \approx J_{(\mathbf{x}_e)} (\mathbf{x} - \mathbf{x}_e) \tag{8.19}$$

which then yields a **linearized approximation** of the original nonlinear system given by

$$\frac{d}{dt}\mathbf{x} = J_{(\mathbf{x}_e)}\left(\mathbf{x} - \mathbf{x}_e\right) \tag{8.20}$$

However, this approximation is true only for small neighborhoods around the equilibrium points. Moreover, if the Jacobian is singular (i.e., it contains some zero eigenvalues) or if it contains pure imaginary eigenvalues, then the truncations may no longer be valid, even for a small neighborhood around the equilibrium points. Thus we need to classify the condition for which linearization would be sufficiently close to the actual flows around the equilibrium points, as given by the following definition:

Definition 8.5. *An equilibrium point* $\mathbf{x}_e$ *of* $d\mathbf{x}/dt = \mathbf{f}(\mathbf{x})$ *is classified as a* **hyperbolic equilibrium point** *if none of the eigenvalues of the Jacobian matrix* $J_{(\mathbf{x}_e)}$ *of* $\mathbf{f}(\mathbf{x})$ *are zero or pure imaginary. Otherwise, it is classified as a* **non-hyperbolic equilibrium point**.

The following theorem, known as the **Hartman-Grobman theorem** or the **linearization theorem**, applies only to hyperbolic equilibrium points:

THEOREM 8.2. *Let* $\mathbf{x}_e$ *be a hyperbolic equilibrium point of* $d\mathbf{x}/dt = \mathbf{f}(\mathbf{x})$. *Then for a small neighborhood around* $\mathbf{x}_e$*, the behavior of the trajectories can be approximated by (8.20).*

This implies that the type and stability of the trajectories surrounding hyperbolic equilibrium points can be determined by simply analyzing the linearized equations.

EXAMPLE 8.5. Consider the following second-order system

$$\frac{d}{dt}\begin{pmatrix} x_1 \\ x_2 \end{pmatrix} = \begin{pmatrix} x_2 \\ \cos(\pi x_1) - x_2 \end{pmatrix} \tag{8.21}$$

The equilibrium points and the corresponding Jacobian matrices at these points are given by

$$(\mathbf{x}_e)_n = \begin{pmatrix} n - \frac{1}{2} \\ 0 \end{pmatrix} \quad \text{and} \quad J_{(\mathbf{x}_e)_n} = \begin{pmatrix} 0 & 1 \\ (-1)^n\pi & -1 \end{pmatrix}$$

where n is an integer. The characteristic equation is $\lambda^2 + \lambda - (-1)^n\pi = 0$, and the eigenvalues are

$$\lambda = -\frac{1}{2} \pm \frac{(i)^n}{2}\sqrt{4\pi + (-1)^n}$$

where $i = \sqrt{-1}$. Thus all the equilibrium points are hyperbolic, and we can use linearized approximations of the trajectories around each of the equilibrium points. Based on the results of Section 8.5, the equilibrium points are saddles for even values of n and stable focuses for odd values of n.

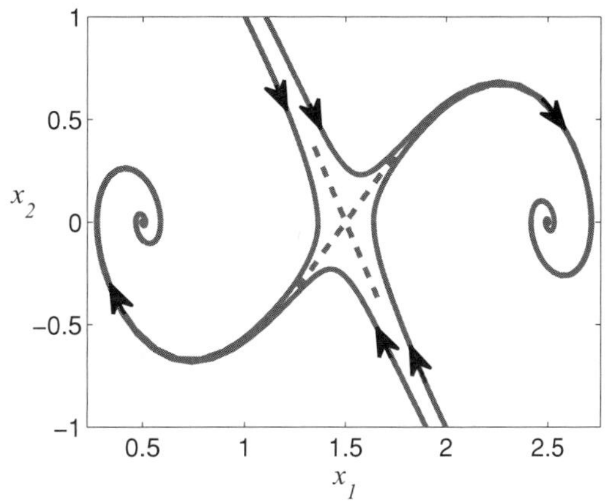

Figure 8.17. The trajectories for (8.21) using the initial conditions given in (8.22). (The dotted lines are in the direction of the eigenvectors corresponding to the Jacobian matrix at $\mathbf{x}_e = (1.5, 0)^T$.)

Using the following initial conditions:

$$(\mathbf{x}_0)_a = \begin{pmatrix} 1 \\ 1 \end{pmatrix} ; (\mathbf{x}_0)_b = \begin{pmatrix} 1.1 \\ 1 \end{pmatrix} ; (\mathbf{x}_0)_c = \begin{pmatrix} 2 \\ -1 \end{pmatrix} ; (\mathbf{x}_0)_d = \begin{pmatrix} 1.9 \\ -1 \end{pmatrix} \tag{8.22}$$

we obtain the phase-plane portrait shown in Figure 8.17. In the figure, we see that the eigenvectors found from the linearized models around the saddle are consistent with the actual trajectories (at least locally). Also, for the linearized models around the focus, we can also note that the $(1, 2)^{\text{th}}$ element of the Jacobian matrix is positive and thus should have a clockwise rotation around the focus. The plots in the figure show that this is consistent with the nonlinear trajectories as well.

We now end this section with an example that the linearization approximation should not be used to approximate the trajectories around non-hyperbolic equilibrium points.

EXAMPLE 8.6. Consider the system

$$\frac{d}{dt}\begin{pmatrix} x_1 \\ x_2 \end{pmatrix} = \begin{pmatrix} x_2 \\ -x_1 - 6x_1^2 x_2 \end{pmatrix} \tag{8.23}$$

The equilibrium point is at the origin, and the linearized equation around the origin is given by

$$\frac{d}{dt}\mathbf{x} = \begin{pmatrix} 0 & 1 \\ -1 & 0 \end{pmatrix}\mathbf{x}$$

The eigenvalues of the Jacobian matrix are $\pm i$, which predicts that the trajectories around the origin should be close to those corresponding to centers. The plot of the trajectory starting at $\mathbf{x}_0 = (-0.15, 0.1)^T$ is shown in Figure 8.18. The trajectory starts out different from trajectories around centers, but at longer times, it does approach a stable focus, with very slow movement toward the origin. It can be shown in the next section that the origin in this example is asymptotically stable (even though it goes very slowly as is nears the origin). Nonetheless, the linearization does predict the clockwise rotation.

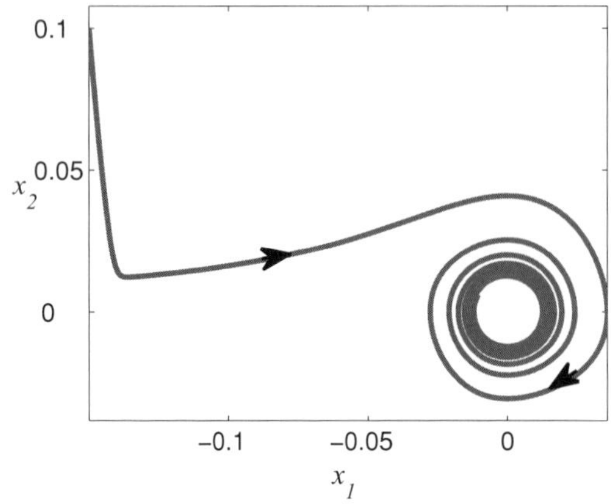

Figure 8.18. The trajectory of (8.23) starting at $\mathbf{x}_0 = (-0.15, 0.1)^T$.

8.7 Method of Lyapunov Functions

In the previous section, we noted that the linearization approach is limited to the local analysis of hyperbolic equilibrium points. A more general approach to assess the stability of equilibrium points is an approach known as **Lyapunov's method** for stability analysis. We assume that the origin will be the equilibrium point. If this is not the case, we could always translate the axes such that, under the new coordinates, the origin is the equilibrium point.

Definition 8.6. *Let the origin be an equilibrium point of $d\mathbf{x}/dt = \mathbf{f}(\mathbf{x})$ and let D be an open neighborhood of the origin. Then for $\mathbf{x} \in D$, a scalar function $V(\mathbf{x})$ is a* **Lyapunov function in** *D for the dynamic system if $V(\mathbf{x})$ is positive definite, that is, $V(\mathbf{x}) > 0$ for $\mathbf{x} \neq 0$, $V(0) = 0$, whereas dV/dt is negative semi-definite, that is,*

$$\frac{dV(\mathbf{x})}{dt} = \frac{dV}{d\mathbf{x}}\mathbf{f} \leq 0$$

THEOREM 8.3. *Let the origin be an equilibrium point of $d\mathbf{x}/dt = \mathbf{f}(\mathbf{x})$ and let D be an open neighborhood of the origin. If there exists a Lyapunov function $V(\mathbf{x})$, then the origin is a Lyapunov stable point. If in addition, dV/dt is negative definite, that is, $dV/dt < 0$ for $\mathbf{x} \neq 0$, then the origin is asymptotically stable. Likewise, if there exists a positive definite function $V(\mathbf{x})$ such that $dV/dt > 0$, then the origin is unstable.*

EXAMPLE 8.7. Consider the system

$$\frac{d}{dt}\mathbf{x} = \begin{pmatrix} x_2 \\ -x_1 - \epsilon x_1^2 x_2 \end{pmatrix}$$

Let $V(\mathbf{x}) = \frac{1}{2}\mathbf{x}^T\mathbf{x}$, then

$$\frac{dV}{dt} = \mathbf{x}^T\left(\frac{d}{dt}\mathbf{x}\right) = x_1x_2 + x_2\left(-x_1 - \epsilon x_1^2 x_2\right) = -\epsilon x_1^2 x_2^2$$

Thus as long as $\epsilon \geq 0$, the origin is Lyapunov stable. If $\epsilon > 0$, the origin is asymptotically stable.

Referring back to Example 8.6, the system there is for the case with $\epsilon = 60$. In that example, we noted that the origin is a non-hyperbolic equilibrium point and that the linearization approach is not applicable. We see that the Lyapunov function approach was still able to assess the stability around the origin. For $\epsilon = 0$, the nonlinear system reduces to a linear system in which the origin will be a center, which is Lyapunov stable but not asymptotically stable.

Although the method of Lyapunov functions is a powerful and general method for the determination of stability of equilibrium points, the biggest problem is the search for these functions. Nonetheless, several candidates are available for one to try. These include the following:

1. $V(\mathbf{x}) = \mathbf{x}^T Q\mathbf{x}$ where Q is positive definite.
2. Krasovskii Forms.

$$V(\mathbf{x}) = \mathbf{f}(\mathbf{x})^T Q\mathbf{f}(\mathbf{x})$$

3. Rosenbrock's Function.

$$V(\mathbf{x}) = \sum_{j=1}^{N} |\mathbf{f}_j(\mathbf{x})|$$

EXAMPLE 8.8. One can also use Lyapunov functions to show whether a linear system given by

$$\frac{d}{dt}\mathbf{x} = A\mathbf{x}$$

is asymptotically stable without having to determine the eigenvalues. Let P be a symmetric positive definite matrix; then by definition, $V = \mathbf{x}^T P\mathbf{x} > 0$. Taking the derivative of V, we have

$$\frac{dV}{dt} = \left(\frac{d}{dt}\mathbf{x}^T\right) P\mathbf{x} + \mathbf{x}^T P\left(\frac{d}{dt}\mathbf{x}\right) = \mathbf{x}^T \left(A^T P + PA\right)\mathbf{x}$$

which will be a Lyapunov function if we can find $P > 0$ such that

$$N = A^T P + PA$$

is a negative definite matrix.

Thus we can choose any N and try to solve for X in

$$A^T X + XA = N \tag{8.24}$$

and if X is positive definite, then the linear system will be asymptotically stable. Thus stability of the linear system can be determined by solving the **Lyapunov matrix equation** (8.24) (which is a special case of Sylvester matrix equation given in (1.23)) and then proving that the solution is positive definite.[6]

For instance, let

$$A = \begin{pmatrix} -2.5 & 4.5 \\ 0.5 & -2.5 \end{pmatrix}$$

then with $N = -I$, we have the solution of (8.24) to be

$$X = \begin{pmatrix} 0.2625 & 0.3125 \\ 0.3125 & 0.7625 \end{pmatrix}$$

Because X can be shown to be positive definite using Sylvester's criterion in Theorem A.2, we can conclude that $\frac{d}{dt}\mathbf{x} = A\mathbf{x}$ will be asymptotically stable. This example may appear to be unnecessary, because one should be able to solve for the eigenvalues of A. Instead, it shows an alternative approach for stability analysis without the need for solving eigenvalues, which may not be easily analyzed if matrix A is large or contains unspecified parameters.[7]

8.8 Limit Cycles

Besides the possibility of having multiple isolated equilibrium points, some nonlinear systems can also produce sustained periodic oscillations called **limit cycles**. An isolated closed trajectory (cycle) of an autonomous system is called a **limit cycle**.

Limit cycles are different from centers in several ways. One difference is that limit cycles can either be stable, unstable, or partially unstable. Another difference is that centers are trajectories that contain the initial point. However, a stable limit cycle is a periodic trajectory that is approached asymptotically from an initial point, yet do not contain the initial point.

EXAMPLE 8.9. Consider the van der Pol's equation

$$\begin{aligned} \frac{dx_1}{dt} &= x_2 \\ \frac{dx_2}{dt} &= x_2(1 - x_1^2) - x_1 \end{aligned} \tag{8.25}$$

The origin is an equilibrium point, and the linearized equation around the origin is given by

$$\frac{d}{dt}\mathbf{x} = \begin{pmatrix} 0 & 1 \\ -1 & 1 \end{pmatrix}\mathbf{x}$$

Because the eigenvalue of the Jacobian is given by $(0.5 \pm 0.866i)$, the behavior around the origin will be an unstable focus. However, the trajectory does not go unbounded. Instead, it settles down into a limit cycle, as shown in Figure 8.19.

8.8.1 Limit Cycle Existence Theorems

There are several methods to assess the existence of limit cycles. For second-order systems, we have two important results: Bendixson's criterion and the Poincaré-Bendixson theorem.

[6] The Lyapunov function approach has also been used to prove Routh-Hurwitz stability criterion for a given characteristic polynomial (see, e.g., C. T. Chen, *Linear System Theory and Design*, Oxford University Press, 1984).

[7] This is also the main utility of the Routh-Hurwitz method for stability analysis.

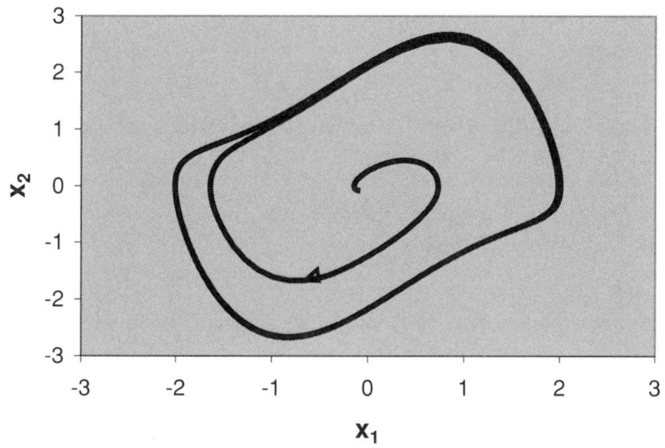

Figure 8.19. Phase-plane plot of the van der Pol system given in (8.25).

THEOREM 8.4. **Bendixson's Criterion.** *For the second-order autonomous system given by*

$$\frac{d}{dt}\mathbf{x} = \mathbf{f}(\mathbf{x})$$

suppose that the divergence

$$\nabla \cdot \mathbf{f} = \frac{\partial f_1}{\partial x_1} + \frac{\partial f_2}{\partial x_2}$$

is not identically zero, nor does it change sign in a simply connected open region D; then no limit cycles exist in D.

PROOF. First assume that there is a limit cycle in D and that the limit cycle is contained in a closed curve C. Let S be the closed region enclosed by C. Then using Green's lemma (cf. Equation 5.1)on the divergence of $\mathbf{f}$ in region S, we have

$$\begin{aligned}\int_S \left(\frac{\partial f_1}{\partial x_1} + \frac{\partial f_2}{\partial x_2}\right) dS &= \oint_C f_1 dx_2 - f_2 dx_1 \\ &= \int_0^T \left(f_1 \frac{dx_2}{dt} - f_2 \frac{dx_1}{dt}\right) dt \\ &= 0\end{aligned}$$

where T is the period of the cycle. However, for the surface integral to be zero, we need the divergence of $\mathbf{f}$ to either be zero or change sign.

EXAMPLE 8.10. Consider the van der Pol system given in (8.25). Calculating the divergence of $\mathbf{f}$, we have

$$\nabla \cdot \mathbf{f} = 1 - x_1^2$$

Let D be a region bounded by a circle centered at the origin having a radius, $r < 1$. The divergence of $\mathbf{f}$ in D is always positive. We then conclude that in this region, there are no limit cycles. This can be verified by observing the plot given in Figure 8.19.

However, note that the criterion does not prevent portions of limit cycles to be in the region that satisfy the conditions of the criterion.

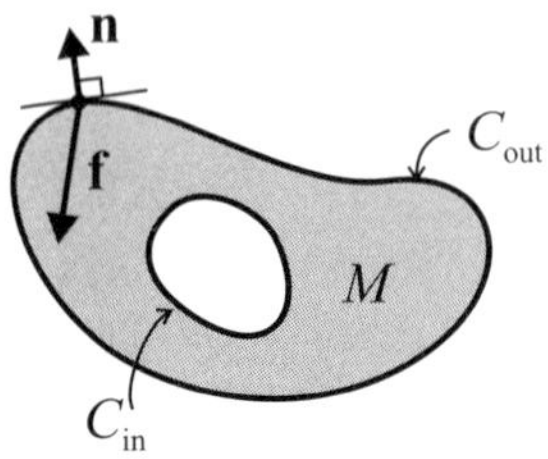

Figure 8.20. The conditions given in Poincaré-Bendixson's theorem.

THEOREM 8.5. **Poincaré-Bendixson's Theorem.** *Let M be a region bounded by two nonintersecting closed curves C_{out} and C_{in}, where C_{in} is inside C_{out}, as shown in Figure 8.20. For the second-order autonomous system given by*

$$\frac{d}{dt}\mathbf{x} = \mathbf{f}(\mathbf{x})$$

If

1. *There are no equilibrium points inside M.*
2. *Along C_{in} and C_{out}*

$$\mathbf{f} \cdot \mathbf{n} \leq 0 \tag{8.26}$$

where **n** *is the outward unit normal vector of region M,*

3. *Inside M*

$$(\mathbf{f})^T \begin{pmatrix} -x_2 \\ x_1 \end{pmatrix} \neq 0 \tag{8.27}$$

then a stable limit cycle exists in M.

This statement of the Poincaré-Bendixson's theorem gives a sufficient condition for the existence of a limit cycle in M. The condition in (8.27) means that the trajectories inside M never point in a radial (in or out) direction.[8] The theorem, however, does not identify whether there is only one or a multiple number of limit cycles in M. Furthermore, this result cannot be extended to third- or higher order autonomous nonlinear systems.

EXAMPLE 8.11. Consider the following autonomous system:

$$\frac{d}{dt}\begin{pmatrix} x_1 \\ x_2 \end{pmatrix} = \begin{pmatrix} x_2 \\ x_1 \end{pmatrix} + \left(1 - (x_1^2 + x_2^2)\right)\begin{pmatrix} x_1 \\ x_2 \end{pmatrix} \tag{8.28}$$

Next, choose C_{in} to be a circle centered around the origin of radius r_{in}, and choose C_{out} to be a circle centered around the origin of radius r_{out}, where $0 < r_{in} < r_{out}$. The outward unit normal vectors at C_{in} are

$$\mathbf{n}_{in} = \begin{pmatrix} -\cos(\theta) \\ -\sin(\theta) \end{pmatrix} \qquad \text{where} \quad \theta = \tan^{-1}\left(\frac{y}{x}\right), \; x^2 + y^2 = r_{in}$$

[8] A more general version of the Poincaré-Bendixson's theorem simply states that if a trajectory never leaves the region M, then either the trajectory is a limit cycle or approaches a limit cycle in M.

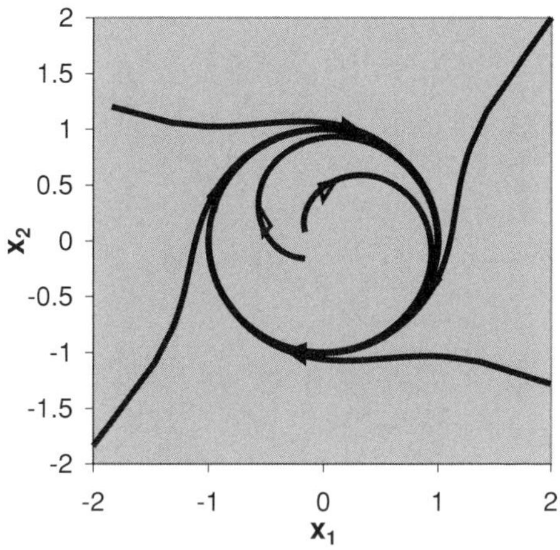

Figure 8.21. A plot of some trajectories of (8.28) showing that the unit circle is a limit cycle.

whereas the outward unit normal vectors at C_{out} are

$$\mathbf{n}_{out} = \begin{pmatrix} \cos(\theta) \\ \sin(\theta) \end{pmatrix} \qquad \text{where} \quad \theta = \tan^{-1}\left(\frac{y}{x}\right), \; x^2 + y^2 = r_{out}$$

Along C_{in}, we find that

$$\mathbf{f} \cdot \mathbf{n} = -r_{in}\left(1 - r_{in}^2\right)\sin(\theta)^2 \qquad 0 \le \theta \le 2\pi$$

whereas along C_{out},

$$\mathbf{f} \cdot \mathbf{n} = r_{out}\left(1 - r_{out}^2\right)\sin(\theta)^2 \qquad 0 \le \theta \le 2\pi$$

and we can see that we need $r_{in} < 1$ and $r_{out} > 1$ to satisfy (8.26).

As for (8.27), we have

$$\mathbf{f}^T \begin{pmatrix} -x_2 \\ x_1 \end{pmatrix} = -\left(x_1^2 + x_2^2\right) \neq 0 \qquad \text{for } (x_1, x_2) \in M$$

A plot of some trajectories of the system is shown in Figure 8.21. From the figure, one can observe that the Poincaré-Bendixson theorem would have predicted a limit cycle inside the annulus region between the outer circle of radius greater than 1 and an inner circle of radius less than 1.

8.8.2 Poincaré Maps and Liènard Systems

The Poincarè-Bendixson theorem is a tool to determine the existence of a stable limit cycle inside a region in a phase-plane, that is, a second-order system. In higher order cases, the theorem may no longer hold because other trajectories may be present, such as quasi-periodic or chaotic trajectories. To determine the presence of limit cycles in these cases, a tool known as **Poincaré maps** (also known as **first-return maps**) can be used. It can also be used to show uniqueness of stable limit cycles.

Let S be an $(n-1)$-dimensional hypersurface that is transverse to the flow of an n-dimensional autonomous system $d\mathbf{x}/dt = \mathbf{f}(\mathbf{x})$ (i.e., none of the trajectories will be parallel to S). A **Poincaré map** P is a mapping obtained by following the trajectory

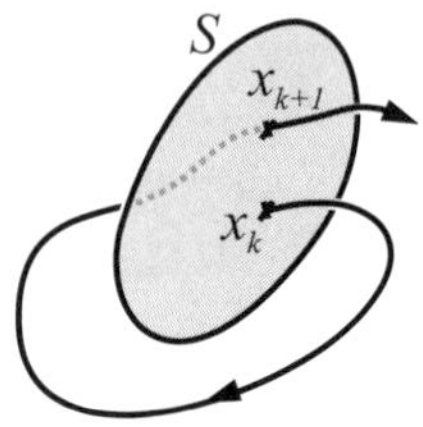

Figure 8.22. Poincaré map based on surface S.

of an intersecting point $\mathbf{x}_k^*$ (of the trajectory with S) to the next intersecting point $\mathbf{x}_{k+1}^*$, that is,

$$\mathbf{x}_{k+1}^* = P\left(\mathbf{x}_k^*\right) \tag{8.29}$$

This is shown diagramatically in Figure 8.22. Note that the hypersurface S is often a bounded or semibounded region in $n-1$ dimension that slices the n dimensional region containing the limit cycle.

If a point $\mathbf{x}_{eq}^*$ in the surface S is the fixed point of the Poincaré map, that is, $\mathbf{x}_{eq}^* = P\left(\mathbf{x}_{eq}^*\right)$, then $\mathbf{x}_{eq}^*$ belongs to a limit cycle.[9] A discrete version of Lyapunov functions can also be used to determine the stability of the limit cycle. This principle will be used in the following example to show the uniqueness of a limit cycle of a particular class of second-order equations known as **Liènard systems**.

EXAMPLE 8.12. Consider the second-order system known as the **Liènard system** given by

$$\frac{d^2x}{dt^2} + f(x)\frac{dx}{dt} + g(x) = 0 \tag{8.30}$$

Instead of the usual state-space representation, the **Liènard coordinates** can be used by defining y as

$$y = \frac{dx}{dt} + F(x) \quad \text{where} \quad F(x) = \int_0^x f(\xi)d\xi \tag{8.31}$$

Then (8.30) becomes

$$\frac{d}{dt}\begin{pmatrix} x \\ y \end{pmatrix} = \begin{pmatrix} y - F(x) \\ -g(x) \end{pmatrix} \tag{8.32}$$

Now assume the following conditions:

1. The function $g(x)$ is an odd function with $g(x) > 0$ if $x > 0$.
2. The integral $F(x)$ is also an odd function and has three roots given by $x = 0$, $x = q > 0$, and $x = -q < 0$, such that $f(0) < 0$ and $f(x) > 0$ at $x = \pm q$.
3. $F(x) \to \infty$ as $x \to \infty$.

[9] We are using a technical distinction between **fixed points** of discrete maps such as Poincaré maps and **equilibrium points** for continuous maps that apply to differential equations.

Figure 8.23. The nullclines of the Liènard system.

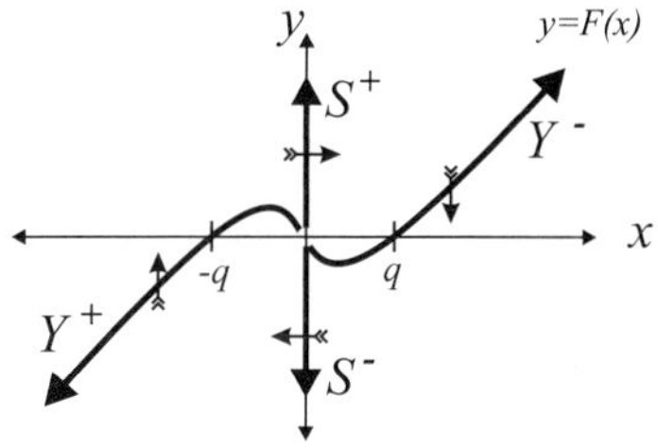

If all these conditions are satisfied, then the Liènard system given by (8.30) will have exactly one stable limit cycle, as we show next. (When we refer to the Liènard system that follows, we are assuming that these conditions are already attached.) The relevance of Liènard systems is that it is a class of autonomous oscillators of which the van der Pol equation is a special case.

The Jacobian matrix resulting from linearization of (8.32) at the origin is given by

$$J_0 = \left.\begin{pmatrix} -f(x) & 1 \\ -dg/dx & 0 \end{pmatrix}\right|_{(x,y)=(0,0)}$$

Because $\mathbf{tr}(J_0) = -f(0) > 0$ and $\mathbf{det}(J_0) = dg/dx > 0$, the origin is hyperbolic and is either an unstable node or unstable focus (see Figure 8.16).

Let S^+ be the strictly positive y-axis (i.e., excluding the origin); then S^+ is a nullcline where the trajectories are horizontal and pointing to the right $(dy/dt|_{(x=0)} = g(0) = 0$ and $dx/dt|_{(x=0)} = y > 0)$. Another nullcline S^- is the strictly negative y-axis where the trajectories are horizontal and pointing to the left. The other two nullclines Y^- and Y^+ will be the graph $y = F(x)$ with $x \neq 0$. The trajectories at Y^- are vertical and pointing down when $x > 0$, whereas the trajectories at Y^+ are vertical pointing up when $x < 0$ ($dy/dt = -g(x) < 0$ for $x > 0$ and $dy/dt = -g(x) > 0$ when $x < 0$). The nullclines are shown in Figure 8.23.

Using all the nullclines, one can apply Poincaré -Bendixson's theorem to prove that a limit cycle exists in a region M whose outer boundary has a distance from the origin that is greater than q (the root of $F(x)$) and inner boundary is a circle of a small radius $\epsilon > 0$ surrounding the origin. However, this does not prove the uniqueness of the limit cycle. To do so, we need to find a unique fixed point of a Poincaré map, which we will choose S^+ to help define the Poincaré map P.

Because we are only given conditions for $g(x)$ and $F(x)$, we cannot evaluate the actual Poincaré map P.[10] Instead, we use a Lyapunov function defined by

$$V(x,y) = \frac{1}{2}y^2 + \int_0^x g(\xi)d\xi$$

whose derivative is given by

$$\frac{dV}{dt} = y\frac{dy}{dt} + g(x)\frac{dx}{dt} = -g(x)F(x)$$

[10] Even if we are given $f(x)$ and $g(x)$, the Liènard equation is an Abel equation, which can be very difficult to solve in most cases.

Furthermore, note that because $F(x)$ and $g(x)$ are odd functions, the Liènard system is symmetric with respect to both x and y; that is, the same equation (8.32) results after replacing x and y by $\widetilde{x} = -x$ and $\widetilde{y} = -y$, respectively. Thus instead of the full Poincaré map P, we can just analyze the map $\widetilde{P}$, which is the map of the trajectory starting at a point in S^+ to a first intersection in S^-, and conclude that y^* is in a limit cycle if and only if $y^* = -\widetilde{P}(y^*)$ due to the special symmetry. The fixed point $(0, y^*)$ at S^+ is unique if the Lyapunov function between Poincaré maps at S^+ is equal, that is,

$$V(0, y^*) = V(0, P(y^*)) \quad \text{or} \quad V(0, y^*) = V\left(0, -\widetilde{P}(y^*)\right)$$

Let $\Delta V_{y_0} = V\left(0, -\widetilde{P}(y_0)\right) - V(0, y_0)$. Starting at $(0, y_0)$ with $y_0 > 0$, the trajectory will intersect the nullcline Y^-. If the intersection with Y^- is at $x \le q$, then $dV/dt > 0$, which means $\Delta V_{y_0} > 0$. However, if the intersection with Y^- is at $x > q$, we can show that the difference $\Delta V_{y_0} > 0$ decreases monotonically with increasing y_0, as sketched here:

1. The value of ΔV_{y_0} can be split into three paths yielding

$$\Delta V_{y_0} = \int_0^{t_1} V(x(t), y(t))dt + \int_{t_1}^{t_2} V(x(t), y(t))dt + \int_{t_2}^{t_3} V(x(t), y(t))dt$$

where t_1 is the first instance that $x = q$, t_2 is the second instance that $x = q$ as the trajectory loops back toward the negative y-axis, and t_3 is the time when $x = 0$ and $y = \widetilde{P}(y_0)$ at the negative y-axis.

2. The first and third integrals can be evaluated as

$$\int_0^{t_1} Vdt = -\int_0^{q} \frac{F(x)g(x)}{y - F(x)}dx \quad \text{and} \quad \int_{t_2}^{t_3} Vdt = -\int_q^{0} \frac{F(x)g(x)}{y - F(x)}dx$$

and will both decrease for an increase in y_0 because the denominator will be a more positive value for the integral from 0 to t_1, and the denominator will be a more negative value for the integral from t_2 to t_3.

3. For the second integral

$$\int_{t_1}^{t_2} Vdt = \int_{y^+(q)}^{y^-(q)} F(x(y))dy$$

the arc from $y^+(q)$ to $y^-(q)$ increases in size to the right and $F(x) > 0$ along this arc as y_0 is increased. However, because $y^-(q) < 0 < y^+(q)$, this integral will also decrease as y_0 increases.

4. Because all three integrals decrease as y_0 increases, we conclude that ΔV_{y_0} decreases monotonically as y_0 is increased. Following the same arguments of the integrals, it can be shown that $\Delta V_{y_0} \to -\infty$ as $y_0 \to \infty$.

   Combining these results, we can see from Figure 8.24 that there is only one fixed point in S^+ (equiv. in S^-); that is, the limit cycle of the Liènard system is stable and unique.

Because the Poincaré map is a discrete map $\mathbf{x}_{k+1} = P(\mathbf{x}_k)$, one could test the stability of the limit cycle passing through a fixed point $\mathbf{x}^*$ in the surface S by introducing a perturbation ϵ_k, that is,

$$\mathbf{x}^* + \epsilon_{k+1} = P(\mathbf{x}^* + \epsilon_k)$$

Figure 8.24. The change in Lyapunov function ΔV_{y_0} as a function initial point y_0, showing only one unique fixed point y^* of the Poincaré map based on S^+ of the Liènard system.

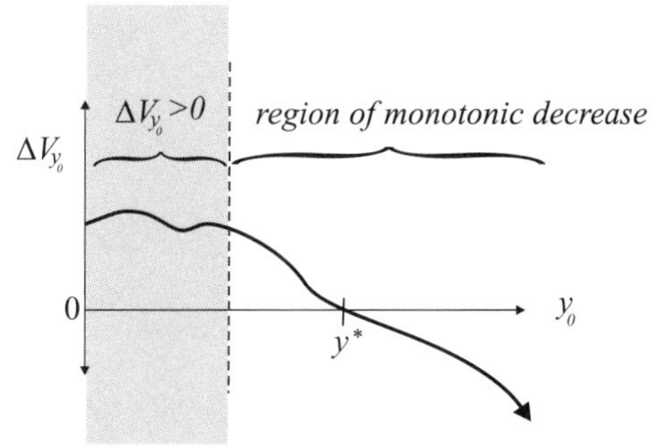

Using Taylor series expansion and truncating the higher order terms, we end up with a linearized Poincaré map J_P for the perturbation

$$\epsilon_{k+1} = J_P \epsilon_k$$

where

$$J_P = \left.\frac{dP}{d\mathbf{x}}\right|_{\mathbf{x}=\mathbf{x}^*} \tag{8.33}$$

Using Theorem 7.2, the perturbations will die out if all the eigenvalues λ_i, $i = 1, \ldots, n$, of the linearized Poincaré map J_P have magnitudes less than 1. These eigenvalues are also known as the **Floquet multipliers** of the Poincaré map P. However, similar to the non-hyperbolic case of an equilibrium point in the continuous case, the stability cannot be determined using linearization if any of the eigenvalues of J_P have a magnitude equal to 1. In those cases, the Lyapunov analysis or other nonlinear analysis methods need to be used.

8.8.3 Nonlinear Centers

If the equilibrium points are non-hyperbolic points, there are still cases in which the surrounding neighborhood will have periodic trajectories (or orbits). If this is the case, the equilibrium point is known as a **nonlinear center**. However, periodic orbits surrounding a nonlinear center are not the same as limit cycles because they are not isolated; that is, any small perturbation will immediately lie in a neighboring periodic orbit.

Two special cases will yield a nonlinear center. The first case involves the **conservative system**, when a Lyapunov function $V(\mathbf{x})$ can be found around the equilibrium point $\mathbf{x}_{eq}$ such that the Lyapunov functions are constant around the neighborhood of the equilibrium point. One class of second-order conservative system is given in Exercise **E8.15**.

The other case is when the system belongs to a class of **reversible systems**, where the trajectories are symmetric with respect to time and one of the coordinates, that is, with $\tau = -t$ and $\widetilde{y} = -y$

$$\frac{d}{dt}\begin{pmatrix} x \\ y \end{pmatrix} = \begin{pmatrix} f_1(x,y) \\ f_2(x,y) \end{pmatrix} \quad \rightarrow \quad \frac{d}{d\tau}\begin{pmatrix} x \\ \widetilde{y} \end{pmatrix} = \begin{pmatrix} f_1(x,\widetilde{y}) \\ f_2(x,\widetilde{y}) \end{pmatrix}$$

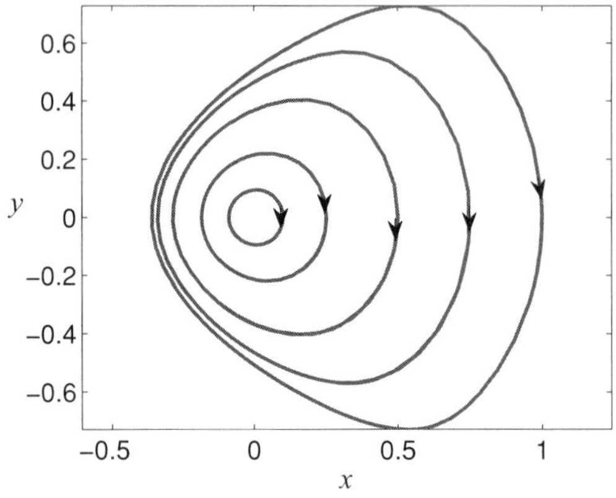

Figure 8.25. Plot of trajectories for (8.34) with initial points at (1, 0), (0.75, 0), (0.5, 0), (0.25, 0), and (0.11, 0).

EXAMPLE 8.13. Consider the reversible system

$$\frac{d}{dt}\begin{pmatrix} x \\ y \end{pmatrix} = \begin{pmatrix} y\left(1-y^2\right) \\ -x+y^2 \end{pmatrix} \tag{8.34}$$

The origin is an equilibrium point that is non-hyperbolic, with the Jacobian matrix at the origin given by

$$J(\mathbf{x}=0) = \begin{pmatrix} 0 & 1 \\ -1 & 0 \end{pmatrix}$$

Various trajectories are shown in Figure 8.25, starting at different initial conditions. Note that the shapes of the periodic trajectories are not circular nor elliptic the farther they are from the origin. Furthermore, the rotation is clockwise, which can be predicted from $J_{12} > 0$ (cf. Section 8.5).

8.9 Bifurcation Analysis

The qualitative analysis discussed thus far has mostly dealt with fixed coefficients. As can be anticipated, if some of the system coefficients are replaced by parameters, the behavior will most likely change. It could mean the addition or loss of equilibrium points or limit cycles, as well as changes in the stability of these behaviors. A brief discussion on bifurcation analysis is also included in Section H.1 as an appendix. These analysis are aimed at how the phase portraits (i.e., collective behavior of the system) are affected by variations in critical parameters.

8.10 EXERCISES

E8.1. Consider a scalar system involving a parameter $0 < q < 1$ given by

$$\frac{dx}{dt} = \mathbf{f}(x) = \begin{cases} 0 & \text{for } x \le 0 \\ x^q & \text{for } x > 0 \end{cases}$$

with initial condition $x(0) = 0$. Show that the following function is a solution:

$$x(t) = \begin{cases} 0 & \text{for } t \le \tau \\ \left[(t-\tau)(1-q)\right]^{1/(1-q)} & \text{for } t > \tau \end{cases}$$

for all $\tau > 0$. This means that the solution is not unique, and this is possible because although $f(x)$ is continuous, its derivative df/dx is not continuous at $x = 0$.

E8.2. Consider the following autonomous system, with $a \neq 0$,

$$\begin{aligned} \frac{dx}{dt} &= -ay \\ \frac{dy}{dt} &= az \\ \frac{dz}{dt} &= -a^2 z \end{aligned}$$

Show that the equilibrium point is Lyapunov stable but not quasi-asymptotically stable.

E8.3. Consider the following system given in Glendinning (1994):

$$\frac{dx}{dt} = x - y + x\left(x^2 + y^2\right) + \frac{xy}{\sqrt{x^2 + y^2}}$$

$$\frac{dy}{dt} = x + y - y\left(x^2 + y^2\right) - \frac{x^2}{\sqrt{x^2 + y^2}}$$

1. Determine the equilibrium points.
2. Obtain a direction field for this system in the domain $-2 \leq x \leq 2$ and $-2 \leq y \leq 2$. (We suggest a 20×20 grid.) Ascertain from the direction field whether the equilibrium is Lyapunov stable, quasi-asymptotic stable, or asymptotically stable.
3. Use the initial value solvers and obtain the trajectories around the equilibrium.

E8.4. For the linear second-order system $d\mathbf{x}/dt = A\mathbf{x}$, once it can be determined to be a focus or center, show that the direction of rotation can be determined to be clockwise or counterclockwise depending on whether $a_{12} > 0$ or $a_{12} < 0$. One approach is given by the following steps:

1. Show that a necessary condition for focus or centers is that $a_{12}a_{21} < 0$.
2. Let $\underline{\vec{\mathbf{r}}} = x_1\underline{\boldsymbol{\delta}}_x + x_2\underline{\boldsymbol{\delta}}_y$ be the position vector of $\mathbf{x}$ and $\underline{\vec{\mathbf{v}}} = y_1\underline{\boldsymbol{\delta}}_x + y_2\underline{\boldsymbol{\delta}}_y$ be the tangent to trajectory at $\mathbf{x}$, where $\mathbf{y} = A\mathbf{x}$. Show that the cross product is given by

$$\underline{\vec{\mathbf{r}}} \otimes \underline{\vec{\mathbf{v}}} = \xi\underline{\boldsymbol{\delta}}_z$$

   where

$$\xi = a_{21}x_1^2 + (a_{22} - a_{11})x_1x_2 - a_{12}x_2^2$$

3. Using the fact that $a_{12}a_{21} < 0$, show that if $a_{21} > 0$, we must have $a_{12} < 0$ then

$$\xi > \frac{1}{-a_{12}}\left[\frac{a_{22} - a_{11}}{2}x_1 - a_{12}x_2\right]^2 \geq 0$$

   and the rotation becomes counterclockwise. Next, you can show the reverse for clockwise rotation.

E8.5. Show that the conditions given in (8.13) will indeed determine whether the improper node is either S-type or Z-type. (Hint: To obtain an S-type improper node, trajectories at either side of $\mathbf{v}$ will have to be counterclockwise for stable equilibrium points and clockwise for unstable equilibrium points. For the Z-type improper nodes, the situations are reversed.)

E8.6. Obtain the trajectories in the phase plane for the degenerate cases with $A \neq 0$ where both eigenvalues are zero.

E8.7. For each of the cases for A that follow, determine the type of equilibrium points and do a phase-plane analysis of each case for

$$\frac{d\mathbf{x}}{dt} = A\mathbf{x}$$

by plotting trajectories at various initial points surrounding the equilibrium point. Also include supplementary information such as bounding ellipse, direction of rotation, shapes, straight line trajectories, and so forth appropriate for the type of equilibrium points.

$$\text{a) } A = \begin{pmatrix} 1 & 2 \\ 4 & 1 \end{pmatrix} \quad \text{b) } A = \begin{pmatrix} -1 & 2 \\ -2 & 1 \end{pmatrix} \quad \text{c) } A = \begin{pmatrix} 2 & 4 \\ -3 & -2 \end{pmatrix}$$

$$\text{d) } A = \begin{pmatrix} 1 & 2 \\ -4 & 1 \end{pmatrix} \quad \text{e) } A = \begin{pmatrix} 0 & 1 \\ -6 & 5 \end{pmatrix} \quad \text{f) } A = \begin{pmatrix} -5 & 0 \\ 0 & -5 \end{pmatrix}$$

$$\text{g) } A = \begin{pmatrix} -2 & 1 \\ 3 & -2 \end{pmatrix} \quad \text{h) } A = \begin{pmatrix} 2 & 1 \\ 4 & 2 \end{pmatrix} \quad \text{i) } A = \begin{pmatrix} 0 & -3 \\ 0 & 0 \end{pmatrix}$$

E8.8. Let x_1 and x_2 be the population of two competitive species 1 and 2. One model for the population dynamics is given by

$$\frac{d}{dt}\mathbf{x} = \begin{pmatrix} x_1 & 0 \\ 0 & x_2 \end{pmatrix} \left(\mathbf{b} - A\mathbf{x} \right)$$

where $\mathbf{b}$ and A are constant, with A nonsingular.

1. Show that there are four equilibrium points given by

$$(\mathbf{x}_e)_1 = \begin{pmatrix} 0 \\ 0 \end{pmatrix} \; ; \; (\mathbf{x}_e)_2 = \begin{pmatrix} \dfrac{b_1}{a_{11}} \\ 0 \end{pmatrix} \; ; \; (\mathbf{x}_e)_3 = \begin{pmatrix} 0 \\ \dfrac{b_2}{a_{22}} \end{pmatrix} \; ; \; (\mathbf{x}_e)_4 = A^{-1}\mathbf{b}$$

2. Prove that if $\mathbf{b}$ is a positive vector and A is a symmetric, positive definite, strictly diagonally dominant matrix with $a_{12} = a_{21} \leq 0$, then all the equilibrium points will be hyperbolic, $(\mathbf{x}_e)_2$ and $(\mathbf{x}_e)_3$ are saddles, $(\mathbf{x}_e)_1$ is an unstable node (or star), and $(\mathbf{x}_e)_4$ is a stable equilibrium point. (Hint: For $(\mathbf{x}_e)_4$, one can use Gershogorin circles.)
3. Consider the specific case:

$$A = \begin{pmatrix} 2 & -1 \\ -1 & 1.5 \end{pmatrix} \quad \text{and} \quad \mathbf{b} = \begin{pmatrix} 2 \\ 3 \end{pmatrix}$$

Predict the local trajectories around the equilibrium points. Obtain a rough sketch of the phase portrait, and then verify using a direction field and using ODE IVP solvers.

E8.9. Consider the following coupled equations describing the dynamics of a non-isothermal continuously stirred tank reactor based on energy and mass balance for a first-order reaction $A \rightarrow B$:

$$\begin{aligned}
\frac{dC}{dt} &= \frac{F}{V}(C_{in} - C) - k_0 C e^{-E/(R(T+460))} \\
\frac{dT}{dt} &= \frac{F}{V}(T_{in} - T) + \frac{(-\Delta H)}{\rho c_p} k_0 C e^{-E/(R(T+460))} - \frac{UA}{V\rho c_p}(T - T_j)
\end{aligned}$$

where C and T are the concentration of A and temperature in the reactor, respectively. Using the following parameter set:

F :	volumetric flow rate	$= 3{,}000\ ft^3/hr$
V :	liquid volume	$= 750\ ft^3$
T_{in} :	temperature of feed	$= 60\ ^oF$
C_{in}	: concentration of A in feed	$= 0.132\ lbmol/ft^3$
ΔH :	molar heat of reaction	$= -45{,}000\ BTU/lbmol$
ρc_p :	heat capacity per volume	$= 33.2\ BTU/(ft^3\ ^oF)$
U :	heat transfer coefficient	$= 75\ Btu/(hrft^2\ ^oF)$
A :	area for heat transfer	$= 1221\ ft^2$
k_o :	Arrhenius frequency factor	$= 15 \times 10^{12}\ hr^{-1}$
E :	activation energy	$= 32{,}400\ Btu/(lbmol\ ^oF)$
R :	universal gas constant	$= 1.987\ Btu/(lbmol\ ^oR)$
T_j:	jacket temperature	$= 55\ ^oF$

1. Find all the equilibrium points and show that they are all hyperbolic.
2. Based on linearization, determine the local behavior of the trajectories surrounding the equilibrium points.
3. Plot a direction field in a phase that contains all three equilibrium points. Then using an ODE IVP solver, obtain the phase portrait that shows the behavior of trajectories using various initial conditions.

E8.10. For the second-order system described by

$$\frac{d}{dt}\mathbf{x} = \begin{pmatrix} h(\mathbf{x}) & k \\ -k & h(\mathbf{x}) \end{pmatrix} \mathbf{x}$$

where $h(\mathbf{x})$ is a scalar function of x_1 and x_2 and k is a real-valued constant.

1. Show that the origin is a non-hyperbolic equilibrium point if $h(\mathbf{0}) = 0$.
2. Show that if $h(\mathbf{x})$ is negative definite (semi-definite), then the origin is asymptotically stable (Lyapunov stable) by using the Lyapunov function $V(\mathbf{x}) = \mathbf{x}^T\mathbf{x}$.
3. Determine the system stability/instability if $h(\mathbf{x})$ is positive definite.

E8.11. Use the Lyapunov function approach as in Example 8.8 to show that the following linear equation is stable:

$$\frac{d}{dt}\mathbf{x} = \begin{pmatrix} -5 & 1 \\ 2 & -4 \end{pmatrix} \mathbf{x}$$

E8.12. Use Bendixon's criterion to show that the system given in Example 8.5,

$$\frac{d}{dt}\begin{pmatrix} x_1 \\ x_2 \end{pmatrix} = \begin{pmatrix} x_2 \\ \cos(\pi x_1) - x_2 \end{pmatrix}$$

will never have limit cycles.

E8.13. For the system given by

$$\frac{d}{dt}\mathbf{x} = \mathbf{f}(\mathbf{x}) = \begin{pmatrix} x_2 \\ -x_1 - x_2\, g(r) \end{pmatrix}$$

where $r = \sqrt{x_1^2 + x_2^2}$.

1. Let $g(r)$ be any monotonic function in the range $r \geq 0$, with a root at $r = r^* > 0$, that is, $g(r) < 0$ for $r < r^*$ and $g(r) > 0$ for $r > r^*$. Use Poincaré-Bendixson's theorem to show that a limit cycle will exist around the origin. Furthermore, show that the limit cycle will have a circular shape. (Hint: Apply the theorem to the annular region between a circle having a radius less than r^* and a circle of radius greater than r^*.) Test this for $g(r) = (r-1)(r+1)$.
2. Let $g(r) = \cos(2\pi r/R)$ for some fixed R. Using Poincaré-Bendixson's theorem, show that there will be multiple circular limit cycles having radius r^* at the roots of $g(r)$, and determine which are stable limit cycles and which are unstable limit cycles.

E8.14. Consider the following system

$$\frac{d}{dt}\mathbf{x} = \mathbf{f}(\mathbf{x}) = \begin{pmatrix} x_2 \\ -\left(\frac{1}{2} - \left(1 - x_1^2\right) e^{-x_1^2}\right) x_2 - x_1 \end{pmatrix}$$

Check that the conditions of the Liènard system given in Example 8.12 are satisfied and thus this equation will have a stable limit cycle. Solve the system using the IVP solvers at different initial conditions to show that the limit cycle is stable and unique.

E8.15. Consider a second-order process given by $\ddot{x} = h(x)$ or, in state-space representation,

$$\frac{d}{dt}\begin{pmatrix} x \\ y \end{pmatrix} = \begin{pmatrix} y \\ h(x) \end{pmatrix}$$

1. Let x_r be any root of $h(x)$ such that
$$\left.\frac{d}{dx}h(x)\right|_{x=x_r} < 0$$
Show that the equilibrium points at $(x, y) = (x_r, 0)$ is non-hyperbolic, having linearized eigenvalues that are pure imaginary.
2. Assume $x_r = 0$; that is, the origin is an equilibrium point where $dh/dx < 0$ (This can be done by a simple translation of coordinates). Then show that the function
$$V(x, y) = \frac{y^2}{2} - \int_0^x h(\xi)d\xi$$
is a Lyapunov function for the origin for which $dV/dt = 0$ for a neighborhood around the origin. Thus $V = constant$ in the neighborhood around the origin, hence a conservative system. Thus the origin will be a nonlinear center.
3. Let $h(x) = -x(x+1)(x+4)$. Then according to the preceding results, the origin should behave as a nonlinear center for this case. Verify by using IVP solvers to obtain a phase-plane portrait using various initial conditions around the origin. What about around the neighborhood of $x = -4$ and $x = -1$?

4. Another special case is the model for an undamped pendulum given by $d^2\theta/dt^2 = -(g/\ell)\sin(\theta)$, where g and ℓ are the gravitational acceleration and length of the pendulum, respectively. Predict, and then simulate, the phase portrait of this conservative system. Show that this system is also a reversible system.

(Note: Problems **E8.16** to **E8.19** involve topics discussed in Appendix H.1.)

E8.16. For the one-dimensional system given by

$$\frac{dx}{dt} = x\left[r + (1-r)\,x\right]$$

where $|r| < 1$, determine the type of bifurcation around $r = 0$ and compare with the criteria given in (H.4), (H.6), or (H.7). Draw the bifurcation diagram and check by solving the system at different values of r and initial conditions.

E8.17. For the one-dimensional system given by

$$\frac{dx}{dt} = (r - x)\left(r + 2x^2\right)$$

determine the type of bifurcation around $r = 0$ and compare using one of the criteria given in (H.4), (H.6), or (H.7). Draw the bifurcation diagram and check by solving the system at different values of $-1 < r < 1$ and different initial conditions. What is the stability of the various branches at $r = -0.5$?

E8.18. Consider the following second-order process

$$\begin{aligned} \frac{dx}{dt} &= -ax + y \\ \frac{dy}{dt} &= \frac{x^2}{1+x^2} - by \end{aligned}$$

with $a, b > 0$.

1. From each equation we obtain two graphs: $y = g_1(x) = ax$ and $y = g_2(x) = x^2/[b(1+x^2)]$. The equilibrium points are at the intersection of both graphs. Plot both graphs under a fixed b to show that with $a_{lo} > 0$ there is a range of $a_{lo} < a < a_{hi}$ such that the number of equilibrium points will change from three to two. Also, for $a > a_{hi}$, there will only be one equilibrium point.
2. Determine the stability of the equilibrium points.
3. Draw a bifurcation diagram for this system for $b = 1$, and classify the type of bifurcations at the bifurcation points.

E8.19. The **Brusselator reaction** is given by following set of reactions:

$$\begin{aligned} A &\rightarrow x \\ B + x &\rightarrow y + C \\ 2x + y &\rightarrow 3x \\ x &\rightarrow D \end{aligned}$$

where the concentrations of C_A, C_B, C_C, and C_D are assumed to be constant. Furthermore, with the reactions assumed isothermal,the specific kinetic rate coefficients k_i for the ith reaction will be constant. The rate of change in the concentrations of x and y will then be given by

$$\begin{aligned}\frac{dC_x}{dt} &= k_1 C_A - k_2 C_B C_x + k_3 C_x^2 C_y - k_4 C_x \\ \frac{dC_y}{dt} &= k_2 C_B C_x - k_3 C_x^2 C_y\end{aligned}$$

1. Obtain the equilibrium point of the reaction system.
2. Choosing C_B as the bifurcation parameter, find the critical point $C_{B,h}$ under which the Brusselator can satisfy the conditions for Hopf bifurcation.
3. Using the IVP solvers to simulate the process, determine whether the Brusselator exhibits supercritical or subcritical Hopf bifurcation under the following fixed parameters: $k_1 = 1.8$, $k_2 = 2.1$, $k_3 = 1.2$, $k_4 = 0.75$, and $C_A = 0.9$.

9 Series Solutions of Linear Ordinary Differential Equations

In this chapter, we focus our attention on obtaining analytical solutions of linear differential equations with coefficients that are not constant. These solutions are not as simple as those for which the coefficients were constant. One general approach is to use a power series formulation.

In Section 9.1, we describe the main approaches of power series solution. Depending on the equation, one can choose to expand the solution around an ordinary point or a singular point. Each of these choices will determine the structure of the series. For an ordinary point, the expansion is simply a Taylor series, whereas for a singular point, we need a series known as a **Frobenius series**.

Although the power series method is straightforward, power series solutions can be quite lengthy and complicated. Nonetheless, for certain equations, solutions can be found based on the parameters of the equations, thus yielding direct solutions. This is the case for two important classes of second-order equations that have several applications. These are the Legendre equations and Bessel equations, which we describe in Sections 9.2 and 9.3, respectively.

We have also included other important equations in the exercises, such as hypergeometric equations, Jacobi equations, Laguerre equations, Hermite equations, and so forth, where the same techniques given in this chapter can be used to generate the useful functions and polynomials. Fortunately, the special functions and polynomials that solve these equations, including Legendre polynomials, Legendre functions and Bessel functions, are included in several computer software programs such as MATLAB.

9.1 Power Series Solutions

Consider the first-order equation

$$\frac{dy}{dx} - ay = 0 \quad \text{with} \quad y(0) = y_0 \tag{9.1}$$

We can apply the successive substitution approach known as **Picard's method** and obtain a series solution. The method begins with an integration of (9.1),

$$y(x) = y_0 + a \int_0^x y(\tau_1) d\tau_1$$

and continued with recursive integrations on $y(\tau_i)$, $i = 1, 2, \ldots$,

$$\begin{aligned} y(x) &= y_0 + a\int_0^x \left(y_0 + a\int_0^{\tau_1} y(\tau_2)d\tau_2 \right) d\tau_1 \\ &= y_0 + axy_0 + a^2\int_0^x\int_0^{\tau_1}\left(y_0 + a\int_0^{\tau_2} y(\tau_3)d\tau_3 \right) d\tau_2 d\tau_1 \\ &= y_0\left(1 + ax + \frac{(ax)^2}{2!} + \frac{(ax)^3}{3!} + \cdots\right) \\ &= y_0 e^{ax} \end{aligned}$$

This example suggests that some of the solutions to linear differential equations could be represented by power series. Furthermore, it shows that, as in the preceding process toward forming e^{ax}, the series solutions themselves can often generate the construction of new functions. For instance, we can show that Legendre polynomials and Bessel functions could be defined from series solutions of certain classes of second-order linear differential equations. However, instead of using Picard's method, we approach the solution by first constructing a power series that contains unknown parameters and coefficients. The parameters are then evaluated after substitution of the power series into the differential equations.

Consider the homogeneous linear N^{th}-order differential equation

$$\psi_N(x)\frac{d^N y}{dx^N} + \cdots + \psi_1(x)\frac{dy}{dx} + \psi_0(x)y = 0 \tag{9.2}$$

Definition 9.1. *A point x^* is an* **ordinary point** *of (9.2) if $\psi_N(x^*) \neq 0$ and $\psi_j(x^*) < \infty$, $j = 0, 1, 2, \ldots, (N-1)$. A point x^* is a* **singular point** *of (9.2) if $\psi_N(x^*) = 0$ or $\psi_j(x^*) = \pm\infty$ for some $0 \le j < N$. Furthermore, if x^* is a singular point, then it is known as a* **regular singular point** *if*

$$\lim_{x\to x^*}\left|(x-x^*)^{N-j}\,\frac{\psi_j(x)}{\psi_N(x)}\right| < \infty \qquad \text{for } j = 0, 1, \ldots, N-1 \tag{9.3}$$

Otherwise, x^* is an **irregular singular point**.

For instance, $x = -2$, $x = -1$, and $x = 0$ are ordinary, regular singular, and irregular singular points, respectively, of the following differential equation:

$$x\frac{d^2y}{dx^2} + \frac{2x}{x+1}\frac{dy}{dx} + \frac{4}{x^2(x+1)^2}y = 0$$

The general form for the power series solution is given by

$$y = \sum_{n=0}^{\infty} a_n\,(x - x_o)^{n+r} \tag{9.4}$$

where x_o is the point of expansion and r is an additional parameter. The choice of x_o will determine the range of x-values where the power series will converge, also known as the **radius of convergence**. This radius turns out to be the distance between x_o and the nearest singular point. For instance, if $x_o = 0$ and $x = \pm 1$ are the only two singular points of (9.2), then the series solution will have a radius of convergence equal to 1. However, suppose $x = 0$ is the only singular point. Then if we choose to expand around $x_o = 0$, the radius of convergence will be infinite. This means that even though expanding around ordinary points are straightforward to evaluate, it is still worthwhile to pursue expansions around singular points if doing so will significantly widen the radius of convergence.

We now list a set of formulas and assumptions that will be used during the development of the power series solutions:

1. A function $f(x)$ is **analytic** at a point x_o if $f(x)$ has a convergent Taylor series expansion around x_o.[1]
2. The product of two infinite series is given by
$$\left(\sum_{i=0}^{\infty} a_i x^i\right)\left(\sum_{j=0}^{\infty} b_j x^j\right) = \sum_{n=0}^{\infty}\sum_{k=0}^{n} a_k b_{n-k} x^n \tag{9.5}$$
3. The **Leibnitz formula** for the nth derivative of a product of functions is given by
$$\frac{d^n}{dx^n} f(x)g(x) = \sum_{k=0}^{n} \binom{n}{k} \frac{d^k f}{dx^k}\frac{d^{n-k} g}{dx^{n-k}} \tag{9.6}$$
where
$$\binom{n}{k} = \frac{n!}{(n-k)!k!}$$
4. Let $X = x - x_o$, then
$$\frac{d^k y}{dX^k} = \frac{d^k y}{dx^k}$$
With $\widehat{\psi}_j(X) = \psi(x)$, we could rewrite (9.2) as
$$\sum_{k=0}^{N} \widehat{\psi}_k(X)\frac{d^k y}{dX^k} = 0$$
Thus, in the next sections, we will assume that x has been shifted such that we could just expand the solution around $x_o = 0$. This will significantly simplify the solutions by avoiding the expansion of $(x - x_o)^k$.
5. The **Gamma Function** is a function defined by
$$\Gamma(x) = \int_0^{\infty} e^{-t} t^{x-1} dt \tag{9.7}$$
The plot of $\Gamma(x)$ is shown in Figure 9.1. Some of the properties of the Gamma functions are:
(a) $\Gamma(1) = 1$ and $\Gamma(1/2) = \sqrt{\pi}$
(b) Using integration by parts, we have
$$\Gamma(x+1) = -e^{-t}t^x\Big|_0^{\infty} + x\int_0^{\infty} e^{-t}t^{x-1}dt = x\Gamma(x) \tag{9.8}$$
(c) For n a non-negative integer, $\Gamma(n+1) = n!$
(d) For m a non-positive integer, $\lim_{x\to m^+}\Gamma(x) = (-1)^m\infty$ and $\lim_{x\to m^-}\Gamma(x) = (-1)^{m+1}\infty$

[1] A more complete discussion of analytic functions and singular points is given in L.2.1.

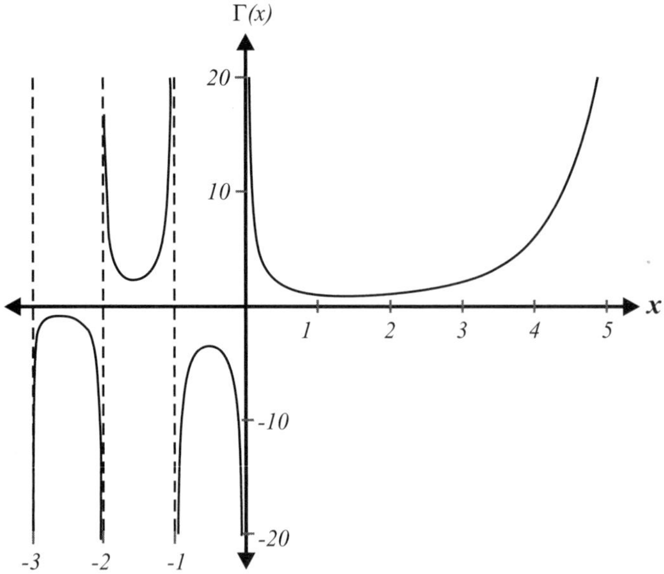

Figure 9.1. A plot of the Gamma function, $\Gamma(x)$.

The gamma function is used later to recast a product of a descending sequence; that is, using (9.8) recursively, one can show that

$$\prod_{k=0}^{m} (x+n-k) = \frac{\Gamma(x+n+1)}{\Gamma(x+n-m)} \tag{9.9}$$

where x is not necessarily an integer. The MATLAB command for $y = \Gamma(x)$ is `y=gamma(x)`.

6. Another product similar to (9.9) but of an ascending sequence is known as the **Pochhammer symbol**, denoted $(x)_n^{\mathcal{P}}$, and is the product defined by

$$(x)_n^{\mathcal{P}} = \begin{cases} 1 & \text{if } n = 0, x \neq 0 \\ (x)(x+1)\cdots(x+n-1) & \text{if } n > 0 \end{cases} \tag{9.10}$$

and is related to the Gamma function by

$$(x)_n^{\mathcal{P}} = \frac{\Gamma(x+n)}{\Gamma(x)} \tag{9.11}$$

7. The **generalized hypergeometric series**, denoted ${}_pF_q$, is an infinite series defined as follows:

$${}_pF_q\left(a_1, \ldots, a_p; b_1, \ldots, b_q; x\right) = \sum_{n=0}^{\infty} \frac{(a_1)_n^{\mathcal{P}} \cdots (a_p)_n^{\mathcal{P}}}{(b_1)_n^{\mathcal{P}} \cdots (b_q)_n^{\mathcal{P}}} \frac{x^n}{n!} \tag{9.12}$$

where $(a_1)_n^{\mathcal{P}}$, $(a_2)_n^{\mathcal{P}}$ and so forth are Pochhammer symbols defined in (9.10)
Special cases includes **Gauss' hypergeometric series** defined as

$${}_2F_1\left(a, b; c; x\right) = \sum_{n=0}^{\infty} \frac{(a)_n^{\mathcal{P}} (b)_n^{\mathcal{P}}}{(c)_n^{\mathcal{P}}} \frac{x^n}{n!} \tag{9.13}$$

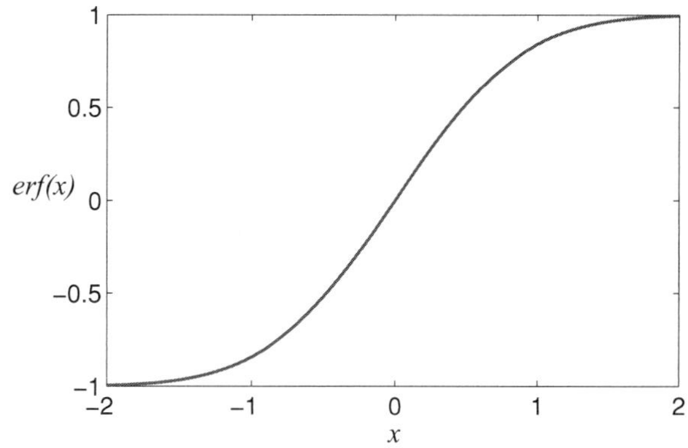

Figure 9.2. A plot of the error function, erf(x).

and the **confluent hypergeometric series** defined as

$$ {}_1F_1\left(a;b;x\right) = \sum_{n=0}^{\infty} \frac{(a)_n^{\mathcal{P}}}{(b)_n^{\mathcal{P}}} \frac{x^n}{n!} \tag{9.14}$$

These functions are useful in evaluating several series solutions. They were originally used to solve differential equations known as **hypergeometric equations** (see Exercise **E9.3**). The MATLAB command for

$$y = {}_pF_q(a_1, \ldots, a_p; b_1, \ldots, b_q, x)$$

is `y=hypergeom([a1,...,ap],[b1,...,bq],x)`.

8. The **error function**, denoted by erf(x), is defined as

$$\mathrm{erf}(x) = \frac{2}{\sqrt{\pi}} \int_0^x e^{-y^2} dy \tag{9.15}$$

A plot of the error function is shown in Figure 9.2.
In the limit as x approaches infinity, we have

$$\lim_{x\to\infty} \mathrm{erf}(x) = 1$$

An associated function known as the **complementary error function**, denoted by erfc(x), is defined as

$$\mathrm{erfc}(x) = 1 - \mathrm{erf}(x) = \frac{2}{\sqrt{\pi}} \int_x^{\infty} e^{-y^2} dy \tag{9.16}$$

In MATLAB, the function `erf` is available for evaluating the error function.

In the sections that follow, we first tackle the series solutions expanded around an ordinary point. Then we follow it with the discussion of the series solutions expanded around a regular singular point. We initially explore the solutions of high-order equations. However, at some point, we must limit it to second-order differential equations to allow for some tractable results.

9.1.1 Series Solutions Around Ordinary Points

Consider the homogenous N^{th}-order linear differential equation given by

$$\sum_{j=0}^{N} p_j(x)\frac{d^j y}{dx^j} = 0 \tag{9.17}$$

then $x = 0$ is an ordinary point of (9.17) if the coefficient functions $p_j(x)$ are all analytic around $x = 0$; that is, the coefficient functions can be expanded as

$$p_j(x) = \sum_{n=0}^{\infty} \rho_{j,n} x^n \qquad j = 0, 1, 2, \ldots, N \tag{9.18}$$

with the additional condition that $\rho_{N,0} \neq 0$.

THEOREM 9.1. *Let $x = 0$ be an ordinary point of the Nth-order linear differential equation described by (9.17) and (9.18). The general series solution is then given by*

$$y = \sum_{n=0}^{\infty} a_n x^n \tag{9.19}$$

with N arbitrary integration constants $a_0, a_1, \ldots, a_{N-1}$, and the remaining constants a_{n+N}, $n = 0, 1, \ldots$, satisfy the following recursion formulas:

$$a_{n+N} = \sum_{k=0}^{n+N-1} \phi_{n,k} a_k \qquad n = 0, 1, \ldots, \infty \tag{9.20}$$

where

$$\phi_{n,k} = (-1)\frac{\rho_{0,n-k} + \displaystyle\sum_{j=1}^{N} \rho_{j,n-k+j} \prod_{i=0}^{j-1}(k-i)}{\rho_{N,0} \displaystyle\prod_{i=1}^{N}(n+i)} \tag{9.21}$$

with

$$\rho_{j,\ell} = 0 \qquad \ell < 0$$

PROOF. (See Section I.6.1 for proof.)

EXAMPLE 9.1. Consider the third-order linear differential equation,

$$\frac{d^3y}{dx^3} + x^2\frac{d^2y}{dx^2} + 3x\frac{dy}{dx} + y = 0 \quad \text{and} \quad y(0) = 2; \frac{dy}{dx}(0) = 1; \frac{d^2y}{dx^2}(0) = 1$$

and we want to obtain a power series solution expanded around the origin. The origin is an ordinary point of the differential equation. The coefficient functions of the derivatives of y are

$$\begin{aligned}
p_3(x) &= 1 &\longrightarrow&\quad \rho_{3,0} = 1\\
p_2(x) &= x^2 &\longrightarrow&\quad \rho_{2,2} = 1\\
p_1(x) &= 3x &\longrightarrow&\quad \rho_{1,1} = 3\\
p_0(x) &= 1 &\longrightarrow&\quad \rho_{0,0} = 1
\end{aligned}$$

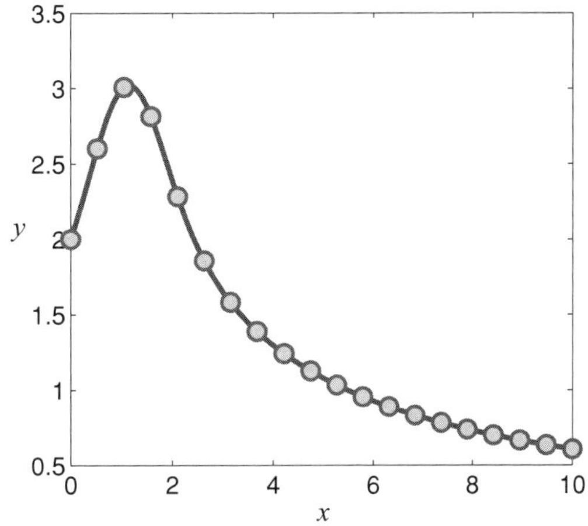

Figure 9.3. Series solution (open circles) together with Runge-Kutta solution.

Based on Theorem 9.1, the coefficients in the recurrence equation (9.20) for $N = 3$ will be given by

$$\phi_{n,k} = (-1)\frac{\rho_{0,n-k} + k\rho_{1,n-k+1} + (k)(k-1)\rho_{2,n-k+2} + (k)(k-1)(k-2)\rho_{3,n-k+3}}{(n+3)(n+2)(n+1)\rho_{3,0}}$$

The nonzero terms occur only for $\rho_{3,0}$ and $\rho_{i,i}$ for $i = 0, 1, 2$. Thus $\phi_{n,k} \neq 0$ only when $n = k$, which yields

$$\phi_{n,n} = (-1)\frac{n+1}{(n+2)(n+3)}$$

The recurrence equation (9.20) is then given by

$$a_{n+3} = \phi_{n,n}a_n = (-1)\frac{n+1}{(n+2)(n+3)}a_n \qquad n = 0, 1, 2, \ldots$$

and a pattern for a_{n+3}, $(n = 0, 1, \ldots)$, in terms of a_0, a_1, and a_2, can be found.

One can show that the solution is then given by

$$y = \sum_{\ell=0}^{2} a_\ell\, x^\ell \left[1 + \ell! \sum_{k=1}^{\infty} \frac{\left(-x^3\right)^k}{(3k+\ell)!} \prod_{j=1}^{k} (3j - 2 + \ell)^2\right]$$

Using the initial conditions, we obtain $a_0 = 2$, $a_1 = 1$ and $a_2 = 1/2$. In terms of generalized hypergeometric functions (cf. (9.12)), we have

$$\begin{aligned} y(x) &= 2\left(1 - \frac{x^3}{6}\,{}_2F_2\left[1, \frac{4}{3}; 2, \frac{5}{3}; \frac{-x^3}{3}\right]\right) + x\left(1 - \frac{x^3}{6}\,{}_2F_2\left[1, \frac{5}{3}; 2, \frac{7}{3}; \frac{-x^3}{3}\right]\right) \\ &\quad + \frac{x^2}{2}\left(1 - \frac{3x^3}{20}\,{}_2F_2\left[1, 2; \frac{8}{3}, \frac{7}{3}; \frac{-x^3}{3}\right]\right) \end{aligned}$$

The series solutions are compared with the numerical solution using Runge-Kutta in Figure 9.3.

In some cases, the series solutions may yield some group of terms that can be replaced by their closed forms. When these occur, the method of order reduction may be useful in finding a closed form for the other solutions.

Let the solution of a second-order homogeneous differential equation be given by $y(x) = Au(x) + Bv(x)$. Suppose $u(x)$ has been found (e.g., using series solution) and has a known closed form. The method of order reduction simply proceeds by letting the second solution take the form of $v(x) = q(x)u(x)$, where $q(x)$ will be solved from an associated equation that will hopefully be from a reduced order equation or an equation that can be solved using other techniques. More details about the method of order reduction can be found in Section I.2 as an appendix.

EXAMPLE 9.2. Consider the following equation[2]

$$a\frac{d^2y}{dx^2} + x\frac{dy}{dx} - y = 0$$

One of the solutions[3] can be quickly determined to be $u(x) = x$. Instead of proceeding completely with the other series solution, we can try to determine whether a second solution in closed form can be found via the order reduction approach.

Let the second solution be given by $v(x) = q(x)u(x) = xq(x)$. Substituting this into the differential equation,

$$\frac{d^2q}{dx^2} + \left(\frac{2}{x} + \frac{x}{a}\right)\frac{dq}{dx} = 0$$

This is a second-order equation with a missing dependent variable. Letting $p = dq/dx$, we have

$$\frac{dp}{dx} + \left(\frac{2}{x} + \frac{x}{a}\right)p = 0 \quad \rightarrow \quad p = C\frac{1}{x^2}e^{-x^2/(2a)}$$

or

$$q = C\int p\, dx = -C\left[\frac{1}{x}e^{-x^2/(2a)} + \sqrt{\frac{\pi}{2a}}\operatorname{erf}\left(\frac{x}{\sqrt{2a}}\right)\right]$$

where $\operatorname{erf}(x)$ is the error function (cf. (9.15)).

Finally, we obtain the alternative form to the series solution given by

$$y = Ax + \bar{C}\left[e^{-x^2/(2a)} + \sqrt{\frac{\pi}{2a}}\, x \operatorname{erf}\left(\frac{x}{\sqrt{2a}}\right)\right]$$

This is relatively easier to evaluate (given that $\operatorname{erf}(x)$ is a function that is available in MATLAB and other programs) than the solution obtained by a complete implementation of Theorem 9.1, which is given by

$$y = Ax + B\left(1 - 2\sum_{k=1}^{\infty}\left(-\frac{1}{2a}\right)^k \frac{(2k-2)!}{(k-1)!(2k)!}x^{2k}\right) \tag{9.22}$$

[2] This equation results from solving the diffusion equation after applying the similarity transform method (cf. Example 11.15).

[3] This first solution derives from Theorem 9.1, where $\phi_{1,1} = 0$. This leads to $a_3 = a_5 = a_7 = \cdots = 0$.

9.1.2 Series Solutions Around Regular Singular Points

The series solutions around regular singular points add significant complexity over those around ordinary points. For the linear N^{th}-order linear differential equation given by

$$\sum_{i=0}^{N} x^i \widetilde{p}_i(x) \frac{d^i y}{dx^i} = 0 \tag{9.23}$$

$x = 0$ is a regular singular point of (9.23) if $\widetilde{p}_i(x)$, $i = 1, \ldots, N$, are analytic around $x = 0$, with

$$\widetilde{p}_i(x) = \sum_{n=0}^{\infty} \widetilde{\rho}_{i,n} x^n \quad ; \quad \widetilde{\rho}_{N,0} \neq 0 \tag{9.24}$$

Then the series solution around the regular singular point $x = 0$ will take a form that is slightly different from the series solution around an ordinary point given in (9.19).

To motivate the form needed for the solution, consider the first-order differential equation,

$$x\frac{dy}{dx} + (q_0 + q_1 x)\, y = 0$$

The differential equation is separable, and the solution can be found to be

$$y = Ax^{-q_0} e^{-q_1 x} = Ax^{-q_0}\left(1 - q_1 x + \frac{1}{2}(q_1 x)^2 + \cdots\right) = A \sum_{n=1}^{\infty} a_n x^{(n-q_0)}$$

where A is an arbitrary constant. Note that the exponents now involve another parameter q_0.

Thus for the series solution around a regular singular point at $x = 0$, we use a more general series known as the **Frobenius series** given by

$$y = \sum_{n=0}^{\infty} a_n x^{n+r} \tag{9.25}$$

where r is an additional parameter, and the solution method is known as the **Frobenius method**. Using (9.25), the product of x_j and jth derivative of y can be evaluated as follows:

$$\begin{aligned} x\frac{dy}{dx} &= \sum_{n=0}^{\infty} (n+r) a_n x^{n+r} \\ &\vdots \\ x^j \frac{d^j y}{dx^i} &= \sum_{n=0}^{\infty} \left(\prod_{\ell=0}^{j-1} n + r - \ell \right) a_n x^{n+r} \end{aligned} \tag{9.26}$$

After substitution of (9.24) and (9.26) into (9.23), with the implementation of (9.5),

$$\sum_{n=0}^{\infty} x^{n+r} \sum_{k=0}^{n} a_k \left[\widetilde{\rho}_{0,n-k} + \sum_{j=1}^{N} \widetilde{\rho}_{j,n-k} \prod_{\ell=0}^{j-1} (k + r - \ell) \right] = 0 \tag{9.27}$$

For $n = 0$, the coefficient of x^r has to be equal to zero and is given by

$$a_0 \left(\widetilde{\rho}_{0,0} + \sum_{j=1}^{N} \widetilde{\rho}_{j,0} \prod_{\ell=0}^{j-1} (r - \ell) \right)$$

By setting a_0 to be an arbitrary constant, the parameter r has to satisfy

$$\widetilde{\rho}_{0,0} + \sum_{j=1}^{N} \widetilde{\rho}_{j,0} \prod_{\ell=0}^{j-1} (r - \ell) = 0 \tag{9.28}$$

Equation (9.28) is known as the **indicial equation** of (9.23), and the N roots of (9.28) $r_1, \ldots, r_N$ are known as the **indicial roots** (also known as **exponents**).

Having set a_0 as arbitrary, (9.27) also implies that the other coefficients can be obtained by using the following recurrence formula:

$$a_n Q_{n,n}(r) = -\sum_{k=0}^{n-1} Q_{n,k}(r) a_k \qquad n = 1, 2, \ldots \tag{9.29}$$

where

$$Q_{n,k}(r) = \widetilde{\rho}_{0,n-k} + \sum_{j=1}^{N} \widetilde{\rho}_{j,n-k} \prod_{\ell=0}^{j-1} (k + r - \ell) \tag{9.30}$$

If $Q_{n,n}(r) \neq 0$,

$$a_n = \widetilde{\phi}_n(r) a_0 \tag{9.31}$$

where

$$\widetilde{\phi}_n(r) = \begin{cases} 1 & \text{if } n = 0 \\ -1 \left(\sum_{k=0}^{n-1} Q_{n,k}(r) \widetilde{\phi}_k(r) \right) / Q_{n,n}(r) & \text{if } n > 0 \end{cases} \tag{9.32}$$

When $Q_{n,n}(r) \neq 0$ for all indicial roots r, the complete solution is immediately given by a linear combination of solutions based on each root,

$$y = \sum_{k=1}^{N} c_k y_k \tag{9.33}$$

where

$$y_k = \left. \sum_{n=0}^{\infty} \widetilde{\phi}_n(r) x^{r+n} \right|_{(r=r_k)} \tag{9.34}$$

$c_1, c_2, \ldots, c_N$ are arbitrary constants needed to satisfy initial and boundary conditions.

The condition under which $Q_{n,n}(r) = 0$ will occur is when some indicial roots differ by an integer. To show this, suppose $r_a - r_b = m$, where r_a and r_b are two of the indicial roots and m is a positive integer. In the process of solving for y_k in (9.34), the mth term in the series will involve the evaluation of $Q_{m,m}(r_b)$. From (9.30),

$$Q_{m,m}(r_b) = \widetilde{\rho}_{0,0} + \sum_{j=1}^{N} \widetilde{\rho}_{j,0} \prod_{\ell=0}^{j-1} (m + r_b - \ell) = \widetilde{\rho}_{0,0} + \sum_{j=1}^{N} \widetilde{\rho}_{j,0} \prod_{\ell=0}^{j-1} (r_a - \ell) \tag{9.35}$$

which is the indicial equation with r replaced by r_a (cf. (9.28)); thus $Q_{m,m}(r_b)$ will be zero. If the numerator for $\widetilde{\phi}_m(r_b)$ in (9.32) happens to be zero, then the recurrence can be continued by setting $\widetilde{\phi}_m(r_b) = 0$. Otherwise, additional procedures will be needed to treat these cases.

There are two cases that yield an integer difference between two indicial roots: when some of the indicial roots are equal and when some indicial roots differ by a nonzero integer. We can partition the indicial roots into equivalence subsets, where in each subset the roots are either repeating or differ by integers. Taking the largest value from each of the equivalence subsets, (9.34) can be used to obtain at least one solution from each of these subsets. To determine the other solutions, we can apply a technique known as the **d'Alembert method of order reduction**. Briefly, this method is used to find another solution $v(x) = u(x)\int z(x)dx$, where $u(x)$ is a known solution and $z(x)$ is a function that solves an associated differential equation whose order is one less than the original differential equation. Details for the method of order reduction are given in Section I.2 as an appendix. Unfortunately, this additional procedure can also get very complicated for high-order differential equations.

We can directly apply the procedure just discussed to second-order equations, and the details for doing so are given in Section I.1 as an appendix. We summarize the results from applying the Frobenius method to the second-order equation, expanded around a regular singular point at $x = 0$, in the following theorem:

THEOREM 9.2. **Frobenius Series Solution of Linear Second Order ODE** *Given the homogenous second-order linear differential equation*

$$x^2\widetilde{P}_2(x)\frac{d^2y}{dx^2} + x\widetilde{P}_1(x)\frac{dy}{dx} + \widetilde{P}_0(x)y = 0$$

where

$$\widetilde{P}_i(x) = \sum_{k=0}^{\infty}\widetilde{\rho}_{i,k}x^k \qquad i = 0, 1, 2 \qquad \textit{and} \;\; \widetilde{\rho}_{2,0} \neq 0$$

The indicial roots r_a and r_b, with $r_a \geq r_b$ are

$$r_a, r_b = \frac{(\widetilde{\rho}_{2,0} - \widetilde{\rho}_{1,0}) \pm \sqrt{(\widetilde{\rho}_{2,0} - \widetilde{\rho}_{1,0})^2 - 4\widetilde{\rho}_{0,0}\widetilde{\rho}_{2,0}}}{2\widetilde{\rho}_{2,0}}$$

and the complete solution is given by $y = Au(x) + Bv(x)$, where A and B are arbitrary coefficients and

$$u(x) = \sum_{n=0}^{\infty} a_n x^{n+r_a} \tag{9.36}$$

$$v(x) = \eta u(x)\ln(x) + \sum_{n=0}^{\infty} b_n x^{n+r_b} \tag{9.37}$$

If the difference between r_a and r_b is an integer, let $m = (r_a - r_b) \geq 0$. Also, define $Q_{n,k}(r)$, $\widetilde{\phi}_n(r)$ and $\sigma_n(r)$ as follows:

$$Q_{n,k}(r) = \begin{cases} \widetilde{\rho}_{0,n-k} + \widetilde{\rho}_{1,n-k}(k+r) \\ \quad + \widetilde{\rho}_{2,n-k}(k+r)(k+r-1) & \text{if } k < n \\ \\ n\left[\widetilde{\rho}_{2,0}n + \widetilde{\rho}_{2,0}\,(2r-1) + \widetilde{\rho}_{1,0}\right] & \text{if } k = n \end{cases} \tag{9.38}$$

$$\widetilde{\phi}_n(r) = \begin{cases} 1 & \text{if } n = 0 \\ \\ -\dfrac{\sum_{k=0}^{n-1} Q_{n,k}(r)\widetilde{\phi}_k(r)}{Q_{n,n}(r)} & \text{if } n > 0 \end{cases} \tag{9.39}$$

$$\sigma_n(r) = \sum_{k=0}^{n} \left[(2r+2k-1)\,\widetilde{\rho}_{2,n-k} + \widetilde{\rho}_{1,n-k}\right]\widetilde{\phi}_k(r) \tag{9.40}$$

then the coefficients a_n, b_n, and η are given by

$$a_n = \widetilde{\phi}_n\,(r_a) \tag{9.41}$$

$$b_n = \begin{cases} \widetilde{\phi}_n\,(r_b) & \text{if } (r_a - r_b) \text{ not an integer} \\ & \quad \text{or} \quad n < m \\ \\ 0 & \text{if } n = m \\ \\ -\dfrac{\sum_{k=0}^{n-1} Q_{n,k}(r_b)b_k + \eta\sigma_{n-m}(r_a)}{Q_{n,n}(r_b)} & \text{if } n > m \end{cases} \tag{9.42}$$

$$\eta = \begin{cases} 0 & \text{if } (r_a - r_b) \text{ not an integer} \\ \\ 1 & \text{if } m = 0 \\ \\ -\dfrac{\sum_{k=0}^{m-1} Q_{m,k}(r_b)\widetilde{\phi}_k(r_b)}{\sigma_0(r_a)} & \text{if } m > 0 \end{cases} \tag{9.43}$$

PROOF. (See Section I.6.2 for proof.)

We have included three examples in Section I.3 where this theorem has been applied. Also, in Section 9.3, we use the result of Theorem 9.2 to obtain the solution of an important class of equations known as **Bessel equations**.

9.2 Legendre Equations

The Legendre equations and the associated Legendre equations are given in Table 9.1. These equations are often obtained during the analytical solutions of

Table 9.1. *Solutions of legendre equations*

Equation	Solution
Legendre equation of order μ:	
$(1-x^2)\frac{d^2y}{dx^2} - 2x\frac{dy}{dx} + \mu(\mu+1)y = 0$	$y = AL_{\text{even}}(x) + BL_{\text{odd}}(x)$
Legendre equation of integer order n:	
$(1-x^2)\frac{d^2y}{dx^2} - 2x\frac{dy}{dx} + n(n+1)y = 0$	$y = A\mathcal{P}_n(x) + B\mathcal{Q}_n(x)$
Associated Legendre equation:	
$(1-x^2)\frac{d^2y}{dx^2} - 2x\frac{dy}{dx} + \left(n(n+1) - \frac{m^2}{1-x^2}\right)y = 0$ (where n, m are integers)	$y = A\mathcal{P}_{n,m}(x) + B\mathcal{Q}_{n,m}(x)$

partial differential equations involving spherical coordinates (see Example 9.3). The series solutions of these equations involve the direct implementation of the results given in Theorem 9.1 of Section 9.1.1. The details of the series solution for these Legendre equations and associated Legendre equations can be found in section I.4 as an appendix. These solutions can be formulated using special functions, and they are summarized in Table 9.1.

The special functions in Table 9.1 result from grouping various terms in the series solutions. These include Legendre polynomials $\mathcal{P}_n(x)$, the Legendre functions $\mathcal{Q}_n(x)$, the Legendre functions of the second kind, L_{even} and L_{odd}, the associated Legendre polynomials $\mathcal{P}_{n,m}$, and the associated Legendre functions $\mathcal{Q}_{n,m}$. The definitions of these functions are given in Table 9.2.

Because of the importance of Legendre polynomials and functions, several computer programs have built-in routines for the evaluation of these functions. In MATLAB, the command for the associated Legendre polynomials is `legendre(n,x)`, which yields a vector $\left(\mathcal{P}_{n,0}(x), \ldots, \mathcal{P}_{n,n}(x)\right)^T$.

The first five Legendre polynomials are given by,

$$\begin{aligned} \mathcal{P}_0 &= 1 & \mathcal{P}_3 &= \frac{5x^3-3x}{2} \\ \mathcal{P}_1 &= x & & \\ \mathcal{P}_2 &= \frac{3x^2-1}{2} & \mathcal{P}_4 &= \frac{35x^4-30x^2+3}{8} \end{aligned}$$

A plot of these are shown in Figure 9.4. Also, the first four Legendre functions are given by

$$\begin{aligned} \mathcal{Q}_0(x) &= \text{atanh}(x) & \mathcal{Q}_2(x) &= \frac{3x^2-1}{2}\text{atanh}(x) - \frac{3}{2}x \\ \mathcal{Q}_1(x) &= x\,\text{atanh}(x) - 1 & \mathcal{Q}_3(x) &= \frac{5x^3-3x}{2}\text{atanh}(x) - \frac{5}{2}x^2 + \frac{2}{3} \end{aligned}$$

A plot of these are shown in Figure 9.5.

Table 9.2. *Legendre polynomials and functions*

Name	Definition
Legendre Polynomial	$\mathcal{P}_n(x) = \sum_{k=0}^{\nu(n)} \frac{(-1)^k\,[2n-2k]!}{2^n k!(n-k)!(n-2k)!} x^{n-2k}$ where $\nu(n) = \begin{cases} n/2 & \text{if } n \text{ is even} \\ (n-1)/2 & \text{if } n \text{ is odd} \end{cases}$
Legendre Function	$\mathcal{Q}_n(x) = \mathcal{P}_n(x) \int \left[\frac{1}{(1-x^2)\left(\mathcal{P}_n(x)\right)^2} \right] dx$
Legendre Functions of 2nd kind	$L_{\text{even}}(x) = 1 + \sum_{n=1}^{\infty} \Lambda_{2n}(\mu) x^{2n}$ $L_{\text{odd}}(x) = x + \sum_{n=1}^{\infty} \Lambda_{2n+1}(\mu) x^{2n+1}$ where, $\Lambda_m = \frac{(-1)^{\nu(m)}}{m!} \prod_{k=0}^{\nu(m)-1} \left[\frac{(\mu-m-2k)}{(\mu+m-2k+1)} \right]$ $\nu(m) = \begin{cases} m/2 & \text{if } m \text{ is even} \\ (m-1)/2 & \text{if } m \text{ is odd} \end{cases}$
Associated Legendre Polynomial	$\mathcal{P}_{n,m} = (-1)^m \left(1-x^2\right)^{m/2} \frac{d^m}{dx^m} \mathcal{P}_n(x)$
Associated Legendre Function	$\mathcal{Q}_{n,m} = (-1)^m \left(1-x^2\right)^{m/2} \frac{d^m}{dx^m} \mathcal{Q}_n(x)$

EXAMPLE 9.3. The Laplace equation in spherical coordinates (cf. (4.95)) is given by

$$\nabla^2 F = 0 = r^2 s_\theta \frac{\partial^2 F}{\partial r^2} + 2rs_\theta \frac{\partial F}{\partial r} + s_\theta \frac{\partial^2 F}{\partial \theta^2} + c_\theta \frac{\partial F}{\partial \theta} + \frac{1}{s_\theta} \frac{\partial^2 F}{\partial \phi^2} \tag{9.44}$$

where $s_\theta = \sin(\theta)$ and $c_\theta = \cos(\theta)$. This differential equation is linear and can be solved using the method of separation of variables. This method is discussed in more detail in Section 11.3. Basically, the solution is first assumed to be a product of three univariable functions, that is, $F = R(r)\Theta(\theta)\Phi(\phi)$. After substituting this product into (9.44), the following equation results

$$A(r) + B(\theta) + \frac{1}{s_\theta^2} C(\phi) = 0 \tag{9.45}$$

where

$$A(r) = \frac{r^2}{R}\frac{d^2 R}{dr^2} + \frac{2r}{R}\frac{dR}{dr}\ ; \qquad B(\theta) = \frac{1}{\Theta}\frac{d^2\Theta}{d\theta^2} + \frac{c_\theta}{s_\theta \Theta}\frac{d\Theta}{d\theta}\ ; \qquad C(\phi) = \frac{1}{\Phi}\frac{d^2\Phi}{d\phi^2}$$

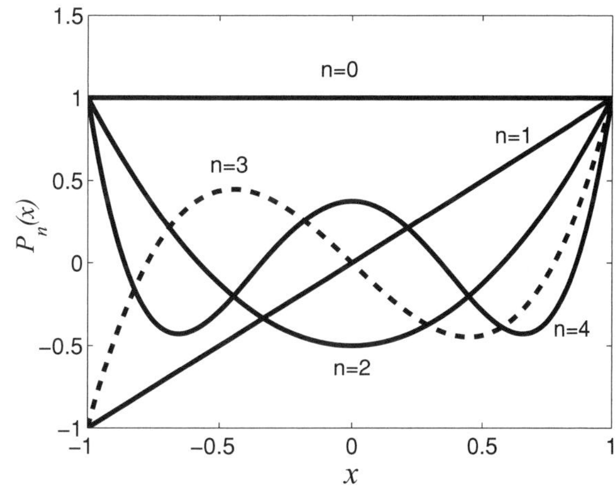

Figure 9.4. Plots of Legendre polynomials of orders 0 to 4.

For (9.45) to be true for arbitrary values of the independent variables, the three functions $A(r)$, $B(\theta)$, and $C(\phi)$ will need to satisfy the following:

$$A(r) = \alpha \quad ; \quad C(\phi) = \beta \quad ; \quad B(\theta) + \alpha + \frac{\beta}{s_\theta^2} = 0$$

where α and β are constants. Focusing on $B(\theta)$, we then have

$$\frac{d^2\Theta}{d\theta^2} + \frac{c_\theta}{s_\theta}\frac{d\Theta}{d\theta} + \left(\alpha + \frac{\beta}{s_\theta^2}\right)\Theta = 0$$

Next, let $x = c_\theta$, then $s_\theta^2 = 1 - x^2$, $dx = -s_\theta\, d\theta$ and

$$\frac{d\Theta}{d\theta} = -s_\theta \frac{d\Theta}{dx} \quad \text{and} \quad \frac{d^2\Theta}{d\theta^2} = -c_\theta \frac{d\Theta}{dx} + s_\theta^2 \frac{d^2\Theta}{dx^2}$$

Switching the independent variable to be $-1 \leq x \leq 1$,

$$\left(1 - x^2\right)\frac{d^2\Theta}{dx^2} - 2x\frac{d\Theta}{dx} + \left(\alpha + \frac{\beta}{1 - x^2}\right)\Theta = 0$$

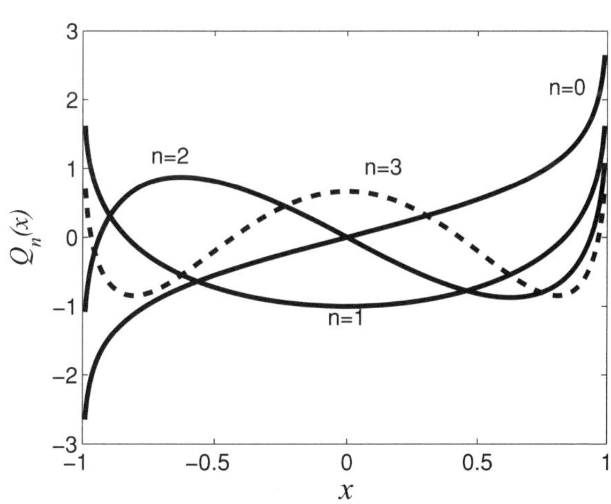

Figure 9.5. Plots of Legendre functions of orders 0 to 4.

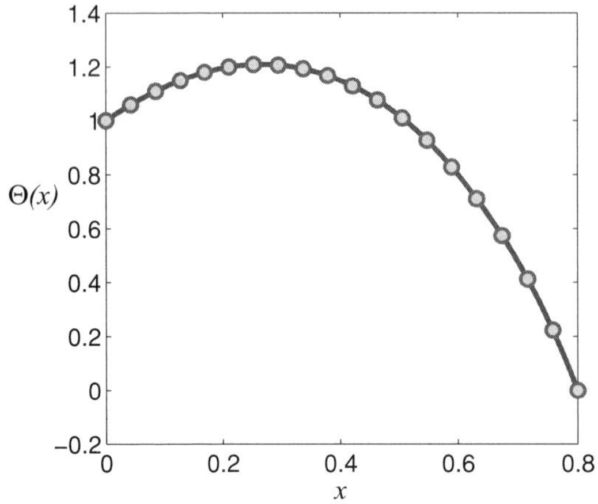

Figure 9.6. The solution to the associated Legendre equation with $n = 2$ and $m = 1$. The solid line is the numerical BVP solution, whereas the open circles are the analytical series solutions.

Furthermore, we set $\alpha = n(n+1)$ and $\beta = -m^2$, which then results in an associated Legendre equation,

$$\left(1 - x^2\right)\frac{d^2\Theta}{dx^2} - 2x\frac{d\Theta}{dx} + \left(n(n+1) - \frac{m^2}{1-x^2}\right)\Theta = 0$$

Using Table 9.1, the solution is $\Theta = A\mathcal{P}_{n,m}(x) + B\mathcal{Q}_{n,m}(x)$. For a numeric example, let $n = 2$ and $m = 1$, and boundary conditions $\Theta(0) = 1$ and $\Theta(0.8) = 0$. Then we have

$$\Theta(x) = 0.503\mathcal{P}_{2,1}(x) + 0.5\mathcal{Q}_{2,1}(x)$$

where

$$\mathcal{P}_{2,1}(x) = -3x\sqrt{1-x^2} \text{ and } \mathcal{Q}_{2,1}(x) = -\sqrt{1-x^2}\left[3x \text{ Atanh}(x) + \frac{3x^2-1}{2(1-x^2)} - \frac{3}{2}\right]$$

A plot of $\Theta(x)$ for this numeric case is shown in Figure 9.6 together with the numerical BVP solution.

Some of the properties of Legendre polynomials include the following:

1. The boundary values are given by $\mathcal{P}_n(1) = 1$ and $\mathcal{P}_n(-1) = (-1)^n$. This can be quickly verified to be the case in Figure 9.4.
2. The polynomials can also be evaluated using the **Rodriguez formula**:

$$\mathcal{P}_n(x) = \frac{1}{2^n n!}\frac{d^n}{dx^n}\left[\left(x^2-1\right)^n\right] \tag{9.46}$$

3. The generating function for Legendre polynomials is

$$g_L(x,t) = \frac{1}{\sqrt{1-2xt+t^2}} = \sum_{n=0}^{\infty}\mathcal{P}_n(x)t^n \tag{9.47}$$

This means that taking the Taylor series of $g_L(x,t)$, based on powers of t while keeping x constant, the coefficients of the series will yield the value of Legendre polynomial at x.

Table 9.3. *Solutions of Bessel equations*

Equation	Solution
Bessel equation of order ν:	
$x^2\dfrac{d^2y}{dx^2}+x\dfrac{dy}{dx}+\left(x^2-\nu^2\right)y=0$	$y=AJ_\nu(x)+BY_\nu(x)$
Bessel equation with parameter λ:	
$x^2\dfrac{d^2y}{dx^2}+x\dfrac{dy}{dx}+\left(\lambda^2x^2-\nu^2\right)y=0$	$y=AJ_\nu(\lambda x)+BY_\nu(\lambda x)$
Modified Bessel equations:	
$x^2\dfrac{d^2y}{dx^2}+x\dfrac{dy}{dx}-\left(x^2+\nu^2\right)y=0$	$y=AI_\nu(x)+BK_\nu(x)$
Generalized Bessel equation:	
$x^2\dfrac{d^2y}{dx^2}+xf_1(x)\dfrac{dy}{dx}+f_2(x)y=0$ $f_1(x)=a+2bx^p$ $f_2(x)=\left[c+bx^p\left(p+a-1\right)+\left(bx^p\right)^2\right]+dx^{2q}$	(see (9.55) for solution)

4. The orthogonality property of the Legendre polynomials is given by

$$\int_{-1}^{1}\mathcal{P}_n(x)\mathcal{P}_m(x)dx=\begin{cases}0 & \text{if } m\neq n\\ \dfrac{2}{2n+1} & \text{if } m=n\end{cases} \tag{9.48}$$

This property allows the Legendre polynomials to be used as a basis for the series approximation of continuous functions in $-1\le x\le 1$, that is, $f(x)=\sum_{n=0}^{\infty}a_n\mathcal{P}_n(x)$.

9.3 Bessel Equations

The Bessel equations and the modified Bessel equations are given in Table 9.3. These second-order equations often result from partial differential equations involving cylindrical coordinates. The series solutions of these equations involve the direct implementation of the results given in Theorem 9.2 of Section F.2. The details of the series solution for the Bessel equations and modified Bessel equations can be found in Section I.5 as an appendix. These solutions can be formulated using special functions, and they are summarized in Table 9.3.

As before, the special functions in Table 9.3 result from grouping various terms in the series solutions. These include Bessel functions of the first kind $J_\nu(x)$, the Bessel function of the second kind $Y_\nu(x)$, the modified Bessel function of the first kind $I_\nu(x)$, and the modified Bessel function of the second kind $K_\nu(x)$. The definitions of these functions are summarized in Table 9.4.

Although the definitions for Bessel functions of the first and second kind appear to be quite complicated, several computer software programs are able to evaluate these functions. In MATLAB, the functions for $J_\nu(x)$ and $Y_\nu(x)$ are given by

Table 9.4. *Bessel and modified Bessel functions*

Name	Definition
Bessel function of 1st kind	$J_\nu(x) = \sum_{n=0}^{\infty}\left(\frac{x}{2}\right)^{2n+\nu}\frac{(-1)^n}{n!\Gamma(n+\nu+1)}$
Bessel function of 2nd kind	$Y_\nu(x) = \frac{J_\nu(x)\cos(\nu\pi) - J_{-\nu}(x)}{\sin(\nu\pi)}$
Modified Bessel function of 1st kind	$I_\nu(x) = \exp\left(-\frac{\nu\pi i}{2}\right)J_\nu(ix)$
Modified Bessel function of 2nd kind	$K_\nu(x) = \exp\left(\frac{(\nu+1)\pi i}{2}\right)\left[J_\nu(ix) + iY_\nu(ix)\right]$

`BESSELJ(nu,x)` and `BESSELY(nu,x)`, respectively. Also, the MATLAB functions for $I_\nu(x)$ and $K_\nu(x)$ are given by `BESSELI(nu,x)` and `BESSELK(nu,x)`, respectively. A plot $J_\nu(x)$, $Y_\nu(x)$, $I_\nu(x)$ and $K_\nu(x)$ can be generated using MATLAB, and these are shown in Figures 9.7 and 9.8.

Instead of using the Frobenius series solution process outlined in Theorem 9.2, it is often worthwhile to determine whether a given equation can be transformed first to either a Bessel or modified Bessel equations, to take advantage of existing solutions (and functions). To find the conditions in which this might be possible, let $X = X(x)$ be the new independent variable and

$$Y = yG(x) \tag{9.49}$$

be the new dependent variable.[4] Evaluating the derivatives dY/dX and d^2Y/dX^2,

$$\begin{aligned} \frac{dY}{dX} &= \frac{yG' + Gy'}{X'} \\ \frac{d^2Y}{dX^2} &= \frac{y\left(G''X' - G'X''\right) + y'\left(2G'X' - GX''\right) + y''GX'}{(X')^3} \end{aligned}$$

where we used the prime notation to denote the operation d/dx, that is, $G' = dG/dx$, $G'' = d^2G/dx^2$, and so forth.

In terms of X and Y, the Bessel equation with parameter λ and order ν is given by

$$X^2\frac{d^2Y}{dX^2} + X\frac{dY}{dX} + \left(\lambda^2X^2 - \nu^2\right)Y = 0 \tag{9.50}$$

Returning to the original variables,

$$x^2\frac{d^2y}{dx^2} + xf_1(x)\frac{dy}{dx} + f_2(x)y = 0 \tag{9.51}$$

[4] A more general approach is to let $X = X(x, y)$ and $Y = Y(x, y)$. As expected, doing so would yield much more complicated results.

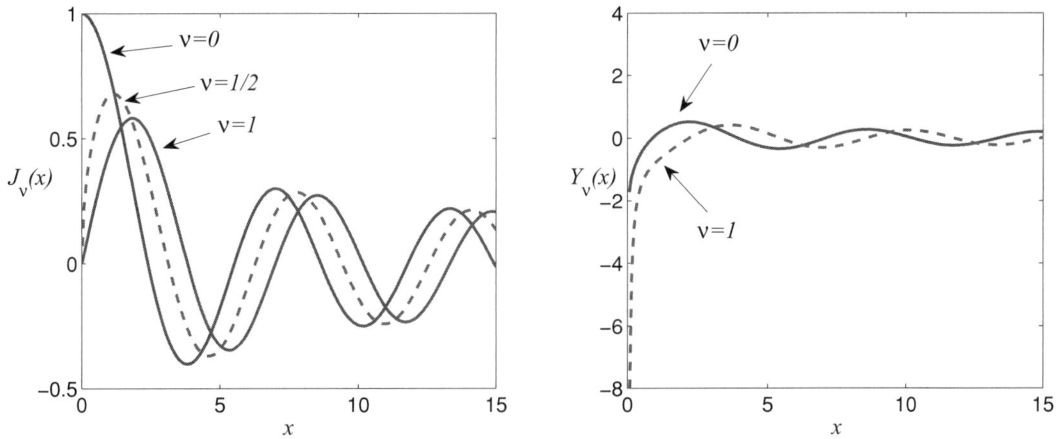

Figure 9.7. Plots of Bessel functions of the first kind, $J_\nu(x)$, and second kind, $Y_\nu(x)$.

where

$$
\begin{aligned}
f_1(x) &= x\left(2\frac{G'}{G}+\frac{X'}{X}-\frac{X''}{X'}\right)\\
f_2(x) &= x^2\left[\frac{G''}{G}+\left(\frac{X'}{X}-\frac{X''}{X'}\right)\frac{G'}{G}+(\lambda X')^2-\left(\nu\frac{X'}{X}\right)^2\right]
\end{aligned}
\qquad (9.52)
$$

Based on choices for $X(x)$ and $G(x)$, both $f_1(x)$ and $f_2(x)$ should be analytic at $x=0$ to guarantee a series solution for (9.51) and, in particular, one that could be written in terms of Bessel or modified Bessel functions.

One set of specifications that have found many applications is to let

$$
\begin{aligned}
X(x) &= x^q\\
G(x) &= x^\alpha \exp\left(\beta x^p\right)
\end{aligned}
$$

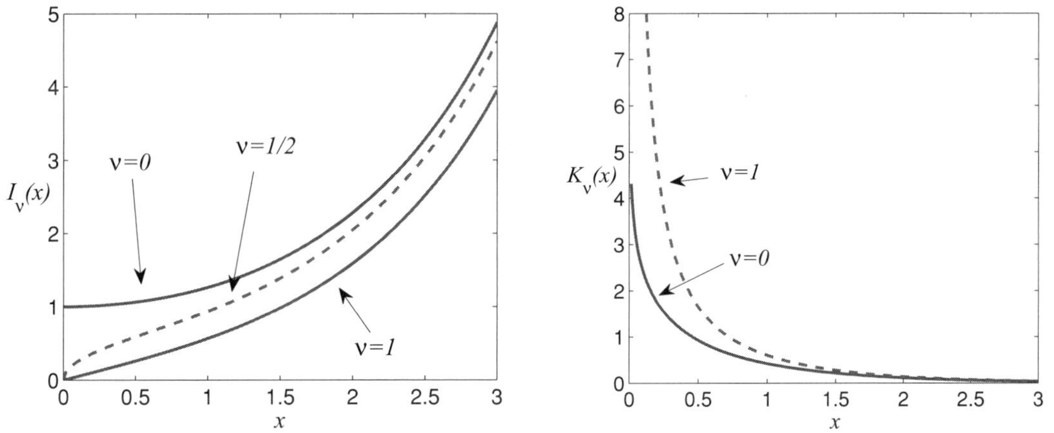

Figure 9.8. Plots of modified Bessel functions of the first kind, $I_\nu(x)$, and second kind, $K_\nu(x)$.

where $q \neq 0$ and $p \neq 0$. The first and second derivatives of $X(x)$ and $G(x)$ are then given by

$$\begin{aligned}
X' &= qx^{q-1} = \frac{q}{x}X \\
X'' &= q(q-1)x^{q-2} = \frac{q-1}{x}X' \\
G' &= \left(\alpha + \beta p x^p\right) x^{\alpha-1} \exp\left(\beta x^p\right) = \frac{\alpha + \beta p x^p}{x} G \\
G'' &= \frac{\alpha + \beta p x^p}{x} G' + \frac{\beta p x^p (p-1) - \alpha}{x^2} G \\
&= \frac{\left(\alpha + \beta p x^p\right)^2 + \beta p x^p (p-1) - \alpha}{x^2} G
\end{aligned}$$

After substituting $X(x)$, $G(x)$, and their derivatives into (9.52), we obtain

$$\begin{aligned}
f_1(x) &= (2\alpha + 1) + 2\beta p x^p \\
f_2(x) &= \left(\alpha^2 - \nu^2 q^2\right) + \beta p x^p (p + 2\alpha) + \left(\beta p x^p\right)^2 + \lambda^2 q^2 x^{2q}
\end{aligned} \tag{9.53}$$

We can summarize the results for this specific form of $X(x)$ and $G(x)$ in the following theorem.

THEOREM 9.3. *For the second-order differential equation, with $q \neq 0$, $p \neq 0$, $d \neq 0$ and $(1-a)^2 \geq 4c$,*

$$x^2 \frac{d^2 y}{dx^2} + x f_1(x) \frac{dy}{dx} + f_2(x) y = 0$$

where

$$\begin{aligned}
f_1(x) &= a + 2bx^p \\
f_2(x) &= dx^{2q} + \left[c + bx^p (p + a - 1) + \left(bx^p\right)^2\right]
\end{aligned} \tag{9.54}$$

The solution is given by

$$y = \begin{cases} \dfrac{AJ_\nu\left(\lambda x^q\right) + BY_\nu\left(\lambda x^q\right)}{G} & \text{if } d > 0 \\[2ex] \dfrac{AI_\nu\left(\lambda x^q\right) + BK_\nu\left(\lambda x^q\right)}{G} & \text{if } d < 0 \end{cases} \tag{9.55}$$

where,

$$\begin{aligned}
G &= x^\alpha \exp\left(\beta x^p\right) & \qquad \nu &= \frac{\sqrt{(a-1)^2 - 4c}}{2q} \\
\alpha &= \frac{a-1}{2} & \\
\beta &= \frac{b}{p} & \qquad \lambda &= \frac{\sqrt{|d|}}{q}
\end{aligned}$$

PROOF. Recall that (9.54) was obtained from the Bessel equation given in (9.50), with

$$X(x) = x^q \qquad \text{and} \qquad Y(x) = yG(x) = yx^{\alpha}\exp\left(\beta x^p\right)$$

The solutions of (9.50) can then be replaced with $X(x)$ and $Y(x)$ to yield (9.55).

EXAMPLE 9.4. The following equation is known as the **Airy equation**:

$$\frac{d^2y}{dx^2} + kxy = 0 \tag{9.56}$$

Multiplying the equation by x^2,

$$x^2\frac{d^2y}{dx^2} + kx^3y = 0$$

By comparing with (9.54), we can identify the parameters to be

$$\alpha = -\frac{1}{2}, \qquad \beta = 0, \qquad q = \frac{3}{2}, \qquad \nu = \frac{1}{3}, \qquad \lambda = \frac{2}{3}\sqrt{|k|}$$

The solution is then given by

$$y = \begin{cases} \sqrt{x}\left[AJ_{1/3}\left(\left[\frac{2}{3}\sqrt{k}\right]x^{3/2}\right) + BJ_{-1/3}\left(\left[\frac{2}{3}\sqrt{k}\right]x^{3/2}\right)\right] & \text{if } k > 0 \\ \sqrt{x}\left[AI_{1/3}\left(\left[\frac{2}{3}\sqrt{-k}\right]x^{3/2}\right) + BI_{-1/3}\left(\left[\frac{2}{3}\sqrt{-k}\right]x^{3/2}\right)\right] & \text{if } k < 0 \end{cases} \tag{9.57}$$

For the case when $k < 0$, a set of functions called the **Airy functions of the first kind**, denoted by $\mathrm{Ai}(x)$, and the **Airy functions of the second kind**, denoted by $\mathrm{Bi}(x)$, are often used instead of the modified Bessel functions. These functions are defined as

$$\begin{aligned} \mathrm{Ai}(x) &= \frac{1}{3}\sqrt{x}\left[-I_{1/3}\left(\frac{2}{3}x^{3/2}\right) + I_{-1/3}\left(\frac{2}{3}x^{3/2}\right)\right] \\ \mathrm{Bi}(x) &= \sqrt{\frac{x}{3}}\left[\,I_{1/3}\left(\frac{2}{3}x^{3/2}\right) + I_{-1/3}\left(\frac{2}{3}x^{3/2}\right)\right] \end{aligned} \tag{9.58}$$

Figure 9.9 shows a plot of both functions. In terms of the Airy functions, the solution of the Airy equation when $k < 0$ is given by

$$y = C_1\mathrm{Ai}\left(|k|^{1/3}x\right) + C_2\mathrm{Bi}\left(|k|^{1/3}x\right) \tag{9.59}$$

In MATLAB, the functions $\mathrm{Ai}(x)$ and $\mathrm{Bi}(x)$ are given by `airy(0,x)` and `airy(2,x)`, respectively.

By looking closer at Figures 9.7 and 9.8, we can immediately note that the Bessel and modified Bessel functions of the first kind, $J_\nu(x)$ and $I_\nu(x)$, are finite

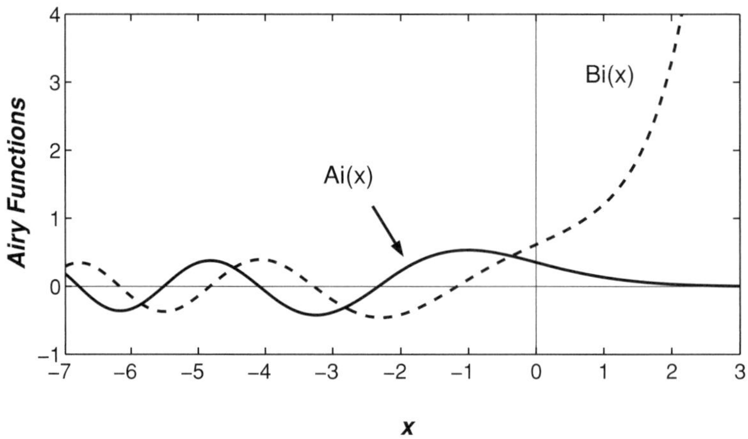

Figure 9.9. Plots of Airy functions A(x) and B(x).

at $x = 0$. However, both $Y_\nu(x)$ and $K_\nu(x)$ approaches $-\infty$ and $+\infty$, respectively, as $x \to 0^+$. These are key properties that can be used to evaluate the arbitrary constants to match the boundary conditions, and we show how these facts are used in the following example. There are several other important properties of Bessel functions, including their derivatives. A summary of the properties are given in Section 9.4.

EXAMPLE 9.5. A model for the one-dimensional steady-state temperature distribution in a triangular cooling fin as shown in Figure 9.10 is given by

$$\xi^2 \frac{d^2U}{d\xi^2} + \xi \frac{dU}{d\xi} - \omega^2 \xi U = 0$$

where U and ξ are the normalized (dimensionless) temperature and distance from the tip,

$$U = \frac{T - T_a}{T_w - T_a} \quad \text{and} \quad \xi = \frac{x}{L}$$

with T_w and T_a being the temperature of the wall and of the surrounding air, respectively. The parameter $\omega = Lh/\left(Wk\cos(\theta)\right)$ is the thermal diffusion constant of the fin, with h and k being the heat transfer coefficient and thermal conductivity of the fin, respectively. The boundary conditions are $U(0) < \infty$ and $U(1) = 1$.

Using the results in Theorem 9.3, we obtain

$$U(\xi) = AI_0\left(2\omega\sqrt{\xi}\right) + BK_0\left(2\omega\sqrt{\xi}\right)$$

Because $U(0) < \infty$ and $K_0(0) = \infty$, we need $B = 0$. Using the other boundary condition, we get $A = \left(I_0\left(2\omega\right)\right)^{-1}$. A plot of the results for different values of ω is shown in Figure 9.11.

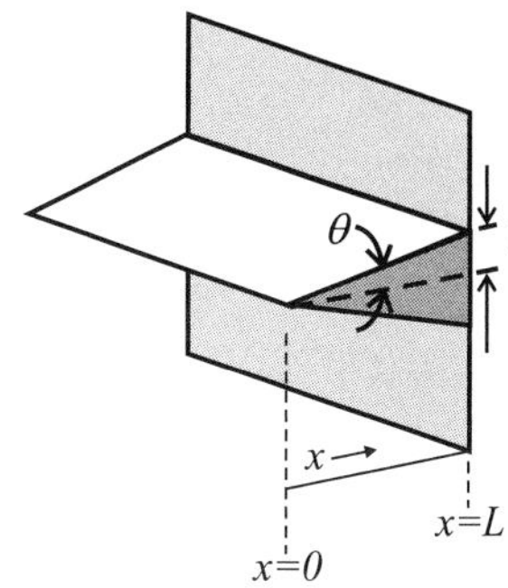

Figure 9.10. Cooling fin.

9.4 Properties and Identities of Bessel Functions and Modified Bessel Functions

The proofs of these properties are included in Section I.6.3 as an appendix.

1. Values near $x = 0$:

$$J_\nu(0) = I_\nu(0) = \begin{cases} 1 & \text{if } \nu = 0 \\ 0 & \text{if } \nu > 0 \\ \pm\infty & \text{if } \nu < 0 \end{cases} \tag{9.60}$$

$$Y_\nu(0) = -\infty \tag{9.61}$$

$$K_\nu(0) = \infty \tag{9.62}$$

2. Derivatives.
   (a) Derivatives of $J_\nu(x)$.

$$\frac{d}{dx}\left(x^\nu J_\nu(x)\right) = x^\nu J_{\nu-1}(x) \tag{9.63}$$

$$\frac{d}{dx}\left(x^{-\nu} J_\nu(x)\right) = -x^{-\nu} J_{\nu+1}(x) \tag{9.64}$$

$$\frac{d}{dx} J_\nu(x) = J_{\nu-1}(x) - \frac{\nu}{x} J_\nu(x) \tag{9.65}$$

$$= -J_{\nu+1}(x) + \frac{\nu}{x} J_\nu(x) \tag{9.66}$$

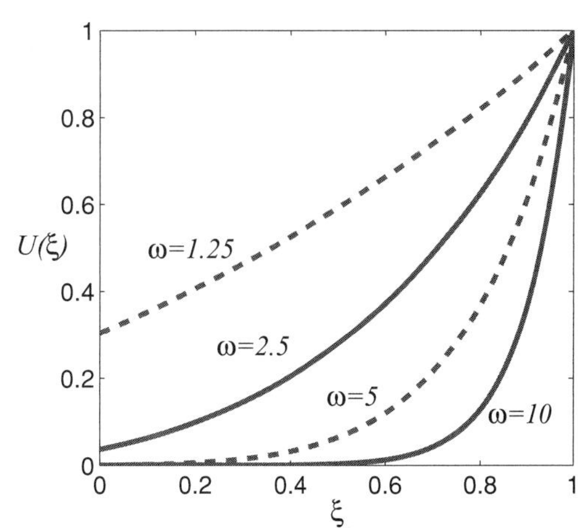

Figure 9.11. Dimensionless temperature profile of the cooling fin for various ω values.

(b) Derivatives of $Y_\nu(x)$.

$$\frac{d}{dx}\left(x^\nu Y_\nu(x)\right) = x^\nu Y_{\nu-1}(x) \tag{9.67}$$

$$\frac{d}{dx}\left(x^{-\nu} Y_\nu(x)\right) = -x^{-\nu} Y_{\nu+1}(x) \tag{9.68}$$

$$\frac{d}{dx}Y_\nu(x) = Y_{\nu-1}(x) - \frac{\nu}{x}Y_\nu(x) \tag{9.69}$$

$$= -Y_{\nu+1}(x) + \frac{\nu}{x}Y_\nu(x) \tag{9.70}$$

(c) Derivatives of $I_\nu(x)$.

$$\frac{d}{dx}\left(x^\nu I_\nu(x)\right) = x^\nu I_{\nu-1}(x) \tag{9.71}$$

$$\frac{d}{dx}\left(x^{-\nu} I_\nu(x)\right) = x^{-\nu} I_{\nu+1}(x) \tag{9.72}$$

$$\frac{d}{dx}I_\nu(x) = I_{\nu-1}(x) - \frac{\nu}{x}I_\nu(x) \tag{9.73}$$

$$= I_{\nu+1}(x) + \frac{\nu}{x}I_\nu(x) \tag{9.74}$$

(d) Derivatives of $K_\nu(x)$.

$$\frac{d}{dx}\left(x^\nu K_\nu(x)\right) = -x^\nu K_{\nu-1}(x) \tag{9.75}$$

$$\frac{d}{dx}\left(x^{-\nu} K_\nu(x)\right) = -x^{-\nu} K_{\nu+1}(x) \tag{9.76}$$

$$\frac{d}{dx}K_\nu(x) = -K_{\nu-1}(x) - \frac{\nu}{x}K_\nu(x) \tag{9.77}$$

$$= -K_{\nu+1}(x) + \frac{\nu}{x}K_\nu(x) \tag{9.78}$$

3. Recurrence equations:

$$J_{\nu-1}(x) = \frac{2\nu}{x}J_\nu(x) - J_{\nu+1}(x) \tag{9.79}$$

$$Y_{\nu-1}(x) = \frac{2\nu}{x}Y_\nu(x) - Y_{\nu+1}(x) \tag{9.80}$$

$$I_{\nu-1}(x) = \frac{2\nu}{x}I_\nu(x) + I_{\nu+1}(x) \tag{9.81}$$

$$K_{\nu-1}(x) = -\frac{2\nu}{x}K_\nu(x) + K_{\nu+1}(x) \tag{9.82}$$

(*The recurrence equations can be obtained directly from the preceding derivative formulas. For instance, equation (9.79) can be obtained by equating (9.65) and (9.66)*).

4. Values for negative integral orders, $\nu = -n, n = 1, 2, \ldots$

$$J_{-n}(x) = (-1)^n J_n(x) \quad (9.83)$$

$$Y_{-n}(x) = (-1)^n Y_n(x) \quad (9.84)$$

$$I_{-n}(x) = I_n(x) \quad (9.85)$$

$$K_{-n}(x) = K_n(x) \quad (9.86)$$

5. Generating functions:

$$\exp\left[\frac{x}{2}\left(t - \frac{1}{t}\right)\right] = \sum_{n=-\infty}^{\infty} J_n(x) t^n \quad (9.87)$$

6. Orthogonalities:

$$\int_0^a x J_\nu(\lambda_n x) J_\nu(\lambda_m x) dx = \begin{cases} 0 & \text{if } n \neq m \\ \frac{1}{2}a^2 \left[J_\nu^2(\lambda_n a) - J_{\nu-1}(\lambda_n a) J_{\nu+1}(\lambda_n a)\right] & \text{if } n = m \end{cases} \quad (9.88)$$

$$\int_0^a x I_\nu(\lambda_n x) I_\nu(\lambda_m x) dx = \begin{cases} 0 & \text{if } n \neq m \\ \frac{1}{2}a^2 \left[I_\nu^2(\lambda_n a) - I_{\nu-1}(\lambda_n a) I_{\nu+1}(\lambda_n a)\right] & \text{if } n = m \end{cases} \quad (9.89)$$

9.5 EXERCISES

E9.1. Obtain the series solution (9.22) given in Example 9.2.

E9.2. The Legendre equations are members of a class of differential equations in which the coefficient of d^2y/dx^2 is $(1 - x^2)$. Other equations in this class of equations are the Jacobi, Chebyshev, and Gegenbauer equations

$$(1 - x^2)\frac{d^2y}{dx^2} + f(x)\frac{dy}{dx} + g(x)y = 0$$

where the functions $f(x)$ and $g(x)$ are shown in Table 9.5.

1. Show that $x = 0$ is an ordinary points for all these cases.
2. For the Chebyshev type 2 equation, show that one of the solutions will be a finite polynomial if n is a positive integer. (These polynomials are used in series approximation of functions due to the desirable othorgonality property of the polynomials.)

Table 9.5. *Differential equations with $(1 - x^2)$ as leading coefficient*

Type	$f(x)$	g(x)
Jacobi	$(\beta - \alpha) - (\alpha + \beta + 2)x$	$n(n + \alpha + \beta + 1)$
Chebyshev type 1	$-x$	n^2
Chebyshev type 2	$-3x$	$n(n + 2)$
Gegenbauer	$-(2\alpha + 1)x$	$n(n + 2\alpha)$

E9.3. The **hypergeometric equation** is defined as

$$x(1-x)\frac{d^2y}{dx^2} + \left(\gamma - (\alpha+\beta+1)x\right)\frac{dy}{dx} - \alpha\beta y = 0 \qquad (9.90)$$

where α, β, and γ are positive real constants.

1. Show that $x = 0$ and $x = 1$ are regular singular points of (9.90).
2. When $x = 1/t$ yields a transformed equation in t where $t = 0$ is a regular singular point, then the original equation is said to have a regular singular point at $x = \infty$. Show that (9.90) has a regular singular point at infinity.
3. Let γ be a non-integer. Use the Frobenius method and show that the complete solution of (9.90) based on a Frobenius series expanded around $x = 0$ will yield the following complete solution:

$$y = A\left[{}_2F_1(\alpha,\beta;\gamma;x)\right] + Bx^{1-\gamma}\left[{}_2F_1(1-\gamma+\alpha, 1-\gamma+\beta; 2-\gamma; x)\right]$$

where A and B are arbitrary coefficients and ${}_2F_1$ is the hypergeometric function defined in (9.12).
4. Let $\alpha = 1$, $\beta = 0.2$ and $\gamma = 0.5$. Find the solution of (9.90) that satisfies the boundary conditions $y(0.1) = 0.1$ and $y(0.8) = -1.0$. Plot this solution for $0.01 \le x \le 0.95$ and compare with the numerical BVP solution.

E9.4. Obtain the analytical solution for the associated Legendre equation given by

$$(1-x^2)\frac{d^2\Theta}{dx^2} - 2x\frac{d\Theta}{dx} + \left(12 - \frac{4}{1-x^2}\right)\Theta = 0$$

with the boundary conditions $\Theta(0) = 0.4$ and $\Theta(0.9) = 0.2$. Plot your solution together with the numerical BVP solution.

E9.5. Consider the steady-state model for the temperature distribution of a transverse fin having a triangular cross-section as shown in Figure 9.12,[5]

$$x(R-x)\frac{d^2u}{dx^2} + (R-2x)\frac{du}{dx} - \beta(R-x)u = 0$$

subject to the following boundary conditions:

$$u(0) < \infty \quad \text{and} \quad u(R-a) = T_p - T_a$$

The independent variable $x = R - r$ is the distance from the outer tip of the fin. The dependent variable is $u = T - T_a$, where $T(x)$ is the temperature at x and T_a is the constant air temperature. As shown in the figure, a and R are the inner and outer radius of the cooling fin, respectively. T_p is the temperature of the contents flowing in the pipe at $x = R - a$. The parameter $\beta = h/(k\sin\theta)$ contains the thermal diffusivity, with θ being the slant angle of the fin. By using the Frobenius series expanding around $x = 0$,

1. Obtain the indicial equation and show that the indicial roots are equal. (According to Theorem 9.2, the second solution will contain a term with $\ln(x)$. Because $u(0) < \infty$, the second solution will not be necessary.)
2. Using Theorem 9.2, show that one of the series solution is given by $u = A\sum_{n=0}^{\infty}\phi_n x^n$, where A is an arbitrary constant and

$$\phi_0 = 1\ ;\ \phi_1 = \beta\ ;\cdots;\ \phi_k = \frac{k(k-1)+\beta R}{k^2R}\phi_{k-1} - \frac{\beta}{k^2R}\phi_{k-2}$$

[5] Adapted from V. G. Jenson and G. V. Jeffreys, *Mathematical Methods in Chemical Engineering*, Academic Press, UK, 1977.

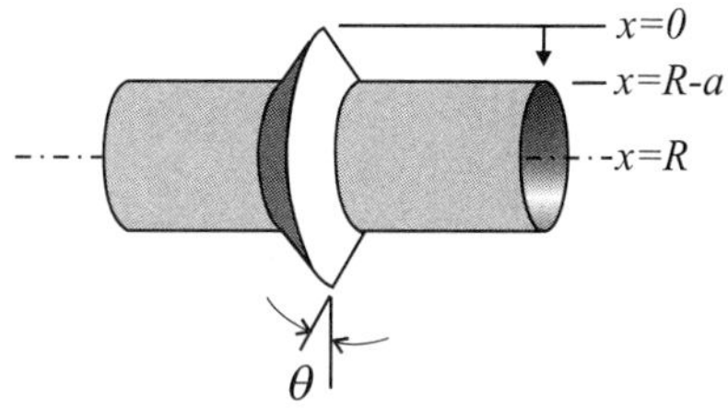

Figure 9.12. A cooling fin with triangular cross-section.

(This series is difficult to reduce to a recursion pattern; however, the convergence is relatively fast; i.e., a few terms of the series may be sufficient.)

3. Using the first ten terms in the series, compare this truncated series solution with the numerical BVP solution for the case where $R = 8\ in$, $a = 4\ in$, $\beta = 3\ in^{-1}$, $T_a = 70^o F$, and $T_p = 300^o F$. Note that for the numerical BVP, instead of starting at $x = 0$, one can start instead at, say, $x = \epsilon = 10^{-4}$. Also, the boundary condition at $x = 0$ can be replaced by setting $x = 0$ for the model and then approximating at $x = \epsilon$, that is,

$$R \left.\frac{du}{dx}\right|_{x=\epsilon} - \beta R\, u(\epsilon) = 0$$

E9.6. Find the series solution expanded around $x = 0$ for the equation given by

$$\frac{d^2y}{dx^2} - bx\frac{dy}{dx} + cy = 0$$

Let $b = 2$ and $c = 2$ and the boundary conditions be given by

$$y(0) = 1 \quad \text{and} \quad y(2) = -3$$

Compare the solution with a numerical BVP solution. (Hint: The series solution can also be reduced to

$$y(x) = A\left(1 - x^2\left[{}_2F_2\left(1, \frac{1}{2}; 2, \frac{3}{2}; x^2\right)\right]\right) + Bx$$

where ${}_2F_2$ is the generalized hypergeometric function, cf. (9.12)).

E9.7. The **Hermite equation** is given by

$$\frac{d^2y}{dx^2} - 2x\frac{dy}{dx} + 2\lambda y = 0$$

Obtain the series solution and show that one of the solutions is a finite order polynomial for λ, a positive integer. These polynomials are known as the **Hermite polynomials**.

E9.8. The **Laguerre equation** is given by

$$x\frac{d^2y}{dx^2} + (1 - x)\frac{dy}{dx} + my = 0$$

Using Theorem 9.2, show that if m is a positive integer, then one of the solutions is given by finite polynomial of order m. Obtain the solution for $m = 5$. (These polynomials are known as the **Laguerre polynomials**.)

E9.9. Derive the following identity using the definition of the Bessel function $J_\nu(x)$ defined in 9.4:

$$J_{1/2}(x) = \sqrt{\frac{2}{\pi x}}\ \sin(x) \tag{9.91}$$

E9.10. The **Helmholtz equation** in polar coordinates is given by

$$\nabla^2 F + k^2 F = \frac{1}{r}\frac{\partial^2 F}{\partial r^2} + \frac{\partial F}{\partial r} + \frac{1}{r^2}\frac{\partial F}{\partial r} + k^2 F = 0$$

Using the method of separation of variables, we let $F = R(r)\Theta(\theta)$, which will then yield one ODE for $\Theta(\theta)$ and another ODE given by

$$r^2\frac{d^2R}{dr^2} + r\frac{dR}{dr} + \left(k^2r^2 - n^2\right)R = 0 \qquad (9.92)$$

where n will be an integer. Let the boundary conditions for R be given by

$$R(0) < \infty \quad \text{and} \quad R(a) = F_0$$

1. Obtain the solution of (9.92).
2. Show that the boundary condition $R(0) < \infty$ implies $R(0) = 0$ if $n > 0$ and $dR/dr = 0$ if $n = 0$. (Note: You could show this in two possible approaches. One approach is to use the analytical solution just obtained. The other approach is to set $r = 0$ in (9.92)).
3. Compare the analytical solution with the numerical BVP solution for the case where $n = 0$, $a = 2$, $k = 1.33$, and $F_0 = 1$. (Note: Start the numerical solution at $r = \epsilon$ instead of $r = 0$, where $0 < \epsilon \ll 1$.)

E9.11. Oxygen diffuses into a spherical microorganism of radius R with diffusitivy $\mathcal{D}_{AB}$ and undergoes a first-order reaction, $\text{rxn}_A = -kC_A$. Assuming spherical symmetry, the steady-state concentration of oxygen C_A inside the microorganism can be modeled by

$$r^2\frac{d^2C_A}{dr^2} + 2r\frac{dC_A}{dr} - \frac{k}{\mathcal{D}_{AB}}r^2C_A = 0$$

with boundary conditions,

$$C_A(R) = C_{A,\text{surr}} \quad \text{and} \quad \left.\frac{dC_A}{dr}\right|_{r=0} = 0$$

where $C_{A,\text{surr}}$ is the concentration of oxygen in the surroundings.[6] For the parameter values given by $R = 10^{-2}\ cm$, $\mathcal{D}_{AB} = 1.32\times10^{-6}\ cm^2/s$, $k = 0.05/s$, determine the minimum value of $C_{A,\text{surr}}$ such that $C_A \geq 2\times 10^{-4} mol/cm^3$ inside the microorganism.

E9.12. For the differential equation given by

$$\frac{d^2y}{dx^2} = \frac{\dfrac{dy}{dx} + \alpha^2 y}{x}$$

Show that the change of variables

$$X = \sqrt{x} \quad \text{and} \quad Y = \frac{y}{x}$$

will transform the original equation to either a Bessel or modified Bessel equation with Y and X as the dependent and independent variables, respectively.

E9.13. Consider the following equation:

$$\frac{d^2y}{dx^2} + \left(\frac{1}{x} + 2\right)\frac{dy}{dx} + \left(1 - \frac{1}{x^2}\right)y = 0$$

[6] Adapted from H. Adidharma and V. Temyanko, *Mathcad for Chemical Engineers*, Trafford Publishing, Canada, 2007.

Determine whether the conditions in Theorem 9.3 can be met, and if so, obtain the general solution. For the boundary condition that $y(1) = 100$ and $y(2) = 20$, obtain the analytical solution and compare with the numerical BVP solution.

E9.14. Consider the following differential equation and the boundary conditions

$$\frac{d^2y}{dx^2} = 2xy \qquad y(-1) = 1,\ y(1) = 0.5$$

Obtain the solution in terms of Airy functions $\text{Ai}(x)$ and $\text{Bi}(x)$. Compare the solution with the numerical BVP solution.

E9.15. Based on (9.50) and (9.51), let $G(x) = 1$, that is, no transformation is needed for the independent variable:

1. For $X(x) = x + a$ show that

$$f_1(x) = \left(\frac{x}{x+a}\right) \quad \text{and} \quad f_2(x) = x^2\left((\lambda)^2 - \left(\frac{\nu}{x+a}\right)^2\right)$$

Thus obtain the solution of

$$\frac{d^2y}{dx^2} + \left(\frac{1}{x+a}\right)\frac{dy}{dx} + \left(\lambda^2 - \frac{\nu^2}{(x+a)^2}\right)y = 0$$

in terms of Bessel functions. (Note: If we multiply this equation by $(x+a)^2$, we could also see immediately that this change of variable, i.e., $X = x + a$, is the appropriate one.)

2. For $X(x) = e^x$ show that

$$f_1(x) = 0 \quad \text{and} \quad f_2(x) = x^2\left(\lambda^2 e^{2x} - \nu^2\right)$$

Thus obtain the solution of

$$\frac{d^2y}{dx^2} + \left(\lambda^2 e^{2x} - \nu^2\right)y = 0$$

in terms of Bessel functions.

PART IV

PARTIAL DIFFERENTIAL EQUATIONS

This part of the book focuses on partial differential equations (PDEs), including the solution, both analytical and numerical methods, and some classification methods. Because the general topic of PDEs is very large, we have chosen to cover only some general methods mainly applicable to linear PDEs, with the exception of nonlinear first-order PDEs.

In Chapter 10, we focus on the solution of first-order PDEs, including the method of characteristics and Lagrange-Charpit methods. The second half of the chapter is devoted to classification of high-order PDEs, based on the factorization of the principal parts to determine whether the equations are hyperbolic, parabolic, or elliptic.

In Chapter 11, we discuss the analytical solutions of linear PDEs. We begin with reducible PDEs that allow for the method of separation of variables. To satisfy various types of initial and boundary conditions, Sturm-Liouville equations are used to obtain orthogonal functions. The techniques can then be extended to the case of nonhomogenous PDEs and nonhomogeneous boundary conditions based on eigenfunction expansions.

In Chapter 12, we discuss integral transform methods such as Fourier and Laplace transforms methods. For the Fourier transforms, we cover the important concepts of the classic transforms, including the use of distribution theory and tempered distributions to find the generalized Fourier transforms of step functions, sine, and cosine functions. A brief but substantial introduction to distribution theory is included in the appendix. For numerical implementation purposes, we have also included a discussion of the fast Fourier transform in the appendix. The Laplace transform methods are then discussed, including the transform of special functions such as gamma functions and Bessel functions. Thereafter, they are applied to solve linear PDEs. To perform the inverse Laplace transform, we employ the theory of residues. A brief but substantial discussion of complex integration and residue theory is also included in the appendix.

The last two chapters cover two of the most popular numerical methods available for solving PDEs. Chapter 13 covers the method of finite differences. Starting with a discussion of various discretization approaches, the finite differences are then applied to the solution of time-independent cases, ranging from the one-dimensional cases to the three-dimensional cases. The time-dependent cases are then handled by the semi-discrete approach, also known as the method of lines, including forward Euler, backward Euler, and Crank Nicholson schemes. Because stability analysis is a key issue for the successful application of the finite difference method, we include a discussion of both the eigenvalue approach and the Von Neumann analysis.

Chapter 14 discusses the other major class of numerical methods, known as the finite element method. The discussion is restricted to the use of triangular finite elements applied to the weak formulation of the PDEs with only two independent variables. After transforming the PDEs to integral equations, they are further reduced to matrix equations.

10 First-Order Partial Differential Equations and the Method of Characteristics

In this chapter, we discuss the solution of first-order differential equations. The main technique that is used is known as the method of characteristics, in which various paths in the domain of the independent variables, known as **characteristics**, are obtained. The solutions are then propagated through these paths, yielding the **characteristic solution curves**. (Note that the two terms are different: **characterics** vs. **characteristic curves**). A solution surface can then be constructed by combining (or "*bundling*") these characteristic curves.

The method of characteristics is applicable to partial differential equations that take on particular forms known as quasilinear equations. Special cases are the semilinear, linear, and strictly linear forms. It can be shown that for the semilinear cases, the characteristics will not intersect each other. However, for the general quasilinear equations, the characteristics can intersect. When they do, the solution will become discontinuous, and the discontinuities are known as **shocks**. Moreover, if the discontinuity is given in the initial conditions, **rarefaction** can occur, in which a fan of characteristics are filled in to complete the solution surface. A brief treatment of shocks and rarefaction is given in Section J.1 as an appendix.

In Section 10.3, we discuss a set of conditions known as Lagrange-Charpit conditions. These conditions are used to generate the solutions of some important classes of nonlinear first order PDEs.

Next, we turn to the classification of higher order differential equations. We show, at least for the second-order case, how the method of characteristics can be used to find canonical forms such as hyperbolic, elliptic, or parabolic. The generalization to higher order cases or cases with more than two independent variables are discussed in Sections J.2 and J.3 as appendices.

Finally, in Section 10.5, we include a brief discussion on how to treat a set of quasilinear first-order partial differential equations. Unfortunately, several issues are omitted because this topic demands a more in-depth treatment. Nevertheless, we show how the method of characteristics can be applied to systems of hyperbolic equations depending on the solution of a corresponding eigenvalue problem, but limited to systems with only two independent variables.

10.1 The Method of Characteristics

The general form of a first-order partial differential equation is given by

$$F\left(\mathbf{x}, u, \frac{\partial u}{\partial x_1}, \ldots, \frac{\partial u}{\partial x_n}\right) = 0 \tag{10.1}$$

where u is the dependent variable, and $\mathbf{x} = \{x_1, \ldots, x_n\}$, are the n independent variables.

Some important special forms, in decreasing generalities, are given by the following:

$$\textbf{Quasilinear}: \quad \sum_{i=1}^{n} \alpha_i(\mathbf{x}, u)\frac{\partial u}{\partial x_i} = f(\mathbf{x}, u) \tag{10.2}$$

$$\textbf{Semilinear}: \quad \sum_{i=1}^{n} \beta_i(\mathbf{x})\frac{\partial u}{\partial x_i} = g(\mathbf{x}, u) \tag{10.3}$$

$$\textbf{Linear}: \quad \sum_{i=1}^{n} \beta_i(\mathbf{x})\frac{\partial u}{\partial x_i} = h(\mathbf{x}) + \gamma(\mathbf{x})u \tag{10.4}$$

$$\textbf{Strictly Linear}: \quad \sum_{i=1}^{n} \beta_i(\mathbf{x})\frac{\partial u}{\partial x_i} = h(\mathbf{x}) \tag{10.5}$$

EXAMPLE 10.1. Recall the conservation laws discussed in Section 5.4.1. When limited to one-dimensional flow with no source term, and assuming sufficient smoothness in the functions, we can obtain the general formula given by

$$\frac{\partial}{\partial t}u(t, z) + \frac{\partial}{\partial z}g\left(u(t, z), t, z\right) = 0 \tag{10.6}$$

where u is the conserved quantities and g is the flux of u. For the special case in which $g = \bar{v}\, u(t, z)$, with $\bar{v}$ as a constant velocity, we obtain

$$\frac{\partial u}{\partial t} + \bar{v}\frac{\partial u}{\partial z} = 0 \tag{10.7}$$

which is known as the **advection equation**, and it can be classified as a strictly linear first-order PDE, that is, (10.5) with $x_1 = t$, $x_2 = z$, $\beta_1(\mathbf{x}) = 1$, $\beta_2(\mathbf{x}) = \bar{v}$ and $h(\mathbf{x}) = 0$.

For the case in which $g = g(u)$, we can rewrite (10.6) by applying the chain rule, to obtain

$$\frac{\partial u}{\partial t} + G(u)\frac{\partial u}{\partial z} = 0 \tag{10.8}$$

where $G(u) = dg/du$. Equation (10.8) can thus be classified as a quasilinear first-order PDE, that is, (10.2) with $\alpha_1(\mathbf{x}, u) = 1$, $\alpha_2(\mathbf{x}, u) = G(u)$ and $f(\mathbf{x}, u) = 0$.

We focus first on the solution of quasilinear forms. These methods can be readily applied to both the linear and semilinear cases. For nonlinear cases that are not

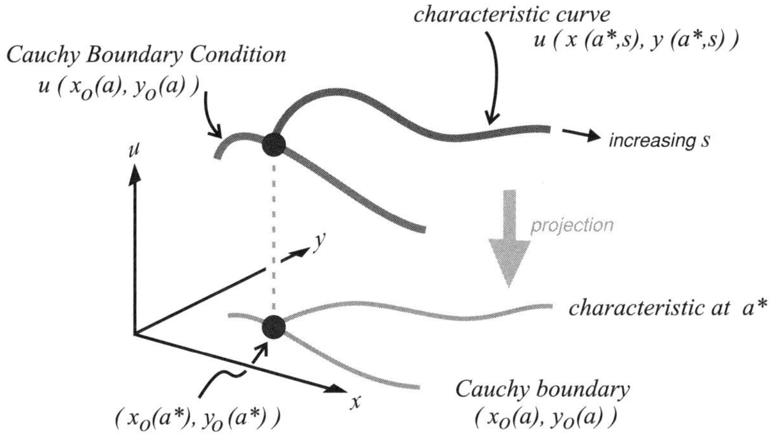

Figure 10.1. The characteristic curve corresponding to $a = a^*$.

quasilinear, additional steps known as the **Lagrange-Charpit methods**, to be discussed later in Section 10.3, can sometimes be used to convert the equations to an associated quasilinear equation.

Consider a quasilinear equation with two independent variables, x and y,

$$\xi(x, y, u) \frac{\partial u}{\partial x} + \eta(x, y, u) \frac{\partial u}{\partial y} = \zeta(x, y, u) \tag{10.9}$$

subject to the boundary condition

$$u = u_o(x_o(a), y_o(a)) \tag{10.10}$$

where $(x_o(a), y_o(a))$ is a boundary curve parameterized by a. This type of boundary condition, where the values are fixed along a continuous curve in the domain of the the independent variables, is known as a **Cauchy condition**.

The **method of characteristics** involves the determination of a solution surface parameterized by a and s in the (x, y, u)-space where x and y are independent variables. The boundary condition is fixed to be the values at $s = 0$ and and is denoted by $x_o(a) = x(a, s = 0)$, $y_o(a) = y(a, s = 0)$, and $u_o(a) = u(a, s = 0)$; that is, the locus of points of the boundary conditions are specified by the parameter a. At any fixed value of a, we then track a curve in the solution surface as s increases from 0. These curves are the integral curves known as the **characteristic curves** of the partial differential system given by (10.9) subject to (10.10). This is illustrated in Figure 10.1.

As the parameter a takes on values of its range, a collection of characteristics curves can be generated. The projections of these curves on the (x, y) plane are known as the **projected characteristic curves**, **characteristic base**, or **characteristics**, which we denote as ϕ_a. As long as the characteristics do not cross each other, we can then take the collection of the characteristic curves and "*bundle*" them together to be the solution of (10.9). This means that the solution surface can be put in terms of parameters a and s. If the problem is well posed, one could transform this solution back in terms of x and y, that is, $u = u(x, y)$.

An incremental movement along the characteristic curve u per incremental change in s is represented by du/ds. Using the chain rule,

$$\frac{\partial u}{\partial x}\frac{dx}{ds} + \frac{\partial u}{\partial y}\frac{dy}{ds} = \frac{du}{ds} \tag{10.11}$$

After comparing (10.11) and (10.9), we obtain the following set of simultaneous ordinary differential equations, known as the **characteristic equations**:

$$\frac{du}{ds} = \zeta(x, y, u) \tag{10.12}$$

$$\frac{dx}{ds} = \xi(x, y, u) \tag{10.13}$$

$$\frac{dy}{ds} = \eta(x, y, u) \tag{10.14}$$

accompanied by the following set of initial conditions,

$$u(0) = u_o(a) \quad ; \quad x(0) = x_o(a) \quad ; \quad y(0) = y_o(a) \tag{10.15}$$

After solving these simultaneous equations, we can then rearrange the results to obtain the solution in terms of x and y. The approach is best illustrated by a simple example.

EXAMPLE 10.2. Consider the partial differential equation given by

$$x\frac{\partial u}{\partial x} - y\frac{\partial u}{\partial y} = 2x^2 - 6y^2 \tag{10.16}$$

subject to the boundary condition:

$$u(x = 1, y) = 3y^2 + 2y - 5 \tag{10.17}$$

The boundary can be parameterized by a by letting $x_o = 1$ and $y_o = a$. The characteristic equations are

$$\frac{dx}{ds} = \xi(x, y, u) = x \tag{10.18}$$

$$\frac{dy}{ds} = \eta(x, y, u) = -y \tag{10.19}$$

$$\frac{du}{ds} = \zeta(x, y, u) = 2x^2 - 6y^2 \tag{10.20}$$

together with the following initial conditions:

$$x_o = 1 \tag{10.21}$$

$$y_o = a = \text{arbitrary constant} \tag{10.22}$$

$$u_o = 3y_o^2 + 2y_o - 5 = 3a^2 + 2a - 5 \tag{10.23}$$

Solving (10.18) and (10.19), while applying (10.21) and (10.22),

$$x = e^s \tag{10.24}$$

$$y = a\, e^{-s} \tag{10.25}$$

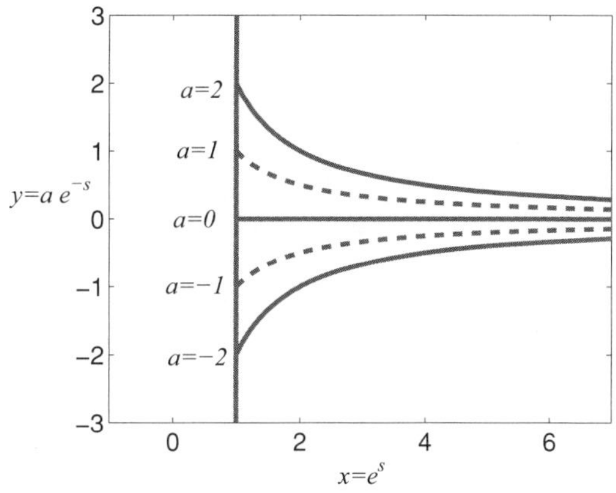

Figure 10.2. The characteristics corresponding to different starting points: $(x, y) = (1, a)$, with $a = -2, -1, 0, 1, 2$.

These two equations define the characteristic $\phi_a(s)$. Plots of the characteristics for $a = -2, -1, 0, 1, 2$ are shown in Figure 10.2. The characteristic curves are then obtained by solving (10.20). Substituting $x = e^s$ and $y = ae^{-s}$,

$$\frac{du}{ds} = 2e^{2s} - 6a^2e^{-2s} \tag{10.26}$$

which can be integrated to yield

$$u = e^{2s} + 3a^2e^{-2s} + C \tag{10.27}$$

where C is an arbitrary constant. Taking the boundary condition (10.23) and equating it to (10.27) at $s = 0$,

$$\begin{aligned} \left(u_o = 3a^2 + 2a - 5\right) &= \left(1 + 3a^2 + C = u_{(s=0)}\right) \\ \rightarrow \quad C &= 2a - 6 \end{aligned}$$

or

$$u = e^{2s} + 3a^2e^{-2s} + 2a - 6 \tag{10.28}$$

Plots of u starting at different values of $(x_o, y_o) = (1, a)$ with $a = -2, -1, 0, 1, 2$ are shown in Figure 10.3. The collection of these curves corresponding to a continuum of a values will be a 3D surface, which forms the solution to partial differential equation.

We could further process the solution to be in terms of the original variables x and y. Substituting $a = ye^s = xy$, (10.28) becomes

$$u = x^2 + 3y^2 + 2xy - 6 \tag{10.29}$$

One can show directly that (10.29) satisfies both the partial differential equation (10.16) and the boundary condition (10.17).

EXAMPLE 10.3. Consider the simplified model for a parametric pump based on the chromatographic process described by the following partial differential equation:

$$(1 + \beta K(t))\frac{\partial c}{\partial t} + v(t)\frac{\partial c}{\partial z} = -v(t)\frac{dK}{dt}c$$

where c, t, and z are the solute concentration, time, and distance from the bottom of the column, respectively. The parameter $\beta = (1 - \epsilon)/\epsilon$ is the ratio

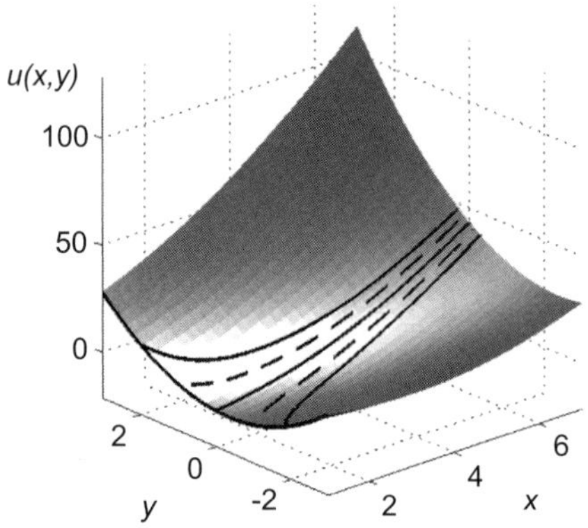

Figure 10.3. The characteristic curves $u = u(x(s), y(s))$ corresponding to $a = -2, -1, 0, 1, 2$.

of solid volume to liquid volume, where ϵ is the porosity, which is assumed constant along the column. The assumption is that the concentration N at the solid phase is proportional to the liquid concentration c given by $N = Kc$, where $K = K(t)$ is a temperature-dependent adsorption parameter. Both the velocity $v(t)$ and adsorption parameter $K(t)$ are assumed to be constant during the upswing (+) and downswing (−) phases of operation, which also assumes that the adsorption equilibrium occurs very fast, with isotherm temperatures T^+ and T^- during the upswing and downswing modes, respectively. This is shown in Figure 10.4.

Let μ^+ and μ^- be defined by

$$\mu^+ = \frac{1}{1+\beta K^+} \quad \text{and} \quad \mu^- = \frac{1}{1+\beta K^-}$$

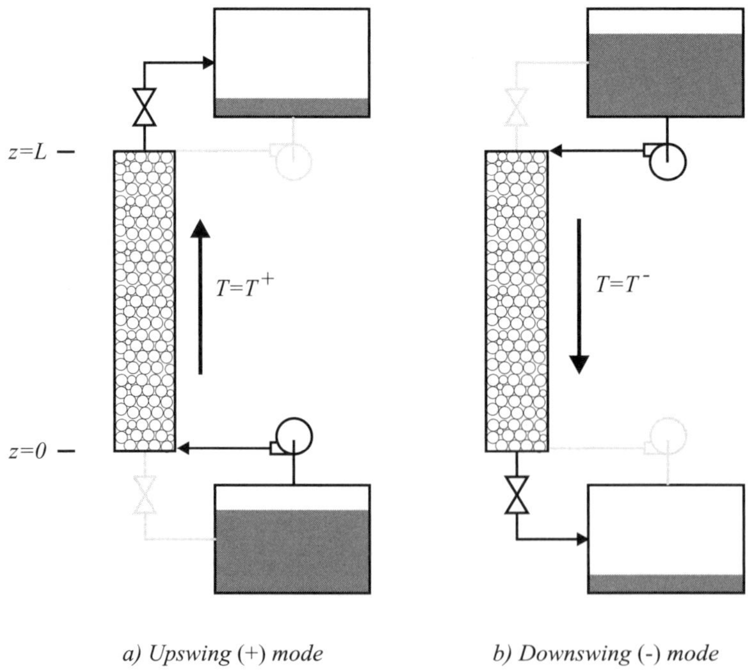

Figure 10.4. The parametric pump process showing the flow configurations during an upswing and downswing mode.

Also, let $c^+(z)$ and $c^-(z)$ denote the concentrations of the fluid in the column, during switches where there are no flow, at temperatures T^+ and T^-, respectively. Then we can model the process separately during each mode as follows:

$$\left.\begin{array}{rcl} \left(\dfrac{1}{\mu^+}\right)\dfrac{\partial c}{\partial t}+v\dfrac{\partial c}{\partial z} &=& 0 \\ c(0,t) &=& c_{\text{in,bottom}}^{(n)} \\ c(z,nP) &=& c^{+(n)}(z) \end{array}\right\} \quad \text{when} \quad nP < t < \left(n+\frac{1}{2}\right)P$$

$$\left.\begin{array}{rcl} \left(\dfrac{1}{\mu^-}\right)\dfrac{\partial c}{\partial t}-v\dfrac{\partial c}{\partial z} &=& 0 \\ c(L,t) &=& c_{\text{in,top}}^{(n)} \\ c(z,(n+\frac{1}{2})P) &=& c^{-(n)}(z) \end{array}\right\} \quad \text{when} \quad \left(n+\frac{1}{2}\right)P < t < (n+1)P \tag{10.30}$$

where v is the constant flow speed, and P is the period for one cycle involving one sequence of upswing mode and downswing mode, with a switch occurring instantaneously at $P/2$. The functions $c^{+(n)}(z)$ (and $c^{-(n)}(z)$) are the concentration profile along the column just prior to upswing flow (downswing flow) at the nth cycle. The model assumes that the column will start with an upswing mode.

Both equations in (10.30) are strictly linear first-order PDEs and can be treated separately. The solutions are constant concentrations traveling along the characteristics. (The derivation of these results is presented as Exercise **E10.2**).

Let $P_n = nP$ and $P_{nh} = (n+0.5)P$. Then for $P_n < t \le P_{nh}$,

$$c(z,t) = \begin{cases} c^{+(n)}\left(z-(t-P_n)v\mu^+\right) & \text{for } z > (t-P_n)v\mu^+ \\ c_{\text{in,bottom}}^{(n)} & \text{for } 0 \le z \le (t-P_n)v\mu^+ \end{cases} \tag{10.31}$$

and for $P_{nh} < t \le P_{n+1}$,

$$c(z,t) = \begin{cases} c^{-(n)}\left(z+(t-P_{nh})v\mu^-\right) & \text{for } z < L-(t-P_{nh})v\mu- \\ c_{\text{in,top}}^{(n)} & \text{for } L-(t-P_{nh})v\mu^- \le z \le L \end{cases} \tag{10.32}$$

where the feed concentrations for $n > 0$ are obtained by mixing the liquid collected from the previous mode, that is, $c_{\text{in,bottom}}^{(n)} = c_{\text{bottom}}(P_n)$ and $c_{\text{in,top}}^{(n)} = c_{\text{top}}(P_{nh})$, where

$$c_{\text{top}}(t) = \begin{cases} \dfrac{1}{t-P_n}\displaystyle\int_{P_n}^{t} c(L,\tau)d\tau & \text{for } P_n < t \le P_{nh} \\ c_{\text{top}}(P_{nh}) & \text{for } P_{nh} < t \le P_{n+1} \end{cases}$$

and

$$c_{\text{bottom}}(t) = \begin{cases} c_{\text{bottom}}(P_n) & \text{for } P_n < t \le P_{nh} \\ \dfrac{1}{t - P_{nh}} \displaystyle\int_{P_{nh}}^{t} c(0,\tau)d\tau & \text{for } P_{nh} < t \le P_{n+1} \end{cases}$$

These equations assume that after a completed upswing or downswing, the bottom reservoir and top reservoir, respectively, are empty.

The relation between the concentrations in the adsorption column at the point of temperature switching can be obtained from mass balance and assuming fast ideal adsorption equilibrium. It is given by

$$\frac{c^-}{c^+} = \frac{\mu^-}{\mu^+} = \lambda$$

We consider the case where $\lambda < 1$; that is, the amount adsorbed in the solids are greater during the downswing mode, making the liquid more dilute with solute as it is being pushed to the bottom reservoir. This implies that the magnitude of the slopes of the characteristics for the downswing mode is less than that of the upswing mode, that is, $v\mu^- < v\mu^+$. Prior to pumping, we have

$$c^{+(n)}(z) = \frac{1}{\lambda}c(z, P_n) \quad \text{and} \quad c^{-(n)}(z) = \lambda c(z, P_{nh})$$

To complete the model, we need to be given the data for the initial conditions, that is, $c^{+(0)}(z)$ and $c^{(0)}_{\text{in,bottom}}$. It can be shown (as part of Exercise **E10.2**) that the top reservoir will keep increasing in solute concentration asymptotically to a maximum level.

For the general case of n independent variables,

$$\sum_{i=1}^{n} \alpha_i(\mathbf{x}, u)\frac{\partial u}{\partial x_i} = f(\mathbf{x}, u) \tag{10.33}$$

the method of characteristics generates characteristic equations composed of $n+1$ ordinary differential equations given by

$$\frac{dx_1}{ds} = \alpha_1(\mathbf{x}, u) \quad ; \quad \cdots \quad ; \quad \frac{dx_n}{ds} = \alpha_n(\mathbf{x}, u) \quad ; \quad \frac{du}{ds} = f(\mathbf{x}, u) \tag{10.34}$$

subject to the following boundary conditions,[1]

$$\begin{aligned} x_1(0) &= x_{o,1}(a_1, a_2, \ldots, a_{n-1}) \\ &\vdots \\ x_n(0) &= x_{o,n}(a_1, a_2, \ldots, a_{n-1}) \\ u(0) &= u_o(x_{o,1}, \ldots, x_{0,n}) \end{aligned} \tag{10.35}$$

The parameters $\{a_1, a_2, \ldots, a_{n-1}\}$ are used to specify the boundary hyper-surface.

[1] The boundary conditions are now specified at an $(n-1)$-dimensional hypersurface that can be parameterized by $\{a_1, \ldots, a_{n-1}\}$.

For the linear and semilinear first-order equations, the characteristic equations become

$$\frac{dx_1}{ds} = \alpha_1(\mathbf{x}, u) = \beta_1(\mathbf{x}) \quad ; \quad \cdots \quad ; \quad \frac{dx_n}{ds} = \alpha_n(\mathbf{x}, u) = \beta_n(\mathbf{x}) \tag{10.36}$$

and the characteristics can be solved independently of u, that is,

$$\begin{aligned} x_1 &= x_1\left(x_{o,1}, \ldots, x_{o,n}, s\right) \\ &\vdots \\ x_n &= x_n\left(x_{o,1}, \ldots, x_{o,n}, s\right) \end{aligned} \tag{10.37}$$

Furthermore, if the boundary surface are parameterized by $\{a_1, \ldots, a_{n-1}\}$, then with a nonsingular Jacobian determinant given by

$$\left|\begin{pmatrix} \dfrac{\partial x_1}{\partial s} & \dfrac{\partial x_1}{\partial a_1} & \cdots & \dfrac{\partial x_1}{\partial a_{n-1}} \\ \vdots & \vdots & \ddots & \vdots \\ \dfrac{\partial x_n}{\partial s} & \dfrac{\partial x_n}{\partial a_1} & \cdots & \dfrac{\partial x_n}{\partial a_{n-1}} \end{pmatrix}\right| \neq 0 \tag{10.38}$$

the solution may be put in terms of $\mathbf{x}$.

It may appear that solving the quasilinear partial differential equations is straightforward; that is, the characteristic curves $u(x_1(s), \ldots, x_n(s))$ simply involve solving simultaneous ordinary differential equations. Unfortunately, for the general case, the characteristics can intersect each other. When this occurs, a discontinuity in the solution surface commences, and we refer to this situation as a "shock." We include a brief discussion of shocks (as well the case of rarefaction) in Section J.1 as an appendix.

10.2 Alternate Forms and General Solutions

By dividing any two equations in (10.34), one can get rid of the dependence on s. For instance, for the first two equations, we have

$$\frac{dx_1}{dx_2} = \frac{\alpha_1}{\alpha_2}$$

which can be rearranged to be

$$\frac{dx_1}{\alpha_1} = \frac{dx_2}{\alpha_2}$$

Thus an alternate formulation of the set of determining equations for the characteristics and characteristic curves given in (10.34) is given as

$$\frac{dx_1}{\alpha_1} = \cdots = \frac{dx_n}{\alpha_n} = \frac{du}{f} \tag{10.39}$$

A limitation with this alternate formulation is that one or more functions α_i may be zero. In those cases, the notation simply implies that $x_i =$ constant, and the term dx_i/α_i can then be removed from (10.39). Likewise, if $f = 0$, we have $u =$ constant, and the last term in (10.39) can also be removed.

The formulation given in (10.39) is usually preferred when one is interested in a **general solution**. A general solution is a solution that satisfies the partial differential equation, without having to specify boundary conditions, and often involves arbitrary functions of the independent and dependent variables.

The set of equations in (10.39) produces n solutions

$$\psi_i(x_1, \ldots, x_n, u) = \text{constant}_i \qquad i = 1, \ldots, n \tag{10.40}$$

These can further be combined in two possible forms, and both are equivalent general solutions. One formulation is to use one of the solutions, say ψ_n, to be an arbitrary function of the other solutions, that is,

$$\psi_n = f(\psi_1, \ldots, \psi_{n-1}) \tag{10.41}$$

where f is an arbitrary function. The other formulation is to simply combine all intermediate equations via another arbitrary function that is then set to a constant, that is,

$$F(\psi_1, \ldots, \psi_n) = \text{constant} \tag{10.42}$$

The following example illustrates the equivalence of (10.41) and (10.42).

EXAMPLE 10.4. Consider the partial differential equation given in Example 10.2,

$$x\frac{\partial u}{\partial x} - y\frac{\partial u}{\partial y} = 2x^2 - 6y^2 \tag{10.43}$$

As prescribed in (10.39),

$$\frac{dx}{x} = \frac{dy}{-y} = \frac{du}{2x^2 - 6y^2} \tag{10.44}$$

By equating the first two terms, the solution is given by

$$\ln(x) = -\ln(y) + \ln(\psi_1) \quad \rightarrow \quad \psi_1 = xy = \text{constant}$$

We could then substitute $y = \psi_1/x$ in the last term of (10.44) and equate the result with the first term,

$$\frac{du}{2x^2 - 6\dfrac{\psi_1^2}{x^2}} = \frac{dx}{x} \quad \rightarrow \quad \begin{aligned} u - \psi_2 &= x^2 + 3\frac{\psi_1^2}{x^2} \\ \psi_2 &= u - \left(x^2 + 3y^2\right) \end{aligned}$$

So the first formulation of the general solution is given by

$$\begin{aligned} \psi_2 &= f(\psi_1) \\ u - \left(x^2 + 3y^2\right) &= f(xy) \\ u &= x^2 + 3y^2 + f(xy) \end{aligned} \tag{10.45}$$

which satisfies (10.43). Next, consider the second formulation of the general solution,

$$F(\psi_1, \psi_2) = F\left(xy, u - x^2 + 3y^2\right) = \text{constant} \tag{10.46}$$

To show that this also satisfies (10.43), take the partial derivatives of F with respect to x and y, respectively,

$$\begin{aligned}\frac{\partial F}{\partial \psi_1}y+\frac{\partial F}{\partial \psi_2}\left(\frac{\partial u}{\partial x}-2x\right) &= 0\\ \frac{\partial F}{\partial \psi_1}x+\frac{\partial F}{\partial \psi_2}\left(\frac{\partial u}{\partial y}-6y\right) &= 0\end{aligned}$$

By multiplying the first equation by x and the second by y, the difference will be

$$\left(\frac{\partial F}{\partial \psi_2}\right)\left(x\frac{\partial u}{\partial x}-2x^2-y\frac{\partial u}{\partial y}+6y^2\right)=0 \tag{10.47}$$

Because F is an arbitrary function, the other factor on the left-hand side of (10.47) must equal zero, which again yields (10.43).

Thus either formulation of the general solution is valid. The advantage of the first formulation is that it can often yield an explicit form, $u = u(x, y)$. In cases in which only implicit solutions are possible, the second formulation is preferred.

10.3 The Lagrange-Charpit Method

The general nonlinear differential equation in n independent variables, $x_1, \ldots, x_n$, is given by

$$F\left(\mathbf{x}, u, \frac{\partial u}{\partial x_1}, \ldots, \frac{\partial u}{\partial x_n}\right)=0 \tag{10.48}$$

To solve (10.48), one approach is the **Lagrange-Charpit method**. First, we assume that the desired solution, $u = u(\mathbf{x})$, also satisfies another nonlinear partial differential equation given by

$$G\left(\mathbf{x}, u, \frac{\partial u}{\partial x_1}, \ldots, \frac{\partial u}{\partial x_n}\right)=0 \tag{10.49}$$

It will turn out later that we do not actually need to specify any particular formulation of $G(\mathbf{x}, u, \ldots)$.

We introduce the following notations to denote the partial derivatives:

$$\begin{aligned}\mu_k &= \frac{\partial u}{\partial x_k}\\ \mu_{k,\ell} &= \frac{\partial^2 u}{\partial x_k \partial x_\ell}\end{aligned}$$

$$\begin{aligned}F_{x_k} &= \frac{\partial F}{\partial x_k} \quad , \quad & G_{x_k} &= \frac{\partial G}{\partial x_k}\\ F_u &= \frac{\partial F}{\partial u} \quad , \quad & G_u &= \frac{\partial G}{\partial u}\\ F_{\mu_k} &= \frac{\partial F}{\partial \mu_k} \quad , \quad & G_{\mu_k} &= \frac{\partial G}{\partial \mu_k}\end{aligned}$$

Taking partial derivatives of (10.48) with respect to x_i, yields

$$\frac{\partial F}{\partial x_i} + \frac{\partial F}{\partial u}\frac{\partial u}{\partial x_i} + \frac{\partial F}{\partial \mu_1}\mu_{i,1} + \cdots + \frac{\partial F}{\partial \mu_n}\mu_{i,n} = 0 \tag{10.50}$$

Similarly, for (10.49),

$$\frac{\partial G}{\partial x_i} + \frac{\partial G}{\partial u}\frac{\partial u}{\partial x_i} + \frac{\partial G}{\partial \mu_1}\mu_{i,1} + \cdots + \frac{\partial G}{\partial \mu_n}\mu_{i,n} = 0 \tag{10.51}$$

Collecting these equations in matrix form,

$$\begin{pmatrix} 0 \\ \vdots \\ 0 \end{pmatrix} = \begin{pmatrix} F_{x_1} \\ \vdots \\ F_{x_n} \end{pmatrix} + F_u \begin{pmatrix} \mu_1 \\ \vdots \\ \mu_n \end{pmatrix} + \begin{pmatrix} \mu_{1,1} & \cdots & \mu_{1,n} \\ \vdots & \ddots & \vdots \\ \mu_{n,1} & \cdots & \mu_{n,n} \end{pmatrix} \begin{pmatrix} F_{\mu_1} \\ \vdots \\ F_{\mu_n} \end{pmatrix} \tag{10.52}$$

$$\begin{pmatrix} 0 \\ \vdots \\ 0 \end{pmatrix} = \begin{pmatrix} G_{x_1} \\ \vdots \\ G_{x_n} \end{pmatrix} + G_u \begin{pmatrix} \mu_1 \\ \vdots \\ \mu_n \end{pmatrix} + \begin{pmatrix} \mu_{1,1} & \cdots & \mu_{1,n} \\ \vdots & \ddots & \vdots \\ \mu_{n,1} & \cdots & \mu_{n,n} \end{pmatrix} \begin{pmatrix} G_{\mu_1} \\ \vdots \\ G_{\mu_n} \end{pmatrix} \tag{10.53}$$

Next, we can premultiply (10.52) by a row vector $(G_{\mu_1}, \ldots, G_{\mu_n})$ and premultiply (10.53) by a row vector $(F_{\mu_1}, \ldots, F_{\mu_n})$. Taking the difference of the two resulting scalar equations, while noting that $\mu_{i,j} = \mu_{j,i}$, we have

$$\sum_{k=1}^{n} \left(F_{\mu_k} G_{x_k} + F_{\mu_k}\mu_k G_u - G_{\mu_k} F_{x_k} - G_{\mu_k}\mu_k F_u\right) = 0$$

or

$$F_{\mu_1} G_{x_1} + \cdots + F_{\mu_k} G_{x_n} + \left(\sum_{k=1}^{n} F_{\mu_k}\mu_k\right) G_u$$
$$- \left(F_{x_1} + F_u\mu_1\right) G_{\mu_1} - \cdots - \left(F_{x_n} + F_u\mu_n\right) G_{\mu_n} = 0 \tag{10.54}$$

which is a quasilinear partial differential equation for G. Thus we can write the $2n$ characteristic equations as follows

$$\frac{dx_1}{F_{\mu_1}} = \cdots = \frac{dx_n}{F_{\mu_n}} = \frac{du}{\sum_{k=1}^{n} F_{\mu_k}\mu_k}$$
$$= \frac{-d\mu_1}{F_{x_1} + F_u\mu_1} = \cdots = \frac{-d\mu_n}{F_{x_n} + F_u\mu_n} \tag{10.55}$$

which we shall refer to as the **Lagrange-Charpit conditions**. Among these $2n$ characteristic equations, we only need n equations that would yield μ_i, $i = 1, 2, \ldots, n$, in terms of the independent variables $x_1, x_2, \ldots, x_n$. Finally, u is obtained by solving the differential equation,

$$du = \mu_1 (x_1, \ldots, x_n)\, dx_1 + \cdots + \mu_n (x_1, \ldots, x_n)\, dx_n \tag{10.56}$$

The solution of (10.56) will involve a set of n arbitrary constants after substituting back into the differential equation (10.48). This solution is known as the **complete solution**. In some cases, the arbitrary constants can be removed to obtain a general solution, where arbitrary functions replace the arbitrary constants.

EXAMPLE 10.5. Consider the following nonlinear first-order differential equation

$$\left(\frac{\partial u}{\partial x_2}\right)\left(1 - \frac{\partial u}{\partial x_1}\right) = x_1 - u \tag{10.57}$$

subject to $u(x_1, 0) = 3x_1 + 4$

The required function F and the partial derivatives are

$$F = \mu_2(1 - \mu_1) \quad + \quad u - x_1$$

$$\frac{\partial F}{\partial x_1} = -1 \quad , \quad \frac{\partial F}{\partial x_2} = 0 \quad , \quad \frac{\partial F}{\partial u} = 1$$

$$\frac{\partial F}{\partial \mu_1} = -\mu_2 \quad , \quad \frac{\partial F}{\partial \mu_2} = 1 - \mu_1$$

Substituting into (10.55),

$$-\frac{dx_1}{\mu_2} = \frac{dx_2}{1 - \mu_1} = \frac{du}{\mu_2 - 2\mu_1\mu_2} = \frac{d\mu_1}{1 - \mu_1} = -\frac{d\mu_2}{\mu_2} \tag{10.58}$$

Of the four equations, we can select only those which can be used to find μ_1 and μ_2 as explicit functions of x_1 and x_2. To do so, we equate the first and the last term, as well as equate the second and the second-to-the-last term.

$$dx_1 = d\mu_2 \quad \text{and} \quad dx_2 = d\mu_1$$

or

$$\mu_2 = \frac{\partial u}{\partial x_2} = x_1 + c_1 \quad \text{and} \quad \mu_1 = \frac{\partial u}{\partial x_1} = x_2 + c_2$$

where c_1 and c_2 are arbitrary constants. Solving, we get

$$u = x_1x_2 + c_1x_2 + c_2x_1 + c_3$$

After substituting back into (10.57), we see $c_3 = c_1c_2 - c_1$. Thus the complete solution is given by

$$u = x_1x_2 + c_1(x_2 - 1 + c_2) + c_2x_1 \tag{10.59}$$

After applying the boundary condition, the solution is then

$$u = x_1x_2 + 2(x_2 + 2) + 3x_1 \tag{10.60}$$

We now list a set of special nonlinear cases in which the Lagrange-Charpit method can be used to obtain the complete solutions:

1. **Case 1: Missing u and x_k: $F(\mu_1, \ldots, \mu_n) = 0$.** We will assume that $F(\mu_1, \ldots, \mu_n) = 0$ can be used to solve for μ_n in terms of $\mu_1, \ldots, \mu_{n-1}$, that is, $\mu_n = f(\mu_1, \ldots, \mu_{n-1})$. Because $F_u = F_{x_1} = \cdots = F_{x_n} = 0$, the last n terms in (10.55) becomes

$$\frac{d\mu_1}{0} = \cdots = \frac{d\mu_n}{0} \quad \rightarrow \quad \mu_k = a_k \, , \; k = 1, \ldots, n$$

where $a_1, \ldots, a_{n-1}$ are arbitrary constants but with $a_n = f(a_1, \ldots, a_{n-1})$. The solution is then given by

$$u = \left(\sum_{k=1}^{n-1} a_k x_k \right) + f(a_1, \ldots, a_{n-1}) x_n + b \tag{10.61}$$

where b is another arbitrary constant.

2. **Case 2: Missing x_k: $F(u, \mu_1, \ldots, \mu_n) = 0$.** Because $F_{x_1} = \cdots = F_{x_n} = 0$, the last n terms in (10.55) becomes

$$\frac{d\mu_1}{\mu_1} = \cdots = \frac{d\mu_n}{\mu_n} \quad \rightarrow \quad \mu_j = a_j \mu_n, \; j = 1, \ldots, n-1$$

where $a_1, \ldots, a_{n-1}$ are arbitrary constants. Next, assume that μ_n can be obtained from $F(u, \mu_1, \ldots, \mu_n) = F(u, a_1\mu_n, \ldots, a_{n-1}\mu_n, \mu_n) = 0$, i.e. $\mu_n = f(u, a_1, \ldots, a_{n-1})$. Then the complete solution is given by

$$\int \frac{du}{f(u, a_1, \ldots, a_{n-1})} = \left(\sum_{k=1}^{n-1} a_k x_k \right) + x_n + b \tag{10.62}$$

where b is another arbitrary constant.

3. **Case 3: Separable terms, missing u: $\sum_{k=1}^{n} f_k(x_k, \mu_k) = 0$.** With $F_u = 0$, equating the $(k \leq n)^{\text{th}}$ term with the $(k+n+1)^{\text{th}}$ term in (10.55) will result in

$$\frac{dx_k}{\partial f_k / \partial \mu_k} = -\frac{d\mu_k}{\partial f_k / \partial x_k} \quad \rightarrow \quad \frac{\partial f_k}{\partial x_k} dx + \frac{\partial f_k}{\partial \mu_k} d\mu_k = 0 \text{ or } f_k(x_k, \mu_k) = a_k$$

where $a_1, \ldots, a_{n-1}$ are arbitrary constants and $a_n = -\sum_{j=1}^{n-1} a_j$. Thus assuming each separate term in the given differential equation can be solved for μ_k, that is,

$$f_k(x_k, \mu_k) = a_k \quad \rightarrow \quad \mu_k = g(a_k, x_k)$$

then the complete solution is given by

$$u = \sum_{k=1}^{n} \left(\int g(a_k, x_k)\, dx_k \right) + b \tag{10.63}$$

where b is another arbitrary constant.

4. **Case 4: Clairaut's PDE: $u = \sum_{k=1}^{n} x_k \mu_k + f(\mu_1, \ldots, \mu_n)$.** Because $F_{x_k} + F_u \mu_k = 0$, the last n terms of (10.55) become

$$\frac{d\mu_1}{0} = \cdots = \frac{d\mu_n}{0} \quad \rightarrow \quad \mu_k = a_k, \; k = 1, \ldots, n$$

where $a_1, \ldots, a_n$ are arbitrary constants (Note that in this case we can set a_n arbitrary and independent of the other constants). Substituting back into the differential equation, we have

$$u = \sum_{k=1}^{n} a_k x_k + f(a_1, \ldots, a_n) \tag{10.64}$$

10.4 Classification Based on Principal Parts

We now generalize the classification of PDEs that are consistent with the first-order case. As is seen next, the method of characteristics is used during the search for canonical forms. For the case of two independent variables, x and y, we use the following notations for the partial derivatives,

$$
\begin{array}{lll}
u_x = \dfrac{\partial u}{\partial x} & ; \quad u_y = \dfrac{\partial u}{\partial y} & \\[2ex]
u_{xx} = \dfrac{\partial^2 u}{\partial x \partial x} & ; \quad u_{xy} = \dfrac{\partial^2 u}{\partial x \partial y} & ; \quad u_{yy} = \dfrac{\partial^2 u}{\partial y \partial y} \\
\vdots & &
\end{array}
\tag{10.65}
$$

However, for a system where u depends on the independent variables $\{x_1, x_2, \ldots, x_n\}$, $n > 2$, we use the following notation for the partial derivatives:

$$
\begin{aligned}
\mu_i &= \frac{\partial u}{\partial x_i} \\
\mu_{i,j} &= \frac{\partial^2 u}{\partial x_i \partial x_j}, \quad 1 \le i, j \le n \\
\mu_{i,j,k} &= \frac{\partial^3 u}{\partial x_i \partial x_j \partial x_k}, \quad 1 \le i, j, k \le n \\
&\vdots
\end{aligned}
\tag{10.66}
$$

The **order** of the partial differential equation is the highest order derivative present in the equation. The most general m^{th}-order form is the nonlinear equation given by

$$F\left(\mathbf{x}, u, \mu_{[1]}, \ldots, \mu_{[m]}\right) = 0 \tag{10.67}$$

where,

$$\mathbf{x} = \{x_1, x_2, \ldots, x_n\} \tag{10.68}$$

$$\mu_{[1]} = \{\mu_1, \mu_2, \ldots, \mu_n\} \tag{10.69}$$

$$\mu_{[2]} = \left\{\mu_{i_1,i_2}, \ 1 \le i_1 < i_2 \le n\right\} \tag{10.70}$$

$$\vdots$$

$$\mu_{[m]} = \left\{\mu_{i_1,\ldots,i_m}, \ 1 \le i_1 < \ldots < i_m \le n\right\} \tag{10.71}$$

To be consistent with the classification we used for the first-order partial differential equations, we decompose the equation into two parts: the principal part F_{prin}, which contains the highest derivatives, and the nonprincipal part F_{non}, in which the highest derivatives are absent,

$$F_{\text{prin}} + F_{\text{non}} = 0 \tag{10.72}$$

The differential equations can then be classified based on special forms that the principal parts may take on.

1. **Quasilinear Equation**

$$F_{\text{prin}} = \sum_{1\le i_1<\cdots<i_m\le n} \alpha_{i_1,\ldots,i_m}\left(\mathbf{x}, u, \mu_{[1]}, \ldots, \mu_{[m-1]}\right)\mu_{i_1,\ldots,i_m} \tag{10.73}$$

   which means the highest derivatives are linearly combined, with coefficients α being functions of the independent variables and lower order derivatives.
2. **Semilinear Equation**

$$F_{\text{prin}} = \sum_{1\le i_1<\cdots<i_m\le n} \beta_{i_1,\ldots,i_m}\left(\mathbf{x}\right)\mu_{i_1,\ldots,i_m} \tag{10.74}$$

   which means the highest derivatives are linearly combined, with coefficients β being functions of only independent variables.
3. **Linear Equation.**

$$F_{\text{prin}} = \sum_{1\le i_1<\cdots<i_m\le n} \beta_{i_1,\ldots,i_m}\left(\mathbf{x}\right)\mu_{i_1,\ldots,i_m} \tag{10.75}$$

$$F_{\text{non}} = \left(\sum_{1\le i_1<\cdots<i_{m-1}\le n} \gamma_{i_1,\ldots,i_{m-1}}\left(\mathbf{x}\right)\mu_{i_1,\ldots,i_{m-1}}\right) + \cdots + \left(\sum_{1\le i\le n}\gamma_i\left(\mathbf{x}\right)\mu_i\right) + \gamma\left(\mathbf{x}\right)u + h(\mathbf{x}) \tag{10.76}$$

We now consider the second-order semilinear equations with u as the dependent variable and with x and y as the two independent variables,

$$A\left(x,y\right)u_{xx} + B\left(x,y\right)u_{xy} + C\left(x,y\right)u_{yy} = -F_{\text{non}}\left(x,y,u,u_x,u_y\right) \tag{10.77}$$

We can always convert this partial differential equation to three canonical forms using a change in coordinates, that is, with $\alpha\left(x,y\right)$ and $\beta\left(x,y\right)$ as the new independent variables. The three canonical forms are[2]:

$$\begin{aligned}
\text{Hyperbolic:}\quad & u_{\alpha\alpha} - u_{\beta\beta} = G_1\left(\alpha,\beta,u,u_\alpha,u_\beta\right)\\
\text{Parabolic:}\quad & u_{\alpha\alpha} = G_2\left(\alpha,\beta,u,u_\alpha,u_\beta\right)\\
\text{Elliptic:}\quad & u_{\alpha\alpha} + u_{\beta\beta} = G_3\left(\alpha,\beta,u,u_\alpha,u_\beta\right)
\end{aligned}$$

For each canonical form, the solution takes on particular behavior and properties. For hyperbolic equations, the solutions are wave-like in nature and are thus usually classified as wave equations. For parabolic equations, the solutions are related to diffusion-like behavior and are thus usually classified as diffusion equations. Lastly, for the elliptic equations, the solutions are related to potential fields and are thus usually classified as potential equations. Therefore, one immediate use of classification is to better identify the appropriate solution approach, either analytically or numerically.

[2] The naming of each canonical form closely matches the functional description of a hyperbola: $(x/a)^2 - (y/b)^2 = f_1(x,y,c)$, a parabola: $(y/b)^2 = f_2(x,y,c)$, and an ellipse: $(x/a)^2 + (y/b)^2 = f_3(x,y,c)$.

Let $\xi(x, y)$ and $\eta(x, y)$ be a pair of new coordinates. Applying the chain rule, we obtain

$$\begin{aligned}
u_x &= \partial_x u = \xi_x u_\xi + \eta_x u_\eta \\
u_y &= \partial_y u = \xi_y u_\xi + \eta_y u_\eta \\
u_{xx} &= \partial_x u_x = \xi_x \left[\xi_x u_{\xi\xi} + \eta_x u_{\eta\xi}\right] + \xi_{xx} u_\xi + \eta_x \left[\xi_x u_{\xi\eta} + \eta_x u_{\eta\eta}\right] + \eta_{xx} u_\eta \\
u_{xy} &= \partial_x u_y = \xi_x \left[\xi_y u_{\xi\xi} + \eta_y u_{\eta\xi}\right] + \xi_{xy} u_\xi + \eta_x \left[\xi_y u_{\xi\eta} + \eta_y u_{\eta\eta}\right] + \eta_{xy} u_\eta \\
u_{yy} &= \partial_y u_y = \xi_y \left[\xi_y u_{\xi\xi} + \eta_y u_{\eta\xi}\right] + \xi_{yy} u_\xi + \eta_y \left[\xi_y u_{\xi\eta} + \eta_y u_{\eta\eta}\right] + \eta_{yy} u_\eta
\end{aligned} \tag{10.78}$$

which converts (10.77) to

$$g u_{\xi\eta} + Q\left(\xi_x, \xi_y\right) u_{\xi\xi} + Q\left(\eta_x, \eta_y\right) u_{\eta\eta} = h\left(\xi, \eta, u, u_\xi, u_\eta\right) \tag{10.79}$$

where,

$$g = 2\left(A\xi_x\eta_x + C\xi_y\eta_y\right) + (B)\left(\xi_x\eta_y + \xi_y\eta_x\right)$$

and $Q(\cdot, \cdot)$ is a quadratic form known as the **characteristic form** of (10.77) defined as

$$Q(w, q) = Aw^2 + Bwq + Cq^2 = \begin{cases} A\left(w - r_1 q\right)\left(w - r_2 q\right) & \text{if } A \neq 0 \\ \\ (Bw + Cq)(q) & \text{if } A = 0 \end{cases} \tag{10.80}$$

with roots r_1 and r_2 given by

$$r_1 = \frac{-B + \sqrt{B^2 - 4AC}}{2A} \quad ; \quad r_2 = \frac{-B - \sqrt{B^2 - 4AC}}{2A} \tag{10.81}$$

To remove the terms involving $u_{\xi\xi}$ and $u_{\eta\eta}$ in (10.79), we can solve for ξ and η such that $Q\left(\xi_x, \xi_y\right) = 0$ and $Q\left(\eta_x, \eta_y\right) = 0$, that is,

$$\begin{aligned}
\left(\xi_x - r_1\xi_y\right)\left(\xi_x - r_2\xi_y\right) = 0 = \left(\eta_x - r_1\eta_y\right)\left(\eta_x - r_2\eta_y\right) \qquad & \text{if } A \neq 0 \\
\left(B\xi_x + C\xi_y\right)\left(\xi_y\right) = 0 = \left(B\eta_x + C\eta_y\right)\left(\eta_y\right) \qquad & \text{if } A = 0
\end{aligned}$$

If $r_1 = r_2 = r$, or equivalently $B^2 = 4AC$, we can solve $\xi_x - r\xi_y = 0\, (= \eta_x - r\eta_y)$ using the method of characteristics and then set $\xi = \eta = \alpha$. This case results in a parabolic canonical form. For $r_1 \neq r_2$, we can solve the following equations using the method of characteristics:

$$\begin{aligned}
\xi_x - r_1\xi_y = 0 \quad ; \quad \eta_x - r_2\eta_y = 0 \qquad & \text{if } A \neq 0 \\
B\xi_x + C\xi_y = 0 \quad ; \quad \eta_y = 0 \qquad & \text{if } A = 0
\end{aligned} \tag{10.82}$$

If $r_1 \neq r_2$ are real, or equivalently $B^2 > 4AC$ (which includes the case when $A = 0$, $B \neq 0$), then ξ and η are also real. Let $\alpha = \xi + \eta$ and $\beta = \xi - \eta$. Then with

$$\partial_\xi = \alpha_\xi \frac{\partial}{\partial\alpha} + \beta_\xi \frac{\partial}{\partial\beta} = \frac{\partial}{\partial\alpha} + \frac{\partial}{\partial\beta} \qquad \text{and} \qquad \partial_\eta = \alpha_\eta \frac{\partial}{\partial\alpha} + \beta_\eta \frac{\partial}{\partial\beta} = \frac{\partial}{\partial\alpha} - \frac{\partial}{\partial\beta}$$

we have $u_{\xi\eta} = u_{\alpha\alpha} - u_{\beta\beta}$, which yields the hyperbolic canonical form. Finally, if r_1 and r_2 are complex conjugates, or equivalently $B^2 < 4AC$, then ξ and η will also be complex conjugates. Let $\alpha = \xi + \eta$ and $\beta = i(\xi - \eta)$. Then with

$$\partial_\xi = \frac{\partial}{\partial\alpha} + i\frac{\partial}{\partial\beta} \qquad \text{and} \qquad \partial_\eta = \frac{\partial}{\partial\alpha} - i\frac{\partial}{\partial\beta}$$

we have $u_{\xi\eta} = u_{\alpha\alpha} + u_{\beta\beta}$ which yields the elliptic canonical form.

In summary, for purposes of classification, the semilinear partial differential equation given in (10.77) can be determined to be hyperbolic, parabolic, or elliptic easily as follows:

$$\begin{aligned} \text{Hyperbolic, if} \quad & B^2 > 4AC \\ \text{Parabolic, if} \quad & B^2 = 4AC \\ \text{Elliptic, if} \quad & B^2 < 4AC \end{aligned}$$

Because the coefficients A, B, and C in (10.77) depend on the values of x and y, an equation can change from elliptic to parabolic, and then to hyperbolic, or the other way around.

EXAMPLE 10.6. Let us classify the linear partial differential equation,

$$xu_{xx} - (x+y)u_{xy} + yu_{yy} = 2xy$$

and find the corresponding canonical form. With $A = x$, $B = -(x+y)$, and $C = y$, which determines the relationship $B^2 - 4AC = (x-y)^2 > 0$, the equation is hyperbolic.

To obtain the canonical form, the roots of the characteristic form in (10.80) are $r_1 = 1$ and $r_2 = y/x$. From (10.82), we determine $\xi = x + y$ and $\eta = xy$. Substituting these into the differential equation using (10.78),

$$-\left(\xi^2 - 4\eta\right) u_{\xi\eta} = \xi u_\eta + 2\eta$$

Next, using $\alpha = \xi + \eta = x + y + xy$ and $\beta = \xi - \eta = x + y - xy$, we obtain the hyperbolic canonical form,

$$u_{\alpha\alpha} - u_{\beta\beta} = -2\frac{(\alpha+\beta)(u_\alpha - u_\beta) + 2(\alpha - \beta)}{(\alpha+\beta)^2 - 8(\alpha-\beta)}$$

One of the advantages of finding the canonical forms is that solutions and solution methods, both analytical and numerical, are readily available for the nonprincipal parts having special reduced forms. For instance, if the canonical hyperbolic form (in terms of ξ and η) is given by

$$u_{\xi\eta} = f(\xi, \eta)$$

then the general solution can be obtained immediately by integration, that is,

$$u(\xi, \eta) = \int\int f(\xi, \eta)\, d\xi d\eta + g(\xi) + h(\eta)$$

The classification of a second-order semilinear equation with more than two independent variables is discussed in Section J.2 as an appendix. Also, the classification of higher order semilinear differential equations with two independent variables is discussed in Section J.3 as an appendix.

10.5 Hyperbolic Systems of Equations

Let us now consider a set of first-order quasilinear equations for a vector of m dependent variables, $\mathbf{u} = (u_1, \ldots, u_m)^T$, and two independent variables, t and x, given by

$$\mathbf{u}_t + A(x, t, \mathbf{u})\,\mathbf{u}_x = \mathbf{b}(x, t, u) \tag{10.83}$$

where

$$\mathbf{u}_t = \begin{pmatrix} \dfrac{\partial u_1}{\partial t} \\ \vdots \\ \dfrac{\partial u_m}{\partial t} \end{pmatrix} \quad \text{and} \quad \mathbf{u}_x = \begin{pmatrix} \dfrac{\partial u_1}{\partial x} \\ \vdots \\ \dfrac{\partial u_m}{\partial x} \end{pmatrix}$$

$A[=]m \times m$ and $\mathbf{b}[=]m \times 1$. One approach is to determine whether the method of characteristics can be used. Let $\mathbf{g}$ be a vector of functions such that

$$\mathbf{g}^T \mathbf{u}_t + \mathbf{g}^T A_{(x,t,\mathbf{u})}\mathbf{u}_x = \mathbf{g}^T \mathbf{b}(x, t, u)$$

which will yield a set of equations in which the dependent variables will propagate along a common characteristic described by

$$\frac{dx}{dt} = \lambda(x, t, \mathbf{u}) \qquad \leftarrow \qquad \mathbf{g}^T(\mathbf{u}_t + \lambda\mathbf{u}_x) = \mathbf{g}^T\mathbf{b} \tag{10.84}$$

This leads to an eigenvalue problem problem given by

$$A^T\mathbf{g} = \lambda\mathbf{g} \tag{10.85}$$

Assuming that the eigenvectors $\mathbf{g}_k$ are real and linearly independent, the partial differential equations can be reduced to a set of m ordinary differential equations,

$$\mathbf{g}_k^T(x, t, \mathbf{u})\frac{d\mathbf{u}}{dt} = \mathbf{g}_k^T(x, t, \mathbf{u})\,\mathbf{b}(x, t, \mathbf{u}) \qquad \text{along} \qquad \frac{dx}{dt} = \lambda_k(x, t, u) \tag{10.86}$$

Note, however, that unlike the eigenvalue problems we have dealt with in the previous chapters, the matrix A is a matrix of functions. Further, note that even though there are m ordinary differential equations, the equations are along different characteristics. This means that the equations cannot be solved simultaneously as they are given in (10.86). Moreover, the system will contain multiple-point boundary value problems.

If the eigenvalues are all real and the eigenvectors are linearly independent, we say that (10.83) is a **hyperbolic system of PDE**. In addition, when the eigenvalues are real and distinct, we say that it is a **strictly hyperbolic system of PDE**. However, if none of the eigenvalues are real, then (10.83) is called an **elliptic system**. If there are fewer than m linearly independent eigenvectors, we have a **parabolic system**.

EXAMPLE 10.7. Consider the special case for (10.83) with $A[=]n \times n$ and $\mathbf{b}[=]n \times 1$ being constant, and A^T having n real and distinct eigenvalues, $\lambda_1 \neq \cdots \neq \lambda_n$ and corresponding eigenvectors $\mathbf{g}_1, \ldots, \mathbf{g}_n$, that is,

$$A^T G = G\Lambda \qquad \text{where} \qquad \Lambda = \begin{pmatrix} \lambda_1 & & 0 \\ & \ddots & \\ 0 & & \lambda_n \end{pmatrix} \quad ; \quad G = \left(\mathbf{g}_1 \,\middle|\, \cdots \,\middle|\, \mathbf{g}_n \right)$$

Suppose also that the initial conditions are given by

$$\mathbf{u}(x, 0) = \mathbf{u}_0(x) \tag{10.87}$$

Then the solution to (10.86) subject to the initial conditions, and following along the characteristics (details of which are left as an exercise in **E10.18**), will be given by

$$\mathbf{u} = \mathbf{b}t + G^{-T} \begin{pmatrix} \mathbf{g}_1^T \mathbf{u}_0(x - \lambda_1 t) \\ \vdots \\ \mathbf{g}_n^T \mathbf{u}_0(x - \lambda_n t) \end{pmatrix} \tag{10.88}$$

One can show by direct substitution that (10.88) will satisfy both the system of differential equations (10.83) and the initial conditions (10.87).

EXAMPLE 10.8. Consider the model for waves in shallow water given by

$$\begin{pmatrix} \dfrac{\partial h}{\partial t} \\ \dfrac{\partial v}{\partial t} \end{pmatrix} + \begin{pmatrix} v & h \\ g & v \end{pmatrix} \begin{pmatrix} \dfrac{\partial h}{\partial x} \\ \dfrac{\partial v}{\partial x} \end{pmatrix} = \begin{pmatrix} 0 \\ 0 \end{pmatrix}$$

where $h(x, t)$ and $v(x, t)$ are the height of the surface and velocity along x, respectively, and g is the gravitational acceleration. The first equation is the mass balance equation, whereas the second equation is the momentum balance (reduced under the constraint of the mass balance equation). The eigenvalues of A^T are the roots of

$$\mathbf{det} \begin{pmatrix} \lambda - v & -g \\ -h & \lambda - v \end{pmatrix} = 0 \quad \rightarrow \quad \lambda_1, \lambda_2 = v \pm \sqrt{hg}$$

and the corresponding eigenvectors are

$$\mathbf{g}_1 = \begin{pmatrix} 1 \\ +\sqrt{\dfrac{h}{g}} \end{pmatrix} \quad \text{and} \quad \mathbf{g}_2 = \begin{pmatrix} 1 \\ -\sqrt{\dfrac{h}{g}} \end{pmatrix}$$

Thus yielding

$$\frac{dh}{dt} + \left(\sqrt{\frac{h}{g}}\right) \frac{dv}{dt} = 0 \quad \text{along} \quad \frac{dx}{dt} = v + \sqrt{hg}$$

and

$$\frac{dh}{dt} - \left(\sqrt{\frac{h}{g}}\right) \frac{dv}{dt} = 0 \quad \text{along} \quad \frac{dx}{dt} = v - \sqrt{hg}$$

Because the right-hand side of both ordinary differential equations are zero, we find that along their respective characteristics the following are constants, that is,

$$R_1 = 2\sqrt{gh} + v = \text{constant} \quad \text{along} \quad \frac{dx}{dt} = v + \sqrt{gh}$$

$$R_2 = 2\sqrt{gh} - v = \text{constant} \quad \text{along} \quad \frac{dx}{dt} = v - \sqrt{gh}$$

The two quantities are known as the **Riemann invariants** along their respective characteristics.

10.6 EXERCISES

E10.1. Classify the following first-order differential equations (i.e., linear, semilinear, nonlinear, etc.)

1. $\dfrac{\partial (u^2)}{\partial t} + \dfrac{\partial (u^2 + 4t)}{\partial z} = ue^{-2z}$
2. $\dfrac{\partial u}{\partial t} + \dfrac{\partial e^{-u}}{\partial z} = -4ze^{-2t}$
3. $\dfrac{\partial u}{\partial t} + \underline{\vec{\mathbf{r}}} \cdot \nabla u = \|\underline{\vec{\mathbf{r}}}\|$, under the rectangular coordinate system, where $\underline{\vec{\mathbf{r}}}$ is the position vector.

E10.2. Obtain the solutions given in (10.31) and (10.32) of the parametric pump process given in Example 10.3. By setting the period $P = 2\dfrac{L}{v\mu^+}$ and $\lambda <$ 1, obtain $c_{\text{bottom}}(t)/c_0$ and $c_{\text{top}}(t)/c_0$, assuming $c^{+(0)}(z) = c^{\text{bottom}} = c_0$. Also, what is $\lim_{t\to\infty} c^{\text{top}}(t)$?

E10.3. For the plug flow reactor undergoing a first-order reaction, a simplified model for the concentration of the reactant in the reactor is given by

$$\frac{\partial c}{\partial t} + v\frac{\partial c}{\partial z} = -kc$$

Assuming v is constant, obtain the solution given the conditions: $c(z, 0) = c_0(z)$ and $c(0, t) = c_{in}(t)$

E10.4. Consider the advection equation with variable coefficients given by

$$\frac{\partial u}{\partial t} + v(t)\frac{\partial u}{\partial z} = 0$$

Obtain the solution given the conditions $u(z, 0) = u_0(z)$ and $u(0, t) = u_{in}(t)$ for a) $v(t) = 2 - e^{-2t}$ and b) $v(t) = 2 + \sin(2t)/2$.

E10.5. Find the solution to the dynamic, 2D advection problem given by

$$\frac{\partial u}{\partial t} + v_x\frac{\partial u}{\partial x} + v_y\frac{\partial u}{\partial y} = 0$$

subject to the conditions:

$$u(x, y, 0) = u_0(x, y) \quad \text{and} \quad u(x, y, t)\Big|_{y=-x} = u_{in}(t)$$

for the domains $t > 0$ and $x + y > 0$, where v_x and v_y are positive constants. Check your solutions by making sure they satisfy the PDE and the boundary and initial conditions.

E10.6. Find the solution to the dynamic, 2D advection problem given by

$$\frac{\partial c}{\partial t} + v_x(t)\frac{\partial c}{\partial x} + v_y(t)\frac{\partial c}{\partial y} = -kc$$

subject to the conditions:

$$c(x, y, 0) = c_0(x, y) \quad ; \quad c(0, y, t) = f(y, t) \quad ; \quad c(x, 0, t) = g(x, t)$$

for the domains $t > 0$, $x > 0$, and $y > 0$, where $v_x(t) = 1 - e^{-t/5}$ and $v_y(t) = 2$. Check your solutions by making sure they satisfy the PDE and the boundary and initial conditions. (A closed solution can be written in terms of the Lambert W-function [cf. (6.34)]. However, a set of implicit equations is often regarded as a sufficient form of the solution.)

E10.7. Consider the 3D advection equation with a generation term $R(c)$ given by

$$\frac{\partial c}{\partial t} + \mathbf{v}(x, y, z) \cdot (\nabla c) = R(c)$$

with the initial condition $c(x, y, z, 0) = f(x, y, z)$. For the special case where $R(c) = -k_0\, c^2$ and

$$\begin{pmatrix} v_x \\ v_y \\ v_z \end{pmatrix} = A \begin{pmatrix} x \\ y \\ z \end{pmatrix}$$

obtain the solution $c(x, y, z, t)$ in terms of f and k_0. (Check your solution by substituting it into the differential equation and the initial conditions.)

E10.8. An nth-order ordinary differential equation is said to admit a similarity transformation $\widetilde{x} = \lambda x$ and $\widetilde{y} = \lambda^\beta y$ (cf. sections 6.2 and 6.4.2) if

$$y^{(n)} = f\left(x, y, \ldots, y^{(n-1)}\right) \quad \longleftrightarrow \quad \widetilde{y}^{(n)} = f\left(\widetilde{x}, \widetilde{y}, \ldots, \widetilde{y}^{(n-1)}\right)$$

where $y^{(k)} = d^k y/dx^k$.

1. Using the conditions for similarity, one obtains

$$\lambda^{\beta-n} y^{(n)} = f\left(\lambda x, \lambda^\beta y, \ldots, \lambda^{\beta-n+1} y^{(n-1)}\right)$$

Show that differentiation with respect to λ followed by the setting of $\lambda = 1$ will yield a linear first-order partial differential equation given by

$$x\frac{\partial f}{\partial x} + \beta y\frac{\partial f}{\partial y} + \cdots + (\beta - n + 1)\, y^{(n-1)}\frac{\partial f}{\partial y^{(n-1)}} = (\beta - n)\, f \qquad (10.89)$$

2. Show that the general solution for (10.89) is given by

$$\frac{f\left(x, y, \ldots, y^{(n-1)}\right)}{x^{\beta-n}} = G\left(u, v_1, \ldots, v_{n-1}\right)$$

where $u = yx^{-\beta}$, $v_k = y^{(k)}x^{-\beta+k}$ for $k = 1, \ldots, n-1$ and G is an arbitrary function to be fitted later to initial and boundary conditions.

3. With u and v_k defined previously, show that a system of $(n-1)$ first-order ordinary differential equations has been obtained that is given by

$$\frac{dv_j}{du} = \frac{v_{j+1} - (\beta - j)\, v_j}{v_1 - \beta u} \qquad \text{for } j = 1, \ldots, n-2$$

$$\frac{dv_{n-1}}{du} = \frac{G\left(u, v_1, \ldots, v_{n-1}\right) - (\beta - n + 1)\, v_{n-1}}{v_1 - \beta u}$$

which means a reduction of one order from the original ordinary differential equation. (Assuming the procedure can be recursively reduced in which the set of equations also admits a similarity transformation, it will end with a separable first-order ordinary differential equation solvable by quadratures.)

E10.9. Consider the following quasilinear equation:

$$\frac{\partial u}{\partial t} + (1+u)^2 x \frac{\partial u}{\partial x} = 0$$

subject to

$$u(x,0) = u_0(x)$$

such that $u_0(x)$ is continuous.

1. Solve for $u(x,t)$ for $u_0(x) = \tanh(x)$. Plot the solution for $-5 \le x \le 5$ at $t = 1,5$ and 10.
2. Show that as long as $du_o/dx > 0$ and $u_0 > -1$, the solution will not contain a shock. (Hint: Show that under this condition, $\partial x/\partial a \neq 0$ for all t where a is parameter for x when $t = 0$; i.e., there will be no break times, and the characteristics will not intersect).

E10.10. Consider the following quasilinear equation:

$$\frac{\partial u}{\partial t} + (2u+1)\frac{\partial u}{\partial x} = -4u + 1$$

subject to

$$u(x,0) = h(x)$$

such that $h(x)$ is continuous.

1. Solve for $u(x,t)$ for $h(x) = 2 - \tanh(x)$ (Note: The solution may need to be in implicit form). Plot the solution for $-5 \le x \le 5$ at $t = 0.01$, 0.05, and 0.1.
2. Show that as long as $dh/dx > -4$, the solution will not contain a shock. (Hint: Show that under this condition, $\partial x/\partial a \neq 0$ for all t where a is parameter for x when $t = 0$; i.e., there will be no break times, and the characteristics will not intersect).

E10.11. Obtain the general solution for the differential equation given by

$$\frac{\partial u}{\partial x} + \left(\frac{x-y}{x+y}\right)\frac{\partial u}{\partial y} = -ku$$

for $x \ge 0$.

E10.12. Consider the following nonlinear first-order differential equation:

$$\left(x\frac{\partial u}{\partial x}\right)^2 + \left(y\frac{\partial u}{\partial y}\right)^2 + \left(z\frac{\partial u}{\partial z}\right)^2 = c^2$$

Obtain the Lagrange-Charpit conditions and find the complete solution.

E10.13. When the nonlinear first-order partial differential equation does not explicitly contain u and when $\partial u/\partial t$ can be separated as a linear term, we obtain the equation known as the **Hamilton-Jacobi equation**:

$$F\left(t, x_1, \ldots, x_n, u_t, u_{x_1}, \ldots, u_{x_n}\right) = u_t + H\left(t, x_1, \ldots, x_n, u_{x_1}, \ldots, u_{x_n}\right) = 0$$

where $u_t = \partial u/\partial t$ and $u_{x_k} = \partial u/\partial x_k$. This equation is very useful in classical mechanics where H is known as the **Hamiltonian**, whereas its solution

$u(t, x_1, \ldots, x_n)$ is known as the **Hamilton principal function**. Obtain the Lagrange-Charpit conditions for this equation, and use it to derive the set of equations known as the **Hamiltonian canonical equations** (of a conservative mechanical system):

$$\frac{dx_k}{dt} = \frac{\partial H}{\partial u_{x_k}} \tag{10.90}$$

$$\frac{du_{x_k}}{dt} = -\frac{\partial H}{\partial x_k} \tag{10.91}$$

E10.14. The equation for propagation of light is given by

$$\left(\frac{\partial u}{\partial x}\right)^2 + \left(\frac{\partial u}{\partial y}\right)^2 = 1$$

Show that using (10.61), the solution is given by

$$u(x, y) = \alpha x + \sqrt{1-\alpha^2}\, y + \beta$$

Further, let $\alpha = \cos\theta$ and $\beta = \beta(\theta)$, show that for constant u, the positions yielding this value are given by

$$\begin{aligned} x &= (u - \beta(\theta))\cos\theta + \frac{d\beta}{d\theta}\sin\theta \\ y &= (u - \beta(\theta))\sin\theta - \frac{d\beta}{d\theta}\cos\theta \end{aligned}$$

E10.15. Use (10.62) to obtain the complete solution of the following equation:

$$\left(\frac{\partial u}{\partial x}\right)^2 + \left(\frac{\partial u}{\partial y}\right)^2 + \left(\frac{\partial u}{\partial z}\right)^2 = ku$$

E10.16. Classify the following second-order linear equation for $y \neq 0$ and $y \neq x$,

$$(x-y)\frac{\partial^2 u}{\partial x^2} + 2x\frac{\partial u}{\partial x}y + (x+y)\frac{\partial^2 u}{\partial y^2} = 0$$

Find the variables ξ and η that would reduce the principal part to $u_{\xi\eta}$.

E10.17. For the linear homogeneous partial differential equation given by

$$\frac{\partial^2 u}{\partial y^2} + (a+b)\frac{\partial^2 u}{\partial x \partial y} + (ab)\frac{\partial^2 u}{\partial x^2} = 0$$

where a and b are real numbers, show that this is a hyperbolic equation and obtain the alternate canonical form given by $u_{\xi\eta} = 0$. Obtain the general solution.

E10.18. Derive the solution $\mathbf{u}(x, t)$ of the system of n first-order partial differential equations given by

$$\frac{\partial \mathbf{u}}{\partial t} + A\frac{\partial \mathbf{u}}{\partial x} = \mathbf{b}$$

subject to $\mathbf{u}(x, 0) = \mathbf{u}_0(x)$, assuming that A and $\mathbf{b}$ are constants and the eigenvalues λ_k, $k = 1, \ldots, n$, of A^T are distinct and real with corresponding eigenvector $\mathbf{g}_k$. Also, show that the solution will satisfy the initial conditions and the system of partial differential equations.

E10.19. From the continuity equation (5.24) and equation of motion (5.25), we can obtain the inviscid gas dynamics in a tube flowing only in the x-direction by assuming viscosity $\mu = 0$ and neglecting gravitational effects due to g,

$$\begin{aligned} \frac{\partial \rho}{\partial t} + \frac{\partial (\rho v)}{\partial x} &= 0 \\ \rho \left(\frac{\partial v}{\partial t} + v \frac{\partial v}{\partial x} \right) + \frac{\partial p}{\partial x} &= 0 \end{aligned}$$

where ρ and v are the gas density and velocity, respectively. Assume that the pressure is given by $p = k\rho^{\gamma}$ where $k > 0$ and $\gamma > 1$. Let $\zeta = \sqrt{dp/d\rho} = \sqrt{\gamma k}\, \rho^{(\gamma-1)/2}$. Show that the Riemann invariants for the system are given by

$$R_1 = \frac{2\zeta}{\gamma - 1} + v = \text{constant} \qquad \text{along} \qquad \frac{dx}{dt} = v + \zeta$$

$$R_2 = \frac{2\zeta}{\gamma - 1} - v = \text{constant} \qquad \text{along} \qquad \frac{dx}{dt} = v - \zeta$$

(Note: Problems **E10.20** through **E10.22** refer to topics covered in Appendix J.1.)

E10.20. For the partial differential equation given by

$$\frac{\partial u}{\partial t} + b(u) \frac{\partial u}{\partial x} = 0$$

subject to

$$u(x, 0) = \begin{cases} u^{\text{left}} & \text{if } x \le a \\ u^{\text{right}} & \text{if } x > a \end{cases}$$

where $b\left(u^{\text{left}}\right) < b\left(u^{\text{right}}\right)$, show that the solution given by

$$u(x, t) = \begin{cases} u^{\text{left}} & \text{if } x \le b\left(u^{\text{left}}\right) t + a \\ b^{-1}\left(\dfrac{x - a}{t} \right) & \text{if } b\left(u^{\text{left}}\right) t + a < x \le b\left(u^{\text{right}}\right) t + a \\ u^{\text{right}} & \text{if } x > b\left(u^{\text{right}}\right) t + a \end{cases}$$

is indeed piecewise continuous. Furthermore, verify this to be the case for $a = 4$, $b(u) = u^2$, $u^{\text{left}} = 2.5$, and $u^{\text{right}} = 3$ at $t = 10$ and $t = 50$.

E10.21. An idealized model for **traffic flow** is given by

$$\frac{\partial u}{\partial t} + \left(1 - \frac{2u}{b} \right) a \frac{\partial u}{\partial x} = 0$$

where u is the car density, a is the maximum car velocity (e.g., at zero car density), and b is the maximum car density when velocity is zero (e.g., at the red light).

1. Suppose the red light at $x = 0$ occurs at $t = 0$. Note that u, a, and b are all positive values. Thus let the initial condition be given by

$$u(x, 0) = \begin{cases} c & \text{for } x \le 0 \\ b & \text{for } x > 0 \end{cases}$$

where $c < b$. Obtain the shock solution, using the Rankine-Hugoniot jump condition (J.18) to determine the shock path.

2. Suppose the green light occurred after all the cars have been stopped at $x \le 0$. So this time, assume the initial condition is given by

$$u(x,0) = \begin{cases} b & \text{for } x \le 0 \\ 0 & \text{for } x > 0 \end{cases}$$

Obtain the rarefaction solution.

E10.22. Consider the Riemann problem involving the inviscid Burger equation given by

$$\frac{\partial u}{\partial t} + u\frac{\partial u}{\partial x} = 0 \qquad \text{subject to} \qquad u(x,0) = \begin{cases} 0 & \text{if } x \le 0 \\ 1 & \text{if } 0 < x \le 1 \\ 0 & \text{if } x > 1 \end{cases}$$

This is an example in which the left discontinuity will involve a rarefaction, whereas the right discontinuity propagates a shock.

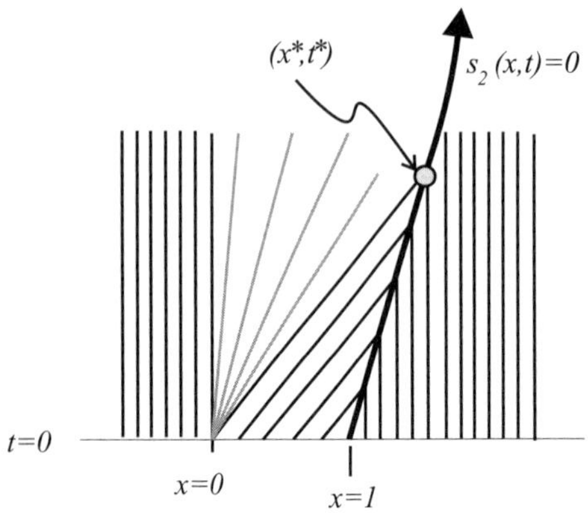

Figure 10.5. The characteristics containing both shock and rarefaction.

1. Based on the Rankine-Hugoniot conditions, let t^* be the time at which the left side jump of the shock stops involving $u^- = 1$ and starts involving the rarefaction solution curves (see Figure 10.5). Determine t^* for this problem.
2. Using the Rankine-Hugoniot conditions, show that the shock $s_2(x,t) = 0$ for $t > t^*$ is given by

$$s_2(x,t) = x - \sqrt{2t}$$

Thus obtain the solution $u(x,t)$ for $t \ge 0$.

11 Linear Partial Differential Equations

In this chapter, we focus on the solution of linear partial differential equations. In general, we consider a partial differential equation to be linear if the partial derivatives together with their coefficients can be represented by an operator L such that it satisfies the property that $L(\alpha u + \beta v) = \alpha L u + \beta L v$, where α and β are constants, whereas u and v are two functions of the same set of independent variables. This linearity property allows for superposition of basis solutions to fit the boundary and initial conditions.

We limit our discussion in this chapter to three approaches: operator factorization of reducible linear operators, separation of variables, and similarity transformations. Another set of solution methods known as integral transforms approach are discussed instead in Chapter 12.

The method using operator factorization is described in Section 11.2, and it is applicable to a class of partial differential equations known as **reducible differential equations**. We show how this approach can be applied to the one-dimensional wave equation to yield the well-known d'Alembert's solutions. In Section K.1 of the appendix, we consider how the d'Alembert solutions can be applied and modified to fit different boundary conditions, including infinite, semi-infinite, and finite spatial domains.

The separation of variables method is described in Section 11.3. It may not be as general at first glance, but it is an important and powerful approach because it has yielded useful solutions to several important linear problems in science and engineering. These include the linear diffusion, wave, and elliptic problems. The solutions based on this approach often involve an infinite series of basis functions that would satisfy the partial differential equations, which is possible because of the linearity property of the differential operators. In the process of fitting the series to satisfy initial conditions and boundary conditions, orthogonality properties of the basis functions are needed. To this end, the theory of Sturm-Liouville systems can be used to find the components needed to obtain the orthogonality properties. A brief discussion of theory of Sturm-Liouville systems is given in Section 11.3.3. However, because Sturm-Liouville systems involve only homogeneous boundary conditions, additional steps are needed to handle both nonhomogeneous differential equations and nonhomogeneous boundary conditions. Fortunately, again due to the linearity properties of the differential operation, we could employ an approach that splits the required solutions to allow the original problem to be divided into one or more additional problems, some of which are homogeneous differential equations, whereas others have homogeneous boundary conditions. These function-splitting techniques are given in Section 11.4. For the case of nonhomogeneous differential equations with homogeneous boundary conditions, a generalization of the separation of variables technique, known as the **eigenfunction expansion approach**, is described in Section 11.4.2 and can be used to solve the nonhomogeneous problem.

The third approach, known as the **method of similarity transformation** (also known as **combination of variables**), is described in Section 11.5. It is a special case of techniques known as **symmetry transformation methods**. It can be applied to equations involving two independent variables to yield an ordinary differential equation based on an independent variable that is a combination of the original pair of independent variables. If the number of independent variables were more than two, and if the approach were applicable, it could be used to reduce the number of independent variables by one.[1] However, it has been deemed by several textbooks to be of limited application, because the boundary conditions may not allow for similarity transformations, even if it applies to the differential equations. Nonetheless, it has been used to derive important solutions such as that for diffusion equations with semi-infinite domains. More importantly, the similarity transformation approach, even more so the general approach of symmetry transformation approach, has been applied to several nonlinear differential equations.

11.1 Linear Partial Differential Operator

The general description of a linear partial differential equation was given in Section 10.4 by (10.75) and (10.76), that is,

$$\sum_{1\le i_1<\cdots<i_m\le n} \beta_{i_1,\ldots,i_m}(\mathbf{x})\,\mu_{i_1,\ldots,i_m} = \sum_{1\le i_1<\cdots<i_{m-1}\le n} \gamma_{i_1,\ldots,i_{m-1}}(\mathbf{x})\,\mu_{i_1,\ldots,i_{m-1}} + \cdots + \sum_{1\le i\le n} \gamma_i(\mathbf{x})\,\mu_i + \gamma(\mathbf{x})\,u + h(\mathbf{x}) \tag{11.1}$$

where

$$\mu_{i_1\ldots i_m} = \frac{\partial^m u}{\partial x_{i_1}\ldots\partial x_{i_m}} \quad ; \quad 1\le i_1,\ldots,i_m\le n$$

Equation (11.1) can be rewritten in operator form as

$$Lu = h(\mathbf{x}) \tag{11.2}$$

[1] Similar principles of similarity transformations methods are discussed in Section 6.2 and Section 6.4.2 for the case of ordinary differential equations. In those cases, first-order differential equations are converted to separable variable types, whereas for second-order or higher order ordinary differential equations, the approach may be used to reduce the order by one.

where L is known as the **linear partial differential operator** given by

$$\begin{aligned} L \;=\; & \sum_{1 \le i_1 < \cdots < i_m \le n} \beta_{i_1,\ldots,i_m}(\mathbf{x}) \frac{\partial^m}{\partial x_{i_1}, \ldots, \partial x_{i_m}} \\ & - \sum_{1 \le i_1 < \cdots < i_{m-1} \le n} \gamma_{i_1,\ldots,i_{m-1}}(\mathbf{x}) \frac{\partial^{m-1}}{\partial x_{i_1}, \ldots, \partial x_{i_{m-1}}} \\ & - \cdots - \sum_{1 \le i \le n} \gamma_i(\mathbf{x}) \frac{\partial}{\partial x_i} - \gamma(\mathbf{x}) \end{aligned} \tag{11.3}$$

L is a linear operator because it satisfies the following property:

$$L(\alpha u_1 + \beta u_2) = \alpha L u_1 + \beta L u_2 \tag{11.4}$$

for any constant α and β.

Following a similar approach to linear ordinary differential equations, the solution of a linear partial differential equation can be taken as the sum of a particular solution and a complementary solution. A function $u_{\text{particular}}$ is called the **particular solution** if it satisfies the partial differential equation given by (11.2), that is,

$$L\, u_{\text{particular}} = h(\mathbf{x}) \tag{11.5}$$

Conversely, a function u_{comp} is called the **complementary homogeneous solution** if it satisfies the homogeneous partial differential equation,

$$L\, u = 0 \tag{11.6}$$

while allowing enough arbitrary functions (or infinite arbitrary constants, as in the case of Fourier series solutions) for the satisfaction of initial or boundary conditions. Any solution that satisfies (11.6) is known as the **homogeneous solution**. Suppose $u_{\text{homog},1}$ and $u_{\text{homog},2}$ are two different homogeneous solutions, then due to (11.4), a linear combination of both solutions will also be a homogeneous solution, that is,

$$L\left(\alpha u_{\text{homog},1} + \beta u_{\text{homog},2}\right) = \alpha L\left(u_{\text{homog},1}\right) + \beta L\left(u_{\text{homog},2}\right) = 0 \tag{11.7}$$

Equation (11.7) can be extended to any number of different homogeneous solutions, including cases in which there are an infinite number of homogeneous solutions. This process is referred to as **superposition of homogeneous solutions**. By including a sufficient number of homogeneous solutions, the complementary homogeneous solution is given by a linear combination

$$u_{\text{comp}} = \sum_{i=1}^{N} \alpha_i u_{\text{homog},i} \tag{11.8}$$

The **complete solution**[2] to $(Lu = h)$ is then given by

$$u = u_{\text{particular}} + u_{\text{comp}} = u_{\text{particular}} + \sum_{i=1}^{N} \alpha_i u_{\text{homog},i} \tag{11.9}$$

[2] Solutions containing arbitrary functions, instead of arbitrary constants, are considered **general solutions**. However, complete solutions and general solutions are often interchangeable.

There are three main types of boundary conditions for partial differential equations:

1. **Dirichlet conditions** (also known as **essential boundary conditions**). These are the conditions in which values of the dependent variables are specified at regions of the boundary. For instance, in a 2D rectangular domain $\Omega : (x, y) \in [0, 1] \times [0, 1]$, then setting $u(x = 0, y) = 10$ is a Dirichlet condition.
2. **Neumann conditions**. These are conditions in which values of the derivatives of the dependent variables are specified at regions of the boundary. For instance, $\partial u/\partial y(x = 1, y) = f(y)$ is a Neumann condition.
3. **Robin conditions** (also known as **mixed boundary condtions**). These are conditions in which a relationship (often a linear combination) of both the derivatives and the values of the dependent variables are specified at regions of the boundary. For instance, $\partial u/\partial y(x = 0, y) + \alpha u(x = 0, y) = g(y)$ is a Robin condition.

Both Neumann and Robin conditions often result from the consideration of specifying the flux behavior at portions of the boundary, because fluxes are usually functions of gradients. For instance, a Neumann condition set to zero is used to specify zero flux. Both Neumann and Robin conditions are also known as **natural boundary conditions**.

11.2 Reducible Linear Partial Differential Equations

In some cases, a linear partial differential equation can be factored into several first-order partial differential operators. Under certain conditions, this process will reduce the homogenous problem into several first-order partial differential equations. These types are known as **reducible linear partial differential equations** and are also known as **factorizable linear partial differential equations**.)

THEOREM 11.1. *Let L be an mth-order linear partial differential operator involving a system with n independent variables. Furthermore, suppose that L can be decomposed into n commutative factors of first-order linear partial differential operators, that is,*

$$L = \left(\prod_{i=1}^{m} L_i\right) \tag{11.10}$$

where

$$L_i = f_{0,i}(\mathbf{x}) + \sum_{q=1}^{n} f_{q,i}(\mathbf{x})\frac{\partial}{\partial x_q} \quad \text{and} \quad L_iL_j = L_jL_i \quad , \quad i \neq j \tag{11.11}$$

with $f_{\ell,i}$ ($\ell = 0, 1, \ldots, n$), being continuous and sufficiently differentiable. If $L_i \neq L_j$ for $i \neq j$ ($i, j = 1, \ldots, n$), then the complete solution of the homogeneous equation, $Lu = 0$, is given by

$$u = \sum_{i=1}^{m} \alpha_i u_i \tag{11.12}$$

where u_i is the solution of $L_iu = 0$ and α_i are arbitrary constants.

In case L_i is repeated k times, then k terms of the complete solution corresponding to L_i^k is given by

$$(L_i)^k v_i = 0 \quad \longrightarrow \quad v_i = \left(\sum_{j=1}^{\ell} \beta_j g_j(\mathbf{x}) \right) u_i \quad \text{where} \quad L_i u_i = 0 \tag{11.13}$$

where β_j are arbitrary constants and g_j are functions such that $L_i^k g_j = 0$.

PROOF. (See Section K.2.1 for proof.)

In general, the search for commuting factors L_i, including the determination of whether L is reducible, is not straightforward. However, there are classes of equations that can be reduced easily, such as operators with constant coefficients. Furthermore, when L_i^k is a kth power of a linear first-order partial differential operator with constant coefficients, that is,

$$L_i = c_0 + \sum_{q=1}^{n} c_\ell \frac{\partial}{\partial x_q} \tag{11.14}$$

where n is the number of independent variables, then the functions g_j in (11.13) can be set up as

$$g_j = x_1^{p_1} x_2^{p_2} \cdots x_n^{p_n} \tag{11.15}$$

where $\sum_{i=1}^{n} p_i \leq k$.

EXAMPLE 11.1. Consider the homogeneous partial differential equation given by

$$\frac{\partial^2 u}{\partial x^2} + 2\frac{\partial^2 u}{\partial x \partial y} + \frac{\partial^2 u}{\partial y^2} + 3\left(\frac{\partial u}{\partial x} + \frac{\partial u}{\partial y} \right) = 0$$

In operator notation we have,

$$Lu = L_1 L_2 u = L_2 L_1 u = 0$$

where

$$L_1 = \frac{\partial}{\partial x} + \frac{\partial}{\partial y} \qquad \text{and} \qquad L_2 = \frac{\partial}{\partial x} + \frac{\partial}{\partial y} + 3$$

Using the method of characteristics for L_1, we have

$$\frac{dx}{1} = \frac{dy}{1} = \frac{du}{0} \qquad \rightarrow \qquad u_1 = \phi\,(x - y)$$

where ϕ is an arbitrary function.

Similarly, for L_2,

$$\frac{dx}{1} = \frac{dy}{1} = \frac{du}{-3u} \qquad \rightarrow \qquad u_2 = e^{-3x}\psi\,(x - y)$$

where ψ is another arbitrary function. The general solution is then given by

$$u = \phi\,(x - y) + e^{-3x}\psi\,(x - y)$$

EXAMPLE 11.2. Consider the case for

$$\left(3\frac{\partial}{\partial x}+2\frac{\partial}{\partial y}+1\right)^3 u=0$$

First, the general solution of

$$\left(3\frac{\partial}{\partial x}+2\frac{\partial}{\partial y}+1\right) w=0$$

via the method of characteristics is

$$w(x,y)=e^{-x/3}\phi\left(2x-3y\right)$$

where $\phi(.)$ is a general function. Because the operator $L=(3\partial_x+2\partial_y+1)$ is repeated $k=3$ times, we can use (11.15) to set

$$g(x,y)=\beta_1+\beta_2 x+\beta_3 y+\beta_4 x^2+\beta_5 xy+\beta_6 y^2$$

Combining, the complete general solution is given by

$$u(x,y)=\left(\beta_1+\beta_2 x+\beta_3 y+\beta_4 x^2+\beta_5 xy+\beta_6 y^2\right)e^{-x/3}\phi\left(2x-3y\right)$$

11.2.1 One-Dimensional Wave Equation

The **one-dimensional wave equation** is given by

$$\frac{\partial^2 u}{\partial x^2}-\frac{1}{c^2}\frac{\partial^2 u}{\partial t^2}=0 \tag{11.16}$$

where c is a real constant. It is hyperbolic, homogeneous, and linear, with constant coefficients. Equation (11.16) can be solved easily after transformation to canonical form, as discussed in Section 10.4. Alternatively, we can approach it as a reducible partial differential equation to show that it also leads to the same solution.

The linear operator for (11.16) can be factored as

$$Lu=L_1L_2u=L_2L_1u$$

where

$$L_1=\frac{\partial}{\partial x}-\frac{1}{c}\frac{\partial}{\partial t}\quad\text{and}\quad L_2=\frac{\partial}{\partial x}+\frac{1}{c}\frac{\partial}{\partial t}$$

For $L_1u=0$, the characteristics equations are

$$\frac{dx}{1}=\frac{dt}{-1/c}=\frac{du}{0}$$

and the general solution is given by

$$u_1=\phi\left(x+ct\right)$$

where ϕ is an arbitrary function. Similarly, for $L_2u=0$, the characteristic equations are

$$\frac{dx}{1}=\frac{dt}{1/c}=\frac{du}{0}$$

and the general solution becomes

$$u_2=\psi\left(x-ct\right)$$

where ψ is an arbitrary function. Superposition of the two solutions,

$$u = \phi(x + ct) + \psi(x - ct) \tag{11.17}$$

For the case in which $-\infty < x < \infty$ and initial conditions given by

$$u(x, 0) = f(x) \qquad \text{and} \qquad \frac{\partial u}{\partial t}(x, 0) = g(x) \tag{11.18}$$

(11.17) reduces to

$$u(x, t) = \frac{1}{2}\left[f(x - ct) + f(x + ct)\right] + \frac{1}{2c}\int_{x-ct}^{x+ct} g(\tau)d\tau \tag{11.19}$$

This result is known as the **d'Alembert solution**. The details for this solution, as well as its extensions to cases with semi-infinite domain and nonhomogeneous Neumann conditions, are given in Section K.1 as an appendix.

11.3 Method of Separation of Variables

Assume that the form of solution $u(\mathbf{x})$ is a product of functions $X_k(x_k)$, each of which are dependent only on one variable, that is,

$$u = X_1(x_1)X_2(x_2)\cdots X_n(x_n) \tag{11.20}$$

We say that the homogeneous partial differential equation $Lu = 0$ is **separable** if, after the substitution of (11.20), we obtain n ordinary differential equations of the form given by

$$G_i\left(x_i, X_i, \frac{dX_i}{dx_i}, \ldots, \frac{d^nX_i}{dx_n^n}\right) = \kappa_i \qquad i = 1, \ldots, n \tag{11.21}$$

These equations may need a sequence of rearrangements, as shown in the next example.

EXAMPLE 11.3. Consider the dynamic temperature distribution of a solid in cylindrical coordinates given by

$$\nabla^2 u = \left[\frac{\partial^2 u}{\partial r^2} + \frac{1}{r}\frac{\partial u}{\partial r} + \frac{1}{r^2}\frac{\partial^2 u}{\partial \theta^2} + \frac{\partial^2 u}{\partial z^2}\right] = a^2\frac{\partial u}{\partial t} \tag{11.22}$$

Let $u = T(t)Z(z)R(r)\Theta(\theta)$. After substitution into (11.22), we could rearrange the resulting equation to be

$$a^2\frac{1}{T}\frac{dT}{dt} = \frac{1}{R}\frac{dR^2}{dr^2} + \frac{1}{rR}\frac{dR}{dr} + \frac{1}{r^2\Theta}\frac{d^2\Theta}{d\theta^2} + \frac{1}{Z}\frac{d^2Z}{dz^2}$$

Because the left-hand side of the equation involves only the variable t, whereas the other side does not involve t, both sides must be equal to a constant, say κ_1, that is,

$$G_1\left(t, T, \frac{dT}{dt}\right) = a^2\frac{1}{T}\frac{dT}{dt} = \kappa_1$$

and

$$\frac{1}{R}\frac{dR^2}{dr^2} + \frac{1}{rR}\frac{dR}{dr} + \frac{1}{r^2\Theta}\frac{d^2\Theta}{d\theta^2} + \frac{1}{Z}\frac{d^2Z}{dz^2} = \kappa_1$$

Next, we can rearrange the last equation to be

$$\left(\kappa_1 - \frac{1}{Z}\frac{d^2Z}{dz^2}\right) = \frac{1}{R}\frac{dR^2}{dr^2} + \frac{1}{rR}\frac{dR}{dr} + \frac{1}{r^2\Theta}\frac{d^2\Theta}{d\theta^2}$$

Setting both sides to another constant κ_2,

$$G_2\left(z, Z, \frac{dZ}{dz}, \frac{d^2Z}{dz^2}\right) = \left(\kappa_1 - \frac{1}{Z}\frac{d^2Z}{dz^2}\right) = \kappa_2$$

and

$$\frac{1}{R}\frac{dR^2}{dr^2} + \frac{1}{rR}\frac{dR}{dr} + \frac{1}{r^2\Theta}\frac{d^2\Theta}{d\theta^2} = \kappa_2$$

or

$$\left(\frac{r^2}{R}\frac{dR^2}{dr^2} + \frac{r}{R}\frac{dR}{dr} - \kappa_2 r^2\right) = -\frac{1}{\Theta}\frac{d^2\Theta}{d\theta^2}$$

Setting both sides to another constant κ_3,

$$G_3\left(r, R, \frac{dR}{dr}, \frac{d^2R}{dr^2}\right) = \left(\frac{r^2}{R}\frac{dR^2}{dr^2} + \frac{r}{R}\frac{dR}{dr} - \kappa_2 r^2\right) = \kappa_3$$

and

$$G_4\left(\theta, \Theta, \frac{d\Theta}{d\theta}, \frac{d^2\Theta}{d\theta^2}\right) = -\frac{1}{\Theta}\frac{d^2\Theta}{d\theta^2} = \kappa_3$$

In summary, we can gather the ordinary differential equations as follows:

$$\begin{aligned}
a^2\frac{1}{T}\frac{dT}{dt} &= \kappa_1 \\
\frac{1}{Z}\frac{d^2Z}{dz^2} &= \kappa_1 - \kappa_2 \\
\frac{r^2}{R}\frac{dR^2}{dr^2} + \frac{r}{R}\frac{dR}{dr} - \kappa_2 r^2 &= \kappa_3 \\
-\frac{1}{\Theta}\frac{d^2\Theta}{d\theta^2} &= \kappa_3
\end{aligned}$$

and conclude that the partial differential equation is indeed separable.

As may be expected, not all linear differential equations are separable. Sometimes, a change of coordinates is needed to attain separability. However, one case is always separable,

$$\sum_{i=1}^{n}\sum_{j=0}^{m}\widehat{\beta}_{ij}\left(x_i\right)\frac{\partial^j u}{\partial x_i^j} + \alpha u = 0 \tag{11.23}$$

where $\widehat{\beta}_{ij}$ is only a function of x_i. After substituting $\left(u = \prod_{\ell=1}^{n} X_\ell(x_\ell)\right)$ and then dividing again by u,

$$\sum_{i=1}^{n}\frac{1}{X_i}\left(\sum_{j=0}^{m}\widehat{\beta}_{ij}\left(x_i\right)\frac{d^j X_i}{dx_i^j}\right) + \alpha = 0$$

resulting in

$$\frac{1}{X_i}\sum_{j=0}^{m}\widehat{\beta}_{ij}\left(x_i\right)\frac{d^j X_i}{dx_i^j} = \kappa_i \qquad i = 1, 2, \ldots, n \tag{11.24}$$

where $\sum_{i=1}^{n}\kappa_i + \alpha = 0$.

11.3.1 One-Dimensional Diffusion Equation

The **one-dimensional diffusion equation** is given by

$$\alpha^2\frac{\partial^2 u}{\partial x^2} = \frac{\partial u}{\partial t} \tag{11.25}$$

where α is a real constant. Substituting $u = T(t)X(x)$ into (11.25),

$$\frac{\alpha^2}{X}\frac{d^2X}{dx^2} = \frac{1}{T}\frac{dT}{dt}$$

yields two ordinary differential equations given by

$$\frac{\alpha^2}{X}\frac{d^2X}{dx^2} = \kappa \quad \text{and} \quad \frac{1}{T}\frac{dT}{dt} = \kappa \tag{11.26}$$

where κ is a constant. The solutions to (11.26) are given by

$$X = Ae^{x\sqrt{\kappa}/\alpha} + Be^{-x\sqrt{\kappa}/\alpha} \qquad \text{and} \qquad T = Ce^{\kappa t}$$

and

$$u = T(t)X(t) = e^{\kappa t}\left(ACe^{x\sqrt{\kappa}/\alpha} + BCe^{-x\sqrt{\kappa}/\alpha}\right) \tag{11.27}$$

where A, B, and C are arbitrary constants. At this point, constraints from the initial and boundary conditions, as well as physical insights, can be used to move the solution forward. For u to remain bounded, we need $\kappa < 0$. As a consequence, the term $\sqrt{\kappa}$ will have imaginary values. Let $\kappa = -\lambda^2$, then (11.27) becomes

$$u = e^{-\lambda^2 t}\left(\beta\sin\left(\frac{\lambda}{\alpha}x\right) + \gamma\cos\left(\frac{\lambda}{\alpha}x\right)\right) \tag{11.28}$$

where β and γ are arbitrary constants.

Next, we consider two cases for the finite domain, $0 \le x \le L$. The first case is when the boundary conditions set both the values of u at $x = 0$ and $x = L$ to be zero. This is known as the case with homogeneous Dirichlet boundary conditions. The second case we consider is the nonhomogeneous Dirichlet boundary conditions.

1. **Homogeneous Dirichlet boundary conditions.** Let the boundary conditions be given by $u(0,t) = u(L,t) = 0$, together with an initial condition given by $u(x,0) = f(x) < \infty$. Then (11.28) becomes

$$u(0,t) = 0 = e^{-\lambda^2 t}\left(\beta\sin(0) + \gamma\cos(0)\right) = \gamma e^{-\lambda^2 t} \qquad \rightarrow \qquad \gamma = 0$$

Applying this together with the boundary condition at $x = L$,

$$u(L,t) = 0 = \beta e^{-\lambda^2 t} \sin\left(\frac{\lambda}{\alpha}L\right)$$

We need $\beta \neq 0$. Otherwise, we obtain the trivial solution $u(x,t) = 0$, which may not satisfy the initial conditions. Thus we need λ to satisfy the following:

$$\sin\left(\frac{\lambda}{\alpha}L\right) = 0 \qquad \Rightarrow \qquad \lambda = \frac{n\pi}{L}\alpha \; ; \; n = \ldots, -2, -1, 0, 1, 2, \ldots$$

Let $k_n = n\pi/L$, then the nth homogeneous solution is given by

$$u_n(x,t) = e^{-(\alpha k_n)^2 t} \sin(k_n x)$$

Using superposition, a complete solution is

$$u = \sum_{n=1}^{\infty} \beta_n u_n = \sum_{n=1}^{\infty} \beta_n e^{-(\alpha k_n)^2 t} \sin\left(k_n x\right) \tag{11.29}$$

where the series involves only positive values of n because $\sin(-k_n x) = -\sin(k_n x)$ (allowing us to combine β_n and β_{-n} into one constant).

To satisfy the initial condition

$$u(x,0) = f(x) = \sum_{n=1}^{\infty} \beta_n \sin\left(\frac{n\pi}{L}x\right) \tag{11.30}$$

the constants β_n can be evaluated using the following orthogonality properties of sinusoids:

$$\int_0^L \sin\left(\frac{n\pi}{L}x\right) \sin\left(\frac{m\pi}{L}x\right) dx = \begin{cases} 0 & \text{if } m \neq n \\ L/2 & \text{if } m = n \end{cases} \tag{11.31}$$

We can multiply (11.30) by $\sin(m\pi x/L)$ and integrate from 0 to L. Using (11.31), we get

$$\int_0^L f(x) \sin\left(\frac{m\pi}{L}x\right) dx = \int_0^L \sin\left(\frac{m\pi}{L}x\right) \sum_{n=0}^{\infty} \beta_n \sin\left(\frac{n\pi}{L}x\right) dx$$

or

$$\beta_m = \frac{2}{L} \int_0^L f(x) \sin\left(\frac{m\pi}{L}x\right) dx \tag{11.32}$$

Table 11.1. *The first eight values of β_m for example 11.4*

m	β_m	m	β_m
1	1.2090×10^{0}	5	4.2028×10^{-7}
2	3.3307×10^{-16}	6	9.4369×10^{-16}
3	2.4586×10^{-1}	7	-9.8859×10^{-2}
4	9.7143×10^{-17}	8	5.9674×10^{-16}

EXAMPLE 11.4. Consider the one-dimensional diffusion problem given by

$$\frac{\partial^2 u}{\partial x^2} = \frac{\partial u}{\partial t}$$

$$\text{subject to:} \quad u(0,t) = u(10,t) = 0$$
$$u(x,0) = f(x) = \varsigma(5,-5,1,x) + \varsigma(-5,45,1,x)$$

where

$$\varsigma(a,b,c,x) = \frac{c}{4}\left(1 + \tanh(ax+b)\right)$$

Based on (11.32), we can evaluate the coefficients of β_m to be

$$\beta_m = 0.2 \int_0^{10} \sin\left(\frac{m\pi}{10}\right)\left[\varsigma(5,-5,1,x) + \varsigma(-5,45,1,x)\right] dx$$

Performing numerical integration, the first eight values of β_m are given in Table 11.1. The approximations to $u(x,0)$ for the series truncated after M term are shown in Figure 11.1. The function approximation is sufficiently acceptable at about $M = 50$ (maximum errors around $\pm 10^{-3}$). Thus using the first 50 terms of the sum given in (11.29), we obtain the surface plot of the solution $u(x,t)$ shown in Figure 11.2.

2. **Nonhomogeneous Dirichlet boundary conditions.** Suppose that the boundary conditions are now given by

$$u(0,t) = U_o \quad \text{and} \quad u(L,t) = U_L$$

with an initial condition

$$u(x,0) = f(x) \qquad \text{bounded}$$

where U_o and U_L are constants that are not both zero. Applying these conditions to (11.28),

$$u(0,t) = U_o = e^{-\lambda^2 t}\left(\beta \sin(0) + \gamma \cos(0)\right)$$
$$u(L,t) = U_L = e^{-\lambda^2 t}\left(\beta \sin\left(\frac{\lambda}{\alpha}L\right) + \gamma \cos\left(\frac{\lambda}{\alpha}L\right)\right)$$

Unfortunately, no values of β, γ or λ values can be found that will satisfy these conditions for all t, unless both U_o and U_L are zero.

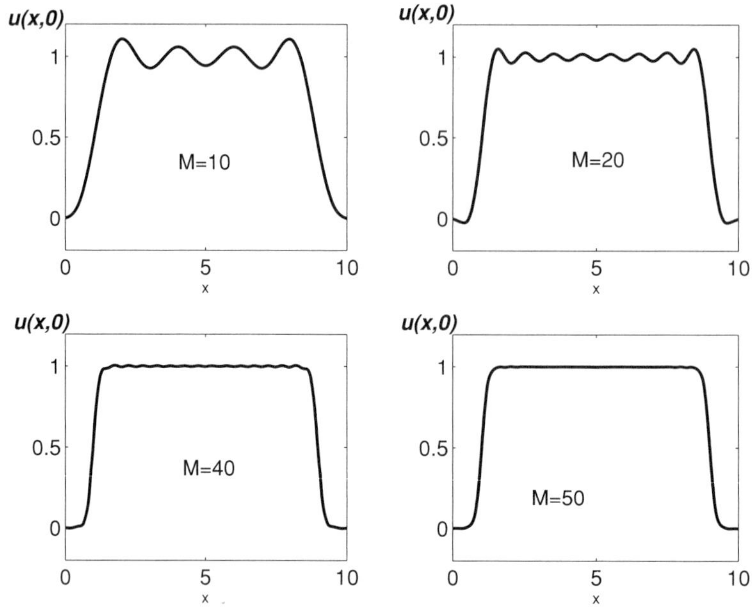

Figure 11.1. Plot of $u(x,0) \approx \sum_{m=1}^{M} \beta_m \sin(m\pi x/10)$ for Example 11.4.

Instead of abandoning the method of separation of variables, we could use superposition to search for an additional solution to handle the nonhomogeneous Dirichlet boundary conditions. This means that we split u into two parts,

$$u = u_{\text{hBC}} + u_{\text{nonhBC}} \tag{11.33}$$

where u_{hBC} and u_{nonhBC} both satisfy the partial differential equation, but u_{hBC} satisfies the homogeneous boundary conditions:

$$u_{\text{hBC}}(0,t) = 0 = u_{\text{hBC}}(L,t) \quad \text{and} \quad u_{\text{hBC}}(x,0) = f(x) - g(x) \tag{11.34}$$

whereas u_{nonhBC} satisfies the nonhomogeneous conditions

$$u_{\text{nonhBC}}(0,t) = U_0 \quad , \quad u_{\text{nonhBC}}(L,t) = U_L \quad \text{and} \quad u_{\text{nonhBC}}(x,0) = g(x) \tag{11.35}$$

The solution of u_{hBC} has already been discussed in the previous case, except that now it depends on an additional function $g(x)$.

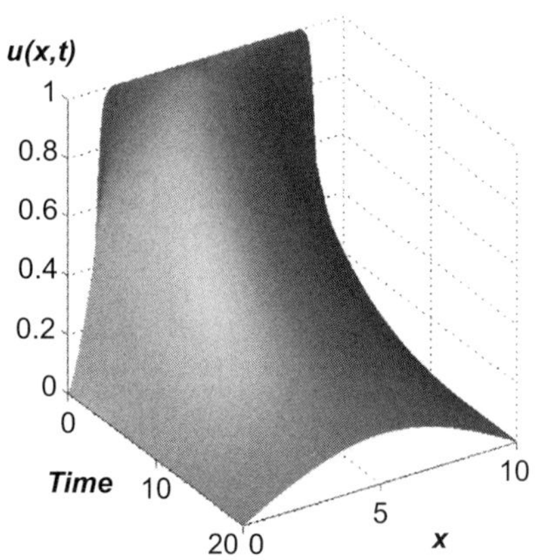

Figure 11.2. Surface plot of the trajectory of the u distribution for Example 11.4.

We address the solution of u_{nonhBC} first. The solution of u_{nonhBC} can take on many possibilities. The simplest approach is to set it independent of t, which implies $u_{\text{nonhBC}} = g(x)$. Substituting $u_{\text{nonhBC}} = g(x)$ in the diffusion equation,

$$\alpha^2 \frac{d^2 g}{dx^2} = 0 \quad \rightarrow \quad g(x) = ax + b$$

To satisfy the boundary conditions given in (11.35), we have $a = (U_L - U_o)/L$ and $b = U_o$. Thus

$$u_{\text{nonhBC}} = (U_L - U_o)\frac{x}{L} + U_o \tag{11.36}$$

Having determined $g(x)$, u_{hBC} can be obtained as a case with homogeneous boundary conditions, that is,

$$u_{\text{hBC}} = \sum_{n=1}^{\infty} \beta_n e^{-(\alpha k_n)^2 t} \sin(k_n x) \tag{11.37}$$

where $k_n = n\pi/L$. To evaluate β_n, we again use the orthogonality property of sinusoids, except this time we apply them on $(f(x) - g(x))$,

$$\beta_n = \frac{2}{L}\int_0^L \left[f(x) - (U_L - U_o)\frac{x}{L} - U_o \right] \sin\left(\frac{n\pi}{L}x\right) dx \tag{11.38}$$

Thus the solution is given by

$$u = (U_L - U_o)\frac{x}{L} + U_o + \sum_{n=1}^{\infty} \beta_n e^{-(\alpha k_n)^2 t} \sin(k_n x) \tag{11.39}$$

where β_n is given by (11.38).

EXAMPLE 11.5. Consider the one-dimensional diffusion equation given by (11.25) with $\alpha = 1$,

$$\frac{\partial^2 u}{\partial x^2} = \frac{\partial u}{\partial t}$$

subject to the conditions: $u(0, t) = 1$, $u(10, t) = 2$ and $u(x, 0) = 1 + \tanh(x)$. Then

$$u_{\text{nonhBC}} = \frac{x}{10} + 1 \quad \text{and} \quad u_{\text{hBC}} = \sum_{n=0}^{\infty} \beta_n e^{-(\alpha k_n)^2 t} \sin(k_n x)$$

where $k_n = n\pi/10$. The coefficients β_n are found as

$$\beta_n = 0.2 \int_0^{10} \left[\tanh(x) - \frac{x}{L} \right] \sin\left(\frac{n\pi}{10}x\right) dx$$

which can be evaluated also by numerical integration. The first eight values of β_n are shown in Table 11.2. The approximations to $f(x) - g(x)$ for the series truncated after N term are shown in Figure 11.3. The function approximation is sufficiently acceptable at about $N = 15$ (maximum errors of around $\pm 10^{-4}$). Thus using only the first $N = 16$ terms of the sum for u_{hBC}, we obtain the surface plot of the solution $u(x, t) = u_{\text{hBC}} + u_{\text{nonhBC}}$ shown in Figure 11.4.

Table 11.2. *The first eight values of β_n for example 11.5*

n	β_n	n	β_n
1	6.1150×10^{-1}	5	5.3671×10^{-2}
2	2.7196×10^{-1}	6	3.2618×10^{-2}
3	1.5077×10^{-1}	7	1.9879×10^{-2}
4	8.8998×10^{-2}	8	1.2129×10^{-2}

The methods discussed thus far are for cases with constant nonhomogeneous boundary conditions. For the more general situations in which nonhomogeneous terms are not constant, homogenization of the boundary conditions will make the differential equations nonhomogeneous. For the treatment of nonhomogeneous differential equations, refer to Section 11.4 for the method of eigenfunction expansion.

11.3.2 Potential Equation in a Circle

The **potential equation**, also known as **harmonic equation** or **Laplace equation**, is given by the following partial differential equation

$$\nabla^2 u = 0 \tag{11.40}$$

regardless of the coordinate system used. Functions that satisfy (11.40) are also known as **harmonic functions**. Using the divergence theorem, several properties can be found for harmonic functions in bounded regions, such as mean-value properties and maximum principles. Our focus is on solving the potential equation (11.40) in a circle of radius $r_{\max} < \infty$ by using the method of separation of variables.

Consider the potential equation in polar coordinates r and θ,

$$\frac{\partial^2 u}{\partial r^2} + \frac{1}{r}\frac{\partial u}{\partial r} + \frac{1}{r^2}\frac{\partial^2 u}{\partial \theta^2} = 0 \tag{11.41}$$

subject to the conditions, $u(r_{\max}, \theta) = f(\theta)$ with $0 \leq \theta \leq 2\pi$. After substitution of $u = R(r)\Theta(\theta)$ into (11.41) and some rearrangements, two ordinary differential equations

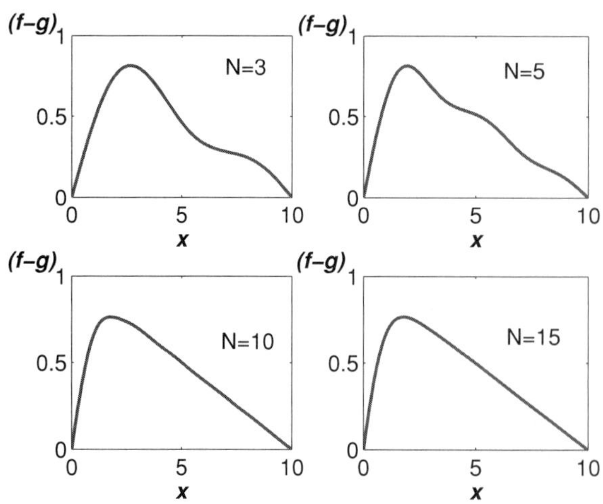

Figure 11.3. Plot of $f(x) - g(x) \approx \sum_{n=1}^{N} \beta_n \sin(n\pi x/10)$ for Example 11.5.

Figure 11.4. Surface plot of the trajectory of the u distribution for Example 11.5.

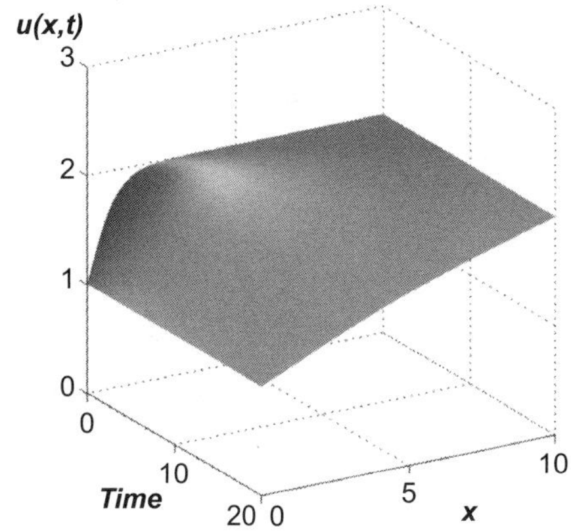

can be obtained,

$$\frac{1}{R}\left(r^2\frac{d^2R}{dr^2}+r\frac{dR}{dr}\right)=\kappa \quad \text{and} \quad \frac{1}{\Theta}\frac{d^2\Theta}{d\theta^2}=-\kappa$$

The first equation is a Euler-Cauchy equation (cf. Section 6.4.3) whose solution is given by

$$R = ar^{\sqrt{\kappa}} + br^{-\sqrt{\kappa}} \tag{11.42}$$

whereas the solution for the second differential equation is given by

$$\Theta = ce^{\theta\sqrt{-\kappa}} + de^{-\theta\sqrt{-\kappa}} \tag{11.43}$$

Because we expect Θ to be periodic, κ needs to be nonnegative. Furthermore, because the period has to be 2π, and θ is already in the range of 0 to 2π, we need $\sqrt{\kappa}=n$, a nonnegative integer. The coefficient b has to be zero to maintain $R(r)<\infty$ at $r=0$. Thus for a fixed n

$$u_n(r,\theta) = r^n\left(\alpha_n\cos(n\theta)+\beta_n\sin(n\theta)\right)$$

and the complete solution is obtained by superposition,

$$u(r,\theta)=\sum_{n=0}^{\infty}u_n(r,\theta) \tag{11.44}$$

To satisfy the boundary conditions at $r=r_{\max}$,

$$f(\theta)=\sum_{n=0}^{\infty}r_{\max}^n\left(\alpha_n\cos(n\theta)+\beta_n\sin(n\theta)\right)$$

Using the orthogonality property of sinusoids and cosinusoids, we obtain

$$\begin{aligned}
\alpha_0 &= \frac{1}{2\pi}\int_0^{2\pi} f(\theta)d\theta \\
\alpha_n &= \frac{1}{\pi r_{\max}^n}\int_0^{2\pi} f(\theta)\cos(n\theta)\,d\theta \\
\beta_n &= \frac{1}{\pi r_{\max}^n}\int_0^{2\pi} f(\theta)\sin(n\theta)\,d\theta
\end{aligned}$$

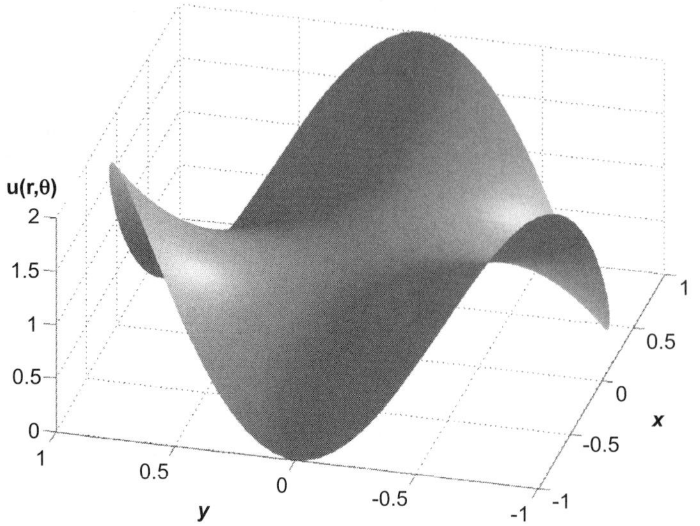

Figure 11.5. A surface plot of $u(r,\theta)$ for Example 11.6.

EXAMPLE 11.6. Consider the potential equation $\nabla^2 u = 0$ in polar coordinates, subject to $u(r_{\max},\theta) = 1 + \cos(3\theta)$, $0 \le \theta \le 2\pi$. Then the solution is given by (11.44), where the only nonzero coefficients are $\alpha_0 = 1$ and $\alpha_3 = r_{\max}^{-3}$. Thus

$$u(r,\theta) = 1 + \left(\frac{r}{r_{max}}\right)^3 \cos(3\theta)$$

One can verify this solution by substitution to $\nabla^2 u = 0$ and $u(r_{\max},\theta) = 1 + \cos(3\theta)$. A plot of $u(r,\theta)$ with $r_{\max} = 1.0$ is shown in Figure 11.5.

Using properties of sine and cosine functions, an alternative form of the solution can be obtained. By substituting the formulas for the coefficients α_k and β_k into (11.44) and then interchanging the order of summation and integration,

$$\begin{aligned} u(r,\theta) &= \frac{1}{\pi}\int_0^{2\pi} f(\phi)\left[\frac{1}{2} + \sum_{n=1}^{\infty}\left(\frac{r}{r_{max}}\right)^n \left(\cos(n\phi)\cos(n\theta) + \sin(n\phi)\sin(n\theta)\right)\right] d\phi \\ &= \frac{1}{\pi}\int_0^{2\pi} f(\phi)\left[\frac{1}{2} + \sum_{n=1}^{\infty}\left(\frac{r}{r_{max}}\right)^n \cos(n(\theta-\phi))\right] d\phi \\ &= \frac{1}{2\pi}\int_0^{2\pi} f(\phi)\left[1 + \sum_{n=1}^{\infty}\xi_{(-)}^n + \sum_{n=1}^{\infty}\xi_{(+)}^n\right] d\phi \end{aligned} \qquad (11.45)$$

where the last line was obtained after using Euler's identity for cosine, with $\xi_{(-)}^n$ and $\xi_{(+)}^n$ defined by

$$\xi_{(-)} = \frac{r}{r_{\max}} e^{-i(\theta-\phi)} \qquad \text{and} \qquad \xi_{(+)} = \frac{r}{r_{\max}} e^{+i(\theta-\phi)}$$

while using the following identities,

$$\xi_{(+)} + \xi_{(-)} = 2\frac{r}{r_{\max}}\cos(\theta - \phi) \qquad \text{and} \qquad \xi_{(+)}\xi_{(-)} = \left(\frac{r}{r_{\max}}\right)^2$$

One can also use the identity

$$\sum_{n=0}^{\infty} \xi^n = \frac{1}{1-\xi}$$

to simplify (11.45) further as

$$\begin{aligned} u(r,\theta) &= \frac{1}{2\pi}\int_0^{2\pi} f(\phi)\left[\frac{1}{1-\xi_{(-)}} + \frac{1}{1-\xi_{(+)}} - 1\right] d\phi \\ &= \frac{1}{2\pi}\int_0^{2\pi} f(\phi)\left[\frac{1-\xi_{(-)}\xi_{(+)}}{1-(\xi_{(+)}+\xi_{(-)})+\xi_{(-)}\xi_{(+)}}\right] d\phi \\ &= \frac{1}{2\pi}\int_0^{2\pi} f(\phi) K_{\text{Poisson}}\left(\rho,\theta,\phi\right) d\phi \end{aligned} \qquad (11.46)$$

where $\rho = r/r_{\max}$ is the normalized radius and $K_{\text{Poisson}}(\rho,\theta,\phi)$ is known as the **Poisson kernel** given by

$$K_{\text{Poisson}}\left(\rho,\theta,\phi\right) = \frac{1-\rho^2}{1-2\rho\cos(\theta-\phi)+\rho^2} \qquad (11.47)$$

Equation (11.46) is known as the **Poisson integral equation**. One advantage of the form given in (11.46) is that the solution is based on a single integration instead of an infinite series.

11.3.3 Sturm-Liouville Systems and Orthogonal Functions

As we observe from the previous example, two features usually result with the application of the method of separation of variables:

1. The solution will usually be in the form of an infinite series.
2. The determination of coefficients of an infinite series will require the use of orthogonality properties.

In Section 11.3.1, we applied the orthogonality property of sinusoids given in (11.31) to find β_n. In general, however, the functions may not involve simple sines or cosines. Instead, they may involve polynomials such as Legendre polynomials or Chebyshev polynomials. In other cases, it may involve Bessel functions. We begin with a general definition of orthogonality.

Definition 11.1. *A set of functions* $\{\phi_k(x)\}$, $k = 0, 1, 2, \ldots,$ *are* **orthogonal with respect to a weighting function** $r(x)$ **within the** (finite or infinite) **interval** (A, B), *if*

$$\int_A^B r(x)\phi_m\phi_n dx = \begin{cases} = 0 & \text{if } m \neq n \\ \neq 0 & \text{if } m = n \end{cases} \qquad (11.48)$$

For $r(x) = 1$, the set of functions are classified as simply **orthogonal**. *Furthermore, if*

$$\int_A^B \phi_n^2(x)dx = 1$$

for all n, then we say that the set of functions are **orthonormal**.

This means that if the set $\{\phi_k\}$ turns out to be orthogonal with respect to $r(x)$ in the interval (A, B), we could find the ℓth coefficient, a_ℓ, in the following series

$$f(x) = \sum_{k=0}^{\infty} a_k \phi_k(x) \tag{11.49}$$

by first multiplying both sides of (11.49) by $r(x)\phi_\ell(x)$ and then integrating from A to B,

$$\begin{aligned} \int_A^B r(x)\phi_\ell(x) f(x)dx &= \sum_{k=0}^{\infty} a_k \int_A^B r(x)\phi_\ell(x)\phi_k(x)dx \\ &= a_\ell \int_A^B r(x)\phi_\ell^2(x)dx \end{aligned}$$

or

$$a_\ell = \frac{\displaystyle\int_A^B r(x)\phi_\ell(x) f(x)dx}{\displaystyle\int_A^B r(x)\phi_\ell^2(x)dx} \tag{11.50}$$

Three items need to be determined: the weighting function $r(x)$ and the limits A and B. The properties of Sturm-Liouville systems described next are useful in helping us obtain these items.

Definition 11.2. *Let $r(x)$, $p(x)$ and $q(x)$ be continuous functions, then a* **Sturm-Liouville system** *is described by the following differential equation*

$$\frac{d}{dx}\left(p(x)\frac{d\phi}{dx}\right) + (q(x) + \lambda r(x))\,\phi = 0 \tag{11.51}$$

subject to the following boundary conditions

$$\begin{aligned} \beta_A \phi(A) + \gamma_A \left.\frac{d\phi}{dx}\right|_{x=A} &= 0 \\ \beta_B \phi(B) + \gamma_B \left.\frac{d\phi}{dx}\right|_{x=B} &= 0 \end{aligned} \tag{11.52}$$

where (β_A, γ_A) are not simultaneously zero and (β_B, γ_B) are not simultaneously zero.

Any homogeneous linear second-order differential equation of the form

$$\alpha_2(x)\frac{d^2\phi}{dx^2} + \alpha_1(x)\frac{d\phi}{dx} + \alpha_0(x)\phi + \lambda\phi = 0 \tag{11.53}$$

where $\alpha_2(x)$ has no zeros in the region of interest, can be recast into the Sturm-Liouville form given in (11.51). To do so, multiply (11.53) by a factor

$$\eta(x) = \exp\left(\int \frac{\alpha_1}{\alpha_2}\, dx\right)$$

and then divide the result by $\alpha_2(x)$. This yields

$$\begin{aligned}
\left(\eta(x)\frac{d^2\phi}{dx^2} + \frac{\alpha_1}{\alpha_2}\eta(x)\frac{d\phi}{dx}\right) + \eta(x)\frac{\alpha_0}{\alpha_2}\phi + \eta(x)\frac{1}{\alpha_2}\lambda\phi &= 0 \\
\frac{d}{dx}\left[\eta(x)\frac{d\phi}{dx}\right] + \left[\eta(x)\frac{\alpha_0(x)}{\alpha_2(x)}\right]\phi + \left[\eta(x)\frac{1}{\alpha_2(x)}\right]\lambda\phi &= 0
\end{aligned}$$

This means that for the given differential given in (11.53), we can identify the terms in (11.51) as

$$\begin{aligned}
p(x) &= \eta(x) = e^{\int(\alpha_1/\alpha_2)dx} \\
q(x) &= \left[e^{\int(\alpha_1/\alpha_2)dx}\right]\frac{\alpha_0(x)}{\alpha_2(x)} \\
r(x) &= \left[e^{\int(\alpha_1/\alpha_2)dx}\right]\frac{1}{\alpha_2(x)}
\end{aligned} \tag{11.54}$$

The following theorem states the important result regarding solutions of Sturm-Liouville systems, which we use to find the items needed for orthogonality, namely the weighting function $r(x)$ and the limits A and B:

THEOREM 11.2. *Let $\phi_n(x)$ be the solutions of the Sturm-Liouville system given by (11.51), (11.52) and (11.52), corresponding to $\lambda = \lambda_n$. Then the set of solutions $\phi = \{\phi_0, \phi_1, \phi_2, \ldots\}$, are orthogonal with respect to $r(x)$ within the interval (A, B), where $r(x)$ is obtained in (11.51), whereas the values of A and B are those given in boundary conditions (11.52).*

PROOF. (See Section K.2.2 for proof.)

For the special case where $\phi_n(x) = e^{i(n\pi x)/p}$, with $p = B - A$ and $r(x) = 1$, where $i = \sqrt{-1}$, the series given by

$$f(x) = \sum_{n=-\infty}^{\infty} a_n e^{i(n\pi x)/p} \tag{11.55}$$

is known as the **Fourier series**.

To summarize, we have the following procedure to find the coefficients in (11.49):

1. Starting with $f(x) = \sum_{k=0}^{\infty} a_k\phi_k(x)$, identify the functions $\phi_k(x)$, which should contain parameters indexed by k.

2. Obtain a homogeneous differential equation such as that given in (11.53),

$$\alpha_2(x)\frac{d^2\phi}{dx^2} + \alpha_1(x)\frac{d\phi}{dx} + \alpha_0(x)\phi + \lambda\phi = 0$$

There are at least two approaches available:
(a) Combine ϕ_k with its first and second derivatives using appropriate coefficients $\alpha_2(x)$, $\alpha_1(x)$, and $\alpha_0(x)$ that would yield zero.
(b) If the series resulted from other procedures such as the method of separation of variables, the required differential equation that generated ϕ_k should already be available.
3. Determine $r(x)$ using the relationship in (11.54),

$$r(x) = \left[e^{\int(\alpha_1/\alpha_2)dx}\right]\frac{1}{\alpha_2(x)}$$

4. Identify points A and B, with $A < B$, such that ϕ_k satifies the boundary conditions in (11.52) and (11.52) for all k. The interval (A, B) should at least cover the region on which $f(x)$ is being approximated.
5. The coefficients a_k can then be obtained by the formula given in (11.50),

$$a_k = \frac{\displaystyle\int_A^B r(x)\phi_k(x)f(x)dx}{\displaystyle\int_A^B r(x)\phi_k^2(x)dx}$$

In Example 11.7 that follows, we show a situation using step 2a, whereas in Example 11.8, we show a situation using step 2b in the context of solving a partial differential equation.

EXAMPLE 11.7. Suppose we want to find the coefficients that would satisfy the following equation:

$$f(x) = \sum_{k=0}^{\infty} a_k \cos\left(k\sqrt{x}\right)$$

We can identify ϕ_k and evaluate its derivatives,

$$\begin{aligned}
\phi_k &= \cos\left(k\sqrt{x}\right) \\
\frac{d\phi_k}{dx} &= -\frac{k}{2\sqrt{x}}\sin\left(k\sqrt{x}\right) \\
\frac{d^2\phi_k}{dx^2} &= -\frac{k^2}{4x}\cos\left(k\sqrt{x}\right) + \frac{k}{4x\sqrt{x}}\sin\left(k\sqrt{x}\right)
\end{aligned}$$

By inspection, one could combine ϕ_k and its derivatives to obtain a homogeneous second-order differential equation,

$$\frac{d^2\phi_k}{dx^2} + \frac{1}{2x}\frac{d\phi_k}{dx} + \frac{k^2}{4x}\phi_k = 0$$

Table 11.3. *The first ten values of a_k for example 11.7*

k	a_k	k	a_k
0	2.4305×10^{-1}	5	-5.7742×10^{-2}
1	-4.2365×10^{-1}	6	-3.8581×10^{-2}
2	-5.6718×10^{-1}	7	1.5017×10^{-2}
3	3.1978×10^{-2}	8	-2.4405×10^{-2}
4	-3.0976×10^{-2}	9	-1.6804×10^{-2}

or, after multiplying by $4x$,

$$4x\frac{d^2\phi_k}{dx^2} + 2\frac{d\phi_k}{dx} + k^2\phi_k = 0$$

We identify the coefficients to be: $\alpha_2(x) = 4x$, $\alpha_1(x) = 2$, $\alpha_0(x) = 0$ and $\lambda_k = k^2$. The weight $r(x)$ is then given by

$$r(x) = \frac{1}{4x}\exp\left(\int \frac{1}{2x}dx\right) = \frac{1}{4\sqrt{x}}$$

Take for instance the function

$$f(x) = \begin{cases} (x-10)/10 & \text{for } x \le 20 \\ (30-x)/10 & \text{for } x \ge 20 \end{cases}$$

that is to be approximated in the region $10 \le x \le 30$.[3] We could arbitrarily set the boundary conditions to be:

$$\frac{d\phi_k}{dx}(A) = -\frac{k}{2\sqrt{A}}\sin\left(k\sqrt{A}\right) = 0 \qquad \text{and} \qquad \frac{d\phi_k}{dx}(B) = -\frac{k}{2\sqrt{B}}\sin\left(k\sqrt{B}\right) = 0$$

with $A = \pi^2 \approx 9.87$ and $B = 4\pi^2 \approx 39.4$. The interval (A, B) covers the needed region: $(10, 30)$. It might appear valid to set $\phi_k(A) = \cos\left(k\sqrt{A}\right) = 0$, which could yield $A = (\pi/2)^2$. Unfortunately, doing so will satisfy $\phi_k(A) = 0$ only for odd integers. Finally, we have the following equations for a_k,

$$a_k = \frac{\displaystyle\int_{\pi^2}^{4\pi^2} \frac{1}{4\sqrt{x}} f(x)\cos\left(k\sqrt{x}\right)dx}{\displaystyle\int_{\pi^2}^{4\pi^2} \frac{1}{4\sqrt{x}}\cos^2\left(k\sqrt{x}\right)dx}$$

The first ten values of a_k are shown in Table 11.3. Figure 11.6 shows the result of using the first N number of terms in the series. Around $N = 50$, the error range at each x is within $\pm 10^{-3}$. Note that the approximations are based on cosines, which are even functions, and thus a more complete orthogonal set should have included sines as well.

[3] Note that we avoided the region $x \le 0$, because $r(x)$ is either unbounded or complex-valued in that region.

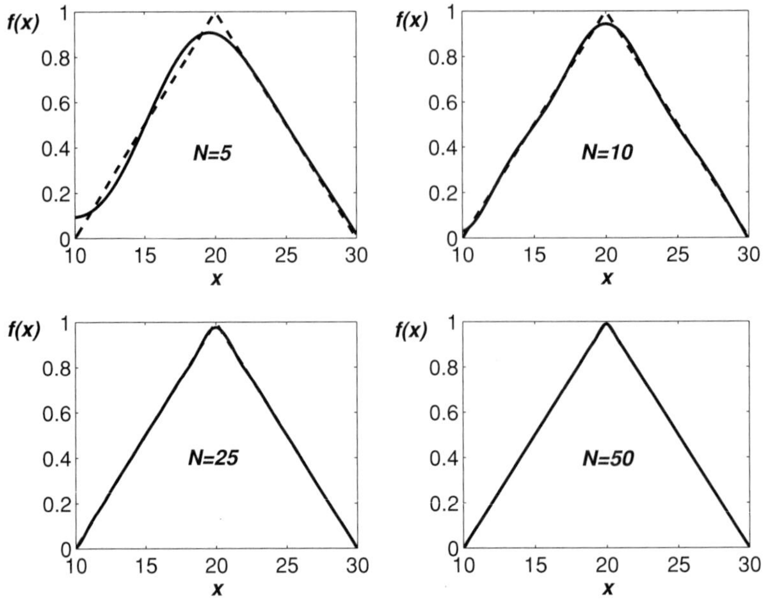

Figure 11.6. Plot of $f(x) \approx \sum_{k=0}^{N} a_k \cos(k\sqrt{x})$ for Example 11.7.

EXAMPLE 11.8. Consider the following heat equation for a circular region,

$$\alpha^2 \nabla^2 u = \alpha^2 \left(\frac{\partial^2 u}{\partial r^2} + \frac{1}{r}\frac{\partial u}{\partial r} + \frac{1}{r^2}\frac{\partial^2 u}{\partial \theta^2} \right) = \frac{\partial u}{\partial t}$$

subject to

$$\text{BC:} \quad u(r_{\text{max}}, \theta, t) = 0 \qquad 0 \le \theta \le 2\pi \;;\; t \ge 0$$

$$\text{IC:} \quad u(r, \theta, 0) = f(r, \theta) \qquad 0 \le \theta \le 2\pi \;;\; 0 \le r \le r_{\text{max}}$$

After substituting $u = T(t)R(r)\Theta(\theta)$ into the differential equation and some rearranging, we obtain

$$\frac{1}{R}\left(\frac{d^2R}{dr^2} + \frac{1}{r}\frac{dR}{dr} \right) + \frac{1}{r^2\Theta}\frac{d^2\Theta}{d\theta^2} = \frac{1}{\alpha^2 T}\frac{dT}{dt}$$

The last equation can be separated into three ordinary differential equations,

$$\frac{dT}{dt} = \kappa_1 \alpha^2 T \qquad ; \qquad \frac{d^2\Theta}{d\theta^2} = -\kappa_2 \Theta$$

$$r^2 \frac{d^2R}{dr^2} + r\frac{dR}{dr} - \left(\kappa_1 r^2 + \kappa_2\right) R = 0$$

whose solutions are

$$T = Ae^{\kappa_1 \alpha^2 t} \qquad ; \qquad \Theta = Be^{\sqrt{-\kappa_2 \theta}} + Ce^{-\sqrt{-\kappa_3 \theta}}$$

$$R = DJ_{\sqrt{\kappa_2}}\left(\sqrt{-\kappa_1} r\right) + EY_{\sqrt{\kappa_2}}\left(\sqrt{-\kappa_1} r\right)$$

where $J_\nu(\lambda r)$ and $Y_\nu(\lambda r)$ are Bessel functions of r of first and second kind, respectively, each of order ν with parameter λ.

Because we need the final solution to be bounded for all t, κ_1 must be negative. So, let $\kappa_1 = -\lambda^2$. For Θ to be periodic with period 2π, we need κ_2 to be a positive integer. So, let $\kappa_2 = n^2$, where n is an integer. Finally, for the solution to be bounded for all r, including the origin, we need the coefficient of $Y_n(\lambda r)$ to be zero. Applying all these into the solutions, and multiplying the values to obtain u, we have

$$u = e^{-(\lambda\alpha)^2 t} J_n(\lambda r)\left(\alpha \cos(n\theta) + \beta \sin(n\theta)\right)$$

To satisfy the homogeneous boundary condition at $r = r_{\max}$ for all $t \geq 0$ and $0 \leq \theta \leq 2\pi$, we need

$$J_n(\lambda r_{\max}) = 0$$

Let ρ_{nm} be the mth root of $J_n(r)$ and let $\lambda_{nm} = \dfrac{\rho_{nm}}{r_{\max}}$. Applying superposition, the complete solution is given by

$$u = \sum_{n=0}^{\infty}\sum_{m=1}^{\infty} e^{-(\lambda_{nm}\alpha)^2 t} J_n(\lambda_{nm} r)\left(\alpha_{nm} \cos(n\theta) + \beta_{nm} \sin(n\theta)\right)$$

To evaluate the coefficients a_{nm} and b_{nm}, we use the initial condition,

$$f(r,\theta) = u(r,\theta,0) = \sum_{n=0}^{\infty}\sum_{m=1}^{\infty} J_n(\lambda_{nm} r)\left(\alpha_{nm} \cos(n\theta) + \beta_{nm} \sin(n\theta)\right)$$

and apply orthogonality properties. The orthogonality property for sine was already given in (11.31). A similar property can be used for cosines. Also, the orthogonality of cosine and sine functions can be shown by direct integration:

$$\int_0^{2\pi} \sin\left(\frac{n\pi}{L}x\right)\cos\left(\frac{m\pi}{L}x\right)dx = 0$$

For the Bessel functions, we can apply the Sturm-Liouville theorem (Theorem 11.2). We begin with the differential equation for $R(r)$, written in such a way that λ_{nm} is the mth eigenvalue,

$$\frac{d^2R}{dr^2} + \frac{1}{r}\frac{dR}{dr} + \left(\lambda_{nm}^2 - \frac{n^2}{r^2}\right)R = 0$$

From this equation, we can immediately identify the weighting function. To avoid confusion, we use $w(r)$ to denote the weighting function,

$$w(r) = \exp\left(\int \frac{1}{r}dr\right) = r$$

Note that the eigenvalues for this equation are indexed by m, not n. The parameters indexed by n are eigenvalues connected with the differential equations involving θ. To obtain the limits needed for the orthogonality, recall some properties of Bessel functions with r as the independent variable (cf. Equations (9.60) and (9.66)):

$$\frac{d}{dr}J_n(\lambda_{nm} r) = J_{n+1}(\lambda_{nm} r) - \frac{n}{r}J_n(\lambda_{nm} r)$$

and

$$\lim_{r\to 0} J_n(\lambda_{nm} r) = 0 \qquad \text{for } n > 0$$

Table 11.4. *The first twenty positive roots of* $J_0(r)$

2.4048	18.0711	33.7758	49.4826
5.5201	21.2116	36.9171	52.6241
8.6537	24.3525	40.0584	55.7655
11.7915	27.4935	43.1998	58.9070
14.9309	30.6346	46.3412	62.0485

This implies that we could set one of the homogeneous boundary condition needed for the Sturm-Liouville conditions at $A = 0$ to be

$$\text{BC 1:} \quad \begin{cases} \dfrac{d}{dr}J_0\left(\lambda_{nm}0\right) = 0 & \text{for } n = 0 \\ J_n\left(\lambda_{nm}0\right) = 0 & \text{for } n > 0 \end{cases}$$

For the second homogenous boundary condition needed for Sturm-Liouville conditions, we set $B = r_{\text{max}}$ and obtain

$$\text{BC 2:} \qquad J_n\left(\lambda_{nm}r_{\text{max}}\right) = 0$$

Combining the preliminary results, the orthogonality property of the Bessel functions is given by

$$\int_0^{r_{\text{max}}} rJ_n\left(\lambda_{nm}r\right)J_n\left(\lambda_{nq}r\right)dr \begin{cases} = 0 & \text{if } m \neq q \\ \neq 0 & \text{if } m = q \end{cases}$$

The coefficients, α_{nm} and β_{nm}, are then evaluated by

$$\alpha_{nm} = \frac{\int_0^{r_{\text{max}}}\int_0^{2\pi} rf\left(r,\theta\right)J_n\left(\lambda_{nm}r\right)\cos\left(n\theta\right)d\theta dr}{\int_0^{r_{\text{max}}}\int_0^{2\pi} rJ_n^2\left(\lambda_{nm}r\right)\cos^2\left(n\theta\right)d\theta dr}$$

$$\beta_{nm} = \frac{\int_0^{r_{\text{max}}}\int_0^{2\pi} rf\left(r,\theta\right)J_n\left(\lambda_{nm}r\right)\sin\left(n\theta\right)d\theta dr}{\int_0^{r_{\text{max}}}\int_0^{2\pi} rJ_n^2\left(\lambda_{nm}r\right)\sin^2\left(n\theta\right)d\theta dr}$$

For a more concrete illustration, consider a specific case where $\alpha = 1$, $r_{\text{max}} = 1$ and $f(r,\theta) = f(r)$ is given by

$$f(r) = 2r^3 - 3r^2 + 1$$

For this case, it turns out that the only nonzero term is for: $n = 0$. Thus the solution for u is given by[4]

$$u = \sum_{m=1}^{\infty} k_m J_0\left(\lambda_m r\right)e^{-\left(\lambda_m\right)^2 t}$$

where

$$k_m = \frac{\int_0^1 rf\left(r\right)J_0\left(\lambda_m r\right)dr}{\int_0^1 rJ_0^2\left(\lambda_m r\right)dr}$$

and $\lambda_m = \rho_m$ is the mth root of $J_0(r)$. The first set of twenty positive roots, ρ_m, of $J_0(r)$ is given in Table 11.4, whereas the values of k_m for $m = 1, 2, \ldots, 20$

[4] It would have been more efficient to use the fact that $f(r,\theta) = f(r)$ and deduce radial symmetry at the outset. Doing so, we could set $\partial^2 u/\partial\theta^2 = 0$ and arrive at the same results.

Table 11.5. *The values of k_m for $m = 1, 2, \ldots, 20$*

m	k_m	m	k_m
1	7.7976×10^{-1}	11	-2.2189×10^{-3}
2	2.5024×10^{-1}	12	1.8426×10^{-3}
3	-5.5448×10^{-2}	13	-1.4560×10^{-3}
4	3.3734×10^{-2}	14	1.2403×10^{-3}
5	-1.6083×10^{-2}	15	-1.0151×10^{-3}
6	1.1263×10^{-2}	16	8.8139×10^{-4}
7	-6.9293×10^{-3}	17	-7.4063×10^{-4}
8	5.2733×10^{-3}	18	6.5260×10^{-4}
9	-3.6803×10^{-3}	19	-5.5986×10^{-4}
10	2.9502×10^{-3}	20	4.9927×10^{-4}

are given in Table 11.5. Truncation of the infinite series for $u(r, 0)$ after $m = 20$ resulted in an approximation of the initial condition $f(r)$ with errors within $\pm 10^{-4}$. Figure 11.7 shows some time-lapse sequence of surface plots of u distribution.

In general, the eigenvalues λ_n for the Sturm-Liouville system, (11.51) and (11.52), may not be directly identifiable in terms of sines, cosines, Bessel functions, Legendre polynomial, and so forth. Instead, the Sturm-Liouville system must be treated as an eigenvalue problem. Just as it was for matrix theory, the eignenvalue problem here is described by

$$L(\phi) = -\frac{1}{r(x)} \left(\frac{d}{dx} p(x) \frac{d}{dx} + q(x) \right) (\phi) = \lambda \phi$$

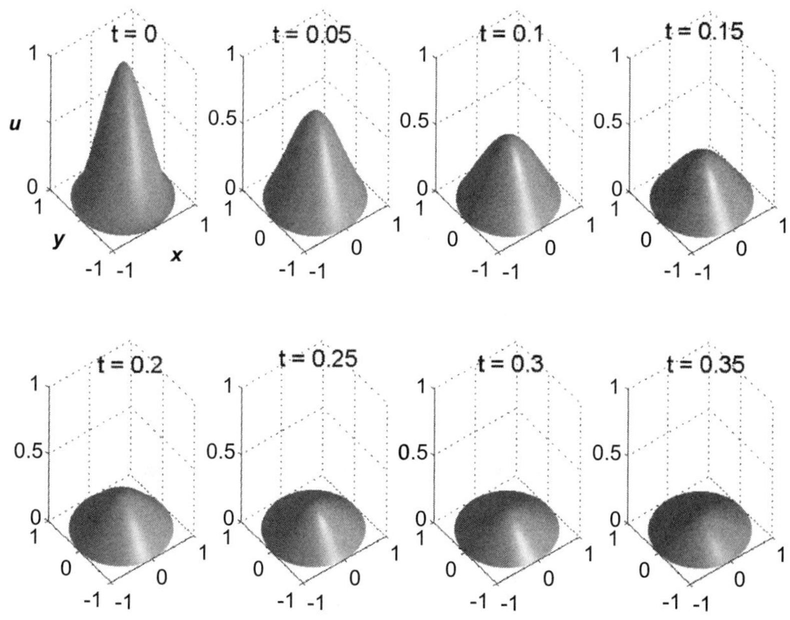

Figure 11.7. Time-lapse sequence of surface plots of u distribution.

such that $\phi \neq 0$. Unfortunately, the solution of the Sturm-Liouville differential equations may not be easily solvable in general, because the coefficients are often functions of the independent variable x. If the solutions are available, the procedure for evaluating the eigenvalues lies in satisfying the boundary conditions such that the solutions are not trivial; that is, ϕ is not identically zero. We show the process in the next example. One identity that is often useful during the determination of the eigenvalues is

$$e^{i2n\pi} = \cos(2n\pi) + i\sin(2n\pi) = 1 \qquad ; \qquad n = \ldots, -1, 0, 1, \ldots \tag{11.56}$$

EXAMPLE 11.9. Consider the eigenvalue problem

$$a_2 x^2 \frac{d^2\phi}{dx^2} + a_1 x \frac{d\phi}{dx} + a_0\phi = \lambda\phi \tag{11.57}$$

subject to

$$\phi(x_0) = 0 \qquad \text{and} \qquad \phi(x_L) = 0$$

with $x_0 \neq 0$ and $x_L \neq 0$.

The differential equation (11.57) is a Euler-Cauchy type and can be reduced to a differential equation with constant coefficient under a new independent variable $z = \ln(x)$ (cf. Section 6.4.3), that is,

$$a_2 \frac{d^2\phi}{dz^2} + (a_1 - a_2)\frac{d\phi}{dx} + (a_0 - \lambda)\phi = 0$$

whose general solution is

$$\phi = Ae^{s_1 z} + Be^{s_2 z} = Ax^{s_1} + Bx^{s_2}$$

where A and B are arbitrary constants, whereas s_1 and s_2 are the roots of the characteristic equation, that is,

$$s^2 + \frac{a_1 - a_2}{a_2}s + \frac{a_0 - \lambda}{a_2} = 0$$

$$\beta = \frac{a_2 - a_1}{2a_2} \qquad \rightarrow \qquad \begin{array}{rcl} s_1 & = & \beta + \gamma \\ s_2 & = & \beta - \gamma \end{array}$$

$$\gamma = \sqrt{\beta^2 + \frac{\lambda - a_0}{a_2}}$$

The boundary conditions are then given by

$$\begin{pmatrix} x_0^{s_1} & x_0^{s_2} \\ x_L^{s_1} & x_L^{s_2} \end{pmatrix} \begin{pmatrix} A \\ B \end{pmatrix} = \begin{pmatrix} 0 \\ 0 \end{pmatrix}$$

For A and B to not be simultaneously zero, we need the matrix involving x_0 and x_L to be singular, that is,

$$0 = x_0^{s_1}x_L^{s_2} - x_0^{s_2}x_L^{s_1} = (x_0 x_L)^{\beta}\left(\frac{x_0}{x_L}\right)^{\gamma}\left[1 - \left(\frac{x_L}{x_0}\right)^{2\gamma}\right]$$

Setting the last factor to zero, that is,

$$\left(\frac{x_L}{x_0}\right)^{2\gamma} = e^{2\gamma\ln(x_L/x_0)} = 1$$

we can then use (11.56) to get the nth eigenvalue λ_n as follows:

$$2\gamma_n \ln(x_L/x_0) = i2n\pi \rightarrow \left[\begin{array}{rcl} \gamma_n & = & i\dfrac{n\pi}{\ln(x_L/x_0)} \\ \beta^2 + \dfrac{\lambda_n - a_0}{a_2} & = & -\left(\dfrac{n\pi}{\ln(x_L/x_0)}\right)^2 \\ \lambda_n & = & a_0 - a_2\left(\beta^2 + \left(\dfrac{n\pi}{\ln(x_L/x_0)}\right)^2\right) \end{array}\right]$$

To find the corresponding eigenfunction ϕ_n, we use the first boundary condition to determine that we need $B_n = -A_n x_0^{2\gamma_n}$. Going back to the general solution,

$$\begin{aligned} A_n x^\beta \left(x^{\gamma_n} - x_0^{2\gamma_n} x^{-\gamma_n}\right) &= \left(A_n x_0^{\gamma_n}\right) x^\beta \left(\left(\frac{x}{x_0}\right)^{\gamma_n} - \left(\frac{x}{x_0}\right)^{-\gamma_n}\right) \\ &= \left(A_n x_0^{\gamma_n}\right) x^\beta \left[\exp\left(i\frac{n\pi\ln(x/x_0)}{\ln(x_L/x_0)}\right)\right. \\ &\qquad \left. - \exp\left(-i\frac{n\pi\ln(x/x_0)}{\ln(x_L/x_0)}\right)\right] \\ &= \left(A_n x_0^{\gamma_n}(2i)\right) x^\beta \sin\left(n\pi\frac{\ln(x/x_0)}{\ln(x_L/x_0)}\right) \end{aligned}$$

After dropping the constant coefficient, the nth eigenfunction is given by

$$\phi_n = x^{(a_2-a_1)/(2a_2)} \sin\left(n\pi\frac{\ln(x/x_0)}{\ln(x_L/x_0)}\right)$$

Furthermore, because the eigenvalue problem satisfies the conditions of a Sturm-Liouville system, we have the following orthogonality property:

$$\int_{x_0}^{x_L} \left(x^{(a_1/a_2)-2}\right) \phi_n \phi_m dx \left\{ \begin{array}{ll} = 0 & \text{if } n \neq m \\ \neq 0 & \text{if } n = m \end{array}\right.$$

11.4 Nonhomogeneous Partial Differential Equations

One technique for solving linear nonhomogeneous partial differential equations is to first solve an associated homogeneous differential equation. Then we can apply an approach similar to the variation of parameters, in which a linear functional combination of the homogeneous solutions are substituted into the nonhomogeneous equations to obtain the final solution. This approach is known as the **method of eigenfunction expansion**.

In the next section, we show a procedure that converts nonhomogeneous boundary conditions to homogeneous boundary conditions. This will, in general, introduce more nonhomogeneous terms into the differential equation. We limit the discussion to second-order linear differential equations for $u = u(x,t)$.

11.4.1 Homogenization of Boundary Conditions

We begin with the partial differential equation,

$$Lu = h(x,t) \tag{11.58}$$

where L is the linear partial differential operator given by

$$L = \alpha_{xx}\frac{\partial^2}{\partial x^2} + \alpha_{xt}\frac{\partial^2}{\partial x\partial t} + \alpha_{tt}\frac{\partial^2}{\partial t^2} + \beta_x\frac{\partial}{\partial x} + \beta_t\frac{\partial}{\partial t} + \gamma$$

subject to initial and boundary conditions given by

$$c_0\, u(x,0) = f_1(x) \qquad \text{and} \qquad c_1\left.\frac{\partial u}{\partial t}\right|_{t=0} = f_2(x)$$

$$a_0\, u(0,t) + a_1\left.\frac{\partial u}{\partial x}\right|_{x=0} = g_1(t) \qquad \text{and} \qquad b_0\, u(1,t) + b_1\left.\frac{\partial u}{\partial x}\right|_{x=1} = g_2(t) \tag{11.59}$$

By splitting the original function u as

$$u(x,t) = S(x,t) + U(x,t) \tag{11.60}$$

and substituting into (11.58) and (11.59), we have

$$LU = h - LS$$

subject to the initial conditions

$$\begin{aligned} c_0\, U(x,0) &= f_1(x) - c_0\, S(x,0) \\ c_1\left.\frac{\partial U}{\partial t}\right|_{t=0} &= f_2(x) - c_1\left.\frac{\partial S}{\partial t}\right|_{t=0} \end{aligned}$$

and boundary conditions

$$\begin{aligned} a_0 U(0,t) + a_1\left.\frac{\partial U}{\partial x}\right|_{x=0} &= g_1(t) - a_0 S(0,t) - a_1\left.\frac{\partial S}{\partial x}\right|_{x=0} \\ b_0 U(1,t) + b_1\left.\frac{\partial U}{\partial x}\right|_{x=1} &= g_2(t) - b_0 S(1,t) - b_1\left.\frac{\partial S}{\partial x}\right|_{x=1} \end{aligned}$$

We want to transform the original problem in $u(x,t)$ to a problem in $U(x,t)$ but with homogeneous boundary conditions; that is, we need

$$a_0 S(0,t) + a_1\left.\frac{\partial S}{\partial x}\right|_{x=0} = g_1(t) \qquad \text{and} \qquad b_0 S(1,t) + b_1\left.\frac{\partial S}{\partial x}\right|_{x=1} = g_2(t) \tag{11.61}$$

One of the simplest choices for $S(x,t)$ is to assume the following form,

$$S(x,t) = S_0(t) + xS_1(t) \tag{11.62}$$

After substitution of (11.62) into (11.61), we can solve S_0 and S_1 to be

$$\begin{pmatrix} S_0(t) \\ S_1(t) \end{pmatrix} = \frac{1}{a_0\left(b_0+b_1\right)-a_1 b_0}\begin{pmatrix} b_0+b_1 & -a_1 \\ -b_0 & a_0 \end{pmatrix}\begin{pmatrix} g_1(t) \\ g_2(t) \end{pmatrix} \tag{11.63}$$

The transformed problem is given by

$$LU = \widetilde{h}(x,t) \tag{11.64}$$

subject to initial conditions

$$c_0\, U(x,0) = \widetilde{f}_1(x) \qquad \text{and} \qquad c_1\, \left.\frac{\partial U}{\partial t}\right|_{t=0} = \widetilde{f}_2(x) \tag{11.65}$$

and boundary conditions

$$a_0 U(0,t) + a_1\, \left.\frac{\partial U}{\partial x}\right|_{x=0} = 0 \qquad \text{and} \qquad b_0 U(1,t) + b_1\, \left.\frac{\partial U}{\partial x}\right|_{x=1} = 0 \tag{11.66}$$

where,

$$\begin{aligned} \widetilde{h}(x,t) &= h(x,t) - L\left(S_0(t) + xS_1(t)\right) \\ \widetilde{f}_1(x) &= f_1(x) - c_0\left(S_0(0) + xS_1(0)\right) \\ \widetilde{f}_2(x) &= f_2(x) - c_1\left.\left(\frac{dS_0}{dt} + x\frac{dS_1}{dt}\right)\right|_{t=0} \end{aligned} \tag{11.67}$$

and S_0 and S_1 are given by (11.63), assuming that $a_0(b_0+b_1) \neq a_1 b_0$. Thereafter, the desired solution for u is given by

$$u(x,t) = U(x,t) + S_0(t) + xS_1(t) \tag{11.68}$$

EXAMPLE 11.10. Consider the following:

$$\frac{\partial u}{\partial t} = \frac{\partial^2 u}{\partial x^2} + e^{-3t}\left(8.7x^2 - 6x + 5.5\right) - 2.8$$

subject to the initial condition, $u(0,x) = -1.5x^2 + x + 0.5$, and boundary conditions

$$2u(t,0) - 0.1\, \left.\frac{\partial u}{\partial x}\right|_{x=0} = 0.9 \qquad \text{and} \qquad 2u(t,1) + 0.5\, \left.\frac{\partial u}{\partial x}\right|_{x=1} = -3.5e^{-3t} + 2.5$$

We can evaluate S_0 and S_1 from (11.63) to be

$$\begin{pmatrix} S_0(t) \\ S_1(t) \end{pmatrix} = \begin{pmatrix} 2 & -0.1 \\ 2 & 2.5 \end{pmatrix}^{-1}\begin{pmatrix} 0.9 \\ -3.5e^{-3t} + 2.5 \end{pmatrix} = \begin{pmatrix} 0.4808 - 0.0673e^{-3t} \\ 0.6154 - 1.3462e^{-3t} \end{pmatrix}$$

With $u = U + S_0 + xS_1$, the transformed problem becomes

$$\frac{\partial U}{\partial t} = \frac{\partial^2 U}{\partial x^2} + e^{-3t}\left(8.7x^2 - 16.313x + 5.298\right) - 2.8$$

subject to the initial condition, $U(x,0) = -1.5x^2 + 1.731x + 0.087$, and boundary conditions

$$2U(0,t) - 0.1\left.\frac{\partial U}{\partial x}\right|_{x=0} = 0 \qquad \text{and} \qquad 2U(1,t) + 0.5\left.\frac{\partial U}{\partial x}\right|_{x=1} = 0$$

11.4.2 Method of Eigenfunction Expansion

With the aid of the homogenization techniques discussed in the previous section, we can now focus on the solution of a nonhomogenous partial differential subject to homogeneous boundary conditions. We also limit our discussion to the partial differential operator L_{sep} given in (11.23), that is,

$$L_{\text{sep}} = \alpha_x(x)\frac{\partial^2}{\partial x^2} + \beta_x(x)\frac{\partial}{\partial x} + \gamma_x(x) + \alpha_t(t)\frac{\partial^2}{\partial t^2} + \beta_t(x)\frac{\partial}{\partial t} + \gamma_t(t) \quad (11.69)$$

Thus consider the nonhomogenous partial differential system given by

$$L_{\text{sep}} U = \widetilde{h}(t,x) \quad (11.70)$$

subject to initial conditions

$$c_0\, U(0,x) = \widetilde{f}_1(x) \qquad \text{and} \qquad c_1\left.\frac{\partial U}{\partial t}\right|_{t=0} = \widetilde{f}_2(x) \quad (11.71)$$

and boundary conditions

$$a_0 U(t,0) + a_1\left.\frac{\partial U}{\partial x}\right|_{x=0} = 0 \qquad \text{and} \qquad b_0 U(t,1) + b_1\left.\frac{\partial U}{\partial x}\right|_{x=1} = 0 \quad (11.72)$$

The **method of eigenfunction expansion** uses the eigenfunctions obtained by first solving the differential equation while setting $\widetilde{h}(t,x)$ first to zero.[5] With $\widehat{U} = \widehat{T}(t)X(x)$,

$$L_{\text{sep}}\left(\widehat{T}(t)X(x)\right) = 0 \quad \rightarrow \quad \begin{array}{rcl} L_T\left(\widehat{T}\right) & = & -\lambda\,\widehat{T} \\ L_X(X) & = & \lambda\, X \end{array} \quad (11.73)$$

where

$$\begin{aligned} L_T &= \alpha_t(t)\frac{d^2}{dt^2} + \beta_t(t)\frac{d}{dt} + \gamma_t(t) \quad \text{and} \\ L_X &= \alpha_x(x)\frac{d^2}{dx^2} + \beta_x(x)\frac{d}{dx} + \gamma_x(x) \end{aligned} \quad (11.74)$$

The solutions X_n to

$$\alpha_x(x)\frac{d^2X}{dx^2} + \beta_x(x)\frac{dX}{dx} + \gamma_x(x)X = \lambda_n X \quad (11.75)$$

[5] This is similar to the step of finding the complementary solutions of ordinary differential equations.

subject to $a_0 X_n(0) + a_1 \left.\frac{dX}{dx}\right|_{x=0} = 0$ and $b_0 X_n(1) + b_1 \left.\frac{dX}{dx}\right|_{x=1} = 0$ are known as the **eigenfunctions** corresponding to $\lambda = \lambda_n$. Based on Theorem 11.2 and (11.54),

$$\int_0^1 r(x) X_n(x) X_m(x) dx = \begin{cases} = 0 & \text{if } n \neq m \\ \neq 0 & \text{if } n = m \end{cases} \tag{11.76}$$

where

$$r(x) = \frac{e^{\int (\beta_x/\alpha_x) dx}}{\alpha_x}$$

Next, we represent $U(t,x)$ and the nonhomogeneous term $\widetilde{h}(t,x)$ as series expansions based on the eigenfunctions

$$U(t,x) = \sum_{n=0}^{\infty} T_n(t) X_n(x) \tag{11.77}$$

$$\widetilde{h}(t,x) = \sum_{n=0}^{\infty} \widetilde{h}_n(t) X_n(x) \tag{11.78}$$

where

$$\widetilde{h}_n(t) = \frac{\int_0^1 r(x) \widetilde{h}(t,x) X_n(x)\, dx}{\int_0^1 r(x) X_n^2(x)\, dx}$$

Substituting these to (11.70),

$$\begin{aligned} L_{\text{sep}} \left(\sum_{n=0}^{\infty} T_n(t) X_n(x) \right) = \sum_{i=0}^{\infty} L_{\text{sep}} \Big(T_n(t) X_n(x) \Big) &= \sum_{n=0}^{\infty} \widetilde{h}_n(t) X_n(x) \\ \sum_{i=0}^{\infty} \Big(T_n(t) \big(L_X X_n(x) \big) + X_n(x) \big(L_T T_n(t) \big) \Big) &= \\ \sum_{i=0}^{\infty} X_n \Big((L_T + \lambda_n)(T_n(t)) - \widetilde{h}_n(t) \Big) &= 0 \end{aligned}$$

because $L_X X_n(x) = \lambda_n X_n(x)$. The terms $T_n(t)$ can then be obtained by solving

$$\alpha_t(t) \frac{d^2 T_n}{dt^2} + \beta_t(t) \frac{dT_n}{dt} + \Big(\gamma_t(t) + \lambda_n \Big) T_n = \widetilde{h}_n(t) \tag{11.79}$$

subject to the initial conditions

$$c_0\, T_n(0) = \frac{\int_0^1 r(x) \widetilde{f}_1(x) X_n(x) dx}{\int_0^1 r(x) X_n^2(x) dx} \quad \text{and} \quad c_1 \left.\frac{dT_n}{dt}\right|_{t=0} = \frac{\int_0^1 r(x) \widetilde{f}_2(x) X_n(x) dx}{\int_0^1 r(x) X_n^2(x) dx}$$

where $\widetilde{f}_1(x)$ and $\widetilde{f}_2(x)$ are the functions specified in (11.71).

EXAMPLE 11.11. Let us start with the problem given in Example 11.10, which, after homogenization of boundary conditions, is given by

$$\frac{\partial U}{\partial t} = \frac{\partial^2 U}{\partial x^2} + e^{-3t}\left(8.7x^2 - 16.313x + 5.298\right) - 2.8$$

subject to

$$U(0,x) = -1.5x^2 + 1.731x + 0.087 \qquad \text{and} \qquad \begin{array}{rcl} 2U(t,0) - 0.1\left.\dfrac{\partial U}{\partial x}\right|_{x=0} & = & 0 \\ 2U(t,1) + 0.5\left.\dfrac{\partial U}{\partial x}\right|_{x=1} & = & 0 \end{array}$$

First, we drop the nonhomogeneous terms in the differential equation to obtain the problem

$$\frac{\partial \widehat{U}}{\partial t} = \frac{\partial^2 \widehat{U}}{\partial x^2}$$

subject to

$$\widehat{U}(0,x) = -1.5x^2 + 1.731x + 0.087 \qquad \text{and} \qquad \begin{array}{rcl} 2\widehat{U}(t,0) - 0.1\left.\dfrac{\partial \widehat{U}}{\partial x}\right|_{x=0} & = & 0 \\ 2\widehat{U}(t,1) + 0.5\left.\dfrac{\partial \widehat{U}}{\partial x}\right|_{x=1} & = & 0 \end{array}$$

Using $\widehat{U} = T(t)X(x)$ and $\lambda = -\kappa^2$,

$$T(t) = Ce^{-\kappa^2 t} \qquad \text{and} \qquad X(x) = A\cos(\kappa x) + B\sin(\kappa x)$$

Applying the boundary conditions, we find $A = 0.05\kappa\, B$ and $B\,\varphi(\kappa) = 0$, where $\varphi(\kappa) = 0.6\kappa\cos(\kappa) + (2 - 0.025\kappa^2)\sin(\kappa)$. The roots of $\varphi(\kappa)$ are used to find the eigenvalues $\lambda_n = -\kappa_n^2$, where $\kappa_n = 0$ is a trivial root in this example. A list of the first ten nonpositive eigenvalues κ_n is given in Table 11.6. Thus the functions

$$X_n(x) = 0.05\kappa_n\cos(\kappa_n x) + \sin(\kappa_n x)$$

are the nth eigenfunctions of the system that has the following orthogonality property:

$$\int_0^1 X_n(x)X_m(x) \left\{ \begin{array}{ll} = 0 & \text{if } n \neq m \\ \neq 0 & \text{if } n = m \end{array} \right.$$

Next, we put back the nonhomogenous terms and expand them in series based on X_n,

$$\widetilde{h}(t,x) = e^{-3t}\left(8.7x^2 - 16.313x + 5.298\right) - 2.8 = \sum_{n=1}^{\infty} \tilde{h}_n(t)X_n(x)$$

where

$$\tilde{h}_n(t) = \frac{\displaystyle\int_0^1 \left(e^{-3t}\left(8.7x^2 - 16.313x + 5.298\right) - 2.8\right) X_n(x)\, dx}{\displaystyle\int_0^1 X_n^2(x)dx}$$

Similarly, with $U(t,x) = \sum_{n=1}^{\infty} T_n(t)X_n(x)$, we generate the nth-ordinary differential equations for T_n given by

$$\frac{dT_n}{dt} + \kappa_n^2 T_n = \widetilde{h}_n(t) \quad \text{subject to} \quad T_n(0) = \frac{\displaystyle\int_0^1 U(0,x)X_n(x)dx}{\displaystyle\int_0^1 X_n^2(x)dx}$$

The explicit calculations can be obtained for this example with the aid of symbolic manipulation software.[6]

Because the original problem in Example 11.10 is for $u(t,x) = U(t,x) + S(t,x)$, where

$$S(t,x) = \left(\frac{25}{52} - \frac{7}{104}e^{-3t}\right) + x\left(\frac{8}{13} - \frac{35}{26}e^{-3t}\right)$$

The final solution is given by

$$u(t,x) = S(t,x) + \sum_{n=1}^{\infty} T_n(t)X_n(x)$$

A plot of the eigenfunction solution using the series truncated after $n = 100$ is shown in Figure 11.8. The exact solution we used to generate the original problem in Example 11.10 is known and is given by

$$u_{\text{exact}}(t,x) = e^{-3t}\left(-\frac{3}{2}x^2 + x + \frac{1}{2}\right) + \left(1 - e^{-3t}\right)\left(\frac{14}{10}x^2 - x + \frac{4}{10}\right)$$

Also included in Figure 11.8 is a plot of the errors between the eigenfunction solution and the exact solution.

[6] Using MathCad, we get $T_n(t) = \mathcal{T}_{n,a}(t) + \mathcal{T}_{n,b}(t) + \mathcal{T}_{n,c}(t)/\mathcal{T}_{n,d}$, where

$$\begin{aligned}
\mathcal{T}_{n,a}(t) &= e^{-3t}\Bigg[\left(\frac{2059}{13}\kappa_n^3 + \frac{67512}{13}\kappa_n\right)\sin(\kappa_n) \\
&\quad + \left(-\frac{3752}{13}\kappa_n^2 + 13920\right)\cos(\kappa_n) + \left(-13920 + 4640\kappa_n^2\right)\Bigg] \\
\mathcal{T}_{n,b}(t) &= e^{-\kappa_n t}\Bigg[\left(\frac{165}{13}\kappa_n^5 - \frac{12738}{13}\kappa_n^3 - \frac{36960}{13}\kappa_n\right)\sin(\kappa_n) \\
&\quad - \left(\frac{3960}{13}\kappa_n^4 + \frac{11088}{13}\kappa_n^2\right)\cos(\kappa_n)\Bigg] \\
\mathcal{T}_{n,c}(t) &= \left(-112\kappa_n^3 + 336\kappa_n\right)\sin(\kappa_n) + \left(2240\kappa_n^2 - 6720\right)\left(\cos(\kappa_n) - 1\right) \\
\mathcal{T}_{n,d}(t) &= \left(-40\kappa_n^5 + 120\kappa_n^3\right)\cos(\kappa_n) + \left(\kappa_n^6 - 403\kappa_n^4 + 1200\kappa_n^2\right)\cos(\kappa_n)\sin(\kappa_n) \\
&\quad + \kappa_n^7 + 437\kappa_n^5 - 1320\kappa_n^3
\end{aligned}$$

Table 11.6. *A list of the first ten positive eigenvalues, λ_n*

n	κ_n	n	λ_n
1	2.4664	6	19.8380
2	5.1244	7	22.8825
3	7.9425	8	25.9425
4	13.8145	9	29.0148
5	16.8133	10	32.0971

11.4.3 Homogenization of the Partial Differential Equation

In the previous sections, we proceeded by first homogenizing the boundary conditions that could then lead to adding nonhomogeneous terms to the partial differential equations, which could later be solved using the eigenfunction approach. However, in some cases, the reverse process may lead to the solution; that is, we could first homogenize the partial differential equation, which would then lead to nonhomogeneous boundary conditions.

We now show how this procedure can be used for solving an equation known as the **Poisson equation** given by

$$\nabla^2 u = -f(x, y, z)$$

where f is a source function. One application of this equation is to model the potential of an electrostatic field, with f representing a constant charge density. Another application is to model the temperature distribution of a domain with $f(x, y, z)$ as the heat generated at that point. We limit our discussion only to the 2D case with Dirichlet boundary conditions.

EXAMPLE 11.12. Consider the Poisson equation

$$\frac{\partial^2 u}{\partial x^2} + \frac{\partial^2 u}{\partial y^2} = -1$$

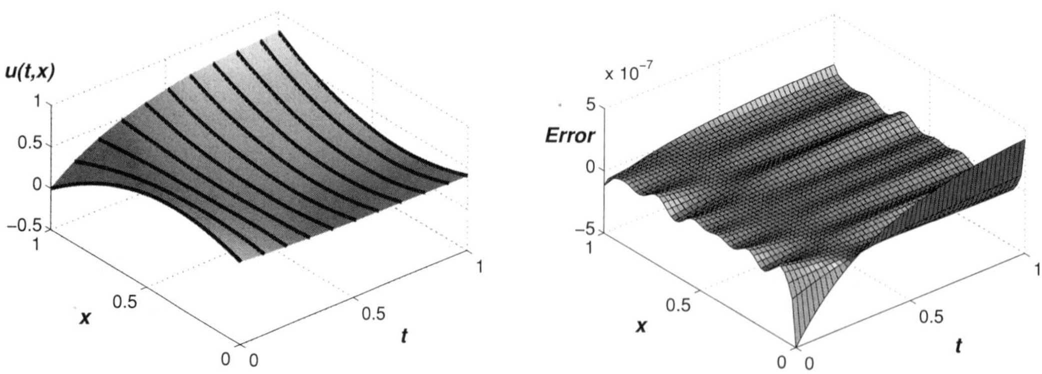

Figure 11.8. A plot of the eigenfunction expansion solution for the system given in Examples 11.10 and 11.11. The plot on the right is the error of the expansion solution from the exact solution.

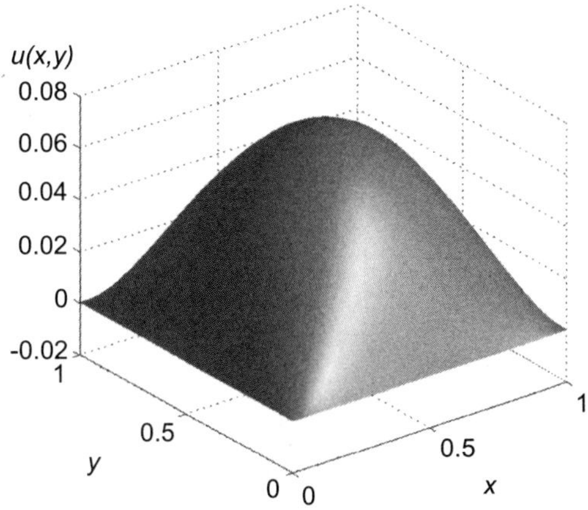

Figure 11.9. A plot of $u(x, y)$ that solves the Poisson equation for $f = 1$ and boundary conditions $u(0, y) = u(1, y) = u(x, 0) = u(x, 1) = 0$.

subject to $u(0, y) = u(1, y) = u(x, 0) = u(x, 1) = 0$. Next, split u as $u(x, y) = U(x, y) + S(x, y)$ such that $\nabla^2 S = -1$. If we choose $S = S(x)$ such that $S(0) = S(1) = 0$, we obtain $S(x) = \dfrac{x(1-x)}{2}$. Then U will have to satisfy

$$\nabla^2 U = \nabla^2 u - \nabla^2 S = 0$$

subject to $U(0, y) = U(1, y) = 0$, $U(x, 0) = U(x, 1) = \dfrac{x(x-1)}{2}$. Using separation of variables, one can show (as in Exercise **E11.16**) that

$$U(x, y) = -\frac{4}{\pi^3} \sum_{n=1,3,5,\ldots}^{\infty} \left(\frac{\sin(n\pi x)}{n^3}\right)\left(\cosh(n\pi y) + \frac{1-\cosh(n\pi)}{\sinh(n\pi)}\sinh(n\pi y)\right) \tag{11.80}$$

A plot of $u = \dfrac{x(1-x)}{2} + U(x, y)$ is given in Figure 11.9.

11.5 Similarity Transformations

Another powerful and general approach to solve partial differential equations, both linear and nonlinear, is based on finding group transformation variables that would achieve a property known as **symmetry**. We can generalize the definition of symmetry transformations given in 6.1 that would apply to partial differential equations. We include this approach in this chapter for linear differential equations because some of the important results in applied mathematics have been obtained using this approach, for example, the solution of diffusion equations.

Definition 11.3. *Let* $\mathbf{x}$ *be a vector of* n *independent variables for a partial differential equation for a single dependent variable* u*,*

$$F\left(\mathbf{x}, u, \frac{\partial u}{\partial \mathbf{x}}, \ldots\right) = 0 \tag{11.81}$$

A set of transformations

$$\widetilde{\mathbf{x}} = \widetilde{\mathbf{x}}(u, \mathbf{x}) \quad \textit{and} \quad \widetilde{u} = \widetilde{u}(u, \mathbf{x}) \tag{11.82}$$

is called a set of **symmetry transformation** *for the differential equation (11.81), if the substitution of these transformed variables result in a differential equation given by*

$$F\left(\widetilde{\mathbf{x}}, \widetilde{u}, \frac{\partial \widetilde{u}}{\partial \widetilde{\mathbf{x}}}, \ldots\right) = 0 \tag{11.83}$$

that is, the same function results, except that $\mathbf{x}$ *and* u *are replaced by* $\widetilde{\mathbf{x}}$ *and* $\widetilde{u}$*, respectively. Furthermore, we say that the new differential equation has attained* **symmetry** *based on the transformations.*

It can be shown that if the differential equation admits a symmetry transformation based on $\widetilde{u}$ and $\widetilde{\mathbf{x}}$, one can reduce the original differential equation to a differential equation with ψ as a new variable that depends on ζ_ℓ, $\ell = 1, \ldots, (n-1)$. This means that a combination of variables from the set $(u, \mathbf{x})$ can yield an ordinary differential equation if $n = 2$ or a partial differential equation with $(n-1)$ independent variables if $n > 2$.

However, in general, the determination of the symmetries of a differential equation can be a very long and difficult process. One type of transformation is easy to check for symmetry. These are the **similarity transformations** (also known as **scaling** or **stretch transformations**), given by

$$\widetilde{u} = \lambda^{-\gamma} u \quad \text{and} \quad \widetilde{x_k} = \lambda^{-\alpha_k} x_k \quad \text{for } k = 1, \ldots, n \tag{11.84}$$

where λ is known as the **similarity transformation parameter**, and at least two of the exponents α_k must be nonzero. To determine whether a given partial differential equation admits a similarity transformation, one would only need to substitute the transformations given in (11.84) into the given differential equation (11.81) and determine whether there exists values of α_k and γ that would yield an equation that does not involve the parameter λ.

EXAMPLE 11.13. Applying the similarity group of transformations (11.84), $t = \lambda^{\alpha}\widetilde{t}$, $x = \lambda^{\beta}\widetilde{x}$ and $u = \lambda^{\gamma}\widetilde{u}$ to the differential equation given by

$$\frac{\partial u}{\partial t} + x\frac{\partial^2 u}{\partial x^2} = A$$

where A is a constant, we obtain

$$\left(\lambda^{\gamma-\alpha}\right)\frac{\partial \widetilde{u}}{\partial \widetilde{t}} + \left(\lambda^{\gamma-\beta}\right)\widetilde{x}\frac{\partial^2 \widetilde{u}}{\partial \widetilde{x}^2} = A$$

Symmetry is achieved if we set $\alpha = \beta = \gamma = 1$. Conversely, one can show that the following differential equation does not admit symmetry based on a similarity transformation:

$$\frac{\partial u}{\partial t} + \frac{\partial^2 u}{\partial x^2} = x^2 + A$$

At this point, we limit the similarity method to handle only partial differential equations for u that depend on two independent variables x and t. For this case, we have the following theorem:

THEOREM 11.3. *The partial differential equation*

$$F\left(x, t, u, \frac{\partial u}{\partial t}, \frac{\partial u}{\partial x}, \frac{\partial^2 u}{\partial x^2}, \frac{\partial^2 u}{\partial x \partial t}, \frac{\partial^2 u}{\partial t^2}, \ldots\right) = 0 \tag{11.85}$$

which admits symmetry for the similarity transformations given by $\widetilde{t} = \lambda^{-\alpha} t$, $\widetilde{x} = \lambda^{-\beta} x$ *and* $\widetilde{u} = \lambda^{-\gamma} u$, *can be reduced to an ordinary differential equation with* ζ *and* ψ *as the independent and dependent variables, respectively, where*

$$\zeta = \frac{\widetilde{x}^{\alpha}}{\widetilde{t}^{\beta}} = \frac{x^{\alpha}}{t^{\beta}} \qquad \text{and} \qquad \psi = \frac{\widetilde{u}^{\alpha}}{\widetilde{t}^{\gamma}} = \frac{u^{\alpha}}{t^{\gamma}}$$

PROOF. (See Section K.2.3 for proof.)

Theorem 11.3 guarantees that a partial differential equation involving two independent variables that admits a similarity transformation can be reduced to an ordinary differential equation. However, one will need to consider additional complexities. In some cases, especially for linear partial differential equations, there can be more than one set of similarity transformations. One will need to determine whether symmetry applies to the initial and boundary conditions as well. Finally, the ordinary differential equations may not necessarily be easy to solve. In some cases, numerical methods might be needed.

EXAMPLE 11.14. Consider the diffusion equation given by

$$\frac{\partial u}{\partial t} = \mu^2 \frac{\partial^2 u}{\partial x^2} \tag{11.86}$$

subject to the conditions: $u(x, 0) = u_i$ and $u(0, t) = u_0$. After substitution of $\widetilde{t} = \lambda^{-\alpha} t$, $\widetilde{x} = \lambda^{-\beta} x$ and $\widetilde{u} = \lambda^{-\gamma} u$, we have

$$\lambda^{\gamma-\alpha} \frac{\partial \widetilde{u}}{\partial \widetilde{t}} = \mu^2 \lambda^{\gamma - 2\beta} \frac{\partial^2 \widetilde{u}}{\partial \widetilde{x}^2}$$

subject to $\lambda^{-\gamma} u\left(\lambda^{-\beta} x, 0\right) = u_i$ and $\lambda^{-\gamma} u\left(0, \lambda^{-\alpha} t\right) = u_0$. For symmetry, we need $\alpha = 2\beta$ and $\gamma = 0$, so we can set $\alpha = 1$, $\beta = 1/2$ and $\gamma = 0$, yielding the invariants $\psi = u = \widetilde{u}$ and $\zeta = x/\sqrt{t} = \widetilde{x}/\sqrt{\widetilde{t}}$. Substituting these into the partial differential equation (11.86) will yield

$$\mu^2 \frac{d^2 u}{d\zeta^2} = -\frac{\zeta}{2} \frac{du}{d\zeta}$$

The general solution is given by

$$u(\zeta) - u(0) = C_1 \int_0^{\zeta} \exp\left(\frac{-\zeta^2}{4\mu^2}\right) d\zeta = A \,\mathrm{erf}\left(\frac{\zeta}{2\mu}\right)$$

where $A = C_1 \mu \sqrt{\pi}$ is an integration constant and $\mathrm{erf}(x)$ is the error function defined by

$$\mathrm{erf}(z) = \frac{2}{\sqrt{\pi}} \int_0^z e^{-q^2} dq$$

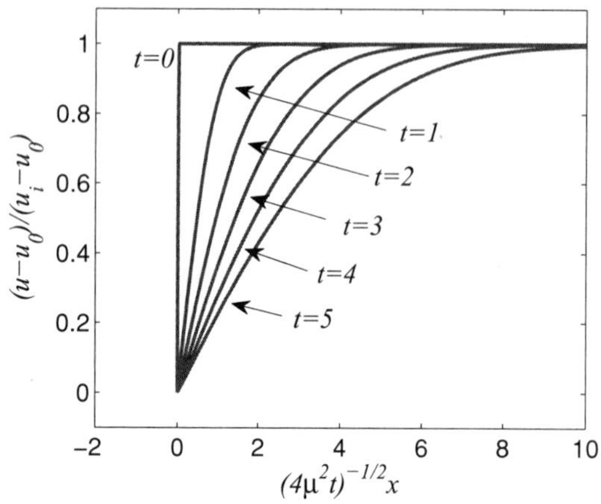

Figure 11.10. A plot of $(u - u_0)/(u_i - u_0)$ distribution along $x/\sqrt{4\mu^2 t}$ at different instants of t.

with $\text{erf}(0) = 0$ and $\text{erf}(\infty) = 1$. After implementing the initial and boundary conditions, that is, $u(0, t) = u(\zeta = 0) = u_0$ and $u(x, 0) = u(\zeta = \infty) = u_i$, the solution is then given by

$$u(x, t) = (u_i - u_0)\,\text{erf}\left(\frac{x}{\sqrt{4\mu^2 t}}\right) + u_0$$

A plot of the ratio $(u - u_0)/(u_i - u_0)$ along the normalized distance $x/\sqrt{4\mu^2 t}$ at different values of t is shown in Figure 11.10.

EXAMPLE 11.15. For the diffusion equation given in (11.86), suppose the conditions are now given by $u(x, 0) = 0$, $\lim_{x\to\infty} u(x, t) = 0$, and $\partial u/\partial x(0, t) = H$. After substitution of $\widetilde{t} = \lambda^{-\alpha} t$, $\widetilde{x} = \lambda^{-\beta} x$ and $\widetilde{u} = \lambda^{-\gamma} u$, the same partial differential equation results as in Example 11.14, except that the initial and boundary conditions are now given by

$$\begin{aligned}\lambda^{\gamma}\widetilde{u}\left(\lambda^{\beta}\widetilde{x}, 0\right) &= 0\\ \lim_{\lambda^{\beta}\widetilde{x}\to\infty}\lambda^{\gamma}\widetilde{u}\left(\lambda^{\beta}\widetilde{x}, \lambda^{\alpha}\widetilde{t}\right) &= 0\end{aligned} \qquad \text{and} \qquad \lambda^{\gamma-\beta}\frac{\partial\widetilde{u}}{\partial\widetilde{x}}(0, \lambda^{\alpha}\widetilde{t}) = H$$

For symmetry, we need $\alpha = 2\beta$ and $\gamma = \beta$, so we can set $\alpha = 1$, $\beta = 1/2$ and $\gamma = 1/2$, yielding the following invariants $\psi = u/\sqrt{t} = \widetilde{u}/\sqrt{\widetilde{t}}$ and $\zeta = x/\sqrt{t} = \widetilde{x}/\sqrt{\widetilde{t}}$. Substituting these together with $u = \sqrt{t}\,\psi(\zeta)$ into the partial differential equation (11.86) yields the following ordinary differential equation:

$$\frac{d^2\psi}{d\zeta^2} + \frac{\zeta}{2\mu^2}\frac{d\psi}{d\zeta} - \frac{1}{2\mu^2}\psi = 0$$

subject to $\lim_{\zeta\to\infty}\psi(\zeta) = 0$ and $\dfrac{d\psi}{d\zeta}(0) = H$. The solution of this equation can be found in Example 9.2. After applying the boundary conditions, we have

$$\psi = H\zeta - \left(\frac{2H\mu}{\sqrt{\pi}}\right)\left(e^{-\zeta^2/(4\mu^2)} + \sqrt{\frac{\pi}{4\mu^2}}\,\zeta\,\text{erf}\left(\frac{\zeta}{\sqrt{4\mu^2}}\right)\right)$$

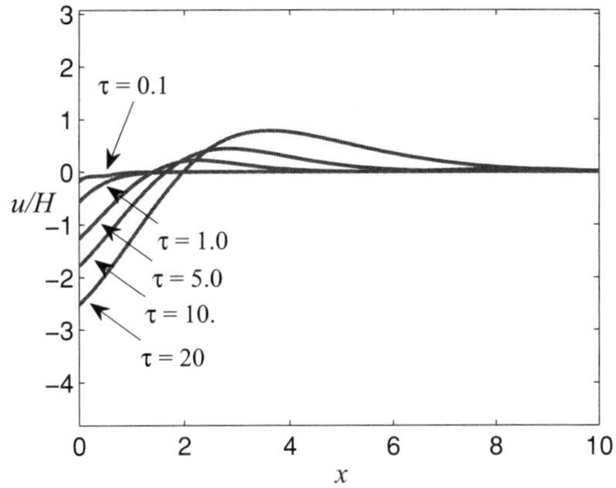

Figure 11.11. A plot of u/H distribution along $x \geq 0$ at different instants of $\tau = 4\mu^2 t$.

or in terms of the original variables,

$$u(x,t) = H\left[x\,\mathrm{erfc}\left(\frac{x}{\sqrt{4\mu^2 t}}\right) - \sqrt{\frac{4\mu^2 t}{\pi}}\exp\left(-\left[\frac{x}{\sqrt{4\mu^2 t}}\right]^2\right)\right]$$

where $\mathrm{erfc}(z) = 1 - \mathrm{erf}(z)$ is known as the **complementary error function**. A plot showing u/H as a function of x at different values of $\tau = 4\mu^2 t$ is shown in Figure 11.11.

Several textbooks consider the similarity transform methods to be limited to a few applications, citing that only special boundary conditions can match the solutions of the resulting ordinary differential equations. However, this should not preclude the fact that symmetry methods, of which similarity transforms is just one possibility, is actually quite general, especially because of its effectiveness in solving several nonlinear partial differential equations. A more serious practical issue is the complexity of the calculations involved in both finding the symmetries and solving the resulting ordinary differential equations. However, with the availability of improved symbolic equation solvers, the symmetry approaches have become more tractable, and they are gaining more acceptance as a viable approach.

11.6 EXERCISES

E11.1. Obtain the general solutions for the following reducible or reduced homogeneous differential equations:

$$6\frac{\partial^2 u}{\partial x^2} + \frac{\partial^2 u}{\partial x \partial y} - 2\frac{\partial^2 u}{\partial y^2} + 8\frac{\partial u}{\partial x} + 3\frac{\partial u}{\partial y} + 2u = 0$$

Hint: One of the factors of L is $\left(2\frac{\partial}{\partial x} - \frac{\partial}{\partial y} + 2\right)$.

E11.2. For the **second-order bivariate hyperbolic equation** given by

$$\frac{\partial^2 u}{\partial x \partial y} + a(x,y)\frac{\partial u}{\partial x} + b(x,y)\frac{\partial u}{\partial y} + c(x,y)u = 0 \tag{11.87}$$

1. Define two linear operators:
$$L_x = \frac{\partial}{\partial x} + b(x,y) \qquad \text{and} \qquad L_y = \frac{\partial}{\partial y} + a(x,y)$$
Show that (11.87) can be rearranged in two forms:
$$L_x L_y(u) = h_a(x,y)u \quad \text{or} \quad L_y L_x(u) = h_b(x,y)u$$
where
$$\begin{aligned} h_a(x,y) &= \frac{\partial a}{\partial x} + a(x,y)b(x,y) - c(x,y) \\ h_b(x,y) &= \frac{\partial b}{\partial y} + a(x,y)b(x,y) - c(x,y) \end{aligned}$$
The functions h_a and h_b are known as the **Laplace invariants** of (11.87).
2. If $h_a = h_b = 0$, (11.87) will be reducible. Solve the case where $a(x,y) = x$, $b(x,y) = y$ and $c(x,y) = 1 + xy$.
3. If only one of the Laplace invariants is zero, one can still proceed by integrating with respect to one independent variable, followed by the integration with respect to the other independent variable. For instance, if $L_x L_y u = 0$, then first solve $L_x z = 0$ followed by solving $L_y u = z$. Using this approach, obtain the general solution of (11.87) for the case where $a(x,y) = y$, $b(x,y) = xy$ and $c(x,y) = xy^2$.

E11.3. For the following initial value problem
$$\frac{\partial^2 u}{\partial x^2} + 3\frac{\partial^2 u}{\partial x \partial t} + 2\frac{\partial^2 u}{\partial t^2} = 0 \quad \text{subject to: } u(x,0) = f(x) \;\; \text{and} \;\; \frac{\partial u}{\partial t}(x,0) = g(x)$$
1. Show that this is a hyperbolic equation, and obtain the canonical variables that would transform this equation to
$$\frac{\partial^2 u}{\partial \xi \partial \eta} = 0$$
whose general solution is given by $u(x,t) = \phi(\xi(x,t)) + \psi(\eta(x,t))$.
2. Show that the same general solution can be obtained by treating the differential equation as a reducible type.
3. Plot the d'Alembert solution for the case where
$$g(x) = \text{sech}(x) \quad \text{and} \quad f(x) = \sum_{i=1}^{4} \zeta(\alpha_i, \beta_i, \gamma_i, x)$$
where
$$\zeta(\alpha, \beta, \gamma, x) = \frac{\gamma}{2}\left[1 + \tanh(\alpha x + \beta)\right] \quad \text{and}$$

i	α_i	β_i	γ_i
1	1	4	1
2	1	−4	−1
3	1	4	−0.5
4	1	10	0.5

for $-15 \le x \le 15$ and $0 \le t \le 2.5$.

E11.4. Solve the wave equation for $x \ge 0$,
$$\frac{\partial^2 u}{\partial t^2} = \frac{\partial^2 u}{\partial x^2}$$
subject to
$$\left.\begin{aligned} u(x,0) &= e^{-2x} \\ \frac{\partial u}{\partial t}(x,0) &= \frac{1}{1+e^{4x}} \end{aligned}\right\} \text{for } x \ge 0 \quad ; \quad u(0,t) = \frac{9 - e^{-4t}}{8} \qquad t \ge 0$$

(Hint: See Section K.1.2 for the solution of the wave equation that includes a Dirichlet boundary condition.)

E11.5. Obtain the displacement $u(x, t)$ for a string fixed at two ends described by the following wave equation

$$\frac{\partial^2 u}{\partial t^2} = \gamma^2 \frac{\partial^2 u}{\partial x^2}$$

subject to $u(x, 0) = f(x)$, $\partial u/\partial t(x, 0) = g(x)$ and $u(0, t) = u(L, t) = 0$. Plot the solution for the case where $L = 1$, $g(x) = 0$ and

$$f(x) = \begin{cases} 0.2x & 0 \le x \le 0.5 \\ 0.2(1-x) & 0.5 < x \le 1 \end{cases}$$

E11.6. Consider the diffusion equation with nonhomogeneous boundary conditions:

$$\frac{\partial u}{\partial t} = \alpha^2 \frac{\partial^2 u}{\partial x^2}$$

subject to

$$u(x, 0) = x + \cos(\pi x) \qquad \text{and} \qquad \begin{aligned} u(0, t) &= 1 \\ 2u(1, t) + \left.\frac{\partial u}{\partial x}\right|_{x=1} &= 1 \end{aligned}$$

1. Show that the approach in Section 11.4 can transform the problem to a homogeneous partial differential equation with homogenous boundary conditions given by

$$\frac{\partial U}{\partial t} = \alpha^2 \frac{\partial^2 U}{\partial x^2}$$

with conditions

$$U(x, 0) = \cos(\pi x) - 1 + \frac{4}{3}x \qquad \text{and} \qquad \begin{aligned} U(0, t) &= 0 \\ 2U(1, t) + \left.\frac{\partial U}{\partial x}\right|_{x=1} &= 0 \end{aligned}$$

where $u(x, t) = U(x, t) + 1 - x/3$.

2. Solve the problem and plot the solutions.

E11.7. Use the Sturm-Liouville approach given in Section 11.3.3 to show that the required weighing functions for the orthogonality of the solutions of the following eigenfunction equations are given by:

1. Legendre polynomials:

$$\frac{d}{dx}\left((1-x)^2 \frac{du}{dx}\right) + n(n+1)u = 0 \rightarrow r(x) = 1$$

2. Associated Legendre polynomials:

$$\frac{d}{dx}\left((1-x)^2 \frac{du}{dx}\right) + \left(n(n+1) - \frac{m^2}{1-x^2}\right)u = 0 \rightarrow r(x) = 1$$

3. Chebyshev polynomials:

$$\left(1-x^2\right)\frac{d^2u}{dx^2} - x\frac{du}{dx} + n^2 u = 0 \rightarrow r(x) = \frac{1}{\sqrt{1-x^2}}$$

4. Laguerre polynomials:

$$x\frac{d^2u}{dx^2} + (1-x)\frac{du}{dx} + nu = 0 \rightarrow r(x) = e^{-x}$$

5. Associated Laguerre polynomials:

$$x\frac{d^2u}{dx^2} + (k+1-x)\frac{du}{dx} + nu = 0 \rightarrow r(x) = e^{-x}x^k$$

6. Hermite polynomials:

$$\frac{d^2u}{dx^2} - 2x\frac{du}{dx} + 2n\,(n+1)\,u = 0 \rightarrow r(x) = e^{-x^2}$$

7. Spherical Bessel functions:

$$\frac{d^2}{dx^2}(xu) + \left(x - \frac{\ell(\ell+1)}{x}\right)u = 0 \rightarrow r(x) = 1$$

E11.8. Let $f(x)$ be a continuous function of x. Show that for a set of functions

$$\phi_n(x) = \cos\left(n\, f(x)\right)$$

a Sturm-Liouville differential eigenfunction equation (in expanded form) for these functions can be constructed as follows

$$\left(\frac{df}{dx}\right)^{-2}\frac{d^2\phi_n}{dx^2} - \left(\frac{d^2f}{dx^2}\right)\left(\frac{df}{dx}\right)^{-3}\frac{d\phi_n}{dx} + n^2\phi_n = 0$$

Using this result, find a series approximation based on $\phi_n(x)$ using $f(x) = x^2$ to approximate

$$h(x) = \begin{cases} (x) & \text{for } 0.25 \leq x \leq 0.5 \\ (1-x) & \text{for } 0.5 < x \leq 0.75 \end{cases}$$

E11.9. A well-known equation for financial option pricing is the **Black-Scholes equation** for $a \leq x \leq b$ and $0 \leq t \leq T$ given by

$$\frac{\partial u}{\partial t} = \frac{1}{2}\sigma^2 x^2 \frac{\partial^2 u}{\partial x^2} + r\left(x\frac{\partial u}{\partial x} - u\right)$$

subject to

$$u(a,t) = u(b,t) = 0 \qquad \text{and} \qquad u(x,0) = u_f(x)$$

where $u(x,t)$ is the value of the option, x is the value of the underlying asset, t is the time from expiry of option, r is the risk-free interest rate, σ is the volatility of the asset, and $u_f(x)$ is the final payoff. The boundaries a and b are barriers under which the option becomes worthless. Obtain the solution and plots for $r = 1$, $\sigma = 2$, $a = 0.5$, $b = 2$ and

$$u_f(x) = \begin{cases} 10x - 5 & \text{if } 0.5 \leq x \leq 1 \\ -5x + 10 & \text{if } 1 \leq x \leq 2 \\ 0 & \text{otherwise} \end{cases}$$

(Hint: Using separation of variables, an eigenvalue problem that is a special case of Example 11.9 will result.)

E11.10. Obtain the solution to the following convective-diffusion equation:

$$\mathcal{D}\frac{\partial^2 u}{\partial x^2} + \frac{\partial u}{\partial x} = \frac{\partial u}{\partial t}$$

subject to:

$$\begin{aligned} u(0,t) &= \sin(\omega t) \\ u(1,t) &= 0 \end{aligned} \qquad \text{and} \qquad u(x,0) = 0$$

Plot the solution for the case where $\mathcal{D} = 0.1$ and $\omega = 2\pi$, $0 \leq t \leq 1$.

E11.11. For the reaction-diffusion equation with first-order reaction given by

$$\frac{\partial u}{\partial t} = \mathcal{D}\frac{\partial^2 u}{\partial x^2} - ku$$

show that this can be converted to a simple diffusion problem for $q(x,t)$ by using a simple transformation $u(x,t) = e^{-kt}q(x,t)$.

E11.12. Consider the **Fokker-Planck equation** with constant coefficients given by

$$\mathcal{D}\frac{\partial^2 u}{\partial x^2} = \frac{\partial u}{\partial t} + \mu\frac{\partial u}{\partial x}$$

where $\mathcal{D}$ and μ are constant diffusion and drift coefficients, respectively. Let $u = e^{Ax}e^{Bt}q(x,t)$. Find the values of A and B that would transform the original equation into a simple diffusion equation for $q(x,t)$, that is, for some α,

$$\frac{\partial q}{\partial t} = \alpha^2\frac{\partial^2 q}{\partial x^2}$$

E11.13. The **Nusselt problem** models the temperature for a laminar flow of a fluid through a pipe. After normalization, we have

$$\frac{\partial u}{\partial z} = \left(\frac{\partial^2 u}{\partial r^2} + \frac{1}{r}\frac{\partial u}{\partial r}\right) \qquad 0 \le z \le L \quad \text{and} \quad 0 \le r \le 1$$

subject to

$$u(1,z) = u_W \qquad \frac{\partial u}{\partial r}(0,z) = 0 \quad \text{and} \quad u(r,0) = u_0$$

Solve this equation using the method of separation of variables. Plot the solution for the case with $u_W = 5$ and $u_0 = 10$.

E11.14. Consider the potential equation for the temperature u of a sphere of radius. Assume cylindrical symmetry for the surface temperature, $u(1,\theta,\phi) = f(\theta)$; then we have only dependence with r and θ. Thus

$$\nabla^2 u = 0 \quad \rightarrow \quad r^2\frac{\partial^2 u}{\partial r^2} + 2r\frac{\partial u}{\partial r} + \frac{\partial^2 u}{\partial \theta^2} + \frac{\cos\theta}{\sin\theta}\frac{\partial u}{\partial \theta} = 0$$

where $0 \le \theta \le \pi$ and $0 \le r \le 1$.

1. Let $u = R(r)\Theta(\theta)$; then show that separation of variables will yield the following ordinary differential equations:

$$r^2\frac{d^2 R}{dr^2} + 2r\frac{dR}{dr} - \mu R = 0 \tag{11.88}$$

$$\frac{d^2\Theta}{d\theta^2} + \frac{\cos\theta}{\sin\theta}\frac{d\Theta}{d\theta} + \mu\Theta = 0 \tag{11.89}$$

2. Letting $x = \cos\theta$ and $\mu = n(n+1)$, show that (11.89) reduces to the Legendre equation (cf. Section 9.2), that is,

$$(1-x^2)\frac{d^2\Theta}{dx^2} - 2x\frac{d\Theta}{dx} + n(n+1)\Theta = 0$$

whose solutions (considering $\mathcal{Q}_n(0) = \pm\infty$) are

$$\Theta_n(\theta) = \mathcal{P}_n(\cos\theta)$$

where $\mathcal{P}_n$ is the Legendre polynomial of order n defined in (I.31).

3. Using $\mu = n(n+1)$, solve for $R(r)$ under the condition that $R(0)$ is finite.
4. Obtain the solution for $u(r,\theta)$ under the additional boundary conditions $u(1,\theta) = 1 - \cos^2(\theta)$, $0 \le \theta \le \pi$ and plot the solution.

E11.15. Consider the heat conduction in a solid sphere of radius $R = 1$ initially at temperature $u(r,0) = f(r)$ that is suddenly immersed in a constant bath temperature of u_s. Assume that the heat transfer is symmetric around the origin, that is, $u = u(r,t)$ then the thermal diffusion equation is given by

$$\frac{\partial u}{\partial t} = \alpha \nabla^2 u = \alpha \left(\frac{\partial^2 u}{\partial r^2} + \frac{2}{r}\frac{\partial u}{\partial r} \right) \tag{11.90}$$

subject to

$$u(r,0) = f(r) \qquad \left.\frac{\partial u}{\partial r}\right|_{r=R} = h\left(u_s - u\right)\Big|_{r=R} \qquad \left.\frac{\partial u}{\partial r}\right|_{r=0} = 0$$

1. Show that with $u = q/r$, (11.90) can be recast in terms of q to yield

$$\frac{\partial q}{\partial t} = \alpha \frac{\partial^2 q}{\partial r^2}$$

2. Solve for $u(r,t)$ for the case where $f(r) = u_0 = 0, u_S = 1, \alpha = 1$ and $h = 5$ and obtain $u(r = 0, t)$.

E11.16. Use separation of variables to derive the solution of

$$\frac{\partial^2 U}{\partial x^2} + \frac{\partial^2 U}{\partial y^2} = 0$$

subject to $U(0,y) = U(1,y) = 0$, $U(x,0) = U(x,1) = \dfrac{x(x-1)}{2}$ to show that

$$U(x,y) = -\frac{4}{\pi^3}\sum_{n=1,3,5,\ldots}^{\infty} \left(\frac{\sin(n\pi x)}{n^3}\right)\left(\cosh(n\pi y) + \frac{1-\cosh(n\pi)}{\sinh(n\pi)}\sinh(n\pi y)\right)$$

E11.17. Solve the following nonhomogeneous partial differential equation using the eigenfunction expansion approach:

$$\frac{\partial u}{\partial t} = \frac{\partial^2 u}{\partial x^2} - 3e^{x-2t} \qquad 0 \le x \le 1 \quad \text{and} \quad t \ge 0$$

subject to

$$u(0,t) = e^{-2t} \quad , \quad u(1,t) = e^{1-2t} \quad \text{and} \quad u(x,0) = e^x$$

Plot the solution and compare with the exact solution $u(x,t) = e^{x-2t}$.
Hint: First, use the approach given in Section 11.4 to obtain a splitting of the solution as $u(x,t) = U(x,t) + S(x,t)$, such that

$$\frac{\partial U}{\partial t} = \frac{\partial^2 U}{\partial x^2} - 3e^{x-2t} - \left(\frac{\partial S}{\partial t} - \frac{\partial^2 S}{\partial x^2}\right)$$

subject to

$$U(0,t) = 0 \quad , \quad U(1,t) = 0 \quad \text{and} \quad U(x,0) = e^x - S(x,0)$$

Then show that the eigenfunction approach solution should yield $U(x,t) = \sum_{n=1}^{\infty} T_n(t)\sin(n\pi t)$ where $T_n(t)$ is the solution of the ordinary differential equation:

$$\frac{dT_n}{dt} + (n\pi)^2 T_n = \xi_n e^{-2t} \quad \text{subject to } T_n(0) = p_n$$

where

$$p_n = \frac{2}{n\pi}\frac{1}{\left[(n\pi)^2 + 1\right]}\left((-1)^n e - 1\right)$$

$$\xi_n = (p_n)\left[(n\pi)^2 - 2\right]$$

E11.18. Another approach to the solution of the Poisson equation

$$\frac{\partial^2 u}{\partial x^2} + \frac{\partial^2 u}{\partial y^2} = f(x, y)$$

with homogeneous boundary conditions: $u(x, 0) = u(x, 1) = u(0, y) = u(1, y) = 0$, is to use eigenfunctions of the form

$$\phi_{mn}(x, y) = \sin(n\pi x)\sin(m\pi y)$$

Thus we can approximate both $u(x, y)$ and $f(x, y)$ by

$$u(x, y) = \sum_{m=1}^{\infty}\sum_{n=1}^{\infty} \alpha_{mn}\phi_{mn}(x, y) \qquad \text{and} \qquad f(x, y) = \sum_{m=1}^{\infty}\sum_{n=1}^{\infty} \beta_{mn}\phi_{mn}(x, y)$$

1. Using these approximations, obtain the equations needed to evaluate the coefficients α_{mn} and β_{mn}.
2. Apply these approximations to the case where $f(x, y) = -1$ and compare them to the results given in Example 11.12.

E11.19. Consider the case of diffusion with linearly varying diffusivity given by

$$x\frac{\partial^2 u}{\partial x^2} = \frac{\partial u}{\partial t}$$

Use the similarity transformation approach to solve the equation and obtain a plot for $x \geq 0$ and $t \geq 0$ based on the following conditions: $u(x, 0) = 1$, $x \geq 0$ and $u(1, t) = 0$.

E11.20. The **chemical vapor deposition reaction** (CVD) that is fed by a diffusion-limited laminar flow between two parallel plates can be modeled approximately by[7]

$$\frac{\partial u}{\partial x} = \frac{Q}{y}\frac{\partial^2 u}{\partial y^2}$$

subject to: $u(x, 0) = 0$, $u(0, y) = u_0$, $\lim_{y\to\infty} u(x, y) = u_0$ and $\lim_{x\to\infty} u(x, y) = 0$, where u is the metal concentration, y is the distance from wall, x is the distance from the entrance along the laminar flow, and $Q = \dfrac{\mathcal{D}B}{2v_{\max}}$, with $\mathcal{D}$, $v_{\max}$, and B as the diffusion coefficient, maximum velocity, and midpoint distance between the two parallel plates, respectively. The solution will be valid only near a plate because the approximation has been obtained via linearization around $y = 0$. Obtain and plot the solution using a similarity transformation approach, with $Q = 10$.

[7] Based on an example in R. G. Rice and D. D. Do, *Applied Mathematics and Modeling for Chemical Engineers*, John Wiley & Sons, New York, 1995, pp. 415–420.

12 Integral Transform Methods

In this chapter, we discuss the integral transform methods for solving linear partial differential equations. Although there are several types of transforms available, the methods that are most often used are the Laplace transform methods and the Fourier transform methods. Basically, an integral transform is used to map the differential equation domain to another domain in which one of the dimensions is reduced from derivative operations to algebraic operations. This means that if we begin with an ordinary differential equation, the integral transform will map the equation to an algebraic equation (cf. Section 6.7). For a 2D problem, the integral transforms should map the partial differential equation to an ordinary differential equation, and so on.

We begin in Section 12.1 with a very brief introduction to general integral transforms. Then, in Section 12.2, we discuss the details of Fourier transforms, its definition, some particular examples, and then the properties. Surprisingly, the initial developments of Fourier transforms are unable to be applied to some of the most useful functions, including step function and sinusoidal functions. Although there were several ad hoc approaches to overcome these problems, it was not until the introduction of the theory of distributions that the various ad hoc approaches were unified and gained a solid mathematical grounding. This theory allows the extension of classical Fourier transform to handle the problematic functions. We have placed the discussion of the theory of distributions in the appendix, and we include the details of the Fourier transform of the various "difficult" functions in the same appendix. Some of properties of Fourier transforms are then explored and collected in Section 12.2.2. We have kept these properties to a minimum, with a focus on solving differential equations. Then, in Section 12.3, we apply Fourier transform to solve partial differential equations. As some authors have noted, Fourier transform methods are most useful in handling infinite domains.

Next, starting in Section 12.4, we switch our attention to the Laplace transforms. We view Laplace transforms as a special case of Fourier transforms. However, with Laplace transform, one can apply the technique on dimensions that are semi-infinite, such as time variables. Thus there are strong similarities between the Laplace and Fourier transforms of functions as well as in their properties. There are also significant differences. For one, the handling of several functions such as step, sinusoidal, exponential, Bessel, and Dirac delta functions are simpler to obtain. The inversion of Laplace transforms, however, can be quite complicated. In several cases, using a table of Laplace transforms, together with separation to partial fractions, are sufficient to obtain the inverse maps. In general, however, one may need to resort to the theory of residues. We have included a review of the residue methods in the appendix. Then, in Section 12.5, we apply Laplace transforms to solve some partial differential equations. Finally, in Section 12.6, we include a brief section to show how another technique known as **method of images** can be used to extend the use of either Fourier or Laplace methods to semi-infinite or bounded domains.

12.1 General Integral Transforms

Along with Fourier and Laplace transforms, other integral transforms include Fourier-Cosine, Fourier-Sine, Mellin, and Hankel transforms. In this section, we discuss some general results about integral transforms before we focus on Fourier and Laplace transforms. For a review of complex integration, we have included a brief discussion in the appendix as Section L.2 to cover the theorems and methods needed for computing the transforms.

Definition 12.1. *For a given function $f(x)$, the integral*

$$\mathcal{I}_{K,a,b}\left[f(x)\right] = \int_a^b K(p,x)\, f(x)\, dx \tag{12.1}$$

is called the **integral transform** *of f, where $K(p,x)$ is called the* **kernel** *of the transform and p is called the* **transform variable** *of the integral transform. The limits of integration a and b can be finite or infinite.*

A list of different useful integral transforms is given in Table 12.1. For our purposes, we take integral transforms simply as a mapping of the original function based on variable x to another function based on a new variable p. The expectation is that in the new domain, convenient properties are obtained, by which the analysis becomes more manageable. For instance, with Laplace transforms, the original problem could be a linear time-invariant linear differential equations. After taking Laplace transforms of these equations, the differential equations are replaced by algebraic equations.

Based on the properties of integrals, integral transforms immediately satisfy the linearity property, that is, with constants α and β,

$$\begin{aligned}
\mathcal{I}_{K,a,b}\left[\alpha f(x) + \beta g(x)\right] &= \int_a^b K(p,x)\left(\alpha f(x) + \beta g(x)\right)\, dx \\
&= \alpha \int_a^b K(p,x)\, f(x)\, dx + \beta \int_a^b K(p,x)\, g(x)\, dx \\
&= \alpha \mathcal{I}_{K,a,b}\left[f(x)\right] + \beta \mathcal{I}_{K,a,b}\left[g(x)\right]
\end{aligned}$$

Once the analytical manipulation in the transform domain is finished, an inverse transformation is needed to obtain a solution in the original domain. Thus another important criteria for the choice of integral transforms is how easy the inverse transformations can be evaluated. In most integral transforms, such as those given

Table 12.1. *Various integral transforms*

Name	Transform
Mellin	$\mathcal{M}\left[f(x)\right] = \int_0^{\infty} f(x)x^{p-1}\, dx$
Fourier Cosine	$\mathcal{F}_c\left[f(x)\right] = \sqrt{\frac{2}{\pi}} \int_0^{\infty} f(x)\cos(xp)\, dx$
Fourier Sine	$\mathcal{F}_s\left[f(x)\right] = \sqrt{\frac{2}{\pi}} \int_0^{\infty} f(x)\sin(xp)\, dx$
Fourier	$\mathcal{F}\left[f(x)\right] = \int_{-\infty}^{\infty} f(x)e^{-ixp}\, dx$
Laplace	$\mathcal{L}\left[f(x)\right] = \int_0^{\infty} f(x)e^{-xp}\, dx$
Hankel transform	$\mathcal{H}_\nu\left[f(x)\right] = \int_0^{\infty} f(x)xJ_\nu(xp)\, dx$

in Table 12.1, the inverse transformation is chosen to be another integral transform, that is,

$$f(x) = \int_{\widehat{a}}^{\widehat{b}} \left(\mathcal{I}_{K,a,b}\left[f\right]\right) \widehat{K}(x, p)\, dp \tag{12.2}$$

where $\widehat{K}(x, p)$ is the kernel of the inverse integral transform. In general, $a \neq \widehat{a}$, $b \neq \widehat{b}$, and $K(x, p) \neq \widehat{K}(x, p)$. In fact, in some cases, a and b could be real numbers, whereas $\widehat{a}$ and $\widehat{b}$ are complex numbers. In case $K(x, p) = \widehat{K}(x, p)$, then we call $K(x, p)$ a **Fourier kernel**. For example, the kernels of the Fourier-cosine transforms and Fourier-sine transforms are Fourier kernels.[1]

12.2 Fourier Transforms

In this section, we first derive the Fourier transform and then obtain some important properties. Later, we apply the operation of Fourier transform to the solution of partial differential equations.

12.2.1 Definition and Notations

The Fourier series of a periodic function $f(x)$ with period T is

$$f(x) = a_0 + \sum_{\ell=1}^{\infty} \left[a_\ell \cos\left(\frac{2\pi}{T}\ell x\right) + b_\ell \sin\left(\frac{2\pi}{T}\ell x\right)\right] \tag{12.3}$$

[1] Ironically, the kernel of the Fourier transform, $K(x, p) = e^{-ixp}$, is not a Fourier kernel; that is, the kernel of the inverse Fourier transformation is $\widehat{K}(x, p) = e^{ixp}/2\pi$.

where a_ℓ and b_ℓ are the Fourier coefficients given by

$$a_\ell = \frac{\displaystyle\int_{-T/2}^{T/2} f(t)\cos\left(2\pi\ell t/T\right)dt}{\displaystyle\int_{-T/2}^{T/2} \cos^2\left(2\pi\ell t/T\right)dt} \qquad \text{and} \qquad b_\ell = \frac{\displaystyle\int_{-T/2}^{T/2} f(t)\sin\left(2\pi\ell t/T\right)dt}{\displaystyle\int_{-T/2}^{T/2} \sin^2\left(2\pi\ell t/T\right)dt}$$

With $T = 2\pi\tau$, the coefficients become

$$\begin{aligned} a_0 &= \frac{1}{2\pi\tau}\int_{-\pi\tau}^{\pi\tau} f(t)dt \\ a_\ell &= \frac{1}{\pi\tau}\int_{-\pi\tau}^{\pi\tau} f(t)\cos\left(\frac{\ell t}{\tau}\right)dt \\ b_\ell &= \frac{1}{\pi\tau}\int_{-\pi\tau}^{\pi\tau} f(t)\sin\left(\frac{\ell t}{\tau}\right)dt \end{aligned} \tag{12.4}$$

An efficient way to solve for the coefficients is through the use of fast Fourier transforms (FFT). The FFT method is described in Section L.1 as an appendix. Substituting (12.4) into (12.3),

$$f(x) = \frac{1}{2\pi\tau}\int_{-\pi\tau}^{\pi\tau} f(t)dt + \sum_{\ell=1}^{\infty}\frac{1}{\pi\tau}\int_{-\pi\tau}^{\pi\tau} f(t)\cos\left(\frac{\ell}{\tau}(x-t)\right)dt$$

Letting $(1/\tau) = \Delta\omega$ and $\ell/\tau = \omega$,

$$f(x) = \frac{\Delta\omega}{2\pi}\int_{-\pi\tau}^{\pi\tau} f(t)dt + \frac{1}{\pi}\sum_{\ell=1}^{\infty}\left[\int_{-\pi/\Delta\omega}^{\pi/\Delta\omega} f(t)\cos\left(\omega(x-t)\right)dt\right]\Delta\omega \tag{12.5}$$

By taking the limit of $f(x)$ as $\Delta\omega$ approaches zero, the summation in (12.5) becomes an integral. Also, with the assumption that $(\int_{-\infty}^{\infty} f(t)dt < \infty)$, the first term in (12.5) will become zero. We end up with the formula known as the **Fourier integral equation**,

$$f(x) = \frac{1}{\pi}\int_0^{\infty}\int_{-\infty}^{\infty} f(t)\cos\left(\omega(x-t)\right)\,dt\,d\omega \tag{12.6}$$

Assuming for now that (12.6) is valid for the given function $f(x)$, we could proceed further by obtaining an alternative form of (12.6) based on Euler's identity. Taking the integral of Euler's identity,

$$\begin{aligned} \int_{-m}^{m} e^{i(x-t)\omega}d\omega &= \int_{-m}^{m}\cos\left((x-t)\omega\right)d\omega + i\int_{-m}^{m}\sin\left((x-t)\omega\right)d\omega \\ &= 2\int_0^{m}\cos\left((x-t)\omega\right)d\omega \end{aligned}$$

or

$$\int_0^{m}\cos\left((x-t)\omega\right)d\omega = \frac{1}{2}\int_{-m}^{m} e^{i(x-t)\omega}d\omega \tag{12.7}$$

Substituting (12.7) into (12.6) while switching the order of integration,

$$\begin{aligned} f(x) &= \frac{1}{\pi}\int_{-\infty}^{\infty} f(t)\left[\int_{0}^{\infty} \cos\left(\omega(x-t)\right)\,d\omega\right] dt \\ &= \frac{1}{2\pi}\int_{-\infty}^{\infty} f(t)\left[\int_{-\infty}^{\infty} e^{i(x-t)\omega}\,d\omega\right] dt \\ &= \frac{1}{2\pi}\int_{-\infty}^{\infty} e^{ix\omega}\left[\int_{-\infty}^{\infty} f(t)e^{-it\omega}\,dt\right] d\omega \end{aligned} \tag{12.8}$$

Equation (12.8) can now be deconstructed to yield the Fourier transform pair.

Definition 12.2. *For a given function $f(x)$, the operator $\mathcal{F}$ acting on $f(x)$ given by*

$$F(\omega) = \mathcal{F}[f] = \int_{-\infty}^{\infty} f(t)e^{-it\omega}\,dt \tag{12.9}$$

is called the **Fourier transform of** $f(t)$. *For a given function $F(\omega)$, the operator $\mathcal{F}^{-1}$ acting on $F(\omega)$ given by*

$$f(x) = \mathcal{F}^{-1}[F] = \frac{1}{2\pi}\int_{-\infty}^{\infty} F(\omega)e^{ix\omega}\,d\omega \tag{12.10}$$

is called the **inverse Fourier transform of** $F(\omega)$.

Thus the kernel of the Fourier transform is $K(x,\omega) = e^{-ix\omega}$, whereas the inverse kernel is given by $\widehat{K}(x,\omega) = (1/2\pi)e^{ix\omega}$; that is, the signs in the exponential power in both kernels are opposites of each other.[2]

The Fourier integral equation, (12.6), is not valid for all types of functions. However, (12.6) is valid for functions that satisfy the Dirichlet conditions and are integrable, that is,

$$\int_{-\infty}^{\infty} \left|f(t)\right| dt \leq \infty \tag{12.11}$$

which in fact generalizes (12.6) to the following formulation:

$$\frac{1}{2}\left[f\left(x^{+}\right) + f\left(x^{-}\right)\right] = \frac{1}{\pi}\int_{0}^{\infty}\int_{-\infty}^{\infty} f(t)\cos\left(\omega(x-t)\right)\,dt\,d\omega \tag{12.12}$$

Details of Dirichlet conditions and the corresponding Fourier integral theorem, Theorem L.8, can be found in an appendix of this chapter, under Section L.3. Included in the discussion is the proof that when Dirichlet and integrability conditions are satisfied, the interchangeability of integration sequence used to derive (12.8) is indeed allowed.

The evaluation of the Fourier and inverse Fourier transforms usually requires the rules and methods of integration of complex functions. A brief (but relatively extensive) review of complex functions and integration methods are included in an appendix under Section L.2. Included in Section L.2 are the theory of residues to

[2] There are several versions for the definition of Fourier transform. One version switches the sign of the exponential. Other versions have different coefficients. Because of the existence of different definitions, it is crucial to always determine which definition was chosen before using any table of Fourier transforms.

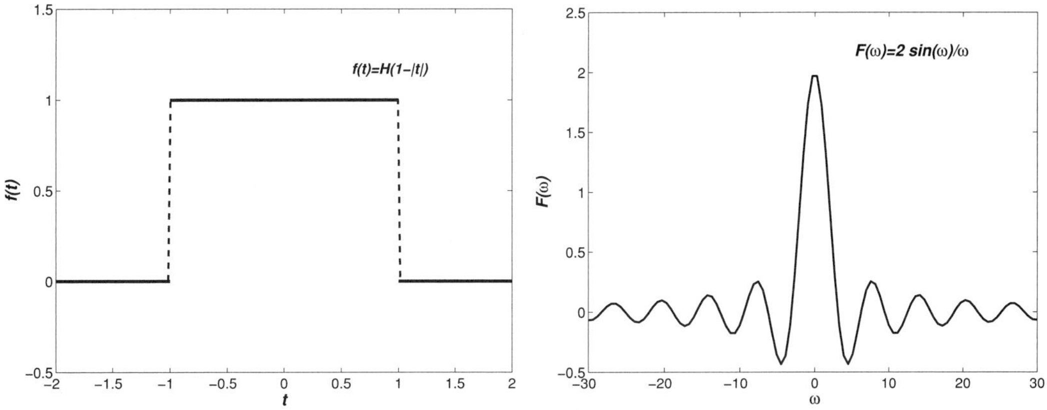

Figure 12.1. A plot of the square pulse function and its corresponding Fourier transform.

evaluate contour integrals and extensions needed to handle integrals with complex paths of infinite limits, regions with infinite number of poles, and functions with branch cuts. Examples addressing Fourier transforms are also included.

EXAMPLE 12.1. Consider the square pulse given by

$$f(t) = \mathcal{H}\left(a - |t|\right) = \begin{cases} 0 & \text{if } |t| > a \\ 1 & \text{if } |t| \leq a \end{cases} \tag{12.13}$$

where $a > 0$ is a constant. Then the fourier transform is given by

$$\begin{aligned} \mathcal{F}\left[\mathcal{H}\left(a - |t|\right)\right] &= \int_{-\infty}^{\infty} \mathcal{H}\left(a - |t|\right) e^{-i\omega t}\, dt \\ &= \int_{-a}^{a} e^{-i\omega t}\, dt \;=\; 2\frac{\sin(a\omega)}{\omega} \end{aligned} \tag{12.14}$$

Plots of (12.13) and (12.14) are shown in Figure 12.1.

The inverse Fourier transform of the square pulse is given by

$$\begin{aligned} \mathcal{F}^{-1}\left[\mathcal{H}\left(b - |\omega|\right)\right] &= \frac{1}{2\pi}\int_{-\infty}^{\infty} \mathcal{H}\left(b - |\omega|\right) e^{i\omega t} d\omega \\ &= \frac{1}{2\pi}\int_{-b}^{b} e^{i\omega t} d\omega = \frac{\sin(bt)}{\pi t} \end{aligned} \tag{12.15}$$

EXAMPLE 12.2. Consider the function, $F(\omega) = e^{-|\omega|}$. Using the definition given in (12.10), the inverse Fourier transform is given by

$$\begin{aligned} \mathcal{F}^{-1}\left[e^{-|\omega|}\right] &= \frac{1}{2\pi}\int_{-\infty}^{\infty} e^{-|\omega|} e^{i\omega x} d\omega \\ &= \frac{1}{2\pi}\left[\int_{-\infty}^{0} e^{(ix+1)\omega} d\omega + \int_{0}^{\infty} e^{(ix-1)\omega} d\omega\right] \\ &= \frac{1}{2\pi}\left(\frac{1}{ix+1} - \frac{1}{ix-1}\right) = \frac{1}{\pi}\frac{1}{x^2+1} \end{aligned} \tag{12.16}$$

Unfortunately, some useful and important functions, such as the unit step function, sine, cosine, and some exponential functions, do not satisfy the required conditions of integrability. The definition of Fourier transforms needed to be expanded to accommodate these important functions.

Before the 1950s, the problematic functions were treated on a case-by-case basis using arguments that involve taking limits of some parameters to approach either 0 or infinity. Many of the results using these approaches were verified by successful physical applications and practice, especially in the fields of physics and engineering. As a result of their successes, these approaches remain valid and acceptable to most practicioners, even at present time. Nonetheless, these approaches lacked crucial mathematical rigor and still lacked generality in their approach. The biggest contention was centered on the fact that the "Dirac delta function" did not satisfy the conditions required of functions.

With the introduction of distribution theory by Laurent Schwartz, most of the mathematical rigor needed was introduced. The theory allowed the definition of the Dirac delta function as a new object called a **distribution**. A subset of distributions called **tempered distributions** was subsequently constructed. Using tempered distributions, a generalized Fourier transform was formulated, and this allowed a general approach to handle Fourier transforms of functions such as sines, cosines, and so forth. Fortunately, the previous approaches using limiting arguments were proven to be equivalent to the methods of distribution theory. Thus the previous methods were validated with more solid mathematical basis. More importantly, a general approach had become available for problems in which the limiting argument may be difficult to perform or justify.

A short introduction to distribution theory and delta distributions is included in Section L.4 as an appendix. Included in Section L.4 are general properties and operations of distributions and specific properties and operations of delta distributions. A discussion of tempered distribution and its application to the formulation of generalized Fourier transform continues in Section L.5 as another appendix. Included in Section L.5 are the definition of the generalized Fourier transforms; the evaluation of Fourier transforms of sines, cosines, unit step functions, and delta distributions using the methods of generalized Fourier transforms; and additional properties such as the Fourier transforms of integrals.

With the formulation of generalized Fourier transforms, we specifically refer to the original definitions given in (12.9) and (12.10) as the classic Fourier transform and the classic inverse Fourier transform, respectively. As shown in Section L.5, the computations used for Fourier transforms of tempered distributions still need both these definitions. Thus we use the term "Fourier transform" to imply generalized Fourier transform, because the classic Fourier transform is already included in the generalized forms. This means that the integral formulas of the classic Fourier and classic inverse Fourier transforms are used for most evaluations, until a problem with integrability or the presence of delta distributions occurs, at which point the methods of generalized Fourier transforms using tempered distributions are applied.

12.2.2 Properties of Fourier Transforms

We now list some of the important properties of Fourier transforms and inverse Fourier transforms. In the following, we use $F(\omega)$ to mean $\mathcal{F}[f(t)]$:

1. **Linearity**. Let a and b be constants,

$$\mathcal{F}\left[af(t)+bg(t)\right] = a\mathcal{F}\left[f(t)\right]+b\mathcal{F}\left[g(t)\right] \tag{12.17}$$

$$\mathcal{F}^{-1}\left[aF(\omega)+bG(\omega)\right] = a\mathcal{F}^{-1}\left[F(\omega)\right]+b\mathcal{F}^{-1}\left[G(\omega)\right] \tag{12.18}$$

2. **Shifting.**

$$\begin{aligned}\mathcal{F}\left[f(t-a)\right] &= \int_{-\infty}^{\infty} f(t-a)e^{-i\omega t}dt \\ &= \int_{-\infty}^{\infty} f(\tau)e^{-i\omega(\tau+a)}d\tau \qquad (\tau = t-a) \\ &= e^{-ia\omega}\int_{-\infty}^{\infty} f(\tau)e^{-i\omega\tau}d\tau \;=\; e^{-ia\omega}\mathcal{F}\left[f(t)\right]\end{aligned} \tag{12.19}$$

$$\begin{aligned}\mathcal{F}^{-1}\left[F(\omega-b)\right] &= \frac{1}{2\pi}\int_{-\infty}^{\infty} F(\omega-b)e^{i\omega t}d\omega \\ &= \frac{1}{2\pi}\int_{-\infty}^{\infty} F(\nu)e^{i(\nu+b)t}d\nu \qquad (\nu = \omega-b) \\ &= e^{ibt}\frac{1}{2\pi}\int_{-\infty}^{\infty} F(\nu)e^{i\nu t}d\nu \;=\; e^{ibt}\mathcal{F}^{-1}\left[F(\omega)\right]\end{aligned} \tag{12.20}$$

3. **Scaling**. Let $a > 0$,

$$\begin{aligned}\mathcal{F}\left[f(at)\right] &= \int_{-\infty}^{\infty} f(at)e^{-i\omega t}dt \\ &= \frac{1}{a}\int_{-\infty}^{\infty} f(\tau)e^{-i\omega(\tau/a)}d\tau \qquad (\tau = at) \\ &= \frac{1}{a}\int_{-\infty}^{\infty} f(\tau)e^{-i(\omega/a)\tau}d\tau = \frac{1}{a}F\left(\frac{\omega}{a}\right)\end{aligned}$$

For $a < 0$,

$$\begin{aligned}\mathcal{F}\left[f(at)\right] &= \int_{-\infty}^{\infty} f(at)e^{-i\omega t}dt \\ &= \frac{1}{a}\int_{\infty}^{-infty} f(\tau)e^{-i\omega(\tau/a)}d\tau \qquad (\tau = at) \\ &= -\frac{1}{a}\int_{-\infty}^{\infty} f(\tau)e^{-i(\omega/a)\tau}d\tau = -\frac{1}{a}F\left(\frac{\omega}{a}\right)\end{aligned}$$

Combining both results,

$$\mathcal{F}\left[f(at)\right] = \frac{1}{|a|}F\left(\frac{\omega}{a}\right) \tag{12.21}$$

In particular, with $a = -1$, we have

$$\mathcal{F}\left[f(-t)\right] = F(-\omega) \tag{12.22}$$

4. **Derivatives.**

$$\begin{aligned}
\mathcal{F}\left[\frac{df}{dt}\right] &= \int_{-\infty}^{\infty} \frac{df}{dt} e^{-i\omega t} dt \\
&= f(t)e^{-i\omega t}\Big]_{-\infty}^{\infty} + i\omega \int_{-\infty}^{\infty} f(t)e^{-i\omega t} dt \quad \text{(after integration by parts)} \\
&= i\omega \mathcal{F}\left[f(t)\right] \quad \text{(after setting values at limits to zero)} \qquad (12.23)
\end{aligned}$$

or in general, we can show by induction that

$$\mathcal{F}\left[\frac{d^n f}{dt^n}\right] = (i\omega)^n \mathcal{F}\left[f(t)\right] \qquad (12.24)$$

For the inverse Fourier transform, we have

$$\begin{aligned}
\mathcal{F}^{-1}\left[\frac{dF}{d\omega}\right] &= \frac{1}{2\pi} \int_{-\infty}^{\infty} \frac{dF}{d\omega} e^{i\omega t} d\omega \\
&= \frac{1}{2\pi} \left(F(\omega)e^{i\omega t}\Big]_{-\infty}^{\infty} - it \int_{-\infty}^{\infty} F(\omega)e^{i\omega t} d\omega \right) \\
&= -it\mathcal{F}^{-1}\left[F(\omega)\right] \\
&= (-it)f(t) \qquad (12.25)
\end{aligned}$$

or in general,

$$\mathcal{F}^{-1}\left[\frac{d^n F}{d\omega^n}\right] = (-it)^n f(t) \qquad (12.26)$$

5. **Integrals.**

$$\mathcal{F}\left[\int_{-\infty}^{t} f(\eta) d\eta\right] = \frac{1}{i\omega} F(\omega) + \pi\delta(\omega)F(0) \qquad (12.27)$$

The derivation of (12.27) is given in the appendix under Section L.5.2 and requires the use of generalized Fourier transforms.

6. **Multiplication and Convolution.** For any two functions $f(\tau)$ and $g(\tau)$, the convolution operation, denoted by '$*$', is defined as

$$\text{Convolution}(f, g) = \left[f * g\right](t) = \int_{-\infty}^{\infty} f(t-\tau)g(\tau)d\tau d\tau \qquad (12.28)$$

The Fourier transform of $\left[f * g\right](t)$ is

$$\mathcal{F}\Big[[f * g](t)\Big] = \int_{-\infty}^{\infty} e^{-i\omega t} \int_{-\infty}^{\infty} f(t-\tau)g(\tau)d\tau dt$$

Let $\eta = t - \tau$, then

$$\begin{aligned}
\mathcal{F}\Big[[f * g](t)\Big] &= \int_{-\infty}^{\infty} e^{-i\omega(\eta+\tau)} \int_{-\infty}^{\infty} f(\eta)g(\tau)d\tau d\eta \\
&= \int_{-\infty}^{\infty} e^{-i\omega\eta} f(\eta)d\eta \int_{-\infty}^{\infty} e^{-i\omega\tau} g(\tau)d\tau \\
&= \mathcal{F}\Big[f(\eta)\Big]\mathcal{F}\Big[g(\tau)\Big] \\
&= F(\omega)G(\omega) \qquad (12.29)
\end{aligned}$$

Note that the convolution operation is commutative, that is, $[f * g](t) = [g * f](t)$.

Going the other way, we can obtain the inverse Fourier transforms of convolutions.

$$\begin{aligned}
\mathcal{F}^{-1}\Big[[F * G](\omega)\Big] &= \frac{1}{2\pi}\int_{-\infty}^{\infty} e^{i\omega t} \int_{-\infty}^{\infty} F(\omega - \nu)g(\nu)d\nu d\omega \\
&\quad \text{let } \sigma = \omega - \nu \\
&= \frac{1}{2\pi}\int_{-\infty}^{\infty} e^{it(\sigma+\nu)} \int_{-\infty}^{\infty} F(\sigma)g(\nu)d\nu d\sigma \\
&= 2\pi\left(\frac{1}{2\pi}\int_{-\infty}^{\infty} e^{-it\sigma}F(\sigma)d\sigma\right)\left(\frac{1}{2\pi}\int_{-\infty}^{\infty} e^{it\nu}G(\nu)d\nu\right) \\
&= 2\pi\mathcal{F}^{-1}[F(\sigma)]\mathcal{F}^{-1}[G(\nu)] \\
&= 2\pi f(t)g(t) \qquad (12.30)
\end{aligned}$$

When it comes to applications of Fourier transform to solve differential equations, the dual versions of (12.29) and (12.30) are more frequently used, that is,

$$\mathcal{F}^{-1}\Big[F(\omega)G(\omega)\Big] = [f * g](t) \qquad (12.31)$$

$$\mathcal{F}\Big[f(t)g(t)\Big] = \frac{1}{2\pi}[F * G](\omega) \qquad (12.32)$$

These six properties are summarized in Table 12.2. Using either direct computation or implementation of the properties of Fourier transforms, a list of the Fourier transforms of some basic functions is given in Table 12.3. In some cases, the Fourier transforms can also be obtained by using known solutions of related differential equations. One such case is the Fourier transform of the Airy function, Ai(x) (see Exercise **E12.1**).

12.3 Solution of PDEs Using Fourier Transforms

In applying Fourier transforms for the solution of partial differential equation, one needs to carefully choose the independent variable used for transformation. For

Table 12.2. *Properties of fourier and inverse fourier transforms*

1. Linearity	$\mathcal{F}\left[af(t)+bg(t)\right] = a\mathcal{F}\left[f(t)\right] + b\mathcal{F}\left[g(t)\right]$ $\mathcal{F}^{-1}\left[aF(\omega)+bG(\omega)\right] = a\mathcal{F}^{-1}\left[F(\omega)\right] + b\mathcal{F}^{-1}\left[G(\omega)\right]$
2. Shifting	$\mathcal{F}\left[f(t-a)\right] = e^{-ia\omega}\mathcal{F}\left[f(t)\right]$ $\mathcal{F}^{-1}\left[F(\omega-b)\right] = e^{ibt}\mathcal{F}^{-1}\left[F(\omega)\right]$
3. Scaling	$\mathcal{F}\left[f(at)\right] = \frac{1}{\lvert a\rvert}F\left(\frac{\omega}{a}\right)$
4. Derivatives	$\mathcal{F}\left[\frac{d^n f}{dt^n}\right] = (i\omega)^n\,\mathcal{F}\left[f(t)\right]$ $\mathcal{F}^{-1}\left[\frac{d^n F}{d\omega^n}\right] = (-it)^n\,\mathcal{F}^{-1}\left[F(\omega)\right]$
5. Integrals	$\mathcal{F}\left[\int_{-\infty}^{t} f(\eta)d\eta\right] = \frac{1}{i\omega}F(\omega) + \pi\delta(\omega)F(0)$
6. Convolution	$\mathcal{F}^{-1}\left[F(\omega)G(\omega)\right] = [f*g](t)$ $\mathcal{F}\left[f(t)g(t)\right] = \frac{1}{2\pi}[F*G](\omega)$

instance, suppose the dependent variable is $u = u(x,t)$, with $-\infty < x < \infty$ and $0 \le t \le \infty$. Then the transform would make sense only with respect to x.

When taking the Fourier transform with respect to x, the other independent variables are fixed during the transformations. For instance, with $u(x,t)$, we define a new variable,

$$U(\omega,t) = \mathcal{F}\left[u(x,t)\right]$$

Some basic rules apply:

1. When taking derivatives with respect to the other independent variables, one can interchange the order of differentiation with the Fourier transform operation, for example,

$$\mathcal{F}\left[\frac{\partial^k}{\partial t^k}u(x,t)\right] = \frac{d^k}{dt^k}\mathcal{F}\left[u(x,t)\right] = \frac{d^k}{dt^k}U(\omega,t) \tag{12.33}$$

   Note that the partial derivative operation has been changed to an ordinary derivative.

2. When taking derivatives with respect to the chosen variable under which Fourier transforms will apply, say x, the derivative property of Fourier transforms can be used, that is,

$$\mathcal{F}\left[\frac{\partial^k}{\partial x^k}u(x,t)\right] = (i\omega)^k\,\mathcal{F}\left[u(x,t)\right] = (i\omega)^k\,U(\omega,t) \tag{12.34}$$

Table 12.3. *Fourier transforms of some basic functions*

	$f(t)$	$\mathcal{F}[f(t)]$	Some remarks
1	$\mathcal{H}(t)=\begin{cases}0 & \text{if } t\le 0\\ 1 & \text{if } t>0\end{cases}$	$\pi\delta(\omega)+[1/(i\omega)]$	See Example L.11
2	$\text{sgn}(t)=\begin{cases}-1 & \text{if } t<0\\ 1 & \text{if } t>0\end{cases}$	$2/(i\omega)$	See Example L.11
3	$\mathcal{H}(a-\lvert t\rvert)=\begin{cases}0 & \text{if } \lvert t\rvert>a\\ 1 & \text{if } \lvert t\rvert\le a\end{cases}$	$2\sin(a\omega)/\omega$	See Example 12.1
4	$\sin(bt)/t$	$\pi\mathcal{H}\left(b-\lvert\omega\rvert\right)$	See Example 12.1
5	$\delta(t-a)$	$e^{ia\omega}$	See Example L.9
6	$a/\left(t^2+a^2\right)$	$\pi e^{-a\lvert\omega\rvert}$	See Example 12.2
7	1	$2\pi\delta(\omega)$	See Example L.10
8	e^{iat}	$2\pi\delta(\omega-a)$	See Example L.10
9	$e^{-\lvert a\rvert t^2}$	$\left(\sqrt{\pi/\lvert a\rvert}\right)e^{-\omega^2/(4\lvert a\rvert)}$	See Example L.4
10	$\cos(at)$	$\pi\left(\delta(\omega+a)+\delta(\omega-a)\right)$	See Example L.10
11	$\sin(at)$	$i\pi\left(\delta(\omega+a)-\delta(\omega-a)\right)$	See Example L.10

However, this relationship is true only if

$$u(x,t)e^{i\omega t}\Big|_{t=-\infty}^{t=\infty}=0$$

One sufficient case is when $u(x,t)\to 0$ as $|x|\to\infty$.

The general approach to solving the partial differential equation for $u(x,t)$ can be outlined as follows:

1. Take the Fourier transform of the partial differential equation with respect to x and reduce it to an ordinary differential equation in terms of $U(\omega,t)$, with t as the independent variable.
2. Take the Fourier transform of the conditions at fixed values of t, for example,

$$U(\omega,0)=\mathcal{F}[u(x,0)] \qquad \text{and} \qquad \frac{d}{dt}U(\omega,0)=\mathcal{F}\left[\frac{\partial u}{\partial t}(x,0)\right]$$

3. Solve the ordinary differential equation for $U(\omega,t)$.
4. Take the inverse Fourier transform of $U(\omega,t)$. In several applications, one could use the table of Fourier transforms in conjunction with the properties of Fourier transforms, such as the shifting theorem and convolution theorem, to obtain a general solution.

EXAMPLE 12.3. Fourier transforms approach to obtain d'Alembert solution. For the one-dimensional wave equation given by

$$\frac{\partial^2 u}{\partial x^2} - \frac{1}{c^2}\frac{\partial^2 u}{\partial t^2} = 0 \qquad -\infty \le x \le \infty$$

subject to initial conditions

$$u(x,0) = f(x) \qquad \text{and} \qquad \frac{\partial u}{\partial t}(x,0) = g(x)$$

We first take the Fourier transform of both sides,

$$\begin{aligned} \mathcal{F}\left[\frac{\partial^2 u}{\partial x^2} - \frac{1}{c^2}\frac{\partial^2 u}{\partial t^2}\right] &= (i\omega)^2 U(\omega,t) - \frac{1}{c^2}\frac{d^2}{dt^2}U &= 0 \\ \frac{d^2 U}{dt^2} + (c\omega)^2 U &= 0 \end{aligned}$$

whose solution can be put in terms of $e^{ic\omega t}$ and $e^{-ic\omega t}$ as

$$U(\omega,t) = Ae^{ic\omega t} + Be^{-ic\omega t}$$

Applying the initial conditions, we have

$$\mathcal{F}[f(x)] = A + B \qquad \text{and} \qquad \mathcal{F}[g(x)] = Aic\omega - Bic\omega$$

or

$$\begin{pmatrix} A \\ B \end{pmatrix} = \frac{1}{2}\begin{pmatrix} 1 & 1/(ic\omega) \\ 1 & -1/(ic\omega) \end{pmatrix}\begin{pmatrix} \mathcal{F}[f] \\ \mathcal{F}[g] \end{pmatrix}$$

Then,

$$U(\omega,t) = \frac{1}{2}\left(e^{ic\omega t} + e^{-ic\omega t}\right)\mathcal{F}[f(x)] + \frac{1}{2ic\omega}\left(e^{ic\omega t} - e^{-ic\omega t}\right)\mathcal{F}[g(x)]$$

Next, take the inverse Fourier transforms of the various terms while applying the shifting theorem (12.19),

$$\begin{aligned} \mathcal{F}^{-1}\left[\frac{1}{2}e^{ic\omega t}\mathcal{F}[f(x)]\right] &= \frac{1}{2}f(x+ct) \\ \mathcal{F}^{-1}\left[\frac{1}{2}e^{-ic\omega t}\mathcal{F}[f(x)]\right] &= \frac{1}{2}f(x-ct) \\ \mathcal{F}^{-1}\left[\frac{1}{2ic\omega}e^{ic\omega t}\mathcal{F}[g(x)]\right] &= \frac{1}{2c}\mathcal{F}^{-1}\left[e^{ic\omega t}\mathcal{F}\left[\int_{-\infty}^{x} g(\eta)d\eta\right]\right] \\ &\quad -\frac{\pi}{2c}\mathcal{F}[g]\bigg|_{\omega=0}\mathcal{F}^{-1}\left[e^{ic\omega t}\delta(\omega)\right] \\ &= \frac{1}{2c}\int_{-\infty}^{x+ct} g(\eta)d\eta - \frac{1}{4c}\mathcal{F}[g]\bigg|_{\omega=0} \\ \mathcal{F}^{-1}\left[\frac{1}{2ic\omega}e^{-ic\omega t}\mathcal{F}[g(x)]\right] &= \frac{1}{2c}\int_{-\infty}^{x-ct} g(\eta)d\eta - \frac{1}{4c}\mathcal{F}[g]\bigg|_{\omega=0} \end{aligned}$$

Adding all the terms together, we obtain d'Alembert's solution,

$$u(x,t) = \frac{1}{2}\Big(f(x+ct) + f(x-ct)\Big) + \frac{1}{2c}\int_{x-ct}^{x+ct} g(\eta)d\eta$$

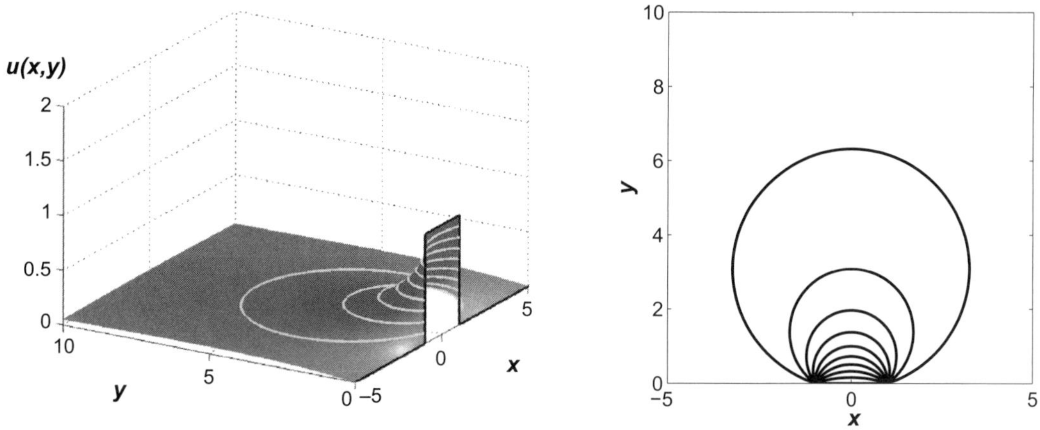

Figure 12.2. A surface plot of $u(x, y)$ given by (12.35) together with the corresponding contour plot.

EXAMPLE 12.4. Consider the Laplace equation for the 2D half-plane, $y \geq 0$,

$$\frac{\partial^2 u}{\partial x^2} + \frac{\partial^2 u}{\partial y^2} = 0$$

subject to

$$\begin{aligned} u(x,0) &= f(x) = \mathcal{H}(1-|x|) \\ u(x,y) &\rightarrow 0 \text{ as } |x| \rightarrow \infty \text{ and } y \rightarrow \infty \end{aligned}$$

Taking the Fourier transform of both sides with respect to x,

$$(-i\omega)^2 U(\omega, y) + \frac{d^2U}{dy^2} = 0$$

whose solution is given by

$$U(\omega, y) = Ae^{|\omega|y} + Be^{-|\omega|y}$$

Because $U(\omega, y) < \infty$ as $y \rightarrow \infty$, we need $A = 0$. Applying the other boundary condition (keeping it in terms of $f(x)$ in the meantime),

$$U(\omega, y) = \mathcal{F}[f]\, e^{-|\omega|y}$$

Using the convolution theorem given by (12.31) and item 6 in Table 12.3, we have

$$\begin{aligned} u(x,y) &= f(x) * \left(\frac{1}{\pi}\frac{y}{x^2+y^2}\right) \\ &= \frac{1}{\pi}\int_{-\infty}^{\infty} \mathcal{H}(1-|\eta|)\,\frac{y}{(x-\eta)^2+y^2}\,d\eta = \frac{1}{\pi}\int_{-1}^{1}\frac{y}{(x-\eta)^2+y^2}\,d\eta \\ &= \frac{1}{\pi}\left(\mathrm{Atan}\left(\frac{x+1}{y}\right) - \mathrm{Atan}\left(\frac{x-1}{y}\right)\right) \end{aligned} \tag{12.35}$$

Figure 12.2 shows a surface plot of $u(x, y)$ given by (12.35), together with some curves at constant u values, which is shown separately as a contour plot. The level curves can be seen to be circles whose centers are located along the line $x = 0$.

12.4 Laplace Transforms

In this section, we derive the Laplace transform as a special case of the Fourier transform. Then we obtain some important properties. We also include a subsection for the use of partial fractions to obtain inverse Laplace transforms of rational polynomial functions. Finally, we apply the Laplace transforms to the solution of partial differential equations, including some examples that combine with the application of Fourier transforms.

12.4.1 Definition and Notations

Consider the function $\breve{f}(x)$ given by

$$\breve{f}(x) = \begin{cases} 0 & \text{if } x \le 0 \\ e^{-ax} f(x) & \text{if } x > 0 \end{cases} \tag{12.36}$$

where a is a positive real number. Assuming $\breve{f}$ is integrable, we can apply it to the Fourier integral equation, (12.8),

$$\begin{aligned} \breve{f}(x) &= \frac{1}{2\pi}\int_{-\infty}^{\infty} e^{ix\omega}\left[\int_{-\infty}^{\infty} \breve{f}(t)e^{-it\omega}\,dt\right] d\omega \\ \mathcal{H}(x)\,e^{-ax} f(x) &= \frac{1}{2\pi}\int_{-\infty}^{\infty} e^{ix\omega}\left[\int_{0}^{\infty} f(t)e^{-(a+i\omega)t}\,dt\right] d\omega \\ \mathcal{H}(x)\,f(x) &= \frac{1}{2\pi}\int_{-\infty}^{\infty} e^{(a+i\omega)x}\left[\int_{0}^{\infty} f(t)e^{-(a+i\omega)t}\,dt\right] d\omega \end{aligned} \tag{12.37}$$

where $\mathcal{H}(x)$ is the unit step function.

From (12.37), and letting $s = a + iw, a \ge 0$, we can extract the integral transform pair called the Laplace transform and inverse Laplace transform.

Definition 12.3. *Let s be a complex variable whose real part is non-negative. For a given function $f(t)$, the operator $\mathcal{L}$ acting on $f(t)$ given by*

$$\widehat{f}(s) = \mathcal{L}[f] = \int_0^{\infty} f(t)e^{-st}\,dt \tag{12.38}$$

is called the **Laplace transform of** f.

For a given function $\widehat{f}(s)$, the operator $\mathcal{L}^{-1}$ acting on $\widehat{f}(s)$ given by

$$\mathcal{H}(t)\,f(t) = \mathcal{L}^{-1}\left[\widehat{f}\right] = \frac{1}{2\pi i}\int_{\gamma-i\infty}^{\gamma+i\infty} \widehat{f}(s)e^{st}\,ds \tag{12.39}$$

is called the **inverse Laplace transform of** $\widehat{f}(s)$. *The value of γ is a non-negative real number chosen to be greater than the real part of any singularities of $\widehat{f}(s)$.*[3]

The kernel of a Laplace transform is given by $K(t,\omega) = e^{-st}$, whereas the inverse kernel is given by $\widehat{K}(t,s) = (1/2\pi i)e^{st}$. During the evaluation of Laplace transforms, one often uses integration by parts and requires that values at the upper limit of $t \to \infty$ are bounded. Alternatively, the direct evaluation of the inverse Laplace

[3] The integral formula given in (12.39) is also known as the **Bromwich integral**.

transforms via (12.39) may be obtained via the method of residues. Details of the residue theorem can be found in Sections L.2.2 and L.2.3.[4] Briefly, we have

$$\frac{1}{2\pi i}\int_{\gamma-i\infty}^{\gamma+i\infty}\widehat{f}(s)e^{st}\,ds=\sum_{\ell=1}^{N}\mathbf{Res}_{z_\ell}\left(\widehat{f}(s)e^{st}\right)\tag{12.40}$$

inside a left semicircle contour composed of a vertical line and an arc large enough to contain the singularities of $\left(\widehat{f}e^{st}\right)$. The residues $\mathbf{Res}_{z_\ell}(g)$, with z_ℓ being a kth order poles of $g(z)$, are given by

$$\mathbf{Res}_{z_o}(g)=\frac{1}{(k-1)!}\lim_{z\to z_o}\frac{d^{k-1}}{dz^{k-1}}\left([z-z_o]^k\,g(z)\right)\tag{12.41}$$

For the special case when the function $g(z)$ is a rational function with a numerator function $num(z)$ and denominator function $den(z)$, that is,

$$g(z)=\frac{num(z)}{den(z)}$$

such that a simple pole at $(z=z_0)$ is a root $den(z)$, then via L'Hospital's rule, we have

$$\mathbf{Res}_{z_o}(g)=\left.\frac{num(z)}{\dfrac{d}{dz}den(z)}\right|_{z=z_0}\tag{12.42}$$

EXAMPLE 12.5. **Laplace transform of** t^ν**.** Consider the function $f(t)=t^\nu$ where $\nu>0$ is a real valued constant. Then

$$\begin{aligned}\mathcal{L}\left[t^\nu\right] &= \int_0^\infty e^{-st}t^\nu dt=\int_0^\infty e^{-y}\left(\frac{y}{s}\right)^\nu d\left(\frac{y}{s}\right)=\frac{1}{s^{\nu+1}}\int_0^\infty e^{-y}y^\nu dy\\ &= \frac{\Gamma(\nu+1)}{s^{\nu+1}}\end{aligned}\tag{12.43}$$

where $\Gamma(x)$ is the gamma function of x (cf. (9.7)).

EXAMPLE 12.6. **Laplace transform of** erfc $\left(1/(2\sqrt{t})\right)$**.** The **error function**, erf (x), and the **complementary error function**, erfc (x) are defined as

$$\text{erf}\,(x)=\frac{2}{\sqrt{\pi}}\int_0^x e^{-\nu^2}d\nu\qquad\text{and}\qquad\text{erfc}\,(x)=\frac{2}{\sqrt{\pi}}\int_x^\infty e^{-\nu^2}d\nu\tag{12.44}$$

Note that erf $(\infty)=1$ and thus

$$\text{erf}\,(x)+\text{erfc}\,(x)=1\tag{12.45}$$

[4] In practice, because the evaluation of the inverse Laplace transforms can be quite complicated, a table of Laplace transforms is usually consulted first.

The Laplace transform of erfc $\left(1/2\sqrt{x}\right)$ is given by

$$\mathcal{L}\left[\operatorname{erfc}\left(\frac{1}{2\sqrt{x}}\right)\right] = \frac{2}{\sqrt{\pi}}\int_0^\infty e^{-st}\int_{1/(2\sqrt{t})}^\infty e^{-\nu}d\nu dt$$

After integration by parts,

$$\begin{aligned}\mathcal{L}\left[\operatorname{erfc}\left(\frac{1}{2\sqrt{x}}\right)\right] &= \frac{1}{2s\sqrt{\pi}}\int_0^\infty \frac{e^{-st}e^{-1/(4t)}}{t\sqrt{t}}dt \\ &= \frac{2}{s\sqrt{\pi}}\int_0^\infty e^{-s/(4q^2)}e^{-q^2}dq \qquad \text{with } q = \frac{1}{2\sqrt{t}}\end{aligned}$$

Let $g(s)$ be equal to the integral term, then

$$\begin{aligned}g(s) &= \int_0^\infty e^{-s/(4q^2)}e^{-q^2}dq \\ \frac{dg}{ds} &= -\int_0^\infty \frac{1}{4q^2}e^{-s/(4q^2)}e^{-q^2}dq \\ \frac{d^2g}{ds^2} &= \int_0^\infty \frac{1}{4^2q^4}e^{-s/(4q^2)}e^{-q^2}dq\end{aligned}$$

By evaluating d^2g/ds^2 through integration by parts,

$$\frac{d^2g}{ds^2} = \frac{1}{s}\int_0^\infty e^{-s/(4q^2)}e^{-q^2}\left(\frac{1}{8q^2}+\frac{1}{4}\right)dq = \frac{1}{s}\left(-\frac{1}{2}\frac{dg}{ds}+\frac{1}{4}g\right)$$

or

$$s^2\frac{d^2g}{ds^2}+\frac{1}{2}s\frac{dg}{ds}-\frac{1}{4}sg = 0$$

This equation is reducible to a Bessel or modified Bessel equation. Using the methods described in Theorem 9.3, we have, after using $g(0) = \sqrt{\pi}/2$ and $|g(\infty)| < \infty$,

$$\begin{aligned}g(s) &= s^{1/4}\left(AI_{1/2}\left(\sqrt{s}\right)+BI_{-1/2}\left(\sqrt{s}\right)\right) = ae^{-\sqrt{s}}+be^{\sqrt{s}} \\ &= \frac{\sqrt{\pi}}{2}e^{-\sqrt{s}}\end{aligned}$$

Combining the results, we have

$$\mathcal{L}\left[\operatorname{erfc}\left(\frac{1}{2\sqrt{x}}\right)\right] = \frac{1}{s}e^{-\sqrt{s}} \qquad (12.46)$$

EXAMPLE 12.7. Let $\widehat{F}(s)$ be given by

$$\widehat{F}(s) = \frac{J_\nu(\xi s)}{J_\nu(s)}$$

where $\xi \neq \{0, 1\}$. To find the inverse Laplace transform, we can use (12.39), that is,

$$\mathcal{L}^{-1}\left[\widehat{F}(x)\right] = \frac{1}{2\pi i}\int_{\gamma-i\infty}^{\gamma+i\infty} e^{st}\frac{J_\nu(\xi s)}{J_\nu(s)}ds$$

and use the theory of residues to evaluate the complex integral. The poles of $\widehat{F}e^{st}$ will be the roots of $J_\nu(s)$, except for $s = 0$, because it a removable singularity as it is also the root of the numerator $J_\nu(\xi s)$. Thus let z_n be the nth root of $J_\nu(z)$. Using (12.40) and (12.42), plus the formula for the derivative of Bessel function given in (9.66), we have

$$\mathcal{L}^{-1}\left[\widehat{F}(x)\right] = \sum_{n=1,\ldots,\infty; z_n \neq 0} e^{z_n t} \frac{J_\nu(\xi z_n)}{\dfrac{\nu}{z_n} J_\nu(z_n) - J_{\nu+1}(z_n)}$$

12.4.2 Properties of Laplace Transforms

We now list some of the important properties of Laplace transforms and inverse Laplace transforms. In the following, we use $\widehat{f}(s)$ to mean $\mathcal{L}[f(t)]$:

1. **Linearity.**

$$\mathcal{L}\Big[\, af(t) + bg(t) \Big] = a\,\mathcal{L}\Big[\, f(t) \,\Big] + b\,\mathcal{L}\Big[\, g(t) \,\Big] \qquad (12.47)$$

$$\mathcal{L}^{-1}\Big[\, a\widehat{f}(s) + b\widehat{g}(s) \,\Big] = a\,\mathcal{L}^{-1}\Big[\, \widehat{f}(s) \,\Big] + b\,\mathcal{L}^{-1}\Big[\, \widehat{g}(s) \,\Big] \qquad (12.48)$$

where a and b are constants. Both (12.47) and (12.48) are immediate consequences of the properties of integrals.

2. **Shifting.**

$$\begin{aligned}
\mathcal{L}\Big[\, \mathcal{H}(t-a)\, f(t-a) \,\Big] &= \int_0^\infty \mathcal{H}(t-a)\, f(t-a) e^{-st} dt \\
&= \int_0^\infty f(\tau) e^{-s(\tau+a)} d\tau \qquad (\tau = t - a) \\
&= e^{-as} \int_0^\infty f(\tau) e^{-s\tau} d\tau \\
&= e^{-as} \mathcal{L}\Big[\, f(t) \,\Big] \qquad (12.49)
\end{aligned}$$

$$\begin{aligned}
\mathcal{L}^{-1}\Big[\, \widehat{f}(s-b) \,\Big] &= \frac{1}{2\pi i} \int_{\gamma-i\infty}^{\gamma+i\infty} \widehat{f}(s-b) e^{st} d\omega \\
&= \frac{1}{2\pi i} \int_{\gamma-i\infty}^{\gamma+i\infty} \widehat{f}(\nu) e^{(\nu+b)t} d\nu \qquad (\nu = s - b) \\
&= e^{bt} \frac{1}{2\pi i} \int_{\gamma-i\infty}^{\gamma+i\infty} \widehat{f}(\nu) e^{\nu t} d\nu \\
&= e^{bt} \mathcal{L}^{-1}\Big[\, \widehat{f}(s) \,\Big] \qquad (12.50)
\end{aligned}$$

3. **Scaling.**

Let Real$(a) \geq 0$, $a \neq 0$

$$\begin{aligned}
\mathcal{L}\Big[\, f(at) \,\Big] &= \int_0^\infty f(at)e^{-st}dtdt \\
&= \frac{1}{a}\int_0^\infty f(\tau)e^{-s(\tau/a)}d\tau \qquad (\tau = at) \\
&= \frac{1}{a}\int_0^\infty f(\tau)e^{-(s/a)\tau}d\tau \\
&= \frac{1}{a}\widehat{f}\left(\frac{s}{a}\right) \qquad (12.51)
\end{aligned}$$

4. **Derivatives.**

$$\begin{aligned}
\mathcal{L}\left[\frac{df}{dt}\right] &= \int_0^\infty \frac{df}{dt}e^{-st}dt \\
&= f(t)e^{-st}\Big]_0^\infty + s\int_0^\infty f(t)e^{-st}dt \quad \text{(after integration by parts)} \\
&= s\mathcal{L}\Big[\, f(t) \,\Big] - f(0) \qquad (12.52)
\end{aligned}$$

or in general, we can show by induction that

$$\mathcal{L}\left[\frac{d^n f}{dt^n}\right] = s^n\mathcal{L}\Big[\, f(t) \,\Big] - \sum_{k=0}^{n-1} s^{n-k-1}\left(\left.\frac{d^k f}{dt^k}\right|_{t=0}\right) \qquad (12.53)$$

5. **Integrals.**

$$\mathcal{L}\left[\int_0^t f(\tau)d\tau\right] = \int_0^\infty e^{-st}\int_0^t f(\tau)d\tau$$

After integration by parts,

$$\mathcal{L}\left[\int_0^t f(\tau)d\tau\right] = \frac{1}{s}\int_0^\infty f(t)e^{-st}dt = \frac{1}{s}\mathcal{L}\Big[\, f(t) \,\Big] \qquad (12.54)$$

6. **Convolution.** For Laplace transforms, we define a different convolution operator:

$$\text{Convolution}(f, g) = [f * g]\,(t) = \int_0^t f(t-\tau)g(\tau)d\tau d\tau \qquad (12.55)$$

Note that the limits for the convolution used in Laplace transform methods are from 0 to t, whereas the convolution for Fourier transform methods are from $-\infty$ to ∞.[5]

The Laplace transform of a convolution is then given by

$$\begin{aligned}
\mathcal{L}\left[f * g\right] &= \int_0^\infty e^{-st}\left[\int_0^t f(t-\tau)g(\tau)d\tau\right] dt \\
&= \int_0^\infty \left[\int_0^t e^{-st} f(t-\tau)g(\tau)d\tau\right] dt \\
&= \int_0^\infty \left[\int_u^\infty e^{-st} f(t-\tau)g(\tau)dt\right] d\tau \\
&= \int_0^\infty \left[\int_0^\infty e^{-s(\eta+\tau)} f(\eta)g(\tau)d\eta\right] d\tau \\
&= \left[\int_0^\infty f(\eta)e^{-s\eta}d\eta\right]\left[\int_0^\infty g(\tau)e^{-s\tau}d\tau\right] \\
&= \mathcal{L}[f]\ \mathcal{L}[g] \qquad (12.56)
\end{aligned}$$

In several applications, the inverse formulation is found to be quite useful, that is,

$$\mathcal{L}^{-1}\left[\mathcal{L}[f]\,\mathcal{L}[g]\right] = f * g \qquad (12.57)$$

EXAMPLE 12.8. Laplace transform of $e^{-\alpha t}$, $\cos(\omega t)$ and $\sin(\omega t)$. Let α be a constant complex number with a non-negative real part. Then the Laplace transform of $f(t) = e^{-\alpha t}$ is given by

$$\begin{aligned}
\mathcal{L}\left[e^{-\alpha t}\right] &= \int_0^\infty e^{-\alpha t}e^{-st}dt = \int_0^\infty e^{-(s+\alpha)t}dt = -\frac{1}{s+\alpha}e^{-(s+\alpha)t}\Big|_{t=0}^{t=\infty} \\
&= \frac{1}{s+\alpha} \qquad (12.58)
\end{aligned}$$

For the special case where $\alpha = i\omega$, we have

$$\mathcal{L}\left[e^{-i\omega t}\right] = \frac{1}{s+i\omega} \qquad (12.59)$$

For cosine and sine, we can use (12.59), Euler's identity, and the linearity property

$$\begin{aligned}
\mathcal{L}\left[\cos(\omega t)\right] &= \frac{1}{2}\mathcal{L}\left[e^{i\omega t} + e^{-i\omega t}\right] = \frac{1}{2}\left(\frac{1}{s-i\omega} + \frac{1}{s+i\omega}\right) \\
&= \frac{s}{s^2+\omega^2} \qquad (12.60)
\end{aligned}$$

[5] Based on the original definition of convolution given in (12.28), observe that (12.55) is simply the result of restricting $f(t) = g(t) = 0$ for $t < 0$, that is,

$$\int_{-\infty}^\infty \left(\mathcal{H}(t-\tau)\, f(t-\tau)\right)\left(\mathcal{H}(\tau)\, g(\tau)\right) d\tau = \int_0^t f(t-\tau)g(\tau)d\tau d\tau$$

$$\begin{aligned}\mathcal{L}\Big[\sin(\omega t)\Big] &= \frac{1}{2i}\mathcal{L}\Big[e^{i\omega t} - e^{-i\omega t}\Big] = \frac{1}{2i}\left(\frac{1}{s-i\omega} - \frac{1}{s+i\omega}\right) \\ &= \frac{\omega}{s^2+\omega^2} \end{aligned} \tag{12.61}$$

EXAMPLE 12.9. Laplace transform of $J_\nu(t)$. Using the definition of $J_\nu(x)$ (cf. (I.43)), the Laplace transform of t^ν given by (12.43) and the linearity property,

$$\begin{aligned}\mathcal{L}\Big[J_\nu(t)\Big] &= \sum_{k=0}^{\infty} \frac{(-1)^k}{k!\, 2^{2k+\nu}\,\Gamma(\nu+k+1)}\,\mathcal{L}\Big[t^{2k+\nu}\Big] \\ &= \sum_{k=0}^{\infty} \frac{(-1)^k\,\Gamma(\nu+2k+1)}{k!\, 2^{2k+\nu}\,\Gamma(\nu+k+1)}\left(\frac{1}{s}\right)^{2k+\nu+1} \\ &= \left(\frac{1}{s(2s)^\nu}\right)\sum_{k=0}^{\infty}\frac{(-1)^k}{k!}\,\frac{\Gamma(\nu+2k+1)}{\Gamma(\nu+k+1)}\left(\frac{1}{4s^2}\right)^k \\ &= \left(\frac{1}{s(2s)^\nu}\right)\left(1+\sum_{k=1}^{\infty}\frac{(-1)^k}{k!}\left[\prod_{j=0}^{k-1}(k+\nu+1)+j\right]\left(\frac{1}{4s^2}\right)^k\right) \end{aligned} \tag{12.62}$$

where we used the Pockhammer product equation given in (9.11). Note further that in this case, to guarantee that the Pockhammer product will be positive, we need to require that $\nu > -1$. Next, define $g(q)$ as

$$g(q) = \left(\frac{\sqrt{4q+1}-1}{2q}\right)^\nu \left(\frac{1}{\sqrt{4q+1}}\right) \tag{12.63}$$

or in terms of $z = \sqrt{4q+1}$,

$$g(z(q)) = \left(\frac{2}{z+1}\right)^\nu \left(\frac{1}{z}\right)$$

Then

$$\begin{aligned} g(q)\Big|_{q=0} &= g(z(q))\Big|_{z=1} = 1 \\ \frac{dg}{dq}\Big|_{q=0} &= \left(\frac{dg}{dz}\right)\left(\frac{dz}{dq}\right)\Big|_{z=1} = 2^{\nu+1}\left(\frac{-(\nu+1)z-1}{(z+1)^{\nu+1}z^3}\right)\Big|_{z=1} = -(\nu+2) \\ &\vdots \\ \left(\frac{d^k g}{dq^k}\right)\Big|_{q=0} &= (-1)^k\prod_{j=0}^{k-1}(k+\nu+1)+j \end{aligned}$$

which means the Taylor expansion of $g(q)$ around $q=0$ is given by

$$g(q) = 1+\sum_{k=1}^{\infty}\frac{(-1)^k}{k!}\left[\prod_{j=0}^{k-1}(k+\nu+1)+j\right]q^k$$

When we let $q = 1/(4s^2)$ in (12.63), then (12.62) reduces to

$$\mathcal{L}\left[J_\nu(t) \right] = \left(\frac{1}{s(2s)^\nu}\right)\left(\frac{\sqrt{1+(1/s^2)}-1}{1/(2s^2)}\right)^\nu \left(\frac{1}{\sqrt{1+(1/s^2)}}\right)$$

$$= \frac{\left(\sqrt{s^2+1}-s\right)^\nu}{\sqrt{s^2+1}} \tag{12.64}$$

Note that this result is valid for $\nu > -1$, including non-integer values.

EXAMPLE 12.10. Laplace transform of $I_\nu(t)$. To determine the Laplace transform of $I_\nu(t)$, we can use the definition given in (I.63)

$$I_\nu(t) = \exp\left(-\frac{\nu\pi i}{2}\right) J_\nu(it) = (-i)^\nu J_\nu(it)$$

Using the scaling property, (12.51), with $a = i$ and the Laplace transform of $J_\nu(t)$ given in (12.64),

$$\mathcal{L}\left[I_\nu(t) \right] = (-i)^\nu \mathcal{L}\left[J_\nu(it) \right] = (-i)^\nu \left[\frac{1}{i} \frac{\left(\sqrt{-s^2+1}-(s/i)\right)^\nu}{\sqrt{-s^2+1}} \right]$$

$$= \frac{\left(s-\sqrt{s^2-1}\right)^\nu}{\sqrt{s^2-1}} \tag{12.65}$$

EXAMPLE 12.11. Laplace transform of $\mathcal{H}(t)$ and $\delta(t)$.

$$\mathcal{L}\left[\mathcal{H}(t) \right] = \int_0^\infty \mathcal{H}(t)\, e^{-st} dt = -\frac{1}{s}\, e^{-st}\Big|_{t=0}^{t=\infty} = \frac{1}{s} \tag{12.66}$$

Using this result plus the fact that $\delta(t)$ is defined as the derivative of $\mathcal{H}(t)$,

$$\mathcal{L}\left[\delta(t) \right] = \mathcal{L}\left[\frac{d}{dt}\mathcal{H}(t) \right] = s\mathcal{L}\left[\mathcal{H}(t) \right] - \mathcal{H}(0) = 1 \tag{12.67}$$

A summary of the Laplace transforms of some basic functions is given in Table 12.4. Note that the table entries could be extended by use of the properties of Laplace transforms, for example, through the combination of scaling, derivation, integration, and convolution properties. A more complete table is available in the literature.

Table 12.4. *Laplace transforms of some basic functions*

	$f(t)$	$\mathcal{L}[f(t)]$	Remarks
1	$\mathcal{H}(t) = \begin{cases} 0 & \text{if } t \le 0 \\ 1 & \text{if } t > 0 \end{cases}$	$\dfrac{1}{s}$	See Example 12.11
2	$\delta(t)$	1	See Example 12.11
3	e^{-t}	$\dfrac{1}{s+1}$	See Example 12.8
4	$\sin(t)$	$\dfrac{1}{s^2+1}$	See Example 12.8
5	$\cos(t)$	$\dfrac{s}{s^2+1}$	See Example 12.8
6	t^{ν}	$\dfrac{\Gamma(\nu+1)}{s^{\nu+1}}$	See Example 12.5
7	$\mathrm{erfc}\left(\dfrac{1}{2\sqrt{t}}\right)$	$\dfrac{1}{s}\, e^{-\sqrt{s}}$	See Example 12.6
8	$\dfrac{1}{\sqrt{t}}\, e^{-1/(4t)}$	$\sqrt{\dfrac{\pi}{s}}\, e^{-\sqrt{s}}$	Left as Exercise **E12.8**
9	$\dfrac{1}{\sqrt{t^3}}\, e^{-1/(4t)}$	$\sqrt{4\pi}\, e^{-\sqrt{s}}$	Left as Exercise **E12.8**
10	$J_{\nu}(t)$	$\dfrac{\left(\sqrt{s^2+1}-s\right)^{\nu}}{\sqrt{s^2+1}}$	See Example 12.9
11	$I_{\nu}(t)$	$\dfrac{\left(s-\sqrt{s^2-1}\right)^{\nu}}{\sqrt{s^2-1}}$	See Example 12.10
12	$t^{\nu/2} J_{\nu}\left(2\sqrt{t}\right)$	$\dfrac{e^{-1/s}}{s^{1+\nu}}$	Left as Exercise **E12.9**

12.4.3 Decomposition to Partial Fractions

For the special case in which the function $\widehat{f}(s)$ is a rational function of polynomials in s given by

$$\widehat{f}(s) = \frac{N(s)}{D(s)} \tag{12.68}$$

where

$$N(s) = \sum_{q=1}^{n-1} \alpha_q s^q \tag{12.69}$$

$$D(s) = \prod_{k=1}^{m} (s - r_k)^{N_k} \qquad \text{with} \qquad \sum_{k=1}^{m} N_k = n \tag{12.70}$$

$\widehat{f}(s)$ can be separated into n additive terms as follows:

$$\widehat{f}(s) = \sum_{k=1}^{m} \left(\sum_{\ell=1}^{N_k} \frac{A_{k\ell}}{(s - r_k)^\ell} \right) \tag{12.71}$$

whose inverse Laplace transform, after using (12.43) together with the linearity, scaling, and shifting properties, is given by

$$f(t) = \sum_{k=1}^{m} \left(\sum_{\ell=1}^{N_k} \frac{A_{k\ell}}{(\ell - 1)! t^{\ell-1}} \right) e^{r_k t} \tag{12.72}$$

To determine A_{KL} (with $1 \le K \le m$ and $1 \le L \le N_K$), first multiply both sides of (12.71) by $(s - r_K)^{N_K}$,

$$(s - r_K)^{N_K} \widehat{f}(s) \; = \; \Omega_{(K)} + \sum_{\ell=1}^{N_K} A_{K\ell}(s - r_K)^{N_K - \ell} \tag{12.73}$$

where,

$$\Omega_{(K)} = \sum_{k \ne K} \left(\sum_{\ell=1}^{N_k} \frac{A_{k\ell}}{(s - r_k)^\ell} (s - r_K)^{N_K} \right) \tag{12.74}$$

Taking the $(N_K - L)^{\text{th}}$ derivative of (12.73) with respect to s,

$$\frac{d^{(N_K - L)}}{ds^{(N_K - L)}} \left[(s - r_K)^{N_K} \widehat{f}(s) \right] \; = \; \frac{d^{(N_K - L)} \Omega_{(K)}}{ds^{(N_K - L)}} + \sum_{\ell=1}^{L} \frac{(N_K - \ell)!}{(L - \ell)!} A_{K\ell}(s - r_K)^{L-\ell} \tag{12.75}$$

Finally, because

$$\lim_{s \to r_K} \frac{d^{(N_K - L)} \Omega_{(K)}}{ds^{(N_K - L)}} = 0$$

the limit of both sides of (12.75) as $s \to r_K$ will yield

$$A_{KL} = \lim_{s \to r_K} \frac{1}{(N_K - L)!} \left(\frac{d^{(N_K - L)}}{ds^{(N_K - L)}} \left[(s - r_K)^{N_K} \widehat{f}(s) \right] \right) \tag{12.76}$$

EXAMPLE 12.12. Consider the function

$$\widehat{f}(s) = \frac{2s^2+s+1}{\left(s+\frac{1}{2}\right)(s+2)^2(s+1-i)^2(s+1+i)^2}$$

Using (12.76) we obtain

k	r_k	N_k	$A_{k\ell}$
1	$-1/2$	1	$A_{11} = -64/225$
2	-2	2	$A_{21} = -19/18$ $A_{22} = -5/6$
3	$-1+i$	2	$A_{31} = (67/100) - (31/100)i$ $A_{32} = -(7/20) - (1/5)i$
4	$-1-i$	2	$A_{41} = (67/100) + (31/100)i$ $A_{42} = -(7/20) + (1/5)i$

After using Euler's identities for $\sin(t)$ and $\cos(t)$, (12.72) can be rearranged to be

$$\begin{aligned} f(t) &= -\frac{64}{225}e^{-t/2} - \left(\frac{5}{6}t + \frac{19}{18}\right)e^{-2t} \\ &\quad + e^{-t}\left[\left(\frac{2}{5}t + \frac{31}{50}\right)\sin(t) + \left(-\frac{7}{10}t + \frac{67}{50}\right)\cos(t)\right] \end{aligned}$$

12.5 Solution of PDEs Using Laplace Transforms

The approach of using Laplace transforms to solve partial differential equation follows closely with the Fourier transforms approach. In contrast to using variables such as $-\infty < x < \infty$, the Laplace transforms need to be applied on semi-infinite variables such $0 \le t \le \infty$. Fortunately, the time variable has this feature, which is mainly why Laplace transforms are usually applied to the time domain. In applications in which one of the variables are defined in a semi-infinite region, such as $z \ge a$, the Laplace transform can also be applied on the translated variable $t = z - a$. The other independent variables may be finite, semifinite, or infinite regions.

As it was in Fourier transforms, when taking the Laplace transform with respect to $t \ge 0$, the other independent variables are fixed during the transformations. For instance, with $u(x,t)$, one could define a new variable,

$$\widehat{U}(x,s) = \mathcal{L}\left[u(x,t)\right] \tag{12.77}$$

Some basic rules apply:

1. When taking derivatives with respect to the other independent variables, one can interchange the order of differentiation with the Laplace transform operation, for example,

$$\mathcal{L}\left[\frac{\partial^k}{\partial x^k}u(x,t)\right] = \frac{d^k}{dt^k}\mathcal{L}\left[u(x,t)\right] = \frac{d^k}{dt^k}\widehat{U}(x,s) \tag{12.78}$$

2. When taking derivatives with respect to the transformed variable, say t, the derivative property of Laplace transforms can be used, that is,

$$\mathcal{L}\left[\frac{\partial^k}{\partial x^k}u(x,t)\right] = s^k\widehat{U}(x,s) - \sum_{n=0}^{k-1} s^{k-n-1} \left.\frac{\partial^n u}{\partial t^n}\right|_{t=0} \tag{12.79}$$

Note that the initial conditions of the partial derivatives need to be specified by the problem for the successful application of Laplace transforms.

The general approach to solving the partial differential equation for $u(x,t)$ can be outlined as follows:

1. Take the Laplace transform of the partial differential equation with respect to $t \geq 0$ and reduce it to an ordinary differential equation in terms of $\widehat{U}(x,s)$, with x as the independent variable.
2. Take the Laplace transform of the conditions at fixed values of x, for example,

$$\widehat{U}(a,t) = \mathcal{L}\left[u(a,t)\right] \tag{12.80}$$

$$\frac{d}{dx}\widehat{U}(a,t) = \mathcal{L}\left[\frac{\partial u}{\partial x}(a,t)\right] \tag{12.81}$$

etc.
3. Solve the ordinary differential equation for $\widehat{U}(x,s)$.
4. Take the inverse Laplace transform of $\widehat{U}(x,s)$.

   In several applications, one could use the table of Laplace transforms together with the properties of Laplace transforms. In other cases, one may need to resort to using the method of decomposition to partial fractions and even the theory of residues to evaluate the inverse Laplace transforms.

EXAMPLE 12.13. **Laplace transform solution of diffusion equation.** Consider the one-dimensional diffusion equation

$$\alpha^2\frac{\partial^2 u}{\partial x^2} = \frac{\partial u}{\partial t} \tag{12.82}$$

with a constant initial condition, $u(x,0) = C_i$. Taking the Laplace transform of (12.82) with respect to t, we obtain

$$\alpha^2\frac{d^2\widehat{U}}{dx^2} - s\widehat{U} = C_i$$

whose general solution is

$$\widehat{U} = Ae^{\lambda x} + Be^{-\lambda x} + \frac{C_i}{s} \tag{12.83}$$

where $\lambda = \left(\sqrt{s}\right)/\alpha$. The solution will depend on the values of the parameters and the boundary conditions.

Let the boundary conditions be given by

$$u(0,t) = C_0 \qquad \text{and} \qquad \lim_{x\to\infty} |u(x,t)| < \infty \tag{12.84}$$

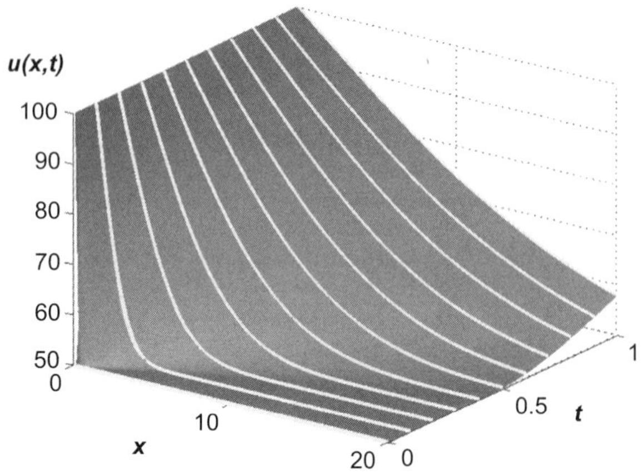

Figure 12.3. A plot of the solution given by (12.85).

Applying these to (12.83), we get $A = 0$ and $B = \dfrac{C_0 - C_i}{s}$. Thus

$$\widehat{U} = \frac{C_0 - C_i}{s} e^{\sqrt{s}/\alpha} + \frac{C_i}{s}$$

Based on item 7 in Table 12.4 and the scaling property given in (12.51),

$$\mathcal{L}\left[\operatorname{erfc}\left(\frac{x}{2\alpha\sqrt{t}}\right)\right] = \frac{1}{s} e^{-\sqrt{(x/\alpha)^2 s}}$$

The solution is then given by

$$u(x,t) = (C_0 - C_i)\operatorname{erfc}\left(\frac{x}{2\alpha\sqrt{t}}\right) + C_i \tag{12.85}$$

A plot of (12.85) with $\alpha = 10$, $C_0 = 100$, and $C_i = 50$ is shown in Figure 12.3.

This preceding case applies to the boundary conditions given in (12.84). For other boundary conditions, the use of residues may be needed to obtain the inverse Laplace transform. In Section L.7, we include a few more examples to show how the solutions under different boundary conditions can still be obtained using Laplace transform methods.

12.6 Method of Images

For a 2D linear partial differential equation $Lu = h(x, y)$ defined in $x > 0$ and $y > 0$ that is subject to two boundary conditions $BC1(x, 0)$ and $BC2(0, y)$ plus the boundedness condition, the Laplace transforms may lack some boundary conditions needed to solve the problem. In this case, the method of Fourier transforms is more appropriate. A technique called the **method of images** can be used to extend the boundary conditions to allow Fourier transforms to apply. This entails breaking down the problem into two problems u_A and u_B, in which one of the boundary conditions is made homogeneous. Depending on whether the homogenized boundary condition is Dirichlet or Neumann, the other boundary condition is extended with an odd function or even function, respectively. Thus the original problem $Lu = h(x, y)$

Table 12.5. *Boundary conditions of subproblems*

Original BC	Subproblem A	Subproblem B
$u(x,0)=f(x)$	$u^A(x,0)=f^{\text{odd}}(x)$	$u^B(x,0)=0$
$u(0,y)=g(y)$	$u^A(0,y)=0$	$u^B(0,y)=g^{\text{odd}}(y)$
$u(x,0)=f(x)$	$u^A(x,0)=f^{\text{even}}(x)$	$u^B(x,0)=0$
$\dfrac{\partial u}{\partial x}(0,y)=g(y)$	$\dfrac{\partial u^A}{\partial x}(0,y)=0$	$\dfrac{\partial u^B}{\partial x}(0,y)=g^{\text{odd}}(y)$

can be split into two subproblems for $Lu^A = h(x, y)$ and $Lu^B = h(x, y)$, and whose boundary conditions are given in Table 12.5, where

$$f^{\text{even}}(x) = \begin{cases} f(x) & \text{for } x \ge 0 \\ f(-x) & \text{for } x < 0 \end{cases} \qquad \text{and} \qquad f^{\text{odd}}(x) = \begin{cases} f(x) & \text{for } x \ge 0 \\ -f(-x) & \text{for } x < 0 \end{cases}$$

EXAMPLE 12.14. Consider the Laplace equation

$$\frac{\partial^2 u}{\partial x^2} + \frac{\partial^2 u}{\partial y^2} = 0 \qquad x \ge 0$$

subject to $u(x,0) = f(x)$ and $\dfrac{\partial u}{\partial x}(0, y) = 0$. Note that using a Laplace transform with respect to x would require another boundary condition because the reduced ordinary differential equation will be second order. Thus we use the method of images and then apply the Fourier transform method. In this case, there is no need for subproblem B.

By extending f to f^{even} and following the same approach in Example 12.4, we obtain

$$\begin{aligned} u(x,y) &= f^{\text{even}}(x) * \left(\frac{1}{\pi}\frac{y}{x^2+y^2}\right) = \frac{1}{\pi}\int_{-\infty}^{\infty} f^{\text{even}}(\eta)\,\frac{y}{(x-\eta)^2+y^2}\,d\eta \\ &= \frac{1}{\pi}\int_0^{\infty} f^{\text{even}}(\eta)\,\frac{y}{(x-\eta)^2+y^2}\,d\eta - \frac{1}{\pi}\int_0^{\infty} f^{\text{even}}(-\eta)\,\frac{y}{(x+\eta)^2+y^2}\,d\eta \\ &= \frac{1}{\pi}\int_0^{\infty} f(\eta)\left(\frac{y}{(x-\eta)^2+y^2} - \frac{y}{(x+\eta)^2+y^2}\right) d\eta \end{aligned}$$

12.7 EXERCISES

E12.1. The Airy equation (9.56) with $k = -1$ is given by

$$\frac{d^2y}{dx^2} = xy \tag{12.86}$$

whose general solution is given by

$$y(x) = C_1\,\mathrm{Ai}(x) + C_2\,\mathrm{B_i}(x)$$

where $\mathrm{Ai}(x)$ and $\mathrm{Bi}(x)$ are the Airy functions of the first and second kind, respectively, defined in (9.58). For $y(x) < \infty$, we need $C_2 = 0$ because

$\text{Bi}(x)$ is unbounded as $x \to \infty$. Using properties of Fourier transforms in Table 12.2, show that

$$x\, y(x) = \frac{d\mathcal{F}\left[y\right]}{d\omega}$$

and when applied to (12.86) should yield

$$\mathcal{F}\left[y\right] = C \exp\left(\frac{i\omega^3}{3}\right)$$

Thus we arrive at the conclusion that

$$\mathcal{F}\left[\text{Ai}(x)\right] = \exp\left(\frac{i\omega^3}{3}\right) \tag{12.87}$$

E12.2. Consider the Laplace equation for an infinite domain in two dimensions given by

$$\frac{\partial^2 u}{\partial x^2} + \frac{\partial^2 u}{\partial y^2} = 0$$

subject to $u(x, 0) = f(x)$ and $u(x, y) < \infty$ for $-\infty \leq (x, y) \leq \infty$.

1. Applying the Fourier transform on x, show that the solution is given by

$$u(x, y) = \frac{1}{\pi} \int_{-\infty}^{\infty} f(x - \tau) \frac{y}{\tau^2 + y^2} d\tau$$

2. Let $\tau = y \tan\theta$, show that an alternative form of the solution is given by

$$u(x, y) = \frac{1}{\pi} \int_{-\pi/2}^{\pi/2} f\left(x - y\tan\theta\right)\, d\theta$$

3. Plot the solution for the case where $f(x) = \dfrac{2}{1 + e^{x^2}}$

E12.3. Consider the third order equation given by

$$\frac{\partial u}{\partial t} + \frac{\partial^3 u}{\partial x^3} = 0$$

subject to $u(x, 0) = f(x)$ and $u(x, t) < \infty$ for $t > 0$ and $-\infty < x < \infty$. Using the Fourier transform approach with respect to x and the transform of Airy function given in (12.87), show that the solution is given by

$$u(x, t) = \frac{1}{2\pi} \int_{-\infty}^{\infty} \text{Ai}(\zeta) f\left(x - (3t)^{1/3}\zeta\right) d\zeta$$

E12.4. Consider the diffusion equation

$$\frac{\partial u}{\partial t} - \mu^2 \frac{\partial^2 u}{\partial x^2} = 0$$

subject to $u(x, 0) = f(x)$ and $u(x, t) < \infty$ for $t > 0$ and $-\infty < x < \infty$. Using the Fourier transform with respect to x, show that the solution is given by

$$u(x, t) = \frac{1}{\sqrt{4\pi\mu^2 t}} \int_{-\infty}^{\infty} \exp\left(-\frac{\xi^2}{4\mu^2 t}\right) f\left(x - \xi\right)\, d\xi$$

However, because the coefficient outside the integral is unbounded when $t \to 0$, let $\zeta = \dfrac{\xi}{\sqrt{4\mu^2 t}}$, thus

$$u(x, t) = \frac{1}{\sqrt{\pi}} \int_{-\infty}^{\infty} e^{-\zeta^2} f\left(x - \zeta\sqrt{4\mu^2 t}\right)\, d\zeta$$

where we assume that $f(x) < \infty$ as $x \to \infty$. Obtain a time-lapse plot of $u(x,t)$ for $-3 \le x \le 3$ and $0 \le t \le 1$ with $\mu = 1$ and

$$f(x) = \begin{cases} 1 & \text{for } |x| \le 1 \\ \\ 0 & \text{otherwise} \end{cases}$$

E12.5. Recall the Black-Scholes equation described in Exercise **E11.9**,

$$\frac{\partial u}{\partial t} = \frac{1}{2}\sigma^2 x^2 \frac{\partial^2 u}{\partial x^2} + r\left(x\frac{\partial u}{\partial x} - u\right)$$

(Note: recall that t is the time from expiry of the option – not actual time.)

1. Taking a hint from the solution of Euler-Cauchy equations (cf. (6.40)), show that with $z = \ln(x)$, the Black-Scholes equation can be transformed to be
$$\frac{\partial u}{\partial t} = \frac{\sigma^2}{2}\frac{\partial^2 u}{\partial z^2} + \left(r - \frac{\sigma^2}{2}\right)\frac{\partial u}{\partial z} - ru$$
2. Next, using $u(z,t) = e^{(Az+Bt)}q(z,t)$, find the values of A and B such that the equation reduces to a diffusion equation given by
$$\frac{\partial q}{\partial t} = \frac{\sigma^2}{2}\frac{\partial^2 q}{\partial z^2}$$
3. Using the solution given in Exercise **E12.4**, show that the solution that satisfies the "European call option": $u(0,t) = 0$, $u(x,t) \to x$ as $x \to \infty$ and $u(x,0) = \max(x - K, 0)$, where K is the strike price, is the **Black-Scholes formula** for European call option given by
$$\begin{aligned} u(x,t) &= x\mathcal{N}(d_1) - Ke^{-rt}\mathcal{N}(d_2) \qquad (12.88) \\ d_1 &= \frac{\left[r + (\sigma^2/2)\right]t + \ln(x/K)}{\sqrt{\sigma^2 t}} \\ d_2 &= \frac{\left[r - (\sigma^2/2)\right]t + \ln(x/K)}{\sqrt{\sigma^2 t}} \end{aligned}$$
where $\mathcal{N}(\cdot)$ is the **cumulative normal distribution function** defined as follows:
$$\mathcal{N}(\xi) = \frac{1}{\sqrt{2\pi}}\int_{-\infty}^{\xi} e^{-\eta^2/2}\,d\eta = \frac{1 + \mathrm{erf}\left(\xi/\sqrt{2}\right)}{2}$$
(Hint: By completing squares in the arguments, it can be shown that
$$\frac{1}{\sqrt{\pi}}\int_{-\infty}^{\xi} \exp\left(-\left[a\eta^2 + b\eta + c\right]\right)d\eta = \frac{1}{\sqrt{a}}\exp\left(\frac{b^2 - 4ac}{4a}\right)\mathcal{N}\left(\frac{b + 2a\xi}{\sqrt{2a}}\right)$$
where $a > 0$.)

E12.6. Obtain the solution for the reaction-diffusion equation with time-dependent coefficient $K(t)$ given by

$$\frac{\partial u}{\partial t} = \mathcal{D}\frac{\partial^2 u}{\partial x^2} - K(t)u$$

subject to $u(x,0) = f(x)$ and $u(x,t) < \infty$ for $t > 0$ and $-\infty < x < \infty$; use the Fourier transform method with respect to x. Compare the solution with one obtained using the transformation suggested in Exercise **E11.11**.

E12.7. Consider the Laplace equation

$$\frac{\partial^2 u}{\partial x^2} + \frac{\partial^2 u}{\partial y^2} = 0 \quad ; \quad -\infty \le x \le \infty \,;\, y \ge 0$$

subject to Neumann condition given by

$$\frac{\partial u}{\partial y}(x,0) = g(x) \quad \text{and} \quad u(x,y) < \infty$$

where $\int_{-\infty}^{\infty} g(x)dx = 0$. Show that the general solution is given by

$$u(x,y) = \frac{1}{2\pi}\int_{-\infty}^{\infty} g(x-\tau)\ln\left(\frac{y^2+\tau^2}{\alpha^2+\tau^2}\right)d\tau$$

where α is an arbitrary constant. (Hint: Let $u = \int_{\alpha}^{y}\phi(x,\eta)d\eta$. Then show that $\nabla^2\phi = 0$ with $\phi(x,0) = g(x)$. After solving for $\phi(x,y)$, obtain $u(x,y)$.)

E12.8. Using the definition of Laplace transforms, prove the following identities:

$$\mathcal{L}\left[\frac{1}{\sqrt{t}}\exp\left(-\frac{1}{4t}\right)\right] = \sqrt{\frac{\pi}{s}}\,e^{-\sqrt{s}} \tag{12.89}$$

$$\mathcal{L}\left[\frac{1}{\sqrt{t^3}}\exp\left(-\frac{1}{4t}\right)\right] = \sqrt{4\pi}\,e^{-\sqrt{s}} \tag{12.90}$$

E12.9. Using a similar approach as Example 12.9, show that

$$\mathcal{L}\left[t^{\nu/2}J_\nu\left(2\sqrt{t}\right)\right] = \frac{e^{-1/s}}{s^{\nu+1}} \tag{12.91}$$

Thus, using the scaling properties, show that for $\gamma > 0$

$$\mathcal{L}^{-1}\left[\frac{1}{s}\exp\left(\frac{\gamma}{s}\right)\right] = I_0\left(2\sqrt{\gamma x}\right) \tag{12.92}$$

E12.10. To obtain the inverse Laplace transform of

$$F(s) = \frac{1}{s}\,\exp\left(\frac{\gamma s}{s+1}\right)$$

First show that

$$\int_0^{\gamma}\left(\frac{1}{s+1}\right)\exp\left(\frac{\xi s}{s+1}\right)d\xi + \frac{1}{s} = \frac{1}{s}\exp\left(\frac{\gamma s}{s+1}\right)$$

Using this identity and (12.92), prove that

$$\mathcal{L}^{-1}\left[\frac{1}{s}\,\exp\frac{\gamma s}{s+1}\right] = \mathcal{H}(t) + \int_0^{\gamma} e^{(\xi-t)}J_0\left(2\sqrt{\xi t}\right)d\xi \tag{12.93}$$

E12.11. The concentration of a reactant A undergoing a first-order reaction $A \to P$ in a plug flow reactor can be modeled by

$$\frac{\partial c}{\partial t} + v\frac{\partial c}{\partial x} = -kc \quad \text{for } 0 \le x \le L, t \ge 0$$

subject to $c(x,0) = f(x)$ and $c(0,t) = g(t)$, where v is the constant velocity of the flow. Use the Laplace transform method to solve this problem. Compare this with the solution obtained using the method of characteristics (cf. Section 10.1). (Note that the method of characteristics can also handle an nth-order reaction, i.e., with $-kc^n$ instead of $-kc$.)

E12.12. Consider a heat exchanger tube immersed in a temperature bath. The temperature of fluid flowing through the tube can be modeled as

$$\frac{\partial u}{\partial t} + v\frac{\partial u}{\partial x} = (H)\left(u_b(t) - u\right)$$

subject to $u(x,0) = f(x)$ and $u(0,t) = g(t)$ for $0 \le x \le L$ and $t \ge 0$, where v, H, and $u_b(t)$ are velocity, heat transfer coefficient, and the time-varying temperature of the bath, respectively. Solve the problem using the Laplace transform method.

E12.13. Consider the system of first-order equations given by

$$\frac{\partial u}{\partial x} = v - u \qquad \text{and} \qquad \frac{\partial v}{\partial t} = u - v$$

subject to the conditions: $u(x,0) = 0$, $v(x,0) = 0$ and $u(0,t) = u_0$, where $x \ge 0$ and $t \ge 0$. Using Laplace transforms, show that the solution for u is given by

$$u(x,t) = u_0 \left[\mathcal{H}(t) - \int_0^x e^{-(\xi+t)} I_0\left(2\sqrt{\xi t}\right) d\xi \right]$$

(Hint: You may need equation (12.93).) Obtain a set of time-lapse plots of $u(x,t)/u_0$.

E12.14. The dynamic equations describing the current and voltage distribution along a thick cable can be modeled by

$$\begin{aligned} \frac{\partial I}{\partial t} + C\frac{\partial V}{\partial t} + GV &= 0 \\ \frac{\partial V}{\partial t} + L\frac{\partial I}{\partial t} + RI &= 0 \end{aligned}$$

where $I(x,t)$ and $V(x,t)$ are the current and voltage values, whereas C, R, L, and G are constant capacitance, resistance, inductance, and leakage, respectively.

1. Show that these equations can be combined to yield

$$\frac{\partial^2 u}{\partial t^2} + (\alpha + \beta)\frac{\partial u}{\partial t} + \alpha\beta u = \gamma^2 \frac{\partial^2 u}{\partial x^2}$$

where u is either $V(x,t)$ or $I(x,t)$, with $\alpha = \dfrac{G}{C}$, $\beta = \dfrac{R}{L}$ and $\gamma = \dfrac{1}{LC}$. These equations are known as the **telegraph equations**.

2. For the special case of $\beta = \alpha$, use the Laplace transform method to solve the equation for the conditions

$$u(0,t) = g(t); \quad u(x,0) = \frac{\partial u}{\partial t}(x,0) = 0 \quad \text{and} \quad u(\infty,t) = 0$$

E12.15. Consider the **Nusselt problem** describing the heat transfer for a plug flow, as given by the following equation:

$$\frac{v}{\alpha}\frac{\partial u}{\partial z} = \frac{\partial^2 u}{\partial r^2} + \frac{1}{r}\frac{\partial u}{\partial r} \qquad 0 \le r \le R, \quad 0 \le z \le L$$

subject to: $u(z,R) = u_R$, $u(0,r) = u_{in}$ and $\dfrac{\partial u}{\partial r}(z,0) = 0$.

1. Applying Laplace transform with respect to z, show that this results in

$$\mathcal{L}[u] = \frac{u_{in}}{s} + \left(\frac{u_R - u_{in}}{s}\right)\frac{J_0\left(ia\sqrt{s}\right)}{J_0\left(ib\sqrt{s}\right)}$$

where $a = r\sqrt{\dfrac{v}{\alpha}}$ and $b = R\sqrt{\dfrac{v}{\alpha}}$ (with $i = \sqrt{-1}$).

2. Using the approach used in Example 12.7 to find the inverse Laplace transform, obtain the solution $u(z,r)$.

E12.16. Consider the **shallow water model** given in Example 10.8. After linearization, we obtain the system described as follows:

$$\begin{pmatrix} \dfrac{\partial h}{\partial t} \\ \\ \dfrac{\partial v}{\partial t} \end{pmatrix} + \begin{pmatrix} 0 & H \\ \\ g & 0 \end{pmatrix} \begin{pmatrix} \dfrac{\partial h}{\partial x} \\ \\ \dfrac{\partial v}{\partial x} \end{pmatrix} = \begin{pmatrix} 0 \\ \\ 0 \end{pmatrix}$$

where $h(x,t)$ and $v(x,t)$ are the height of the surface and velocity along x, respectively. The constants g and H are the gravitational acceleration and mean depth, respectively. Assume that the initial and boundary conditions are $h(x,0) = v(x,0) = 0$ and $v(0,t) = f(t)$, with $v(x,t) < \infty$. Using the Laplace transform method, show that the solution is given by

$$v(x,t) = f\left(t - \frac{x}{\sqrt{Hg}}\right) \qquad \text{and} \qquad h(x,t) = \sqrt{\frac{H}{g}}\, f\left(t - \frac{x}{\sqrt{Hg}}\right)$$

E12.17. Consider the diffusion equation given by

$$\frac{\partial u}{\partial t} = \mathcal{D}\frac{\partial^2 u}{\partial x^2}$$

subject to $u(x,0) = f(x)$ and $u(0,t) = g(t)$ for $x \geq 0$ and $t \geq 0$. Based on the discussion in Section 12.6, the problem can be split into two subproblems:

$$\frac{\partial u^A}{\partial t} = \mathcal{D}\frac{\partial^2 u^A}{\partial x^2} \qquad \text{subject to } u^A(x,0) = f^{\text{odd}}(x), \;\; u^A(0,t) = 0$$

$$\frac{\partial u^B}{\partial t} = \mathcal{D}\frac{\partial^2 u^B}{\partial x^2} \qquad \text{subject to } u^B(x,0) = 0, \;\; u^B(0,t) = g(t)$$

where f^{odd} is the odd function extension of $f(x)$. Solve for u^A using the Fourier transform method and u^B using the Laplace transform method. (Note that because the time derivative is first order, only one initial condition is needed.) Thus obtain the final solution as $u(x,t) = u^A + u^B$.

13 Finite Difference Methods

In this chapter, we discuss one powerful approach to obtain a numerical solution of partial differential equations. The basic approach is to replace derivatives by discrete formulas called **finite difference approximations**. After these approximations are applied to the given differential equations, the boundary conditions are included by modifying the equations that involve the boundary points. This often results in a large and sparse matrix equation, in which the desired values at the chosen grid points are combined as a vector or a matrix, depending on whether the problem involves one dimension or two dimensions. For steady-state cases, the unknown vector can be obtained by solving an algebraic equation. Conversely, for non–steady-state cases, algebraic iteration is used to obtain a time-marching solution.

Throughout the chapter, we limit our discussion to a discretization based on uniform grids. Under this assumption, different finite difference approximations can be formulated using the Taylor series expansions as discussed in Section 13.1. Several formulations are possible depending on the choice of neighboring points for different order of derivatives. Formulas for various approximations of first-order and second-order derivatives, including their order of accuracy, are given in Tables 13.1 and 13.2, respectively.

Once the derivatives in the differential equations are replaced with their finite difference approximations, the resulting formulas can be recast as matrix algebraic equations. We limit our applications to second-order linear partial differential equations. We first discuss the time-independent cases before moving on to time-dependent cases. For the time-independent cases, we start with the solutions of one-dimensional cases. Although this is technically an ordinary differential equation, it is an appropriate point to develop the basic formulas for handling either Dirchlet, Neumann, or Robin conditions. Several of the matrices are used when extending the method to two and three dimensions. For the 2D and 3D cases, additional matrix operations such as Kronecker products, vectorization, and Haddamard products (cf. Table 1.3), including their properties (cf. Table 1.4), are used extensively in our formulas. Although the finite difference method is used most often in rectangular domains, the case of cylindrical or spherical coordinates can be handled with minimal modifications, as discussed in Section 13.2.3. Ultimately, the matrix equations reduce to the linear equation of the form $A\mathbf{x} = \mathbf{b}$. Thus the techniques covered in Chapter 2 can be implemented to take advantage of the sparseness and structure of the equations.

Having developed matrix equations for one, two, or three-dimensional equations, the transition to time-dependent cases is straightforward. This is achieved by using an approach called the **semi-discrete method**, which essentially reduces the partial differential equations first into initial value problems. This means that the techniques covered in Chapter 7 can be used directly. However, due to size of the problems, the Euler methods (forward and backward difference), as well as the averaged version known as **Crank Nicholson methods**, are often the methods of choice. As it was in Chapter 7, the issue of stability becomes important, where implicit methods have the advantages of much larger stability regions. In Section 13.4, we discuss the stability analysis based on the spectral radius, but we also include another method based on Fourier analysis known as the **von Neumann analysis**.

We include the use of finite difference equations for handing hyperbolic equations in Section M.3 as an appendix. Recall that the challenge for handling these types of equations is that the solutions are expected to travel as waves. Thus discontinuities will be propagated in the path of characteristics (in nonlinear cases, these could also involve shock formations). Some of the methods are simple substitution of finite difference formulas such as the **upwind formulas**. However, other methods, such as the Lax-Wendroff method, which uses the Taylor series approximation directly on the differential equation, will arrive at a different scheme altogether.

We do not cover the solution of nonlinear partial differential equations. Instead, we just note that the extension to some nonlinear problems, such as for semilinear or quasilinear partial differential equations, are often straightforward. The additional complexity will be the convergence of the linearized formulas, either by successive substitution or Newton-type approaches.

13.1 Finite Difference Approximations

As mentioned in the introduction, the main steps in a finite difference scheme involve the following steps:

1. Discretize the domain.
2. Approximate the derivatives.
3. Include the boundary conditions.
4. Solve the algebraic equations.

In this section, we discuss the first and second steps in more detail, that is, how to obtain various finite difference approximations of derivatives. Specifically, we develop the formulas for first- and second-order derivatives. The third step, that is, the inclusion of boundary conditions, is discussed in the next section.

Let us first set some specific index notations. We will assume that t, x, y, and z domains have been discretized by uniform increments of Δt, Δx, Δy, and Δz, respectively. Thus let $u_{k,n,m}^{(q)}$ be defined as

$$u_{k,n,m}^{(q)} = u\left(q\Delta t, k\Delta x, n\Delta y, m\Delta z\right) \tag{13.1}$$

Throughout this chapter, we reserve the integers q, k, n, and m as indices for t, x, y, and z, respectively.

We also assume that each spatial dimension has already been scaled to be in the range of $[0, 1]$. For instance, the range of discretized x will be $(0, \Delta x, \ldots, K\Delta x,$

$(K+1)\Delta x)$, where $\Delta x = 1/(K+1)$. The points $x = 0$ and $x = (K+1)\Delta x$ are at the boundaries.

When some of the dimensions are fixed, the corresponding superscripts or subscripts will be suppressed. For instance, in a time-varying system that is dependent on only one spatial dimension, we use $u_k^{(q)} = u(q\Delta t, k\Delta x)$, whereas for a time-independent system in two spatial dimensions, we have $u_{k,n} = u(k\Delta x, n\Delta y)$.

Let us now start with the approximation of a first-order derivative. The Taylor series expansion of $u_{k+1} = u(x + \Delta x)$ around $x = x_k$ is given by

$$u_{k+1} = u_k + \left.\frac{du}{dx}\right|_{x_k} \Delta x + \frac{1}{2}\left.\frac{d^2u}{dx^2}\right|_{x_k} \Delta x^2 + \ldots \tag{13.2}$$

Rearranging,

$$\left.\frac{du}{dx}\right|_{x_k} = \frac{u_{k+1} - u_k}{\Delta x} + \mathcal{O}(\Delta x)$$

then dropping the $\mathcal{O}(\Delta x)$ terms,

$$\left.\frac{du}{dx}\right|_{x_k} \approx \frac{u_{k+1} - u_k}{\Delta x} \tag{13.3}$$

which is a first-order approximation known as the **forward difference approximation**. We could also expand $u_{k-1} = u(x - \Delta x, t)$ around $x = x_k$,

$$u_{k-1} = u_k - \left.\frac{du}{dx}\right|_{x_k} \Delta x + \frac{1}{2}\left.\frac{d^2u}{dx^2}\right|_{x_k} \Delta x^2 + \ldots \tag{13.4}$$

Rearranging,

$$\left.\frac{du}{dx}\right|_{x_k} = \frac{u_k - u_{k-1}}{\Delta x} + \mathcal{O}(\Delta x)$$

then dropping the $\mathcal{O}(\Delta x)$ terms,

$$\left.\frac{du}{dx}\right|_{x_k} \approx \frac{u_k - u_{k-1}}{\Delta x} \tag{13.5}$$

which is a first-order approximation known as the **backward difference approximation**.

Alternatively, we can subtract (13.4) from (13.2),

$$u_{k+1} - u_{k-1} = 2\left.\frac{du}{dx}\right|_{x_k} \Delta x + 2\frac{\Delta x^3}{6}\left.\frac{d^3u}{dx^3}\right|_{x_k} + \ldots \tag{13.6}$$

from which we get

$$\left.\frac{du}{dx}\right|_{x_k} = \frac{u_{k+1} - u_{k-1}}{2\Delta x} + \mathcal{O}\left(\Delta x^2\right) \tag{13.7}$$

and upon dropping the $\mathcal{O}\left(\Delta x^2\right)$ terms,

$$\left.\frac{du}{dx}\right|_{x_k} \approx \frac{u_{k+1} - u_{k-1}}{2\Delta x} \tag{13.8}$$

This is a second-order approximation known as the **central difference approximation**.

The finite difference approximations of higher order derivatives can likewise be obtained as a linear combination of the neighboring points as follows:

$$\left.\frac{d^P u}{dx^P}\right|_{x_k} \approx D_P = \frac{1}{\Delta x^P}\sum_{j=\ell_1}^{\ell_2} \alpha_j u_{k+j} \tag{13.9}$$

The choice for the indices ℓ_1 and ℓ_2 will determine both the order of approximation as well as the bandwidth of the matrices used to solve the finite difference equations once they are applied to the differential equations. For instance, the forward, backward, and central difference formula given in (13.3), (13.5), and (13.7) uses the limits $(\ell_1, \ell_2) = (0, 1)$, $(\ell_1, \ell_2) = (-1, 0)$, and $(\ell_1, \ell_2) = (-1, 1)$, respectively. Both forward and backward approximation formulas have bandwidths of 2 yielding a first-order approximation, whereas the central approximation formula has a bandwidth of 3 but yields a second-order approximation.

One approach to find the coefficients α_j is known as **method of undetermined coefficients** and is based on the Taylor series of u or of its Mth derivative at $x = x_{k+j}$, which we can rewrite in the following form:

$$\left.\frac{d^M u}{dx^M}\right|_{x_{k+j}} = \frac{1}{\Delta x^M}\sum_{\iota=0}^{\infty} \varphi_\iota \, \gamma_{(\iota-M),j} \tag{13.10}$$

where

$$\varphi_r = \Delta x^r \left.\frac{d^r u}{dx^r}\right|_{x_k}, \qquad \gamma_{r,s} = \begin{cases} s^r/r! & \text{if } r > 0 \\ 1 & \text{if } r = 0 \\ 0 & \text{if } r < 0 \end{cases} \tag{13.11}$$

The form given in (13.9) does not include derivatives at neighboring points. Thus substituting (13.10) with $M = 0$, $\ell_1 \le 0$, $\ell_2 \ge 0$ and $P \le (\ell_2 - \ell_1)$ into (13.9),

$$\begin{aligned} \left.\frac{d^P u}{dx^P}\right|_{x_k} &= \sum_{j=\ell_1}^{\ell_2} \frac{1}{\Delta x^P}\alpha_j u_{k+j} \\ \frac{\varphi_P}{\Delta x^P} &= \left[\sum_{\iota=0}^{(\ell_2-\ell_1)} \frac{\varphi_\iota}{\Delta x^P}\sum_{j=\ell_1}^{\ell_2}\gamma_{\iota j}\alpha_j\right] + \left[\sum_{\upsilon=(\ell_2-\ell_1)+1}^{\infty} \frac{\varphi_\upsilon}{\Delta x^P}\sum_{j=\ell_1}^{\ell_2}\gamma_{\upsilon j}\alpha_j\right] \end{aligned}$$

By setting the second sum on the right side to be the truncation error, we have the following lemma:

LEMMA 13.1. *The coefficients α_j of the finite difference approximation of a Pth-order derivative under the form given by (13.9), with $\ell_1 \le 0$, $\ell_2 \ge 0$ and $P \le (\ell_2 - \ell_1)$ is given by*

$$\begin{pmatrix} \alpha_{\ell_1} \\ \vdots \\ \alpha_{\ell_2} \end{pmatrix} = \begin{pmatrix} \gamma_{0,\ell_1} & \cdots & \gamma_{0,\ell_2} \\ \vdots & \ddots & \vdots \\ \gamma_{\ell_2-\ell_1,\ell_1} & \cdots & \gamma_{\ell_2-\ell_1,\ell_2} \end{pmatrix}^{-1} \mathbf{e}_{P+1} \tag{13.12}$$

where $\mathbf{e}_{P+1}$ *is the* $(P+1)^{\text{th}}$ *unit vector of length* $(\ell_2 - \ell_1 + 1)$*, yielding a truncation error given by*

$$\text{Error} = \sum_{\nu=(\ell_2-\ell_1)+1}^{\infty} \left[\sum_{j=\ell_1}^{\ell_2} \gamma_{\nu j} \alpha_j \right] \left(\left. \frac{d^\nu u}{dx^\nu} \right|_{x_k} \right) \Delta x^{\nu - P} \tag{13.13}$$

EXAMPLE 13.1. Approximation of Second-Order Derivatives.
For the finite difference approximation of the second-order partial derivative with respect to x, let $P = 2$, $\ell_1 = -1$ and $\ell_2 = 1$. Then (13.12) becomes

$$\begin{pmatrix} \alpha_{-1} \\ \alpha_0 \\ \alpha_1 \end{pmatrix} = \begin{pmatrix} 1 & 1 & 1 \\ -1 & 0 & 1 \\ 1/2 & 0 & 1/2 \end{pmatrix}^{-1} \begin{pmatrix} 0 \\ 0 \\ 1 \end{pmatrix} = \begin{pmatrix} 1 \\ -2 \\ 1 \end{pmatrix}$$

Thus

$$\left. \frac{d^2 u}{dx^2} \right|_{x_k} \approx \frac{1}{\Delta x^2} \left(u_{k-1} - 2u_k + u_{k+1} \right) \tag{13.14}$$

Looking at the truncation error, we see that the leading term in the summation found in (13.13) is zero, that is,

$$\gamma_{3,-1}\, \alpha_{-1} \;+\; \gamma_{3,0}\, \alpha_0 \;+\; \gamma_{3,1}\, \alpha_1 = \frac{(-1)^3}{3!} - 0 \cdot 2 + \frac{1^3}{3!} = 0$$

This means that the level of approximation in (13.14) is second-order, which is a fortunate case because the leading terms of the truncation errors do not vanish in general.

The finite difference form given in (13.9) can be extended to include derivatives at the neighboring points:

$$\left. \frac{d^P u}{dx^P} \right|_{x_k} \approx D_p = \sum_{j \in [\ell_1, \ell 2],\ \lambda \in [0, P]} \left(\frac{1}{\Delta x^{P-\lambda}} \right) \alpha_{j,\lambda} \left. \left(\frac{d^\lambda u}{dx^\lambda} \right) \right|_{x_{k+j}} \tag{13.15}$$

One reason for including derivatives of neighboring points is to conveniently apply Neumann and Robin boundary conditions at the points near the boundary. Another reason is to obtain higher order accuracy while balancing smaller matrix bandwidths against increased complexity. By applying the general Taylor series given by (13.10) to (13.15), we have the following generalization of Lemma 13.1, which we now refer to as the **finite difference approximation lemma**

LEMMA 13.2. *The* $(Q+1)$ *coefficients* $\alpha_{j,\lambda}$ *for the finite difference approximation (13.15) can be obtained by solving the following set of simultaneous equations:*

$$\sum_{j,\lambda} \gamma_{(\iota-\lambda),j} \alpha_{j,\lambda} = \delta_{P\iota} \qquad ; \qquad j \in [\ell_1, \ell_2],\ \lambda \in [0, P],\ \iota = 0, \ldots, Q \tag{13.16}$$

where δ_{rs} is the Kronecker delta (i.e., $\delta_{rr} = 1$ and 0 otherwise). The truncation error is given by

$$\text{Error} = \sum_{\nu=Q+1}^{\infty} \left[\sum_{j\in[\ell_1,\ell_2],\lambda\in[0,P]\ell_1} \gamma_{(\nu-\lambda),j}\alpha_{j,\lambda} \right] \left(\frac{d^\nu u}{dx^\nu} \right)\bigg|_{x_k} \Delta x^{\nu-P} \tag{13.17}$$

EXAMPLE 13.2. Let J and Λ be the list of indices for j and λ, respectively. For instance, set $J = \{-1, 0, 1\}$ and $\Lambda = \{1, 0, 0\}$ for the approximation given by

$$\frac{d^2u}{dx^2} \approx \frac{1}{\Delta x^2}\left(\alpha_{-1,1}\Delta x \left.\frac{du}{dx}\right|_{x_{k-1}} + \alpha_{0,0}u_k + \alpha_{1,0}u_{k+1} \right)$$

Lemma 13.2 yields the following equation:

$$\begin{pmatrix} 0 & 1 & 1 \\ 1 & 0 & 1 \\ -1 & 0 & 1/2 \end{pmatrix} \begin{pmatrix} \alpha_{-1,1} \\ \alpha_{0,0} \\ \alpha_{-1,0} \end{pmatrix} = \begin{pmatrix} 0 \\ 0 \\ 1 \end{pmatrix} \quad \rightarrow \quad \begin{pmatrix} \alpha_{-1,1} \\ \alpha_{0,0} \\ \alpha_{-1,0} \end{pmatrix} = \frac{2}{3}\begin{pmatrix} -1 \\ -1 \\ 1 \end{pmatrix}$$

or

$$\frac{d^2u}{dx^2} \approx \frac{2}{3\Delta x^2}\left(-\Delta x \left.\frac{du}{dx}\right|_{x_{k-1}} - u_k + u_{k+1} \right) \tag{13.18}$$

Furthermore, the lemma also predicts Error $= \mathcal{O}(\Delta x)$ for this approximation.

Remarks: A MATLAB code to obtain the coefficients of a finite difference approximation of a Pth-order derivative based on a given list of indices J and Λ is available on the book's webpage as `eval_pade_gen_coef.m`. The function is invoked by the statement `[v,ord]=eval_pade_gen_coef(J,Lambda,P)` to obtain `v` as the vector of coefficients and `ord` as the order of approximation.

A list of common finite difference approximations for the first- and second-order derivatives based on the method of undetermined coefficients is given in Tables 13.1 and 13.2, respectively. These are used later in formulating the numerical solution of differential equations. As expected, more terms are needed for increased precision. A balance has to be struck between a smaller Δx or using more terms in the finite difference approximations. Having more terms in the approximation formula will increase the number of computations due to a matrix with larger bandwidths. However, a smaller Δx will also increase the number of computations because the size of the matrix will have to be enlarged to achieve a similar precision.

Note that items 7 through 12 of Table 13.1 and items 5 through 12 of Table 13.2 involve derivatives at the leftmost or rightmost neighbor. These approximation formulas are used when dealing with Neumann boundary conditions.

For mixed derivatives, the same method of undetermined coefficients can be used with the Taylor series expansion of multivariable functions but would require much more complex formulation. For the case of $\dfrac{\partial^2 u}{\partial t \partial x}$, we have

$$\left.\frac{\partial^2 u}{\partial x \partial t}\right|_{(x_k,t_q)} \approx \frac{1}{4\Delta x \Delta t}\left[\left(u_{k+1}^{(q+1)} - u_{k-1}^{(q+1)} \right) - \left(u_{k+1}^{(q-1)} - u_{k-1}^{(q-1)} \right) \right] \tag{13.19}$$

Table 13.1. *Finite difference approximations of first-order derivatives,* $\left(\left.\dfrac{du}{dx}\right|_{x_k}\right)$

Item	Approximation formula	Error order
1	$\dfrac{1}{\Delta x}\left[-u_k + u_{k+1}\right]$	$\mathcal{O}\left(\Delta x\right)$
2	$\dfrac{1}{\Delta x}\left[-u_{k-1} + u_k\right]$	$\mathcal{O}\left(\Delta x\right)$
3	$\dfrac{1}{2\Delta x}\left[-u_{k-1} + u_{k+1}\right]$	$\mathcal{O}\left(\Delta x^2\right)$
4	$\dfrac{1}{12\Delta x}\left[+u_{k-2} - 8u_{k-1} + 8u_{k+1} - u_{k+2}\right]$	$\mathcal{O}\left(\Delta x^4\right)$
5	$\dfrac{1}{12\Delta x}\left[-3u_{k-1} - 10u_k + 18u_{k+1} - 6u_{k+2} + u_{k+3}\right]$	$\mathcal{O}\left(\Delta x^4\right)$
6	$\dfrac{1}{12\Delta x}\left[-u_{k-3} + 6u_{k-2} - 18u_{k-1} + 10u_k + 3u_{k+1}\right]$	$\mathcal{O}\left(\Delta x^4\right)$
7	$\dfrac{1}{3}\left[\left(\dfrac{du}{dx}\right)_{k-1} - \dfrac{2}{\Delta x}u_k + \dfrac{2}{\Delta x}u_{k+1}\right]$	$\mathcal{O}\left(\Delta x^2\right)$
8	$\dfrac{1}{3}\left[-\dfrac{2}{\Delta x}u_{k-1} + \dfrac{2}{\Delta x}u_k + \left(\dfrac{du}{dx}\right)_{k+1}\right]$	$\mathcal{O}\left(\Delta x^2\right)$
9	$\dfrac{1}{75}\left[-3\left(\dfrac{du}{dx}\right)_{k-2} - \dfrac{38}{\Delta x}u_{k-1} - \dfrac{9}{\Delta x}u_k + \dfrac{54}{\Delta x}u_{k+1} - \dfrac{7}{\Delta x}u_{k+2}\right]$	$\mathcal{O}\left(\Delta x^4\right)$
10	$\dfrac{1}{150}\left[18\left(\dfrac{du}{dx}\right)_{k-1} - \dfrac{197}{\Delta x}u_k + \dfrac{279}{\Delta x}u_{k+1} - \dfrac{99}{\Delta x}u_{k+2} + \dfrac{17}{\Delta x}u_{k+3}\right]$	$\mathcal{O}\left(\Delta x^4\right)$
11	$\dfrac{1}{75}\left[\dfrac{7}{\Delta x}u_{k-2} - \dfrac{54}{\Delta x}u_{k-1} + \dfrac{9}{\Delta x}u_k + \dfrac{38}{\Delta x}u_{k+1} - 3\left(\dfrac{du}{dx}\right)_{k+2}\right]$	$\mathcal{O}\left(\Delta x^4\right)$
12	$\dfrac{1}{150}\left[-\dfrac{17}{\Delta x}u_{k-3} + \dfrac{99}{\Delta x}u_{k-2} - \dfrac{279}{\Delta x}u_{k-1} + \dfrac{197}{\Delta x}u_k + 18\left(\dfrac{du}{dx}\right)_{k+1}\right]$	$\mathcal{O}\left(\Delta x^4\right)$

with Error $= \mathcal{O}\left(\Delta t^2, \Delta x^2\right)$. Likewise, we have

$$\left.\frac{\partial^2 u}{\partial x \partial y}\right|_{(x_k, y_n)} \approx \frac{1}{4\Delta x \Delta y}\left[\left(u_{k+1,n+1} - u_{k+1,n-1}\right) - \left(-u_{k-1,n+1} + u_{k-1,n-1}\right)\right] \qquad (13.20)$$

with Error $= \mathcal{O}\left(\Delta x^2, \Delta y^2\right)$. The details for this result can be found in example M.1 contained in the appendix as Section M.1. For higher order approximations, a simpler

Table 13.2. *Finite difference approximations of second order derivatives,* $\left(\left.\dfrac{d^2u}{dx^2}\right|_{x_k}\right)$

Item	Approximation formula	Error order
1	$\dfrac{1}{\Delta x^2}\left[u_{k-1} - 2u_k + u_{k+1}\right]$	$\mathcal{O}\left(\Delta x^2\right)$
2	$\dfrac{1}{12\Delta x^2}\left[-u_{k-2} + 16u_{k-1} - 30u_k + 16u_{k+1} - u_{k+2}\right]$	$\mathcal{O}\left(\Delta x^4\right)$
3	$\dfrac{1}{12\Delta x^2}\left[10u_{k-1} - 15u_k - 4u_{k+1} + 14u_{k+2} - 6u_{k+3} + u_{k+4}\right]$	$\mathcal{O}\left(\Delta x^4\right)$
4	$\dfrac{1}{12\Delta x^2}\left[u_{k-4} - 6u_{k-3} + 14u_{k-2} - 4u_{k-1} - 15u_k + 10u_{k+1}\right]$	$\mathcal{O}\left(\Delta x^4\right)$
5	$\dfrac{1}{3\Delta x^2}\left[-2\Delta x\left(\dfrac{du}{dx}\right)_{k-1} - 2u_k + 2u_{k+1}\right]$	$\mathcal{O}\left(\Delta x\right)$
6	$\dfrac{1}{3\Delta x^2}\left[2u_{k-1} - 2u_k + 2\Delta x\left(\dfrac{du}{dx}\right)_{k+1}\right]$	$\mathcal{O}\left(\Delta x\right)$
7	$\dfrac{1}{11\Delta x^2}\left[-6\Delta x\left(\dfrac{du}{dx}\right)_{k-1} - 4u_k + 2u_{k+1} + 2u_{k+2}\right]$	$\mathcal{O}\left(\Delta x^2\right)$
8	$\dfrac{1}{11\Delta x^2}\left[2u_{k-2} + 2u_{k-1} - 4u_k + 6\Delta x\left(\dfrac{du}{dx}\right)_{k+1}\right]$	$\mathcal{O}\left(\Delta x^2\right)$
9	$\dfrac{1}{822\Delta x^2}\left[\begin{array}{c} 30\Delta x\left(\dfrac{du}{dx}\right)_{k-2} + 946u_{k-1} - 1905u_k \\ + 996u_{k+1} - 31u_{k+2} - 6u_{k+3}\end{array}\right]$	$\mathcal{O}\left(\Delta x^4\right)$
10	$\dfrac{1}{1644\Delta x^2}\left[\begin{array}{c} -600\Delta x\left(\dfrac{du}{dx}\right)_{k-1} + 945u_k - 3548u_{k+1} \\ + 3918u_{k+2} - 1572u_{k+3} + 257u_{k+4}\end{array}\right]$	$\mathcal{O}\left(\Delta x^4\right)$
11	$\dfrac{1}{822\Delta x^2}\left[\begin{array}{c} -6u_{k-3} - 31u_{k-2} + 996u_{k-1} - 1905u_k \\ +946u_{k+1} - 30\Delta x\left(\dfrac{du}{dx}\right)_{k+2}\end{array}\right]$	$\mathcal{O}\left(\Delta x^4\right)$
12	$\dfrac{1}{1644\Delta x^2}\left[\begin{array}{c} +257u_{k-4} - 1572u_{k-3} + 3918u_{k-2} - 3548u_{k-1} \\ + 945u_k + 600\Delta x\left(\dfrac{du}{dx}\right)_{k+1}\end{array}\right]$	$\mathcal{O}\left(\Delta x^4\right)$

but slightly limited approach for mixed partial derivatives is to build up the finite difference approximation by using the fact that

$$\frac{\partial^2 u}{\partial x \partial y} = \frac{\partial}{\partial x}\left(\frac{\partial u}{\partial y}\right), \quad \frac{\partial^3 u}{\partial x \partial^2 y} = \frac{\partial}{\partial x}\left(\frac{\partial^2 u}{\partial y^2}\right), \ldots, \text{ etc.}$$

Thus (13.20) could also have been derived by applying central difference approximations on the variable y followed by another central difference approximation on the variable x, that is,

$$\frac{\partial^2 u}{\partial x \partial y} = \left(\left.\frac{u_{i,n+1} - u_{i,n-1}}{2\Delta y}\right|_{i=k+1} - \left.\frac{u_{i,n+1} - u_{i,n-1}}{2\Delta y}\right|_{i=k-1} \right) \left(\frac{1}{2\Delta x} \right) + \mathcal{O}\left(\Delta x^2, \Delta y^2\right)$$

In case one needs to accommodate a Neumann boundary condition at $y = 0$, another useful form based on item 7 in Table 13.1 is given by

$$\begin{aligned} \frac{\partial^2 u}{\partial x \partial y} &= \frac{1}{2\Delta x \Delta y} \left[\frac{\Delta y}{3} \left(\left.\frac{\partial u}{\partial y}\right|_{k+1,n-1} - \left.\frac{\partial u}{\partial y}\right|_{k-1,n-1} \right) - \frac{2}{3} \left(u_{k+1,n} - u_{k-1,n} \right) \right. \\ &\qquad \left. + \frac{2}{3} \left(u_{k+1,n+1} - u_{k-1,n+1} \right) \right] \end{aligned} \tag{13.21}$$

The derivation of (13.21) is included as exercise **E13.2**.

13.2 Time-Independent Equations

We first apply the finite-difference approach to solve partial differential equations that are independent of time. The finite difference approximations are expressed in matrix forms. This reduces the problem to the solution of a set of simultaneous equations for u at the grid points.

13.2.1 One-Dimensional Case

A general second-order linear differential equation is given by

$$\frac{d^2u}{dx^2} + \beta(x)\frac{du}{dx} + \zeta(x)u + \eta(x) = 0 \tag{13.22}$$

Note that the problem is a two-point boundary value problem that can also be approached using the techniques given in Section 7.5 based on shooting methods. Nonetheless, the matrix formulations in this section are used when they are extended to the 2D and 3D cases.

Define the following vectors and matrices as:

$$\mathbf{u} = \begin{pmatrix} u_1 \\ \vdots \\ u_K \end{pmatrix} \quad \underline{\underline{\beta}} = \begin{pmatrix} \beta_1 & & 0 \\ & \ddots & \\ 0 & & \beta_K \end{pmatrix} \quad \underline{\underline{\zeta}} = \begin{pmatrix} \zeta_1 & & 0 \\ & \ddots & \\ 0 & & \zeta_K \end{pmatrix} \quad \underline{\eta} = \begin{pmatrix} \eta_1 \\ \vdots \\ \eta_K \end{pmatrix}$$

where $u_k = u(k\Delta x)$, $\beta_k = \beta(k\Delta x)$, $\zeta_k = \zeta(k\Delta x)$, and $\eta_k = \eta(k\Delta x)$. The terms connected with the derivatives are given in matrix form as

$$\frac{du}{dx} \rightarrow \mathcal{D}_1\mathbf{u} + \mathfrak{b}_1 \quad \text{and} \quad \frac{d^2u}{dx^2} \rightarrow \mathcal{D}_2\mathbf{u} + \mathfrak{b}_2$$

The elements and structure of the $K \times K$ matrix $\mathcal{D}_P$ are formed using the formulas in Tables 13.1 and 13.2, depending on the choice of precision and types of boundary conditions, whereas the elements of the $K \times 1$ vector $\mathfrak{b}_P$ contain the given boundary values of the problem. Thus the finite difference approximation of (13.22) can be represented by a matrix equation

$$(\mathcal{D}_2\mathbf{u} + \mathfrak{b}_2) + \underline{\underline{\beta}}(\mathcal{D}_1\mathbf{u} + \mathfrak{b}_1) + \underline{\zeta}\mathbf{u} + \underline{\eta} = 0 \quad \rightarrow \quad A\mathbf{u} = \mathbf{b} \tag{13.23}$$

where

$$A = \mathcal{D}_2 + \underline{\underline{\beta}}\mathcal{D}_1 + \underline{\zeta} \qquad \text{and} \qquad \mathbf{b} = -\left(\mathfrak{b}_2 + \underline{\underline{\beta}}\mathfrak{b}_1 + \underline{\eta}\right)$$

and the solution can be found using techniques found in Chapter 2 that take advantage of the band structure and sparsity of matrix A. What remains at this point is to specify $\mathcal{D}_1$, $\mathcal{D}_2$, $\mathfrak{b}_1$, and $\mathfrak{b}_2$.

We limit our discussion to two cases. In the first case, the boundary conditions are Dirichlet type at both $x = 0$ and $x = 1$, that is, $u(0) = Di_0$ and $u(1) = Di_1$. Using the formulas in Tables 13.1 and 13.2, we have for Error $= \mathcal{O}\left(\Delta x^2\right)$ (where we denote the Dirichlet or Neumann boundary condition by (Di) or (Neu), respectively as a superscript for $x = 0$ and as a subscript for $x = 1$)

$$\left[\mathcal{D}_1\right]_{(Di)}^{(Di)}\Bigg|_{\mathcal{O}(\Delta x^2)} = \frac{1}{2\Delta x}\begin{pmatrix} 0 & 1 & & \\ -1 & \ddots & \ddots & \\ & \ddots & \ddots & 1 \\ & & -1 & 0 \end{pmatrix} \text{ and } \left[\mathfrak{b}_1\right]_{(Di)}^{(Di)}\Bigg|_{\mathcal{O}(\Delta x^2)} = \frac{1}{2\Delta x}\begin{pmatrix} -Di_0 \\ 0 \\ \vdots \\ 0 \\ Di_1 \end{pmatrix} \tag{13.24}$$

$$\left[\mathcal{D}_2\right]_{(Di)}^{(Di)}\Bigg|_{\mathcal{O}(\Delta x^2)} = \frac{1}{\Delta x^2}\begin{pmatrix} -2 & 1 & & \\ 1 & \ddots & \ddots & \\ & \ddots & \ddots & 1 \\ & & 1 & -2 \end{pmatrix} \text{ and } \left[\mathfrak{b}_2\right]_{(Di)}^{(Di)}\Bigg|_{\mathcal{O}(\Delta x^2)} = \frac{1}{\Delta x^2}\begin{pmatrix} Di_0 \\ 0 \\ \vdots \\ 0 \\ Di_1 \end{pmatrix} \tag{13.25}$$

For higher precision, that is, Error $= \mathcal{O}\left(\Delta x^4\right)$, we have

$$\left[\mathcal{D}_1\right]_{(Di)}^{(Di)}\Bigg|_{\mathcal{O}(\Delta x^4)} = \frac{1}{12\Delta x}\begin{pmatrix} -10 & 18 & -6 & 1 & & & \\ -8 & 0 & 8 & -1 & & & \\ 1 & -8 & 0 & 8 & -1 & & \\ & \ddots & \ddots & \ddots & \ddots & \ddots & \\ & & 1 & -8 & 0 & 8 & -1 \\ & & & 1 & -8 & 0 & 8 \\ & & & -1 & 6 & -18 & 10 \end{pmatrix}$$

$$\left[\mathfrak{b}_1\right]_{(Di)}^{(Di)}\Bigg|_{\mathcal{O}(\Delta x^4)} = \frac{1}{12\Delta x}\begin{pmatrix} -3Di_0 \\ Di_0 \\ 0 \\ \vdots \\ 0 \\ -Di_1 \\ 3Di_1 \end{pmatrix} \tag{13.26}$$

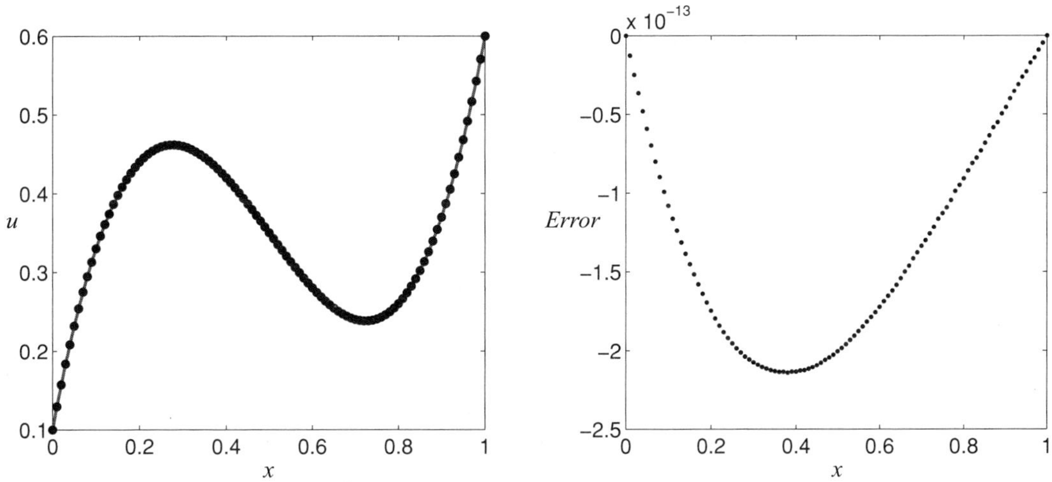

Figure 13.1. A plot of the exact solution for u (solid line) and finite difference solution (points). On the right is a plot of the errors of the approximate solution from the exact solution.

$$\left[\mathcal{D}_2\right]^{(Di)}_{(Di)}\Big|_{\mathcal{O}(\Delta x^4)} = \frac{1}{12\Delta x^2}\begin{pmatrix} -15 & -4 & 14 & -6 & 1 & & \\ 16 & -30 & 16 & -1 & & & \\ -1 & 16 & -30 & 16 & -1 & & \\ & \ddots & \ddots & \ddots & \ddots & \ddots & \\ & & -1 & 16 & -30 & 16 & -1 \\ & & & -1 & 16 & -30 & 16 \\ & & 1 & -6 & 14 & -4 & -15 \end{pmatrix}$$

$$\left[\mathfrak{b}_2\right]^{(Di)}_{(Di)}\Big|_{\mathcal{O}(\Delta x^4)} = \frac{1}{12\Delta x^2}\begin{pmatrix} 10Di_0 \\ -Di_0 \\ 0 \\ \vdots \\ 0 \\ -Di_1 \\ 10Di_1 \end{pmatrix} \tag{13.27}$$

EXAMPLE 13.3. Consider the differential equation

$$\frac{d^2u}{dx^2} + (3-x)\frac{du}{dx} + 5u = 10x^3 + 22.5x^2 - 3x - 5.5$$

subject to Dirichlet conditions $u(0) = 0.1$ and $u(1) = 0.6$. The exact solution is known to be: $u(x) = 5x^3 - 7.5x^2 + 3x + 0.1$.

Using $\Delta x = 0.01$ plus the matrices defined in (13.26) and (13.27) for high precision, we obtain the plots for the solution and errors shown in Figure 13.1. Note that the errors are within $\pm 10^{-12}$. It can be shown that using (13.24) and (13.25) instead would have yielded errors within $\pm 10^{-3}$.

For the second case, we consider the boundary condition at $x = 0$ to be a Neumann type and the boundary condition at $x = 1$ to remain a Dirichlet type, that is,

$du/dx(0) = Neu_0$ and $u(1) = Di_1$. Then we can use item 7 in Table 13.1 and item 7 in Table 13.2 for Error $= \mathcal{O}\left(\Delta x^2\right)$,

$$\left[\mathcal{D}_1\right]_{(Di)}^{(Neu)}\Bigg|_{\mathcal{O}(\Delta x^2)} = \frac{1}{2\Delta x}\begin{pmatrix} -\frac{4}{3} & \frac{4}{3} & & & \\ -1 & 0 & 1 & & \\ & \ddots & \ddots & \ddots & \\ & & -1 & 0 & 1 \\ & & & -1 & 0 \end{pmatrix}$$

$$\left[\mathfrak{b}_1\right]_{(Di)}^{(Neu)}\Bigg|_{\mathcal{O}(\Delta x^2)} = \frac{1}{2\Delta x}\begin{pmatrix} \frac{2}{3}\Delta x\, Neu_0 \\ 0 \\ \vdots \\ 0 \\ Di_1 \end{pmatrix} \tag{13.28}$$

$$\left[\mathcal{D}_2\right]_{(Di)}^{(Neu)}\Bigg|_{\mathcal{O}(\Delta x^2)} = \frac{1}{\Delta x^2}\begin{pmatrix} \frac{-4}{11} & \frac{2}{11} & \frac{2}{11} & & \\ 1 & -2 & 1 & & \\ & \ddots & \ddots & \ddots & \\ & & 1 & -2 & 1 \\ & & & 1 & -2 \end{pmatrix}$$

$$\left[\mathfrak{b}_2\right]_{(Di)}^{(Neu)}\Bigg|_{\mathcal{O}(\Delta x^2)} = \frac{1}{\Delta x^2}\begin{pmatrix} \frac{-6}{11}\Delta x\, Neu_0 \\ 0 \\ \vdots \\ 0 \\ Di_1 \end{pmatrix} \tag{13.29}$$

The matrices for the second case above corresponding to Error $= \mathcal{O}\left(\Delta x^4\right)$ is left as an exercise in **E13.3**.

Another approach to handling Neumann boundary conditions is to apply a central difference approximation of the derivative itself. For instance, let the Neumann boundary condition be given as $du/dx(0) = Neu_0$. Then this can be approximated first as

$$\left.\frac{du}{dx}\right|_{x_0} \approx \frac{-u_{-1} + u_1}{2\Delta x} \quad \longrightarrow \quad u_{-1} = u_1 + 2\Delta x Neu_0 \tag{13.30}$$

Because u_{-1} is a value at a *fictitious or ghost point*; that is, it is not in the actual problem domain, the approach is sometimes known as **method of ghost points**. To complete this method, the solution domain will be extended to include the point at x_0; that is, we extend the previous definitions as

$$\mathbf{u}^* = \begin{pmatrix} u_0 \\ u_1 \\ \vdots \\ u_K \end{pmatrix} \quad \underline{\underline{\beta}}^* = \begin{pmatrix} \beta_0 & & & \\ & \beta_1 & & 0 \\ & & \ddots & \\ 0 & & & \beta_K \end{pmatrix} \quad \underline{\underline{\zeta}}^* = \begin{pmatrix} \zeta_0 & & & \\ & \zeta_1 & & 0 \\ & & \ddots & \\ 0 & & & \zeta_K \end{pmatrix} \quad \underline{\eta}^* = \begin{pmatrix} \eta_0 \\ \eta_1 \\ \vdots \\ \eta_K \end{pmatrix}$$

The matrix formulation becomes

$$\left[\mathcal{D}_1^*\right]_{(Di)}^{(Neu)}\Bigg|_{\mathcal{O}(\Delta x^2)} = \frac{1}{2\Delta x}\begin{pmatrix} 0 & 0 & & & \\ -1 & 0 & 1 & & \\ & \ddots & \ddots & \ddots & \\ & & -1 & 0 & 1 \\ & & & -1 & 0 \end{pmatrix}$$

$$\left[\mathfrak{b}_1^*\right]_{(Di)}^{(Neu)}\Bigg|_{\mathcal{O}(\Delta x^2)} = \frac{1}{2\Delta x}\begin{pmatrix} 2\Delta x Neu_0 \\ 0 \\ \vdots \\ 0 \\ Di_1 \end{pmatrix} \tag{13.31}$$

$$\left[\mathcal{D}_2^*\right]_{(Di)}^{(Neu)}\Bigg|_{\mathcal{O}(\Delta x^2)} = \frac{1}{\Delta x^2}\begin{pmatrix} -2 & 2 & & & \\ 1 & -2 & 1 & & \\ & \ddots & \ddots & \ddots & \\ & & 1 & -2 & 1 \\ & & & 1 & -2 \end{pmatrix}$$

$$\left[\mathfrak{b}_2^*\right]_{(Di)}^{(Neu)}\Bigg|_{\mathcal{O}(\Delta x^2)} = \frac{1}{\Delta x^2}\begin{pmatrix} -2\Delta x Neu_0 \\ 0 \\ \vdots \\ 0 \\ Di_1 \end{pmatrix} \tag{13.32}$$

and the solution can be obtained by solving

$$A^*\mathbf{u}^* = \mathbf{b}^* \tag{13.33}$$

where

$$A^* = \mathcal{D}_2^* + \underline{\underline{\beta}}^*\mathcal{D}_1^* + \underline{\underline{\zeta}}^* \qquad \text{and} \qquad \mathbf{b}^* = -\left(\mathfrak{b}_2^* + \underline{\underline{\beta}}^*\mathfrak{b}_1^* + \underline{\eta}^*\right)$$

EXAMPLE 13.4. Consider again the differential equation of Example 13.3

$$\frac{d^2u}{dx^2} + (3-x)\frac{du}{dx} + 5u = 10x^3 + 22.5x^2 - 3x - 5.5$$

except this time we have a Neumann condition $du/dx(0) = 3$ and a Dirichlet condition $u(1) = 0.6$. The exact solution is still $u(x) = 5x^3 - 7.5x^2 + 3x + 0.1$.

Using $\Delta x = 0.01$, the plots shown in Figure 13.2 compare the result using the method of ghost points and the result using the matrices in (13.28) and (13.29). For both methods the errors are within $\pm 10^{-2}$, with the direct method slightly better for this example.

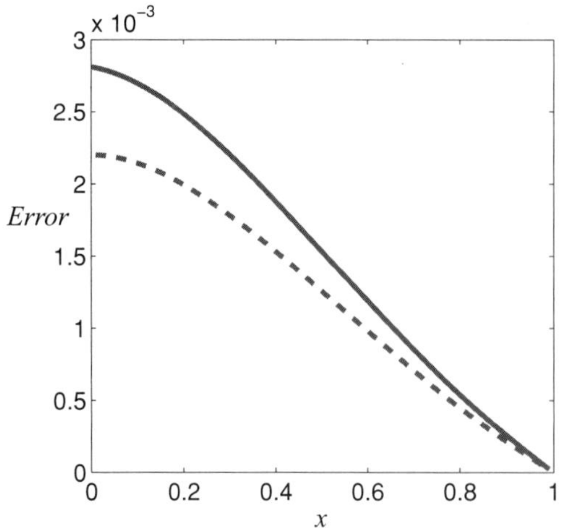

Figure 13.2. A plot of the errors from exact solution using the method of ghost points (solid line) and the direct method using matrices defined in (13.28) and (13.29) (dashed line).

13.2.2 Two-Dimensional Case

To formulate the problem for the 2D case, we can collect the values of u in a $K \times N$ matrix $U = (u_{ij})$ where the row indices refer to the x positions and the column indices refer to the y positions, that is,

$$u_{ij} = u\left(i\Delta x, j\Delta x\right) \qquad 1 \le i \le K, 1 \le j \le N$$

and $u_{0,j}$, $u_{K+1,j}$, $u_{i,0}$ and $u_{i,N+1}$ refer to the boundary values. With this convention, we can put the terms corresponding to finite difference approximations of the partial derivatives in (13.36) in the following matrix forms:

$$\frac{\partial u}{\partial x} \to \mathcal{D}_{(1,x)}U + \mathcal{B}_{(1,x)} \quad ; \quad \frac{\partial^2 u}{\partial x^2} \to \mathcal{D}_{(2,x)}U + \mathcal{B}_{(2,x)}$$

$$\frac{\partial u}{\partial y} \to U\mathcal{D}_{(1,y)}^T + \mathcal{B}_{(1,y)}^T \quad ; \quad \frac{\partial^2 u}{\partial y^2} \to U\mathcal{D}_{(2,y)}^T + \mathcal{B}_{(2,y)}^T \tag{13.34}$$

Depending on boundary conditions and order of approximations, the matrices $\mathcal{D}_{(1,x)}$, $\mathcal{D}_{(2,x)}$, $\mathcal{D}_{(1,y)}$, and $\mathcal{D}_{(2,y)}$ can take the forms of $\mathcal{D}_i$'s given in (13.24) to (13.27) or (13.28) and (13.29), where the additional subscripts x and y simply indicate the sizes $K \times K$ and $N \times N$, respectively. Conversely, the matrices $\mathcal{B}_{(1,x)}$, $\mathcal{B}_{(2,x)}$, $\mathcal{B}_{(1,y)}$, and $\mathcal{B}_{(2,y)}$ are formed by augmenting the vectors $\mathfrak{b}_i$'s given in (13.24) to (13.27) or (13.28) and (13.29), where each column is evaluated at fixed values of the other variable. For instance, if the problem specifies Dirichlet conditions at both $x = 0$ and $x = 1$ while using the central difference formula, then

$$\mathcal{B}_{(1,x)} = \left(\left(\mathfrak{b}_{(1,x)}\right)_{y_1} \Big| \cdots \Big| \left(\mathfrak{b}_{(1,x)}\right)_{y_N} \right) = \frac{1}{2\Delta x}\left(\begin{array}{c|c|c} -u_{0,1} & & -u_{0,N} \\ 0 & & 0 \\ \vdots & \cdots & \vdots \\ 0 & & 0 \\ u_{K+1,1} & & u_{K+1,N} \end{array}\right)$$

For the mixed derivative $\partial^2 u/(\partial x \partial y)$, the matrix formulation can be obtained by applying the partial derivatives in sequence, that is,

$$\frac{\partial^2 u}{\partial x \partial y} \to \mathcal{D}_{(1,x)}U\mathcal{D}_{(1_y)}^T + \mathcal{B}_{(1,x,1,y)} \tag{13.35}$$

where

$$\mathcal{B}_{(1,x,1,y)} = \mathcal{D}_{(1,x)}\mathcal{B}_{(1,y)}^T + \mathcal{B}_{(1,x)}\mathcal{D}_{(1,y)}^T + \mathcal{C}_{(1,x,1,y)}$$

and $\mathcal{C}_{(1,x,1,y)}$ introduces data from extreme corner points of the domain. For instance, if the boundary conditions are all of Dirichlet types, while using the central difference formulas of first-order derivatives, we have

$$\mathcal{C}_{(1,x,1,y)} = \frac{1}{4\Delta x \Delta y}\begin{pmatrix} u_{0,0} & 0 & \cdots & 0 & u_{0,N+1} \\ 0 & & & & 0 \\ \vdots & & 0 & & \vdots \\ 0 & & & & 0 \\ u_{N+1,0} & 0 & \cdots & 0 & u_{N+1,N+1} \end{pmatrix}$$

With these matrix representations, we can now formulate the finite difference solution of a linear second-order linear differential equation given by

$$\mu_{xx}(x,y)\frac{\partial^2 u}{\partial x^2} + \mu_{xy}(x,y)\frac{\partial^2 u}{\partial x \partial y} + \mu_{yy}(x,y)\frac{\partial^2 u}{\partial y^2}$$
$$+\beta_x(x,y)\frac{\partial u}{\partial x} + \beta_y(x,y)\frac{\partial u}{\partial y} + \zeta(x,y)u + \eta(x,y) = 0 \tag{13.36}$$

Similar to matrix U, we can evaluate the coefficients μ_{xx}, μ_{xy}, … at different spatial positions to construct the following $K \times N$ matrices: $\underline{\underline{\mu_{xx}}}$, $\underline{\underline{\mu_{xy}}}$, $\underline{\underline{\mu_{yy}}}$, $\underline{\underline{\beta_x}}$, $\underline{\underline{\beta_y}}$, $\underline{\underline{\zeta}}$, and $\underline{\underline{\eta}}$, where the row and column indices also refer to the x and y positions, respectively. Thus we have

$$\underline{\underline{\mu_{xx}}} \circ \left(\mathcal{D}_{(2,x)}U + \mathcal{B}_{2,x}\right) + \underline{\underline{\mu_{xy}}} \circ \left(\mathcal{D}_{(1,x)}U\mathcal{D}_{(1,y)}^T + \mathcal{B}_{1,x,1,y}\right) + \underline{\underline{\mu_{yy}}} \circ \left(U\mathcal{D}_{(2,y)}^T + \mathcal{B}_{2,y}^T\right)$$
$$+\underline{\underline{\beta_x}} \circ \left(\mathcal{D}_{(1,x)}U + \mathcal{B}_{1,x}\right) + \underline{\underline{\beta_y}} \circ \left(U\mathcal{D}_{(1,y)}^T + \mathcal{B}_{1,y}^T\right)$$
$$+\underline{\underline{\zeta}} \circ U + \underline{\underline{\eta}} = 0 \tag{13.37}$$

where ($\circ$) is the Hadammard product (or element-wise product). These can be further rearranged to formulate a linear equation by recalling the properties of matrix vectorization (cf. Table 1.4). Specifically,

$$\begin{aligned} \text{vec}\,(A+B) &= \text{vec}\,(A) + \text{vec}\,(B) \\ \text{vec}\,(BAC) &= \left(C^T \otimes B\right)\text{vec}\,(A) \\ \text{vec}\,(A \circ B) &= \text{vec}(A) \circ \text{vec}(B) = \left(A^{\text{dv}}\right)\text{vec}(B) \end{aligned} \tag{13.38}$$

where we use the superscript symbol "dv" to denote the diagonalized-vectorization operation, that is,

$$[\ \diamond\ ]^{\text{dv}} = \text{diag}\left(\text{vec}\left(\diamond\right)\right)$$

With (13.38), equation (13.37) can be reduced to the following matrix equation:

$$\mathfrak{R}_{2D}\mathbf{v} = -\mathfrak{f}_{2D} \tag{13.39}$$

where $\mathbf{v} = \text{vec}(U)$ and

$$\begin{aligned}\mathfrak{R}_{2D} &= \underline{\underline{\mu_{xx}}}^{\text{dv}}\left(I_N \otimes \mathcal{D}_{(2,x)}\right) + \underline{\underline{\mu_{xy}}}^{\text{dv}}\left(\mathcal{D}_{(1,y)} \otimes \mathcal{D}_{(1,x)}\right) + \underline{\underline{\mu_{yy}}}^{\text{dv}}\left(\mathcal{D}_{(2,y)} \otimes I_K\right)\\ &\quad + \underline{\underline{\beta_x}}^{\text{dv}}\left(I_N \otimes \mathcal{D}_{(1,x)}\right) + \underline{\underline{\beta_y}}^{\text{dv}}\left(\mathcal{D}_{(1,y)} \otimes I_K\right) + \underline{\underline{\zeta}}^{\text{dv}}\end{aligned}$$

$$\begin{aligned}\mathfrak{f}_{2D} &= \underline{\underline{\mu_{xx}}}^{\text{dv}}\left(\text{vec}\left(\mathcal{B}_{(2,x)}\right)\right) + \underline{\underline{\mu_{xy}}}^{\text{dv}}\left(\text{vec}\left(\mathcal{B}_{(1,x,1,y)}\right)\right) + \underline{\underline{\mu_{yy}}}^{\text{dv}}\left(\text{vec}\left(\mathcal{B}_{(2,y)}^T\right)\right)\\ &\quad + \underline{\underline{\beta_x}}^{\text{dv}}\left(\text{vec}\left(\mathcal{B}_{(1,x)}\right)\right) + \underline{\underline{\beta_y}}^{\text{dv}}\left(\text{vec}\left(\mathcal{B}_{(1,y)}^T\right)\right) + \text{vec}\left(\underline{\eta}\right)\end{aligned}$$

Remarks: A MATLAB code that implements (13.39) based on Dirichlet boundary conditions and central difference approximation formulas is available on the book's webpage as `lin2D_centralfd_dirich.m`. The function is invoked by the statement `[u,x,y]=lin2D_centralfd_dirich(K,N)` to obtain the solution u of size `K` × `N` at grid points x and y. The program will need to be be edited to customize its application with user-defined function coefficients $\mu_{xx}, \ldots$, and so forth and user-defined boundary conditions.

EXAMPLE 13.5. Consider the 2D Laplace equation

$$\frac{\partial^2 u}{\partial x^2} + \frac{\partial^2 u}{\partial y^2} = 0$$

subject to $u(0,y) = g_a(y)$, $u(1,y) = g_b(y)$, $u(x,0) = h_c(x)$ and $u(x,1) = h_d(x)$. In this case, $\mu_{xx}(x,y) = \mu_{yy}(x,y) = 1$ and the other coefficients are zero. Let us now use the central difference formulas

$$\mathcal{D}_{(2,x)} = \frac{1}{\Delta x^2}\begin{pmatrix} -2 & 1 & & 0\\ 1 & -2 & \ddots & \\ & \ddots & \ddots & 1\\ 0 & & 1 & -2\end{pmatrix} \quad\text{and}\quad \mathcal{D}_{(2,y)} = \frac{1}{\Delta y^2}\begin{pmatrix} -2 & 1 & & 0\\ 1 & -2 & \ddots & \\ & \ddots & \ddots & 1\\ 0 & & 1 & -2\end{pmatrix}$$

$$\mathcal{B}_{(2,x)} = \left(\begin{array}{c|c|c} (\hat{g}_a)_1 & & (\hat{g}_a)_N\\ 0 & & 0\\ \vdots & \cdots & \vdots\\ 0 & & 0\\ (\hat{g}_b)_1 & & (\hat{g}_b)_N\end{array}\right) \quad\text{and}\quad \mathcal{B}_{(2,y)} = \left(\begin{array}{c|c|c} \left(\hat{h}_c\right)_1 & & \left(\hat{h}_d\right)_1\\ 0 & & 0\\ \vdots & \cdots & \vdots\\ 0 & & 0\\ \left(\hat{h}_c\right)_K & & \left(\hat{h}_d\right)_K\end{array}\right)$$

with $\mathcal{D}_{(2,x)}$ and $\mathcal{D}_{(2,y)}$ having sizes $K \times K$ and $N \times N$, respectively, and $(\hat{g}_a)_i = g_a(i\Delta y)/\Delta x^2$, $\left(\hat{h}_c\right)_j = h_c(j\Delta x)/\Delta y^2, \ldots$, and so forth. The matrices in (13.39) then become

$$\mathfrak{R}_{2D} = \left(\begin{array}{c|c|c|c} A_K & sI_K & & 0\\ \hline sI_K & A_K & \ddots & \\ \hline & \ddots & \ddots & sI_K\\ \hline 0 & & sI_K & A_K\end{array}\right) [=] (KN) \times (KN)$$

and

$$\mathfrak{f}_{2D} = \text{vec}\left(\begin{array}{c|ccc|c} (\hat{g}_a)_1 + \left(\hat{h}_c\right)_1 & (\hat{g}_a)_2 & \cdots & (\hat{g}_a)_{N-1} & (\hat{g}_a)_N + \left(\hat{h}_d\right)_1 \\ \hline \left(\hat{h}_c\right)_2 & & & & \left(\hat{h}_d\right)_2 \\ \vdots & & 0 & & \vdots \\ \left(\hat{h}_c\right)_{K-1} & & & & \left(\hat{h}_d\right)_{K-1} \\ \hline (\hat{g}_b)_1 + \left(\hat{h}_c\right)_K & (\hat{g}_b)_2 & \cdots & (\hat{g}_b)_{N-1} & (\hat{g}_b)_N + \left(\hat{h}_d\right)_K \end{array}\right)$$

where $A_K = \begin{pmatrix} q & r & & 0 \\ r & \ddots & \ddots & \\ & \ddots & \ddots & r \\ 0 & & r & q \end{pmatrix}$, $s = \Delta y^{-2}$, $r = \Delta x^{-2}$ and $q = -2\,(r+s)$.

Note that $\mathfrak{R}_{2D}$ has a block tri-diagonal structure and A_K has a tri-diagonal structure. These special structures allow for efficient methods such as the Thomas algorithm and the block Thomas algorithms discussed in Section 2.2.2 and Exercise **E2.16**, respectively, to be used.

EXAMPLE 13.6. Consider the nonhomogeneous Helmholtz differential equation

$$\nabla^2 u + 2u = g(x,y) \quad ; \quad 0 \le x, y \le 1 \tag{13.40}$$

where

$$g(x,y) = \left(2x^2 - 2x + \frac{5}{2}\right) + \cos\left(\frac{\pi x}{2}\right)\left(-2\pi\cos(\pi y) + \left(5\pi^2 - 8\right)\frac{y}{4}\sin(\pi y)\right)$$

subject to

$$\begin{aligned} u(0,y) &= \frac{1}{4} - y\sin(\pi y) & \qquad u(x,0) &= \frac{1}{4}(2x-1)^2 \\ u(1,y) &= \frac{1}{4} & \qquad u(x,1) &= \frac{1}{4}(2x-1)^2 \end{aligned} \tag{13.41}$$

For this example, we know that the exact solution is given by

$$u(x,y) = \left(x - \frac{1}{2}\right)^2 - y\cos\left(\frac{\pi x}{2}\right)\sin(\pi y)$$

Solving the matrix equations formulated in (13.39) with $K = N = 99$, that is, $\Delta x = \Delta y = 0.01$ (as implemented using the MATLAB function `lin2D_centralfd_dirich.m`), we have the plot of $u(x,y)$ shown in Figure 13.3 together with a plot of the errors from the exact solution. Note that the errors are within $\pm 10^{-4}$.

The extension to three dimensions is straightforward but quite an increase in complexity. The formulation for the steady-state solutions of these types of problems, including examples, are given in Section M.2 as an appendix.

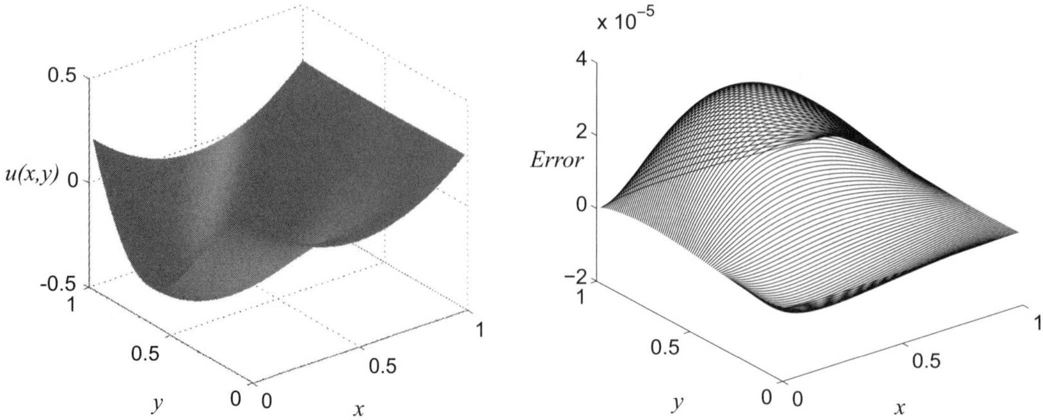

Figure 13.3. The finite difference solution to (13.40) subject to conditions (13.41) is shown on the left plot, whereas the error from the exact solution in shown on the right plot.

13.2.3 Polar and Spherical Coordinates

The finite difference methods are often applied to rectangular-shaped domains or domains that can be mapped into a rectangular domain. However, if the domain is chosen under polar coordinates (r, θ), the approach will require some modifications. First, the angular variable θ is limited to $0 \leq \theta \leq 2\pi$, where the values become periodic beyond this principal range. Second, the differential operators such as the Laplacian will yield formulas with r^n in the denominators, that is,

$$\nabla^2 u = \frac{\partial^2 u}{\partial r^2} + \frac{1}{r}\frac{\partial u}{\partial r} + \frac{1}{r^2}\frac{\partial^2 u}{\partial \theta^2}$$

and complications arise at $r = 0$.

To handle the issue of periodicity, we can incorporate the constraint $u(\theta = 0) = u(\theta = 2\pi)$ to modify the matrices $\mathcal{D}_1$ and $\mathcal{D}_2$ in (13.24) and (13.25). Let $n = 1, \ldots, N+1$ with $\Delta\theta = 2\pi/(N+1)$ and $\theta = (n-1)\Delta\theta$. Then

$$\begin{aligned} u(r, 0) = u(r, 2\pi) \quad &\rightarrow \quad u_{k,1} = u_{k,N+1} \\ u(r, -\Delta\theta) = u(r, 2\pi - \Delta\theta) \quad &\rightarrow \quad u_{k,0} = u_{k,N} \end{aligned} \tag{13.42}$$

Thus, with $U = \left(u_{k,n}\right) [=] K \times N$, $u_{k,n} = u\left(r_k, \theta_n\right)$, the central difference formulas will yield

$$\frac{\partial u}{\partial \theta} = U\left(\mathcal{D}_1^{\text{per}}\right)^T \qquad \text{and} \qquad \frac{\partial^2 u}{\partial \theta^2} = U\left(\mathcal{D}_2^{\text{per}}\right)^T \tag{13.43}$$

where

$$\mathcal{D}_1^{\text{per}} = \frac{1}{2\Delta\theta}\begin{pmatrix} 0 & 1 & & -1 \\ -1 & \ddots & \ddots & \\ & \ddots & \ddots & 1 \\ 1 & & -1 & 0 \end{pmatrix} \quad \text{and} \quad \mathcal{D}_2^{\text{per}} = \frac{1}{\Delta\theta^2}\begin{pmatrix} -2 & 1 & & 1 \\ 1 & \ddots & \ddots & \\ & \ddots & \ddots & 1 \\ 1 & & 1 & -2 \end{pmatrix} \tag{13.44}$$

which are tri-diagonal matrices with additional terms at positions $(1, N)$ and $(N, 1)$. Note that both $\mathcal{D}_1^{\text{per}}$ and $\mathcal{D}_2^{\text{per}}$ are circulant matrices.

There are many approaches to handling the complications at the origin. We discuss a simple approach that will stretch the domain by introducing fictitious points at $r < 0$ and a discretization that would bypass the origin yet make the fictitious points naturally disappear from the finite difference approximation equations.[1] Assume that the radius has been normalized to be $0 \le r \le 1$. Let $k = 1, \ldots, K$ with $\Delta r = \dfrac{2}{2K+1}$ and $r_k = (k - \frac{1}{2})\Delta r$. Thus $r_{K+1} = 1$ and the fictitious points occur at $r_0 = -\Delta r/2$. With this choice and under the second-order central difference approximation of Laplacian at $k = 1$, the formula becomes

$$\nabla^2 u\Big|_{r=\Delta r/2} \approx \frac{u_{2,n} - 2u_{1,n} + u_{0,n}}{\Delta r^2} + \frac{1}{\Delta r/2}\frac{u_{2,n} - u_{0,n}}{2\Delta r} + \frac{1}{\Delta r^2/4}\frac{u_{1,n+1} - 2u_{1,n} + u_{1,n-1}}{\Delta\theta^2}$$

in which the terms involving $u_{0,n}$ do cancel out to yield

$$\nabla^2 u\Big|_{r=\Delta r/2} \approx \frac{u_{2,n} - 2u_{1,n}}{\Delta r^2} + \frac{1}{(\Delta r/2)}\frac{u_{2,n}}{2\Delta r} + \frac{1}{\Delta r^2/4}\frac{u_{1,n+1} - 2u_{1,n} + u_{1,n-1}}{\Delta\theta^2}$$

Assuming Dirichlet conditions at $r = 1$ given by $u(1, \theta) = g(\theta)$, we have

$$\nabla^2 u \quad \rightarrow \quad \mathcal{D}_2 U + \mathcal{B}_2 + V\mathcal{D}_1 U + V\mathcal{B}_1 + WU\left(\mathcal{D}_2^{\text{per}}\right)^T \tag{13.45}$$

where $U = \left(u_{k,n}\right)$, $u_{k,n} = u(r_k, \theta_n)$, $\mathcal{D}_2$ and $\mathcal{D}_1$ are $K \times K$ matrices given by (13.24) and (13.25), respectively, $\mathcal{D}_2^{\text{per}}$ is given in (13.44), V and W are $K \times K$ diagonal matrices given by

$$V = \text{diag}\left(\frac{1}{r_1}, \ldots, \frac{1}{r_K}\right) \qquad W = \text{diag}\left(\frac{1}{r_1^2}, \ldots, \frac{1}{r_K^2}\right)$$

$\mathcal{B}_2 = \dfrac{1}{\Delta r^2}\mathcal{H}$ and $\mathcal{B}_1 = \dfrac{1}{2\Delta r}\mathcal{H}$, with

$$\mathcal{H} = \left(\begin{array}{c|c|c} 0 & & 0 \\ \vdots & \cdots & \vdots \\ 0 & & 0 \\ g(\theta_1) & & g(\theta_N) \end{array}\right)$$

Thus for a Poisson equation

$$\nabla^2 u = f(r, \theta)$$

with $F = \left(f_{k,n}\right)$, $f_{k,n} = f(r_k, \theta_n)$, the finite difference equation reduces to a linear equation given by

$$A\mathbf{u} = \mathbf{b} \tag{13.46}$$

where $\mathbf{u} = \text{vec}\,(U)$ and

$$\begin{aligned} A &= \Big[I_N \otimes (\mathcal{D}_2 + V\mathcal{D}_1)\Big] + \Big[\mathcal{D}_2^{\text{per}} \otimes W\Big] \\ \mathbf{b} &= \text{vec}\,(F) - \text{vec}\,(\mathcal{B}_2 + V\mathcal{B}_1) \end{aligned}$$

[1] This method is based on M. C. Lai, "A note on finite difference discretizations for Poisson equation on a disk," *Numerical Methods for Partial Difference Equations*, vol. 17, 2001, pp. 199–203.

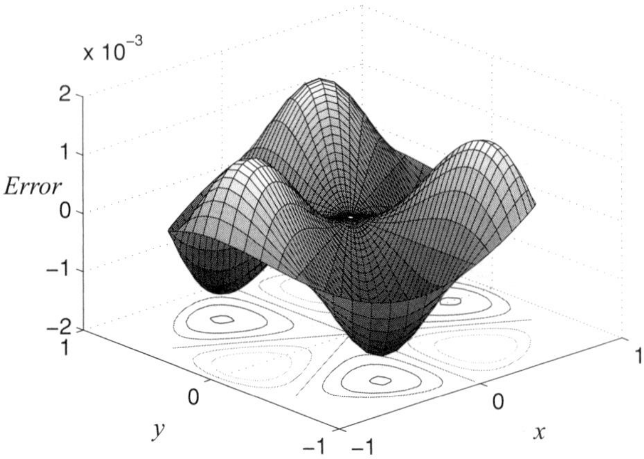

Figure 13.4. The errors of the finite difference solution from the exact solution for the Laplace equation in polar coordinates given in Example 13.7.

These results can be extended to handle the more general case of nonhomogeneous Helmholtz equation given by

$$\nabla^2 u + h(r,\theta)u = f(r\theta)$$

The finite difference solution for this case is included as an exercise in **E13.7**.

The structure of A is a $KN \times KN$, block-circulant matrix, which is a block tri-diagonal matrix with additional blocks at the corners, that is,

$$A = \left(\begin{array}{c|c|c|c} G & H & & H \\ \hline H & G & \ddots & \\ \hline & \ddots & \ddots & H \\ \hline H & & H & G \end{array}\right)$$

and there are efficient algorithms for solving problems with this special structure. (See, e.g., Section B.6.1, where the matrices can be split to take advantage of the block tri-diagonal inner structure.)

EXAMPLE 13.7. Consider the Laplace equation in polar coordinates $\nabla^2 u = 0$ with boundary condition $u(1,\theta) = 1 + \cos(3\theta)$. The exact solution was found in Example 11.6 to be

$$u = 1 + r^3 \cos(3\theta)$$

Based on (13.46), the errors of the finite difference solution from the exact solution are shown in Figure 13.4 while using $K = 30$ and $N = 60$ or $\Delta r = 0.328$ and $\Delta\theta = 0.1047$. The errors are within $\pm 2 \times 10^{-3}$.

For the case of spherical coordinates, the Laplacian is given by

$$\nabla^2 u = \frac{\partial^2 u}{\partial r^2} + \frac{2}{r}\frac{\partial u}{\partial r} + \frac{1}{r^2}\frac{\partial^2 u}{\partial \theta^2} + \frac{\cos\theta}{r^2 \sin\theta}\frac{\partial u}{\partial \theta} + \frac{1}{r^2 \sin^2\theta}\frac{\partial^2 u}{\partial \phi^2} \tag{13.47}$$

The same matrices $\mathcal{D}_1^{\text{per}}$ and $\mathcal{D}_2^{\text{per}}$ defined in (13.44) are needed to satisfy the periodicity along the θ and ϕ variables. However, to handle the issue of the origin, there is no need to change the discretization along r as was done for the case of polar or cylindrical coordinates. This means that we can set $r_k = k\Delta r$, $k = 1, \ldots, K$, where $\Delta r = 1/(K+1)$. To see this, take the finite difference approximation of the terms in (13.47) that involve partial derivatives with respect to r at $r = r_1 = \Delta r$

$$\left.\left(\frac{\partial^2 u}{\partial r^2} + \frac{2}{r}\frac{\partial u}{\partial r}\right)\right|_{r=r_1=\Delta r} \approx \frac{u_{2,n,m} - 2u_{1,n,m} + u_{0,n,m}}{\Delta r^2} + \frac{u_{2,n,m} - u_{0,n,m}}{\Delta r^2}$$
$$= \frac{2u_{2,n,m} - 2u_{1,n,m}}{\Delta r^2}$$

Thus, for the case of azimuthal symmetry, that is, $u = u(r,\theta)$, we can use the regular discretizations $k = 1, \ldots, K$, $\Delta r = 1/(K+1)$, $r_k = k\Delta r$ and $n = 1, \ldots, N$, $\Delta\theta = 2\pi/(N+1)$, $\theta = (n-1)\Delta\theta$ to obtain a matrix equation similar to (13.45) that approximates the Laplacian operation in spherical coordinates, that is,

$$\nabla^2 u \quad \rightarrow \quad \mathcal{D}_2 U + \mathcal{B}_2 + 2V\mathcal{D}_1 U + 2V\mathcal{B}_1 + WU\left(\mathcal{D}_2^{\text{per}}\right)^T + WU\left(\mathcal{D}_1^{\text{per}}\right)^T Q \quad (13.48)$$

where the same definitions for $\mathcal{D}_1$, $\mathcal{D}_2$, $\mathcal{B}_2$, $\mathcal{B}_1$, $\mathcal{D}_1^{\text{per}}$, $\mathcal{D}_2^{\text{per}}$, V, and W used in (13.45) still apply, except for the main difference that $r_k = k\Delta r$. The additional matrix Q is given by

$$Q = \text{diag}\left(\frac{\cos\theta_1}{\sin\theta_1}, \quad \cdots \quad, \frac{\cos\theta_N}{\sin\theta_N}\right)$$

As a final detail, note that because we have $(\cos\theta_1/\sin\theta_1)$ as the first term in Q, division by zero should be avoided by shifting the θ-grids by $\Delta\theta/2$, that is, set $\theta_n = (n+1/2)\Delta\theta$.

Thus for a Poisson equation in spherical coordinates under azimuthal symmetry, $\nabla^2 u = f(r,\theta)$, and under a Dirichlet condition, $u(1,\theta) = g(\theta)$, we can reduce the equation to a linear equation given by

$$A\mathbf{u} = \mathbf{b} \quad (13.49)$$

where $\mathbf{u} = \text{vec}(U)$, $F = \left(f_{k,n}\right)$, $f_{k,n} = f(r_k, \theta_n)$,

$$A = \left[I_N \otimes \left(\mathcal{D}_2 + 2V\mathcal{D}_1\right)\right] + \left[\left(\mathcal{D}_2^{\text{per}} + Q\mathcal{D}_1^{\text{per}}\right) \otimes W\right]$$
$$\mathbf{b} = \text{vec}(F) - \text{vec}(\mathcal{B}_2 + 2V\mathcal{B}_1)$$

Remarks: MATLAB codes that implement (13.46) and (13.49) for the 2D Poisson equation under polar and spherical (with azimuthal symmetry), respectively, are available on the book's webpage as `poisson2d_polar_dirich.m` and `poisson2d_polar_sphere.m`, respectively. The function is invoked by the statement

```
[U,r,th,x,y]=poisson2d_polar_dirich(K,N)
```

(or `[U,r,th,x,y]=poisson2d_sphere_dirich(K,N)`) to obtain the solution U of size `K` $\times$ `N` at grid points x and y (or polar coordinates `r`=r and `th`=θ). The program will need to be be edited to customize its application with user-defined forcing function $f(r,\theta)$ and user-defined boundary conditions $u(1,\theta) = g(\theta)$.

13.3 Time-Dependent Equations

For partial differential equations that are dependent on time, the general form of finite-difference schemes is given by

$$\mathbf{u}^{(q+1)} = f\left(\mathbf{u}^{(q+1)}, \mathbf{u}^{(q)}, \mathbf{u}^{(q-1)}, \ldots, \mathbf{u}^{(q-p)}\right) \tag{13.50}$$

where $\mathbf{u}^{(q)} = \mathbf{u}(q\Delta t)$ and contains the values of $u(t, \mathbf{x})$ at the grid points. Equation (13.50) is called the **time-marching** scheme; that is, from an initial condition $\mathbf{u}^{(0)} = \mathbf{u}_0$, (13.50) is evaluated iteratively as q is incremented. If the function f in (13.50) is not dependent on $\mathbf{u}^{(q+1)}$, the scheme is classified as **explicit**; otherwise, it is classified as **an implicit**. The integer p determines how many past values are used. For $p > 0$, we end up with a scheme that is known as a **multilevel scheme**.[2]

We limit our discussion only on linear time-marching schemes, that is,

$$\mathbf{u}^{(q+1)} = \sum_{i=-1}^{p} \alpha_i^{(q)} \mathbf{u}^{(q-i)} + g^{(q)} \tag{13.51}$$

Thus, if $\alpha_{-1}^{(q)} = 0$ for all $q \geq 0$, the scheme is explicit.

13.3.1 The Semi-Discrete Approach

By first replacing only the spatial partial derivatives with their corresponding finite-difference approximations, the original partial differential equation can be reduced to an initial value problem in the form of an ordinary differential equation given by

$$\mathbf{M}(t)\frac{d}{dt}\mathbf{u}(t) = \widehat{\mathbf{F}}(t)\mathbf{u}(t) + \widehat{\mathbf{B}}(t) \tag{13.52}$$

or assuming $\mathbf{M}(t)$ is nonsingular for $t \geq 0$,

$$\frac{d}{dt}\mathbf{u}(t) = \mathbf{F}(t)\mathbf{u}(t) + \mathbf{B}(t) \tag{13.53}$$

When (13.53) is discretized with respect to time, this yields the time-marching equations of the form (13.51). This approach is known as the **semi-discrete** approach, also known as the **method of lines**.

For instance, consider the following linear, time-varying, second-order equation in 2D space:

$$\alpha_t \frac{\partial u}{\partial t} + \sum_{s=t,x,y} \varphi_{ts} \frac{\partial^2 u}{\partial t \partial s} + \sum_{p,s=x,y} \mu_{ps} \frac{\partial^2 u}{\partial p\, \partial s} + \sum_{s=x,y} \beta_s \frac{\partial u}{\partial s} + \zeta u + \eta \quad = \quad 0 \tag{13.54}$$

Using the procedures discussed in Section 13.2.2, the dependent variable u and the coefficients α_t, φ_{ts}, μ_{ps}, β_s, γ and η, with $p, s = x, y$, can be represented in matrix

[2] More specifically, for $p \geq 0$, we have a $(p+2)$-level scheme in which two of the levels are for $t = (q+1)\Delta t$ and $t = q\Delta t$.

form, for example,

$$U = \begin{pmatrix} u_{1,1} & \cdots & u_{1,N} \\ \vdots & \ddots & \vdots \\ u_{K,1} & \cdots & u_{K,N} \end{pmatrix} ; \quad \underline{\underline{\mu_{xx}}} = \begin{pmatrix} (\mu_{xx})_{1,1} & \cdots & (\mu_{xx})_{1,N} \\ \vdots & \ddots & \vdots \\ (\mu_{xx})_{K,1} & \cdots & (\mu_{xx})_{K,N} \end{pmatrix} \quad ; \text{ etc.}$$

The terms with partial derivative in times then become

$$\begin{aligned} \frac{\partial u}{\partial t} &= \frac{d}{dt}U \\ \frac{\partial^2 u}{\partial t^2} &= \frac{d^2}{dt^2}U \\ \frac{\partial^2 u}{\partial t \partial x} &= \frac{d}{dt}\left(\mathcal{D}_{(1,x)}U + \mathcal{B}_{(1,x)}\right) = \mathcal{D}_{(1,x)}\left(\frac{d}{dt}U\right) + \frac{d}{dt}\mathcal{B}_{(1,x)} \\ \frac{\partial^2 u}{\partial t \partial y} &= \frac{d}{dt}\left(U\mathcal{D}_{(1,y)}^T + \mathcal{B}_{(1,y)}\right) = \left(\frac{d}{dt}U\right)\mathcal{D}_{(1,y)}^T + \frac{d}{dt}\mathcal{B}_{(1,y)} \end{aligned} \tag{13.55}$$

After substitution to (13.54), we obtain

$$\mathcal{M}_2 \frac{d^2}{dt^2}(\text{vec}(U)) + \mathcal{M}_1 \frac{d}{dt}(\text{vec}(U)) + \mathcal{M}_0 (\text{vec}(U)) + \mathcal{N} = 0 \tag{13.56}$$

where

$$\begin{aligned} \mathcal{M}_2 &= \left[\underline{\underline{\varphi_{tt}}}\right]^{\text{dv}} \\ \mathcal{M}_1 &= \left[\underline{\underline{\alpha_t}}\right]^{\text{dv}} + \left[\underline{\underline{\varphi_{tx}}}\right]^{\text{dv}}\left(I_N \otimes \mathcal{D}_{(1,x)}\right) + \left[\underline{\underline{\varphi_{ty}}}\right]^{\text{dv}}\left(\mathcal{D}_{(1,y)} \otimes I_K\right) \\ \mathcal{M}_0 &= \mathfrak{R}_{2\text{D}} \\ \mathcal{N} &= \mathfrak{f}_{2\text{D}} + \left[\underline{\underline{\mu_{tx}}}\right]^{\text{dv}}\frac{d}{dt}\mathcal{B}_{(1,x)} + \left[\underline{\underline{\mu_{ty}}}\right]^{\text{dv}}\frac{d}{dt}\mathcal{B}_{(1,y)} \end{aligned}$$

and the terms $\mathfrak{R}_{2\text{D}}$ and $\mathfrak{f}_{2\text{D}}$ were given in (13.39). Next, let

$$\mathbf{v} = \begin{pmatrix} \text{vec}(U) \\ \\ \dfrac{d}{dt}(\text{vec}U) \end{pmatrix} \tag{13.57}$$

For the special case of nonsingular $\mathcal{M}_2$, we could rearrange (13.56) into semi-discrete form (13.53), that is,

$$\frac{d}{dt}\mathbf{v}(t) = \mathbf{F}(t)\mathbf{v}(t) + \mathbf{B}(t) \tag{13.58}$$

where

$$\mathbf{F} = \left(\begin{array}{c|c} I & 0 \\ \hline -\mathcal{M}_2^{-1}\mathcal{M}_1 & -\mathcal{M}_2^{-1}\mathcal{M}_0 \end{array}\right) \quad \text{and} \quad \mathbf{B} = \left(\begin{array}{c} 0 \\ \hline -\mathcal{M}_2^{-1}\mathcal{N} \end{array}\right)$$

Another simple case occurs for diffusion problems. In this case, $\alpha_t = -1$ and $\varphi_{tt} = \varphi_{tx} = \varphi_{ty} = 0$. Then $\mathcal{M}_2 = 0$, $\mathcal{M}_1 = -1$, $\mathcal{M}_0 = \mathfrak{R}_{2D}$, and $\mathcal{N} = \mathfrak{f}_{2D}$. With $\mathbf{v} = \text{vec}(U)$, (13.56) reduces to

$$\frac{d}{dt}\mathbf{v} = \mathfrak{R}_{2D}\,\mathbf{v} + \mathfrak{f}_{2D} \tag{13.59}$$

EXAMPLE 13.8. Consider the linear reaction-diffusion differential equation given by

$$\frac{\partial u}{\partial t} = \mu\nabla^2 u + \zeta u + \eta(t, x, y) \tag{13.60}$$

with boundary conditions

$$\begin{aligned} u\,(t, 0, y) &= v_0(t, y) & \qquad u\,(t, x, 0) &= w_0(t, x) \\ u\,(t, 1, y) &= v_1(t, y) & \qquad u\,(t, x, 1) &= w_1(t, x) \end{aligned} \tag{13.61}$$

and initial condition

$$u\,(0, x, y) = \xi(x, y) \tag{13.62}$$

where μ and ζ are constants.

Let

$$U = \begin{pmatrix} u_{1,1} & \cdots & u_{1,N} \\ \vdots & \ddots & \vdots \\ u_{K,1} & \cdots & u_{K,N} \end{pmatrix} \qquad \underline{\underline{\eta}}(t) = \begin{pmatrix} \eta_{1,1}(t) & \cdots & \eta_{1,N}(t) \\ \vdots & \ddots & \vdots \\ \eta_{K,1}(t) & \cdots & \eta_{K,N}(t) \end{pmatrix}$$

with $u_{kn} = u(t, k\Delta x, n\Delta y)$ and $\eta_{k,n} = \eta(t, k\Delta x, n\Delta y)$.

Using the finite difference approximations used in (13.39), we obtain

$$\frac{d}{dt}\mathbf{v} = \mathfrak{R}\,\mathbf{v} + \mathfrak{f}(t) \tag{13.63}$$

where

$$\begin{aligned} \mathbf{v} &= \text{vec}\,(U) \\ \mathfrak{R} &= \mu\left[I_N \otimes \mathcal{D}_{(2,x)} + \mathcal{D}_{(2,y)} \otimes I_K\right] \;+\; \zeta\, I_{NK} \\ \mathfrak{f} &= \text{vec}\left(\,\underline{\underline{\eta}}\,\right) \;+\; \mu\;\text{vec}\left(\mathcal{B}_{(2,x)} + \mathcal{B}_{(2,y)}^T\right) \end{aligned}$$

which is an ordinary differential equation with an initial condition,

$$\mathbf{v}(0) = \text{vec}\left(\,\underline{\underline{\xi}}\,\right)$$

In this example, $\mathfrak{R}$ is a constant matrix. However, in general, it can be time-varying.

Once the problem has been reduced to an initial-value problem, the methods discussed in Chapter 7, including Runge-Kutta and multistep methods can be employed. One concern, however, is that the size of the state vector can be enormous as the spatial-grid resolution increases to meet some prescribed accuracy. Moreover, matrix $\mathbf{F}$ in (13.53) will become increasingly ill-conditioned as the matrix size grows,

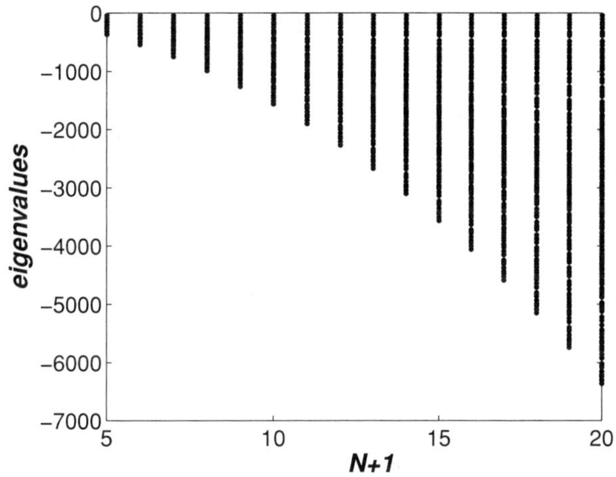

Figure 13.5. The eigenvalues of $\mathfrak{R}$ at different values of $N+1$.

as shown in Example 13.9 that follows. As discussed in Section 7.4.2, stiff differential equations favor implicit time-marching methods.

EXAMPLE 13.9. Consider the problem given in Example 13.8. For $\mu = 2$, $\zeta = -3$ and $N = M$ (or $\Delta x = 1/(N+1) = \Delta y$), matrix $\mathfrak{R}$ becomes a block tri-diagonal matrix given by

$$\mathfrak{R} = \begin{pmatrix} R_a & R_b & & 0 \\ R_b & \ddots & \ddots & \\ & \ddots & \ddots & R_b \\ 0 & & R_b & R_a \end{pmatrix} \tag{13.64}$$

where,

$$R_a = \begin{pmatrix} \zeta & \mu & & 0 \\ \mu & \ddots & \ddots & \\ & \ddots & \ddots & \mu \\ 0 & & \mu & \zeta \end{pmatrix} \; ; \; R_b = \begin{pmatrix} \mu & & 0 \\ & \ddots & \\ 0 & & \mu \end{pmatrix} \; ; \; \zeta = -\frac{8}{\Delta x^2} - 3 \; ; \; \mu = \frac{2}{\Delta x^2}$$

Figure 13.5 shows a plot of the eigenvalues of $\mathfrak{R}$ at different numbers of grid points $N+1 (= M+1)$. Note also that all the eigenvalues are negative. Recall that the condition number ρ is given by the ratio

$$\rho = \frac{\max_i(|\lambda_i|)}{\min_j(|\lambda_j|)}$$

where λ_i is an eigenvalue of $\mathfrak{R}$. Then $\rho = 5.47$ at $N = 4$ and grows to $\rho = 150.1$ at $N = 19$, which is often considered "mildly stiff".

Instead of exploring various types of time-marching algorithms, we focus on the simplest one-step time-marching methods, known as the weighted-average Euler schemes.

13.3.2 Weighted-Average Euler Schemes

Although it is true that the size of matrix $\mathbf{F}$ in (13.53) will grow as the number of grid points increases, the finite-difference approximations will still involve only the neighboring points. This means that the matrices are quite sparse. Several algorithms are available to take advantage of the sparsity properties. They usually involve iterative procedures for the evaluation of the time-marching schemes. Because of this, the most popular methods are the three types of one-step Euler methods: the forward-Euler (explicit), the backward Euler (implicit), and the midpoint-Euler (implicit). The last type is also known as the **Crank-Nicholson method**.

From the initial-value problem (13.53) that was obtained from the semi-discrete approach:

$$\frac{d}{dt}\mathbf{v}(t) = \mathbf{F}(t)\mathbf{v}(t) + \mathbf{B}(t)$$

subject to $\mathbf{v}(0) = \mathbf{v}^{(0)}$, we have

1. **Forward-Euler Schemes.**

$$\begin{aligned} \frac{\mathbf{v}^{(q+1)} - \mathbf{v}^{(q)}}{\Delta t} &= \mathbf{F}^{(q)}\mathbf{v}^{(q)} + \mathbf{B}^{(q)} \\ \mathbf{v}^{(q+1)} &= \left(I + \Delta t\mathbf{F}^{(q)} \right)\mathbf{v}^{(q)} + \Delta t\mathbf{B}^{(q)} \end{aligned} \tag{13.65}$$

This scheme is an explicit type and is one of the easiest time-marching schemes to implement. Starting with the initial condition, $\mathbf{v}^{(0)}$, the values of $\mathbf{v}^{(q)}$ are obtained iteratively. However, as with most explicit methods, stability is limited by the size of time steps used. Sometimes, the time steps required to maintain stability can be very small, resulting in a very slow time-marching process.[3] Nonetheless, due to the ease of implementation, forward schemes are still used in several applications, with the caveat that stability may be a problem under certain parametric conditions.

2. **Backward Euler Schemes.**

$$\begin{aligned} \frac{\mathbf{v}^{(q+1)} - \mathbf{v}^{(q)}}{\Delta t} &= \mathbf{F}^{(q+1)}\mathbf{v}^{(q+1)} + \mathbf{B}^{(q+1)} \\ \mathbf{v}^{(q+1)} &= \left(I - \Delta t\mathbf{F}^{(q+1)} \right)^{-1}\left(\mathbf{v}^{(q)} + \Delta t\mathbf{B}^{(q+1)} \right) \end{aligned} \tag{13.66}$$

This is an implicit method and it requires inversion of $(I - \Delta t\mathbf{F}^{(q+1)})$. Due to the sparsity in $\mathbf{F}$, procedures such as LU factorizations, Thomas algorithms, or GMRES can be used to take advantage of the matrix structures (cf. Section 2.2).

[3] In some cases, the required time steps may be too small such that round-off errors become very significant. In this case, the explicit scheme is impractical.

For equations in two- and three-spatial dimensions, other modifications such as ADI schemes are also used (see the appendix given as Section M.4).

Compared with the forward Euler schemes, the backward schemes are more stable. In some cases, the scheme will be **unconditionally stable**. This means that the scheme will be stable for any time step chosen. Of course, the solution accuracy still demand small time steps.[4]

3. **Crank-Nicholson Schemes.**

$$\frac{\mathbf{v}^{(q+1)} - \mathbf{v}^{(q)}}{\Delta t} = \frac{1}{2}\left(\left[\mathbf{F}^{(q+1)}\mathbf{v}^{(q+1)} + \mathbf{B}^{(q+1)}\right] + \left[\mathbf{F}^{(q)}\mathbf{v}^{(q)} + \mathbf{B}^{(q)}\right]\right)$$

$$\mathbf{v}^{(q+1)} = \left(I - \frac{\Delta t}{2}\mathbf{F}^{(q+1)}\right)^{-1}\left[\left(I + \frac{\Delta t}{2}\mathbf{F}^{(q)}\right)\mathbf{v}^{(q)} + \frac{\Delta t}{2}\left(\mathbf{B}^{(q+1)} + \mathbf{B}^{(q)}\right)\right] \tag{13.67}$$

This is also an implicit method and again requires the inversion of $(I - (\Delta t/2)\mathbf{F}^{(q+1)})$. One advantage of Crank-Nicholson schemes over Euler backward schemes is an increase in the order of accuracy. It can be shown that the accuracy of a Crank-Nicholson will be $\mathcal{O}\left(\Delta t^2, \Delta x^n\right)$, compared with the accuracy of a backward Euler scheme, which is $\mathcal{O}\left(\Delta t, \Delta x^n\right)$. However, for discontinuous or non-smooth boundary conditions, the Crank-Nicholson method can introduce undesired oscillations unless the value Δt is small enough.[5]

For a simple exploration of the three methods applied to a one-dimensional diffusion equation, see Exercise **E13.9**. For equations in two- and three-spatial dimensions, other modifications such as ADI (alternate-dimension implicit) schemes are also used (see the appendix given as Section M.4).

All three methods are special cases of an umbrella scheme called the Euler-θ method, which is also known as **weighted-average Euler method**. The weighted-average Euler method is given by

$$\frac{\mathbf{v}^{(q+1)} - \mathbf{v}^{(q)}}{\Delta t} = \theta\left(\mathbf{F}^{(q+1)}\mathbf{v}^{(q+1)} + \mathbf{B}^{(q+1)}\right) + \left(1 - \theta\right)\left(\mathbf{F}^{(q)}\mathbf{v}^{(q)} + \mathbf{B}^{(q)}\right) \tag{13.68}$$

From (13.68), we see that $\theta = 0$, $\theta = 1/2$, and $\theta = 1$ yield the Euler-forward, Crank-Nicholson, and Euler-backward schemes, respectively.

EXAMPLE 13.10. Consider the following time-dependent scalar field in the (x, y)-plane,

$$u(t, x, y) = e^{-2t}\varphi(x, y) + \left(1 - e^{-2t}\right)\psi(x, y) \tag{13.69}$$

[4] If the required solution is only the steady-state profiles, time accuracy may not be as important. In this case, large time steps can be used to speed up the convergence.

[5] Other additional techniques such as smoothing via averages can be used to reduce the amount of oscillations.

where,

$$\varphi(x,y) = \frac{8}{1+20r(x,y)} \quad ; \quad \psi(x,y) = \frac{1}{1+5s(x,y)}$$

$$r(x,y) = e^{2-2x-y} \quad ; \quad s(x,y) = e^{8\left[(x-0.8)^2-y\right]}$$

A linear reaction-diffusion type differential equation that has (13.69) as the solution is

$$\frac{\partial u}{\partial t} = 2\nabla^2 u - 3u + g(t,x,y) \tag{13.70}$$

where g can be set by substituting (13.69) into (13.70).

$$\begin{aligned}
g(t,x,y) &= f\left(\varphi + 2h(x,y) - \psi - 2q(x,y)\right) + \left(3\psi - 2h(x,y)\right) \\
q(x,y) &= \frac{32000\, r^2}{(1+20r)^3} - \frac{800r}{(1+20r)^2} \\
h(x,y) &= \left[3200 + 50\left(16x - \frac{64}{5}\right)^2\right]\frac{s^2}{(1+5s)^3} \\
&\quad - \left[400 + 5\left(16x - \frac{64}{5}\right)^2\right]\frac{s}{(1+5s)^2}
\end{aligned}$$

Let the boundary conditions given by

$$\begin{aligned}
u(t,0,y) = \alpha_0(t,y) &= e^{-2t}\varphi(0,y) + \left(1-e^{-2t}\right)\psi(0,y) \\
u(t,1,y) = \alpha_1(t,y) &= e^{-2t}\varphi(1,y) + \left(1-e^{-2t}\right)\psi(1,y) \\
u(t,x,0) = \beta_0(t,x) &= e^{-2t}\varphi(x,0) + \left(1-e^{-2t}\right)\psi(x,0) \\
u(t,x,1) = \beta_1(t,x) &= e^{-2t}\varphi(x,1) + \left(1-e^{-2t}\right)\psi(x,1)
\end{aligned} \tag{13.71}$$

and initial condition

$$u(0,x,y) = \varphi(x,y) \tag{13.72}$$

This initial value-boundary condition problem satisfies the same situation given in Example 13.8. Thus the matrix formulation is given by

$$\frac{d}{dt}\mathbf{v} = \mathfrak{R}\,\mathbf{v} + \mathfrak{f}(t)$$

Using the Crank-Nicholson method,

$$\left(I - \frac{\Delta t}{2}\mathfrak{R}\right)\mathbf{v}^{(q+1)} = \left(I + \frac{\Delta t}{2}\mathfrak{R}\right)\mathbf{v}^{(q)} + \frac{\Delta t}{2}\left(\mathfrak{f}^{(q+1)} + \mathfrak{f}^{(q)}\right)$$

$$\mathbf{v}^{(q+1)} = \left[\left(I - \frac{\Delta t}{2}\mathfrak{R}\right)^{-1}\left(I + \frac{\Delta t}{2}\mathfrak{R}\right)\right]\mathbf{v}^{(q)} + \frac{\Delta t}{2}\left(I - \frac{\Delta t}{2}\mathfrak{R}\right)^{-1}\left(\mathfrak{f}^{(q+1)} + \mathfrak{f}^{(q)}\right)$$

With $\Delta x = \Delta y = 0.05$ ($K = N = 19$) and $\Delta t = 0.001$, the plots of the approximate solutions at different time slices are shown in Figure 13.6, together with the exact solution. The plots of the error distribution at different time slices are shown in Figure 13.7. The errors are within $\pm 2 \times 10^{-3}$.

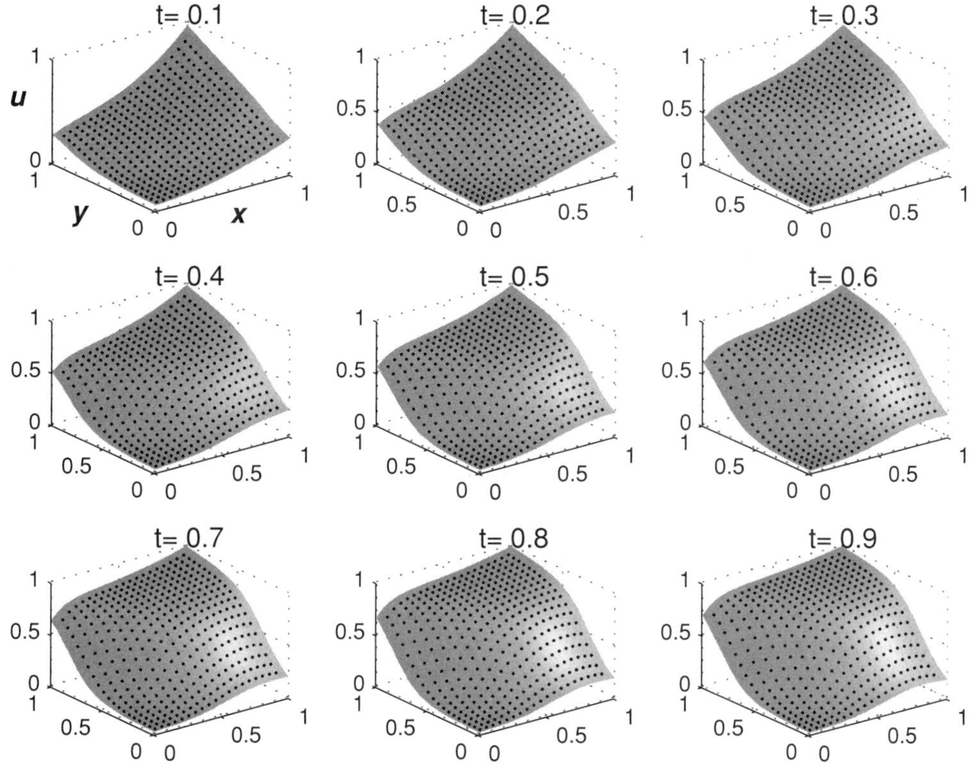

Figure 13.6. The finite difference solution to (13.70) at different slices of time, subject to conditions (13.71) and (13.72), using the Crank-Nicholson time-marching method. The approximations are shown as points, whereas the exact solutions, (13.69), at the corresponding t values are shown as a surface plots.

The backward Euler scheme is given by

$$\mathbf{v}^{(q+1)} = \left(I - \Delta t \mathfrak{R}\right)^{-1} \left(\mathbf{v}^{(q)} + \Delta t \mathfrak{f}^{(q+1)}\right)$$

For this example, the results of the backward Euler are similar to those of the Crank-Nicholson. This just means that even though one would expect the Crank-Nicholson scheme to increase the accuracy from $\mathcal{O}\left(\Delta t^2, \Delta x^2\right)$ to $\mathcal{O}\left(\Delta t^2, \Delta x^2\right)$, the accuracy of both schemes were still limited by the chosen Δx.

The forward Euler scheme is given by

$$\mathbf{v}^{(q+1)} = \left(I + \Delta t \mathfrak{R}\right) \mathbf{v}^{(q)} + \Delta t \mathfrak{f}^{(q)}$$

However, time marching with the same step size $\Delta t = 0.001$, which was stable for Crank-Nicholson and backward schemes, was unstable for the forward scheme. As shown in Figure 13.8, the maximal absolute errors (near the edges) were greater than 4.0 at $t = 0.014$. The errors grow unbounded at further time steps. However, using a time step size $\Delta t \leq 0.00025$ will produce a stable Euler-forward scheme.

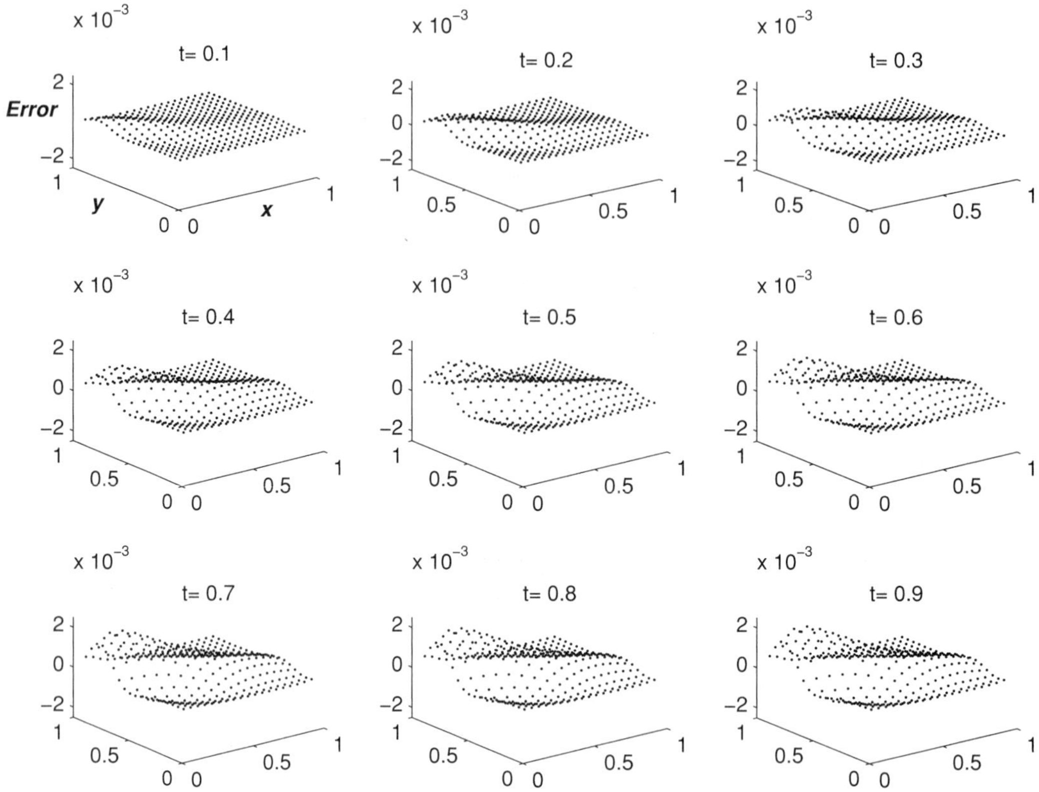

Figure 13.7. The error distribution between the finite difference approximation (using central difference formulas for spatial derivatives and Crank-Nicholson time-marching method) and the exact solutions, (13.69), at different t values.

13.4 Stability Analysis

Three major issues that often need to be addressed by any numerical scheme are consistency, convergence, and stability. Although these three properties are closely related, they are strictly different.

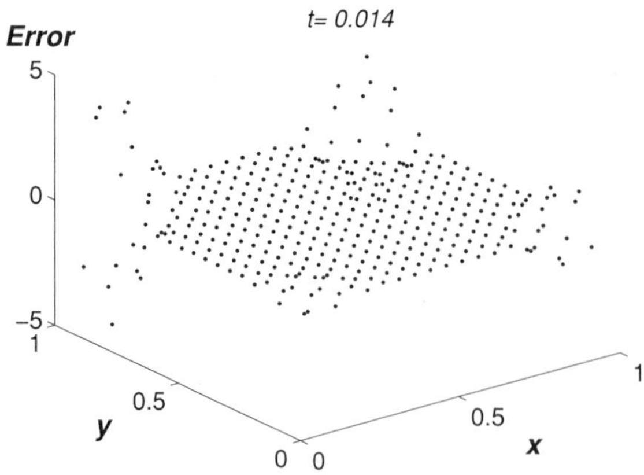

Figure 13.8. The error distribution for finite difference approximation of u (using central difference formulas for spatial derivatives and Euler forward time-marching scheme with $\Delta t = 0.001$) from the exact solutions, (13.69), at $t = 0.014$.

In this section, we limit our discussion to linear equations. Let the partial differential equation be

$$L\mathbf{v} = \mathbf{f} \tag{13.73}$$

where L is a linear partial differential operator and f is the nonhomogenous term. Based on (13.73), let the approximation schemes be described by

$$L_\Delta \mathbf{v}_\Delta = \mathbf{f}_\Delta \tag{13.74}$$

in which the subscript Δ is attached with a refinement path that relates how the step size Δt and the other grid sizes such as Δx, Δy, and Δz are to be reduced.[6]

Definition 13.1. *Let $\mathbf{v}_p$ be the solution of (13.73) at point p, where p belongs to the discretized domain. If*

$$\lim_{\Delta \to 0} \left(L_\Delta \mathbf{v}_p - \mathbf{f}_\Delta \right) = 0 \tag{13.75}$$

then scheme (13.74) is said to be **consistent** *with (13.73).*

Definition 13.2. *Let $\mathbf{v}_p$ and $\mathbf{v}_{p,\Delta}$ be the solution of (13.73) and (13.74), respectively, at the same point p, where p belongs to the discretized domain. If*

$$\lim_{\Delta \to 0} \left(\left| \mathbf{v}_p - \mathbf{v}_{p,\Delta} \right| \right) = 0 \tag{13.76}$$

for all points p in the discretized space, then the scheme (13.74) is said to be **convergent** *with (13.73).*

Definition 13.3. *Suppose the homogeneous part of (13.74), that is, $L_\Delta u_{p,\Delta} = 0$, is rearranged into a two time-level formula given by*

$$\mathbf{v}_\Delta^{(q+1)} = C^{(q)} \mathbf{v}_\Delta^{(q)} \tag{13.77}$$

where $\mathbf{v}^{(q)}$ is a vector containing $u_{p,\Delta}$ at all the points $p = (q\Delta t, k\Delta x, n\Delta y, m\Delta z)$ in the discretized space.

If for any initial condition $\mathbf{v}_\Delta^{(0)} = \mathbf{v}^{(0)}$, the vector $\mathbf{v}_\Delta^{(q)}$ remains bounded as $\Delta \to 0$, then (13.77) is said to be **Lax-Richtmyer stable**.

Convergence is the most desirable property among the three because it aims for an accurate result for the unknown: $u(t, x, y, z)$. Unfortunately, except for simpler equations, convergence of a proposed scheme can be very difficult to prove. However, consistency of a scheme is quite straightforward to show. Using Taylor series approximations, a proposed scheme will normally reveal the mismatch $(L_\Delta u_p - f_\Delta)$ as a truncation error in terms of Δt, Δx, Δy and Δz. Thus, along the refinement path where $\Delta \to 0$, the truncation errors will usually disappear if finite difference approximations of partial derivatives are used during the construction of the schemes.

[6] For instance, one refinement path could be one that maintains the ratio $\Delta t/(\Delta x \Delta y \Delta z)$ constant.

For Lax-Richtmyer stability, there exist several tests for necessary conditions, including spectral radius and von Neumann methods, as well as other sufficient conditions. In general, stability properties are still easier to prove than convergence properties.

Fortunately, a crucial theorem called the **Lax equivalence theorem** states that for two-level linear consistent schemes, Lax-Richtmyer stability is both a necessary and sufficient condition for the convergence of a given scheme.[7] Thus we describe next some of the tools of stability analysis. From these analyses, we can determine the range of time steps that maintains boundedness of the approximate solutions.

13.4.1 Eigenvalue Method for Stability Analysis

Regardless of whether the semi-discrete approach was used or not, the finite difference time-marching schemes for a linear system can be represented by

$$\mathbf{v}^{(q+1)} = C^{(q)}\mathbf{v}^{(q)} + \Omega^{(q)} \tag{13.78}$$

where $\Omega^{(q)}$ contains information from boundary conditions and nonhomogenous terms of the differential equation.

For instance, for the Euler schemes discussed in Section 13.3.2, the ordinary differential equation resulting from discretization of spatial coordinates that yielded

$$\frac{d}{dt}\mathbf{v}(t) = \mathbf{F}(t)\mathbf{v}(t) + \mathbf{G}(t)$$

can be put into the form given by (13.78), where the matrix $C^{(q)}$ is

$$C^{(q)} = \begin{cases} \left(I + \Delta t\mathbf{F}^{(q)}\right) & \text{for forward Euler scheme} \\ \\ \left(I - \Delta t\mathbf{F}^{(q+1)}\right)^{-1} & \text{for backward Euler scheme} \\ \\ \left(I - \frac{\Delta t}{2}\mathbf{F}^{(q+1)}\right)^{-1}\left(I + \frac{\Delta t}{2}\mathbf{F}^{(q)}\right) & \text{for Crank-Nicholson scheme} \end{cases}$$

whereas for matrix $\Omega^{(q)}$,

$$\Omega^{(q)} = \begin{cases} \Delta t\mathbf{G}^{(q)} & \text{for forward Euler scheme} \\ \\ \left(I - \Delta t\mathbf{F}^{(q+1)}\right)^{-1}\Delta t\mathbf{G}^{(q+1)} & \text{for backward Euler scheme} \\ \\ \left(I - \frac{\Delta t}{2}\mathbf{F}^{(q+1)}\right)^{-1}\frac{\Delta t}{2}\left(\mathbf{G}^{(q+1)} + \mathbf{G}^{(q)}\right) & \text{for Crank-Nicholson scheme} \end{cases}$$

As long as matrix $\Omega^{(q)}$ is bounded, the stability of the finite difference scheme will only depend on matrix $C^{(q)}$. For the case in which C is stationary, that is, not dependent on q, we have the following result.

THEOREM 13.1. *Let $\Omega^{(q)}$ be bounded for all $q \geq 0$. Let $\lambda = \{\lambda_1, \ldots, \lambda_m\}$ be the set of distinct eigenvalues of C, with s_i copies of λ_i. A necessary condition for the scheme given in (13.78) to be* **numerically stable** *(i.e., $|\mathbf{v}^{(q)}| < \infty$) is that*

$$\max_{i=1,\ldots,m}\left\{\ |\lambda_i|\ \right\} \leq 1 \tag{13.79}$$

[7] For a proof of the Lax equivalence theorem, see Morton and Myers (2005).

PROOF. Based on Section 3.6, a square matrix C can be factored to Jordan canonical forms,

$$C = T^{-1} J T$$

where T is a nonsingular matrix, whereas J is a block diagonal matrix composed of m Jordan blocks,

$$J = \begin{pmatrix} J_1 & & 0 \\ & \ddots & \\ & & J_m \end{pmatrix} \qquad \text{with} \qquad J_i = \begin{pmatrix} \lambda_i & 1 & & 0 \\ & \ddots & \ddots & \\ & & \lambda_i & 1 \\ 0 & & & \lambda_i \end{pmatrix} [=] \, s_i \times s_i$$

Also, recalling formulas (3.36)-(2), the powers of J_i are given by

$$J_i^q = P_{(q,s_i)} \lambda_i^q$$

where

$$P_{(q,s)} = \begin{pmatrix} 1 & \beta_{[q,q-1]}\lambda_i^{-1} & \cdots & \beta_{[q,q-s+1]}\lambda_i^{-s+1} \\ 0 & 1 & \cdots & \beta_{[q,q-s+2]}\lambda_i^{-s+2} \\ \vdots & \vdots & \ddots & \vdots \\ 0 & 0 & \cdots & 1 \end{pmatrix} [=] \, s \times s$$

$$\beta_{k,j} = \begin{cases} \dfrac{k!}{(k-j)!j!} & \text{if } j \geq 0 \\ 0 & \text{otherwise} \end{cases}$$

The homogeneous solution of (13.78) is then given by

$$\begin{aligned} \mathbf{v}_{\text{homog}}^{(q)} &= \left(T^{-1} J T\right)^q \mathbf{v}^{(0)} \\ &= T^{-1} \begin{pmatrix} P_{(q,s_1)}\lambda_1^q & & 0 \\ & \ddots & \\ 0 & & P_{(q,s_m)}\lambda_m^q \end{pmatrix} T \, \mathbf{v}^{(0)} \end{aligned}$$

Thus

$$\max_i \left(|\lambda_i|\right) > 1$$

is sufficient to produce an instability for $\mathbf{v}^{(q)}$.

Condition (13.79) becomes both a necessary and sufficient condition for numerical stability when all the eigenvalues are distinct. The maximum absolute eigenvalue is also known as the **spectral radius**. Thus condition (13.79) is also known as the spectral radius condition.

Note that numerical stability is only a necessary condition for Lax-Richtmyer stability. For Lax-Richtmyer stability, additional conditions are needed for the refinement paths; that is, numerical stability should be maintained as $\Delta \to 0$. Nonetheless, for practical purposes, (13.79) usually satisfies the needs of most finite difference schemes. In fact, it can be shown that when C is a normal matrix, numerical stability becomes necessary and sufficient for Lax-Richtmyer stability.

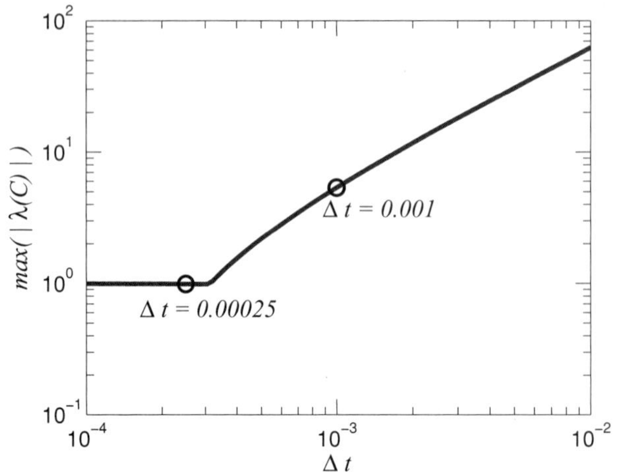

Figure 13.9. A plot of spectral radius of C in (13.80) at different values of Δt using forward Euler method.

Because condition (13.79) depends on the eigenvalues, it is important to note that the calculation of eigenvalues usually contains inaccuracies. Thus one should still apply tolerance levels on the stability ranges for Δt, especially for large C matrices, and test the algorithms directly. For some banded matrices, such as tri-diagonal matrices, there exist explicit formulas for the eigenvalues.[8]

EXAMPLE 13.11. In Example 13.10, the forward Euler method was unstable for $\Delta t = 0.001$ but was stable for $\Delta t = 0.00025$. The spectral radius for this scheme can be evaluated for

$$C = I + \Delta t\, \mathfrak{R} \tag{13.80}$$

where $\mathfrak{R}$ was defined in (13.64). With $\Delta x = \Delta y = 0.05$, $K = N = 19$, as used in Example 13.10, the spectral radius is plotted in Figure 13.9. From the figure, we can see that the spectral radius for $\Delta t = 0.001$ is greater than 1, whereas the spectral radius for $\Delta t = 0.00025$ is 1. Therefore, $\Delta t = 0.001$ should be unstable and $\Delta t = 0.00025$ should be stable because all the eigenvalues are distinct and C is a normal matrix. The curve in Figure 13.9 has a spectral radius greater than 1 when $\Delta t > 3.20 \times 10^{-4}$.

Conversely, by plotting the spectral radius of C for the backward Euler and the Crank-Nicholson methods, we see from Figure 13.10 that either of these methods will be stable for a wide range of Δt. In fact, from the plots, both the backward Euler and the Crank-Nicholson schemes appear to be unconditionally stable.

13.4.2 Von Neumann Method for Stability Analysis

Although methods exists for the calculation of spectral radius of large sparse matrices, these methods are still computationally intensive. An alternative for stability analysis is to use a method known as the **von Neumann method** (also known as Fourier analysis method). Instead of taking the full matrix C into account, the von

[8] See for instance, J. W. Thomas (1995).

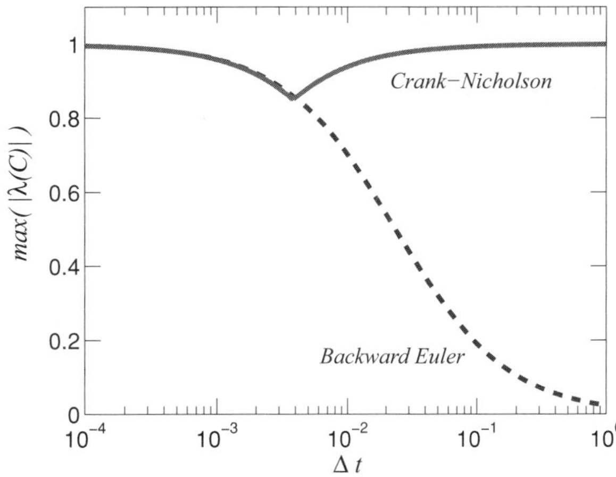

Figure 13.10. A plot of spectral radius of C at different values of Δt for backward Euler and Crank-Nicholson methods.

Neumann approach takes the point formulas, that is, the time-marching difference equation for $u(q\Delta t, k\Delta x, n\Delta y, m\Delta z)$, and determines whether the difference scheme will amplify the magnitude of u at $t + \Delta t$ at some points in the (x, y, z)-space.

Specifically, the method proceeds as follows:

1. Set up the time-marching difference formula for $u\left((q+1)\Delta t, k\Delta x, n\Delta y, m\Delta z\right)$ and consider only the homogeneous terms of the partial differential equation.
2. Set u to be

$$u\left(q\Delta t, k\Delta x, n\Delta y, m\Delta z\right) = \zeta^q \, e^{i\rho_x k\Delta x} \, e^{i\rho_y n\Delta y} \, e^{i\rho_z m\Delta z} \tag{13.81}$$

   where ζ, ρ_x, ρ_y and ρ_z are arbitrary real numbers. ($i = \sqrt{-1}$).
3. Substitute u given by (13.81) into the homogeneous time-marching scheme obtained from step 1.
4. Solve for ζ in terms of Δt, Δx, Δy and Δz.
5. Determine Δt such that

$$|\zeta| \leq 1 \tag{13.82}$$

   for all possible values of $\rho_x \Delta x$, $\rho_y \Delta y$ and $\rho_z \Delta z$

The value of $\left|\zeta\right|$ is known as the **amplification factor**. This method gives a necessary and sufficient condition for the stability of initial value problems or initial-boundary value problems with periodic boundary conditions. However, it is only a necessary condition for problems with general nonperiodic boundary conditions. (See Exercise **E13.11** for an example of a scheme in which the von Neumann condition is satisfied yet the scheme is found to be unstable for a process with nonperiodic boundary conditions.) Nonetheless, this method is one of the more straightforward approaches, and it is relatively easy to evaluate. It also provides a very good estimate of stable range for Δt, even in cases with nonperiodic boundary conditions.

EXAMPLE 13.12. Let us now use the von Neumann method to estimate the range of Δt that would yield a stable scheme for Example 13.10. With the forward Euler scheme, the time-marching scheme for (13.70)

$$\frac{\partial u}{\partial t} = 2\nabla^2 u - 3u + g(t, x, y)$$

at point $u_{k,n,m}^{(q)}$ is

$$\begin{aligned}\frac{1}{\Delta t}\left(u_{k,n,m}^{(q+1)} - u_{k,n,m}^{(q)}\right) &= \frac{2}{\Delta x^2}\left(u_{k+1,n,m}^{(q)} - 2u_{k,n,m}^{(q)} + u_{k-1,n,m}^{(q)}\right) \\ &\quad + \frac{2}{\Delta y^2}\left(u_{k,n+1,m}^{(q)} - 2u_{k,n,m}^{(q)} + u_{k,n-1,m}^{(q)}\right) \\ &\quad - 3u_{k,n,m}^{(q)} + g\left(q\Delta t, k\Delta x, n\Delta y, m\Delta z\right)\end{aligned}$$

To consider only the homogeneous part, we remove g from the equation. Next, we substitute (13.81) and then divide out $u_{k,n,m}^{(q)}$ from both sides to obtain

$$\begin{aligned}\frac{1}{\Delta t}(\zeta - 1) &= \frac{4}{\Delta x^2}\left(\cos\left(\rho_x \Delta x\right) - 1\right) \\ &\quad + \frac{4}{\Delta y^2}\left(\cos\left(\rho_y \Delta y\right) - 1\right) - 3 \\ \zeta &= 1 - (3 + 8\mu)\Delta t\end{aligned}$$

where

$$\mu = \frac{1}{\Delta x^2}\sin^2\left(\rho_x \frac{\Delta x}{2}\right) + \frac{1}{\Delta y^2}\sin^2\left(\rho_y \frac{\Delta y}{2}\right) \geq 0$$

Because the stability requirement should hold for the worst case situation, the values of Δt needed to keep $|\zeta| \leq 1$ can be determined by setting

$$\left(3 + 8\max_{\rho_x, \rho_y}(\mu)\right)\Delta t \leq 2$$

or with $\max(\mu) = (1/\Delta x^2) + (1/\Delta y^2)$,

$$\Delta t \leq \frac{2\Delta x^2 \Delta y^2}{3\Delta x^2 \Delta y^2 + 8\left(\Delta x^2 + \Delta y^2\right)}$$

By setting $\Delta x = \Delta y = 0.05$, we find that $\Delta t \leq 3.123 \times 10^{-4}$. This is comparable to the stability range found using the eigenvalue method (see Example 13.11), which is $\Delta t \leq 3.20 \times 10^{-4}$.

For the backward Euler method, we have

$$\begin{aligned}\frac{1}{\Delta t}\left(u_{k,n,m}^{(q+1)} - u_{k,n,m}^{(q)}\right) &= \frac{2}{\Delta x^2}\left(u_{k+1,n,m}^{(q+1)} - 2u_{k,n,m}^{(q+1)} + u_{k-1,n,m}^{(q+1)}\right) \\ &\quad + \frac{2}{\Delta y^2}\left(u_{k,n+1,m}^{(q+1)} - 2u_{k,n,m}^{(q+1)} + u_{k,n-1,m}^{(q+1)}\right) \\ &\quad - 3u_{k,n,m}^{(q+1)} + g\left((q+1)\Delta t, k\Delta x, n\Delta y, m\Delta z\right)\end{aligned}$$

After removing g and substituting (13.81), the amplification $|\zeta|$ can be evaluated to be

$$\left|\zeta\right| = \left|\frac{1}{1+(3+8\mu)\,\Delta t}\right|$$

which means unconditional stability, because $\mu \geq 0$ and $\Delta t > 0$.

Finally, for Crank-Nicholson schemes, the amplification formula can be found to be

$$\left|\zeta\right| = \left|\frac{1-(3+8\mu)\,(\Delta t/2)}{1+(3+8\mu)\,(\Delta t/2)}\right|$$

Because $\mu \geq 0$, we see that $|\zeta| \leq 1$ for $\Delta t > 0$. So once again, we have unconditional stability.

13.5 EXERCISES

E13.1. Obtain the coefficients for the finite difference formulas of the following and determine the order:

1. $$\left.\frac{d^2u}{dx^2}\right|_{x_k} = a\left(\frac{d^2u}{dx^2}\right)_{k-1} + \frac{1}{\Delta x^2}\left(bu_{k-1} + cu_k + du_{k+1}\right) + e\left(\frac{d^2u}{dx^2}\right)_{k+1}$$
2. $$\left.\frac{d^2u}{dx^2}\right|_{x_k} = a\frac{1}{\Delta x}\left(\frac{du}{dx}\right)_{k-1} + \frac{1}{\Delta x^2}\left(bu_k + cu_{k+1}\right) + d\left(\frac{d^2u}{dx^2}\right)_{k+1}$$

E13.2. Derive the approximation formula given in (13.21) for $\dfrac{\partial^2 u}{\partial x \partial y}$, where $\partial u/\partial y$ was approximated using the formula in item 7 of Table 13.1.

E13.3. Find matrices $\mathcal{D}_2$, $\mathcal{D}_1$, $\mathfrak{b}_2$, and $\mathfrak{b}_1$ (based on notations given in Section 13.2.1) for a fourth-order finite difference approximation that would handle the Neumann condition at $x = 0$ and Dirichlet condition at $x = 1$, that is, $(du/dx)(0) = Neu_0$ and $u(1) = Di_1$.

E13.4. Consider the following second-order differential equation:

$$\nabla^2 u + 2\frac{\partial^2 u}{\partial x \partial y} + \frac{\partial u}{\partial x} - 2\frac{\partial u}{\partial y} - 9y = 0 \quad ; \quad 0 \leq x, y \leq 1$$

subject to $u(0, y) = -2y^2$, $u(1, y) = 1 + y - 2y^2$, $u(x, 0) = x^2$ and $u(x, 1) = x^2 + x - 2$. Obtain the finite difference solution using the central difference approximations of all the partial derivatives, that is, using the formulation given in (13.39). Try using $K = N = 100$. Note: The exact solution for this problem is $u = (x - y)(x + 2y)$.

E13.5. Develop a finite difference solution of the steady-state linear 2D equation given in (13.34) using the matrix formulation of (13.39) that would handle the case where a Neumann boundary condition is fixed at $y = 0$, that is, $\dfrac{\partial u}{\partial y}(y = 0) = \psi_0(x)$. (Suggestion: One could modify the matrices for $\mathcal{D}_i$'s and $\mathcal{B}_i$'s in the program `lin2D_centralfd_dirich.m`.) Test this program on the same problem given in Exercise **E13.4**, except change the boundary condition at $y = 0$ to be $\left.\dfrac{\partial u}{\partial y}\right|_{y=0} = x$. The same exact solution applies, that is, $u = (x - y)(x + 2y)$.

E13.6. Use the method given in Section 13.2.3 to obtain the finite difference solution of the following Poisson equation in polar coordinates:

$$\nabla^2 u = 4 - 21\cos(5\theta) \qquad 0 \le r \le 1\ ,\ 0 \le \theta \le 2\pi$$

subject to $u(1,\theta) = 2 + \cos(5\theta)$. Noting that the exact solution is given by $u = 1 + r^2(\cos(5\theta) + 1)$, plot the solution and the error distribution for the case with $K = 30$ grid points along r and $N = 200$ grid points along θ.

E13.7. Modify the formulation of matrix A in (13.46) to handle the generalization of the Poisson equation to the nonhomogeneous Helmholtz equation in polar coordinates given by

$$\nabla^2 u + h(r,\theta)u = f(r,\theta)$$

Create a program (or modify `poisson2d_polar_dirich.m`) to accommodate the change and test it on the case with $h = 3$, $f = 18 - 15r^3\sin(3\theta)$ and boundary condition $u(1,\theta) = 6 - 5\sin(3\theta)$. The exact solution for this case is given by $u = 6 - 5r^3\sin(3\theta)$. Plot the error distribution for the case using $K = 100$ grid points along r and $N = 100$ grid points along θ.

E13.8. Use the method given in Section 13.2.3 to obtain the finite difference solution of the following Poisson equation in spherical coordinates under azimuthal symmetry:

$$\nabla^2 u = \frac{5 - 4\cos^2(\theta)}{\sin(\theta)} \qquad 0 \le r \le 1\ ,\ 0 \le \theta \le 2\pi$$

subject to $u(1,\theta) = 3 + \sin(\theta)$. Noting that the exact solution is given by $u = 3 + r^2(\sin(\theta))$, plot the solution and the error distribution for the case with $K = 30$ grid points along r and $N = 100$ grid points along θ.

E13.9. Consider the diffusion equation given by

$$\frac{\partial u}{\partial t} = \frac{\partial^2 u}{\partial x^2}$$

subject to the conditions: $u(x,0) = 1$, $u(0,t) = 0$, and $u(1,t) = 0$.

1. Recall the method of separation of variables and show that the analytical solution is given by
$$u(x,t) = 2\sum_{k=1}^{\infty}\left(\frac{1-\cos(k\pi)}{k\pi}\right)e^{-(k^2\pi^2 t)}\sin(k\pi x)$$
2. Obtain the time-marching equation based on the central difference formula for $\partial^2 u/\partial x^2$; that is, find **F** and **B** in (13.51) for this problem.
3. Using $\Delta x = 0.01$, try to obtain the finite difference solution from $t = 0$ to $t = 0.1$ using the three weighted-Euler methods, that is, the forward Euler, backward Euler, and Crank-Nicholson. First try $\Delta t = 0.005$ and then try $\Delta t = 5 \times 10^{-5}$. Plot the time lapse solutions and error distribution of the solutions (if they are bounded).
4. Using the von Neumann stability method, show that the maximum time increment allowed for a stable marching of the forward Euler method for this problem will be $\Delta t = \Delta x^2/2$ (thus $\Delta t_{max} = 5 \times 10^{-5}$ will be stable.)
5. For the Crank-Nicholson method, the value of Δt need not be as small as the one needed for the stability of the forward Euler method to remove the oscillations. Try $\Delta t = 5 \times 10^{-4}$ and show that the oscillations are absent.

E13.10. Consider the one-dimensional advection-diffusion equation

$$\frac{\partial^2 u}{\partial x^2} + \alpha \frac{\partial u}{\partial x} = \frac{\partial u}{\partial t}$$

1. Using the von Neumann stability method, obtain the amplification factor for the difference equation based on the central difference formulas for the spatial derivatives and forward Euler method for the time derivatives, in terms of α, Δt, and Δx.
2. Determine the maximum time increment Δt such that the forward-Euler method will be stable for $\alpha = -2$ and $\Delta x = 0.05$. Verify your prediction for the initial and boundary conditions given by $u(x, 0) = 0$ and $u(0, t) = u(1, t) = 1$.

E13.11. Consider the following scheme known as the **Wendroff scheme**:

$$(1+\eta)\, u_{k+1}^{(q+1)} + (1-\eta)\, u_k^{(q+1)} = (1-\eta)\, u_{k+1}^{(q)} + (1+\eta)\, u_k^{(q)} \tag{13.83}$$

where $\eta = \lambda \Delta t / \Delta x$.

1. Show that, based on the von Neumann method, the amplification factor is given by $|\zeta| = 1$ for all real η.
2. Use this scheme for $\lambda = 0.5$, $\Delta t = 0.01$ for a finite domain $x_k = k\Delta x$, $\Delta x = 0.01$ $k = 0, \ldots, 100$, with Dirichlet conditions $u_0^{(q)} = u_{100}^{(q)} = 0$ and initial condition

$$u_k^{(0)} = \begin{cases} 1 & \text{for } 0.2 \le x_k \le 0.4 \\ 0 & \text{otherwise} \end{cases}$$

   Note that this scheme will turn out to be unstable. Prove this instability using the eigenvalue analysis.
3. Change the previous case to have a periodic boundary condition and note that the system will become stable (although with growing oscillations). Show that the eigenvalue analysis for this case predicts stability. (This exercise shows that the von Neumann condition is only a necessary condition of stability for nonperiodic boundary conditions, but it becomes both necessary and sufficient for systems having only periodic boundary conditions. This is because the von Neumann method is based on Fourier analysis.)

E13.12. Discuss how the matrices in (13.63) for Example 13.8 will have to be modified if the boundary conditions were changed to become

$$\frac{\partial u}{\partial x}(t, 0, y) = 0 \qquad\qquad u(t, x, 0) = w_0(t, x)$$

$$\frac{\partial u}{\partial x}(t, 1, y) = 0 \qquad\qquad u(t, x, 1) = w_1(t, x)$$

after using the method of ghost points.

E13.13. Write a general program that would solve the dynamic 2D Poisson equation in polar coordinates given by

$$\frac{\partial u}{\partial t} + f(r, \theta) = \frac{\partial^2 u}{\partial r^2} + \frac{1}{r}\frac{\partial u}{\partial r} + \frac{1}{r^2}\frac{\partial^2 u}{\partial \theta^2}$$

subject to static initial and boundary conditions given by

$$u(r, \theta, 0) = U_i(r, \theta) \qquad u(1, \theta, t) = U_R(\theta)$$

where $f(r,\theta)$ is a forcing function. Test the program with the problem given in Example 11.8 and determine whether you get similar results to those shown in Figure 11.7.

E13.14. Consider the one-dimensional time-dependent heat equation for a sphere that is symmetric around the origin given by

$$\frac{\partial T}{\partial t} = \nabla^2 T = \left(\frac{\partial^2 T}{\partial r^2} + \frac{2}{r}\frac{\partial T}{\partial r}\right)$$

with initial and boundary conditions given by $T(r,0) = 0$, $(\partial T/\partial r)_{r=0} = 0$ and $T(0,t) = 1$. Based on the separation of variables method, the analytical solution is given by

$$T(r,t) = 1 + \frac{2}{\pi r}\sum_{n=0}^{\infty}\left(\frac{(-1)^n}{n}\right) e^{-\alpha(n\pi)^2 t}\sin(n\pi r) \tag{13.84}$$

1. Obtain the time-dependent temperature at the origin, $T_0(t)$, based on (13.84).
2. Use the central difference approximation and the Crank-Nicholson scheme to obtain a numerical solution of the temperature distribution using $\Delta r = 0.05$ and $\Delta t = 0.001$ for $0 \le r \le 1$ and $0 \le t \le 0.5$. Compare the values for $T(0,t_k)$ to the function $T(0,t)$ found in the previous problem and plot the errors.
3. Obtain the numerical solution using the Crank-Nicholson scheme under the same settings and problem as before, except modify the boundary condition at $r = 1$ to a Robin condition given by

$$-\beta \left.\frac{\partial T}{\partial r}\right|_{r=1} + T(1,t) = 1$$

with $\beta = 0.1$. (Hint: You can use the method of ghost points to introduce this boundary condition while adding $T(r_{K+1}, t_q)$ as unknowns.)

E13.15. Use the von Neumann method to determine the stability region of the six basic finite difference schemes given in Table M.1.

E13.16. Apply the same six finite difference schemes on the system given in Example M.3 but instead using an initial condition that contains a triangular pulse given by:

$$u(x,0) = \begin{cases} 10x - 2 & \text{for } 0.2 \le x \le 0.3 \\ -10x + 4 & \text{for } 0.3 < x \le 0.4 \\ 0 & \text{otherwise} \end{cases}$$

Observe whether significant oscillations will occur when using the Crank-Nicholson, leapfrog, or Lax-Wendroff schemes.

14 Method of Finite Elements

In this chapter, we discuss the finite element method for the solution of partial differential equations. It is an important solution approach especially when the shape of the domain (including possible holes inside the domains) cannot be conveniently transformed to a single rectangular domain. This includes domains whose boundaries cannot be formulated easily under existing coordinate systems.

In contrast to finite difference methods which are based on replacing derivatives with discrete approximations, finite element (FE) methods approach the problem by piecewise interpolation methods. Thus the FE method first partitions the whole domain Ω into several small pieces Ω_n, which are known as the **finite elements** represented by a set of nodes in the domain Ω. The sizes and shapes of the finite elements do not have to be uniform, and often the sizes may need to be varied to balance accuracy with computational efficiency.

Instead of tackling the differential equations directly, the problem is to first recast it as a set of integral equations known as the **weak form** of the partial differential equation. There are several ways in which this integral is formulated, including least squares, collocation, and weighted residual. We focus on a particular weighted residual method known as the **Galerkin method**. These integrals are then imposed on each of the finite elements. The finite elements that are attached to the boundaries of Ω will have the additional requirements of satisfying the boundary conditions. As is shown later, these integrals can be reduced to matrix equations in which the unknowns are the nodal values. Because neighboring finite elements share the same nodes, the various "local" matrix equations have to be assembled to form the "global" matrix equation. Basically, the result is either a large linear matrix equation for steady-state problems or a large matrix iteration scheme for transient problems.

There are several implementations of finite element method. In this chapter, we limit our discussion to the simplest approaches. First, we only tackle 2D linear second-order partial differential equations. The construction of the weak form for these problems is discussed in Section 14.1. We opted to tackle the 2D case because the 1D case can already be easily handled by finite-difference methods, whereas there are several 2D domains that are difficult to solve using only finite difference methods. With a good basic understanding of the 2D case, the extensions to three dimensions should be straightforward.

Second, we limit our finite element meshes to be composed of triangles and apply only the simplest shape function (also known as interpolation function) based on three vertices of the triangle. The various properties and integrals based on triangular elements are discussed in Sections 14.2 and 14.2.1, respectively. We include much later in Section 14.4 a brief description of mesh construction called the **Delaunay triangulation**. The use of triangles, although not the most accurate, simplifies the calculation of the integrals significantly compared with high-order alternatives. The inclusion of the boundary conditions depends on the types of conditions. The application of Neumann conditions and Robin conditions involves line integrals, which are discussed in Section 14.2.2. This will require another set of one-dimensional shape functions.

After the weak form has been applied to the local elements to obtain the various matrix equations, the assembly of these local matrix equations to a global equation is discussed in Section 14.3. Once the assembly process has finished, only then are the Dirichlet conditions included. There are two approaches available for doing this: the matrix reduction approach and the overloading approach. Following most implementations of finite element methods, we focus on the overloading approach. Afterward, the various steps of the finite element method based on triangular elements are summarized in Section 14.5.

Having a basic description of the particular implementation of the finite element method using triangles for a linear second-order differential equation, we also include three extensions in this chapter. One is the improvement for convection-dominated cases known as the streamline upwind Petrov-Galerkin (SUPG) method, discussed briefly in Section N.2 as an appendix. Another extension is the treatment of axisymmetric cases discussed in Section 14.6, in which the techniques of the 2D finite element method can be used almost directly but with the inclusion of r as a factor inside the integrals. Finally, in Section 14.7, we discuss the use of the finite element method to handle unsteady state problems via the Crank-Nicholson method.

14.1 The Weak Form

The general steps for the method of finite elements for the numerical solution of partial differential equations can be summarized as follows:

1. Reformulate the partial differential equation as an equivalent integral equation known as the **weak form**.
2. Decompose the domain into non-overlapping finite elements.
3. Evaluate the integral equations locally for each finite element, applying Neumann and Robin conditions when appropriate.
4. Assemble the local elements together and implement any Dirichlet boundary conditions to generate a global linear matrix equation.
5. Solve the matrix equation for the nodal values.

We first limit our discussion to the solution of the following linear, second-order, elliptic partial differential equation:

$$F\left(x, y, u, \frac{\partial u}{\partial x}, \frac{\partial u}{\partial y}, \frac{\partial^2 u}{\partial x^2}, \frac{\partial^2 u}{\partial y^2}\right) = \Big[\nabla \cdot (\mathbf{M}(x,y) \cdot \nabla u)\Big] + \Big[\mathbf{b}(x,y) \cdot \nabla u\Big] + g(x,y)u + h(x,y) = 0 \qquad (14.1)$$

$$\text{for } (x,y) \in \Omega$$

Figure 14.1. Surface approximation using triangular finite elements.

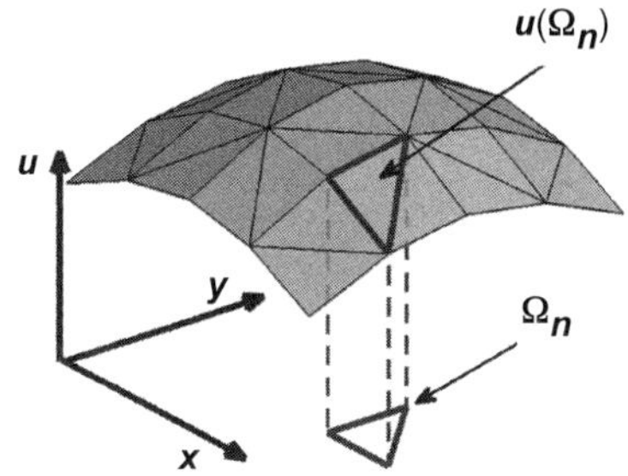

subject to the following boundary conditions:

$$\left[\mathbf{M}(x,y)\cdot\nabla u\right]\cdot\underline{\mathbf{n}} = \overline{q}(x,y) \qquad \text{on boundary }(\Omega)_{\text{NBC}} \tag{14.2}$$

$$\left[\mathbf{M}(x,y)\cdot\nabla u\right]\cdot\underline{\mathbf{n}} = \mu(x,y)u+\overline{q}(x,y) \qquad \text{on boundary }(\Omega)_{\text{RBC}} \tag{14.3}$$

or

$$u = \overline{u}(x,y) \qquad \text{on boundary }(\Omega)_{\text{DBC}} \tag{14.4}$$

where Ω is a region in the (x, y) plane, with the subscripts NBC, RBC, and DBC denoting Neumann, Robin, and Dirichlet boundary conditions, respectively, and $\underline{\mathbf{n}}$ is the unit normal vector pointing outward of Ω. The $\mathbf{M}$ tensor is assumed to be symmetric and positive definite. The functions $\overline{u}(x, y)$, $\overline{q}(x, y)$ and $\mu(x, y)$ are all assumed to be continuous functions along their respective boundary.

As mentioned in the introduction, we limit our approach to the use of triangular finite elements and linear piecewise approximations. This means that the domain will be partitioned into triangular elements, each defined by the three vertices known as **nodes**. By assuming a linear-piecewise approximation of the surface in each element, a stitching of the neighboring elements will provide a continuous, albeit non-smooth, surface solution to the partial differential equation. To illustrate, consider Figure 14.1. The finite element domain, Ω_n, is a triangular region in the (x, y)-plane defined by three vertices (or nodes), $\mathbf{p}_1$, $\mathbf{p}_2$, and $\mathbf{p}_3$. The collection of u-values at each node, that is, $u(\mathbf{p}_j)$, will form the desired numerical solution to the partial differential equation. The linear approximation will then be a plane, denoted by $u(\Omega_n)$, as shown in Figure 14.1. Doing so, we see that the triangular planes will stitch continuously with their neighboring elements. Obviously, the approximation can be improved by increasing the order of approximation, that is, allowing for curved surfaces. When doing so, one needs to make sure that the surface elements will form a continuous surface when combined. In our case, we assume that increasing the resolution of the mesh will be sufficient for improving accuracy.

In committing to use first-order approximations for each finite element, the first-order derivatives $\partial u/\partial x$ and $\partial u/\partial y$ will in general not be continuous across the shared edges of neighboring finite elements. This means that the second-order partial derivatives can no longer be evaluated around these neighborhoods. We need to reformulate the original partial differential equation to proceed with the first-order approximation.

One such formulation is called the **weak form**. It transforms the original second-order partial differential equation as given in (14.1) into an integral equation in which the integrands will involve only first-order partial derivatives. We begin by including a weighing function on the right-hand side of (14.1). We can choose the variation of u, denoted by δu, as the weighing function,[1]

$$\begin{aligned} 0 &= \int_{\Omega} \delta u \, F \, dA \\ &= \int_{\Omega} \delta u \left[\nabla \cdot (\mathbf{M}(x,y) \cdot \nabla u) \right] dA + \int_{\Omega} \delta u \left[\mathbf{b}(x,y) \cdot \nabla u \right] dA \\ &\quad + \int_{\Omega} \delta u \left[g(x,y) u + h(x,y) \right] dA \end{aligned} \tag{14.5}$$

Thus the solution is to find the values of u at the nodes that would make the sum of integrals in (14.5) as close to zero as possible, based on the chosen forms of the weights.

An exact solution $u(x, y)$ has the property that the values and derivatives of u will satisfy $F = 0$ as given in (14.1), while satisfying the boundary conditions. These are known as the **strong solutions**. By fixing the solutions to be composed of flat triangles, we are immediately setting the objective toward obtaining an approximate solution that is defined by the node values, and we do not expect $F = 0$, except in special circumstances. Instead, the factor inside the integrand that is weighted by δu can be taken instead as a "residual error" or "residual." Thus the approach based on the weak form given in (14.5) is also known as the **weighted residual method**. And the solution to (14.5) is known as a **weak solution**.

The first integrand in (14.5) contains the second-order partial derivatives. As we had already mentioned, this causes a problem because the second derivatives of a surface composed of flat triangles will be zero inside each element, while becoming indeterminate at the edges of the elements. To alleviate this problem, we use two previous results: the divergence of a scalar product of a vector (cf. (4.55)) and the divergence theorem (cf. (5.5)), which we repeat below:

$$\nabla \cdot \psi \underline{\mathbf{v}} = \psi \nabla \cdot \underline{\mathbf{v}} + \underline{\mathbf{v}} \cdot \nabla \psi \tag{14.6}$$

$$\int_S \mathbf{F} \cdot \underline{\mathbf{n}} \, dS = \int_V \nabla \cdot \mathbf{F} \, dV \tag{14.7}$$

Applying (14.6) to the integrand of the term in (14.5) that contains **M**,

$$\begin{aligned} \delta u \left[\nabla \cdot (\mathbf{M}(x,y) \cdot \nabla u) \right] &= \nabla \cdot (\delta u \mathbf{M}(x,y) \cdot \nabla u) \\ &\quad - \nabla (\delta u) \cdot \left[\mathbf{M}(x,y) \cdot \nabla u \right] \end{aligned} \tag{14.8}$$

Next, using (14.7) to the first term on the right-hand side of equation (14.8),

$$\int_{\Omega} \nabla \cdot (\delta u \, \mathbf{M}(x,y) \cdot \nabla u) \, dA = \int_{\text{Bound}(\Omega)} (\delta u \, \mathbf{M}(x,y) \cdot \nabla u) \cdot \underline{\mathbf{n}} \, ds \tag{14.9}$$

[1] Instead of using the concept of "weight," a similar approach is to treat δu as a "test function."

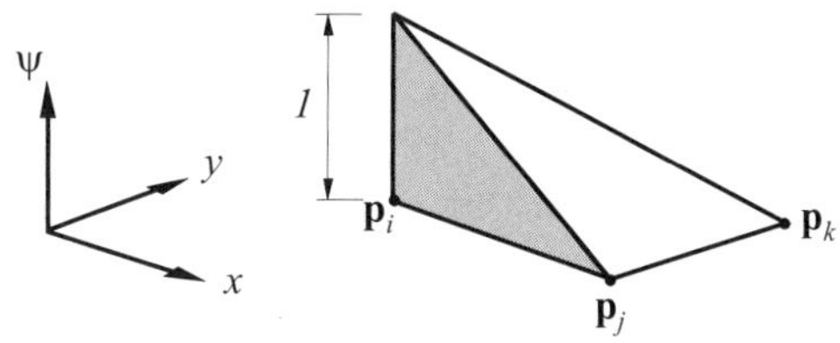

Figure 14.2. The shape function $\psi_i(\mathbf{p})$.

where Bound (Ω) is the boundary of Ω and $\underline{\mathbf{n}}$ is the outward unit normal at this boundary. Substituting (14.8) and (14.9) to (14.5), and then after some rearranging, we get

$$\begin{pmatrix} \int_\Omega \nabla(\delta u)\cdot\left[\mathbf{M}(x,y)\cdot\nabla u\right]dA \\ -\ \int_\Omega \delta u\left[\mathbf{b}(x,y)\cdot\nabla u\right]dA \\ -\ \int_\Omega\left[\delta u\, g(x,y)\,u\right]\,dA \end{pmatrix} = \begin{pmatrix} \int_{\text{Bound}(\Omega)}\left(\delta u\,\mathbf{M}(x,y)\cdot\nabla u\right)\cdot\underline{\mathbf{n}}\,ds \\ +\ \int_\Omega \delta u\, h(x,y)\,dA \end{pmatrix} \tag{14.10}$$

We refer to (14.10) as our weak-form working equation for the finite element method.

14.2 Triangular Finite Elements

Consider the 2D finite element domain Ω_n, shown in Figure 14.1. This triangular domain is fixed by three distinct nodes, denoted by vectors $\mathbf{p}_1 = (x_1, y_1)^T$, $\mathbf{p}_2 = (x_2, y_2)^T$, and $\mathbf{p}_3 = (x_3, y_3)^T$, with $x_i, y_i \in \Omega,\ i = 1, 2, 3$. Corresponding to each node, we can build three linear **shape functions**, $\psi_1(x, y)$, $\psi_2(x, y)$, and $\psi_3(x, y)$, given by

$$\psi_i(\mathbf{p}) = \frac{\mathbf{det}\Big(\ (\mathbf{p}_j - \mathbf{p})\ \Big|\ (\mathbf{p}_k - \mathbf{p})\ \Big)}{\mathbf{det}\Big(\ (\mathbf{p}_j - \mathbf{p}_i)\ \Big|\ (\mathbf{p}_k - \mathbf{p}_i)\ \Big)} \tag{14.11}$$

where $i, j, k = \{1, 2, 3\}$, $i \neq j \neq k$ and $\mathbf{p} = (x, y)^T$ is a point inside the triangle Ω_n formed by the three nodes $\mathbf{p}_i$ $\mathbf{p}_j$ and $\mathbf{p}_k$. Figure 14.2 shows the shape function $\psi_i(\mathbf{p})$.

EXAMPLE 14.1. Let the Ω be the triangle whose vertices are given by

$$\mathbf{p}_1 = \begin{pmatrix} 1 \\ 1 \end{pmatrix} \qquad \mathbf{p}_2 = \begin{pmatrix} 2 \\ 1.5 \end{pmatrix} \qquad \mathbf{p}_3 = \begin{pmatrix} 1.5 \\ 2 \end{pmatrix}$$

then

$$\psi_1 = \frac{\mathbf{det}\begin{pmatrix} 2-x & 1.5-x \\ 1.5-y & 2-y \end{pmatrix}}{\mathbf{det}\begin{pmatrix} 2-1 & 1.5-1 \\ 1.5-1 & 2-1 \end{pmatrix}} = \frac{7}{3} - \frac{2}{3}x - \frac{2}{3}y \qquad \psi_2 = \frac{\mathbf{det}\begin{pmatrix} 1-x & 1.5-x \\ 1-y & 2-y \end{pmatrix}}{\mathbf{det}\begin{pmatrix} 1-2 & 1.5-2 \\ 1-1.5 & 2-1.5 \end{pmatrix}} = -\frac{2}{3} + \frac{4}{3}x - \frac{2}{3}y$$

$$\psi_3 = -\frac{2}{3} - \frac{2}{3}x + \frac{4}{3}y$$

From this point on, we assume that the indexing of the points follow a cyclic sequence, that is, $(i, j, k) = (1, 2, 3), (2, 3, 1)$, or $(3, 1, 2)$ should trace a counter-clockwise path. Let $D/2$ be the area of the triangle formed by $\mathbf{p}_i$, $\mathbf{p}_j$ and $\mathbf{p}_k$, that is, (cf. (1.76) in Exercise **E1.32**),

$$D = \mathbf{det}\left(\mathbf{p}_i \mid \mathbf{p}_j\right) + \mathbf{det}\left(\mathbf{p}_j \mid \mathbf{p}_k\right) + \mathbf{det}\left(\mathbf{p}_k \mid \mathbf{p}_i\right) = \mathbf{det}\left(\begin{array}{c|c|c} 1 & 1 & 1 \\ \hline \mathbf{p}_i & \mathbf{p}_j & \mathbf{p}_k \end{array}\right) \tag{14.12}$$

Using the multilinearity property of determinants (cf. item 7 in Table 1.6), one can show that

$$D = \mathbf{det}\Big(\ (\mathbf{p}_j - \mathbf{p}_i) \ \Big| \ (\mathbf{p}_k - \mathbf{p}_i) \ \Big)$$

which is the denominator in (14.11). Thus with $\mathbf{p}$ constrained to be inside or on the triangle formed by $\mathbf{p}_i$, $\mathbf{p}_j$ and $\mathbf{p}_k$, we have

$$\psi_i = \frac{2\text{Area}\left(\Delta_{\mathbf{p}\mathbf{p}_j\mathbf{p}_k}\right)}{2\text{Area}\left(\Delta_{\mathbf{p}_i\mathbf{p}_j\mathbf{p}_k}\right)} \qquad \text{and} \qquad 0 \le \psi_i\left(\mathbf{p}\right) \le 1$$

We now list some of the properties of the shape functions (with $i, j, k \in \{1, 2, 3\}$ and $i \neq j \neq k$). These results can be obtained by direct application of (14.11).

1. $\psi_i(\mathbf{p})$ has a maximum value of 1 at $\mathbf{p} = \mathbf{p}_i$.

$$\psi_1(\mathbf{p}_1) = 1 \quad ; \quad \psi_2(\mathbf{p}_2) = 1 \quad ; \quad \psi_3(\mathbf{p}_3) = 1 \tag{14.13}$$

2. ψ_i has a minimum value of 0 at $\mathbf{p} = \lambda\mathbf{p}_j + (1-\lambda)\mathbf{p}_k$ where $0 \le \lambda \le 1$.

$$\psi_i\left(\lambda\mathbf{p}_j + (1-\lambda)\mathbf{p}_k\right) = 0 \qquad i \neq j \neq k \quad ; \quad 0 \le \lambda \le 1 \tag{14.14}$$

3. The sum of all the three shape functions is 1.

$$\psi_1\left(\mathbf{p}\right) + \psi_2\left(\mathbf{p}\right) + \psi_3\left(\mathbf{p}\right) = 1 \tag{14.15}$$

4. ψ_i is an affine function of x and y.

   Applying the multilinearity property of determinants to (14.11),

$$\psi_i(\mathbf{p}) = \frac{\left(p_{jy} - p_{ky}\right)}{D}x + \frac{\left(p_{kx} - p_{jx}\right)}{D}y + \frac{\mathbf{det}\Big(\ \mathbf{p}_j \ \Big| \ \mathbf{p}_k \ \Big)}{D} \tag{14.16}$$

5. The value of all shape functions is equal to $1/3$ at $\mathbf{p} = \frac{1}{3}\left(\mathbf{p}_1 + \mathbf{p}_2 + \mathbf{p}_3\right)$.

$$\psi_1\left(\mathbf{p}^*\right) = \psi_2\left(\mathbf{p}^*\right) = \psi_3\left(\mathbf{p}^*\right) = \frac{1}{3} \qquad \text{with } \mathbf{p}^* = \frac{1}{3}\left(\mathbf{p}_1 + \mathbf{p}_2 + \mathbf{p}_3\right) \tag{14.17}$$

6. The surface integral of ψ_i with respect to Ω_n is equal to $\left(\dfrac{D}{6}\right)$.

$$\int_{\Omega_n} \psi_i \, dA = \frac{D}{6} \tag{14.18}$$

   Equation (14.18) is just the volume of the pyramid formed by the shape function ψ_i, as shown in Figure 14.2. The altitude of the pyramid is 1, whereas the area

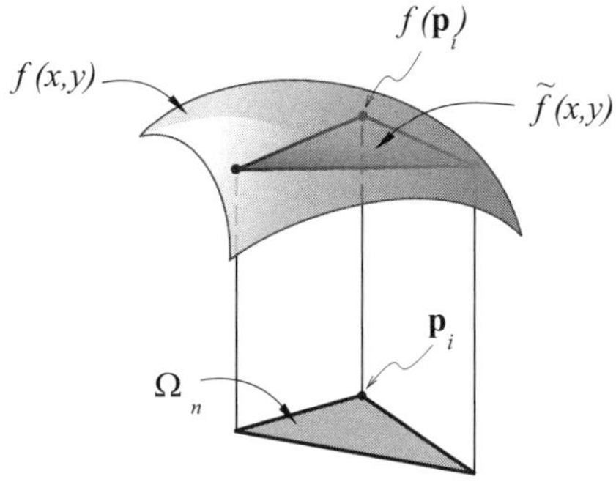

Figure 14.3. A linear (planar) approximation of $f(x, y)$ with x and y in the region Ω_n.

of the triangular base has been determined earlier to be $D/2$. Thus because the volume of the pyramid is one-third the product of base area and the altitude, we obtain the value of $D/6$, where D is given by (14.12)

14.2.1 Surface Integrals of the Finite Elements

Using the shape functions for each node in Ω_n, we now obtain formulas for evaluating the surface integrals in the working equation (14.10). For a continuous function $f(x, y)$, a first-order approximation $\tilde{f}$ in Ω_n is a plane, as shown in Figure 14.3. This plane can be defined by the values of $f(x, y)$ at the three vertices, that is, let $f_i = f(\mathbf{p}_i)$, then

$$\tilde{f} = f_1\psi_1 + f_2\psi_2 + f_3\psi_3 \tag{14.19}$$

Based on (14.19), the surface integral in Ω_n is approximated as follows:

$$\begin{aligned} \int_{\Omega_n} f(x, y)dA &\approx \int_{\Omega_n} \tilde{f}(x, y)dA = f_1 \int_{\Omega_n} \psi_1\, dA + f_2 \int_{\Omega_n} \psi_2\, dA + f_3 \int_{\Omega_n} \psi_3\, dA \\ &= \frac{D}{2}\left(\frac{f_1 + f_2 + f_3}{3}\right) \end{aligned} \tag{14.20}$$

where we used (14.18) to evaluate the surface integrals of ψ_1, ψ_2, and ψ_3 .

Note that in (14.20), each f_i represents one function evaluation. An alternative is to use the Gaussian quadrature approach and replace the three function evaluations with just one function evaluation. This means that instead of vertex values f_1, f_2, and f_3, we need only one constant value, say f^*, evaluated at some point inside Ω_n. This point turns out to be $\mathbf{p}^* = (1/3)(\mathbf{p}_1 + \mathbf{p}_2 + \mathbf{p}_3)$. Based on (14.17),

$$f^* = \tilde{f}(\mathbf{p}^*) = f_1\psi_1(\mathbf{p}^*) + f_2\psi_2(\mathbf{p}^*) + f_3\psi_3(\mathbf{p}^*) = \frac{f_1 + f_2 + f_3}{3} \tag{14.21}$$

and the surface integral of f^*, held constant in Ω_n, yields

$$\int_{\Omega_n} f^* dA = f^* \int_{\Omega_n} dA = \frac{D}{2} f^* = \frac{D}{2}\left(\frac{f_1 + f_2 + f_3}{3}\right) \tag{14.22}$$

Thus the surface integrals given by (14.20) and by (14.22) are equal. The constant value, f^*, acts like a mean value of $\tilde{f}$ in Ω_n as shown in Figure 14.4.

We will be replacing $f(\mathbf{p})$ with several objects. In some cases, we will substitute $f(\mathbf{p})$ with the elements of $\mathbf{M}(x, y)$ and $\mathbf{b}(x, y)$, the functions $g(x, y)$ and $h(x, y)$, as

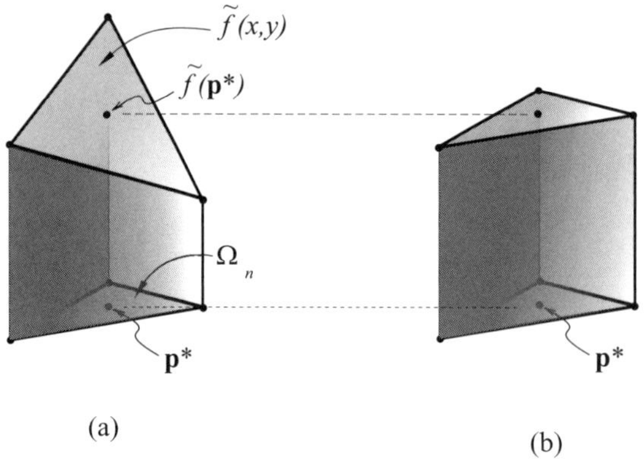

Figure 14.4. The volume in (a) is the same as the volume in (b). The volume in (a) is generated by the surface integral of $\tilde{f} = \sum_{i=1}^{3} f_i \psi_i$. The volume in (b) is a prism of constant height and is generated by the surface integral of $\tilde{f}^* = \sum_{i=1}^{3} \tilde{f}(\mathbf{p}^*)\psi_i$

given in (14.10), even u and δu. Furthermore, we will be using $f^* = f(\mathbf{p}^*)$ – not $\tilde{f}(\mathbf{p}^*)$, that is,

$$\int_{\Omega_n} f(x, y)dA \approx f(\mathbf{p}^*) \frac{D}{2} \tag{14.23}$$

Strictly speaking, this is not exactly the same as (14.22) except when $f(x, y)$ is flat in Ω_n, for example, in the limit where the nodes of Ω_n approach each other. Nonetheless, because $\tilde{f}(x, y)$ is just a linear approximation of $f(x, y)$ inside Ω_n to begin with, the integral approximation in (14.23) may not be worse than the integral approximation in (14.20).

Letting $f(x, y)$ be u in (14.19), we have

$$\tilde{u} = u_1\psi_1 + u_2\psi_2 + u_3\psi_3 \tag{14.24}$$

where $u_i = u(\mathbf{p}_i)$. And at $\mathbf{p}^* = \frac{1}{3}(\mathbf{p}_1 + \mathbf{p}_2 + \mathbf{p}_3)$, we can use (14.17) to evaluate $\tilde{u}(\mathbf{p}^*)$,

$$\tilde{u}(\mathbf{p}^*) = u_1\psi_1(\mathbf{p}^*) + u_2\psi_2(\mathbf{p}^*) + u_3\psi_3(\mathbf{p}^*) = \zeta^T [u] \tag{14.25}$$

where

$$[u] = \begin{pmatrix} u_1 \\ u_2 \\ u_3 \end{pmatrix} \quad \text{and} \quad \zeta = \frac{1}{3}\begin{pmatrix} 1 \\ 1 \\ 1 \end{pmatrix} \tag{14.26}$$

Similarly, for δu,

$$\delta\tilde{u} = \delta u_1\psi_1 + \delta u_2\psi_2 + \delta u_3\psi_3 \tag{14.27}$$

$$\delta\tilde{u}(\mathbf{p}^*) = \delta u_1\psi_1(\mathbf{p}^*) + \delta u_2\psi_2(\mathbf{p}^*) + \delta u_3\psi_3(\mathbf{p}^*) = \zeta^T [\delta u] \tag{14.28}$$

where

$$[\delta u] = \begin{pmatrix} \delta u_1 \\ \delta u_2 \\ \delta u_3 \end{pmatrix}$$

A finite element approach in which the particular choice of the weight δu is formulated by (14.27) is known as the **Galerkin method**.[2] It can be shown that for $\mathbf{b} = 0$

[2] Other weights can be chosen, which may indeed produce a more accurate solution, for example, the Petrov-Galerkin approach to be discussed later.

and **M** that is symmetric, the Galerkin method yields an optimal solution of the weak form (14.5) for a given mesh.

The gradients ∇u and $\nabla(\delta u)$ will be approximated by $\nabla \tilde{u}$ and $\nabla(\delta \tilde{u})$ given below. Substituting (14.16) into (14.24) and rearranging,

$$\begin{aligned}
\tilde{u} &= \frac{1}{D}\left[\begin{pmatrix} x, & y \end{pmatrix}\left(\begin{array}{c|c|c} (p_{2y}-p_{3y}) & (p_{3y}-p_{1y}) & (p_{1y}-p_{2y}) \\ (p_{3x}-p_{2x}) & (p_{1x}-p_{3x}) & (p_{2x}-p_{1x}) \end{array}\right)\right. \\
&\quad \left. +\left(\mathbf{det}\left(\mathbf{p}_2 \,|\, \mathbf{p}_3 \right), \ \mathbf{det}\left(\mathbf{p}_3 \,|\, \mathbf{p}_1 \right), \ \mathbf{det}\left(\mathbf{p}_1 \,|\, \mathbf{p}_2 \right) \right)\right]\begin{pmatrix} u_1 \\ u_2 \\ u_3 \end{pmatrix} \\
&= \left[\begin{pmatrix} x, & y \end{pmatrix}\mathbf{T} + \eta\right][u]
\end{aligned} \tag{14.29}$$

where,

$$\mathbf{T} = \frac{1}{D}\begin{pmatrix} 0 & -1 \\ 1 & 0 \end{pmatrix}\begin{pmatrix} \mathbf{p}_1 \,|\, \mathbf{p}_2 \,|\, \mathbf{p}_3 \end{pmatrix}\begin{pmatrix} 0 & 1 & -1 \\ -1 & 0 & 1 \\ 1 & -1 & 0 \end{pmatrix}$$

$$\eta = \frac{1}{D}\left(\mathbf{det}\left(\mathbf{p}_2 \,|\, \mathbf{p}_3 \right), \ \mathbf{det}\left(\mathbf{p}_3 \,|\, \mathbf{p}_1 \right), \ \mathbf{det}\left(\mathbf{p}_1 \,|\, \mathbf{p}_2 \right) \right)$$

The gradient of $\tilde{u}$ then becomes

$$\nabla \tilde{u} = \mathbf{T}[u] \tag{14.30}$$

Similarly, for $\delta\tilde{u}$,

$$\nabla(\delta\tilde{u}) = \mathbf{T}[\delta u] \tag{14.31}$$

In summary, we have the following evaluations of the surface integrals in (14.10) for the nth finite element:

$$\begin{aligned}
\int_{\Omega_n} \nabla(\delta u)\cdot[\mathbf{M}(x,y)\cdot\nabla u]\,dA &\approx \left\{\frac{D}{2}[\delta u]^T\left(\mathbf{T}^T\,\mathbf{M}_{(\mathbf{p}^*)}\mathbf{T}\right)[u]\right\}_n \\
\int_{\Omega_n} \delta u\,[\mathbf{b}(x,y)\cdot\nabla u]\,dA &\approx \left\{\frac{D}{2}[\delta u]^T\left(\zeta\,\mathbf{b}^T_{(\mathbf{p}^*)}\mathbf{T}\right)[u]\right\}_n \\
\int_{\Omega_n} [\delta u\, g(x,y)\, u]\,dA &\approx \left\{\frac{D}{2}[\delta u]^T\left(g_{(\mathbf{p}^*)}\,\zeta\zeta^T\right)[u]\right\}_n \\
\int_{\Omega_n} \delta u\, h(x,y)\,dA &\approx \left\{\frac{D}{2}[\delta u]^T\left(h_{(\mathbf{p}^*)}\zeta\right)\right\}_n
\end{aligned} \tag{14.32}$$

14.2.2 Line Integrals Along the Boundary

For the line integral in (14.10), we need to first identify whether two or more nodes of Ω_n belong to the boundary of the global domain Ω. For cases (a) and (b) in Figure 14.5, that is, only one or none of the nodes of Ω_n belong to the boundary, the line integral will be set to zero. For case (c), the two collinear points that belong to boundary(Ω) will have values that contribute to the global line integral. In case

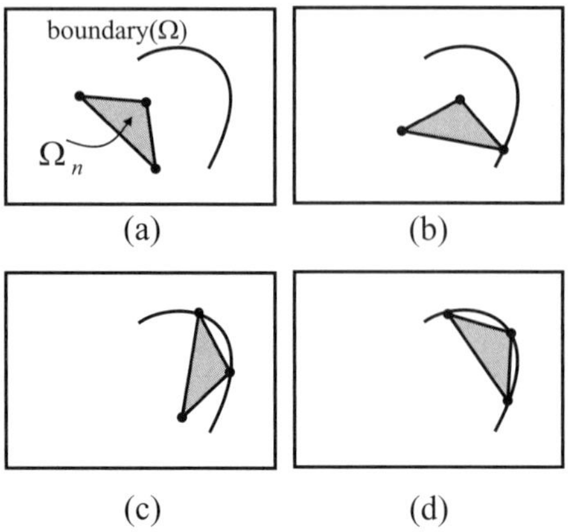

Figure 14.5. Possible attachments of Ω_n with boundary(Ω).

(d), all three vertices belong to boundary(Ω). However, for simplicity, we avoid this situation by constraining the mesh triangulation of Ω to only allow cases (a), (b) or (c).[3]

Let $\mathbf{p}_a$ and $\mathbf{p}_b$ be two points of Ω_n that belong to boundary(Ω). As before, we can define shape functions for each point that would yield a linear approximation of the segment in the boundary. This time, the shape function for point $\mathbf{p}_a$ and $\mathbf{p}_b$ will be given by $\chi_a(\mathbf{p})$ $\chi_b(\mathbf{p})$, respectively, defined by

$$\chi_a(\mathbf{p}) = \frac{\mathbf{det}\left(\mathbf{p}_b \mid \mathbf{p} \right)}{\mathbf{det}\left(\mathbf{p}_b \mid \mathbf{p}_a \right)} \qquad \text{and} \qquad \chi_b(\mathbf{p}) = \frac{\mathbf{det}\left(\mathbf{p}_a \mid \mathbf{p} \right)}{\mathbf{det}\left(\mathbf{p}_a \mid \mathbf{p}_b \right)} \tag{14.33}$$

where $\mathbf{p} \in \overline{\mathbf{p}_a\mathbf{p}_b}$ with $\overline{\mathbf{p}_a\mathbf{p}_b}$ being the line segment connecting $\mathbf{p}_a$ and $\mathbf{p}_b$ given by

$$\overline{\mathbf{p}_a\mathbf{p}_b} = \left\{ \begin{pmatrix} x \\ y \end{pmatrix} = \left(\lambda\mathbf{p}_a + (1-\lambda)\,\mathbf{p}_b\ ,\ 0 \le \lambda \le 1 \right) \right\}$$

Figure 14.6 shows the shape function χ_b.

Some of the properties of these shape functions are:

$$\chi_a(\mathbf{p}_a) = \chi_b(\mathbf{p}_b) = 1 \quad ; \quad \chi_a(\mathbf{p}_b) = \chi_b(\mathbf{p}_a) = 0 \quad ; \quad \chi_a(\mathbf{p}) + \chi_b(\mathbf{p}) = 1$$

$$\chi_a\left(\frac{\mathbf{p}_a + \mathbf{p}_b}{2}\right) = \chi_b\left(\frac{\mathbf{p}_a + \mathbf{p}_b}{2}\right) = \frac{1}{2} \tag{14.34}$$

$$\int_{\overline{\mathbf{p}_a\mathbf{p}_b}} \chi_a\, ds = \int_{\overline{\mathbf{p}_a\mathbf{p}_b}} \chi_b\, ds = \frac{\left\|\mathbf{p}_b - \mathbf{p}_a\right\|}{2}$$

Along points in $\overline{\mathbf{p}_a\mathbf{p}_b}$, a continuous function $f(x, y)$ can be approximated linearly by

$$f(\mathbf{p}) \approx f(\mathbf{p}_a)\,\chi_a(\mathbf{p}) + f(\mathbf{p}_b)\,\chi_b(\mathbf{p}) \tag{14.35}$$

The line integral can then be approximated by

$$\int_{\overline{\mathbf{p}_a\mathbf{p}_b}} f(x, y)ds \approx \frac{f(\mathbf{p}_a) + f(\mathbf{p}_b)}{2} \left\|\mathbf{p}_b - \mathbf{p}_a\right\| \tag{14.36}$$

[3] For instance, when case (d) occurs, we can split the triangle into two triangles along the middle vertex such that the two smaller triangle elements satisfy case (c).

Figure 14.6. The shape function χ_b.

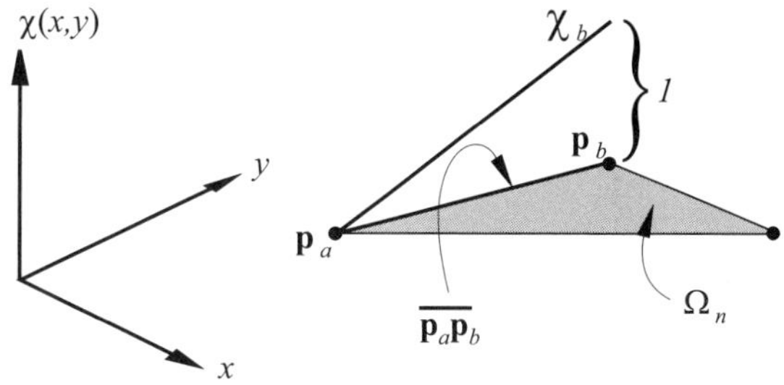

where

$$\left\| \mathbf{p}_b - \mathbf{p}_a \right\| = \sqrt{(\mathbf{p}_b - \mathbf{p}_a)^T (\mathbf{p}_b - \mathbf{p}_a)}$$

Recalling the terms involving a line integral in (14.10), we can apply (14.36),

$$\begin{aligned} \int_{\text{Bound}(\Omega_n)} \left(\delta u \, \mathbf{M}(x,y) \cdot \nabla u \right) \cdot \underline{\mathbf{n}} \, ds &\approx \frac{\delta u_a \, \overline{q}_a + \delta u_b \, \overline{q}_b}{2} \left\| \left(\mathbf{p}_b - \mathbf{p}_a \right) \right\| \\ &= \left\{ [\delta u]^T \left(\mathbf{e}_a \mid \mathbf{e}_b \right) \begin{pmatrix} \overline{q}_a/2 \\ \\ \overline{q}_b/2 \end{pmatrix} L \right\}_n \end{aligned} \quad (14.37)$$

where $\delta u_a = \delta u(\mathbf{p}_a)$, $\delta u_b = \delta u(\mathbf{p}_b)$,

$$\overline{q}_a = \left. (\mathbf{M}(x,y) \cdot \nabla u) \cdot \underline{\mathbf{n}} \right|_{\mathbf{p}=\mathbf{p}_a} \quad ; \quad \overline{q}_b = \left. (\mathbf{M}(x,y) \cdot \nabla u) \cdot \underline{\mathbf{n}} \right|_{\mathbf{p}=\mathbf{p}_b}$$

$$L = \sqrt{\mathbf{v}^T \mathbf{v}} \quad \text{with} \quad \mathbf{v} = \left(\mathbf{p}_1 \mid \mathbf{p}_2 \mid \mathbf{p}_3 \right) \left(\mathbf{e}_b - \mathbf{e}_a \right)$$

and $\mathbf{e}_i$ is the ith unit vector, that is,

$$\mathbf{e}_1 = \begin{pmatrix} 1 \\ 0 \\ 0 \end{pmatrix} \quad ; \quad \mathbf{e}_2 = \begin{pmatrix} 0 \\ 1 \\ 0 \end{pmatrix} \quad ; \quad \mathbf{e}_3 = \begin{pmatrix} 0 \\ 0 \\ 1 \end{pmatrix}$$

In (14.37), we are assuming that the Neumann boundary conditions will supply the values of $\overline{q}_a$ and $\overline{q}_b$. If this is not the case, the Dirichlet conditions will set the values of u at the boundaries – making the line integral calculations unnecessary.

This result can be generalized easily for the Robin (i.e., mixed) conditions. The only change necessary is to replace $\overline{q}_a$ and $\overline{q}_b$ by

$$\left. (\mathbf{M}(x,y) \cdot \nabla u) \cdot \underline{\mathbf{n}} \right|_{\mathbf{p}=\mathbf{p}_a} = \mu_a u(\mathbf{p}_a) + \overline{q}_a \quad ; \quad \left. (\mathbf{M}(x,y) \cdot \nabla u) \cdot \underline{\mathbf{n}} \right|_{\mathbf{p}=\mathbf{p}_b} = \mu_b u(\mathbf{p}_b) + \overline{q}_a$$

where μ_a, μ_b, $\overline{q}_a$ and $\overline{q}_b$ should all be supplied at the boundary points with the Robin conditions.

14.3 Assembly of Finite Elements

We now combine the results of the two previous sections. For the nth-finite element, the approximations of (14.32) and (14.37) can be grouped together based on the weak-form (14.10) as follows:

$$\begin{aligned}
&\begin{pmatrix} \int_{\Omega_n} \nabla(\delta u)\cdot[\mathbf{M}(x,y)\cdot\nabla u]\,dA \\ -\int_{\Omega_n}\delta u\,[\mathbf{b}(x,y)\cdot\nabla u]\,dA \\ -\int_{\Omega_n}[\delta u\, g(x,y)\, u]\,dA \end{pmatrix} \\
&\approx \left\{\frac{D}{2}(\delta u)^T\left(\mathbf{T}^T\mathbf{M}_{(\mathbf{p}^*)}\mathbf{T}-\zeta\mathbf{b}_{(\mathbf{p}^*)}^T\mathbf{T}-g_{(\mathbf{p}^*)}\zeta\zeta^T\right)[u]\right\}_n \\
&= [\delta u]_n^T K_n [u]_n
\end{aligned} \tag{14.38}$$

and

$$\begin{aligned}
\begin{pmatrix} \int_{\text{Bound}(\Omega_n)}(\delta u\mathbf{M}(x,y)\cdot\nabla u)\cdot\underline{\mathbf{n}}\,ds \\ +\int_{\Omega_n}\delta u\, h(x,y)\,dA \end{pmatrix} &\approx \left\{\frac{D}{2}(\delta u)^T\left(h_{(\mathbf{p}^*)}\zeta+Q\right)\right\}_n \\
&= [\delta u]_n^T\Gamma_n
\end{aligned} \tag{14.39}$$

where

$$K_n = \left\{\left(\mathbf{T}^T\mathbf{M}_{(\mathbf{p}^*)}\mathbf{T}-\zeta\mathbf{b}_{(\mathbf{p}^*)}^T\mathbf{T}-g_{(\mathbf{p}^*)}\zeta\zeta^T\right)\frac{D}{2}\right\}_n \tag{14.40}$$

$$\Gamma_n = \left\{\left(h_{(\mathbf{p}^*)}\zeta+Q\right)\frac{D}{2}\right\}_n \tag{14.41}$$

The constant matrices ζ and $\mathbf{T}$ were defined in (14.26) and (14.29), respectively, and

$$Q = \begin{cases} \left(\ \mathbf{e}_a \mid \mathbf{e}_b\ \right)\begin{pmatrix}\overline{q}_a \\ \overline{q}_b\end{pmatrix}\left(\dfrac{L}{D}\right) & \text{if } \mathbf{p}_a,\mathbf{p}_b\in\text{Bound}(\Omega)_{\text{NBC}} \\ 0 & \text{otherwise}\end{cases} \tag{14.42}$$

The formulas for D and L were defined in (14.12) and (14.37), respectively. Note that the sizes of matrices K_n and Γ_n are 3×3 and 3×1, respectively.

For the special case in which $\mathbf{p}_a$ and $\mathbf{p}_b$ both belong to the boundary points specified by Robin conditions, a term will have to be added to K_n as follows:

$$K_n = \left\{\left(\mathbf{T}^T\mathbf{M}_{(\mathbf{p}^*)}\mathbf{T}-\zeta\mathbf{b}_{(\mathbf{p}^*)}^T\mathbf{T}-g_{(\mathbf{p}^*)}\zeta\zeta^T-\Lambda\right)\frac{D}{2}\right\}_n \tag{14.43}$$

$$\Gamma_n = \left\{\left(h_{(\mathbf{p}^*)}\zeta+Q+Q^{(rbc)}\right)\frac{D}{2}\right\}_n \tag{14.44}$$

where

$$\Lambda = \left(\left(\ \mathbf{e}_a \,\middle|\, \mathbf{e}_b\ \right)\begin{pmatrix}\mu_a & 0 \\ 0 & \mu_b\end{pmatrix}\begin{pmatrix}\mathbf{e}_a^T \\ \hline \mathbf{e}_b^T\end{pmatrix}\right)\frac{L}{D}$$

and

$$Q^{(rbc)} = \begin{cases} \left(\mathbf{e}_a \mid \mathbf{e}_b \right) \begin{pmatrix} \overline{q}_a^{(rbc)} \\ \overline{q}_b^{(rbc)} \end{pmatrix} \left(\frac{L}{D} \right) & \text{if } \mathbf{p}_a, \mathbf{p}_b \in \text{Bound}(\Omega)_{\text{RBC}} \\ 0 & \text{otherwise} \end{cases}$$

What we have done so far is obtain the surface and line integrals of the finite elements Ω_n in which ($\Omega_i \bigcap \Omega_j = 0$ for $i \neq j$) and $\bigcup_{n=1}^{N_{\text{elements}}} \Omega_n = \Omega$. We now combine the various integrals as follows:

$$\int_{\Omega} [\diamond]\, dA = \sum_{n=1}^{N_{\text{elements}}} \int_{\Omega_n} [\diamond]\, dA$$

and

$$\int_{\text{Bound}(\Omega)} [\diamond]\, ds = \sum_{n=1}^{N_{\text{elements}}} \int_{\text{Bound}(\Omega_n)} [\diamond]\, ds$$

where $[\diamond]$ stands for the appropriate integrands. Using (14.38) and (14.39), applied to the weak-form working equation given in (14.10), we obtain

$$\sum_{n=1}^{N_{\text{elements}}} [\delta u]_n^T K_n [u]_n = \sum_{n=1}^{N_{\text{elements}}} [\delta u]_n^T \Gamma_n \tag{14.45}$$

However, because $[u]_i$ and $[u]_j$, with $i \neq j$, refer to different vectors, they cannot be added together directly. The same is true for $[\delta u]_i$ and $[\delta u]_j$, $i \neq j$. Instead, we need to first represent (14.45) using $\mathbf{u}$, the global vector of node values.

To proceed at this point, assume that a mesh of triangular elements has been generated for domain Ω. One particular method for mesh generations that would yield a triangulation is known as the **Delaunay triangulation**, which is discussed in Section 14.4. Regardless of the triangulation approach, we assume that the mesh is represented by two sets of data. One set of data is the collection of the node positions given by **node matrix P**,

$$\mathbf{P} = \left(\begin{array}{c|c|c|c} x_1 & x_2 & \cdots & x_{N_{\text{nodes}}} \\ y_1 & y_2 & & y_{N_{\text{nodes}}} \end{array} \right) \tag{14.46}$$

where N_{nodes} is the number of nodes in Ω. The second set of data is represented by a matrix of indices (of the nodes) that make up each triangular finite element, given by **index matrix** $\mathcal{I}$,

$$\mathcal{I} = \begin{pmatrix} \mathcal{I}_{1,1} & \mathcal{I}_{1,2} & \mathcal{I}_{1,3} \\ \mathcal{I}_{2,1} & \mathcal{I}_{2,2} & \mathcal{I}_{2,3} \\ \vdots & & \\ \mathcal{I}_{N_{\text{elements}},1} & \mathcal{I}_{N_{\text{elements}},2} & \mathcal{I}_{N_{\text{elements}},3} \end{pmatrix} \tag{14.47}$$

where N_{elements} are the number of elements in Ω and $\mathcal{I}_{i,j} \in \{1, 2, \ldots, N_{\text{nodes}}\}$ are the indices of the vertices of the ith triangular element. To illustrate,consider the mesh

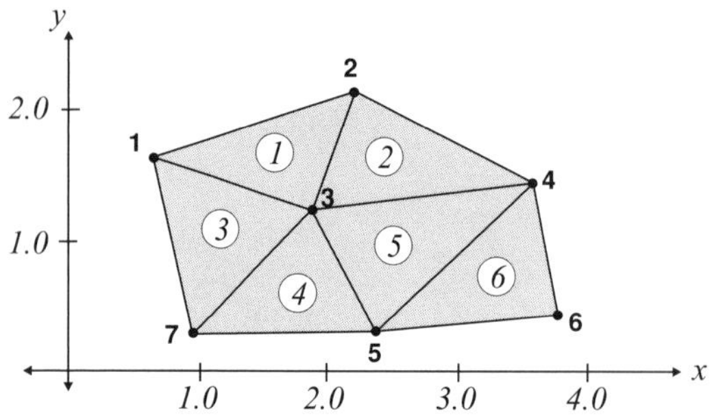

Figure 14.7. Mesh triangulation involving 7 nodes and 6 elements. (The elements are identified by circled numbers.)

triangulation shown in Figure 14.7; then the corresponding matrices $\mathbf{P}$ and $\mathcal{I}$ are given by

$$\mathbf{P} = \begin{pmatrix} 0.67 & 2.22 & 1.90 & 3.60 & 2.38 & 3.78 & 0.96 \\ 1.63 & 2.13 & 1.25 & 1.43 & 0.31 & 0.45 & 0.29 \end{pmatrix} \quad \text{and} \quad \mathcal{I} = \begin{pmatrix} 1 & 3 & 2 \\ 2 & 3 & 4 \\ 1 & 7 & 3 \\ 3 & 7 & 5 \\ 3 & 5 & 4 \\ 4 & 5 & 6 \end{pmatrix}$$

Note that the sequence of the nodes in each row of $\mathcal{I}$ is such that it follows a counterclockwise path.

For the nth element, we can use the rows of $\mathcal{I}$ to define a matrix operator E_n of size $(N_{\text{nodes}} \times 3)$ whose elements are given by

$$E_{n(i,j)} = \begin{cases} 1 & \text{if } i = \mathcal{I}_{n,j} \\ 0 & \text{otherwise} \end{cases} \tag{14.48}$$

When the transpose of E_n premultiplies a vector $\mathbf{u}$ of length N_{nodes}, the result is a (3×1) vector, $[u]_n$, whose elements are extracted from positions indexed by $\{\mathcal{I}_{n,1}, \mathcal{I}_{n,2}, \mathcal{I}_{n,3}\}$.[4]

To illustrate, suppose $N_{\text{nodes}} = 6$ and $\mathcal{I}_n = \{\mathcal{I}_{n,1}, \mathcal{I}_{n,2}, \mathcal{I}_{n,3}\} = \{6, 1, 3\}$, then

$$E_n^T \mathbf{u} = \begin{pmatrix} 0 & 0 & 0 & 0 & 0 & 1 \\ 1 & 0 & 0 & 0 & 0 & 0 \\ 0 & 0 & 1 & 0 & 0 & 0 \end{pmatrix} \begin{pmatrix} u_1 \\ u_2 \\ u_3 \\ u_4 \\ u_5 \\ u_6 \end{pmatrix} = \begin{pmatrix} u_6 \\ u_1 \\ u_3 \end{pmatrix}$$

Returning to the main issue of summing the various terms in (14.45), we see that the (3×1) vector $[u]_n$ given by

$$[u]_n = \begin{pmatrix} u_{\mathcal{I}_{n,1}} \\ u_{\mathcal{I}_{n,2}} \\ u_{\mathcal{I}_{n,3}} \end{pmatrix} \tag{14.49}$$

[4] We can also use E_n to generate the matrix $[\mathbf{p}]_n = (\mathbf{p}_1, \mathbf{p}_2, \mathbf{p}_3)_n$ from matrix $\mathbf{P}$ given in (14.46), that is,

$$[\mathbf{p}_1, \mathbf{p}_2, \mathbf{p}_3]_n = \mathbf{P}\, E_n = \begin{pmatrix} x_{[\mathcal{I}_{n,1}]} & x_{[\mathcal{I}_{n,2}]} & x_{[\mathcal{I}_{n,3}]} \\ y_{[\mathcal{I}_{n,1}]} & y_{[\mathcal{I}_{n,2}]} & y_{[\mathcal{I}_{n,3}]} \end{pmatrix}$$

The local matrix $[\mathbf{p}]_n$ can then be used to evaluate matrices K_n and Γ_n in (14.38) and (14.39).

can be extracted from the global vector

$$\mathbf{u} = \begin{pmatrix} u_1 \\ u_2 \\ \vdots \\ u_{N_{\text{nodes}}} \end{pmatrix} \tag{14.50}$$

using the matrix operator E_n, that is,

$$[u]_n = E_n^T \mathbf{u} \tag{14.51}$$

The same is true for $[\delta u]_n$,

$$[\delta u]_n = E_n^T \delta\mathbf{u} \tag{14.52}$$

Substituting (14.51) and (14.52) into (14.45), we get

$$\begin{aligned} \sum_{n=1}^{N_{\text{elements}}} [\delta\mathbf{u}]^T E_n K_n E_n^T \mathbf{u} &= \sum_{n=1}^{N_{\text{elements}}} [\delta\mathbf{u}]^T E_n \Gamma_n \\ [\delta\mathbf{u}]^T \left(\sum_{n=1}^{N_{\text{elements}}} E_n K_n E_n^T \right) \mathbf{u} &= [\delta\mathbf{u}]^T \left(\sum_{n=1}^{N_{\text{elements}}} E_n \Gamma_n \right) \\ [\delta\mathbf{u}]^T \hat{K} \mathbf{u} &= [\delta\mathbf{u}]^T \widehat{\Gamma} \end{aligned} \tag{14.53}$$

where

$$\begin{aligned} \widehat{K} &= \sum_{n=1}^{N_{\text{elements}}} K_n^{(G)} \\ \widehat{\Gamma} &= \sum_{n=1}^{N_{\text{elements}}} \Gamma_n^{(G)} \\ K_n^{(G)} &= E_n K_n E_n^T \\ \Gamma_n^{(G)} &= E_n \Gamma_n \end{aligned}$$

Because the vector of variations $\delta\mathbf{u}$ is allowed to vary arbitrarily, we could remove it from both sides of (14.53). The assembled form is then given by,

$$\widehat{K}\mathbf{u} = \widehat{\Gamma} \tag{14.54}$$

One final detail still needs to be addressed. The Dirichlet conditions in (14.4) have not yet been included in (14.54). To address this issue, let the nodes that are attached to the Dirichlet conditions be indexed by vector $\mathcal{D}$,

$$\mathcal{D} = (\mathcal{D}_1, \mathcal{D}_2, \ldots, \mathcal{D}_{N_{\text{DBC}}}) \tag{14.55}$$

where $\mathcal{D}_i \in \{1, 2, \ldots, N_{\text{nodes}}\}$ and N_{DBC} is the number of nodes that are involved in the Dirichlet boundary conditions. We now describe two possible approaches.

1. **Approach 1: Reduction of unknowns.** In this approach, $\mathbf{u}$ is split into two vectors: $\mathbf{u}_{\text{Dirichlet}}$, which contains the known u-values at the Dirichlet boundary, and

$\mathbf{u}_{\text{nonDirichlet}}$, which contains the unknown u-values. Let vector $\mathcal{D}^{\text{non}}$ be the vector of indices that remain after the removal of $\mathcal{D}$, that is,

$$\mathcal{D}^{\text{non}} = \left\{\mathcal{D}_1^{\text{non}}, \ldots, \mathcal{D}_{(N_{\text{nodes}}-N_{\text{DBC}})}^{\text{non}}\right\} \tag{14.56}$$

where

$$\mathcal{D}_i^{\text{non}} \notin \mathcal{D} \qquad \text{and} \qquad \mathcal{D}_i^{\text{non}} < \mathcal{D}_j^{\text{non}} \quad \text{if } i < j$$

Then

$$\left[\mathbf{u}_{\text{Dirichlet}}\right]_i = \mathbf{u}_{(\mathcal{D}_i)} \qquad i = 1, 2, \ldots, N_{\text{DBC}} \tag{14.57}$$

and

$$\left[\mathbf{u}_{\text{nonDirichlet}}\right]_j = \mathbf{u}_{(\mathcal{D}_j^{\text{non}})} \qquad j = 1, 2, \ldots, (N_{\text{nodes}} - N_{\text{DBC}}) \tag{14.58}$$

With reference to (14.54), let $\widehat{K}_{\mathcal{D}\downarrow}$ be a matrix obtained from $\widehat{K}$ but with the rows and columns that were indexed by $\mathcal{D}$ removed. Likewise, let $\widehat{\Gamma}_{\mathcal{D}\downarrow}$ be the vector obtained from $\widehat{\Gamma}$ but with elements that were indexed by $\mathcal{D}$ removed.

Next, construct vector $\Upsilon = \left[\Upsilon_1, \ldots, \Upsilon_{(N_{\text{nodes}}-N_{\text{DBC}})}\right]^T$ using the known values of $\mathbf{u}_{\text{Dirichlet}}$,

$$\Upsilon_i = \sum_{\ell=1}^{N_{\text{DBC}}} \widehat{K}_{(i,\mathcal{D}_\ell)} \left[\mathbf{u}_{\text{Dirichlet}}\right]_\ell \tag{14.59}$$

Then we get the reduced set of simultaneous equations given by

$$\left(\widehat{K}_{\mathcal{D}\downarrow}\right)\left[\mathbf{u}_{\text{nonDirichlet}}\right] = \left(\widehat{\Gamma}_{\mathcal{D}\downarrow} - \Upsilon\right) \tag{14.60}$$

2. **Approach 2: Overloading matrix $\widehat{K}$.** Let $\mathbf{K}$ be the new matrix obtained from $\widehat{K}$ whose elements are given by

$$\mathbf{K}_{ij} = \begin{cases} (1/\epsilon)\,\widehat{K}_{ii} & \text{if } i = j = \mathcal{D}_\ell \ ; \ \ell = 1, \ldots, N_{DBC} \\ \widehat{K}_{ij} & \text{otherwise} \end{cases} \tag{14.61}$$

and Γ be the new vector obtained from $\widehat{\Gamma}$ whose elements are given by

$$\Gamma_i = \begin{cases} (1/\epsilon)\,\widehat{K}_{ii}\mathbf{u}_{\mathcal{D}_\ell} & \text{if } i = \mathcal{D}_\ell \ ; \ \ell = 1, \ldots, N_{DBC} \\ \widehat{\Gamma}_i & \text{otherwise} \end{cases} \tag{14.62}$$

The parameter ϵ is chosen to be sufficiently small such that the $\mathcal{D}_\ell^{\text{th}}$ diagonal element dominates the other elements in matrix $\mathbf{K}$. This will effectively set $\mathbf{u}_i = \mathbf{u}_i$ if $i \in \mathcal{D}$. For instance, one could fix a very small value such as $\epsilon = 10^{-10}$. The final equation then becomes

$$\mathbf{K}\mathbf{u} = \Gamma \tag{14.63}$$

Remarks: In the first approach, the number of unknowns is reduced by a count of N_{DBC}. The tradeoff to these reductions is the additional bookkeeping of the new indices in $\mathbf{u}_{\text{nonDirichlet}}$ needed to map both $\mathbf{u}_{\text{nonDirichlet}}$ and $\mathbf{u}_{\text{Dirichlet}}$ back to $\mathbf{u}$ for post-processing. The second approach of overloading the matrix avoids the additional re-mapping procedures, but lacks the advantage of a reduced number of unknowns. Moreover, the overloading approach preserves any symmetry present in $\widehat{K}$, for example, when $\mathbf{b} = 0$. Because the solution of a linear equation is easier

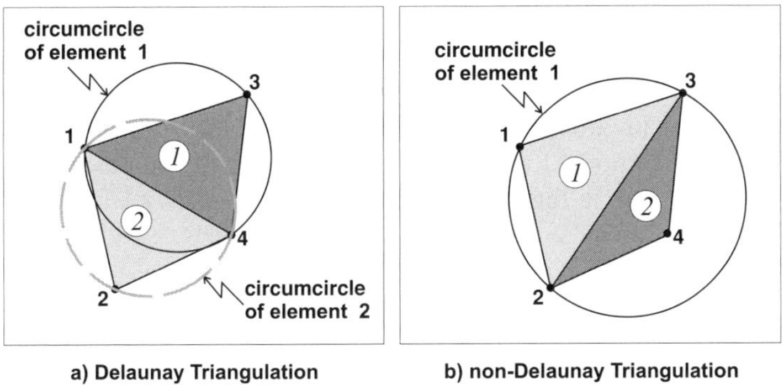

Figure 14.8. An example of two alternative triangulations.

with a symmetric matrix, the overloading approach has become the preferred route in several finite element solutions. For our discussion, we use the overloading approach due to its simpler form. Specifically, in (14.63), **u** refers to the same vector in (14.54).

14.4 Mesh Generation

In this section, we discuss one approach of partitioning Ω into non-overlapping triangular subdomains or finite elements, Ω_n, where $n = 1, 2, \ldots, N_{\text{elements}}$. Specifically, we discuss an "unstructured" approach of constructing Delaunay triangles from a given set of nodes.[5]

A **Delaunay triangulation** is a mesh that has the property that the **circumcircle** of each finite element; that is, the circle formed by the vertices of each triangle will not enclose nor touch any other node of the domain. For example, Figure 14.8 shows one mesh as a Dalaunay triangulation and another that is not.

One reason for the choice of the Delaunay criteria is that these meshes often minimize the occurrence of "spear-shaped" triangles. Also, several algorithms exist that use the **circumcenters**, that is, the centers of the circumcircles, to refine meshes to obtain guaranteed limits on both maximum and minimum angles of each triangles. However, we just focus on one procedure to achieve the Delaunay triangulation without the refinement methods.

Basically, the procedure involves two major steps:

1. <u>**Generate nodes in the domain.**</u>
   (a) Boundary points are selected such that when the lines connect adjacent points, the resulting shapes are sufficiently close to those of the boundaries.
   (b) Nodes are then introduced inside the boundaries. Around the regions where solution surfaces are expected to be sharply varying, more points are needed. These include points near curved boundary regions as well. For regions that are flatter, fewer points may suffice.

[5] The classical approach is the "structured" method of mapping quadrilateral subregions that make up the domain. The structured approach often yields more uniform patterns. However, these procedures sometimes demand several inputs and setup time from the users. Some versions also require the solution of associated partial differential equations.

(c) Internal nodes, that is, excluding the boundary points, may need to be moved around locally so that the nodes are mostly equidistant to each other. This step is called **smoothing**. We do not discuss these smoothing techniques, but some of the more popular methods include the Laplacian smoothing and the force-equilibrium method.[6] Instead, for simplicity, we just generate the points as close to equal size in the desired domain.

The result of the first part of the procedure is the matrix of node positions P, that is,

$$P = \left(\begin{array}{c|c|c|c} x_1 & x_2 & \cdots & x_{N_{\text{nodes}}} \\ y_1 & x_2 & & y_{N_{\text{nodes}}} \end{array} \right) \tag{14.64}$$

2. **Identify the Delaunay triangles.** One of the simpler methods for finding the Delaunay triangles is to use a process known as **node lifting**. Let (x_o, y_o) be a point near the center of domain Ω. A paraboloid function defined by

$$z(x, y) = (x - x_o)^2 + (y - y_o)^2 \tag{14.65}$$

is used to generate a set of 3D points

$$\widehat{P} = \left\{ \widehat{P}_1, \widehat{P}_2, \ldots, \widehat{P}_{N_{\text{nodes}}} \right\} \qquad \text{where} \qquad \widehat{P}_i = \left(\begin{array}{c} x_i \\ y_i \\ z_i(x_i, y_i) \end{array} \right) \tag{14.66}$$

as shown in Figure 14.9. Next, a tight polygonal cover of the lifted nodes can be generated to form a **convex hull** of these nodes using triangular facets. One particular method, which we refer to as the simplified-**QuickHull algorithm**, can be used to obtain such a convex hull of the lifted points. Details of this algorithm are given in the appendix as Section N.1.[7]

When the nodes of each facet are projected down onto the (x, y) plane, the triangular mesh that is generated will satisfy the conditions of a Delaunay triangulation, thus identifying the nodes of a finite element from each projected facet. This can be seen in Figure 14.9, where a vertical circular cylinder enclosing a facet of the triangulated paraboloid generates a circumcircle of the finite element in the (x, y)-plane. Because this cylinder that cuts through the paraboloid will include only the three points of the triangular facet, the projected triangle in the (x, y)-plane will have a circumcircle that should not contain other nodes.

[6] See, for example, P. O. Persson and G. Strang, A Simple Mesh Generator in MATLAB, *SIAM Review*, vol. 46, pp. 329–345, June 2004.

[7] The facets belonging to the top of the paraboloid will need to be removed. These facets are distinguished from the bottom facets using the property that their outward normal vectors are pointing upwards, that is, having a positive z-component.

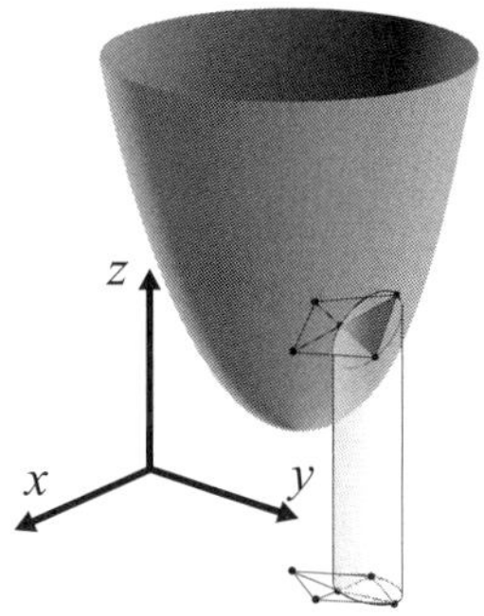

Figure 14.9. Projection of the triangular facets of the paraboloid will yield a Delaunay triangulation in the (x, y) plane.

Using the nodes identified with each facet, the matrix $\mathcal{I}$ can now be given by

$$\mathcal{I} = \begin{pmatrix} \mathcal{I}_{11} & \mathcal{I}_{12} & \mathcal{I}_{13} \\ \mathcal{I}_{21} & \mathcal{I}_{22} & \mathcal{I}_{23} \\ \cdots & \cdots & \cdots \\ \mathcal{I}_{N_{\text{elements}},1} & \mathcal{I}_{N_{\text{elements}},2} & \mathcal{I}_{N_{\text{elements}},3} \end{pmatrix} \qquad (14.67)$$

As cited earlier, to have a consistent sign for the area calculations of D in (14.12), we need to order the indices of each finite element in such a way that it follows the same counterclockwise direction.

Remarks: In MATLAB, a Delaunay triangular mesh can be obtained easily using the command `ind=delaunay(px,py)` or `ind=delaunayn([px',py'])`, where `px` and `py` are row vectors for the x and y positions, respectively, of the boundary and internal nodes, whereas `ind` will yield the index matrix for $\mathcal{I}$ in the same structure given in (14.67).[8] In addition, another function `triplot(ind,px,py)` can be used to generate a plot of all the mesh of triangles associated with index matrix `ind` and points defined by `px` and `py`.

EXAMPLE 14.2. Consider the triangulation of a rectangular plate with a circular hole. The triangulation begins with the assignment of points at the boundaries (see Figure 14.10(a)). Then other points are added to fillin the region (see Figure 14.10(b)). Next, the convex hull algorithm was used to determine the triangles (see Figure 14.10(c)). Finally, the triangles inside the holes are removed (see Figure 14.10(d)). (A MATLAB file `squared_annulus_mesh_example.m` is available on the book's webpage. The results shown in Figure 14.10 are generated by `[px,py,ind]= squared_annulus_mesh_example(0.5,1.0,10)`, where the inner and outer radius are 0.5 and 1.0, respectively.)

14.5 Summary of Finite Element Method

We now gather all the results obtained thus far. This summary is based on solving the linear partial differential equation given in (14.1), repeated here:

$$[\nabla \cdot (\mathbf{M}(x, y) \cdot \nabla u)] + [\mathbf{b}(x, y) \cdot \nabla u] + g(x, y)u + h(x, y) = 0$$

[8] We have found that the function `delaunayn` is more robust than `delaunay`.

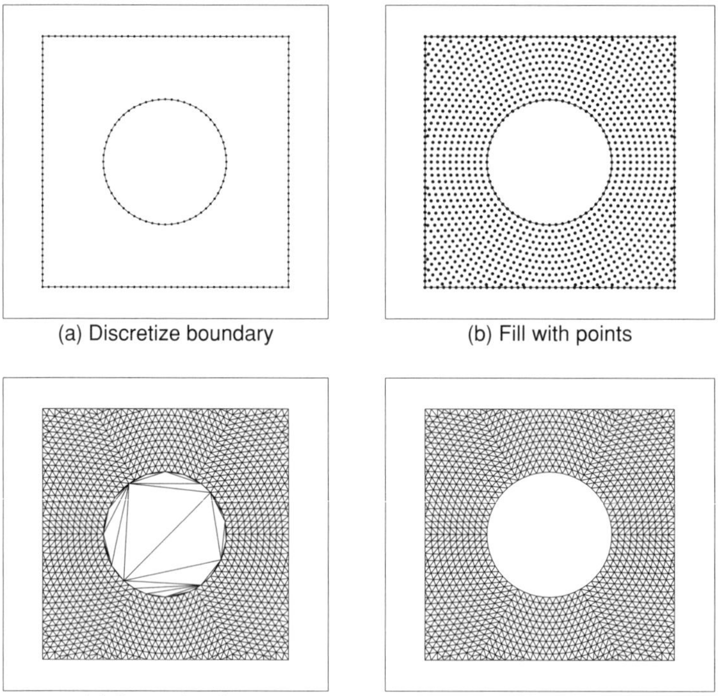

(a) Discretize boundary (b) Fill with points

(c) Form triangles (d) Clear holes

Figure 14.10. Delaunay triangulation of a square with circular hole.

The methods are limited to linear approximations using triangular elements only. This summary also refers only to the use of the overloading technique for incorporating the Dirichlet boundary conditions.

1. **<u>Generate the mesh data</u>**.

P^{Global}: matrix of node positions

$$= \left(\mathbf{p}_1^{\text{Global}} \;\middle|\; \mathbf{p}_2^{\text{Global}} \;\middle|\; \cdots \;\middle|\; \mathbf{p}_{N_{\text{nodes}}}^{\text{Global}} \right) \quad \text{where} \quad \mathbf{p}_i^{\text{Global}} = \begin{pmatrix} x_i \\ y_i \end{pmatrix}$$

$\mathcal{I}$: matrix of node indices of each element

$$= \begin{pmatrix} \mathcal{I}_1 \\ \mathcal{I}_2 \\ \vdots \\ \mathcal{I}_{N_{\text{elements}}} \end{pmatrix} \quad \text{where} \quad \mathcal{I}_n = (\mathcal{I}_{n1}, \mathcal{I}_{n2}, \mathcal{I}_{n3})$$

$\mathcal{D}$: set of indices of Dirichlet boundary nodes $= \left\{ \mathcal{D}_1, \mathcal{D}_2, \ldots, \mathcal{D}_{N_{\text{DBC}}} \right\}$

$\mathcal{N}$: set of indices of Neumann boundary nodes $= \left\{ \mathcal{N}_1, \mathcal{N}_2, \ldots, \mathcal{N}_{N_{\text{NBC}}} \right\}$

$\mathcal{R}$: set of indices of Robin boundary nodes $= \left\{ \mathcal{R}_1, \mathcal{R}_2, \ldots, \mathcal{R}_{N_{\text{RBC}}} \right\}$

2. **Calculate local matrices of each element.** Let $\zeta = \frac{1}{3}\begin{pmatrix} 1, & 1, & 1 \end{pmatrix}^T$. For the nth finite element, with $n = 1, 2, \ldots, N_{\text{elements}}$,

$$P_n^{\text{Local}} = \left(\left[\mathbf{p}_1^{\text{Local}}\right]_n \middle| \left[\mathbf{p}_2^{\text{Local}}\right]_n \middle| \left[\mathbf{p}_3^{\text{Local}}\right]_n \right)$$

$$\text{where} \quad \left[\mathbf{p}_i^{\text{Local}}\right]_n = \mathbf{p}_{\mathcal{I}_{ni}}^{\text{Global}}, \quad i = 1, 2, 3$$

$$D_n = \mathbf{det}\begin{pmatrix} 1 \;|\; 1 \;|\; 1 \\ \hline P_n^{\text{Local}} \end{pmatrix}$$

$$\mathbf{T}_n = \frac{1}{D_n}\begin{pmatrix} 0 & -1 \\ 1 & 0 \end{pmatrix}\left(P_n^{\text{Local}}\right)\begin{pmatrix} 0 & 1 & -1 \\ -1 & 0 & 1 \\ 1 & -1 & 0 \end{pmatrix}$$

$$\mathbf{p}_n^* = \left(P_n^{\text{Local}}\right)(\zeta)$$

$$Q_n = \begin{cases} \left(\mathbf{e}_a \,|\, \mathbf{e}_b\right)\begin{pmatrix} \overline{q_a} \\ \overline{q_b} \end{pmatrix}\left(\dfrac{L_n}{D_n}\right) & \text{if} \quad \mathcal{I}_{n,a} \text{ and } \mathcal{I}_{n,b} \in \mathcal{N}_n \\ 0 & \text{otherwise} \end{cases}$$

$$Q_n^{RBC} = \begin{cases} \left(\mathbf{e}_a \,|\, \mathbf{e}_b\right)\begin{pmatrix} \overline{q_a}^{(RBC)} \\ \overline{q_b}^{(RBC)} \end{pmatrix}\left(\dfrac{L_n}{D_n}\right) & \text{if} \quad \mathcal{I}_{n,a} \text{ and } \mathcal{I}_{n,b} \in \mathcal{R}_n \\ 0 & \text{otherwise} \end{cases}$$

$$\Lambda_n = \begin{cases} \left(\mathbf{e}_a \,|\, \mathbf{e}_b\right)\begin{pmatrix} \mu_a & 0 \\ 0 & \mu_b \end{pmatrix}\begin{pmatrix} \mathbf{e}_a^T \\ \hline \mathbf{e}_b^T \end{pmatrix}\left(\dfrac{L_n}{D_n}\right) & \text{if} \quad \mathcal{I}_{n,a} \text{ and } \mathcal{I}_{n,b} \in \mathcal{R}_n \\ 0 & \text{otherwise} \end{cases}$$

$$\text{where} \quad L_n = \sqrt{\mathbf{v}^T\mathbf{v}}\;, \quad \mathbf{v} = P_n^{\text{Local}}\left(\mathbf{e}_b - \mathbf{e}_a\right)$$

$$\mathbf{e}_1 = \begin{pmatrix} 1 \\ 0 \\ 0 \end{pmatrix}, \mathbf{e}_2 = \begin{pmatrix} 0 \\ 1 \\ 0 \end{pmatrix}, \mathbf{e}_3 = \begin{pmatrix} 0 \\ 0 \\ 1 \end{pmatrix}$$

$$K_n = \left(\mathbf{T}_n^T \mathbf{M}_{(\mathbf{p}_n^*)}\mathbf{T}_n - \zeta\,\mathbf{b}_{(\mathbf{p}_n^*)}^T\mathbf{T}_n - g_{(\mathbf{p}_n^*)}\zeta\zeta^T - \Lambda_n\right)\frac{D_n}{2}$$

$$\Gamma_n = \left(h_{(\mathbf{p}_n^*)}\zeta + Q_n + Q_n^{(RBC)}\right)\frac{D_n}{2}$$

3. **Form the global matrices**
   (a) Assemble finite elements

$$\widehat{K} = \sum_{n=1}^{N_{\text{elements}}} E_n K_n E_n^T \qquad \text{and} \qquad \widehat{\Gamma} = \sum_{n=1}^{N_{\text{elements}}} E_n \Gamma_n$$

where
$$(E_n)_{ij} = \begin{cases} 1 & \text{if } i = \mathcal{I}_{nj} \\ 0 & \text{otherwise} \end{cases}$$

(**Remarks:** In MATLAB, once the local matrices K_n and Γ_n have been found, instead of using E_n, we could use the simpler command lines such as

```
K(In,In)   = K(In,In) + K_n ;
Gamma(In)  = Gamma(In) + Gamma_n ;
```

where `In` is the nth row of $\mathcal{I}$; thus $\widehat{K}$ and $\widehat{\Gamma}$ are updated iteratively from $n = 1$ to N_{elements}.)

   (b) Include Dirichlet boundary conditions using the "overloading" approach

$$\mathbf{K} = \{\mathbf{K}_{ij}\} \qquad \text{where} \qquad \mathbf{K}_{ij} = \begin{cases} (1/\epsilon)\widehat{K}_{ii} & \text{if } i = j = \mathcal{D}_\ell \\ & \quad \ell = 1, \ldots, N_{DBC} \\ & \quad 0 < \epsilon \ll 1 \\ \widehat{K}_{ij} & \text{otherwise} \end{cases}$$

$$\Gamma = \{\Gamma_i\} \qquad \text{where} \qquad \Gamma_i = \begin{cases} (1/\epsilon)\widehat{K}_{ii}\mathbf{u}_{\mathcal{D}_\ell} & \text{if } i = \mathcal{D}_\ell \\ & \quad \ell = 1, \ldots, N_{DBC} \\ & \quad 0 < \epsilon \ll 1 \\ \widehat{\Gamma}_i & \text{otherwise} \end{cases}$$

4. **Solve the linear equation**

$$\mathbf{K}\mathbf{u} = \Gamma$$

Remarks:

1. In MATLAB, a plot of the solution mesh or surface can be obtained by using the command: `trimesh(Ind,x,y,u)` or `trisurf(Ind,x,y,u)`, where `Ind` is the node index matrix $\mathcal{I}$. Alternatively, one could generate a separate mesh grid using the command `meshgrid` and then use `griddata` to obtain the interpolated data on the new mesh. The latter approach will allow one to apply `surfl` to generate a surface plot with lighting or to generate contour plots using `contour` or `contour3`.
2. A MATLAB function `linear_2d_fem.m` is available on the book's webpage that implements the steps in the summary described previously. This function is used in the example that follows.

EXAMPLE 14.3. To test the finite element method as summarized, consider the differential equation

$$[\nabla \cdot (\mathbf{M}(x,y) \cdot \nabla u)] \;+\; [\mathbf{b}(x,y) \cdot \nabla u] \;+\; g(x,y)u \;+\; h(x,y) \;=\; 0$$

with

$$\mathbf{M} = \begin{pmatrix} (3 + x^2y^2/2) & 1/4 \\ 1/4 & 2 \end{pmatrix} \;;\; \mathbf{b} = \begin{pmatrix} -x+1 \\ -y+1 \end{pmatrix} \;;\; g = (x-y)^2$$

and

$$h = -\left(\alpha\cos(2\pi x) + \beta\sin(2\pi x) + \gamma\right)$$

where

$$\begin{aligned} \alpha &= 4\pi\left(1 + xy^2 - x\right) \\ \beta &= -24\pi^2 - (2\pi xy)^2 + 2\,(x-y)^2 \\ \gamma &= 21.2x^2y^3 - 10.6y^2x^3 + y\left(5.3x^4 - 15.9x^2 + 10.6x + 31.8\right) + 5.3x(1+x) \end{aligned}$$

We now fix the domain to be a unit square domain with a hole similar to the one given in Example 14.2 but with a circular hole of radius 0.2 located in the center of the square and with the lower left corner situated at the origin. For our test, we set the Dirichlet conditions the left, top, and bottom sides of the square, plus at the circular edge of the hole given by

$$u = 5.3x^2y + 2\sin(2\pi x) \quad \text{for} \quad \begin{cases} x = 0 & , \quad 0 \le y \le 1 \\ 0 \le x \le 1 & , \quad y = 0 \\ 0 \le x \le 1 & , \quad y = 1 \\ \sqrt{x^2+y^2} = 0.2 & \end{cases}$$

For the right side of the square, we set the Neumann conditions given by

$$\text{at } x = 1,\; 0 \le y \le 1 \;: \quad \left(\begin{aligned} [\mathbf{M} \cdot \nabla u] \cdot \mathbf{n} \;&=\; \left(6 + x^2y^2\right) 2\pi\cos(2\pi x) \\ &\quad + 31.8xy + 5.3x^3y^3 + 1.325x^2 \end{aligned} \right)$$

The exact solution of this problem is known (which was in fact used to set h and the boundary conditions) and given by

$$u(x,y) = 5.3x^2y + 2\sin(2\pi x)$$

After applying the finite element method based on a Delaunay triangulation generated by the same method used in Example 14.2, we obtain the solution shown in Figure 14.11 together with a plot of the errors between the numerical the exact solution. The errors are within $\pm 10^{-2}$, where the larger errors occur in this case around the boundary with Neumann conditions, suggesting potential areas for improvement of the method. It is expected that increasing the resolution may improve the accuracy. However, there are high-order accuracy improvements that can be implemented (which are not discussed here). The results shown here can be reproduced by running the MATLAB m-file `fem_sqc_test1.m` that is available on the book's webpage.

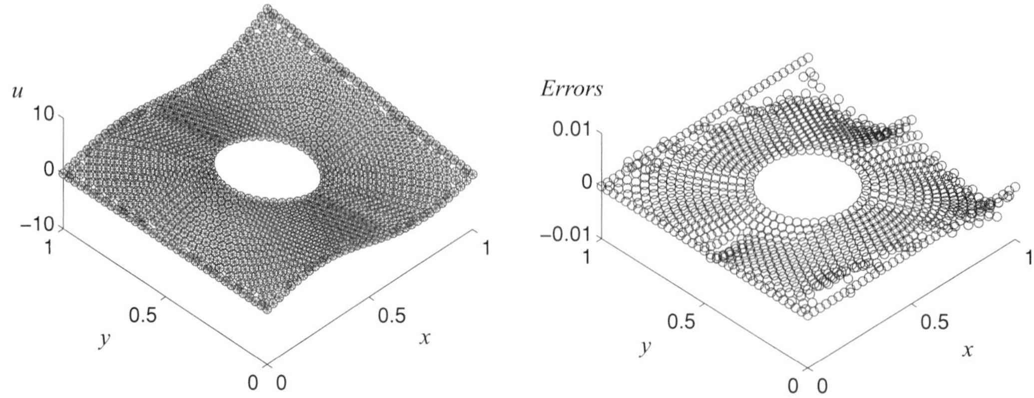

Figure 14.11. Results using finite element method in Example 14.3.

As we have noted in the introduction to this chapter, there are cases where convection dominates over the diffusion terms; that is, the effects of the coefficients **b** are much larger than those of **M**. In these cases, a slight modification known as the Streamline-Upwind Petrov-Galerkin (SUPG) method can handle some of the problems. A brief introduction to this method is given in Section N.2 as an appendix.

14.6 Axisymmetric Case

The extension of the finite element method to three dimensions is straightforward, albeit more complex and demanding of more computational resource. Instead of areas and arcs, we have volumes and surface areas. Instead of triangular finite elements, we could have triangular tetrahedral finite elements. The weak formulation given by (14.10) becomes

$$\begin{pmatrix} \int_\Omega \nabla(\delta u)\cdot\left[\mathbf{M}(x,y,z)\cdot\nabla u\right]dV \\ -\ \int_\Omega \delta u\left[\mathbf{b}(x,y,z)\cdot\nabla u\right]dV \\ -\ \int_\Omega \left[\delta u\, g(x,y)\,u\right]\,dV \end{pmatrix} = \begin{pmatrix} \int_{\text{Bound}(\Omega)}\left(\delta u\,\mathbf{M}(x,y,z)\cdot\nabla u\right)\cdot\underline{\mathbf{n}}\,dS \\ +\ \int_\Omega \delta u\, h(x,y,z)\,dV \end{pmatrix} \tag{14.68}$$

We do not discuss the general finite element method for the 3D case. Instead, we just focus on the special case known as the **axisymmetric case**; that is, we assume symmetry about the z-axis, which can be formalized as $\partial(\bullet)/\partial\theta = 0$ in the cylindrical coordinate system. Under this condition, the terms in (14.68) reduce to

$$\int_{\widehat{\Omega}(r,z)}\begin{pmatrix}\dfrac{\partial(\delta u)}{\partial r}, & \dfrac{\partial(\delta u)}{\partial z}\end{pmatrix}\begin{pmatrix} m_{11}(r,z) & m_{12}(r,z)\\ m_{21}(r,z) & m_{22}(r,z)\end{pmatrix}\begin{pmatrix}\dfrac{\partial u}{\partial r}\\ \dfrac{\partial u}{\partial z}\end{pmatrix}2\pi r\,dr\,dz$$

$$-\int_{\widehat{\Omega}(r,z)}(\delta u)\begin{pmatrix} b_1(r,z), & b_2(r,z)\end{pmatrix}\begin{pmatrix}\dfrac{\partial u}{\partial r}\\ \dfrac{\partial u}{\partial z}\end{pmatrix}2\pi r\,dr\,dz - \int_{\widehat{\Omega}(r,z)}(\delta u)g(r,z)u2\pi r\,dr\,dz$$

$$= \int_{\text{Bound}\left(\widehat{\Omega}(r,z)\right)} q(r,z)2\pi r ds + \int_{\widehat{\Omega}(r,z)}(\delta u)h(r,z)2\pi r\,dr\,dz \tag{14.69}$$

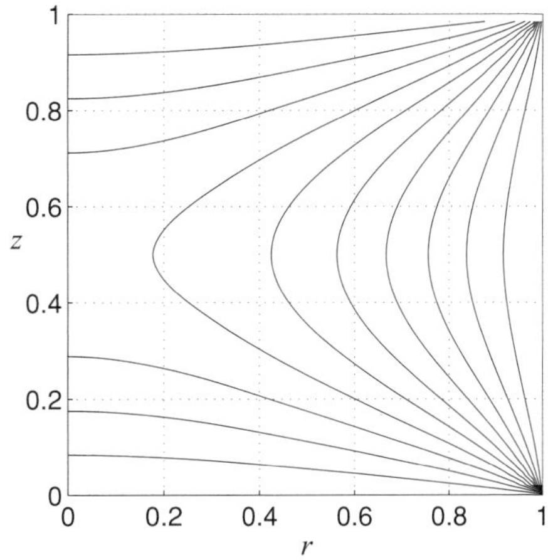

Figure 14.12. A contour plot for a half of the cross-section of a cylinder modeled by a Poisson equation subject to Dirichlet conditions at top, bottom, and radial surface.

where $q(r, z) = \left(\delta u\, \mathbf{M}(r, z) \cdot \nabla u\right) \cdot \underline{\mathbf{n}}$. By close observation, this is the same as (14.10) after division by 2π, with the exception that, other than replacing x and y by r and z, respectively, the coefficients are now multiplied by r, that is, $\mathbf{M}$ is replaced by $r\mathbf{M}$, and so forth. Thus the procedure will be similar as before; however, note that the integrands must now be multiplied by r evaluated at $\mathbf{p}^*$, that is, at the centroids of the triangular finite elements.

EXAMPLE 14.4. As a simple example, consider the Poisson equation

$$\nabla^2 u = 0$$

applied to a cylinder of radius $R = 1$ and height $L = 1$. The boundary conditions are given by $u(R, z) = 100$, $u(r, 0) = u(r, 1) = 30$ and

$$\nabla u \cdot \underline{\mathbf{n}}\Big|_{r=0} = 0$$

Applying (14.69), the finite element solution is given as a contour plot in Figure 14.12 for a half of a cross-section attached to the central axis.

Remarks: A MATLAB function `fem_cylinder.m` is available on the book's webpage and was used to generate the solution. However, to generate the contour plots, one should use the function `griddata` to first interpolate the values on a new mesh generated using `meshgrid`. Afterward, the function `contour` can be used to generate the contour plots.

14.7 Time-Dependent Systems

The finite element methods discussed in the previous sections can be extended to handle time dependence. We limit the discussion to the linear partial differential equation given by

$$\widehat{F}\left(t, x, y, u, \frac{\partial u}{\partial t}, \frac{\partial u}{\partial x}, \frac{\partial u}{\partial y}, \frac{\partial^2 u}{\partial x^2}, \frac{\partial^2 u}{\partial y^2}\right) = 0$$

$$\begin{aligned} &= [\nabla \cdot (\mathbf{M}(t,x,y) \cdot \nabla u)] + [\mathbf{b}(t,x,y) \cdot \nabla u] \\ &\quad + g(t,x,y)u + h(t,x,y) - c(t,x,y)\frac{\partial u}{\partial t} \\ &\qquad \text{for } (x,y) \in \Omega \end{aligned} \tag{14.70}$$

with boundary conditions

$$[\mathbf{M}(t,x,y) \cdot \nabla u] \cdot \mathbf{n} = \overline{q}(t,x,y) \qquad \text{on boundary } (\Omega)_{\text{NBC}} \tag{14.71}$$

$$[\mathbf{M}(t,x,y) \cdot \nabla u] \cdot \mathbf{n} = \mu(t,x,y)u + \overline{q}(t,x,y) \qquad \text{on boundary } (\Omega)_{\text{RBC}} \tag{14.72}$$

or

$$u = \overline{u}(t,x,y) \qquad \text{on boundary } (\Omega)_{\text{DBC}} \tag{14.73}$$

and initial conditions

$$u = \widehat{u}(x,y) \qquad \text{for } t = 0 \tag{14.74}$$

Compared with the steady-state version given by (14.1), the new system described by (14.70) now includes an additional term: $[c(t,x,y)\ \partial u/\partial t]$. Furthermore, the coefficients: $\mathbf{M}$, $\mathbf{b}$, g and h, have all been given time dependence.

As discussed previously, when the equation becomes convection-dominated, the Galerkin method may become increasingly inaccurate. In fact, in the limit of $\mathbf{M} = 0$ with $\mathbf{b} \neq 0$, we have hyperbolic equations, which contain propagated wave solutions. These convection-dominated cases may need to be handled differently, most likely moving the solution along the characteristics. We defer the discussion of these characteristics-based solutions to other sources. Instead, we only treat equations that can be consider to be **diffusion-dominated**, that is, where the effects of **b** are much less that the effects due to **M**.

We can use the same **semi-discrete method**, also known as **method of lines**, introduced in Section 13.3.1. As before, we start with a spatial-discretization of the problem and then convert the original partial differential equation into an initial value problem, that is, an ordinary differential equation in time. First, we let $\dot{u} = \partial u/\partial t$ and treat it as another independent variable. Using the approach in Section 14.1, while keeping t fixed, we obtain a weak formulation of (14.70),

$$\begin{pmatrix} \int_\Omega \nabla(\delta u) \cdot [\mathbf{M}(t,x,y) \cdot \nabla u]\, dA \\ -\ \int_\Omega \delta u\, [\mathbf{b}(t,x,y) \cdot \nabla u]\, dA \\ -\ \int_\Omega [\delta u\, g(t,x,y)\, u]\, dA \\ +\ \int_\Omega [\delta u\, c(t,x,y)\, \dot{u}]\, dA \end{pmatrix} = \begin{pmatrix} \int_{\text{Bound}(\Omega)} (\delta u \mathbf{M}(t,x,y) \cdot \nabla u) \cdot \mathbf{n}\, ds \\ +\ \int_\Omega \delta u\, h(t,x,y)\, dA \end{pmatrix} \tag{14.75}$$

For the nth element, we can use the linear approximation given in (14.19),

$$\{\dot{u}\}_n \approx \{\dot{u}_1\psi_1 + \dot{u}_2\psi_2 + \dot{u}_3\psi_3\}_n \tag{14.76}$$

where $\{\dot{u}_i\}_n$ is the value of $\dot{u}$ at node i of element n. Consequently,

$$\int_{\Omega_n} [\delta u\, c(t,x,y)\, \dot{u}]\; dA \approx \left\{ \frac{D}{2} [\delta u]^T \left(c_{(t,\,\mathbf{p}^*)}\, \zeta\zeta^T \right) [\dot{u}] \right\}_n \tag{14.77}$$

where

$$[\dot{u}]_n = \left\{ \begin{pmatrix} \dot{u}_1 \\ \dot{u}_3 \\ \dot{u}_3 \end{pmatrix} \right\}_n$$

Using the same procedures to assemble the elements, we end up with

$$\mathbf{K}(t)\mathbf{u} + \mathbf{C}(t)\dot{\mathbf{u}} = \Gamma(t) \tag{14.78}$$

where

$$\mathbf{C}(t) = \sum_{n=1}^{N_{\text{elements}}} E_n C_n(t) E_n^T \qquad \text{and} \qquad C_n(t) = \left\{ \frac{D}{2} c_{(t,\,\mathbf{p}^*)}\, \zeta\zeta^T \right\}_n$$

and both $\mathbf{K}(t)$ and $\Gamma(t)$ already include the handling of Dirichlet and/or Neumann conditions, which are possibly time-dependent as well. Matrix $\mathbf{C}$ is often also referred to as a **mass matrix**.

Equation (14.78) is now a set of simultaneous ordinary differential equation. If the resulting system is linear with constant coefficients, then the solution can be obtained analytically using the techniques discussed in Section 6.5.1. If $\mathbf{C}$ is singular with rank r, then the equation could be transformed into a set of r ordinary differential equations and a set of $N_{\text{nodes}} - r$ algebraic equations.

More generally, the ODE solvers discussed in Chapter 7 can be used. In the event that $\mathbf{C}(t)$ becomes singular, the system will be a set of **differential-algebraic equations** (DAE) and would then require special treatments. Among the ODE solvers, due to the large, albeit sparse, matrices involved, the usual preferred choice is the Euler method based on central difference approximation of the time derivatives. A slight, yet significant, variation to the central difference approximation is the **Crank-Nicolson** method, which is also used in finite difference methods (cf. Section 13.3.2). In this approach, the equations are modified to locate the points at the middle of two consecutive time points used in the time march. This results in a method with good stability properties and accuracy (as long as the boundary data are sufficiently smooth).

Thus the Crank-Nicolson method uses two approximations:

$$\left.\frac{d}{dt}\mathbf{u}\right|_{\left(t+\frac{\Delta t}{2}\right)} \approx \frac{\mathbf{u}(t+\Delta t) - \mathbf{u}(t)}{\Delta t} \tag{14.79}$$

$$\mathbf{u}\left(t + \frac{\Delta t}{2}\right) \approx \frac{\mathbf{u}(t+\Delta t) + \mathbf{u}(t)}{2} \tag{14.80}$$

where (14.79) is actually an Euler central-difference approximation of the time derivative when one considers the evaluation to be at $t_k + \Delta t/2$ and time increment $\Delta t/2$. Equation (14.80) sets the value of $\mathbf{u}$ at the half-time increment to be the average of the values at t and $t + \Delta t$. Applying these approximations to (14.78) at $t + \Delta t/2$ will yield, after some rearrangements,

$$\mathbf{L}_k \mathbf{u}_{k+1} = \mathbf{S}_k \mathbf{u}_k + \mathbf{H}_k \tag{14.81}$$

where

$$\mathbf{L}_k = \mathbf{C}(\tau_k) + \frac{\Delta t}{2}\mathbf{K}(\tau_k)$$

$$\mathbf{S}_k = \mathbf{C}(\tau_k) - \frac{\Delta t}{2}\mathbf{K}(\tau_k) \qquad \text{and} \qquad \tau_k = \left(k + \frac{1}{2}\right)\Delta t$$

$$\mathbf{H}_k = \Delta t\, \Gamma(\tau_k)$$

If $\mathbf{L}_k$ is nonsingular, we can premultiply (14.81) by $\mathbf{L}_k^{-1}$,

$$\mathbf{u}_{k+1} = \left[\mathbf{L}_k^{-1}\mathbf{S}_k\right]\mathbf{u}_k + \left[\mathbf{L}_k^{-1}\mathbf{H}_k\right] \tag{14.82}$$

Note that instead of using the “overloading" approach used in solving steady-state equations, a simpler solution for the transient case is to substitute the values of u at the boundary nodes by the boundary values specified by the Dirichlet conditions at each t_k.

EXAMPLE 14.5. Consider the linear reaction-diffusion type differential equation equation and boundary conditions given in Example 13.10,

$$\frac{\partial u}{\partial t} = 2\nabla^2 u - 3u + h(t, x, y) \tag{14.83}$$

where

$$h(t, x, y) = f\left(\varphi + 2\eta_b(x, y) - \psi - 2\eta_a(x, y)\right) + \left(3\psi - 2\eta_b(x, y)\right)$$

$$\eta_a(x, y) = \frac{32000\, r^2}{(1+20r)^3} - \frac{800r}{(1+20r)^2}$$

$$\eta_b(x, y) = \left[3200 + 50\left(16x - \frac{64}{5}\right)^2\right]\frac{s^2}{(1+5s)^3} - \left[400 + 5\left(16x - \frac{64}{5}\right)^2\right]\frac{s}{(1+5s)^2}$$

$$\varphi(x, y) = \frac{8}{1 + 20r(x, y)} \quad ; \quad \psi(x, y) = \frac{1}{1 + 5s(x, y)}$$

$$r(x, y) = e^{2-2x-y} \quad ; \quad s(x, y) = e^{8\left[(x-0.8)^2 - y\right]}$$

on the domain $0 \le x \le 1$ and $0 \le y \le 1$ and subject to Dirichlet conditions on all four edges and whose solution is given by

$$u(t, x, y) = e^{-2t}\varphi(x, y) + \left(1 - e^{-2t}\right)\psi(x, y)$$

Implementing the Crank-Nicholson finite element method using a Delaunay mesh shown in the left plot in Figure 14.13, we obtain a set of time-lapse plots shown in the right side of Figure 14.13. The corresponding time-lapse plots of the errors between the finite element approximation and the exact solution are shown in 14.14. The errors were within ± 0.03, which is larger than those obtained using finite differences based on central difference formulas. This is

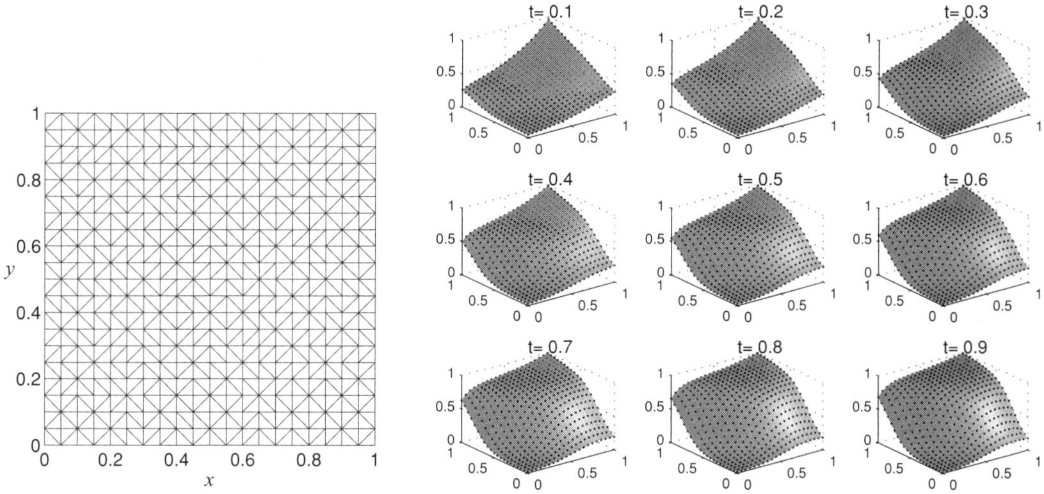

Figure 14.13. The left plot shows the Delaunay triangulation used for the finite element solution. The plots on the right are the finite element solution at different time instants using the Crank-Nicholson method. The approximations are shown as points, whereas the exact solutions are shown as surface plots.

expected because we have restricted our approach to only using first-order approximations.

Remark: A MATLAB function `dynamic_2d_fem.m` is available on the book's webpage and applies the Crank-Nicholson method. Another MATLAB function `dynamic_fem_test1.m` is available on the book's webpage and generates the solution at nine time instants, as shown in this example. One can look at this function to see how to include the specific functions for the various coefficients **M**, **b**, g, h, and c, as well as how to include time-varying Dirichlet or Neumann boundary conditions.

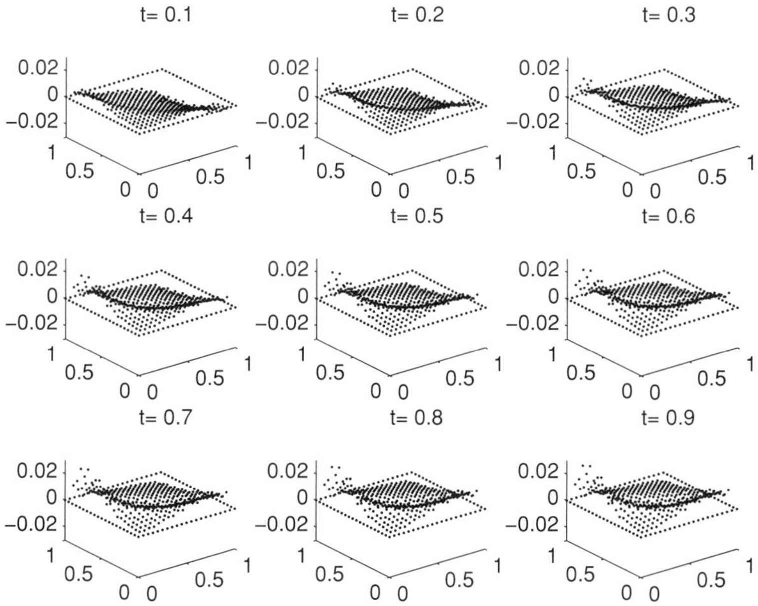

Figure 14.14. The error distribution between the exact solution and the finite element solution using the Crank-Nicholson method at different time instants.

14.8 EXERCISES

E14.1. Generate a triangulation, that is, the set of points P and a corresponding index set $\mathcal{I}$ defined in (14.64) and (14.67), respectively, for the domains given in Figure 14.15 such that the edge (or characteristic length) of the triangles are at most 0.1. (Note: Once the positions have been assigned for both boundary and internal points, you may use the MATLAB function `delaunay` to generate both P and $\mathcal{I}$ as well as the MATLAB function `triplot` to obtain a mesh plot.)

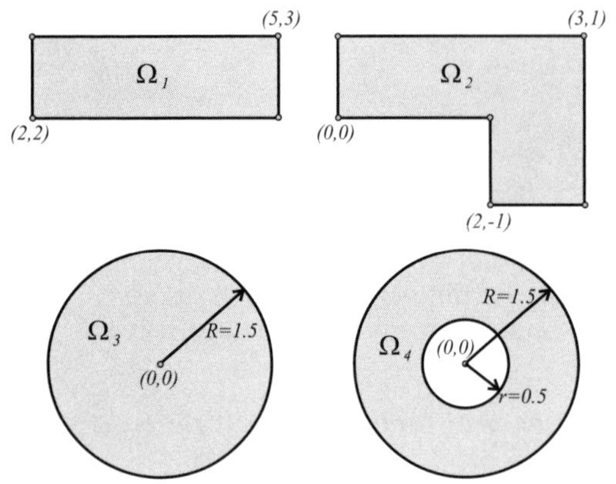

Figure 14.15. Various domains to be triangulated.

E14.2. As a small tutorial to help understand the procedure given in Section 14.5, consider the following differential equation

$$2\nabla^2 u - \frac{\partial u}{\partial x} + \frac{\partial u}{\partial y} - u + (2x - 3y + 7) = 0$$

subject to

$$u(1, y) = -3y + 4 \quad ; \quad u(x, -1) = 2x + 5 \quad , \quad u(x, 0.5) = 2x + 0.5$$

and $\left.\dfrac{\partial u}{\partial x}\right|_{x=2} = 2.0$ in the domain $1 \le x \le 2$ and $-1 \le y \le 0.5$. Let the mesh be described based on the node position matrix P and element index matrix $\mathcal{I}$ be given by

$$P = \begin{pmatrix} 1.0 & 1.0 & 1.0 & 1.0 & 1.5 & 1.5 & 1.5 & 1.5 & 2.0 & 2.0 & 2.0 & 2.0 \\ -1.0 & -0.5 & 0.0 & 0.5 & -1.0 & -0.5 & 0.0 & 0.5 & -1.0 & -0.5 & 0.0 & 0.5 \end{pmatrix}$$

$$\mathcal{I} = \begin{pmatrix} 5 & 2 & 1 \\ 6 & 2 & 5 \\ 6 & 5 & 10 \\ 10 & 5 & 9 \\ 6 & 10 & 7 \\ 7 & 10 & 11 \\ 7 & 2 & 6 \\ 3 & 2 & 7 \\ 4 & 3 & 7 \\ 7 & 8 & 4 \\ 11 & 12 & 7 \\ 12 & 8 & 7 \end{pmatrix}$$

1. Identify the following: functions $\mathbf{M}$, $\mathbf{b}$, g, and h, the set of Dirichlet nodes $\mathcal{D}$ and Neumann nodes $\mathcal{N}$.
2. Show that element matrices K_n and Γ_n for elements 6 and 7 can be computed to be

$$K_6 = \frac{1}{72}\begin{pmatrix} 67 & 7 & -71 \\ -5 & 79 & -71 \\ -77 & -65 & 145 \end{pmatrix} \quad \text{and} \quad \Gamma_6 = \frac{1}{144}\begin{pmatrix} 67 \\ 211 \\ 211 \end{pmatrix}$$

$$K_7 = \frac{1}{72}\begin{pmatrix} 67 & -5 & -59 \\ -5 & 67 & -59 \\ -77 & -77 & 157 \end{pmatrix} \quad \text{and} \quad \Gamma_7 = \frac{1}{9}\begin{pmatrix} 4 \\ 4 \\ 4 \end{pmatrix}$$

3. Show that after assembly, the first five columns of the global matrix $\widehat{K}$ and $\widehat{\Gamma}$, prior to any overloading, are given by

$$\widehat{K}_{\bullet,(1,\ldots,5)} = \frac{1}{72}\begin{pmatrix} 145 & -77 & 0 & 0 & -65 \\ -71 & 280 & -83 & 0 & 14 \\ 0 & -65 & 278 & -77 & 0 \\ 0 & 0 & -71 & 134 & 0 \\ -71 & -10 & 0 & 0 & 304 \\ 0 & -154 & 0 & 0 & -130 \\ 0 & 2 & -154 & -10 & 0 \\ 0 & 0 & 0 & -77 & 0 \\ 0 & 0 & 0 & 0 & -77 \\ 0 & 0 & 0 & 0 & 2 \\ 0 & 0 & 0 & 0 & 0 \\ 0 & 0 & 0 & 0 & 0 \end{pmatrix} ; \ \widehat{\Gamma} = \frac{1}{144}\begin{pmatrix} 71 \\ 264 \\ 112 \\ 105 \\ 294 \\ 276 \\ 480 \\ 108 \\ 223 \\ 576 \\ 416 \\ 261 \end{pmatrix}$$

4. After overloading, solve for the finite element solution u. Compare this with the exact solution given by $u = 2x - 3y + 2$. Note that because the solution is a plane, the finite element solution should also be quite close to the exact solution.

E14.3. Consider the Laplace equation $\nabla^2 u = 0$ with a unit circle domain $R = 1$ subject to the Dirichlet boundary condition $u(R = 1) = 1 + 3\cos\theta, 0 \leq \theta < 2\pi$. Obtain the finite element solution using the rectangular coordinate system and compare the solution with the exact solution given by $u(r, \theta) = 1 + r^3 \cos(3\theta)$ (cf. Example 11.6). (Hint: For the triangulation of the unit circle, you can use the command `[x,y,Ind]=circle_mesh_example(1,20)` that is available on the book's webpage for a circle of radius 1 and 20 concentric circles.)

E14.4. Consider the Poisson equation $\nabla^2 u = -1$ in the domain $0 \leq x \leq 1$, and $0 \leq y \leq 1$, subject to the Dirichlet boundary conditions $u(0, y) = u(1, y) = u(x, 0) = u(x, 1) = 0$. Obtain the finite element solution and compare it with the series solution given in (11.80); that is, plot the difference.

E14.5. A cross-section of a cooling fin is shown in Figure 14.16 with the base located at $y = 0$. The steady-state temperature distribution u inside the fin satisfies the Laplacian equation $\nabla^2 u = 0$, subject to the Dirichlet boundary condition: $u(y = 0) = T_0$, and a mixed boundary condition at $y > 0$:

$$\left(\kappa_T \nabla u \cdot \underline{\mathbf{n}}\right)\Big|_{\text{Boundary}, y>0} = (h_T)\left(T_{\text{surr}} - u\right)$$

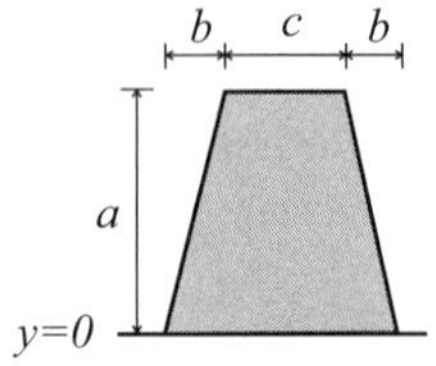

Figure 14.16. The dimensions for a trapezoid-shaped fin.

1. Build a program (under MATLAB or other platforms) that would generate plots of $u(x, y)$ that takes the following as parameters: a, b, c, κ_T, h_T, T_{surr}, and T_0, as well as ℓ, which is a characteristic length of the triangles. (Hint: You can use the program `trapezoid_mesh.m`, or the algorithm within, that is available on the book's webpage, for the triangulation of the domain.)
2. Try the following values: $a = 4, b = 1, c = 1, , T_{surr} = 25, T_0 = 10$, and $\ell = 0.1$. Use the following cases $\kappa_T/h_T = 0.5,\ 1,\ 10$ and observe what happens to the temperature distribution. Discuss the trend of the finite element solution obtained here at the various ratios of k_T/h_T, and compare them with the trend obtained in Example 9.5, where it was assumed that the temperature at each cross-sectional slice of the fin is fixed, that is, $u = u(y)$.

E14.6. Consider the nonhomogeneous Helmholtz equation in polar coordinates given by

$$\nabla^2 u + 3u = 18 - 15r^3 \sin(3\theta)$$

in a unit circle with a Dirichlet boundary condition given by $u = 6 - 5\sin(3\theta)$ at $r = 1$. Obtain a finite element solution using the rectangular coordinates. (You may use the triangulation using the command `[x,y,Ind]=circle_mesh_example(1,20)`). Compare the finite element solution with those found using the finite difference method (cf. Exercise **E13.7**). (Note: The exact solution is given by $u(r, \theta) = 6 - 5r^3 \sin(3\theta)$.)

E14.7. For the equation

$$\alpha\nabla^2 u - v\frac{\partial u}{\partial y} - ku = h(x, y) \quad \text{where } 0 \le x \le 1; 0 \le y \le 10$$

with

$$h(x, y) = e^{-x^2}\left(A + Be^{-x/2}\right); \quad A = 2\alpha + k - 4\alpha x^2 ;$$

$$B = \frac{-35}{4}\alpha + \frac{5}{2}v - 5k + 20\alpha x^2$$

subject to $u(x, 0) = 4e^{-x^2}$, $u(1, y) = e^{-1}\left(5e^{-y/2} - 1\right)$ and

$$\left(\frac{\partial u}{\partial y}\right)_{y=10} = 0 \quad \text{and} \quad \left(\frac{\partial u}{\partial x}\right)_{x=0} = 0$$

Obtain the finite element solution and compare it with the exact solution, which is given by $u = e^{-x^2}\left(5e^{-y/2} - 1\right)$. Try it for the following parameters: $\alpha = 10^{-3}$, $v = 10$, and $k = 0.2$.

E14.8. For an irrotational flow, that is, $\nabla \times \underline{\mathbf{v}} = 0$, we could use the fact that the curl of a gradient is the zero vector to specify a potential function Φ such that $\nabla\Phi = \underline{\vec{\mathbf{v}}}$. If we further assume incompressible flow, that is, $\nabla \cdot \underline{\vec{\mathbf{v}}} = 0$, this will result in a Laplacian equation for Φ.[9]

$$\nabla^2\Phi = 0$$

This means that the velocity profile, as well as the streamline of an irrotational, incompressible flow, can be obtained by first solving a scalar function Φ. Thereafter, the two components of the velocity can be evaluated by taking the gradient of Φ. Consider this to be the case for the fluid flowing between two plates of length L units that are separated Y units apart and flowing past a horizontal cylinder of radius r with center at (x_c, y_c) and located in the middle of the two plates (see Figure 14.17.) The boundary conditions are:

$$-\nabla\Phi \cdot \underline{\mathbf{n}}\Big|_{x=0} = \nabla\Phi \cdot \underline{\mathbf{n}}\Big|_{x=L} = \underline{\vec{\mathbf{v}}}_0$$

that is, the flow entering contains only an x-component, and the velocity is uniform and equal at the entrance and at the exit. Also, because the flow cannot penetrate the walls of the plates or the cylinder, we have,

$$\nabla\Phi \cdot \underline{\mathbf{n}}\Big|_{y=0} = \nabla\Phi \cdot \underline{\mathbf{n}}\Big|_{y=Y} = \nabla\Phi \cdot \underline{\mathbf{n}}\Big|_{(x-x_c)^2+(y-y_c)^2=r^2} = 0$$

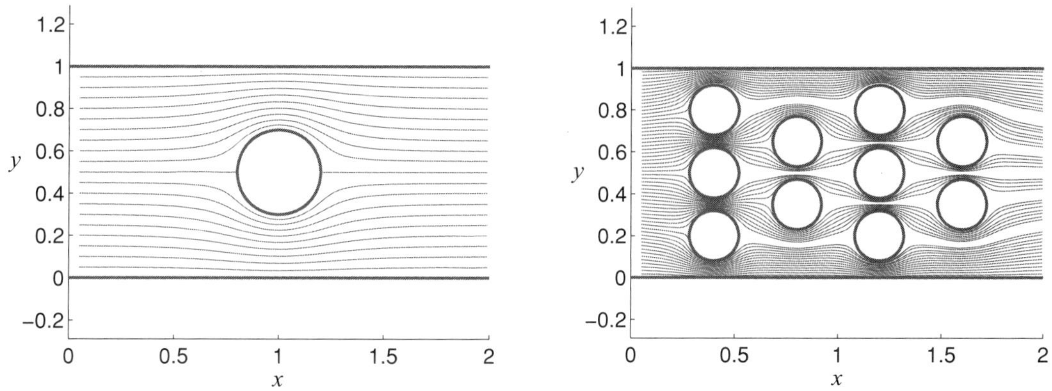

Figure 14.17. The streamlines for a potential flow around one cylinder or an array of cylinders.

1. Using the finite element method, solve the Laplacian equation for Φ subject to the boundary conditions given above. Specifically, let $\underline{\vec{\mathbf{v}}}_0 = 3$, $L = 2$, $Y = 1$, $(x_c, y_c) = (1, 0.5)$ and $r = 0.2$. (Hint: You can use the MATLAB file `rect_circ_mesh.m` to build the mesh).
2. Using (14.30), write a program that would find the gradients $\nabla\Phi$ of each finite element, to be located at the centroid of the triangle; that is, using the results found for P, $\mathcal{I}$, and $\mathbf{u}$, evaluate the vectors:

$$\mathbf{x}_{grad} = \begin{pmatrix} x_{grad,1} \\ \vdots \\ x_{grad,N} \end{pmatrix}; \mathbf{y}_{grad} = \begin{pmatrix} y_{grad,1} \\ \vdots \\ y_{grad,N} \end{pmatrix}; \mathbf{v}_x = \begin{pmatrix} v_{x,1} \\ \vdots \\ v_{x,N} \end{pmatrix}; \mathbf{v}_y = \begin{pmatrix} v_{y,1} \\ \vdots \\ v_{y,N} \end{pmatrix}$$

[9] For an inviscid and incompressible fluid flowing in two dimensions, the Helmholtz vorticity equation given in (4.106) implies irrotational flow at steady state because $(\nabla \times \underline{\vec{\mathbf{v}}}) \cdot \underline{\vec{\mathbf{v}}} = 0$ in a 2D flow.

where $\mathbf{x}_{grad}$ and $\mathbf{y}_{grad}$ are positions of the centroids of the triangles, and $\mathbf{v}_x$ and $\mathbf{v}_y$ are the x- and y- components of the velocitiy, and N is the number of finite elements.

3. With the vectors, $\mathbf{x}_{grad}$, $\mathbf{y}_{grad}$, $\mathbf{v}_x$ and $\mathbf{v}_y$, try to reproduce the streamline plot shown in the left plot in Figure 14.17. (Hint: Once all the necessary vectors have been found, you can use the MATLAB file `fem_exer_streamline.m` that is available on the book's webpage.)
4. Apply the same solution approach to a path involving an array of horizontal cylinders as shown in the right plot in Figure 14.17. Hint: You can use the following MATLAB commands that are available on the book's webpage as follows:

```
[choles,rholes]=circ_arrays(3,4,0.4,0.3,0.12,1,0.5);

[x,y,Ind]=rect_circ_mesh([0;0],[2;1],choles,rholes,0.03);
```

to obtain the array of cylinders as well as the triangulation of the fluid region. Then after the velocities have been found, you can again use the command `fem_exer_streamline` to draw the streamlines.

E14.9. For the potential velocity flow past one cylinder as given in Exercise **E14.8** (cf. Figure 14.17), assume that the velocities are independent of temperature. Applying the thermal energy balance given by (5.32) to an incompressible and inviscid fluid at steady state, we have

$$\underline{\vec{\mathbf{v}}} \cdot \nabla \widehat{u} = -\nabla \cdot \mathbf{q}$$

With internal energy $\widehat{u} = \rho C_v T$, together with Fourier's law, $\mathbf{q} = -\kappa \nabla T$, this reduces to

$$\underline{\vec{\mathbf{v}}} \cdot \nabla T = \alpha \nabla^2 T$$

with $\alpha = \kappa/(\rho C_v)$. Assuming that the two plates are insulated and that a constant heat flux Q occurs at the surface of the cylinder, we can consider the boundary conditions to be given by $T(x=0) = 60$,

$$\nabla T \cdot \underline{\mathbf{n}}\Big|_{y=0} = \nabla T \cdot \underline{\mathbf{n}}\Big|_{y=Y} = \nabla T \cdot \underline{\mathbf{n}}\Big|_{x=L} = 0$$

and

$$\nabla T \cdot \underline{\mathbf{n}}\Big|_{(x-x_c)^2+(y-y_c)^2=r^2} = \frac{Q}{\rho C_v} = 40$$

Let the thermal diffusivity $\alpha = 1$. Using the velocity field found in Exercise **E14.8**, let the entrance velocity $\underline{\vec{\mathbf{v}}}_0 = 4$, and use the finite element method to obtain the temperature field and a tomographic plot of T. Note: Once the values for $x, y, \mathcal{I}$, and T (which are the x and y node positions, nodes indices of the elements, and temperature values at the nodes, respectively) have been found, a temperature tomography plot can be obtained in MATLAB, with Ind=$\mathcal{I}$, using the following commands:

```
trisurf( Ind, x, y, T);
axis equal;  view([0 90]);  shading interp;
```

E14.10. The steady-state temperature of a fluid flowing through a cylindrical pipe is modeled by the following equation

$$v\frac{\partial u}{\partial z} = \alpha\nabla^2 u + \beta\left(u_R - u\right)$$

subject to $u(R, \theta, z) = u_R$, $u(r, \theta, 0) = u_0$

$$\left.\left(\frac{\partial u}{\partial r}\right)\right|_{r=0} = 0 \quad \text{and} \quad \left.\left(\frac{\partial u}{\partial z}\right)\right|_{z=L} = 0$$

Use the finite element method under the axisymmetric assumption to find the distribution $u(r, z)$ for $L = 10$, $R = 1$, $\alpha = 0.2$, $v = 3$, $\beta = 0.3$, and $u_R = 80$. Generate a tomographic with contour map of $u(r, z)$ for $-R \leq r \leq R$ and $0 \leq z \leq L$.

E14.11. At steady state, a parallel, Newtonian viscous flow in the z-direction, through a pipe having a fixed cross-section domain Ω, will have $v_x = v_y = 0$, and v_z modeled by

$$\left(\frac{\partial^2 v_z}{\partial x^2} + \frac{\partial^2 v_z}{\partial x^2}\right) = \beta$$

where $\beta = -(1/\mu)d(p + \mathbf{g}z)/dz$, with μ, p, and $\mathbf{g}$ being the viscosity, pressure, and gravitational acceleration, respectively. The domain Ω can either be simply connected (without holes) or multiply connected (with holes). Either way, the no-slip boundary condition will have to apply; that is, we have the Dirichlet conditions, $v_z = 0$ at Boundary(Ω).

1. Let the cross-section domain Ω be formed by a circle of radius R but with a hole formed by a smaller circle of radius $r_0 = \kappa R$ ($\kappa < 1$) whose center coincides with the larger circle. The exact solution of this flow is known,[10] and it is given by

$$v_z(r) = \frac{\beta R^2}{4}\left[1 - \left(\frac{r}{R}\right)^2 - \frac{1-\kappa^2}{\ln(1/\kappa)}\ln\left(\frac{R}{r}\right)\right] \tag{14.84}$$

and the average velocity is given by

$$v_{z,\text{ave}} = \frac{\int_\Omega v_z d\Omega}{\int_\Omega d\Omega} = \frac{\beta R^2}{8}\left[\frac{1-\kappa^4}{1-\kappa^2} - \frac{1-\kappa^2}{\ln(1/\kappa)}\right] \tag{14.85}$$

For simplicity, let $\beta = 1$, $\kappa = 0.3$, and $R = 1$. Using the finite element method, compare the numerical solution using a mesh based on rectangular coordinates with the exact solution given in (14.84). (Hint: You can use the triangulation obtained by the following MATLAB command that is available on the book's webpage:

```
[x,y,Ind]=circle_annulus(c,r,choles,rholes,dL);
```

where `c` and `r` are the center and radius of the outer circle, respectively, whereas `choles` and `rholes` are the centers and radii of the holes, respectively, and `dL` is the approximate characteristic length of the elements.)

Furthermore, using the same triangulation and the numerical solution of v_z for the annular flow found above, use the approximation of the integrals based on the shape functions (c.f. (14.22)) to determine the

[10] See, for example, R. B. Bird et al., *Transport Phenomena*, 2nd edition, J. Wiley & Sons, New York, 2002, pp. 54–55.

average velocity $v_{z,\text{ave}}$, and compare it with the exact value given by (14.85).

2. Instead of one hole at the center, consider the case as shown in Figure 14.18, where there are three holes with centers and their corresponding radii given by:

$$c_{\text{holes}} = \left(\begin{array}{c|c|c} c_1 & c_2 & c_3 \end{array} \right) = \begin{pmatrix} \frac{1}{2} & \frac{1}{2}\cos(2\pi/3) & \frac{1}{2}\cos(4\pi/3) \\ 0 & \frac{1}{2}\sin(2\pi/3) & \frac{1}{2}\sin(4\pi/3) \end{pmatrix}$$

and $r_{\text{holes}} = \begin{pmatrix} 0.2, & 0.2, & 0.2 \end{pmatrix}$. Obtain a tomographic plot of v_z under viscous flow through Ω for this case, as well as the average velocity. Finally, for density $\rho = 1$, determine the mass flow rate $Q = \rho\, v_{z,\text{ave}} A_\Omega$, where $A_\Omega = \int_\Omega d\Omega$ is the area of Ω.

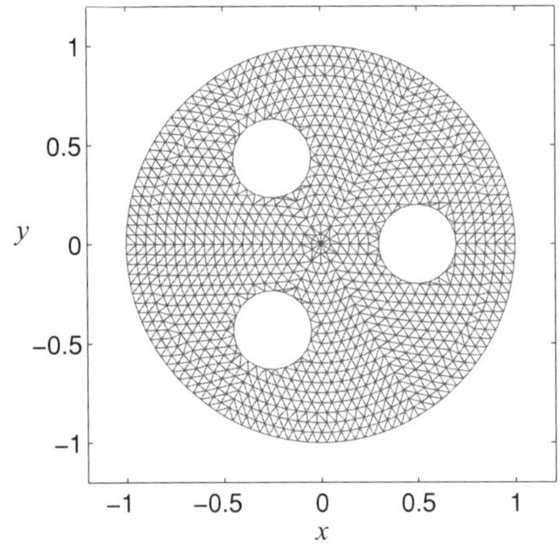

Figure 14.18. A circular cross-section domain Ω containing three holes.

E14.12. Consider the torus shown in Figure 14.19, with $a = 2$ and $R = 1$. Assume the temperature to be initially at $u = 60$. At time $t > 0$, the surface of the torus was set to 120. The conduction inside the torus can be modeled by the heat diffusion equation with $\alpha = 0.5$.

$$\frac{\partial u}{\partial t} = \alpha \nabla^2 u$$

Obtain a time-lapse temperature (surface plot or tomograph plot) of a circular slice of the torus at $t = 0.0, 0.1, 0.2, \ldots, 0.8$.

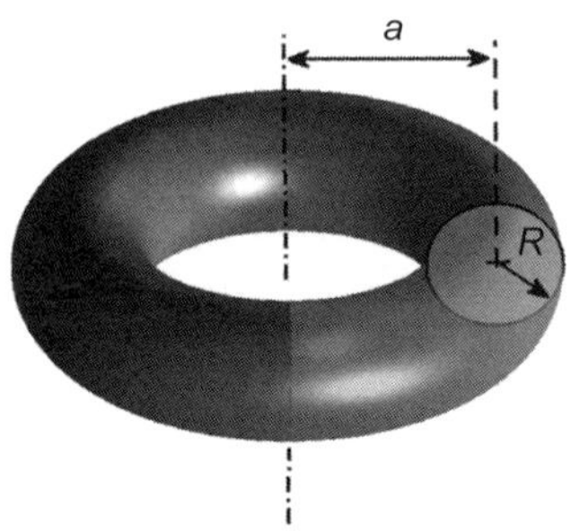

Figure 14.19. A solid torus with major radius a and minor radius R.

E14.13. A cylindrical tank with a hemispherical bottom contains a fluid that is sealed from the top. A cylindrical rod is then inserted at the center. (See Figure 14.20). The temperature distribution inside the tank can be modeled using the diffusion equation

$$\frac{\partial T}{\partial t} = \alpha \nabla^2 u$$

subject to $T(r, \theta, z, 0) = T_w$, $T_{\text{along wall}} = T_w$, $T_{\text{along rod}} = T_{\text{rod}}$ and

$$\left.\frac{\partial T}{\partial r}\right|_{r=0} = 0 \qquad \text{and} \qquad \left.\frac{\partial T}{\partial z}\right|_{\text{along top surface}} = 0$$

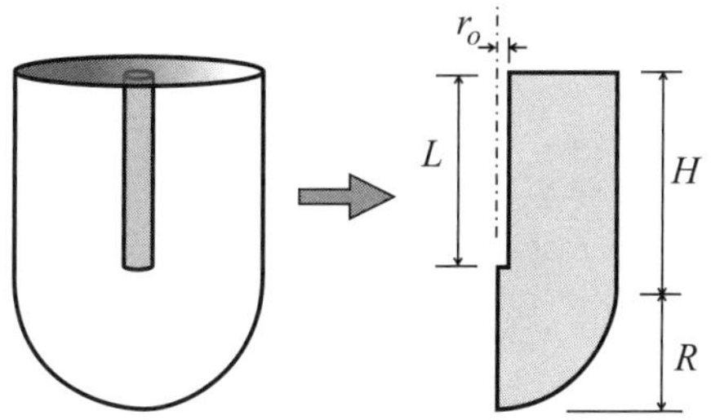

Figure 14.20. A cylindrical tank with hemisphere bottom of radius R, including a solid cylindrical heating rod at the center. The right plot shows the half cross-section from the cental vertical axis.

Let $\alpha = 1.2$, $L = 0.8$, $R = 1$, $H = 1$, and $r_o = 0.5$. Assuming axisymmetric distribution, obtain a time-lapse plot of the temperature distribution at $t = 0.0, 0.01, 0.02, \ldots, 0.08$. Hint: You can use the triangulation obtained using a MATLAB function that is available on the book's webpage:

```
[x,y,tri] = tank_mesh(R,H,L,r0,dL)
```

where dL is the maximum edge of the triangular elements.

Bibliography

[1] N. R. Amundson, *Mathematical Methods in Chemcial Engineering: Matrices and Their Applications*, vol. 1, Prentice Hall, New Jersey, 1966.

[2] N. R. Amundson and R. Aris, *Mathematical Methods in Chemcial Engineering*, vol. 2, Prentice Hall, New Jersey, 1973.

[3] G. Arfken, *Mathematical Methods for Physicists*, Academic Press, New York, third ed., 1985.

[4] V. I. Arnold, *Ordinary Differential Equations*, Springer-Verlag, Berlin Heidelberg, third ed., 1992.

[5] O. Axelsson and V. A. Barker, *Finite Element Solution of Boundary Value Problems*, Society for Industrial and Applied Mathematics, Philadelphia, 2001.

[6] P. Bamberg and S. Sternberg, *A Course in Mathematics for Students of Physics*, vol. 1 and 2, Cambridge University Press, Cambridge, UK, 1990.

[7] K. J. Beers, *Numerical Methods for Chemical Engineering*, Cambridge University Press, Cambridge, UK, 2007.

[8] R. B. Bird, W. E. Stewart, and E. N. Lightfoot, *Transport Phenonomena*, John Wiley & Sons, second ed., 2007.

[9] W. E. Boyce and R. C. DiPrima, *Elementary Differential Equations and Boundary Value Problems*, John Wiley & Sons, New York, third ed., 1977.

[10] K. E. Brenan, S. L. Campbell, and L. R. Petzold, *Numerical Solution of Initial Value Problems in Differential Algebraic Equations*, North-Holland, New York, 1989.

[11] D. N. Burghes and M. S. Borrie, *Modeling with Differential Equations*, Ellis Horwood, West Sussex, England, 1981.

[12] G. Cain and G. H. Meyer, *Separation of Variables for Partial Differential Equations, An Eigenfunction Approach*, Chapman & Hall/CRC, Boca Raton, FL, 2006.

[13] B. J. Cantewell, *Introduction to Symmetry Analysis*, Cambridge University Press, Cambridge, UK, 2002.

[14] H. S. Carslaw and J. C. Jaeger, *Conduction of Heat in Solids*, Oxford University Press, London, second ed., 1959.

[15] C. T. Chen, *Linear System Theory and Design*, Oxford University Press, USA, 1984.

[16] C. R. Chester, *Techniques in Partial Differential Equations*, McGraw-Hill, New York, 1970.

[17] R. Courant and D. Hilbert, *Methods of Mathematical Physics*, vol. 1 and 2, John Wiley & Sons, New York, 1962.

[18] G. Dahlquist and Å. Björck, *Numerical Methods*, Dover Publications, New York, 1974.

[19] L. Debnath, *Nonlinear Partial Differential Equations for Scientists and Engineers*, Birkhäuser, Boston, 1997.

[20] A. S. Deif, *Advanced Matrix Theory for Scientists and Engineers*, Abacus Press, Kent, England, 1982.

[21] J. E. Dennis and R. B. Schnabel, *Numerical Methods for Unconstrained Optimization and Nonlinear Equations*, Prentice Hall, New Jersey, 1983.

[22] J. Donea and A. Huerta, *Finite Element Methods for Flow Problems*, John Wiley & Sons, New York, 2003.

[23] L. Dresner, *Similarity Solutions of Nonlinear Partial Differential Equations*, Pitman Publishing, London, 1983.

[24] P. DuChateau and D. Zachmann, *Applied Partial Differential Equations*, Dover Publications, New York, 1989.

[25] L. Edelstein-Keshet, *Mathematical Models in Biology*, Society for Industrial and Applied Mathematics, Philadelphia, 2005.

[26] D. K. Faddeev and V. N. Faddeeva, *Computational Methods of Linear Algebra*, W.H. Freeman, San Francisco, 1963.

[27] J. D. Faires and R. L. Burden, *Numerical Methods*, Brook/Cole Publishing Company, Pacific Grove, CA, third ed., 2002.

[28] S. J. Farlow, *Partial Differential Equations for Scientists and Engineers*, Dover Publications, New York, 1993.

[29] J. H. Ferziger and M. Perić, *Computational Methods of Fluid Dynamics*, Springer Verlag, Berlin, 2002.

[30] B. A. Finlayson, *The Method of Weighted Residuals and Variational Principles*, Academic Press, New York, 1972.

[31] G. B. Folland, *Fourier Analysis and Its Applications*, Brooks/Cole Publishing Company, Pacific Grove, CA, 1992.

[32] G. Friedlander and M. Joshi, *Introduction to the Theory of Distributions*, Cambridge University Press, Cambridge, UK, second ed., 1998.

[33] J. C. Friedly, *Dynamic Behaviour of Processes*, Prentice-Hall, New Jersey, 1972.

[34] G. F. Froment and K. B. Bischoff, *Chemical Reactor Analysis and Design*, John Wiley & Sons, New York, first ed., 1979.

[35] F. R. Gantmacher, *Matrix Analysis*, vol. 1 and 2, Chelsea Publishing Company, New York, 1977.

[36] C. W. Gear, *Numerical Initial Value Problems in Ordinary Differential Equations*, Prentice-Hall, New Jersey, 1971.

[37] N. E. Gibbs, J. William, G. P̃oole, and P. K. Stockmeyer, *An algorithm for reducing the bandwidth and profile of a sparse matrix*, SIAM J. Numer. Anal., 13 (1976), pp. 236–250.

[38] P. Glendinning, *Stability, Instability, and Chaos: an Introduction to the Theory of Nonlinear Differential Equations*, Cambridge University Press, Cambridge, UK, 1994.

[39] G. H. Golub and C. F. V. Loan, *Matrix Computations*, John Hopkins University Press, Baltimore, third ed., 1996.

[40] M. D. Greenberg, *Foundations of Applied Mathematics*, Prentice Hall, New Jersey, 1978.

[41] J. Guckenheimer and P. Holmes, *Nonlinear Oscillations, Dynamical Systems, and Bifurcations of Vector Fields*, Springer Verlag, Berlin, second ed., 1983.

[42] W. Hahn, *Stability of Motion*, Springer-Verlag, Berlin, 1968.

[43] E. Hairer, S. P. Nørsett, and G. Wanner, *Solving Ordinary Differential Equations I: Nonstiff Problems*, Springer-Verlag, Berlin Heidelberg, second ed., 1993.

[44] E. Hairer and G. Wanner, *Solving Ordinary Differential Equations II: Stiff and Differential-Algebraic Problems*, Springer-Verlag, Berlin Heidelberg, second ed., 1996.

[45] F. B. Hildebrand, *Methods of Applied Mathematics*, Dover Publications, New York, second ed., 1965.

[46] L. Hogben, ed., *Handbook of Linear Algebra*, Chapman & Hall/CRC, Boca Raton, FL, 2007.

[47] R. A. Horn and C. R. Johnson, *Matrix Analysis*, Cambridge University Press, Cambridge, UK, 1985.

[48] ——, *Topics in Matrix Analysis*, Cambridge University Press, Cambridge, UK, 1991.

[49] T. J. R. Hughes, *The Finite Element Method: Linear Static and Dynamic Finite Element Analysis*, Dover Publications, New York, 2000.

[50] M. Humi and W. Miller, *Second Course in Ordinary Differential Equations for Scientists and Engineers*, Springer Verlag, New York, 1987.

[51] P. E. Hydon, *Symmetry Methods for Differential Equations, A Beginner's Guide*, Cambridge University Press, Cambridge, MA, 2000.

[52] E. L. Ince, *Ordinary Differential Equations*, Dover Publications, New York, 1956.

[53] E. Isaacson and H. B. Keller, *Analysis of Numberical Methods*, John Wiley & Sons, New York, 1966.

[54] A. Iserles, *A First Course in the Numerical Analysis of Differential Equations*, Cambridge University Press, Cambridge, UK, 1996.

[55] V. G. Jenson and G. V. Jeffreys, *Mathematical Methods in Chemical Engineering*, Academic Press, London, second ed., 1977.

[56] C. Johnson, *Numerical Solution of Partial Differential Equations by the Finite Element Method*, Dover Publications, New York, 2009.

[57] T. Kailath, *Linear Systems*, Prentice-Hall, New Jersey, 1980.

[58] A. C. King, J. Billigham, and S. R. Otto, *Differential Equations: Linear, Nonlinear, Ordinary, Partial*, Cambridge University Press, Cambridge, UK, 2003.

[59] R. Knobel, *An Introduction to the Mathematical Theory of Waves*, American Mathematical Society, Providence, RI, 1999.

[60] E. Kreyszig, *Advanced Engineering Mathematics*, John Wiley & Sons, ninth ed., 2006.

[61] M. C. Lai, *A note on finite difference discretizations for poisson equation on a disk*, Numerical Methods for Partial Difference Equations, 17 (2001), pp. 199–203.

[62] P. D. Lax, *Hyperbolic Systems of Conservation Laws and the Mathematical Theory of Shock Waves*, Society of Industrial and Applied Mathematics, Philadephia, 1987.

[63] E. S. Lee, *Quasilinearization and Invariant Imbedding*, Academic Press, New York, 1968.

[64] R. J. LeVeque, *Numerical Methods for Conservation Laws*, Birkhäuser Verlag, Switzerland, 1992.

[65] ——, *Finite Volume Methods for Hyperbolic Problems*, Cambridge University Press, Cambridge, UK, 2002.

[66] R. W. Lewis, P. Nithiarasu, and K. N. Seethataramu, *Fundamentals of the Finite Element Method for Heat and Fluid Flow*, John Wiley & Sons, New York, 2004.

[67] H. Lomax, T. H. Pulliam, and D. W. Zingg, *Fundamentals of Computational Fluid Dynamics*, Springer Verlag, Berlin, 2001.

[68] H. S. Mickley, T. K. Sherwood, and C. E. Reed, *Applied Mathematics in Chemical Engineering*, McGraw-Hill, Company, New York, 1957.

[69] K. W. Morton and D. F. Mayers, *Numerical Solution of Partial Differential Equations*, Cambridge University Press, Cambridge, UK, second ed., 2005.

[70] G. M. Murphy, *Ordinary Differential Equations and Their Solutions*, D. Van Nostrand Company, Princeton, NJ, 1960.

[71] P. V. O'Neil, *Advanced Engineering Mathematics*, Cengage Learning – Engineering, Stanford, CT, seventh ed., 2011.

[72] P. O. Persson and G. Strang, *A simple mesh generator in* `matlab`, SIAM Review, 46 (2004), pp. 329–345.

[73] W. H. Press, S. A. Teukolsky, W. T. Vetterling, and B. P. Flannery, *Numerical Recipes: The Art of Scientific Computing*, Cambridge University Press, New York, third ed., 2007.

[74] H.-K. Rhee, R. Aris, and N. R. Amundson, *First-Order Partial Differential Equations: Theory and Applications of Single Equations*, vol. 1, Dover Publications, New York, 2001.

[75] R. G. Rice and D. D. Do, *Applied Mathematics and Modeling for Chemical Engineers*, John Wiley & Sons, New York, 1995.

[76] J. I. Richards and H. K. Youn, *Theory of Distributions, A Nontechnical Introduction*, Cambridge University Press, Cambridge, UK, 1990.

[77] K. F. Riley, M. P. Hobson, and S. J. Bence, *Mathematical Methods for Physics and Engineering*, Cambridge University Press, Cambridge, UK, third ed., 2006.

[78] Y. Saad, *Iterative Methods for Sparse Linear Systems*, Society for Industrial and Applied Mathematics, Philadelphia, second ed., 2003.

[79] H. M. Schey, *Div, Grad, Curl and All That: An Informal Text on Vector Calculus*, W. W. Norton & Company, New York, 1992.

[80] J. H. Seinfeld, *Mathematical Methods in Chemcial Engineering*, vol. 3, Prentice Hall, New Jersey, 1974.

[81] I. N. Sneddon, *Fourier Transforms*, McGraw-Hill, New York, 1951.

[82] I. P. Stavroulakis and S. A. Tersian, *Partial Differential Equation: An Introduction with Mathematica and MAPLE*, World Scientific Publishing Company, Singapore, 1999.

[83] H. Stephani, *Differential Equations, Their Solution Using Symmetries*, Cambridge University Press, Cambridge, UK, 1989.

[84] H. H. Storey and C. Rosenbrock, *Computational Techniques for Chemical Engineers*, Pergamon Press, New York, 1966.

[85] G. Strang, *Introduction to Applied Mathematics*, Wellesley-Cambridge Press, Wellesley, MA, 1986.

[86] S. H. Strogatz, *Nonlinear Dynamics and Chaos: With Applications to Physics, Biology, Chemistry, and Engineering*, Westview Press, Cambridge, MA, 2001.

[87] J. W. Thomas, *Numerical Partial Differential Equations: Finite Difference Methods*, Springer Verlag, New York, 1995.

[88] E. G. Thompson, *Introduction to the Finite Element Method*, John Wiley & Sons, New York, 2005.

[89] N. Tufillaro, T. Abbott, and J. Reilly, *An Experimental Approach to Nonlinear Dynamics and Chaos*, Addison-Wesley Publishing Company, Redwood City, CA, 1992.

[90] C. R. Wiley and L. C. Barrett, *Advanced Engineering Mathematics*, McGraw-Hill Book, New York, fifth ed., 1982.

[91] O. C. Zienkiewicz, R. L. Taylor, and P. Nithiarasu, *The Finite Element Method for Fluid Dynamics*, Elsevier Butterworth-Heinemann, Amsterdam, sixth ed., 2005.

[92] O. C. Zienkiewicz, R. L. Taylor, and J. Z. Zhu, *The Finite Element Method, Its Basis and Fundamentals*, Elsevier Butterworth-Heinemann, Amsterdam, sixth ed., 2005.

[93] D. Zwillinger, *Handbook of Integration*, Jones and Bartlett Publishers, Boston, 1992.

[94] ——, *Handbook of Differential Equations*, Academic Press, San Diego, third ed., 1997.

Index

Affine operator, 106–107
definition, 106
Airy equation and functions, 367, 477
Alternating direction implicit (ADI) schemes. *See* Finite difference methods–ADI schemes
Ampere's law, 222
Analytic functions
definition, 799
branch points, 816
Cauchy's theorem, 801
Jordan's lemma, 808
Morera's theorem, 803
poles, k^{th} order, and simple poles, 800
properties, 800
residues
definition, 803
residue theorem, 804
special applications, 811–819
singular and regular points, 800
types, 800
Analytical functions
Cauchy integral representation theorem, 803
Arnoldi's method, 629, 632

Balance equations. *See* (Conservation laws)
Bernoulli equation of motion, 229
Bessel equations, 363–368
equations reducible to Bessel equations, 366
Plots, Bessel functions of the first and second kind, 365
Plots, modified Bessel functions of the first and second kind, 365
properties and identities of Bessel functions, 369–371
Table, Bessel and modified Bessel functions, 364
Table, types and solutions, 363
Bifurcation analysis, 738–742
cusp points, 741
Hopf bifurcation, 742
hysteresis, 742
Table, normal forms for 2D systems, 743
Table, one-dimensional systems, 739
Black-Scholes formula, 479
Boundary value problems (BVP)
boundary conditions (separated/mixed), 299
linear BVP, 301
Ricatti equation method (invariant imbedding method), 734
shooting method, 301
MATLAB BVP solver, 716
nonlinear BVP, 731
flowchart, nonlinear shooting based on Newton's method, 733
Bromwich integral, 464
Butterworth filter, 269

Cauchy equation of motion, 218
Cauchy principal values, 807
Cauchy's theorem, 801
Cauchy-Riemann conditions, 800
Cayley-Hamilton theorem, 122
Characteristic polynomial, 108
Danilevskii method, 660
Chebyshev equation, 371
Chemical vapor deposition reaction (CVD), 449
Complementary error function, 443, 465
Component balance, 220
binary component balance equation, 220
Conjugate gradient method, 78, 620–625
algorithm, 621
Conservation laws, 216–220
continuity equation, 217
Continuant equation, 50
Convex hull, 867–870
simplified algorithm, 869
Coordinate systems
base vector/reciprocal base vector, 190
coordinate curve, 190
coordinate surface, 190
cylindrical coordinate system, 184–187

Coordinate systems (*cont.*)
rectangular coordinate system, 149
spherical coordinate system, 187–189
Table, relationship between rectangular and cylindrical, 186
Table, relationship between rectangular and spherical, 188
torroidal coordinate system, 229
vector differential operation in cylindrical and spherical coordinates, 194
Courant number, 857
Crank-Nicholson method
See (Finite difference method-)
Cubic spline interpolation, 89
Curl
See also (Vector differential operation)
of a vector field, 176
vorticity, 177
Curvature of path/radius of curvature, 166
Cuthill-Mckee algorithm, 601

d'Alembert method of order reduction, 357
procedure, 748–750
d'Alembert solutions, 786–791
Dahlquist stability test, 291, 296
Danielevskii method, 660–663
Danielson-Lanczos equation, 796
Darboux's inequality, 806
Delaunay triangulation. *See* (Finite element methods)
Delta functions. *See* (Distributions)
Difference equation
linear, with constant coefficients
characteristic equation, 292
complementary solution, 292
stability, 294
Differential algebraic equation (DAE)
index, 304
mass matrix form, 305
MATLAB DAE solver, 717
semi-explicit, 303
Digraph, 595
strongly connected subdigraphs, 596
Directional derivative, 171
Dirichlet integral theorem, 836
Discrete Fourier transforms. *See* (Fast Fourier transforms)
Distribution theory
definition, 821
delta distribution (delta function), 820
in higher dimensions, 828–830
limit identities, 826–827
properties and identities, 825–826
derivative (theorem) of distributions, 824
properties, 823–824
tempered distributions, 830–831
definition, 831
generalized Fourier transform. *See* Fourier transforms-generalized Fourier transforms
test functions, 821
Divergence
See also (Vector differential operation)
of a vector field, 174
Divergence theorem, 208–210
Duhamel's principle, 258

Eigenfunctions, 435
Eigenvalues and eigenvectors, 107–115
definition, 107
left eigenvectors, 113
list of properties, 113
modified QR method, 651–655
of companion matrices, 112
power method, 649
QR method, 650
spectral radius, 125
spectrum, 109
Electromagnetics
Table, terms and relationships, 221
Energy balance, 218–220
mechanical energy balance, 219
thermal energy balance, 219
total energy balance, 219
Error function, 465
Euler equation of motion, 218

Faraday's law, 222
Fast Fourier transforms
algorithm, 798
discrete Fourier transforms, 796
Fick's law, 220
Finite difference equations
ADI schemes, 863–866
Finite difference method
consistency, 513
convergence, 513
finite difference approximations
backward difference, 485
central difference, 485
forward difference, 485
method of undetermined coefficients, 486–491, 851
finite difference approximations
finite difference approximation lemma, 487
Table, for first-order derivatives, 489
Table, for second-order derivatives, 490
hyperbolic equations, 855–862
Courant number, 857
Crank-Nicholson scheme, 860
Lax-Friedrichs scheme, 859
Lax-Wendroff scheme, 860
leapfrog scheme, 859
Table, basic schemes, 858
upwind schemes, 856
Wendroff scheme, 521
Lax-Richmyer stabilty, 513
stability analysis
amplification factor, 517
eigenvalue method, 514–516
Von Neumann method, 516–519

time-dependent
backward Euler, 508
Crank-Nicholson, 509
forward Euler, 508
semi-discrete approach (or method of lines), 504–507
weighted Euler methods, 509
time-independent
one dimension, 491–496
polar and spherical coordinates, 500–503
three dimensions, 852–854
two dimensions, 496–499
Finite element methods
streamlined-upwind-Petrov-Galerkin (SUPG), 870
Shakib formula, 870
stabilization parameter, 870
assembly
method of reduction of unknowns, 537
overloading method, 538
axisymmetric problems, 546–547
Delaunay triangulation, 539–541
node lifting, 540
quick-hull algorithm, 540, 867–870
Galerkin method, 530
summary, main steps, 542–544
time-dependent
Crank-Nicholson method, 549
mass matrix, 549
semi-discrete approach, 548
triangular finite elements, 527–533
index matrix, 535
line integrals, 531–533
node matrix, 535
properties of shape functions, 528–529
shape functions, 527
surface integrals, 529–531
weak solution, 526
weighted residual method, 526
First-order PDE
Cauchy condition, 381
characteristic curves, 381
characteristics
characteristics, 381
Clairaut's equation, 392
general solution, form, 388
Lagrange-Charpit conditions, 390
Lagrange-Charpit method, 389
method of characteristics, 380–387
special forms (quasilinear, semilinear, etc.), 380
Floquet multipliers, 339
Fourier integral equation, 453
Fourier integral theorem, 819
technical lemma, 838
Fourier kernel, 452
Fourier series, 423, 452
Fourier transforms
definition, Fourier/inverse Fourier transform, 454
convolution, 458
Fourier transforms. *See* (Fast Fourier transforms)
generalized Fourier transforms, 831–836
definition, 833
of integrals, 835
Table, properties, 460
Table, transforms of basic functions, 461
Frenet formulas, 198
Frobenius series, 355

Gamma function, 349
Gauss' laws, 222
Gauss-Jordan elimination, 55–57
algorithm, 591
SVD alternative method, 594
Gauss-Legendre quadrature method, 684, 687
Gegenbauer equation, 371
Generalized functions. *See* (Distributions)
Generalized inverse (Moore-Penrose), 128
Gershgorin circles, 113, 143
Givens operator, 101
GMRES method, 79, 629–634
algorithm, 634
Gradient
See also (Vector differential operation)
ascent/descent method, 200
of a scalar, 170
operator (del operator), 170
Gradient-vector dyad operation, 179
Gram-Schmidt orthogonalization, 616
Green's Lemma, 205–208
Gronwall's inequality, 313

Hamiltonian canonical equations, 402
Hamiltonian/Hamiltonian principal function, 402
Harmonic equation. *See* (Partial differential equations–Laplace equation)
Harmonic functions, 179
Hartman-Grobman theorem (linearization theorem), 328
Helmholtz vorticity equation, 202
Hermite equation and polynomials, 373
Holomorphic function. *See* (Analytic functions)
Householder operator, 103
Hypergeometric equation, 372
Hypergeometric series
confluent hypergeometric series, 351
Gauss' hypergeometric series, 350
general series, 350

Initial value problems (IVP)
error control, 724
Dormand-Prince 5/4 tableau, 727
embedded Runge-Kutta methods, 726
Fehlberg 4/5 tableau, 727
flowchart, 726
step doubling, 728
Euler methods (forward, backward), 275
multistep, 282
Adams predictor-corrector method, 286
Adams-Bashforth (explicit multistep), 284

Initial value problems (IVP) (*cont.*)
Adams-Moulton (implicit multistep), 285
backward difference formula (BDF) method (Gear's method), 287
BDF with variable step size, 723
Milne-Simpson method, 308
trapezoidal method, 285
Runge-Kutta methods, 276
explicit fourth order, 279
explicit second order (Huen's method), 307
implicit fourth order (Gauss-Legendre), 280
Runge-Kutta tableau, 278
stability
Dahlquist stability test, 296
principal roots/spurious roots, 297
stability regions
Adams-Moulton method, 299
backward Euler method, 297
explicit Runge-Kutta, 297
fourth-order Gauss-Legendre, 298
types of numerical stability, 300
stiff differential equation, 298
Table, MATLAB IVP solvers, 716
Integral theorems
divergence theorem (Gauss-Ostrogradski-Green), 208
Gauss' theorem, 210
Green's Lemma, 205
Green's theorem (Green's identities), 209
Stokes' theorem, 210
Integral transforms
definition, 451
Dirichlet conditions, 819
Table, examples, 452

Jacobi equation, 371
Jordan block, 119
Jordan canonical form (Jordan decomposition), 118

Kronecker delta, 154
Krylov subspace, 631, 706

Laguerre equation and polynomials, 373
Lambert's W function (Omega function), 247
Laplace invariants, 444
Laplace transforms
definition, Laplace/inverse Laplace transforms, 464
convolution operation, 468
inverse transformation via partial fractions, 472–474
List, properties, 467–469
Table, transforms of basic functions, 472
Laplacian
See also (Vector differential operation)
of a scalar field, 178
of vector fields, 181
operator, 178
Lax entropy conditions, 779
Least-squares solution, 71–77
forgetting factor, 93
Levenberg-Marquardt method, 639
algorithm, 641
Moré method, 641
normal equation, 71
recursive least squares, 93
weighted least squares, 92
with linear constraints, 76
Legendre equations, 358–363
Plots, Legendre functions of orders 0 to 4, 361
Plots, Legendre polynomials of orders 0 to 4, 361
properties of Legendre polynomials, 362
Table, Legendre polynomials and functions, 360
Table, types and solutions, 359
Leibnitz derivative formula (Leibnitz rule), 224–225
for one dimension, 224
for three dimensions, 224
Leibnitz formula (n^{th} derivative of products), 349
Levenberg-Marquardt
See (Least-squares solution)
Levi-Civita symbol, 155
Liènard systems, 336
Limit cycle, 332
Bendixson's criterion, 333
Poincaré-Bendixson's theorem, 334
Line integral, 673
Linear PDE
boundary conditions (Dirichlet, Neumann and Robin), 408
complementary solution, 407
d'Alembert solution, 411
linear partial differential operator, 407
non-homogeneous PDE
homogenization of boundary conditions, 432
homogenization of PDE, 438
particular solution, 407
reducible PDEs, 408–409
similarity transformation and similarity transformation parameter, 440
solution method
Fourier transform method, 459–463
Laplace transform methods, 474–476
method of eigenfunction expansion, 434–437
method of images, 476–477
separation of variables, 411–428
similarity transformation, 439–443
superposition, 407
Lipschitz condition, 312
Logistic solution, 265
LU decomposition, 59–65
block LU decomposition, 605
Choleski's method, 61
Crout's method, 61
Doolittle's method, 61
Thomas algorithm, 63
Lyapunov function, 330
Krasovskii form, 331
Rosenbrock function, 331
Lyapunov matrix equation, 331

Mass balance, 216–217
Mass matrix. *See* (Finite element methods–time dependent)
Matrix
  definition, 5
  adjugate, 14
  asymptotically stable, 124
  bandwidth, 62
  block matrix inverse, 30
  block matrix product, 30
  Boolean, 595
  characteristic equation, 108
  circulant, 116
  classes
    Table, based on operational properties, 562–563
    Table, based on structure and composition, 567
  cofactor, 13
  companion (Frobenius), 112
  condition number, 136
  Cramer's rule, 29
  cross-product operator, 141
  derivative, multivariable
    gradient, 35
    Hessian, 37
    Jacobian, 36
  derivative, univariable function
    definition, 32
    Table of properties, 33
  determinant
    definition, 13
    block matrix formulas, 30
    Table of properties, 25
    via row/column expansion, 13
  diagonalizable, 117
  diagonally dominant, 61, 143
  elementary row/column operators, 11
  exponential, 32, 253
  Fourier, 43
  grammian, 71
  Hadamard product, 12
  Hermitian/skew-Hermitian, 6
  Hessenberg, 652
    algorithm based on Householder operations, 653
  idempotent, 42, 104
  ill-conditioned, 137
  integral, univariable function
    definition, 32
    Table of properties, 34
  inverse
    definition, 12
    block matrix formulas, 30
    Moore-Penrose generalized inverse, 128
    of diagonal matrices, 27
    of triangular matrices, 27
    via adjugates, 14
    Woodbury matrix formula, 28
  Jordan canonical basis, 119
  Kronecker product, 12
  modal, 119
  negative definite, 38
  nilpotent, 43
  nondefective, 117
  normal, 116
  norms, 135–138
  operations, algebraic, 10–12
    Table, 9
    Table of properties, 19
  operations, restructuring, 6–8
    Table, 7–8
  operators, 100–107
    affine, 106
    permutation, 101
    projection, 104
    reflection (Householder), 103
    rotation (Givens), 101
  orthogonal, 101
  permutation, 11, 101
  positive definite, 38, 587
    Sylvester's criterion, 588
  projection, 42, 104
  pseudo-inverse, 72
  quadratic form, 18
    gradient and Hessian, 39
  rank, 56
  redact, 8
  reducible/irreducible, 598–600
  reshape, 6
  semisimple, 117
  sparse matrix, 39–40
    coordinate format, 40
    Table of MATLAB commands, 40
  spectrum, 109
  square root, 132
  stable, 124
  submatrix, 7
  symmetric/skew-symmetric, 6
  unitary, 43, 101
  vectorization, 6
Matrix diagonalization, 117–118
Matrix iterative methods
  conjugate gradient method, 78
  Gauss-Seidel method, 68
  GMRES method, 79
  Jacobi method, 67
  succesive over-relaxation (SOR), 68
Matrix norms, 135–138
  Frobenius, 135
  induced norm, 135
Matrix splitting
  diakoptic method, 605
  Schur complements method, 608
Matrix, functions of, 120–124
  evaluation
    diagonalization, 120
    finite sums, 122
    Jordan decomposition, 121
    Sylvester's theorem, 659
  well-defined functions, 120
Maxwell's equations, 222

Method of ghost points, 494
Method of lines
  See (Finite difference method-time-dependent-semidiscrete approach)
Michaelis-Menten kinetics, 272, 728
Molecular momentum flux, 217
Momentum balance, 217–218
Multiply connected regions, 213, 801

Navier-Stokes equation, 201
Newton's method
  basic algorithm, 81
  Broyden method, 81
  double-dogleg algorithm, 637
  line-search algorithm, 84
  secant method, 81
Nonlinear center, 339–340
  conservative systems, 339
  reversible systems, 339
Nusselt problem, 447, 481

Ordinary differential equation
  autonomous, 250, 313
  Clairaut's equation, 704
  contact transformation, 701
  decoupled system description, 258
  equilibrium point, 313
    hyperbolic equilibrium point, 328
  Euler-Cauchy, 249
  existence and uniqueness, 312
  finite series solution, 705
  first order
    Bernouli, 245
    differential form (Pfaffian form), 237
    exact differential equation, 237
    homogeneous type, 241
    integrating factors, 242
    isobaric, 241
    linear equation, 244
    method of separation of variables, 237
    ratio of affine terms, 241
    similarity transformation, 239
    symmetry transformation, 238
  general Ricatti equation, 700
  instability theorem, 259
  Laplace transform methods, 262
  Legendre transformation, 701
  matrix exponentials, 253
    matrizant (fundamental matrix), 257, 301
    method of finite sums, 255
    properties, 253
  Ricatti transformation, 700
  second order
    Emden-Fowler equation, 269
    missing explicit dependence on x or y, 246
    order reduction via similarity transformation, 247
  singular solutions, 703
  stability
    asymptotic stability, 319
    Lyapunov stability, 317
    quasi-asymptotic stability (attractive property), 317
  state space, 250
  state vector, 250
Orthogonal curvilinear coordinates, 192–194
  List, vector differential operations, 193–194
  scaling factors, 192

Parametric pump, 383
Partial differential equations
  reaction-diffusion equation, 479
Partial differential equations
  Black-Scholes equation, 446, 479
  Clairaut's equation, 392
  classification
    based on principal parts, 393
    characteristic forms, 395
    semilinear, second order, 394, 781–784
    semilinear, higher order, 784–785
  diffusion equation (one dimensional), 413, 478
  Fokker-Planck equation, 447
  Hamilton-Jacobi equations, 401
  Helmholtz equation, 374
    in polar coordinates, 554
    linear PDEs. *See* (Linear PDE)
  Helmholz equation, 499
  hyperbolic/strictly hyperbolic systems, 397
  inviscid Burger equation, 772, 778, 779, 781
  Laplace equation, 224, 418, 478
    polar coordinates, 502
  linear PDEs. *See* (Linear PDE)
  Poisson's equation, 223, 438
    polar coordinates, 501
  reaction-diffusion equation, 220
  telegraph equations, 481
  wave equation (one dimensional), 410
    d'Alembert solutions, 786
Path-independent line integral, 211
Permutation sign function, 12
Permutation symbol, 155
  identities, 155
Phase-space analysis, 315
  degenerate points, 326
  direction field, 316
  flows, 316
  focus and center, 324
  improper nodes, 323
  isoclines, 316
  nodes, 322
  nullclines, 316
  saddles, 323
  stars, 323
  trajectories, 317
Picard's method, 256, 347
Pochhammer symbol, 350
Poincaré maps (first return maps), 335
Poisson kernel, 421
Polar decomposition, 132–135
Positive definite function, 586

Potential equation. *See* (Partial differential equations–Laplace equation)
Principal axes, 163
Principal component analysis, 131–132
Projection operator, 104–106
Pursuit path, 266

QR decomposition, 77
  algorithm, 649

Rarefaction. *See* (Shocks and rarefaction)
Rayleigh quotient, 143
Residues. *See* (Analytic functions)
Riemann invariants, 399
Riemann space, 199
Rodriguez formula (for Legendre polynomials, 362

Schur triangularization, 116–117
  algorithm, 658
Schwartz class (of functions), 830
Series solution (ODE)
  around ordinary points, 352
  around regular singular points
    Frobenius series, 355
  around regular singular points (Frobenius method), 355
    indicial equation and roots, 356
  linear second-order ODE, Frobenius series, 357
  ordinary point, 348
  radius of convergence, 348
  singular point, 348
Shallow water model, 482
Shocks and rarefaction, 771
  break times, 772
  jump notation, 778
  Rankine-Hugonoit jump condition, 778
  rarefaction, 780
  Riemann problems, 779
  shock-fitting scheme, 776
  shock speed, 778
  shocks/shock paths, 771
  weak solutions, 774
Similarity transformation
  in ODEs, 239
  in PDEs, 440
  of matrices, 113
Simply connected regions, 213, 801
Singular value decomposition (SVD), 127–132
  algorithm, 659
  application to principal component analysis, 131
  reduced (economical) SVD, 130
  to obtain Gauss-Jordan factors, 594
  to obtain polar decomposition, 133
Singular values, 127
Spectral radius, 66, 125, 515
Spring and dashpot system, 270
Stokes' theorem, 210–215
Streamlines, 169
  potential flow around cylinders, 555
Sturm-Liouville systems
  definition, 421
  orthogonality of (eigen-)solutions, 423
Substantial time derivative, 183
surface integral, 678
Sylvester matrix equation, 24
Sylvester's criterion (for positive definite matrix)V, 588
Sylvester's theorem, 121, 659
Symmetry transformation
  in ODEs, 238
  in PDEs, 440

Taylor series expansion
  definition, 572
  linearization, 573
  second-order approximation, 573
Tensors
  See also (Vector (and tensors))
  definition, n^{th} order, 159
  metric tensors (fundamental tensor), 199
  operations, 162
  stress tensor, 160, 218
  Table, correspondence with matrix operations, 163
  unit tensor, 160
Thermal conductivity coefficient, 220
Thermal diffusivity coefficient, 220
Torsion of path/radius of torsion, 167
Traffic flow (idealized model), 403

Unit vector
  for matrices, 5
  for vectors and tensors, 154

van der Pol equation, 332, 717
Vector and scalar fields, 169
Vector differential operation
  curl, 176–178
  divergence, 174–175
  gradients, 170–173
  Laplacian, 178–179
  list of identities, 182
  List, spherical and cylindrical, 194
  miscellaneous operations, 179–182
Vector field
  Beltrami, 178, 183
  complex lamellar, 178
  gradient vector field, 170
  irrotational, 177
  solenoidal, 174
Vectors (and tensors)
  acceleration vector, 167
  basis unit vectors, 154
  binormal unit vector, 166
  derivative, 164–165
    List of identities, 164
  dimension, 152
  dyad, 157
  linearly dependent/independent, 152
  polyads, 159
  position vector, 165

Vectors (and tensors) (*cont.*)
  Table, correspondence with matrix operations, 163
  Table, fundamental operations, 151
  Table, operations based on unit vectors, 156
  Table, properties, 153
  traction vector, 160
  unit normal vector (to curve), 166
  unit normal vector (to surface), 168, 173
  unit tangent vector, 166
  velocity vector, 165
Vinograd system, 318
Volume integral, 684
Von Neumann series, 126
Von Neumann stability analysis, 516–519
Vorticity, 177

Wei-Prater kinetics, 260